MARINE ELECTRICAL AND ELECTRONICS BIBLE

Fourth Edition

ABOUT THE AUTHOR

John C. Payne is a professional marine electrical engineer and surveyor. During his long maritime and offshore oil career, he has served on a variety of commercial ships, from general cargo, reefers to oil tankers and dive support vessels, both as an Electrical Officer and Marine Engineer. He has served as a Diving Technician and is a Merchant Marine war veteran. In the offshore oil drilling industry, he is still actively involved with the world's most technologically advanced deepwater drilling rigs as a Construction and Commissioning Manager. John has been employed by one of the leading maritime consultancy companies as a surveyor on a wide variety of vessels, and acted as electrical expert witness on several large marine litigation cases.

As a professionally qualified technical author and writer, John has been responsible for defining and writing technical and operations manuals on a diverse range of air force, marine and naval projects including naval research vessels and submarine sonar systems. He is a licensed electrical contractor and a Certified Maritime Safety Auditor under the ISM Code. Previously he has been a Fellow of the Institute of Diagnostic Engineers and The Institute of Scientific and Technical Communicators.

John is the author of *The Motorboat Electrical and Electronics Manual*, the Understanding Boat series, *The Fisherman's Electrical Manual*, and *Piracy Today*. As his passion for cruising is local cuisine, he also authored *The Great Cruising Cookbook*, or *Cruisine©*, as he likes to call it. He has been widely published around the world in various boating and yachting magazines and regularly lectures on the subject.

His sailing career started in high performance racing dinghies. He has restored, cruised, and lived aboard a classic 37-foot Herreshoff ketch and a 34-foot wooden sloop cruising Europe, the UK, the Pacific, Australia and the Mediterranean. After several years living aboard his 115-year-old Dutch barge navigating European rivers and canals, he is now back sailing and cruising aboard a 36-foot ketch, and is a member of the Westerly Owners Association. He is also a member of the American Boat and Yacht Council (ABYC) and the Cruising Association (CA). His website is www.fishingandboats.com. Follow him on Twitter @JohnCPayne55579.

PRAISE FOR THE
PREVIOUS EDITION

"Laid out like an extended outline, this is, perhaps, the most easy-to-follow electrical reference to date . . ." —*Cruising World Magazine* (US)

"On electrics, the best book by far we have come across . . ." —*Motor Boat and Yachting* magazine (UK)

"This book ranks with the very best technical works ever produced for yachtsmen. . . . Any serious cruising yachtsman will probably buy two copies, one for the boat and one for the home. If I was undertaking a long cruise on a modern yacht, I would rank this book an essential part of my equipment. If you are installing equipment aboard a new or current vessel, then again this book is for you. It is very good indeed." —*The Island*

"A concise, useful and thoroughly practical guide. . . . It's a 'must have on board' book." —*Sailing Inland & Offshore*

"Everything a sailor could possibly want to know about marine electrics and electronics is here, [in] very sensibly signposted chapters . . . as a reference book on the subject it is outstanding." —*Classic Boat*

"All in all, this book makes an essential reference manual for both the uninitiated and the expert." —*Yachting Monthly* (UK)

"An ideal reference book that every professional and serious amateur fitter should have on hand. Even the non-technical yachtsman will find vital guidance and numerous valuable tips." —*Cruising*

"If you are mystified by all those wires, fuses, buttons, switches, etc., this could be your ticket to understanding not the last word, but the latest." —*WoodenBoat*

"John Payne has put together a concise, useful and thoroughly practical guide explaining in detail how to select, install, maintain and troubleshoot all the electrical and electronic systems on a boat. As the author rightly points out, there are no 24-hour road services offshore, so this book is essential reading for practical sailors. . . . It's well illustrated too, so for those unsure of the subject but willing to learn, the clear and concise illustrations will help immensely. It's a 'must have on board' book." —*Sailing Magazine* (South Africa)

"This comprehensive volume tells you just about everything you need to know about your boat's electric and electronics systems, and is invaluable if you ever intend to leave your marina berth!" —*Lifeboat* (UK)

ALSO BY JOHN C. PAYNE

MARINE ELECTRICAL AND ELECTRONICS BIBLE

Fourth Edition

A Practical Handbook for Cruising Sailors

A Sheridan House Guide to Boat Maintenance

JOHN C. PAYNE

ADLARD COLES

LONDON • OXFORD • NEW YORK • NEW DELHI • SYDNEY

For Jane, the best first mate ever!

And for Ian, future electronics engineer.

ADLARD COLES
Bloomsbury Publishing Plc
50 Bedford Square, London, WC1B 3DP, UK
29 Earlsfort Terrace, Dublin 2, Ireland

BLOOMSBURY, ADLARD COLES and the Adlard Coles logo are trademarks of Bloomsbury Publishing Plc

Reprinted from the English Language edition of *Marine Electrical and Electronics Bible*, 4th Edition,
By John C. Payne, originally published by Sheridan House, an imprint of The Rowman & Littlefield
Publishing Group, Inc. Copyright © 2024 by the author(s). Published by arrangement with the
Rowman & Littlefield Publishing Group, Inc.

John C. Payne has asserted his right under the Copyright, Designs and Patents Act, 1988, to be identified
as Author of this work

A catalogue record for this book is available from the British Library

ISBN: HB: 978-1-3994-1418-0; ePDF: 978-1-3994-1416-6; eBook: 978-1-3994-1417-3

10 9 8 7 6 5 4 3 2 1

Typeset in Times
Printed and bound in Great Britain by CPI Digital (UK) Ltd, Croydon CR0 4YY

MIX
Paper | Supporting
responsible forestry
FSC
www.fsc.org   FSC® C013604

To find out more about our authors and books visit www.bloomsbury.com
and sign up for our newsletters

CONTENTS

SECTION TWO   MARINE ELECTRONICS SYSTEMS

FOREWORD

Wow! Thirty-three years after completing the BOC Challenge and eleven boats (ranging from 36m to 5.8m) later, I am still dealing with boat electrics! My latest refit is a 1978 Nautor Swan 57 Ketch *Explorer* for the 2023 Ocean Globe Race, a celebration of the first 1975 Whitbread Round the World Race. I just sold *Trekka*, my Class Globe 5.80 that I raced solo across the Atlantic in 2021 in twenty-eight days. This little plywood "home-built" Mini was exclusively solar-powered. It had all the gear, including electric outboard, satellite communications, VHF, tiller pilot, echo sounder, chart plotter etc., and a wiring diagram.

When I sailed my Talisker *Bounty* boat (a 24-foot open whaleboat) 4,000 miles and forty-eight days across the Pacific in the wake of William Bligh following the mutiny on the *Bounty*, we had nothing! No toilet paper, no charts, only two weeks of water, and virtually no food (I lost 18kg!).

But we had an electrical system (and wiring diagram) to run the camera, computer, emergency satellite phones, navigation lights, and a VHF radio. *You just cannot get away from electricity on boats!* If and when your life depends on it—as it often does in my voyages—you have to do the power systems only one way: the right way!

I still have my original, well-fingered copy of John Payne's great book *Marine Electrical and Electronics Bible*! It's still relevant, and now the new fourth edition is sure to be another hit around the world. Knowledge is everything, and it is all right here! Thanks, John.

—Don McIntyre
Founder and organizer of the Ocean Globe Race,
Golden Globe Race, Mini Globe Race
"Adventure is any activity with an unknown outcome."
www.McIntyreAdventure.com

INTRODUCTION

Marine electrics and electronics on any yacht requires a holistic approach, from power generation and electrical wiring distribution to energy storage and end user electrical and marine electronics equipment. There are many older boats out on the water that have both older technology systems as well as upgraded and new-generation equipment. This fourth edition encompasses much of the older existing technologies along with the newer ones.

The *Marine Electrical and Electronics Bible* had its genesis around a saloon table over a few ice-cold beers with some fellow cruising sailors lamenting the lack of clear guidance on how to approach the myriad electrical and electronics systems on a yacht. As a result, it has been written to meet the practical and real-world situations of most yachts. While there is a sail and cruising bias in the book for 30- to 65-foot yachts, the majority of the information is directly relevant and applicable to the majority of trawler yachts, power- and motorboats, and even superyachts.

In this fourth edition, the Marine Electrical chapters have been reviewed and updated to reflect the many changes in battery technology, such as lithium-ion batteries and associated charging. We have witnessed the emergence of inverters and new DC motor technologies and are at the threshold of an era where electric propulsion is starting to enter mainstream boating. This has major electrical system ramifications. Marine electrical systems are often poorly planned, installed, and maintained. Electrical theory is explained to a level for informed equipment selection, installation, operation, maintenance, and troubleshooting. Within these chapters I have infused my own professional and lived cruising experiences along with lessons learned with decades of continual input from fellow sailors. I have deliberately attempted to correct the dangerous illusion that vessel and automotive battery systems and charging are alike except for the voltage levels.

This fourth edition has updated and enhanced Marine Electronics chapters to reflect the significant advances in everything from the Global Maritime Distress and Safety System (GMDSS) to communications and position fixing. This, along with radar, the automatic identification system (AIS), integrated instrument systems, and electronic charting, has dramatically enhanced marine safety. For many, the rapid development of smartphone technology with an app for almost everything has changed the game for many boaters. Notebook computers, smartphones, and tablets are fairly standard devices on most boats today, along with the supporting software.

One of the inherent dangers in an age of automation and electronic devices is the loss or surrendering of essential seamanship skills. Far too many people sail away without the knowledge or ability to survive a loss of electronic navigation aids or the ability to maintain equipment and systems. The principal goal of this book is to provide support in keeping your systems alive and healthy. Given the sobering annual boating fatality and incident statistics, this book is designed to help keep you and your crew safe.

This book encapsulates nearly 40 years of professional experience as a marine electrical engineer and surveyor on everything from merchant vessels to offshore oil drilling rigs as well as several of my own cruising yachts and power vessels. In this latest edition I have attempted to answer many of the frequently asked questions and requests I get through my website (fishingandboats.com) and address criticisms, both constructive and otherwise. I cannot overstress the importance of adopting a keep-it-simple approach to the reliability of marine electrical and electronics systems.

ABBREVIATIONS AND ACRONYMS

ABBREVIATIONS AND SYMBOLS

μF	microfarad	km/hr	kilometers per hour
A	ampere	kPa	kilopascal
AC	alternating current	kt	knot
Ah	amp-hour	kVA	kilovolt-amp
cd	candela	kW	kilowatt
cfh	cubic feet per hour	l/h	liters per hour
cfm	cubic feet per minute	l/min	liters per minute
cm	centimeter	lb	pound/pounds
cm²	square centimeter	Li-air	lithium-air (battery)
cp	candlepower	m	meter
cu ft	cubic foot	m/s	meters per sec
dB	decibel	m²	square meter
dBa	Decibel A weighted	mA	milliampere
DC	direct current	mbar	millibar
F	farad	Mbps	megabits per second
fpm	feet per minute	MHz	megahertz
ft	foot	ml	milliliter
ft²	square foot	mm	millimeter
g/h	gallons per hour	mm²	square millimeter
g/min	gallons per minute	mph	miles per hour
GHz	gigahertz	ms	millisecond
hp	horsepower	mV	millivolt
Hz	hertz	nm	nautical mile
in	inch	Nm	Newton-meter
K	kelvin	oz	ounce/ounces
kA	kiloampere	ppm	parts per million
Kcal	kilocalorie	psi	pounds per square inch
kg	kilogram	rpm	revolutions per minute
kgF	kilogram-force	VA	volt-amp
kgm	kilogram-meter	vpc	volts per cell
kHz	kilohertz	Ω	ohm

ACRONYMS

ABS	American Bureau of Shipping
ABYC	American Boat and Yacht Council
ACB	activated carbon block
ACU	actuator control unit
AES	advanced encryption standard
AFCI	arc fault circuit interrupter
AGM	absorbed glass mat (battery)
AHRS	attitude heading reference sensor

AHS   Australian Hydrographic Society
AI   artificial intelligence
AIS   Automatic Identification System
AIS-SART   AIS search and rescue transponder
ALRS   Admiralty List of Radio Signals UK
ANSI   American National Standards Institute
ANT   adaptive network technology
API   American Petroleum Institute
APM   alternator protection module
ARPA   automatic radar plotting aid
ARRL   American Radio Relay League
ASM   application specific messages
AST   advanced steering technology
ASTM   American Society for Testing and Materials
ASU   automatic start-up
ATU   automatic tuner unit
AV   audiovisual
AVR   automatic voltage regulator
AWG   American wire gauge
AWQ   Australian Waters Qualification

BBI   battery bank integrator
BDC   bottom dead center
BER   bit error rate
BLE   Bluetooth Low Energy
BMEA   British Marine Electronics Association
BMS   battery management system
BOM   Bureau of Meteorology (Australia)
BPG   personal bathymetric generator
BTU   British thermal unit
BVI   British Virgin Islands

C/A   course and acquisition
CAN   controller area network
CAPE   Convective Available Potential Energy
CCA   cold cranking amps
CEF   charging efficiency factor
CFC   chlorofluorocarbon
CFR   Code of Federal Regulations
CFS   Climate Forecast Model
CHIRP   compressed high-intensity radiated pulse
CM   condition monitoring
CMB   continuous marine broadcast
CME   coronal mass ejection
COG   course over ground
COLREGs   Convention on the International Regulations for Preventing Collisions at Sea, 1972
CPA   closest point of approach
CPR   cardiopulmonary resuscitation
CPU   central processing unit

CSA	cross-sectional area
CSP	concentrated solar power
CSTDMA	carrier-sense time-division multiple access
CT	current transformer
CTV	Color Thermal Vision
DGPS	Differential GPS
DI	direct injection/directly injected
DMM	digital multimeter
DNV	Det Norske Veritas
DOD	US Department of Defense
DOL	direct online
DOP	dilution of precision
DSC	digital selective calling
DZR	dezincification-resistant
EB	equipotential bond
EBL	electronic bearing line
ECB	electronic circuit breaker
ECDIS	Electronic Chart Display and Information System
ECM	engine control module
ECMWF	European Centre for Medium-Range Weather Forecasts
ECS	electronic chart system
ECU	electronic control unit
EEPROM	electrically erasable programmable read-only memory
EGC	enhanced group calling
EGNOS	European Geostationary Navigation Overlay Service
ELCI	equipment leakage circuit interrupter
ELT	emergency locator transmitter
EMC	electromagnetic compatibility
EMF	electromotive force
EMI	electromagnetic interference
EMP	electromagnetic pulse
ENC	electronic navigational chart
EPA	US Environmental Protection Agency
EPDM	ethylene propylene diene monomer
EPR	ethylene propylene rubber
ES	Edison screw
ESA	Electrical Signature Analysis; European Space Agency
ESD	electrostatic discharge
ETA	estimated time of arrival
ETFE	ethylene tetrafluoroethylene
EUV	extreme ultraviolet
EVA	ethylene vinyl acetate
FCC	Federal Communications Commission
FDA	US Food and Drug Administration
FMCW	frequency modulated continuous wave
FMEA	Failure Mode and Effect Analysis
FMECA	Failure Mode and Effect Criticality Analysis

FMOP	frequency modulation on pulse
FRN	FCC registration number
FRP	fiberglass-reinforced polymer
FTA	fault tree analysis
FTC	fast time constant
FW	freshwater
GAC	granular activated carbon
GAGAN	GPS Aided GEO Augmented Navigation
GDOP	geometric dilution of precision
GDT	gas discharge tube
GEOSAR	Geostationary Orbiting Search and Rescue
GFCI	ground fault circuit interrupter
GFS	Global Forecast System
GMDSS	Global Maritime Distress and Safety System
GNSS	global navigation satellite system
GOCP	General Operators Certificate of Proficiency
GOM	Gulf of Mexico
GPS	Global Positioning System
GRP	glass-reinforced plastic
GRT	gross registered tonnage
HamSCI	Ham Radio Science Citizen Investigation
HAZOP	hazard and operability study
HCA	hot cranking amps
HCFC	hydrochlorofluorocarbon
HDOP	horizontal dilution of precision
HEUI	hydraulically actuated, electronic controlled, unit injector
HIRLAM	High Resolution Limited Area Model
HOH	hard-over to hard-over
HPU	hydraulic power unit
HRC	high rupturing capacity
HRRR	high-resolution rapid refresh
HV	high voltage
HVAC	heating, ventilation, and air-conditioning
I/O	input/output
IC	Industry Canada
ICAO	International Civil Aviation Organization
ICCP	impressed current cathodic protection
ICON	Icosahedral Nonhydrostatic Model
ICW	Intracoastal Waterway
IDI	indirect injection/indirectly injected
IEC	International Electrotechnical Commission
IEEE	Institute of Electrical and Electronics Engineers
IGBT	insulated gate bipolar transistor
IMU	inertial measurement unit
IOC	Index of Cooperation
IoT	Internet of Things
IP	ingress protection

IPS	in-plane switching
IPv6	Internet Protocol Version 6
IR	infrared; insulation resistance; interference rejection
IRNSS	Indian Regional Navigation Satellite System
Isc	short-circuit current
ISO	International Standards Organization
ITU	International Telecommunication Union
JPL	Jet Propulsion Laboratory
KDF	kinetic degradation fluxion
LAN	local area network
LCD	liquid crystal display
LDPE	low density polyethylene
LED	light emitting diode
LEL	lower explosion limit
LEN	load equivalency number
LEO	low-earth orbit
LEOSAR	Low Earth Orbiting Search and Rescue
LF	low frequency
LFM	linear frequency modulation
LFP	lithium iron phosphate (battery)
LIDAR	light detecting and ranging
LOA	length overall
LPG	liquefied petroleum gas
LPM	lines per minute
LR	Lloyd's Register
LRC	Long Range Certificate
LROCP	Long-Range Operator Certificate of Proficiency
LSB	lower side band
LUT	local user terminal
MARPA	mini automatic radar plotting
MARPOL	International Convention for the Prevention of Pollution from Ships
MCA	marine cranking amp; Motor Circuit Analysis
MCB	miniature circuit breaker
MCC	mission control center; Motor Control Center
MCOV	maximum continuous operating voltage
MCU	motor control unit
MEO	medium-earth-orbit
MEOSAR	Medium-altitude Earth Orbit Search and Rescue
MFD	multifunction display
MPP	maximum power point
MMSI	maritime mobile service identity
MOB	man overboard
MOS	metal oxide semiconductor
MOSFET	metal-oxide-semiconductor field-effect transistor
MOV	metal oxide varistor
MPPT	maximum power point tracker/tracking

MRCC	Maritime Rescue Coordination Center
MRU	motion reference unit
MSAS	Multifunction Satellite Augmentation System
MSD	marine sanitation device
MSI	maritime safety information
MSW	modified sine wave
MVOC	Maritime VHF Operator's Certificate
NAM	North American Mesoscale Forecast System
NAMS	National Association of Marine Surveyors
NASPA	Nevis Air and Sea Ports Authority
NAVTEX	Navigational Telex
NBDP	narrow band direct printing
NC	normally closed
NCEP	National Centers for Environmental Prediction
NDGPS	Nationwide Differential GPS
NEC	National Electrical Code
NFPA	National Fire Protection Association
NiCad	nickel-cadmium (battery)
NiFe	nickel-iron (battery)
NiMH	nickel metal hydride (battery)
NIST	National Institute of Standards and Technology
NMEA	National Marine Electronics Association
NO	normally open
NOAA	National Oceanic and Atmospheric Administration
NSCV	National Standard for Commercial Vessels
NSW	New South Wales
NTC	negative temperature coefficient
NWP	numerical weather prediction
NWS	National Weather Service
OS	Open Skiron Model
PCB	printed circuit board
PCU	pedestal control unit
PDOP	position dilution of precision
PDT	pressure diffusion technology
PE	polyethylene
PEEK	polyether ether ketone
PEM	proton exchange membrane
PEMFC	proton exchange membrane fuel cell
PF	power factor
PIC	programmable intelligent computer
PIR	passive infrared
PLB	personal locator beacon
PLC	programmable logic controller
PM	permanent magnet
PP	polypropylene
PPE	personal protective equipment
PPI	plan position indicators

PPS	Precise Positioning Service
PRV	pressure relief valve
PSOC	partial state of charge
PSU	polysulfone
PSU	power supply unit
PSV	pressure safety valve
PTC	positive temperature coefficient
PTFE	polytetrafluoroethylene
PTO	power takeoff
PTT	push-to-talk
PVC	polyvinyl chloride
PWM	pulse width modulation/modulated
RAM	random access memory
RAMN	Radio Aids to Marine Navigation
RCC	rescue coordination center
RCCB	residual current circuit breaker
RCD	Recreational Craft Directive; residual current device
RCM	reliability-centered maintenance
RCS	radar cross section
RDF	radio direction finding
RED	Radio Equipment Directive
RF	radio frequency
RFI	radio frequency interference
RFID	radio frequency identification
RISC	reduced instruction set computing
RLS	return link service
RMS	root-mean-square
ROT	rate of turn
RPN	risk priority number
RROP	Restricted Radiotelephone Operator Permit
RTCM	Radio Technical Commission for Maritime Services
RTD	resistance temperature devices
RTE	radar target enhancer
RV	recreational vehicle
S/N	signal/noise
SA	selective availability
SAE	Society of Automotive Engineers
SAR	search and rescue
SART	search and rescue radar transponder
SBAS	satellite-based augmentation system
SCPD	short-circuit protective device
SCR	silicon controlled rectifier
SCU	self-contained unit
SD	secure digital
SELV	safety or separated extra low voltage
SG	specific gravity
SHE	standard hydrogen electrode

SI	Système International
SID	sudden ionosphere disturbance
SIM	subscriber identity module
SMC	Smart Macerator Control
SMS	short messaging services
SOC	state-of-charge
SOG	speed over ground
SOLAS	International Convention for Safety of Life at Sea
SOTDMA	self-organized time-division multiple access
SP	self-pumping
SPD	surge protection device
SPP	single point positioning
SPS	Standard Positioning Service
SRC	Short Range Certificate
SRF	surge reduction filter
SROCP	Short-Range Operator Certificate of Proficiency
SSB	single-sideband modulation
SSL	ship's station license
STC	Smart Toilet Control
SVGA	super video graphics array
TCPA	time to closest point of approach
TDC	top dead center
TDMA	time-division multiple access
TDOP	time dilution of precision
TEFC	totally enclosed fan cooled
TEU	twenty-foot equivalent unit
TFSC	thin-film solar cell
TFT	thin film transistor
THDi	total harmonic current distortion
TNC	terminal node controller
TSR	tip-speed ratio
TSW	true sine wave
TVS	transient voltage suppression
TWA	true wind angle
TWS	true wind speed
TX	thermostatic expansion
UK	United Kingdom
UKHO	UK Hydrographic Office
UKMO	UK Meteorological Office
UL	Underwriters Laboratories
UPS	uninterruptible power supply
URE	user range error
US	United States
USB	universal serial bus
USB	upper side band
USCG	United States Coast Guard
USL	Uniform Shipping Law

USVI	US Virgin Islands
UTC	Coordinated Universal Time
UTP	unshielded twisted pair
UV	ultraviolet
VAC	voltage alternating current
VCI	volatile corrosion inhibitor
VDC	volts direct current
VDES	VHF data exchange system
VDL	VHF data link
VDOP	vertical dilution of precision
VDR	voltage dependent relay/resistor; voyage-data recorder
VDSMS	VHF Digital Small Message Services
VFD	variable frequency drive
VG	viscosity grade
VGA	video graphics array
VHF	very high frequency
VMG	velocity made good
VOC	volatile organic compound
VoIP	voice over internet protocol
VpCI	vapor phase corrosion inhibitor
VRLA	valve regulated lead-acid (battery)
VRM	variable range maker
VSD	variable-speed drive
Vsi	pulse breakdown voltage
VSR	voltage sensitive relay
VSWR	voltage standing wave ratio
VTS	vessel traffic system/service
WA	Western Australia
WAAS	Wide Area Augmentation System
WGS84	World Geodetic Spheroid 1984
WHO	World Health Organization
WRF	Weather Research and Forecasting
WSC	World Shipping Council
WSVGA	wide super video graphics array
WWNWS	Worldwide Navigational Warning Service
WXGA	wide extended graphics array
XLPE	cross-linked polyethylene
XPS	extruded polystyrene
XTE	cross-track error
ZOV	zinc oxide varistor

SECTION ONE

MARINE ELECTRICAL SYSTEMS

Marine Electrics

Marine electrics have continued to evolve over the last decade in tandem with marine electronics. When the first edition of the *Marine Electrical and Electronics Bible* was published, equipment was a lot less sophisticated and is now mostly obsolete. Marine electrics have not undergone the same developmental revolutions that marine electronics have, but they form the basis for reliability for those systems and devices. Like the electronics space, smartphones and apps have emerged for everything from charging systems to lithium-ion batteries and more. Electrical systems used to be a series of both discrete and interconnected systems, but the holistic and integrated approach has emerged. I regularly visit and assist people who have this older technology on board and wish to continue with it, so this section will cover both the old and the new.

1.1   Technology Advances. The advances continue within the marine electrics space. Electric propulsion is never far from the spotlight, enabled by big advances in battery technology and a step up in voltage levels. Microprocessors are in everything, and smart battery charging systems have developed to keep pace with the changing battery technologies. While wiring and circuit protection have remained relatively unaltered, DC electric motors have evolved, with permanent magnet technology replacing brush types. These are rapidly replacing motors on windlass and deck winch applications. Lighting has dramatically altered, with light-emitting diode (LED) lights rapidly replacing the century-old incandescent and fluorescent light technologies.

1.2   Marine Systems Companies. There have been many mergers and acquisitions by companies, and many of these companies are now vertically integrated so as to offer total systems solutions. Brunswick, known for its ownership of Mercury Marine, and the Navico Group now have under their umbrella many household names in marine systems, including Ancor, Attwood, B&G, C-Map, BEP, Blue Sea Systems, Lenco, Garelick, Marinco, Mastervolt, ProMariner, RELiON, CZone, Whale, Lowrance, Simrad, Progressive Industries, ProMariner, Motorguide, and even more in the marine electronics space. The various companies remain leaders in their respective fields, and technology developments tend to have a halo effect and spill over into all other adjacent technology areas. Industry leaders such as Victron Marine have superb power products and are commonplace on yachts, powerboats, and commercial shipping. Vetus is owned by Yanmar. Xylem owns Jabsco, Rule, Flojet, and many others.

Batteries

2.1   About Batteries. This chapter covers battery technology and principles, and the important task of selection, maintenance, and installation of batteries in boats. This is for batteries used in service or house power roles and for engine-starting duty for the main propulsion diesel engine. The majority of electrical power problems arise from improper battery selection along with deficient battery charging. Battery bank capacities are either seriously underrated, with resultant power shortages, or overrated so that the charging system cannot properly recharge them. The result is a premature failure of the batteries due to sulfation in flooded-cell lead-acid batteries. Initially, it is essential that all the equipment on board be listed, along with power consumption ratings. The holistic approach is where the battery bank is selected to supply a calculated load, and then a battery charging source to properly restore charge levels. Battery technology has continued to evolve, and we have come a long way from the traditional flooded-cell lead-acid battery to the lithium-ion battery technologies available today. Additionally, we have developments in carbon foam batteries as well as major developments in foam and graphene technologies along with the lithium-air (Li-air) battery technology front that employs an oxygen reagent. Other research is looking at magnesium as a major battery element along with sodium-ion. Technology marches on in this space and will continue to enhance boat electrical power systems.

2.2   Battery Rating Selection. You cannot plan your power system without quantifying what the load is, so please do an energy audit for your boat by listing all your equipment and devices and then precisely calculating your vessel loads. The following are my own personal minimum recommendations for battery installation.

   a.   **House Battery Capacity.** The house or service battery capacity should be based on calculation of the boat power consumption for an autonomous 24-hour period. The power calculations should include all the equipment and systems that will run continuously or intermittently during the calculated period. Equipment in this category includes lighting, navigation instruments, communications, radar, autopilots, inverters, refrigeration, entertainment systems, and more. The deep-cycle battery is used for these applications. Calculations are based on the maximum power consumption over the longest probable period between battery recharging—typically a period of 24 hours. I have met those who factor this for 36 to 48 hours, which is prudent for long-distance cruising.

   b.   **Start Batteries.** A separate battery should be provided for the diesel propulsion engine. The engine-start battery capacity should be based on the provision of ten consecutive start attempts of 5 seconds' duration, with a 30-second period between each attempt, at an ambient temperature of 41°F (5°C). The engine manufacturer's recommendations should be the minimum battery specification for a starting battery. Starting loads require large current levels for relatively short time periods. Loads in this category include the engine starter motor, the diesel engine preheating system, the anchor wind-lass, electric sail handling deck winches, bow thrusters, and electric toilets. The starting battery is normally used for all these applications, although they are frequently connected to the house battery. The battery rating should always allow for a worst-case starting scenario. Battery efficiency is low-ered in cold temperatures, and engine starting requires greater power due to increased oil viscosity. The rating should allow for problems where multiple start attempts are required for cranking, diesel fuel system bleeding, or trou-bleshooting. (See the section on lithium-ion batteries).

2.3   **Battery Rating Calculation.** The calculation tables illustrate the typical equipment and device power consumptions, with space provided to insert and calculate your own vessel equipment and device-specific data.

   a.   **Load Calculation Table.** To calculate the total system loading, multiply the total current values by the number of hours to get the amp-hour (Ah) rating. If equipment uses 1A over 24 hours, then it consumes 24Ah.

   b.   **Capacity Calculation.** Depending on the frequency between charging peri-ods, select the column that suits your vessel activity. The most typical sce-nario is one of the boat lying at anchor or on a mooring and operating the engine every 24 hours to pull down refrigeration temperatures and also charge the battery. If you do a lot of voyaging, that may be the prime scenario:

   e.g.   Total consumption is 100Ah over 24 hours = 4.16A per hour.

   c.   **Capacity Derating in Lead-Acid Batteries.** As we wish to keep our dis-charge capacity to 50% of nominal battery capacity, we can assume that a battery capacity of 200Ah is the basic minimum level. In an ideal world, this would be a minimum requirement, but certain frightening realities must now be introduced into the equation. The figures below typify a common system, with alternator charging and a standard regulator. Maximum charge deficiency is based on the premise that boat batteries are rarely above 70% charge and cannot be fully recharged with normal regulators, and there is reduced capacity due to sulfation, which is typically a minimum of 10% of capacity. The key to maintaining optimum power levels and avoiding this common and potentially concerning set of numbers is the charging system.

   d.   **Amp-hour (Ah) Capacity.** It is important to discuss a few more relevant points regarding amp-hour capacity, as it has significant ramifications for the selection of capacity and discharge characteristics:

      (1)   **Fast Discharge (Peukert's Equation).** The faster a battery is dis-charged over the nominal rating (either 10-hour or 20-hour rate), the less the real amp-hour capacity the battery has. This effect is defined by Peukert's equation, which has a logarithmic characteristic. This equation is based on the high and low discharge rates and discharge times for each to derive the Peukert coefficient 'n.' Average values

are around 1.10 to 1.20. If we discharge a 250Ah battery bank, which has nominal battery discharge rates for each identical battery of 12A per hour at a rate of 16A, we will actually have approximately 10% to 15% less capacity.

(2)   **Slow Discharge.** The slower the discharge over the nominal rate, the greater the real capacity. If we discharge our 240Ah battery bank at 6A per hour, we will actually have approximately 10% to 15% more capacity. The disadvantage here is that slowly discharged batteries are harder to charge if deep cycled below 50%.

Table 2-1. Battery Capacity Derating

Capacity Status	Capacity Deduction	Battery Capacity
Nominal Capacity		240Ah
Maximum Cycling Level	(50%) Deduct	120Ah
Maximum Charge Deficiency	(30%) Deduct	72Ah
Lost Capacity	(10%) Deduct	24Ah
Available Battery Capacity		**24Ah**

2.4   Sailing Load Calculations. To start power system program or simply understand what you have, it is essential that all equipment on board be recorded along with power consumption ratings. Ratings can usually be found on equipment nameplates or in equipment manuals. Perform a complete audit and insert your own values into the "Actual Amps" column. I always recommend, as a minimum, basing the calculation on a 24-hour period. If your onboard lifestyle and routine is 12 hours, then divide the outcome by 2. To convert power in watts into current in amps, simply divide the power value by your system voltage. Add up all the current figures relevant to your vessel and multiply by hours to get an average Ah consumption rate. Space is reserved to add in specific values. Most of these items will be energized when sailing or at anchor, but many will not be relevant if you are tied up in a marina connected to a battery charger. The calculation assumes that your engine will not be in operation and no generator with battery charging will be operational.

Table 2-2. DC Load Calculation: Table 1—Sailing Loads

Equipment	Typical Amps	Actual Amps	12 hours	24 hours
Radar Transmit	5			
Radar Standby	0.5			
HF/SSB Receive	0.5			
VHF Receive	0.5			
AIS	0.5			
Satcom Receive	1			

(continued)

5

Table 2-2. *Continued*

Equipment	Typical Amps	Actual Amps	12 hours	24 hours
Weatherfax	0.5			
NAVTEX	0.5			
MFD/GPS	0.5			
Instruments	0.5			
Fish-finder	0.5			
Stereo	0.5			
Gas Detector	0.25			
Chart Plotter	0.25			
Refrigeration	4			
Inverter Standby	0.5			
Deck Winch	20			
Tricolor Nav Light	0.25			
Tablet Charger	0.25			
Phone Charger	0.25			
Computer Charger	0.5			
Subtotal Table 1				

2.5   Additional Load Calculations. Other basic load characteristics have to be factored into load calculations. Add up all the current figures relevant to your vessel and multiply by expected run times to get an average Ah consumption rate.

a. **Intermittent Loads.** It is often hard to quantify actual real current demands with intermittent loads. My suggestion is simply to use a baseline of 6 minutes per hour, which is one-tenth of 1 hour.

b. **Device Loads.** These are all the devices we have and can be called parasitic loads if left on all the time; cumulatively they drain battery power, so always monitor them.

c. **Motoring Loads.** Certain loads are only applicable when motoring. Loads must be subtracted from charge current values and actually may impact charging system efficiency at low speeds. Loads include navigation lights, refrigeration electric clutch, water-maker clutch, and ventilation fans.

Table 2-3. DC Load Calculation: Table 2—Intermittent Loads

Equipment	Typical Amps	Actual Amps	12 Hours	24 Hours
Intermittent Loads				
Bilge Pump	3.5			
Shower Pump	3.5			
Water Pump	4.0			
Washdown Pump	2.5			
Toilet	15			
Macerator	14			
MSD	15			
HF/SSB Transmit	4			
VHF Transmit	3			
Spotlight	1			
Extraction Fan	1			
Inverter	30			
Cabin Lights	2			
Coffee Machine	5			
Deck Winch	20			
Device Loads				
Drone Charger				
Battery Tool Charger				
Phone Charger				
Computer Charger				
Tablet Charger				

(continued)

Table 2-3. *Continued*

Equipment	Typical Amps	Actual Amps	12 Hours	24 Hours
Spotlight Charger				
Subtotal Table 2				
Subtotal Table 1				
Total Load				

d. **Battery Load Matching.** The principal aim is to match the discharge characteristics of the battery bank to that of our calculated load of 10A per hour over 12 hours. Assume that we have a modified charging system so that we can recharge batteries to virtually 100% of nominal capacity. The factors affecting matching are as follows:

(1) **Discharge Requirement.** The nominal required battery capacity of 240Ah has been calculated as that required to supply 10A per hour over 12 hours to 50% of battery capacity. In most cases, the discharge requirements are worst for the night period, and this is the 12-hour period that should be used in calculations. What is required is a battery bank with similar discharge rate to the boat electrical current consumption rate. This will maximize the capacity of the battery bank with respect to the effect defined in Peukert's coefficient.

(2) **Battery Requirements.** As the consumption rate is based on a 12-hour period, a battery bank that is similarly rated at the 10-hour rate is required. In practice you will not match the precise required capacity; therefore you should go to the next battery size up. This is important, as the battery will be discharged longer and faster over 12 hours, so a contingency margin is required. If you choose a battery that has 240Ah at the 20-hour rate, in effect you will be installing a battery that in the calculated service has 10% to 15% less capacity than that stated on the label, which will then be approximately 215Ah, so you are below the required capacity.

e. **Battery Capacity Formulas.** A range of formulas are frequently used to determine battery capacity:

(1) **Four-Day Consumption Formula.** One of the more unrealistic formulas states that you should be able to supply all electrical

needs over four complete 24-hour periods without recharging batteries. Given that an average 10A per hour is or was a typical consumption rate, the 4-day formula tells us we'll use 960Ah. If we only discharge to 50%, that translates to an incredible 2000Ah battery capacity. In addition, the recharging period, which requires an additional 20%, must replace about 1200Ah. Even with a fast-charge device, a 160A alternator, the finite charge acceptance rate of the battery will require at least 12 hours charging. Lithium-ion batteries are changing this scenario.

(2)   **75/400 System.** This was included in a magazine article as one of three formulas for various sized vessels and was for a 40ft to 45ft (12m to 14m) yacht. This was the nearest I have seen to a rational set of numbers, based on a 75A consumption over 24 hours. Though perhaps too conservative, the formula is based on a 130A or 150A alternator with fast-charge device to recharge half of a 400Ah battery bank. In reality, there are no perfect formulas. Each vessel has different requirements, and systems must be tailored to suit.

2.6   Battery Ratings. In any sailing yacht, the battery has a primary role as a power storage device and a secondary one as a "buffer," absorbing power surges and disturbances that arise during charging and discharging. The foundation of a reliable and efficient power system is a correctly specified and rated battery. The following chapters explain the factors essential to the installation of a reliable power system. Manufacturers use a range of ratings figures to indicate battery performance levels. When selecting a battery, it is essential to understand the ratings and how they apply to your own requirements. Data sheets sometimes state the number of plates, and this is defined as the number of positive and negative plates within a cell. The more plates installed, the greater the plate material surface area. This increases the current during high current rate discharges, and cranking capacity and cold weather performance are improved. Battery casings are mostly plastic, and it is now rare to see rubber-cased units.

a.   **Amp-Hour Rating.** The amp-hour (Ah) rating refers to the available current over a nominal period until a specified final voltage is reached. Rates are normally specified at the 10- or 20-hour rate. This rating is normally only applicable to deep-cycle batteries. For example, a battery is rated at 84Ah at the 10-hour rate, with a final voltage of 1.7 volts per cell (vpc). The battery is capable of delivering 8.4A for 10 hours, when a cell voltage of 1.7V will be attained (Battery Volts = 10.2V). Where a battery is discharged faster than the nominal rating, the available capacity also decreases. This is the Peukert effect, and the decline follows a logarithmic curve.

b.   **Reserve Capacity Rating.** This rating specifies the number of minutes a battery can supply a nominal current at a nominal temperature without the voltage dropping below a nominal level. This rating is normally applied to automotive applications. It indicates the power available to operate ignition and auxiliaries when an alternator fails. Typically, the rating is specified for a 30-minute period at 77°F (25°C) with a final battery voltage of 10.2V.

c. **Cold Cranking Amps (CCAs).** This rating defines the current available at 0°F (–18°C) for a period of 30 seconds, while being able to maintain a cell voltage that exceeds 1.2vpc. This rating is applicable for engine starting purposes. The higher the battery rating, the more power available, especially in cold weather conditions.

d. **Marine Cranking Amps (MCAs).** This rating defines the current available at 32°F (0°C) for a period of 30 seconds while being able to maintain a cell voltage exceeding 1.2vpc. Again, this rating is only applicable for engine starting purposes. In cold-climate areas such the UK, northern Europe, United States, and Canada, the CCA is more relevant.

e. **Hot Cranking Amps (HCAs).** This rating defines the current available at 80°F (27°C) for a period of 30 seconds while being able to maintain a cell voltage exceeding 1.2vpc (7.2V for 12V battery).

f. **Marine Battery.** Plates may be thicker than normal, or there may be more of them. Internal plate supports are used for vibration absorption. Cases may be manufactured with a resilient rubber compound and have carry handles fitted. Filling caps may be of an anti-spill design.

2.7   Deep-Cycle Batteries. Service loads require a battery that can withstand cycles of long continuous discharge and repeated recharging. This deep cycling requires using the suitably named deep-cycle battery. Top of the range Rolls (Surrette) and Trojan batteries typify the high specification deep-cycle battery. The deep-cycle lead-acid battery has the following characteristics.

a. **Construction.** The deep cycle lead-acid battery is typified by the use of thick, high-density flat-pasted plates, or a combination of flat and tubular. The plate paste materials may contain small proportions of antimony to help stiffen them. Porous, insulating separators are used between the plates, and these have proprietary designs for electrochemical performance improvements. Glass matting is used to assist in retaining active material on the plates that may break away as plates expand and contract during charge and recharge cycles. If material accumulates at the cell base, a cell short circuit may occur, although this is less common in modern batteries. If material is lost, the plates will have reduced capacity or insufficient active material to sustain the chemical reaction, resulting in cell failure. Leading manufacturers have done much to develop stronger and more efficient plates. The grid design has fewer heavier sections to hold the high-density active material. This is due to the dynamic forces that normally cause expansion and contraction, with subsequent warping and cracking. Separator design has evolved, and some use double-insulated, thick glass woven separators that totally encase the positive plate, along with a microporous polyethylene envelope. This retains any material shed from the plates that can cause cells to short-circuit.

b.   **Cycling.** The number of available cycles varies between individual battery makes and models. Typically, it is within the range of 800 to 1,500 cycles of discharge to 50% of nominal capacity and complete recharging. Battery life is a function of the number of cycles and the depth of cycling. Batteries discharged to only 70% of capacity will last appreciably longer than those discharged to 40% of capacity. In practice, you should plan your system so that discharge is limited to 50% of battery capacity. The typical life of batteries when they are properly recharged and cycle capabilities maximized can be from 5 to 10 years.

c.   **Charging.** The recommended charging rate for a deep-cycle battery is often given as 15% of capacity. In vessel operations, it is not possible to apply these criteria accurately. Essentially, the correct charge voltage, corrected for temperature, should be used. Deep-cycle battery charging characteristics are as follows:

    (1)   **Counter Voltage.** During charging, a phenomenon called "counter voltage" occurs. Primarily, this is caused by the inability of the electrolyte to percolate at a sufficiently high rate into the plate material pores and subsequently convert both plate material and electrolyte. As the battery resists charging, plate surface voltage rises artificially high and "fools" the regulator into prematurely reducing charging.

    (2)   **Charging Voltage.** To properly charge a deep-cycle battery, a charge voltage of around 14.5V is required, corrected for temperature. Contrary to assertions by some "experts," a charge level of approximately 80% does not represent a fully-charged battery. It is not acceptable if you want a reliable electrical power system and reasonable battery life. If you do not fully recharge the battery, it will rapidly deteriorate and sustain permanent damage.

d.   **Equalization Charge.** An equalization charge consists of applying a higher voltage level at a current rate of 5% of battery capacity. This is done to "reactivate" the plates. There is a mistaken belief that this will completely reverse the effects of sulfation. Although there may be an improvement following the process, it will not reverse long-term permanent damage. Equalization at regular intervals can increase battery longevity by ensuring complete chemical conversion of plates, but care must be taken.

2.8   **Starting Batteries.** The starting battery must be capable of delivering the engine starter motor, or other consumers, with sufficient current to turn and start the diesel engine. This starting load can be affected by engine compression, oil viscosity, and engine-driven loads. Some loads, such as an inverter, thruster, or an anchor windlass operating full load, require similarly high values of current. Starting batteries have the following characteristics.

    a.   **Construction.** The starting battery is characterized by thin, closely spaced porous plates, which give maximum exposure of active plate material to the electrolyte and offer minimal internal resistance. This enables maximum electrochemical reaction rates to take place and maximum current availability. Physical construction is similar to deep-cycle batteries.

b.   **Cycling.** Starting batteries cannot withstand deep cycling and, if deep cycled or completely discharged, will have an extremely short service life. Ideally, batteries should be maintained within 95% of full charge.

c.   **Sulfation.** In practice, sulfation is not normally a problem, as batteries are generally fully charged if used for engine starting applications only. If improperly used for deep-cycle applications and undercharged, batteries will sulfate.

d.   **Self-Discharge.** Starting batteries have low self-discharge rates; this is generally not a problem in normal engine installations.

e.   **Efficiency.** Cold temperatures dramatically affect battery performance. Engine lubricating oil viscosities are also affected by low temperatures and further increase starting loads on the battery. If the reduction in battery capacity at low temperatures is combined with the increased starting current requirements, the importance of fully charged batteries is amplified. Table 2-4 illustrates the typical cranking power loss when temperature decreases from 80°F to 32°F (27°C to 0°C) using a 10W-30 multi-viscosity lubricating oil and the increased percentage of power required to turn over and start an engine.

Table 2-4. Battery Power Table

Temperature	Battery Level	Power Required
80°F (27°C)	100%	100%
32°F (0°C)	65%	155%
0°F (−18°C)	40%	210%

f.   **Charging.** Recharging of engine starting batteries is identical to deep-cycle batteries. Additional factors to consider are as follows:

(1)   Discharged current must be restored quickly to avoid damage. Similarly, temperature compensation must be made.

(2)   Normally, after a high current discharge of relatively short duration, there is no appreciable decrease in electrolyte density. The battery is quickly recharged, as the counter voltage phenomenon does not have time to build up and consequently has a negligible effect on the charging.

g.   **Battery Ratings.** Starting batteries are normally specified on the basis of an engine manufacturer's recommendations, although I have found these to be imprecise; the following is given as a guide only. Table 2-5 shows the recommended battery ratings for various marine diesel engines, as well as typical starter motor currents:

(1) **Start Capability.** It is practical to calculate a good safety margin, allowing for a multistart capability. Some classification societies specify a minimum of six consecutive starts, and that should be the absolute minimum value.

(2) **Temperature Allowance.** Additional allowances should be made for the decreased efficiency in cold climates, as a greater capacity and greater load current are required.

h. **Additional Start Battery Loadings.** The start battery should be used to supply short-duration, high-current loads. Check with your engine supplier for the recommended battery rating, and then add a good margin for safety. Depending on the equipment you have installed, consideration may require a second battery bank for high current loads. Additionally, factor in the following loads that may be added to your load analysis and calculations:

(1) **Anchor Windlass.** The very heavy current loadings that electric windlasses demand require a much higher battery bank rating. The battery banks should be doubled up so that two identical batteries are then parallel connected, although engines should be operating when anchor handling.

(2) **Deck Winches.** The electric deck winch has very heavy current loadings and will require a much higher battery bank rating. The battery banks should be doubled up so that two identical batteries are then parallel connected. Consult your manufacturer's manual for precise power requirements.

(3) **AC Generator.** In some cases, the engine battery can be used for starting; if it is, use caution if starting the engine while the generator is running. It is common to see installed 10A to 15A alternators suffer damage from the high current load of the engine starting motor. It is better to have a separate battery, which gives some redundancy.

Table 2-5. Minimum Battery Ratings Table

Voltage	Engine Rating	Current Load	Battery CCA
12	10hp 7.5kW	59A	375
12	15hp 11kW	67A	420
12	20hp 15kW	67A	420
12	30hp 22kW	75A	450
12	40hp 30kW	85A	500
12	50hp 37kW	115A	500
12/24	100hp	115/60A	500

(4)   **Bow Thrusters.** Direct current (DC) powered thrusters are generally powered directly from the engine start batteries. As the engine should be running, much of the load is supplied directly from the alternator. It must be noted that, as the engine is often in slow speed or idle, the full output is often not available, and considerable load can be taken from the battery if engine speed is too low.

2.9   **Battery Technologies.** Battery technology has evolved considerably over the past decade or two. The traditional lead-acid flooded-cell battery is being displaced by more advanced and efficient battery types. Due to cost constraints and budgets, many boat owners are still choosing to use flooded-cell batteries. This is covered in detail, as maintenance becomes one of the main drivers for reliability.

a.   **Lead-Acid Batteries.** The lead-acid flooded-cell battery has been used in the majority of marine installations for many years, and the principles are explained.

b.   **Low-Maintenance Batteries.** These are a variation of and improvement on the standard flooded-cell battery.

c.   **Gel Cell Batteries.** Gel cell batteries are widely installed on boats, and these are explained.

d.   **AGM Batteries.** Absorbed glass mat (AGM) batteries have been used extensively on boats and continue to be used; the advantages will be discussed.

e.   **Lithium-Ion Batteries.** These batteries are starting to revolutionize boat battery power supplies and are now a real alternative to lead-acid technologies. All major battery manufacturers now have lithium-ion battery options.

f.   **Carbon Foam Batteries.** This emerging battery technology is offering an alternative to lithium-ion batteries.

g.   **Graphene Batteries.** The latest technology entrant has vastly improved performance characteristics. Not mainstream yet but will be here soon.

h.   **Lead Carbon Batteries.** This innovation is rapidly gaining acceptance in boat applications and is worth monitoring.

i.   **NiCad and NiFe Batteries.** Alkaline batteries were usually found on larger cruising vessels and are an alternative to lead-acid batteries, although less common now.

2.10   **Lead-Acid Batteries.** The fundamental theory of the lead-acid battery is that a voltage is developed between two electrodes of dissimilar metal when they are immersed in an electrolyte. In the typical lead-acid cell, the nominal generated voltage is 2.1V. The typical 12V battery consists of six cells, which are internally connected in series to make up the battery.

a. **Cell Components.** The principal cell components are as follow:

    (1)   **Lead Dioxide (PbO_2).** This is the positive plate active material.

    (2)   **Sponge Lead (Pb).** This is the negative plate material.

    (3)   **Sulfuric Acid (H_2SO_4).** This is the electrolyte.

b. **Discharge Cycle.** Discharging of the battery occurs when an external load is connected across the positive and negative terminals. That load is any electrical device or equipment on your boat. An electrochemical reaction takes place between the two plate materials and the electrolyte. During the discharge reaction, the plates interact with the electrolyte to form lead sulfate and water. This reaction dilutes the sulfuric acid electrolyte, reducing the density. As both plates become similar in composition, the cell loses the ability to generate a voltage. As will be discussed elsewhere, the discharge cycle is impacted by the load size, the ambient temperature, and the condition of the battery when discharged. The formula for the discharge cycle is $PbO_2 + Pb + 2H_2SO_4 \rightarrow 2PbSO_4 + 2H_2O$.

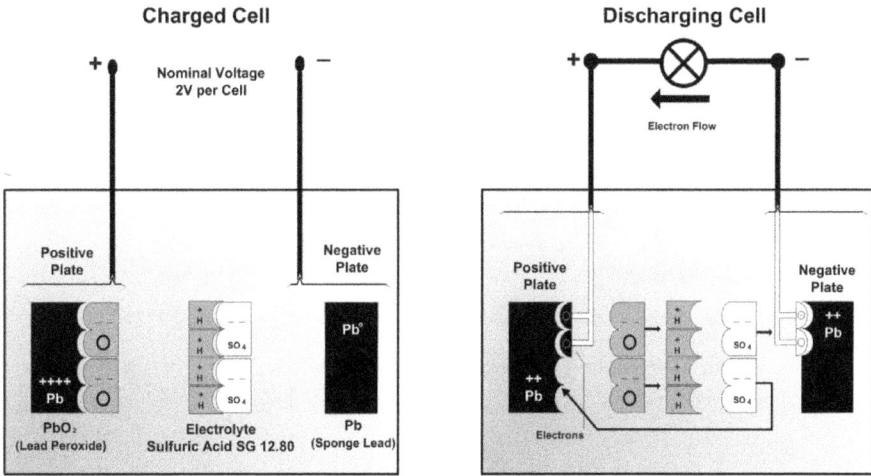

Figure 2-1. Lead-Acid Battery Charge and Discharge Reaction

c. **Charge Cycle.** Charging the battery reverses the electrochemical reaction described in the discharge cycle. The water decomposes to release hydrogen and oxygen. The two plate materials are reconstituted to the original material. When the plates are fully restored and the electrolyte is returned to the nominal density, the battery is completely recharged. As described in the battery charging and battery charger sections, while often based on a new battery, charging efficiency is subject to many factors. The battery depth of discharge is the first factor. The second factor is the degree of sulfation of plates, as this directly affects charge acceptance and the ability of the battery to be fully recharged. The internal resistance of the battery impacts the charge cycle. If it is excessive and the battery charge voltage is too high, the battery will lose energy through heating. The initial bulk charging level is dependent on the charge voltage and is when maximum charge current will flow. The level of charge that a battery can accept is known as the natural absorption of the battery. The last 20% of charging takes much longer than the initial bulk charging stage. The float stage to bring the battery to full charge occurs around 85% to 95% of full charge level. The complete reversal of the discharge electromechanical process will depend on the condition of plates. The formula for the charging cycle is $2PbSO_4 + 2H_2O \rightarrow PbO_2 + Pb + 2H_2SO_4$.

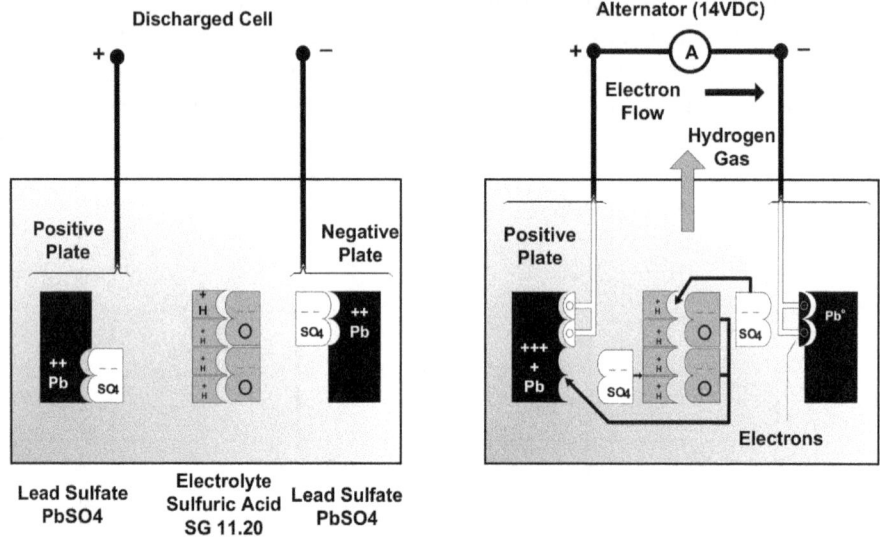

Figure 2-2. Lead-Acid Battery Discharge and Charge Reaction

2.11   Battery Electrolyte. The cell electrolyte is a dilute solution of sulfuric acid (H_2SO_4) and pure water. Specific gravity (SG) is a measurement defining electrolyte acid concentration. A fully charged cell has an SG typically in the range 1.240 to 1.280, corrected for temperature. This is an approximate volume ratio of acid to water of 1:3. Sulfuric acid has an SG of 1.835; water, a nominal 1.0. The following factors apply to battery electrolytes.

    a.   **Temperature Effects.** For accuracy, all hydrometer readings should be corrected for temperature. Ideally, actual cell temperatures should be used, but in practice ambient battery temperatures are sufficient. Hydrometer floats have the reference temperature printed on them, and this should be used for calculations. As a guide, the following should be used for calculation purposes:

        (1)   For every 1.8°F (1.0°C) the cell temperature is ABOVE the reference value, ADD 1 point (0.001) to the hydrometer reading.

(2) For every 1.8°F (1.0°C) the cell temperature is BELOW the reference value, SUBTRACT 1 point (0.001) from the hydrometer reading.

b. **Nominal Electrolyte Densities.** Recommended densities are normally obtainable from battery manufacturers. In tropical areas it is common to have battery suppliers put in a milder electrolyte density. This does not deteriorate the separators and grids as quickly as electrolyte densities used in a temperate climate.

c. **Battery Gas.** Battery cells contain an explosive mixture of hydrogen and oxygen gas at all times. An explosion risk exists whenever naked flames, sparks, or cigarettes are introduced into the immediate vicinity. Always use insulated tools, and cover the terminals with an insulating material to prevent an accidental short circuit. Note that watchbands, bracelets, and neck chains can accidentally cause a short circuit when working with batteries.

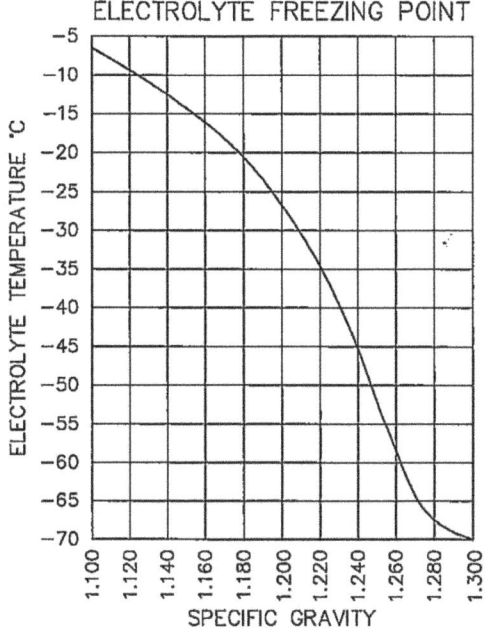

Figure 2-3.
Battery Electrolyte
Temperature Effects

2.12   Battery Water. When topping up the cell electrolyte, always use distilled, demineralized, or deionized water. Rainwater is acceptable, but under no circumstances use faucet or tap water. Faucet or tap water generally has an excessive mineral content or other impurities that may pollute and damage the cells. Impurities introduced into the cell will remain, and concentrations will accumulate at each top-up, reducing service life. Long and reliable service life is essential, so the correct water must always be used. Water purity levels are defined in various national standards.

**Table 2-6. Electrolyte Correction
at 68°F (20°C)**

Temperature	Correction Value
23°F (–5°C)	deduct 0.020
32°F (0°C)	deduct 0.016
41°F (5°C)	deduct 0.012
50°F (10°C)	deduct 0.008
59°F (15°C)	deduct 0.004
77°F (25°C)	add 0.004
86°F (30°C)	add 0.008
95°F (35°C)	add 0.012
104°F (40°C)	add 0.016

2.13   Battery Plate Sulfation. Sulfation is the single greatest cause of flooded-cell battery failure. The destructive and battery life shortening process is as follows.

a.   **Discharge.** During discharge, the chemical reaction causes both plates to convert to lead sulfate ($PbSO_4$). If recharging is not carried out within a couple of hours, the lead sulfate starts to harden and crystallize. This is characterized by white crystals on the typically brown plates and is almost nonreversible. If a battery is only 80% charged, this does not mean that only 20% is sulfating; the entire plate material has not fully converted and subsequently sulfates. Lead sulfate that is formed in the cell discharge reaction crystalizes to create larger surface areas of lead sulfate crystals, which are called hard lead sulfate. This inhibits and blocks the conductive path required for cell recharging. Sulfation has a tendency to deform the plates.

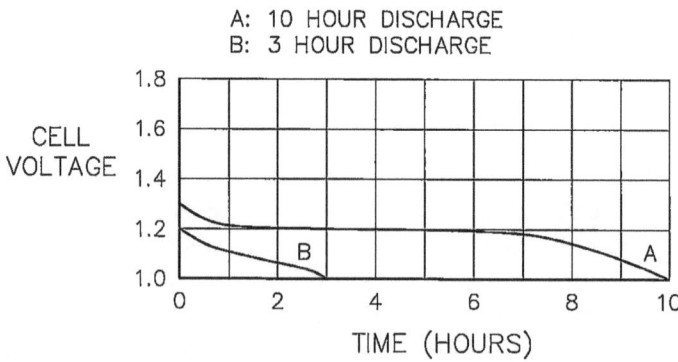

Figure 2-4. Battery Discharge Rates

18

b.   **Sulfation.** The immediate effect of sulfation is partial and permanent loss of capacity as the active materials are reduced. Electrolyte density also partially decreases, as the chemical reaction during charging cannot be fully reversed. This sulfated material introduces higher resistances within the cell and inhibits charging. As the level of sulfated material increases, the ability of the cell to retain a charge is reduced and the battery fails. The deep-cycle battery has unfairly gained a bad reputation, but the battery is not the problem; improper and inadequate charging is. As long as some charging is taking place, even from a small solar panel, a chemical reaction is taking place and sulfation will not occur.

c.   **Efficiency.** Battery efficiency is affected by temperature. At 32°F (0°C), efficiency falls by 60%. Batteries in warm tropical climates are more efficient, but they may have reduced life spans. Batteries in cold climates have increased operating lives but are less efficient.

d.   **Self-Discharge.** During charging, a small quantity of antimony (Sb) or other impurities dissolve out of the positive plates and deposit on the negative ones. Other impurities are introduced with impure topping-up water and deposit on the plates. A localized chemical reaction then takes place, slowly discharging the cell. Self-discharge rates are affected by temperature, with the following results:

   (1)   At 32°F (0°C), self-discharge rates are minimal.

   (2)   At 86°F (30°C), self-discharge rates are high; the specific gravity can decrease by as much as 0.002 per day, typically up to 4% per month.

   (3)   The use of a solar panel, wind generator, or regular and complete recharging will prevent permanent damage, as it can equal or exceed the self-discharge rate.

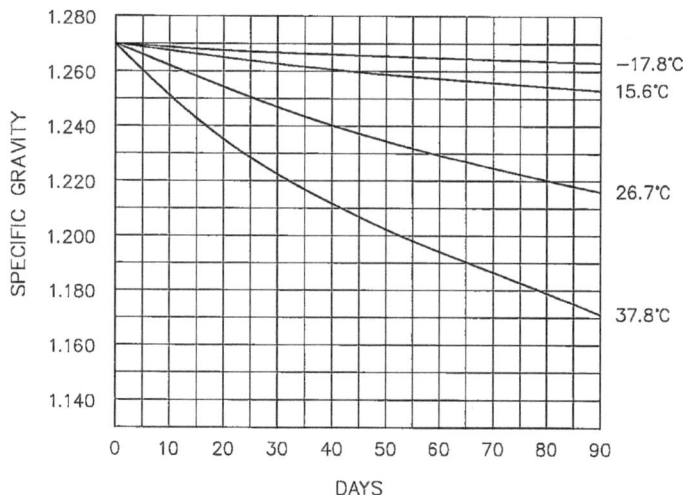

Figure 2-5. Battery Self-Discharge Rates

19

2.14   Battery Voltages and Installation. Batteries must be installed correctly. There are a number of important criteria to consider when installing battery banks to make up the required voltage and capacity.

a.   **Cell Size.** Battery banks may be installed in cell multiples of either 1.2V, 6V, 12V, 24V, 32V, 36V, or 48V. Each configuration has both physical and operational advantages:

(1)   **1.2V.** This is generally impractical from an overall size aspect. The battery plates are generally more robust and thicker. This leads to increased service life, but it is an expensive option, with installation space challenges.

(2)   **6V.** This is the ideal arrangement. The cells are far more manageable to install and remove. Large-capacity batteries are simply connected in series. Electrically, they are often better than 12V batteries and generally have thicker and more durable plates. Contrary to some opinions, a series arrangement does not reduce the available power range. If one battery requires replacement, the other should also be replaced simultaneously. The one proviso is that batteries must be of the same make, model, and age.

(3)   **12V.** This is the most common arrangement. Physically batteries up to around 115Ah are easily managed, and paralleled in banks of two or three is the most common arrangement. It is common to see traction or truck batteries of very large dimensions such as 8D sizes installed, and this is impractical from any service standpoint. If the battery space is constructed to take a three-battery arrangement, it is relatively easy to replace one unit. If you have a multiple bank and lose one with cell failure, two will remain.

(4)   **24V.** This is prevalent on larger vessels and is a standard voltage in commercial shipping control and backup power supply systems. This is simply any of the above battery or cell sizes connected in series to obtain the 24 volts.

(5)   **48V.** This voltage level is now being used for electric propulsion and many other systems. This entails a series battery connection of four 12V batteries to get 48V.

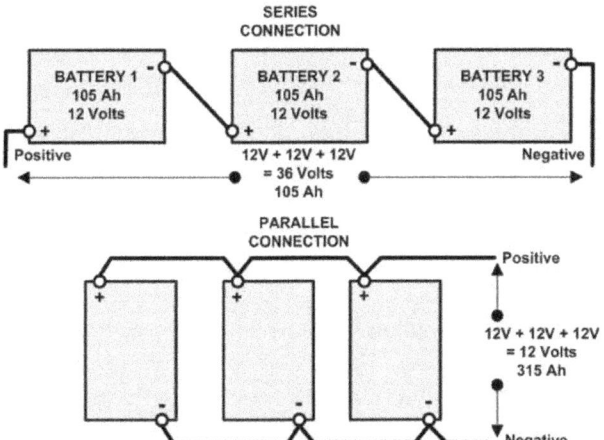

Figure 2-6. Cell and Battery Arrangements

2.15 **Battery Commissioning.** After installation, the following commissioning procedures should be carried out.

a. **Battery and DC System Safety.** Prior to performing any work and maintenance on a battery system, perform a basic risk assessment. Identify the hazards and assess the risks associated with the proposed task. These include electric shock, chemicals, arc flash, and thermal impacts.

b. **Battery Electrolyte Level.** Check the electrolyte level in each cell:

(1) Cells with separator guard, fill to the top of the guard.

(2) Cells without guard, fill to 2mm above plates.

c. **Battery Electrolyte Filling.** If the level is low, and evidence suggests a loss of acid in transit, refill with an electrolyte of similar density. Specific gravity is normally in the range of 1.240 to 1.280 at 59°F (15°C). If no evidence of spillage is apparent, top up electrolyte levels with demineralized, deionized, or distilled water to the correct levels.

d. **Battery Terminals.** Battery terminals are a simple piece of equipment, yet they cause an inordinate number of problems. When servicing the battery isolate so no loads are on the battery to prevent sparking when disconnecting terminal cables:

(1) **Install terminals.** Install heavy-duty marine-grade brass terminals. Do not use the cheaper plated brass terminals—they are not robust and fail quickly. Ensure that terminal posts do not have any raised sections and are not deformed, as a poor surface contact and high resistance connection will result.

(2) **Clean terminals.** Remove battery terminals and ensure that terminal posts are clean and free of corrosion deposits. Baking soda is

a good cleaner. You can clean terminals with a battery brush. Refit and tighten terminals and coat with petroleum jelly, not grease; see the terminal protection notes below.

(3)   **Replace connections.** Replace the standard wing nuts on these terminal types with stainless steel nuts, washers, and spring washers. The wing nuts are very difficult to tighten properly without deformation and breakage. I have encountered many installations where the wing castings or ears are broken.

(4)   **Terminal protection.** There are special battery terminal coatings such as CRC Battery Terminal Protector and a range of similar products. There are also wax based battery terminal sealants. It is a good idea to install PVC insulative battery terminal covers or color-coded protectors. These are available in a wide variety of configurations to suit most terminals and help protect against accidental short circuit. There are also felt terminal protectors that are installed on the posts under the terminals for corrosion protection and are impregnated with corrosion-inhibiting chemicals.

e.   **Battery Cleaning.** Cleaning involves the following tasks:

(1)   **Clean surfaces.** Regularly clean the battery surfaces with a clean, damp cloth. A solution of baking soda and water will neutralize acid spray residues. Moisture and salt can allow tracking across the top from the positive to the negative pole, slowly discharging the battery. This is a common cause of dead and discharged batteries, and those mysterious but untraceable system "leaks." Do this for all battery types.

(2)   **Remove grease and oil.** Grease and oil can be removed with a mild detergent and cloth.

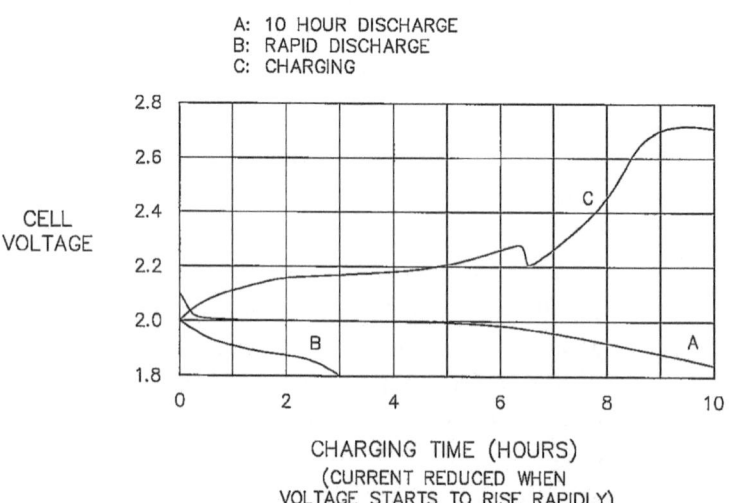

Figure 2-7. Lead-Acid Battery Characteristics

f. **Battery Charging.** After taking delivery of a new battery, perform the following:

   (1) **Initial Charge.** Give a freshening charge immediately.

   (2) **Routine Charging.** Give a charge every week if the vessel is incomplete or not in service.

2.16 **Battery Routine Testing.** The following tests can be made on a weekly basis to monitor the condition of the battery. Battery status can be measured by checking the electrolyte density and the voltage as follows:

a. **Stabilized Voltage Test.** Voltage readings should be taken with an accurate voltmeter. Switchboards should incorporate a high-quality meter, not a typical engine gauge charge indicator. The difference between fully charged and discharged is less than 1 volt, so accuracy is essential. A digital voltmeter is the ideal. Battery voltage readings should only be taken a minimum of 30 minutes after charging or discharging. Turn off all electrical loads before measuring. Typical values at 59°F (15°C) are shown in Table 2-7. Manufacturers have slightly varying densities, so check with your supplier.

Table 2-7. Typical Open Circuit Voltages and Densities

Charge Level	SG Temperate	SG Tropical	Voltage
100%	1.250	1.240	12.75
90%	1.235	1.225	12.65
80%	1.220	1.210	12.55
70%	1.205	1.195	12.45
60%	1.190	1.180	12.35
50%	1.175	1.165	12.25
40%	1.160	1.150	12.10
30%	1.145	1.135	11.95
20%	1.130	1.120	11.85
10%	1.115	1.105	11.75
0%	1.100	1.090	11.65

b. **Battery Electrolyte Specific Gravity.** A hydrometer should be used weekly to check acid density. The hydrometer is essentially a large syringe with a calibrated float. The calibration scale is corrected to a nominal temperature value, which is normally marked on the float. The following points should be observed during testing with a hydrometer:

   (1) **Testing.** Never test immediately after charging or discharging. Wait at least half an hour until the cells stabilize; this is because it takes time for the pockets of varying electrolyte densities to equalize. Never test immediately after topping up the electrolyte. Wait until after a charging period, as it takes time for the water and acid to evenly mix.

(2)   **Inspection.** Ensure that the float is clean and not cracked and the rubber has not deteriorated. Keep the hydrometer vertical. Ensure that the float does not contact the side of the barrel, which may give a false reading. Draw sufficient electrolyte into the barrel to raise the float. Ensure that the top of the float does not touch the top of the barrel. Observe the level on the scale. Disregard the liquid curvature caused by surface tension. Adjust your reading for temperature to obtain the actual value.

(3)   **Cleaning.** Wash the hydrometer with clean water when finished.

c.   **Battery Load Test.** The load test is carried out only if the batteries are suspect. The load tester consists of two probes connected by a resistance and a meter. The tester is connected across the battery terminals, effectively putting a heavy load across it. A load of approximately 275A at 8V is normal. Take your battery to your nearest automotive vehicle electrician or battery service center for a test.

2.17   Battery Additives. There are a number of chemical additives on the market. The claims made by manufacturers offer significant performance enhancement. The chemical compounds magnesium sulfate ($MgSO_4$), Epsom salts, and caustic soda are specifically designed to prevent sulfation or dissolve buildups from the plate surfaces on aging lead-acid batteries to boost declining performance. Battery plates are shedding material throughout their life cycle, and this often settles at the bottom between separators and eventually short-circuits the cell. Chemical additives are not able to replace active plate material or repair cracked or split plates. Ultimately it is a stopgap measure and may buy you a couple of months in a seriously sulfated battery. As flooded lead-acid batteries are gradually headed the way of the dodo, with newer battery technologies and maintenance-free units, there is less demand for this methodology.

Table 2-8. Lead-Acid Battery Troubleshooting

Symptom	Probable Fault
Will not accept charge	Plates heavily sulfated Maximum battery life reached Cell plates shorting Incorrect charge voltage Excessive deep discharge Electrolyte low specific gravity High resistance at battery terminals
Low cell electrolyte SG	Cell sulfated Cell failure Heavy cell sulfation

Symptom	Probable Fault
Low battery SG value	Low charge level (regulator failure) Plates sulfated Temperature affected Cell failing
Will not support load	Low charge level (undercharging problem) Plates sulfated One cell failed Excessive load High resistance at terminals
Cell failure	Electrolyte contaminated (impure water) Overcharging problem (regulator failure) Undercharging problem (regulator failure) Excess vibration and plate damage
Cell internal short circuit	Plates structurally damaged Plates sulfated Excessive charge current (regulator failure) Cells damaged

2.18   Low-Maintenance Batteries. Sealed low-maintenance batteries are not generally suited to most boat or cruising vessel applications. Frequently, they are installed without considering their performance characteristics or their various advantages and disadvantages.

a.   **Low-Maintenance Principles.** Basic chemical reactions are similar to the conventional lead-acid cell. The differences are as follows:

(1)   **Lead-Acid Batteries.** In a normal lead-acid battery, water loss occurs when water is decomposed into oxygen and hydrogen close to the end of charging. In any battery during charging, oxygen will develop at the positive plate at approximately 75% of full charge level. Hydrogen is generated at the negative plate at approximately 90% of full charge. These are the bubbles seen in the cells during charging. In normal batteries, the gases disperse to the atmosphere, resulting in electrolyte loss that requires periodic water replacement.

(2)   **Low-Maintenance Batteries.** The low-maintenance recombinational battery has different characteristics. The plates and separators are held under pressure. During charging, oxygen is able to move through the separator pores only from positive to negative, reacting with the lead plate. The negative plate charge is then effectively maintained below 90%, inhibiting hydrogen generation.

b.   **Low-Maintenance Battery Safety.** Batteries are totally sealed but incorporate a safety valve. Each cell is sealed, with a one-way vent. When charging commences, oxygen generation exceeds the recombination rate, and the vents release excess pressure within the battery. Excessive charge rates can create internal pressure buildup. If the pressure exceeds the safety vent's discharge rate, an explosion can occur.

c.   **Charging.** Low-maintenance batteries must be charged only at recommended charging rates and charge starting currents. The result of any overcharging may be excessive gas generation and a possible explosion.

d.   **Advantages.** The following are advantages of low-maintenance batteries:

    (1)   **Low Water Loss.** Low water loss is the principal advantage; however, performing a routine monthly inspection and occasional topping up of a lead-acid battery is neither labor intensive nor inconvenient. This factor is the main advantage put forward for these batteries. If you are continually topping up, you have a charging problem or a high ambient temperature.

    (2)   **Inversion, Heel, and Self-Discharge.** The batteries are safe at inversion or excessive heel angles without acid spilling, and they have a low self-discharge rate.

e.   **Disadvantages.** Two major disadvantages make low-maintenance batteries unsuitable for cruising applications:

    (1)   **Overvoltage Charging.** Low-maintenance batteries are incapable of withstanding any overvoltage during charging. If they are subjected to high charging voltages above 13.8V, water will vent out, and they have been known to explode. This means no fast-charging devices should be installed to charge them.

    (2)   **Cycle Availability.** Cycle availability is restricted, and an approximate lifespan of 500 cycles to 50% of nominal capacity is typical. Any discharge to 40% of capacity or less makes recharging extremely difficult, if not impossible, and requires special charging techniques.

2.19   **Gel Cell Batteries.** These are dryfit batteries manufactured by Sonnenschein. A quality deep-cycle lead-acid battery can have a life exceeding 2,500 cycles of charge and discharge to 50%. A gel cell has a life of approximately 800 to 1,000 cycles. They have a much greater cycling capability than normal starting batteries but not good deep-cycle or AGM batteries.

a.   **Gel Battery Electrolyte.** Unlike normal lead-acid cells, the gel cell has a solidified thixotropic gel as an electrolyte, which is locked into each group of plates. The gel electrolyte has a high viscosity and develops voids and cracks during the charge and discharge process. These can impede the flow of acid and cause capacity loss. During charging the gel liquefies due to its thixotropic properties. Solidification after charging can exceed an hour, as thixotropic gels have a reduced viscosity under stress. The newer types use phosphoric acid (H_3PO_4) in the gel to retard the sulfation hardening rates. Water loss can occur in valve regulated lead acid (VRLA) batteries, and the oxygen recombination cycle is used to minimize electrolyte loss. Water loss is small for each source; however, cumulatively they can cause failure. The process is called dry-out. Water loss causes include reduced recombination efficiency. This is due to high charge voltages, corrosion of the positive grid, and transpiration through the cell casing. Typically, this is in temperatures exceeding 104°F (40°C), and self-discharge in temperatures greater than 68°F (20°C). Batteries should be installed in cool areas wherever possible.

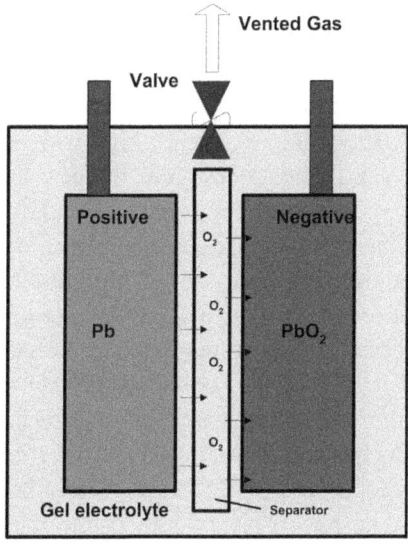

Figure 2-8.
Gel Battery
Characteristics

b. **Construction.** The plates are reinforced with calcium (Ca) rather than antimony (Sb), which reduces self-discharge rates, and they are relatively thin. This facilitates gel diffusion and improves the charge acceptance rate, as diffusion problems are reduced. The separator provides electrical and mechanical isolation of the plates, and must have a high porosity to facilitate ion migration and electrolyte acceptance. Each cell has a safety valve to relieve excess pressure if the set internal pressure is exceeded. Typical values are 114psi (100mbar). The valve recloses tightly to prevent oxygen from entering the cell. Lead-acid batteries suffer from corrosion of the current collector grid. Typical grids in VRLA batteries are manufactured from lead-calcium-tin (Pb-Ca-Sn), lead-tin alloys, or lead-antimony-cadmium (Pb-Sb-Cd). Positive plate grids corrode due to the conversion of lead (Pb) into lead dioxide (PbO_2). This corrosion effect effectively doubles for every 18°F (10°C), as well as charge voltages and electrolyte density. As the lead dioxide requires a substantially higher volume, this causes mechanical stressing and deformation of the grid. The effect is to degrade the active plate material contact with the grid, causing capacity loss, internal short circuits, and even cell case ruptures, which increases with service time. The process consumes water, reducing the electrolyte and causing lowered performance. VRLA battery cells are prone to a chemical degradation of the negative plate lugs and strap surfaces called sulfate rot. This is caused by the oxygen recombination reaction and electrolyte inorganic sulfate salts.

2.20   Gel Cell Charging. Batteries have a much higher charge acceptance rate, and therefore a more rapid charge rate is possible. Gel cells cannot tolerate having any equalizing charge applied, and this overcharge condition will seriously damage them. During charging, the current causes decomposition of the water and the evolvement of oxygen at the positive plate. The oxygen diffuses through the unfilled glass mat separator pores to the negative plate and chemically reacts to form lead oxide, lead sulfate, and water. The charge current then reduces and does not evolve hydrogen gas. The end of charge voltage is typically 2.22vpc (volts per cell) to 2.28vpc. If recombination of hydrogen is incomplete during overcharge conditions, the gases may vent to the battery locker, creating an explosion risk. Although accepting a higher charge rate than a lead-acid deep-cycle battery, and consequentially charging to a higher value, there is, at a certain point, the problem of attaining full charge and therefore capacity usage of the battery bank. As no fast-charge devices can be safely used, a longer engine run time is required for complete recharging. While these batteries will accept some 30% to 40% greater current than an equivalent lead-acid battery, they are restricted in the voltage levels allowed. Typical open-circuit voltage at 100% charge is 12.85V, 75% is 12.65V, 50% is 12.35V, and 25% is 12V; 11.8V is completely discharged. Gel cells are intolerant to overvoltage charge conditions and can be damaged in overcharge situations. The normal optimum voltage tolerance on dryfit units is 14.4V. There are some minimal heating effects during charging, caused by the recombination reaction. Where batteries are not kept within reasonable temperature ranges, thermal runaway can occur. This normally occurs during charging when the temperature of the battery and charge current create a cumulative increase in temperatures that leads to battery destruction. Continuous over- or undercharging of gel cells is the most common cause of premature failure.

2.21   Absorbed Glass Mat (AGM) or Valve Regulated Lead-Acid (VRLA) Batteries. I have installed these on my own boat, and the performance has been excellent. There are variations to flat plate manufacturing techniques, and the Optima AGM batteries have a spiral cell, dual plate construction. Another important feature is a greater shock and vibration resistance than gel or flooded cell batteries. They are available with very high CCA values of up to 800A at 0°F (–18°C).

    a.   Recombinant Gas Absorption Principles. In a normal lead-acid battery, water loss will occur when it is electrochemically decomposed into oxygen and hydrogen near the end of charging. In a battery during charging, oxygen will evolve at the positive plate at approximately 75% of full charge level. Hydrogen evolves at the negative plate at approximately 90% of full charge. In normal batteries, the evolved gases disperse to the atmosphere, resulting in electrolyte loss and periodic water replacement. These are the bubbles seen in the cells during charging. During charging the current causes decomposition of the water, and oxygen is evolved on the positive plate. The oxygen then migrates through the unfilled pores of the separator matting to react with the negative plate and form lead oxide (PbO), lead sulfate ($PbSO_4$), and water (H_2O). The charge current reduces and does not generate hydrogen. The plates and separators are held under pressure. During charging, the evolved oxygen is only able to move through the separator pores from positive to negative, reacting with the lead plate. The negative plate charge is then effectively maintained below 90%, inhibiting hydrogen generation. These batteries emit less than 2% hydrogen gas during severe overcharge, where 4.1% is the flammable level.

b. **AGM Electrolyte.** The electrolyte is held within a very fine microporous, boron-silicate glass matting that is placed between the plates. This absorbs and immobilizes the acid while still allowing plate interaction. They are also called starved electrolyte batteries, as the mat is only 95% soaked in electrolyte.

c. **AGM Charging.** Charging of AGM cells has few limitations, and no special charge settings are required. Typical charge voltages are in the range 14.4V to 14.6V at 68°F (20°C). The batteries have a very low internal resistance, which results in minimal heating effects during heavy charge and discharge. They can be bulk charged at very high currents, typically by a factor of 5 over flooded cells and by a factor of 10 over gel batteries. They allow 30% deeper discharges and recharge 20% faster than gel batteries, and they have good recovery from full-discharge conditions. Self-discharge rates are only 1% to 3%. If you are a weekend, harbor, or river cruiser and have limited motoring periods or leave the boat unattended for long periods, the AGM battery is a viable proposition; it has very low self-discharge rates and very high recovery rates from deep discharges. If a solar panel or wind generator is left connected with a suitable regulator, these batteries will recoup the annual cost of replacing deep-cycle batteries by lasting several seasons, with the more important improvements in reliability. Typical charge voltage levels are 12.8 to 12.9V at 100%, 12.6V at 75%, 12.3V at 50%, and 12V at 25%; 11.8V is discharged. At high temperatures, AGM batteries and gel cells are unable to dissipate the heat generated by oxygen and hydrogen recombination. This can create thermal runaway and will lead to gassing and the drying out of cells. A premature loss of capacity can occur when the positive plate and grids degrade as a result of higher operating temperatures due to the exothermic recombination process and higher charge currents. In addition, negative plates degrade because of inadequate plate conversion. The main failure modes are cell shorting and pressure vent malfunctions caused by manufacturing faults, the latter being relatively uncommon.

2.22   Optima Batteries. These batteries deserve their own chapter. Among the most advanced absorbed glass mat (AGM) batteries, they are used in commercial maritime and offshore systems. The Optima AGM battery is constructed with spirally wound cells. The Spiralcell Technology utilizes a series of individual spiral-wound cells composed of two pure (99.99%) lead plates with a precise coating of lead oxide. This contributes to a lower internal resistance, allowing the Optima batteries to handle high discharge and recharge rates. This means increased cranking amps and much faster recharge rates. The lead plates and separator are wound and then tightly compressed into a cell tube. These tightly-coiled AGM separators, which are set between each cylindrical cell, hold the electrolyte fluid similar to a sponge, which eliminates acid spillages. This prevents any movement, plate shedding, and breakage or fractures, even when subjected to severe shock and vibration. I have read test results where an Optima battery was subjected to 5G vibrations for 12 hours and it still stayed operational. One of the positive factors of this battery type is that there is no free-flowing acid that can leak or spill from the cells, so operation at any angle of heel, even inverted, has no effect on performance. Battery charging offers significant advantages over the slow charging of flooded-cell batteries. When charging an Optima battery using a 100A alternator charging into a 100% discharged battery (my worst nightmare) that is nominally at 10.5V, complete recharging will take approximately 35 minutes. This test was for an Optima D27F deep-cycle battery. Optima deep-cycle batteries are rated in amp-hours (C20 rate). Many deep-cycle batteries are often rated at 100Ah, but in reality, it is possible only to discharge 70% of this value. With Optima deep-cycle batteries, you can discharge 100%, although I do not recommend doing that.

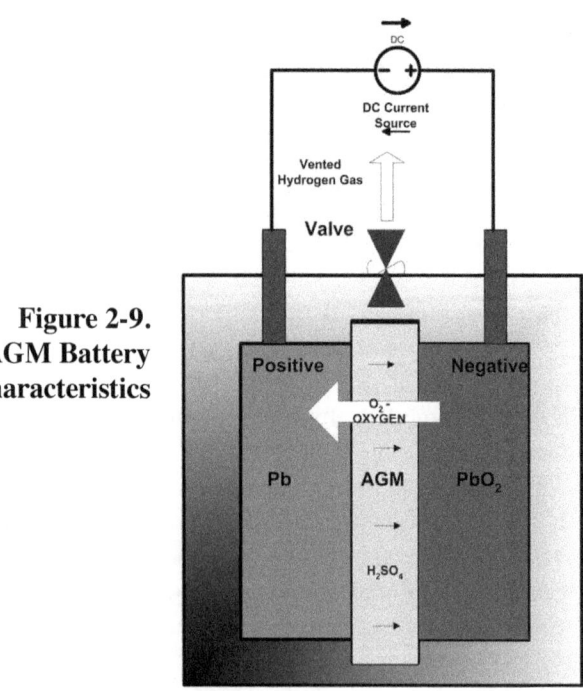

**Figure 2-9.
AGM Battery
Characteristics**

2.23 **Lithium-ion Batteries.** These batteries are rapidly entering mainstream service on boats. Lithium-ion–based battery technology is revolutionizing the mobile power market. They are a viable alternative to lead acid, AGM, and gel batteries. The batteries are available in a variety of voltages, including 12V, 24V, 36V, and 48VDC. Lithium batteries quote peak current ratings. Typically, this is 68°F (20°C) for 5 to 10 seconds. There are many advantages, including almost no power loss when discharging to near 100%; the entire battery capacity is available. Lithium batteries possess a very high energy and power density. Other advantages include much lighter weights and having approximately 500% greater cycling capability than an AGM. Drawbacks are that these batteries are not generally suitable for engine starting and that there is a cost premium of about 100% to 300% or more. This is easily amortized by significantly longer life, much greater cycle life, and a power discharge and charge profile that is much friendlier to boating applications. In late 2022 the UK company Aceleron launched the Essential 2.0 LFP battery, which is a Group 32 standalone LFP battery. This incorporates a battery management system (BMS) that has Bluetooth and controller area network (CAN bus) connectivity that allows network connection and remote monitoring of charge and discharge, cycle numbers, and internal cell temperatures. They have designed ergonomically correct carry handles, rubber feet to cushion against vibration, and cobalt-free lithium–iron phosphate technology. It has a peak power rating of more than 200A and is guaranteed for 5,000 cycles. The American Boat and Yacht Council (ABYC) published revised lithium-ion battery recommendations in Standard E-13 in 2023. In 2022 the United States Coast Guard (USCG) requested that the ABYC test lithium-ion batteries and to try to replicate many of the reported issues and failures. They were unable to re-create any failures, regardless of how aggressive the testing was and have reportedly stated that lithium-ion batteries are safe. Following several major incidents, the jury is still out within the commercial maritime space.

a. **Lithium-ion Battery Construction.** Lithium-ion batteries use lithium cobalt dioxide ($LiCoO_2$) or lithium manganese oxide ($LiMn_2O_4$) as a cathode. A lithium-iron battery uses lithium (Li), iron (Fe), phosphate (PO_4), or $LiFePO_4$, as the cathode. Like most battery technologies, the battery cell comprises two electrodes—the cathode and anode—with an insulative separator or barrier between them. The anode stores the lithium and is made from graphite carbon with a metallic backing. The cathode stores lithium and is manufactured with a metal oxide chemical compound ($LiFePO_4$).

b. **Lithium-ion Electrolyte.** The electrolyte has a key role within the battery cell, transporting positive lithium ions between the cathode and anode. The electrolyte used within lithium battery technologies comprises a very high purity lithium salt, lithium hexafluorophosphate ($LiPF_6$). This white crystalline powder is dissolved within an organic carbonate solvent, or nonaqueous solution. Several other chemical additives are used in the electrolyte to attain the required electrolyte properties.

c. **Lithium-ion Battery Discharging.** How much power can be delivered throughout the discharge cycle is the key to power supply quality and reliability. It should be noted that batteries can be damaged when overdischarged. This is usually a result of cumulative small residual or parasitic loads, such as devices that run all the time.

Table 2-9 illustrates typical voltage levels at various discharge levels of different battery technologies.

Table 2-9. Battery Discharge Comparisons

State of Charge %	Lead Acid Voltage	AGM/Gel Voltage	Lithium-ion Voltage	Comments
100	12.7	12.80	14.40	Effective power in bold
90	12.6	12.64	13.40	Effective power in bold
80	12.42	12.50	13.30	Effective power in bold
70	12.34	12.37	13.25	Effective power in bold
60	12.15	12.27	13.20	End of effective capacity
50	11.96	12.01	13.10	End of effective capacity
40	11.81	11.81	13.10	Serious damage level
30	11.65	11.66	13.00	Serious damage level
20	11.39	11.12	12.90	Permanent damage level
10	10.94	10.50	12.00	Permanent damage level

2.24 Lithium-ion Battery Charging. The lithium ions pass through the separator and block the passage of the electrons. In the charging cycle, the ions pass from positive to negative; conversely, in the discharge cycle, they move from negative to positive. This movement of ions creates a potential difference and what we know as voltage. Lithium-ion batteries have a nominal voltage of 3.7vpc. Like other battery types, cells are connected in series to achieve the required battery bank voltage. That means four cells for a 14.8V battery. The cells are placed in parallel to increase the amp-hour capacity required. These batteries have a capacity rating of 1C, which means that a fully charged battery with a nominal capacity of 100A can discharge 100A in 1 hour, or 10Ah is 10A in 1 hour. Maximum discharge rates are usually rated at 2C. Like all batteries, regular charging and discharging is the best path to reliability. The average cell voltage across the discharge range is 3.6V to 3.7V; full charge is 4.2vpc, 3V when at minimum level. Running them down to 2.5vpc will damage them. A fully charged LiFePO$_4$ battery will have a stable voltage of 13.3V to 13.4V; the lead-acid battery will be around 12.6V.

2.25 Lithium-ion Battery Installation. Do not install these batteries with other battery types, and do not be tempted to connect one in parallel with an AGM or other battery type, as there is an explosion and fire risk. Like all batteries, they do not tolerate rough mistreatment; absolutely do not short-circuit one. Do not overcharge them or subject them to reverse polarity. Do not crush or fracture the casing, and do not attempt to disassemble one—you may experience chemical burns as well fire and explosion. Lithium-ion battery age and life expectancy is subject to the temperature and the state-of-charge (SOC). The higher the temperature, the faster these batteries will age; ideally, the cooler they are kept, the better they will age.

2.26   Lithium-ion Battery Storage. These batteries are intolerant to temperatures exceeding 140°F (60°C) and should always be stored in a cool place. The nominal range is –4°F to 140°F (–20°C to 60°C); storage at temperatures below 68°F (20°C) will result in permanent capacity loss. If you are laying up your boat—as is common through many US and European winters in countries that experience subzero or temperatures well below 68°F (20°C)—it would be advisable to take the batteries out and store them somewhere at the house. There are SOC and voltage advisories as well. Short-term storage is recommended in the range of 3.0V to 4.2vpc in series. For long-term winter layups, this is at about 70% to 80% of battery capacity or less with a voltage level of 3.85V to 4.0V. The battery will lose storage capacity if it is maintained at 100%. They do not like being run down below the minimum voltage, which is between 2.4V and 3.0vpc.

2.27   Thermal Runaway. A frequent cautionary or advisory note is that lithium-ion batteries have a flammable electrolyte and are at risk of what is known as "thermal runaway." In certain situations, a lithium-ion battery can suffer very rapid internal heating, and once this exothermic reaction is initiated, it is hard to extinguish, contain, or stop. The initial cause of thermal runaway is an internal short circuit. Another potential cause is very fast charging and discharging, which generates heat. Thermal runaway can also be caused by damage to a lithium-ion battery due to careless and rough handling. Maintaining the battery bank within its nominal temperature range is important; therefore, effective ventilation is necessary. Avoid overcharging by only using charging systems, fast charge regulators, and battery chargers that are suitable for use with lithium-ion batteries; many are not. This is one of the reasons that a BMS should be installed at all times. The BMS monitors and shuts off when it detects either high or low voltage limits, high or low temperature limits, and when current charge limits are exceeded, both charging and discharging. The BMS should monitor these parameters at cell level. Only install batteries that are from manufacturers with integral cell-level BMS monitoring systems.

2.28   High Load Applications. Some equipment manufacturers specify only certain lithium-ion battery makes and models. Andersen is one of them for its compact deck winches. They only accept Super B, Mastervolt, and Victron Energy (MG Energy Systems) batteries. These batteries have undergone compatibility testing and have integral protection to prevent battery, motor, and systems damage. The use of these batteries in start battery applications has not been recommended. In late 2022, Mercury Marine approved the use of one battery type for starting outboard engines. That battery was the RELiON Model RB100-HP. This is a LiFePO$_4$ battery with a minimum cranking amp rating of 800A for 8 seconds minimum at 20°F (–7°C). It has a peak acceptance charge of 165A for 1 minute, a maximum alternator charge rating of 150A, and a maximum charge voltage of 14.8V. Many engine makers advise that using a lithium-ion battery that is not approved will void the warranty.

2.29   Battery Management System (BMS). Many lithium-ion batteries incorporate an integrated BMS, and many are very comprehensive and sophisticated. The term "management" is the operative one, as these systems are also built-in protection devices. These systems are able to carry a 100A continuous load with a 200A surge capacity for up to 30 seconds. The systems also have high and low voltage protection along with short-circuit protection, as well as high and low temperature protection and automatic cell balancing. BMS modules may incorporate cell balancing circuits that balance series-connected sections during charging and discharge. Victron Energy has the Lynx Smart Battery Management System. The Lynx BMS is designed for lithium-ion batteries and incorporates state of charge monitoring, alarms, Bluetooth connectivity, and remote monitoring capability. Integrated battery management systems are now being offered by many manufacturers. BEP uses a system called Smart Battery Hub, which is an intelligent battery management system for multiengine installations. Smart Battery Hub operates with remotely activated switches, automatic voltage sensitive switching, and emergency parallel functionality. The Lithionics NeverDie BMS includes battery state-of-health, status, fault codes, and SOC monitoring. The patent-pending BMS utilizes customized microprocessors and firmware that enable customization of the BMS to perform as a programmable logic controller (PLC). The NeverDie BMS is standard on all Lithionics Battery systems to guarantee that the lithium-ion batteries perform within rated specifications. Optional Bluetooth monitors battery voltage, SOC, temperature, current, and status codes from a mobile device. The optional Bluetooth features include a live telemetry data feed and much more. A free app is available at the Google Play and Apple app stores. The Navico Group has introduced the Fathom e-power system, an integrated lithium-ion power-management system that facilitates the operation of multiple onboard systems. The system package, which includes lithium-ion batteries along with sensors, switches, and controllers, is a product of Navico's vertical integration that includes companies within the group, such as Mastervolt, BEP, CZone, Ancor, and Blue Sea Systems. The system has an intuitive user interface that allows monitoring and control of power consumption using multifunction displays or an app for smart mobile devices. Check out lithionicsbattery.com, www .victronenergy.com, and www.navico.com/fathom.html.

2.30   Carbon Foam Batteries. Carbon foam batteries, another relative newcomer in the battery technology revolution, are a viable alternative to lithium-ion batteries in some applications. They offer several advantages, including far greater energy density, increased cycle life, and more efficient charging. Manufacturers include Ocean Planet Energy (Firefly Battery Company), a leading innovator in this space (fireflyenergy.com).

    a.   Carbon Foam Construction. Carbon foam batteries are constructed using a composite grid made of carbon-based porous foam. This foam consists of hundreds of thousands of highly porous spherical microcells. The structure allows enhanced efficiency of the lead acid reaction, as each microcell increases the active material area given the increased surface area and has 70% or greater porosity. This results in very high levels of thermal and electrical conductivity due to the high surface area–to-volume ratio. The cell anode is made of antimonide—copper blended with antimony—and this is electroplated onto the carbon foam. The carbon foam has a significant amount of air, but each bubble has a very large surface area. These discrete 3-D microcells form an electrolyte structure not unlike a beehive's honeycomb lattice. The cathode is made of nickel oxyhydroxide ($NiO(OH)$).

    b.   Carbon Foam Chemistry. Carbon foam batteries utilize a rather unique working principle. They are in fact half battery and half capacitor. A capacitor is able to store a charge and, similar to a battery cell, has two metal plates separated by a dielectric. The chemical reaction creates energy and stores this energy in the capacitor section of the cell in the form of electrolyte ions, which are attracted to the carbon anode. The negative plate is resistant to sulfation, which improves longevity; the drawback is that batteries will not charge to full capacity.

c. **Carbon Foam Discharging and Charging.** Charging efficiency is increased, and this is quoted as around 50%. One leading manufacturer recommends a complete 100% charge weekly as well as a periodic restoration charge. These batteries have a low discharge rate, which is an advantage. Carbon foam batteries are able to withstand occasional fast charging; however, they will degrade when charged at high current levels. The carbon foam battery is able to use conventional battery charging sources.

d. **Cycling Ability.** There are claims that these battery types have a much longer lifespan than a standard flooded-cell lead-acid battery. The most common factor quoted is 200% to 400%, although this is hard to substantiate. Another claim is that they can be discharged to 80% and 100% of capacity without sulfation or capacity reductions. In my opinion, I don't think any battery should be discharged to that depth; deep discharges of any battery result in reduced life.

2.31   Graphene Batteries. Graphene is a form of carbon composed of two dimensional planar monolayer sheets just a single atom thick. These carbon atoms are configured in a honeycomb, hexagonal-shaped lattice arrangement. Graphene is the strongest, lightest, most electrically conductive substance there is. Grabat Energy, a Spanish company, has developed an innovative graphene polymer battery they have named the Grabat. They claim it is one-third the weight of a lithium-ion battery, with a 400% increase in energy density. Even more significant, these batteries can be recharged at a speed some 3,000% faster and have a significantly higher cycle ability. Watch this space!

2.32   Lead Carbon Batteries. Replacing the active material of the negative plate by a lead carbon composite potentially reduces sulfation and improves the charge acceptance of the negative plate. The advantages of lead carbon batteries include decreased sulfation when partially discharged. They have a lower charging voltage, which results in increased efficiency and reduced positive plate corrosion. The overall benefit is increased cycle life. Testing has shown that lead carbon batteries have a cycle life of 500 100% discharge cycles. Some manufacturers have claimed a cycle life of 2,000 90% discharge cycles. Real-world results are as yet unknown; however, the advances here are significant and make these batteries a viable option. The chemistry is interesting, and these battery types employ a standard lead positive electrode and a supercapacitor negative electrode. Theoretically, the lead carbon battery has an unlimited discharge and fast charge rate. This innovative and complex technology uses activated carbon and nano carbon (graphene) to optimize the electrochemical reaction between the electrolyte and active electrode material and prevent sulfation. Carbon has very good conductivity and capacitance characteristics. This combination of asymmetric supercapacitors and lead-acid cell processes results in significantly enhanced performance due to low internal resistance and energy conduction. Another advantage is that partial state of charge (PSOC) is no longer an issue with destructive plate sulfation. Most batteries must be fully recharged, which is not the case with this technology.

2.33   Alkaline Batteries. Alkaline cells are typified by nickel cadmium (NiCad) and nickel iron (NiFe) batteries. The components of the NiCad cell are nickel-hydroxide ($2Ni(OH)_2$), the positive plate; cadmium hydroxide ($Cd(OH)_2$), the negative plate; and potassium hydroxide (KOH), the electrolyte. The principal factors are cost (typically 500% greater), greater weight, and physically larger bank size. For these reasons, these batteries normally will be found only in larger motor and sailing vessels. They have completely different operating characteristics to the lead-acid cell. The obvious difference is the use of an alkaline electrolyte instead of an acid. Unlike lead-acid cells, plates undergo changes in their oxidation state, altering very little physically. As the active materials do not dissolve in the electrolyte, plate life is very long. The electrolyte is a potassium hydroxide solution with a specific gravity of 1.3. The electrolyte transports ions between the positive and negative plates, and the alkaline solution is chemically more stable than lead-acid cell electrolytes. Unlike lead-acid cells, the density does not significantly alter during charge and discharge, and hydrometer readings cannot be used to determine the state of charge. Electrolyte loss is relatively low in operation.

2.34   NiCad Battery Characteristics. The open-circuit voltage of a vented cell is around 1.28 volts. This depends on the temperature and time interval from the last charge period. Unlike a lead-acid cell, the voltage does not indicate the state of charge. The nominal voltage is 1.2V. This voltage is maintained during discharge until approximately 80% of the 2-hour rated capacity has been discharged. This is also affected by temperature and rate of discharge. The closed-circuit voltage is measured immediately after load connection. Typically, it is 1.25V to 1.28V volts per cell. The working voltage is the voltage observed on the level section of the discharge curve of a NiCad cell, voltage plotted against time. Typically, the voltage averages 1.22 volts per cell. Capacity is specified in amp-hours and is normally quoted at the 5-hour rate. The nominal rating is the amp-hour delivery rate over 5 hours to a nominal voltage of 1.0 volt per cell. Internal resistance values are typically very low. This is due to the large plate surface areas used, and is why the cells can deliver and accept high current values. A NiCad battery accepts high charge currents and will not be damaged by them. At 1.6 volts per cell, a NiCad can absorb up to 400% of capacity from a charging source. In most cases, it will accept whatever the alternator can supply.

2.35   Battery Installation. Batteries should be installed within a separate space or compartment that is located above the maximum bilgewater level and protected from mechanical damage. Flooded-cell lead-acid batteries should be installed in a lined box protected from temperature extremes. The preferred temperature range is 50°F to 80°F (10°C to 27°C). The box should be located as low down as possible in the vessel for weight distribution reasons, but high enough to avoid bilgewater or flooding. After a bad knockdown, and with water over the sole, many boats using flooded-cell batteries have compounded their problems by having the batteries contaminated with salt water; spillage occurs at angles exceeding 20 degrees. In the 1998 Sydney to Hobart yacht race when several yachts foundered with multiple fatalities, the inquest had several findings. One concerned flooded-cell lead-acid batteries. When some of the yachts capsized, rolled, or were knocked down, the sulfuric acid electrolyte spilled from the batteries. This mixed with saltwater and the evolved Chlorine (Cl_2) and Hydrogen Chloride (HCl) gas vapors adversely affected the crew. In addition, the power supply for communications and engine starting was also adversely affected.

This is less of an issue with today's widespread use of sealed batteries. On multihulls the batteries should either be located centrally in the mast area to center the weight or be divided into two banks, with one bank in each hull. This effectively gives two separate house banks plus two engine-start batteries. Sufficient natural light should be available for testing or servicing. If this is not possible, an ignition-protected light can be installed. Allow sufficient clearance to install and remove batteries. Ensure there is sufficient vertical clearance to allow hydrometer testing if you have flooded-cell batteries. Ideally, batteries should not be located within machinery spaces where they might be exposed to high ambient temperatures.

a. **Battery Location.** Batteries should not be installed adjacent to any fuel tank, fuel pipe, fuel filters, separators and valves, or other parts of the fuel system. Any leak or accumulation of fuel represents a serious hazard that can ignite. Battery compartments should not contain any electrical equipment capable of causing ignition of any generated gases or vapor. Sparks can be generated and cause ignition of any accumulated hydrogen gas generated after battery charging. Be aware that sealed batteries can emit gas in certain overcharge and overpressure situations.

b. **Battery Enclosure.** Batteries should be installed within an enclosure or have a tray that will contain any spills of electrolyte at all angles of heel or inversion. The box should be made of plastic, fiberglass, or lead lined to prevent acid spills from contacting wood or water. Boxes should be at least the full height of the battery so that any spills will be contained at all times. Polyvinyl chloride (PVC) battery boxes are acceptable alternatives. Sealed maintenance-free batteries can still be a hazard if a casing is damaged or cracked. Batteries should be protected from high temperatures. In the majority of boats, high temperature is not an issue, although insulating the battery box area is a good idea. This is very important in very cold temperature areas during the winter layup.

c. **Movement Protection.** Batteries should be protected from vibration and any movement. It is good practice to install a battery on a thick piece of neoprene rubber to absorb any vibrations. Insert rubber spacers around and under the batteries to stop any minor movements and vibrations. This is highly recommended when installing lithium-ion batteries. Batteries should be secured so that they do not move at any angle of heel or inversion. Physically secure batteries with either straps or a removable restraining clamp or rod across the top. Make sure you use a nonconductive clamp across the battery. The ideal configuration is to arrange flooded cell batteries athwartships. This offers marginally better protection against acid spilling under excess heel if you are using open-cell batteries. A friend had a gimballed tray on his steel cruising yacht to prevent electrolyte spills when heeling, although this is probably excessive. What is important is that all batteries are installed so they don't move under any angle of heel or a rollover, knockdown, and capsize event. Be careful not to overtighten any hold-down clamps, particularly with lithium-ion batteries, so the battery case is not distorted, cracked, or damaged. Overtightening can lead to internal short circuit and thermal runaway, as electrolytes may leak out. Note that lead-acid batteries can explode as well in overcharge situations with hydrogen gas generation and ignition. The color insert image shows improperly installed batteries, with no support, no restraints, degraded terminals, and other issues.

d.　**Battery Terminals.** Battery terminals and connections should be installed or protected against any accidental contact with metallic objects. Battery box lids should be in place at all times and secured. PVC or other connection covers should be installed where accidental contact by metallic tools or other items is possible. The prospective short-circuit current of a standard vehicle battery can be in the range 1,000A up to 20,000A. This is particularly important with lithium-ion batteries. Battery terminals can be coated with petroleum jelly or an equivalent compound to prevent corrosion or interaction with electrolyte spray. There are terminal torque recommendations, and these are typically in the range of 5 foot-pounds (7Nm) to 7.5 foot-pounds (10Nm). The color insert image shows a dry and tight battery terminal.

e.　**Ventilation.** Because batteries generate explosive hydrogen gas, the battery compartment should have adequate ventilation to vent to the atmosphere all evolved gases. An extraction fan is rarely required, but it should be considered if natural convection methods are insufficient. If a fast-charging device is installed, ensure that the ventilation is sufficient to remove any generated gases and prevent them from accumulating. Note that sealed batteries are still able to emit gas in overpressure situations, so the need for ventilation holds true for all battery types.

f.　**Battery Electrical Connections.** Where start and service batteries have an interconnecting switch for emergency power supply, the switch should be in the normally open position. I frequently have come across incidents where both the engine starting battery and the house power batteries were all completely discharged as a result of the switch not being in the proper position. Attach a warning label or tag to the switch as a reminder.

g.　**Battery Segregation.** Starting and house power service batteries should be electrically separate and arranged so that service loads cannot discharge the engine start battery. In most cases, the battery negatives are bridged, and a separate negative for each should be installed. This requires a separate alternator, charging negative and grounding negative to the same grounding point as the other battery. Any arrangement should ensure that the start battery cannot not be accidentally discharged. Where a solenoid, voltage dependent relay (VDR), or battery integrator system is used to parallel the batteries for charging, it must always open when the charging ceases. Battery interconnection cables should have the same rating as the main supply circuit cables. In a dual battery system, the cables connecting each battery negative or positive should be rated the same as the main supply cables. I have observed installations where battery interconnections are much smaller than the main cables. Equipment with high current ratings, such as thrusters and windlass systems, should be installed to limit the disturbances or effects on the stability of the electrical system. Where high-current equipment can cause system disturbances, such as large load surges and voltage droops, consideration should be given to installing a separate battery bank with the required characteristics to power the equipment. Alternatively, ensure that if they are powered off the engine starting battery, the battery rating is increased to supply the additional loads.

2.36   Device Batteries. Standard disposable batteries are zinc carbon, long-life alkaline, and super alkaline, lithium, silver-oxide, and zinc-air batteries. Rechargeable batteries are far more economical. The small nickel cadmium (NiCad) battery can be recharged several hundred times. A rechargeable battery should be completely discharged before recharging. The nickel metal hydride (NiMH) cell used in many cellular phones is an example of more recent battery technology. These cells can withstand recharging up to 1,000 cycles. They do not suffer with partial discharge and charge, although it is still good practice. Please do not dispose of batteries over the side; it is environmentally unsafe.

2.37   Lithium-Ion Portable Devices. There has been a spate of yacht fires associated with these batteries, resulting in a total loss of the boat. Lithium-ion battery-powered devices include rechargeable power tools, laptops, tablets, drones, smart watches, mobile phones, power banks, electric toothbrushes, portable speakers, e-bikes, and scooters (a popular travel mode on my large marina), rechargeable appliances, toys, tender and dinghy outboards, and so on. Some of these fires have initiated due to non-original charging devices, and most fires appear to occur when the device is left on charge. Recently a major fire started on an oil tanker when a handheld VHF radio with lithium-ion battery erupted while left on charge. For safety, I would recommend you do not leave your devices on charge when you leave the boat unattended or at night. Consider a timer on the charger power supply set to 3 hours maximum.

Battery Charging Systems

An efficient battery charging system is essential for optimum battery and electrical system performance. The principal charging systems on cruising yachts, trawler yachts, and powerboats consist of the following.

a. **Engine-Mounted Alternators.** The standard alternator along with the optional second alternator is the principal charging source on the majority of boats, including sailing yachts. In many cases, it is the only source utilized, even in a marina slip, due to the alternator's higher available charging currents. High-output battery chargers are expensive.

b. **Alternative Energy Charging Systems.** Alternative energy charging is available as an option to supplement engine charging sources. These include solar panels, wind generators, hydrogenerators, and fuel cell chargers. Read about these in chapter 4, "Alternative Energy Charging Systems."

c. **Battery Chargers.** When a boat is in a marina slip, particularly in liveaboard situations where the main power source is via a shore-powered charger, the battery charger has an important role in the power system. When paired with a diesel genset, battery chargers have a very crucial role in many powerboats, motorboats, and trawler yachts.

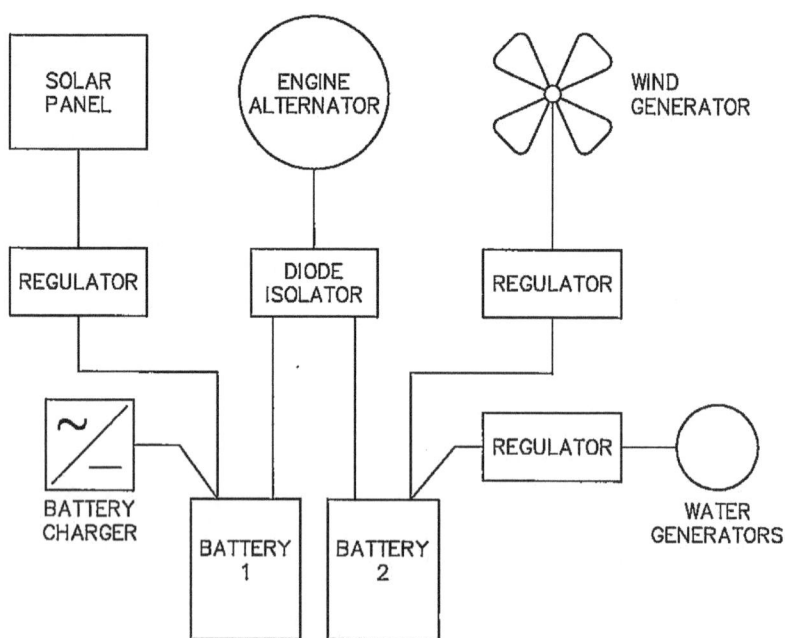

Figure 3-1. Battery Charging Systems

3.1   **Charging Cycles.** There are four recognized components of any charging cycle. Understanding these phases is crucial to understanding charging systems and problems.

a.   **Bulk Charge.** The bulk-charge phase is the initial charging period before the gassing point is reached. This is typically in the range 14.4V to 14.6V, corrected for temperature, although with a traditional alternator and regulator, output is fixed at around 14.2V. The bulk charge rate can be anywhere between 25% and 40% of rated amp-hour capacity at the 20-hour rate, as long as temperature rises are limited.

b.   **Absorption Charge.** After attaining the gassing voltage, the charge level should be maintained at 14.4V until the charge current falls to 5% of battery capacity. This level normally should equate to 85% of capacity. In a typical 300Ah battery bank, this will be 15A.

c.   **Float Charge.** The battery charge rate should be reduced to a float voltage of approximately 13.2V to 13.8V to maintain the battery at full charge. In this phase, any system loads are supplied by the charger.

d.   **Equalization Charge.** A periodic charge rated at 5% of the installed battery capacity should be applied for a period of 3 to 4 hours until a voltage of 16V is reached. This is generally once a month or greater.

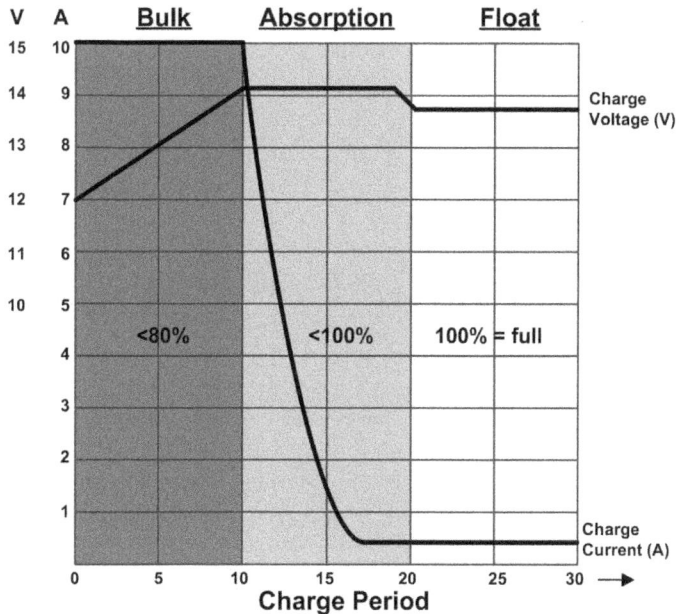

Figure 3-2. Battery Charging Cycles

41

3.2 **Charging Efficiency.** Before any charging systems can be considered, a number of factors must be summarized and taken into account. Nominal capacities of batteries are specified by manufacturers, and the total capacity of the bank must be taken into consideration. Older batteries have reduced capacities due to normal in-service aging and plate sulfation. Sulfation increases internal resistance and therefore inhibits the charging process. The electrolyte is temperature dependent, and temperature is a factor in setting maximum charging voltages.

a. **State-of-Charge (SOC).** The state of charge when charging commences can be checked using the open-circuit voltage test and electrolyte density. The level of charge will affect the charging rate, and the temperature is critical to the state of charge. Temperature has a dramatic effect on charge voltages, as indicated in the State of Charge and Temperature Characteristics illustration.

b. **Charging Voltage and Regulation.** Charging voltage is defined as the battery voltage plus the internal cell voltage drops. Cell voltage drops are due to internal resistance, plate sulfation, electrolyte impurities, and gas bubble formation that evolves on the plates during charging. These resistances oppose charging and must be exceeded to effectively recharge the battery. Resistance to charging increases as a battery reaches a fully-charged state and decreases with discharge. A battery is self-regulating in terms of the current it can accept under charge. Overcurrent charging at excessive voltages simply generates heat and damages the plates.

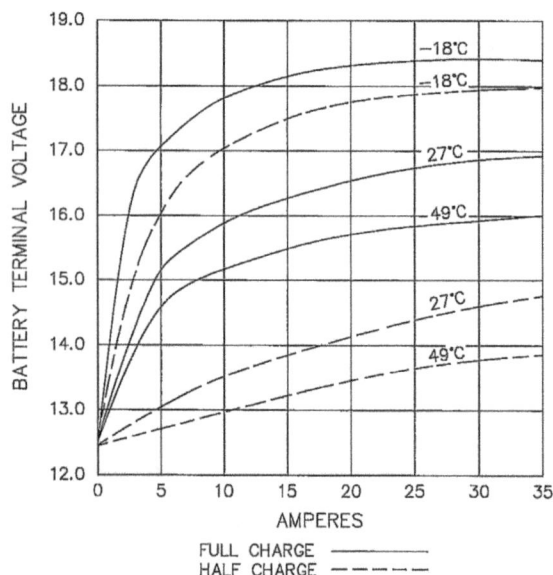

Figure 3-3. State of Charge and Temperature Characteristics

3.3 **Charging System Configurations.** There are three principal charging systems in use. These are the changeover switch, the relay (solenoid, battery combiner or integrator), and the diode isolator. The charging system on most engines uses the same cabling as the engine starter circuit. Basically, it consists of a switch with three positions and "off." The center position parallels both battery banks. It is not uncommon to see both batteries left accidentally paralleled under load, with complete discharge of both. Paralleling of a heavily discharged battery and a fully charged one during charging can cause some instability in the charging as they both equalize.

a. **Switch Operation Under Load.** If a changeover switch is operated under load, the spike and surge will probably destroy the alternator diodes. Most switches incorporate an auxiliary make-before-break contact for connection of field. This advanced field switching disconnects the field and therefore de-energizes the alternator fractionally before opening the main circuit. In reality, this is rarely connected; most alternators have integral regulators, and it is difficult to connect the switch into the field circuit.

b. **Voltage Surges.** If both battery banks are paralleled during an engine start, sensitive electronics can be damaged by the resultant voltage surge or spike. Read the section on electronic interference.

c. **Circuit Resistance.** In most cases, the cables must run from the batteries to the switch location and back to the starter motor, introducing voltage drops. Switches are notoriously unreliable and can introduce voltage drops into the circuit and total alternator or switch failure.

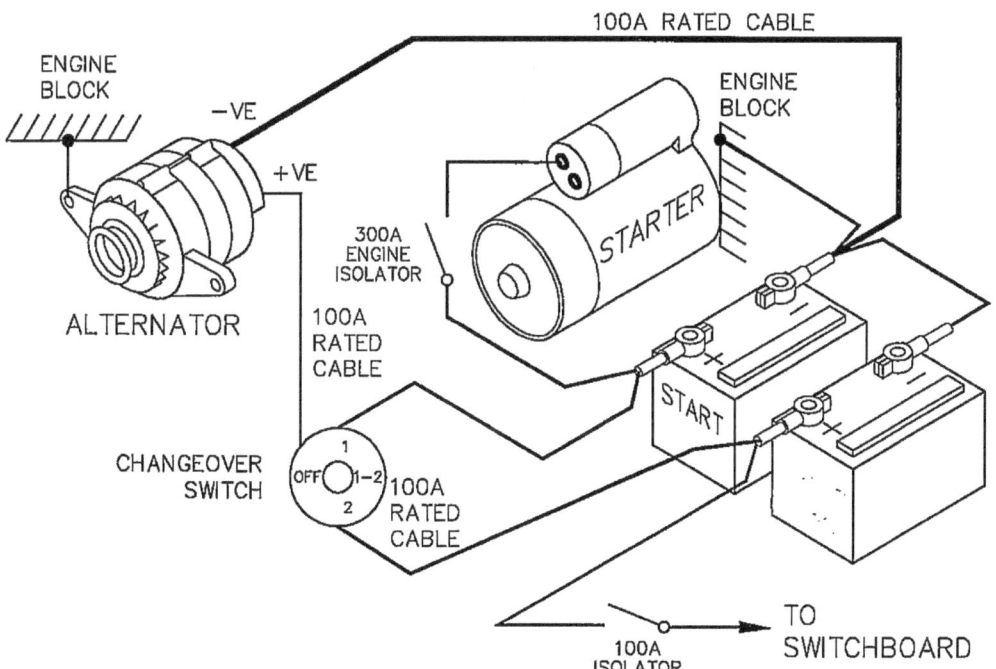

Figure 3-4. Single Engine Changeover Switch Charging System

3.4   Relay/Solenoid Configuration. This system is able to improve on the switch system and enables separation of the charging system from starting circuits. The relay or solenoid does offer a point of failure if incorrectly rated for the task. The relay interconnects both batteries during charging and separates them when off. This prevents discharge between the batteries. The relay-operating coil is interlocked with the ignition and energizes when the key is turned on. When modifying the system, it is necessary to separate the charging cable from the alternator to starter motor main terminal, where it is usually connected. A cable is taken directly from the alternator output terminal to the relay, as illustrated below. Relay ratings should at least match the maximum rated output of the alternator. It is prudent to overrate the relay. Relays are marketed in various forms, the most common being automotive solenoid types. Another device is the voltage sensitive relay (VSR). The relay is open when the engine is started, and when the voltage rises to 13.7V, it closes to parallel the two batteries, which then charge together. When the engine stops, and the voltage decreases, the relay opens to split the two batteries again.

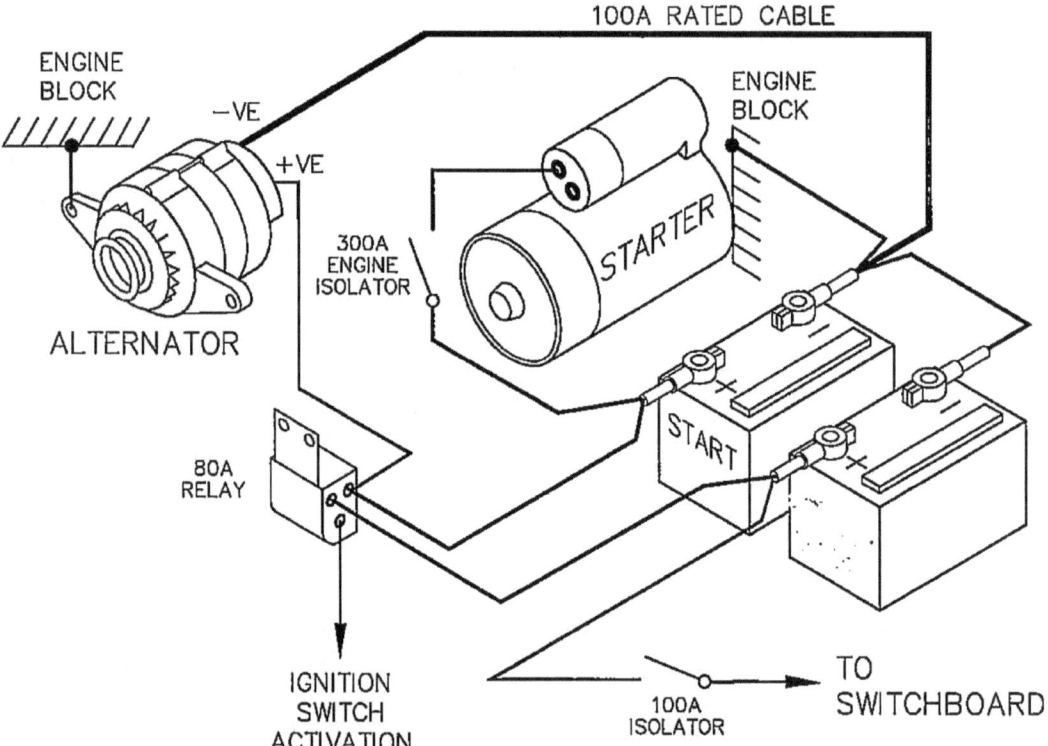

Figure 3-5. Single Engine Relay Charging System Configuration

3.5   Diode Configuration. The diode system is the simplest configuration and the most reliable. A diode has an inherent voltage drop of typically 0.7V to 0.8V. This is totally unacceptable in a normal charging circuit. If the alternator is machine sensed and does not have any provision for increasing the output in compensation, the diode should not be used. Essentially, a diode isolator consists of two diodes with their inputs connected. They allow voltage to pass one way only so that each battery has an output. This prevents any back feeding between the batteries. They are mounted on heat sinks specifically designed for the maximum current-carrying capacity and maximum heat dissipation. The diode isolators must be rated for at least the maximum rating of the alternator and, if mounted in the engine compartment, must be overrated to compensate for the derating effect caused by engine heat. Heat sink units should have their cooling fins in the vertical position to ensure maximum convection and cooling. Do not install switches in the cables from each output of the diode to the batteries. Newer devices are now available such as the Victron Argodiode and Argofet splitters, which have very low forward voltage drops of just 0.1V and use Schottky diodes. They come in two- and three-battery options and are a viable alternative to voltage sensitive relays and older diode splitters. I have installed these on my own boat.

3.6   Diode Isolator Testing. With the engine running, the diode output terminal voltages should be identical, and should read approximately 0.75V higher if a non–battery sensed regulator is being used. The input terminal from the alternator should be zero when the engine is off. Test with power off and batteries disconnected.

a.   Select the resistance setting on a multimeter or set the meter scale to ohms (Ω) x1 on older meters, and then connect the red positive probe to the input terminal. Connect the black negative probe to the output terminal 1 or 2.

b.   If the diode is good, the meter will indicate minimal or no resistance.

c.   Reverse the probes and repeat the test. The reading should indicate high resistance, or overrange.

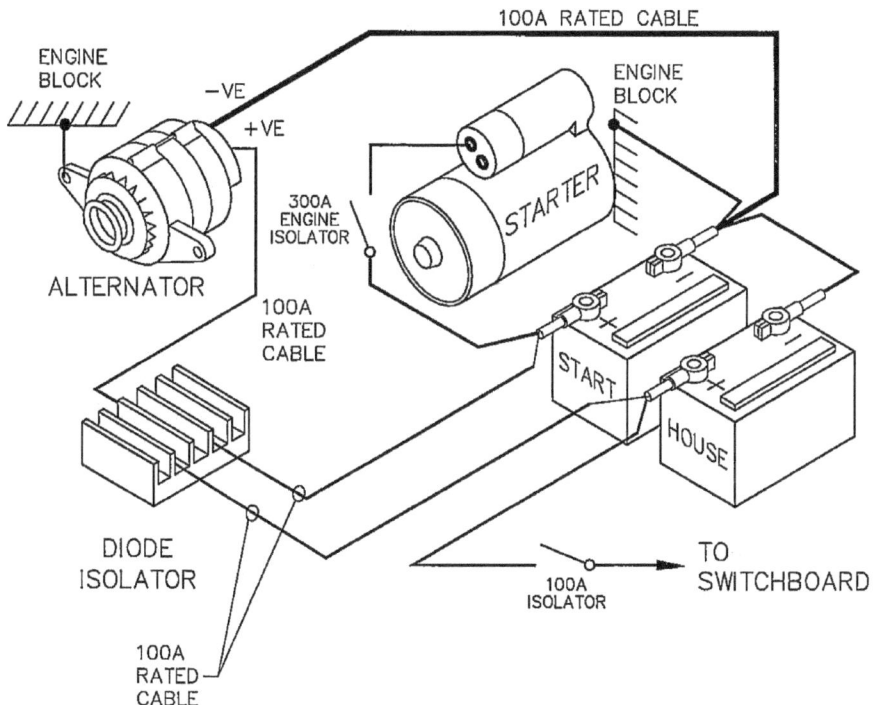

Figure 3-6. Single Engine Diode Charging System Configuration

45

3.7   Electronic Battery Switches. These are also known as charge distributors or battery integrators and are characterized by the NewMar Battery Bank Integrator (BBI). When a charge voltage is detected that exceeds 13.3V, the unit switches on. The unit consists of a low-contact resistance relay that closes to parallel the batteries for charging. When charging ceases and voltage falls to 12.7V, the relay opens, isolating the batteries. The unit incorporates a voltage comparator and time delay circuit, which prevents the unit from cycling in the event of a voltage transient or load droop on the circuit reducing the voltage below the cutout level.

3.8   Battery Charging Capacity. The primary battery charging source should be calculated with a minimum output of 30% of the total installed battery capacity. From the power analysis in chapter 2, we have calculated the maximum current consumption. Added to this is a 20% margin for battery losses, giving a final charging value. A battery requires replacement of 120% of the discharged current to restore it to full charge. This value is required to overcome losses within the battery due to battery internal resistances during charging. A popular benchmark is that the alternator rating should be approximately 30% of battery capacity. An 80A alternator used to be the maximum recommended alternator, but 150A alternators are now being installed on engines. With a suitable regulator system, this is usually adequate for most charging and load requirements; it really depends on your lifestyle. As a battery is effectively self-limiting in terms of charge acceptance levels, we cannot simply push in the discharged value and hope the battery will recharge. During charging, the battery is reversing the chemical reaction of discharge, and this can occur only at a finite rate. The alternator and the regulator therefore must be selected, if possible, to recharge at the battery optimum-charge rate, as specified. Charging, by necessity, has a tapered characteristic, which is why start and finishing rates are specified. The point at which charging reaches close to full capacity is sometimes referred to as the end amps, tail current, or return amps, and the charge current decreases. This can be in the range of 2% to 4% of capacity rating.

3.9   Charging Support. Alternative charging sources such as wind, solar, and water should not be included within power calculations. These systems are to be classified as supplementary charging sources. Other charging sources should be viewed as additional and not be used in the primary calculations, as they are reliant on weather and other conditions. Where two alternators are installed to provide redundancy or improved charging capability, one should be dedicated to starting battery duty. Both alternators should not be charging in parallel to the same battery bank.

3.10   Charging Circuits. The positive cable from the alternator to the battery, or charge distribution device, such as a diode or relay, should be rated at the maximum rated alternator current and for a maximum voltage drop of 3%. All charge circuit cables must be rated for maximum current capacity of the alternator, with minimal voltage drop, and allow for high ambient temperatures. Many installations are underrated and therefore overheat and fail. A negative cable should be installed, equivalent in size to the positive cable, from the negative terminal or case of the alternator to the battery. Where more than one alternator is installed, the negative cables should be connected to the respective battery negatives to maintain system separation and minimize voltage drops in the charging circuit, which normally includes the engine block. These should go to the corresponding battery under charge. All charging system cable terminations and connectors should be rated for the maximum alternator current.

All charging system cable connectors should be crimped. Many charging system terminations are underrated for the current capacity of the cable. Ensure that crimp terminal connections of the correct current capacity are used with rings of the correct size for the termination bolts on alternator and battery. Many are oversized and make poor electrical contact. Soldered connections frequently fail or are high-resistance points in the circuit. See the image in the color insert showing undersized cable and crimp terminal on an alternator.

3.11   Charging Isolation and Protection. No alternator output cable should have any isolation switch or fuse installed within the circuit so that opening of the circuit during operation could cause damage to or failure of the alternator. Alternator failures caused by inadvertent operation of changeover switches are common. When a switch is opened, a voltage spike is generated as the field collapses, and this normally destroys the alternator rectifier diodes. The majority of alternators have a fixed output of around 14.4V to 14.6V, with some models having the option of regulator adjustment up to around 14.8V for isolation diode voltage drop compensation. The regulator should not be able to create high voltages that cause excessive gassing of batteries or in excess of normal equipment voltage input ranges. In the case of flooded cells, AGM, or gel batteries, this might cause catastrophic damage to the batteries and the venting of dangerous gases and explosion.

3.12   Charging System Failure Mode and Effect Analysis (FMEA). To understand this process, read the "FMEA" chapter. In general, many people look at the charging system in terms of discrete components. A charging system must not be viewed as simply a collection of series-connected components but as a complete system. The typical charging system comprises a considerable number of functional elements.

    a.   **The Charging System.** It is recommended to trace out each circuit on your boat, and draw in each component and mark each connection on it. At a minimum, you will have four main positive circuit connections, four main negative circuit connections, four control circuit connections, two changeover switch contacts, a meter shunt, the alternator, the regulator, and the battery:

        (1)   The alternator, which includes several components such as brushes, brush gear, slip rings, bearings, diodes, and windings.

        (2)   The regulator, which may be integral or separate.

        (3)   The DC positive circuit, which includes connections at alternator and battery, and the changeover switch.

        (4)   The DC negative circuit, which includes connections at alternator and battery, the cable back to the battery, and the meter shunt if fitted. In addition, the engine block becomes part of the negative circuit, along with the alternator support bracket, holding bolts, etc.

        (5)   The engine start battery.

    b.   **Charging System Failure Mode Analysis.** The math of this simple analysis is that there is a total of fourteen connection points plus the alternator, regulator, and battery that can impact the starting system, plus the boat owner. Each point represents a single point of failure with subsequent total system failure, with no apparent redundancy. For this exercise, wind, water, and solar panels are considered extra or supplementary charge sources, as are generators with chargers. These, however, can be factored into redundancy provisions. The operational factors must be considered. If a changeover switch is opened or fails during operation, the alternator diodes can be destroyed. The result is no charging capability.

    c.   **Support Systems Failure Mode Analysis.** A similar analysis should be carried out on the engine starting system, the diesel fuel system, the engine cooling water systems, and the air intake system.

3.13   **Charging System Redundancy.** Relatively simple modifications can be carried out on the charging system to improve efficiency and reliability.

    a.   **Second Alternator.** Install a second alternator on the engine; this will require adding a second pulley. The second alternator is for the house battery charging circuit, with the existing alternator being used for charging the start battery and a second house battery bank. Each alternator will have a separate positive circuit without any switches or other devices in it. This will eliminate changeover switch problems on alternators that commonly destroy the alternator rectifier diodes. This reduces connections to just two. It eliminates accidental (human error) switch operation under load, or switch contact failures, which are very common. Each alternator will have a separate negative circuit cable running back to the respective battery from the alternator. This provides separation from the starter motor to the battery negative, with the main starter negative serving as a backup. This reduces connections to just two. It also takes the engine block out of the circuit and generally reduces voltage drop in the circuit as the circuit resistance is reduced.

    b.   **Separate Charging System.** Separate the charging system from the starting circuit; in the long term, this will considerably reduce problems and increase reliability. This process entails the deletion of battery selection changeover switches and the installation of a separate charging circuit, which may include charge splitting diodes or relays. An emergency crossover switch between battery banks can be installed; however, this does not affect the circuit during operations.

    c.   **Install Separate Negative Cables.** Install a separate negative conductor that equals or exceeds the maximum output of your alternator from each alternator case or negative terminal directly back to the corresponding battery negative. Given that some new alternators are rated at 160A or more, this means a substantial conductor size of 00AWG to 2/0AWG ($67mm^2$) or more when allowing for voltage drop. This bypasses the engine block and all the cumulative resistances of mountings and brackets. It offers a good low-resistance path and reduces stray currents through the block, which can cause pitting of bearings. It eliminates a single-point failure of the main negative connection to the engine block. If the main battery negative on an engine comes loose, as they do, you can lose everything.

d. **Replace Positive Cable.** Most installed positive charge cables are under-rated, especially if a fast-charge device is installed. Given that some new alternators are rated at 160A or more, this means a substantial conductor size of 00AWG to 2/0AWG (67mm²) or more when allowing for voltage drop. One problem is that besides having a maximum current going through it with fast-charge devices or when heavily discharged batteries are recharged, the heat of the engine compartment derates the current capacity of the cable. In most cases, a significant voltage drop develops across the cable under full output conditions.

3.14 **Failure Mode Mitigation Results.** There is now a significant reduction in single-point failures. With two separate charging circuits, there is full redundancy and now only a total of four connections in both the positive and negative circuits, the negative having a backup with the starter motor negative. In a typical system using a changeover switch, that is a reduction of up to 75% in possible failure points. The human element has been engineered out of the system, as the possibility of accidentally switching off the changeover switch is removed. There is a significant improvement in charging efficiency, with a gain of up to or exceeding 0.5V due to lower circuit resistance in both the positive and negative circuits. This reduces alternator loads and can shorten charging time, reduce charge current, and extend alternator life. The starting system is more efficient, with the negative at the starter reducing voltage drops, lowering current, reducing run time, and improving starting times. Coupled with a spare starter motor and solenoid, there is a reasonable chance of being operational within an hour. Separation of start and charging systems eliminates the many problems of voltage spikes, surges, and transients. There are now two redundant power systems, each one being capable of powering the vessel; any single failure of one system will not affect the other.

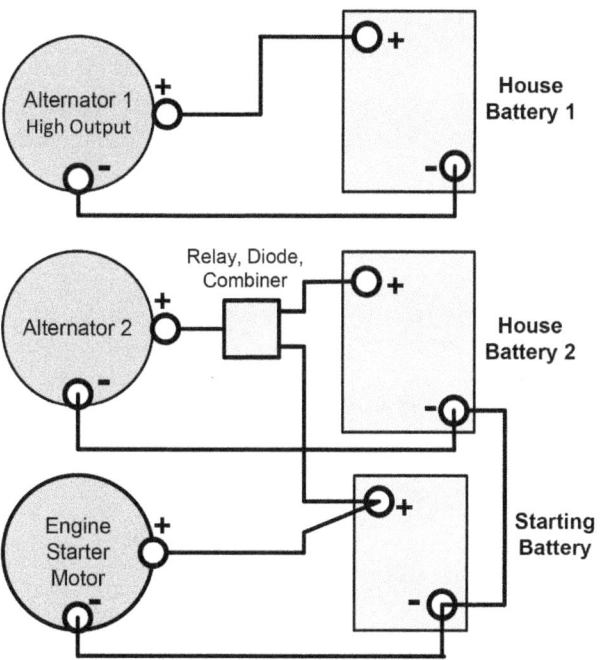

Figure 3-7. Charging and Start Circuit Separation

3.15   Charging System Maintenance. Maintenance strategies and principles are explained in the "Maintenance and Troubleshooting" chapter. Perform the maintenance on all critical battery charging system elements.

a.   **Alternators.** Alternators have a relatively low failure rate, as actual operating hours are relatively low. Failures are generally caused by diode failures or overheating, particularly with fast-charge regulators and oversized battery banks. Alternators should be cleaned and overhauled on a regular basis, ideally not exceeding 2 years. In high-hours sailing, look at bearing replacements prior to each major cruise. Consider a higher rated alternator to reduce overloading and heating. If changing to other battery technologies, such as lithium-ion, review charging system compatibility.

b.   **Batteries.** Batteries have the highest maintenance-dependent failure rates. This is generally due to either inadequate charging, with resultant sulfation; lost capacity and failure; or complete discharge of the battery, with subsequent damage. The second-highest failure is inadequate inspection and topping up of electrolytes, with resultant plate damage. Consider AGM, carbon foam, and lithium-ion battery types with lower failure rates and less maintenance.

c.   **Connections.** All connections on alternators, starters, engine blocks, and batteries should be checked and tightened every 6 months. This easy task results in fewer intermittent and complete failures.

d.   **Spares.** It is rare to see a boat with a spare starter motor or alternator, and these should be a prerequisite on an extended voyage. While some boats may carry spare bearings, diode plates, brush-gear, and so on, good skill sets are required. It is quicker to change out the entire alternator as failures tend to occur at the worst possible times. Invest in a spare starter motor and alternator, and store them properly so they are in good condition when you need them.

e.   **Configuration.** Assess whether your current charging system works for you. The image shows how complex it can get for a twin engine interconnected system.

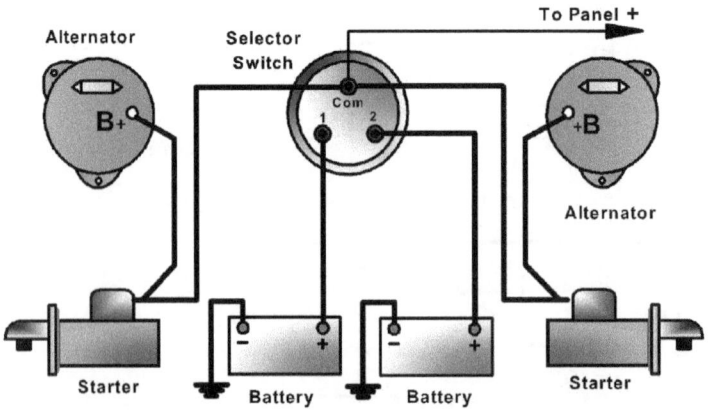

Figure 3-8. Dual Engine Switch Arrangement

3.16 **Multiple Alternator Charging System Configurations.** Twin-engine vessels such as catamarans have, by default, two charging systems. In single-engine boats such as monohulls and trimarans, the fitting of a second alternator is a useful option where redundancy is required. There are several different system configurations for multiple alternator installations.

a. **Discrete Systems.** These systems usually have the original engine alternator charging the engine start battery only. The additional alternator, usually a higher rated unit, charges the house batteries only. If there is more than one bank, this may be split through either a diode isolator or a switch. Ideally the start battery alternator should be used to charge a third battery bank, as the alternator is underutilized since start batteries require very little charging.

b. **Cross-Feed Systems.** These systems usually have each alternator charging a primary battery bank, except that each alternator cross-feeds to the other battery bank via a diode isolator. Although initially this looks complicated, it is relatively simple, and the advantage of such a layout is that the arrangement allows charging of both battery banks even if one alternator should fail, providing some redundancy. It is easier to balance loads between battery banks in order to achieve similar discharge levels of the same periods. This allows both batteries to be charged at a similar rate, which is faster overall, assuming that alternators and regulators are the same.

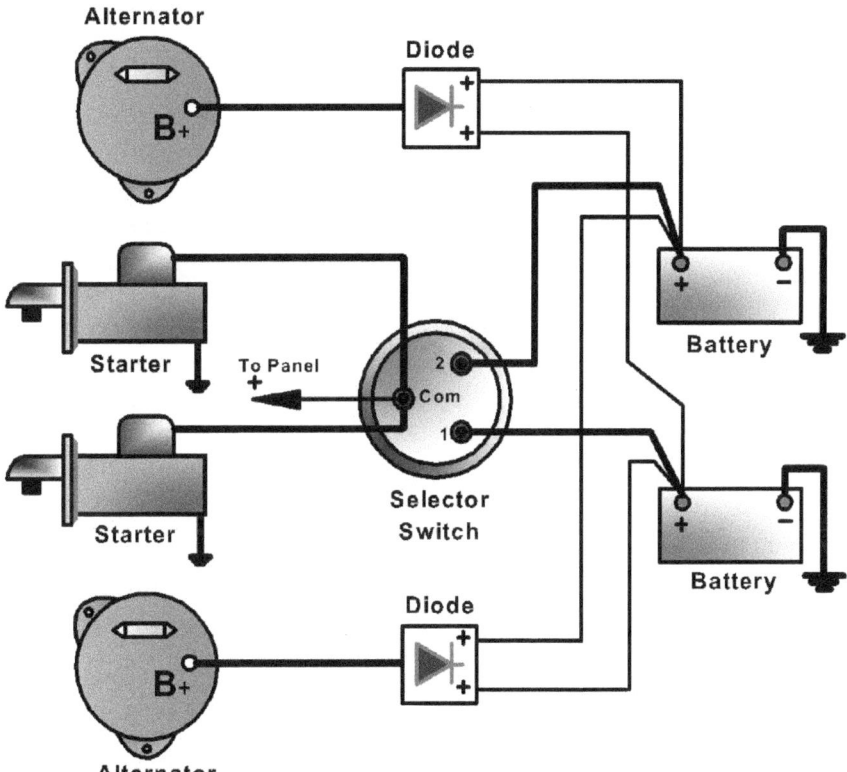

Figures 3-9. Dual Engine Charging Arrangements

Alternative Energy Charging Systems

More misconceptions exist about the capabilities of alternative energy systems on cruising yachts than virtually any other equipment. In most cases, expectations are overly optimistic and the realities are at best disappointing. Some absolute truths must be recognized before embarking on projects that entail large expenditures, and often quite a lot of engineering. They must be faced in spite of the philosophical and environmental arguments. Sustainability is possible, carbon neutrality is possible, and low to zero emissions is possible. The important factors are outlined below for consideration in the decision-making process.

    a.   **Secondary Power Sources.** Alternative energy sources are auxiliary charging sources. They should be integrated into the power system as a secondary power source when no further charging capacity can be derived from the engine alternator. In most cases, alternative combined generation sources significantly reduce dependence on engine-based hydrocarbon systems. A battery can lose as much as 14% of its charge per month; an alternative energy charging system would be ideal in this situation.

    b.   **Primary Power Sources.** Many people, for a variety of reasons, choose to rely solely on renewable power sources to supply electrical power. It can be done and is a combination of water, wind, and solar. In many cases there is a significant lack of understanding of basic electrical design with respect to charging and charging requirements, along with unrealistic expectations. Consider these points:

        (1)   **Design Considerations.** Alternative energy systems once required considerably more stringent design criteria. They also require a sailing philosophy that excludes a large amount of electrical and electronic equipment, as well as a very disciplined lifestyle while cruising. If you want all the comforts and technologies, you are going to need to adopt a holistic approach that integrates solar panels, wind generators, and probably hydrogenerators as well when voyaging. Regrettably, the natural forces that control alternative sources are far from predictable, which is why many have had to adjust their cruising behavior to one dominated by the search for ways to conserve battery power and recharge batteries.

        (2)   **Output Data.** The quoted output data of devices is almost always achieved in ideal laboratory conditions. In real-world situations, you will require a safety factor to get acceptable results.

        (3)   **The Downward Spiral.** In practice, battery charge levels tend to slowly spiral downward as the battery charging fails to keep up with real power demand. The real trick in getting the most out of alternative energy systems is to fully charge the battery to 100%.

This involves a properly calculated charging regime that includes a fast-charge engine alternator system before the batteries discharge low enough to be damaged. That enables the alternative systems to keep up. There has been a drastic improvement in solar power, wind power generation, and, lately, very efficient water-based charging systems. It is possible to reduce engine run time requirements and end the battery power spiral.

4.1   Solar Systems. Many cruising yachts use solar panels and wind generators to charge batteries when on an anchorage, mooring, or marina slip, as well as under sail. In an age of sustainability, solar forms an important plank in the overall power supply system, not as a backup system but as a primary system. Solar energy concepts are not new, dating back to 1839 when the French scientist Henri Becquerel discovered the photovoltaic phenomenon. Solar systems, the most commonly used alternative energy source, offer renewable and nearly maintenance-free energy. The fundamental process of a solar cell is that a voltage is generated when light falls onto the cell. This is called the photovoltaic principle. The basic solar cell consists of two silicon semiconductor layers, one positive and one negative. When light energy photons enter the cell, the silicon atoms absorb some photons. This frees electrons in the negative layer, which then flow through the external circuit (the battery) and back to the positive layer. When manufactured, the cells are electronically matched and connected into an array by connecting in series to form complete solar panels, with typical peak power outputs of 16V to 18V. There are a number of solar cell types, based on the cell material or structure used.

a.   **Monocrystalline.** Pure, defect-free silicon slices from a single grown crystal are used for these structures. The cell atomic structure is rigid and ordered and, unlike amorphous cells, cannot be easily bent. These panels generally possess the highest efficiency and power output and are approximately 17% to 20% efficient. The thin pure silicon wafers are etched within a caustic solution to create a textured surface. This textured surface consists of millions of four-sided pyramids, which act as efficient light traps, reducing reflection losses. Panels are made by interconnecting sixty or seventy-two cells placed onto a glass back and encapsulated.

b.   **Polycrystalline.** These high power output cell types use high-purity silicon 0.2mm-thick wafers from a single block. The wafers are bonded to an aluminum substrate. Some cells are covered with a tempered low-iron glass and a titanium dioxide antireflective coating to improve light absorption. The polycrystalline cell has an efficiency in the range of 16% to 18%.

c.   **Amorphous Silicon.** These cells are formed from several layers applied to a substrate. They have a characteristic black appearance, and some cells have a tin oxide–based coating to improve conductivity and light absorption. Unlike crystalline cells, these thin film panels have a loosely arranged atomic structure and are much less efficient at 9% to 10%. They do have the advantage that the cells can be applied to flexible plastic surfaces to make flexible panels. Additionally, unlike crystalline cells, they are capable of generating a voltage under low-light conditions. The big disadvantage is that power outputs are nearly one-quarter of crystalline cells of the same size.

d. **Thin-Film Solar Cells (TFSCs).** These cells are made using a single layer or multiple layers over a surface of glass, metal, or plastic. Depending on which technology is used, they can be as high as 19% efficient and in laboratory conditions as high as 22%. They are often used on boats because they can be made flexible.

e. **Solar Cell Construction.** Cell arrays are normally laminated under ethylene vinyl acetate (EVA). Antireflection coatings based on titanium dioxide are used, and some are characterized by a blue coloring. This increases the gathering of light at the blue end of the spectrum. Panels are constructed to be moisture and ultraviolet resistant. Glass surfaces are tempered and sometimes textured to reduce reflection, increase surface area, and improve light gathering at low lighting angles. Solar arrays often utilize front and rear connections to improve faulty cell redundancy.

f. **Flexible Solar Panels.** These are becoming very common as technology advances. Companies such as SunPower Maxeon have developed some very efficient solar cells. Representative of these are the Sunpower Maxeon cells used in the SpectraLite Pro range, which have an output range of 50W, 100W, 150W, and 200W. The cell construction technique uses a solid copper foundation overlaid with an ultrapure silicon layer and topped with an antireflective glass. They are flexible up to 30%, and the cell is able to flex in variable temperatures. The efficiency improvements result in power generation at lower irradiance levels. The SpectraLite SemiFlex Pro features the same Maxeon cells and is coated with DuPont ETFE (ethylene tetrafluoroethylene), which further enhances the flexible and nonslip characteristics. DuPont Tefzel ETFE film is chemically inert and solvent resistant to virtually all chemicals, has a high dielectric strength exceeding 160kV/mm for 0.025mm film, does not suffer from electric tracking, and has several other properties. These cells are monocrystalline. To clarify a misconception, walking on these cells is not recommended; cell cracking can still occur and, if damaged, can affect output. Each solar panel includes standard MC4 connectors, which makes connection and disconnection easy. The image in the color insert shows bimini-mounted Sunware Textile flexible panels and rigid panels installed in a stern-mount arrangement on a catamaran.

4.2   Solar Ratings and Efficiency. Efficiency is optimum when a solar panel is angled directly toward the sun, and manufacturers rate panels at specific test standards. Output ratings are normally quoted to a standard, typically $1000W/m^2$ at 77°F (25°C) cell temperature; the level of irradiance is measured in watts per square meter. The irradiance value is multiplied by time duration to give watt-hours per square meter per day. Location and seasonal factors affect the amount of energy available. Cells are approximately 15% efficient and start producing a voltage as low as 5% of full sunlight value. Solar angles are important to the efficiency

Table 4-1. Peak Solar Level Table

Location	Winter Hours	Summer Hours	Average
California	4.0	5.0	4.5
Miami	3.6	6.2	4.9
Central Pacific	4.5	6.0	5.3
Caribbean	5.5	5.5	5.5
Azores	2.2	6.0	4.1
Northern Europe	1.5	4.0	2.7
Southern England	0.6	5.0	2.8
South France	2.5	7.5	5.0
Greece	2.4	7.4	4.9
Southeast Asia	4.0	5.5	4.7
Cape Town	4.0	5.0	4.5
Red Sea	6.0	6.5	6.3
Indian Ocean	5.0	5.5	5.3
Eastern Australia	4.5	5.5	5.0

of panels. With the sun at 90 degrees overhead, panels give 100% output. When angled at 75 degrees, the output falls to approximately 95%; at 50 degrees output falls to 75%; and a lower light angle of 30 degrees gives a reduction to 50%. Most solar panels will give some output on dull days. The table shows typical seasonal hours and yearly averages based on a solar array tilted toward the sun at an angle equal to latitude of the location plus 15 degrees.

4.3   Panel Regulation. In any panel larger than a small 12W to 15W unit, a regulator is required to limit voltage to a safe level. It is not uncommon to have a solar panel output rise to 15V to 18V and boil the batteries dry over an extended unsupervised period. There are solar control devices in use, which must not be confused. One simply limits voltage to safe levels, and regulators and others, such as an MPPT charge controller, optimize power under all solar conditions. Solar systems have inherent losses, and the tendency is to oversize an array to compensate for low-irradiation conditions in poor weather. In addition, there are dust and dirt accumulations, shading from spars and so on, as well as improper solar alignment, which is a challenge on most boats. It is very easy to accumulate charging losses that exceed 30% to 40% on yachts. It should be noted that you should never exceed the solar charge controller maximum input voltage or current rating.

      a.   **Regulators.** The regulator serves to limit panel output to a safe level and prevent damage to a battery. Some units simply limit voltage to 13.8V, the maximum float level, and dissipate heat through a heat sink. More sophisticated regulators get more from the panel. These units incorporate an automatic boost level of 14.2V and a float setting of 13.8V. The regulator float charges the battery until a lower limit of approximately 12.5V is reached before switching to boost. The units normally eliminate the need for an additional blocking diode. Check the manufacturer's data sheet first. Some regulators have temperature compensation and must be installed adjacent to the batteries. In practice, the charge controller output current rating should be about 10% to 20% of the battery Ah rating. A 150W panel is able to generate a 10A charge current for a 100Ah battery to attain the adsorption charge voltage subject to optimal orientation and no shading.

b.   **Linear Current Booster.** These electronic devices boost current from the solar module. They are designed to prevent permanent magnet motors from stalling, but effectively they are constant current devices. Such units are used primarily in applications where panels directly supply a load. They are not useful on boats where the panel is used to charge a battery.

c.   **Pulse Width Modulation (PWM) Controllers.** These relatively simple controller types modulate the charging using an electronic circuit. The circuit output opens and closes hundreds of times every second to maintain a constant battery voltage and modulate the current. This allows the solar panel output voltage to equal the battery voltage level.

d.   **Maximum Power Point Tracker (MPPT) Charge Controller.** These are more sophisticated than PWM units. They allow the solar panel to operate at the maximum and optimum power point, which is more efficient. The MPPT is essentially a DC-DC converter. The reality is that the level of sunlight or irradiance level is constantly changing due to clouds or angle, so the voltage and current flow also changes. The MPPT unit is able to calculate the best voltage and current ratio to output the most power. For yachts with two or more panels, this is essential given the way a yacht moves and changes angles. MPPT devices from many manufacturers are similar. I have installed a Victron MPPT unit on my own boat.

4.4   **Diodes.** Most panels have diodes installed. There is a rather flawed argument that the use of a diode reduces charging voltage. This is true, as a diode reduces voltage by approximately 0.75V. But if you are installing a couple of 3A panels, which is typical, you will need a regulator to reduce the voltage to avoid overcharging and damaging your batteries. If the regulator is a good unit, the control will float between 14.5V and 13.8V, so the small voltage drop will probably not be a problem. If the regulator has the appropriate reverse-current protection diode, the panel-installed diode can be removed to increase the input voltage to the regulator, which gives a marginally higher output. If you do not regulate the solar supply, failing to install, or removing, the diode will result in a dead battery overnight. Schottky barrier diodes are used due to their low forward voltage drop and very fast switching action. There are two functional uses of diodes.

a.   **Bypass Diodes.** Bypass diodes, normally factory installed in module junction boxes, reduce power losses that might occur if a module within the array is partially shaded. For 12V systems, these offer sufficient circuit protection without the use of a blocking diode. A 24V array requires two 12V panels in series. An array for larger current outputs requires the parallel connection of these series arrangements. If one module of a parallel array is shaded, reverse current flow may occur.

b.   **Blocking Diodes.** Blocking diodes are often connected in series with the solar panel output to prevent the battery from discharging back to the array at night, but not all manufacturers install them as standard. If the panels do not have a diode, a diode rated to 1.5 times the maximum output should be installed at the regulator input. Most solar regulators will have the diode incorporated. Generally, all panels with a bypass diode installed in the connection box do not require another diode.

4.5   Charging System Interaction. There is often an interaction between solar panels and alternator charging regulators during engine charging. This can also cause problems when wind and solar are installed together. If the solar panels are not regulated, it is quite common to see a voltage of up to 16V or more across the battery. When an alternator's regulator senses this high a voltage level, it simply registers it as a fully charged battery; as a result, the alternator does not charge the battery, or does so at a minimal rate. When installing panels and regulators, consider the following features if not using an MPPT regulator or any other more advanced regulator type.

a.   **Isolation Switch.** Install an isolation switch on the incoming line to the panel so that it can be switched out of circuit. Open-circuit voltages can get quite high.

b.   **Engine Interlock.** This circuit automatically disconnects the solar panel via a relay so that the solar panel does not impress a higher voltage and "confuse" the alternator regulator.

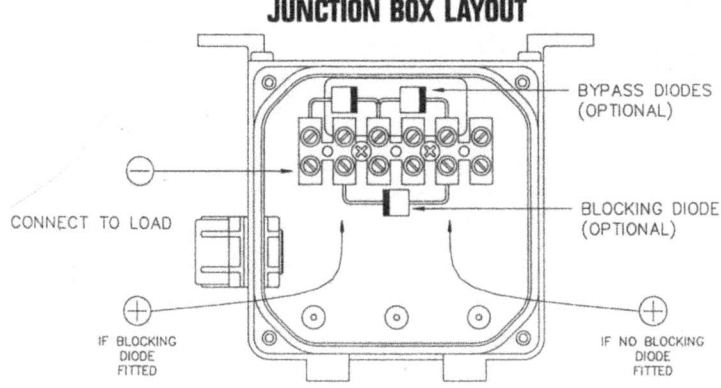

Figure 4-1. Solar Panel Diodes

4.6   **Solar Panel Site Selection.** Solar-panel siting is largely dependent on the physical space available for installation. The following options are the most common and most efficient. In all cases, it is essential to ensure that panels are not shadowed by sails, spars, or any other equipment. Ideally, panels should be angled toward the sun, if at all possible, but this is not always practical on a cruising yacht. Generally, flat-mounted panels offer the best compromise, which is why the stern arch configuration is becoming so popular. Flexible panel technology has opened up some new possibilities. Images of the various options are shown in the color insert section.

a.   **Coach House.** Panels can be mounted on coach-house tops, but one panel will often be shaded and the other irradiated, depending on the tack you are on.

b.   **Stern Arches.** This is the most popular method, as it allows the easy installation of at least two unobstructed panels. Depending on the vessel type, this can be somewhat challenging though, and the arch structure is also home to wind generators, radomes, radio aerials, GPS and AIS antenna and aerials, as shown in the color insert section.

c.   **Stern Pulpit (Pushpit) Rails.** This arrangement uses two panels mounted on swing-up brackets on each side of the vessel, normally close to and on the pulpit rails. Depending on tack or direction of sun, the panels can be put into service and then folded down if not used.

d.   **Bimini.** The availability of flexible panels allows easy installation on the canvas covers, as shown in the color image section.

e.   **Deck Mountings.** The availability of flexible panels enables relatively easy installation on deck areas.

f.   **Multihulls.** The greater deck area of a multihull and a nearly level sailing attitude make site selection much easier and offers increased efficiency. In most cases, a large coach house can be utilized, and on trimarans, arrays can be mounted on the outer hulls well clear of shadows. The wider stern allows installation of an array, which is a popular solution.

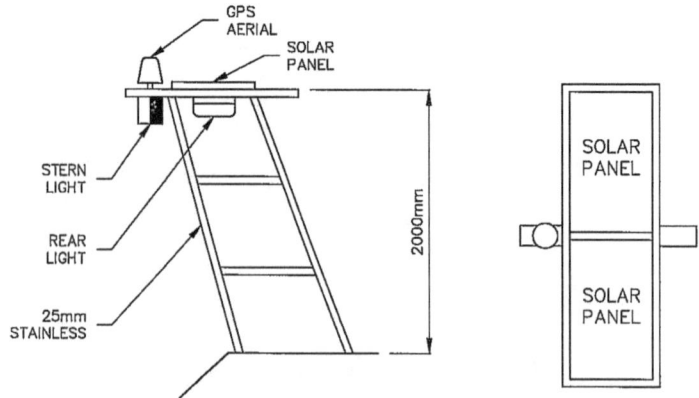

Figure 4-2. Stern Arch Arrangement

4.7 **Solar Panel Installation.** Solar panels are manufactured in either rigid or flexible form. Cabling should be properly rated to avoid voltage drop. To cope with two 65W panels, 14AWG (2.5mm²), 16A cable is the minimum size; bigger is better. Use only tinned copper, double-insulated marine cable. Most panels have weatherproof connection boxes, and connections can be simply twisted and terminated in terminals. Do not use connectors or solder the wire ends. Manufacturers specify grounding the solar array or module metallic frames.

    a. **Panel Safety.** Cover solar panels to prevent voltage from being generated during installation or removal to avoid accidental short-circuiting of terminals or cables.

    b. **Panel Mounting.** Each panel should be securely mounted and able to withstand mechanical loads. Ideally, panels should be aligned to provide unrestricted sunlight from 09:00 to 15:00 solar time. Not an easy thing on a yacht.

    c. **Panel Standoffs.** Allow sufficient ventilation under the panel. Excessive heat levels will reduce output and damage cells. Most panels in frames have sufficient clearance incorporated into them.

4.8 **Solar Panel Maintenance.** Maintenance requirements for solar panels are minimal. The essential tasks are as follows.

    a. **Panel Cleaning.** Panels should be cleaned periodically to remove salt deposits, dirt, and seagull droppings. Use water and a soft cloth or sponge. Mild, nonabrasive cleaners may be used. Do not use scouring powders or similar materials.

    b. **Panel Connections.** Make sure the terminal box connections are secure and dry. Fill the box with silicon compound if they are exposed to a lot of water.

4.9 **Solar Panel Troubleshooting.** Faults are normally the result of catastrophic mechanical damage. A single cell failure will not seriously reduce performance, as multiple cell interconnections provide some redundancy. Reliability is very high, and manufacturers supply 10-year warranties to support this. Faults can be virtually eliminated by proper mounting and regular maintenance. As with all electrical systems, the most common faults are cable connections. The following checks should be carried out if charging is below normal levels.

    a. **Regulator Output.** Check regulator output for rated voltage, typically 13.2V.

    b. **Regulator Input.** Check regulator input; voltage will typically be above 14V. Disconnected from battery, it can be 17V up to 30V, depending on the panels and arrangement.

    c. **Panel Connections.** Check panel junction boxes for moisture or corroded connections.

4.10 **Wind Charging Systems.** Wind generators, or micro wind turbines, are the second most used alternative energy source. As with all charging systems, there are important factors to consider when deciding whether to install a unit as part of a balanced power system. This chapter outlines the various factors to consider.

a. **Cruising Patterns.** Wind generators are more effective in some areas than others. They are very effective in the Caribbean, with its consistent trade winds. In the Mediterranean, solar power is often considered more efficient. If you sail downwind following the trades, wind generators are not effective because the apparent wind speed is reduced, along with charging capability. If your cruising takes you primarily to sheltered anchorages, wind generators may not be an economical or practical proposition. It is at anchorages that wind generators can be most useful and provide 24-hour charging, albeit with some variation as you swing on the hook to both tide and wind.

b. **Generator Types.** Essentially, a wind generator is either a DC generator (on older models) or an AC alternator driven by a propeller. The trend has been for smaller diameter multiblade units, although there are several high-performance three-blade units. Most wind generators now have a permanent magnet rotor, with neodymium rare earth magnets and up to twelve poles. A three-phase alternating current is generated and rectified to DC within a rectifier, similar to engine driven alternators.

4.11 **Wind Generator Features.** Wind generators from all manufacturers have protection systems installed to prevent wind generator damage or battery overcharge situations.

a. **Low Cut-in Speed.** Wind generators work best when the blades are humming along in 10–15 knots of wind, but that's not always the case, especially when you are tucked away in some idyllic anchorage. Low cut-in speed capability means your batteries are getting a charge in the lightest of airs, if only a very low-level trickle charge. Most manufacturers are constantly trying to improve this, from low inertia bearings, motor design, and blade efficiency.

b. **Auto Brake.** If you happen to be ashore when the wind kicks up, you don't want your turbine to spin wildly out of control. Some generators feature a type of automatic speed governor that slows the blades before they self-destruct. Back in the day, you had to pull down on a cord to apply the brakes, then climb up the mast or mounting pole and manually tie off the blades.

c. **Overcharge Protection.** To prevent overcharging, a "brake" kicks in when the batteries are nearing or at full capacity. Some controllers slow the rotation of the blades; others "dump" the excess power generated through a resistor and heat sink. Wiring an in-line resettable fuse can prevent damage to your batteries should a malfunction occur within the charging system.

d. **About Performance Curves.** Most manufacturers base outputs on wind tunnel testing or theoretical analysis or predicted outputs based on calculations that are usually based on a steady-state wind speed in clear and unobstructed wind flows. In real life, we know that boats are heeling, pitching, heaving, and yawing with gusts, wind bullets, and other weather conditions. This all affects the way wind impacts the blade and how the generator performs. Exercise expectation management when selecting a wind generator; while this does not detract from a published set of data and the ability of a device to perform, be aware that there are factors when installed. It is almost impossible for designers to consider all situations when afloat.

4.12   **Wind Speed Equivalents.** Wind speed numbers and values can be confusing, whether you use meters per sec (m/s), knots (kts), kilometers per hour (km/hr), miles per hour (mph), or the Beaufort wind scale as a reference value. Table 4-2 shows the different values and equivalents used by manufacturers. Most wind generator manufacturers quote the generator's maximum wind speed rating. Many commence output limitation in the range of 25kts to 40kts depending on the model. Survival wind speeds are 78kts, 90mph (40m/s) for an AirX Marine unit. The Superwind 350-II is 97kts or 112mph (50m/s) in shutdown mode.

Table 4-2. Wind Speed Equivalents

Kts	m/s	mph	km/hr	Force
1.9	1	2.2	3.6	1
3.9	2	4.5	7.2	2
5.8	3	6.7	10.8	2
7.7	4	8.9	14.4	3
9.7	5	11.2	18	3
11.7	6	13.4	21.6	4
13.6	7	15.7	25.2	4
15.6	8	17.9	28.8	4
17.5	9	20.1	32.4	5
19.4	10	22.4	36	5
21.4	11	24.6	39.6	5
22.3	12	26.8	43.2	6
25.3	13	29.1	46.8	6
27.2	14	31.3	50.4	6
29.2	15	33.5	54	7
31.1	16	35.8	57.6	7
33.1	17	38.1	61.2	7
34.9	18	40.1	64.8	8
36.9	19	42.5	68.4	8
38.9	20	44.7	72	8
40.8	21	46.9	75.6	9
42.7	22	49.2	79.2	9
44.7	23	51.4	82.8	9

(*continued*)

Table 4-2. *Continued*

Kts	m/s	mph	km/hr	Force
46.6	24	53.7	86.4	9
48.6	25	55.9	90	10
50.5	26	58.2	93.6	10
52.5	27	60.1	97.2	10
54.4	28	62.6	100.8	10
56.4	29	64.9	104.4	11
58.3	30	67.1	108	11
60.3	31	69.3	111.6	11
62.2	32	71.6	115.2	11
64.1 >	33	73.8	118.8	12
66.1 >	34	76.1	122.4	12

4.13   Marlec Rutland 914i and 1200 Wind Generator. Marlec has been developing wind generators for many years, and they are common among the cruising community. Visit marlec.co.uk.

a.   **Generator Design.** The Rutland 1200 has a cut-in wind speed of 2.5m/s. The generator is a low-friction, brushless, three-phase alternator with high-specification rare earth magnets. It has a slip ring and brush gear for 360-degree free rotation. Protection features include electronic stalling to prevent overcharging.

b.   **MPPT Controller.** MPPT technology is an intelligent microprocessor-controlled unit used to optimize power generation from solar and wind sources. This enables charging at very low rotor speeds and for solar cells to start charging at relatively low irradiance levels. The charge controller incorporates a temperature sensor to adjust charge levels to match the battery temperature. The Marlec unit has a serial data socket to enable connection to an optional remote display, data collection, or with a Marlec controller interface cable and app for programming of voltage and other parameters required for battery types, such as lithium-ion. The device integrates a run and stall button, and this is used to manually start or soft stall the generator. Soft stalling allows controlled deceleration of the rotor until it spins at slow idle. The unit has a bicolor LED to indicate charging status: standby, charging, and regulating. The protection function automatically activates in high-current situations to reduce electrical load. This function stalls the turbine for 5 minutes and will cycle in high-wind environments. In the case of high controller internal temperatures, the turbine will be temporarily stalled. An algorithm embedded within the overcharge control provides multistage charging to maintain it through the bulk, absorption, and float phases. An image of an MPPT controller is shown in the color insert section.

4.14   Eclectic D400 Wind Generator. This is a horizontal-axis upwind turbine with a diameter of 43in (1.1m). The published output rated power is stated as 235W @ 11m/s (22 kt), 420W @ 14m/s (28kt). The maximum output power is 600W, the rotational speed is 1,100rpm @ 14m/s, and the cut-in speed is 2.5m/s (5kt). The D400 is designed to reduce vibration by running the turbine at a slower speed for any given electrical output. This equates to around a 75% speed reduction compared to other units. Visit eclectic-energy.co.uk.

a.   **Blade Design.** The D400 has reengineered the air blades with precision injection molding manufactured from glass-reinforced nylon. The design accounts for taper and twist and incorporates a low Reynolds variable camber airfoil section, which has a variation between root to blade tip. This design detail is straight out of aeronautical engineering. The Reynolds number for an airfoil is stated as a dimensionless number used to determine whether a fluid (air) is exhibiting laminar flow or turbulent flow; the Reynolds number is denoted as "Re." The swept area is $0.95m^2$, with a tip speed ratio of 4. The unique tail design is to impart stiffness and prevent resonance. Noise reduction is stated as 2dBA to 6dBA over background noise, which is very low or whisper quiet. The yaw system is classed as passive low resonance, and this relates to the tail design. Image of the Eclectic D400 mounted on a stern arch is shown in the color insert section.

b.   **Regulator.** The AC-rectified DC output voltages are 12V, 24V, and 48V, which will be useful for those transitioning to electric or hybrid propulsion systems. The unit has a heavy-duty slip ring assembly, with saddle spring loaded output brushes to transfer power from generator to output cables to regulator and control.

c.   **Braking Switch.** The system has a braking switch that enables turbine shutdown. The switch is a double throw "break-before-make" type that is rated at 40A or more. On operation, the switch disconnects the D400 from the batteries prior to short-circuiting the turbine. The application of a short circuit results in the turbine slowing to facilitate rotating the unit out of the wind and securing the blades.

d.   **Alternator.** For the electrical side, the D400 has a highly efficient direct-driven twelve-pole, three-phase AC axial field permanent magnet alternator, which is passed through a rectifier to derive the DC output. This consists of two large annular magnet rotors with stator coils positioned between them. The stator coils are wound from heavy-gauge copper and are encapsulated in a heat-conductive resin to facilitate heat dissipation. This alternator design is "ironless," which eliminates "cogging" and provides very low friction rotation. Cogging is the name applied to the alternators where the magnets on a rotor attract the poles within the laminated iron core stator. If you turn a wind generator alternator with this construction, it will feel stiff and have some resistance. This impacts performance, as the static torque inhibits blade rotation at low wind speeds. Alternators that do not have iron core design do not suffer from this and therefore start more easily. They do not have iron losses, so they are more efficient. To minimize inertia and friction losses, the unit uses high-quality bearings and twin-lipped radial shaft seals to protect them.

4.15   Watt and Sea Racing Wind Generator. This is a relatively new system arising out of high-speed mono- and multihull offshore races, which include catamarans and trimarans. They are designed with high stability and a rear-facing turbine propellor with no tail fin. The alternator is a permanent magnet type; it is oil lubricated and naturally has a high wind range. The unit has a storm mode that is selectable on the converter and is able to withstand wind speeds of up to 50kts. The blades are of a low drag design. The start-up wind speed is 17kts, with the working wind speed range of 13kts to 35kts. It is not suitable for the cruising space; however, for high-speed multihull and monohull vessels, it has a place.

4.16   TESUP Master940 Wind Generator. The TESUP Master940 wind turbine has a permanent magnet generator that incorporates strong rare earth N42 neodymium ($Nd_2Fe_{14}B$) magnets with steel slots inside. When the wind speed exceeds 3m/s, the turbine will rotate freely by exceeding the holding or cogging torque. The charge controller incorporates dissipation resistors for overcharging and load dumping, and maximum charge levels are adjustable via a potentiometer. The wind turbine blades have been designed for lower noise and vibration, greater efficiency, and low wind speed start-up. This uses a lift and drag type blade design.

4.17   Primus Air Silent X and Air-X Marine. These are common on cruising yachts. The Air-X marine turbine unit comprises three rotor blades of injection-molded composite construction. The alternator is a permanent magnet brushless type. It has a microprocessor-based smart controller and electronic torque control. The start-up wind speed is 8mph (3.58m/s). The wind speed range is 8mph to 49mph (3.6m/s to 22m/s), and the optimum wind range is 25mph to 32mph (11m/s to 15m/s). The swept area is 11.5 ft² (1.07m²). The rotor diameter is 46in (1.17m). The Silent X is designed to reduce noise and has carbon fiber blades and hub. The start-up wind speed is 8mph (3.59m/s). The Air Breeze generator is part of the Primus range. These units have composite blades and an internal smart microprocessor regulator that incorporates peak power tracking capability. The alternator is a brushless neodymium type. The rotor diameter is 46in (1.17m) and a start-up wind speed of 6mph (2.68m/s). Rated output is 200W at 28mph (12.5m/s), and overspeed protection is through electronic torque control. Visit primuswindpower.com.

4.18   Leading Edge. This is a UK generator with two marine models: the LE-300 and LE-400. The three-bladed LE-300 uses a direct-drive, axial flux, eight-pole neodymium rare earth permanent magnet three-phase DC-rectified generator. It has a 3ft (1m) rotor diameter. It features zero cogging and has low TSR Whispower blades. The blades are constructed of glass-reinforced, UV-resistant nylon. The rotor speed range is 0 to 2,000rpm. The swept area is 0.785m², the tip speed ratio is 6, and the swept area is 0.785m². Generator regulation utilizes a dump-type regulator. The unit has a cut-in speed of 6.7mph (3m/s), has a peak output of 30W, and will output 85W at wind speeds of 18mph (8m/s). Visit leadingedgepower.com.

4.19   SilentWIND Pro. These turbines use a three-phase permanent magnet generator that uses hand-laminated carbon fiber blades. They have low wind start-up and low cogging torque at 2.2m/s wind speed and charging at 2.8m/s (5.4kts). They have a stop switch and electronic braking. Rated output is 420W. They come in 12V, 24V, and 48V models. Visit silentwindgenerator.com.

4.20   MarineKinetix MK4. This innovative generator goes back to some crucial design basics. The unit has a 5.8-knot start-up speed and a 6.7-knot cut-in speed, which is the charging start point. I have extracted some of the design philosophy behind their generators, as it is very informative. An image of the MarineKinetix MK4 is provided in the color insert. Visit marinekinetix.com.

a.   **Blades and Tail.** The carbon fiber–reinforced blade set has an extremely low rotational inertia, with a strong and lightweight blade. The 20% carbon polymer blades have a very quiet noise footprint as well as improved aerodynamic performance. The quoted audible noise level is 35dB at 17ft (5m) at 10kts. The starting torque on a wind turbine is generated in the blade area closest to the hub, while the power-producing torque is produced in the blade area closer to the tips. Improved hub design and precise tight tolerance blade fastening are used for perfect blade alignment. Lightweight blades have a low rotational inertia, which is critical. Low rotational inertia allows the blades to accelerate more quickly, which provides faster rotation in low wind speeds, and this keeps the tip-speed ratio (TSR), which is the blade tip speed relative to wind speed, more constant. When the wind generator operational envelope is close to the optimum tip-speed ratio during wind gusts, this allows the turbine to "capture" more wind. The available power to a wind generator is a function of the square of the diameter, or swept area of the blades, and the cube of the wind speed. Efficiency is dependent on blade length, or swept area, and the TSR. When the blades rotate too fast relative to the wind, they appear as a solid disk; air accumulates, blocking or piling up at the blade front, interrupting the wind flow behind it. This forms a pressure bubble or wave that creates turbulence and interrupts airflow. In the case of low air speed, the wind flow passes through the slow spinning blade gaps, so no power is generated. The MK4+ has blade lengths some 9in (23cm) longer than typical blades, attaining a swept area some 40% higher than other designs. Wind speed is a vital factor in wind energy capture, and given the cube relationship, there is twenty-seven times more power within a 15-knot wind than a 5-knot wind.

b.   **Tail and Yaw Error.** Yaw error is the difference between the direction the wind turbine is facing and the actual direction of the wind itself. As this yaw error increases, power decreases geometrically. Tail design, and the reduction or elimination of yaw error, is another very important element in the MK4+'s design. The upward-facing fin is not blocked by the mounting pole, and the large self-tracking tail is wind-tunnel designed for minimal yaw error and maximum tracking efficiency.

c.   **Alternator.** The alternator uses a direct-drive, three-phase, dual-bearing AC permanent magnet synchronous generator. This is made from a rare earth neodymium iron boron ($Nd_2Fe_{14}B$) permanent magnet synchronous design.

This comprises a twelve-pole rotor with a bread-loaf magnet profile and asymmetrical pole-shifted magnet placement for low cogging torque performance. Earlier wind generators used a skewed rotor design that improved low speed performance due to low rotational inertia.

d. **Braking.** The brake comprises an automatic back-EMF (electromotive force) principle, and braking activates at full charge, or 40-knot overspeed protection. When an electromotive force or voltage is applied to a motor armature, current will flow; this creates a magnetic force, and the armature starts to rotate. A counterforce of eddy currents is generated by the rotating magnetic field; this known as a back electromotive force.

e. **Marine Kinetix Regulator/Controller.** The MK4+ system uses an integrated charging control concept. This is a hybrid controller that can control both wind and solar inputs. This is microprocessor controlled, with automatic set points for AGM, gel, VRLA, flooded, and lithium-iron phosphate (LiFePO$_4$) batteries. The three-phase AC generator output is rectified within the microprocessor-based controller by use of an IU charging profile, which uses a patented two-stage pulse width modulation (PWM) controller with hysteresis braking. IU is a two-step charge process in which the I-phase is a constant current step and the U-phase is constant voltage. The unit has an integrated battery monitor, stop control ammeter, and watt meter. What really sets this controller apart is the RISC CPU 8-bit programmable intelligent computer (PIC) microprocessor-based circuit that is the heart of the charge controller. Reduced instruction set computing (RISC) is a type of microprocessor architecture that uses small and highly optimized sets of instructions. RISC processors are able to perform complex instructions by combining several simpler instructions, making them very efficient. The central processing unit (CPU) is able to execute these instructions very rapidly. The controller has three set points, and when the MK4+ controller senses that the batteries are close to full charge, it switches voltage-control charge mode. When it enters this mode, excess power is dumped or shed using the PWM load-dumping sequence, which comprises thousands of steps. At this point, the controller dumps power it doesn't need to complete the final charging stage of the battery. When the controller senses the battery has reached full charge, all power is shed and no further charging occurs until voltage drops below the set point. When full charge is attained in high wind conditions, the controller activates the back-EMF brake; this magnetically adds load to the generator, resulting in decreased blade speed. As this prevents overcharge conditions, it is suitable for any battery chemistry, including AGM, gel cell, and lithium-ion batteries. The final set point is a safety braking set point; it is activated only if there is excess current production during charging, which could result in generator winding damage. The safety brake stays activated for up to 20 minutes and will release automatically when the temperature decreases to a set value.

4.21   **Superwind 350 II Wind Generator.** This innovative unit comes from a company that road tests wind generators in some of the harshest environments on the planet. It features a patented autonomous furling pitch blade mechanism or automatic-feathering overspeed-avoidance system. When the wind speed reaches 25kts, the blades will start to feather, or dump air; this reduces efficiency and reduces the turbine speed to avoid overspeed operation. As the wind speed reduces, the blades re-pitch and increase efficiency. The optimized rotor blades were developed to achieve the desired airfoil design and match the specially engineered rotors. See the image of wind generator in the color insert. Visit superwind.com.

a.   **Blade Design.** Starting with wind tunnel testing simulations, the company factored in real-world parameters such as turbulence and crosswinds. The design has relatively broad rotor blades that are matched with a special pitch angle that results in high start-up torque with a rotor start-up wind speed of 6.8kts (3.5m/s). The principal innovation is the unique aerodynamic rotor control system. This will adjust and alter the rotor blade pitch angles in a short time period of milliseconds to prevent damage. The precision engineered pitch control mechanism is completely integrated within the hub. The mechanical control mechanism, and resultant rotor blade adjustment, is activated by aerodynamic wind forces on the blades and centrifugal force. The mechanical overspeed control limits how much energy is input to the charge control and regulation unit. The wind force moves the turbine into rated power speeds, and the blades then change the pitch automatically into an angle for reducing their aerodynamic lift. This feathering is almost instantaneous; the blades then recover automatically to normal position as the wind speed decreases. The blades are able to alter pitch to a 40-degree angle. Another feature on the blades is what are called turbulators. These turn the laminar boundary layer into a turbulent boundary layer. When air flows over the blade, there is a layer of air known as the boundary layer between the blade surface and where the air is undisturbed. The turbulators eliminate the separation bubble and the drag it creates. See the image of a hub assembly in the color insert.

b.   **Generator.** Like many other manufacturers, the system employs a brushless three-phase AC system using a neodymium permanent magnet construction. Some background on the neodymium magnet, also known as a NdFeB magnet. Because it has really good magnetic properties, it is the main type of permanent magnet motor. The permanent magnet is manufactured from an alloy comprising neodymium, iron, and boron.

c.   **Regulation.** The company has a system called Ever-Loaded Generator Control. This is a temperature-compensated and diversion control system. This system uses the SCR 12 Marine charging regulator for control. The regulator has a secondary function, which is to maintain the wind generator under load. When the battery's full charge level is reached, the PWM circuit automatically starts to divert excess power to a dump resistor. There are two resistors rated at 0.35Ω and 40A. The state of charge is indicated by an LED. See the image of a dump regulator system in the color insert.

4.22 **Wind Generator Ratings.** Ratings curves are always a function of wind speed and are quoted at rated 12V or 24V output voltages. Some performance curve wind speeds are denoted in m/sec or watts; I have converted back to wind speed in knots and current in amps to get the values for comparison. The data is taken from published output curves; if it is slightly incorrect, my apologies.

Table 4-3. Wind-Generator Output Table

Make and Model	Output Current (amps)	Wind Speed (kts)
Aerogen 6	2.5	10
	10	20
	20	30
Air Silent X Marine	4.2	13
	9.2	17
	22.5	21
	35.8	24
Rutland 914i	2.25	10
	11.6	20
	21.6	30
	35.8	24
MarineKinetix Mk4+	2.4	10
	7	15
	15	20
	27	25
Superwind 350-II	3.3	12
	12	15
	15	19
	25	22
	29	25
Eclectic D400	2.5	10
	7.5	15
	15	20
	27	25
	40	30
Rutland 1200	10	2.9 (5 m/s)
	21	21.3 (11 m/s)
	29	35.5 (15 m/s)
Watt&Sea Racing	15	4.2
	20	10.4
	25	20.8
	35	33.3

4.23   Wind Generator Charging Regulation. A number of features are incorporated into wind generators to protect batteries and generators.

   a.   Charge Controllers/Regulators. A regulator is required to limit normal charging voltages to a safe level of 14.5V and to limit output at high wind speeds. On older-technology generators, a shunt regulator was the preferred technique over a normal solar panel regulator, as it is more suited to constant loads. Shunt regulators divert excess current to a resistor, which functions as a heater and dissipates heat through a heat sink. Although shunt or dump controllers are still used, they are being superseded by MPPT devices in some wind generators.

   b.   MPPT Controllers. Maximum power point tracking (MPPT) is a control technique commonly used with variable power sources like solar and wind. It is an electronic DC-DC converter. The basic concept is that an MPPT controller maximizes or optimizes the power output or charging efficiency by matching the generator impedance to the load or battery. The maximum power point (MPP) of a solar array or wind turbine is the operating point at which maximum output power occurs.

   c.   PWM Controllers. Pulse width modulation (PWM) is a mature technology for regulating charging. The concepts are in some cases combined into the latest-generation controller.

   d.   Regulator Interaction. Like solar panel installations, interaction may occur with alternator charging systems. The output or battery input charging voltages confuse one of the regulators when they are running together; the engine alternator, when applied to a battery with wind, water, and solar inputs, can create interaction. The charging should be switched out of circuit or diverted to a battery other than the sensed one (e.g., the start battery).

   e.   Chokes. Some older units incorporate a choke to limit the charge produced at high wind speeds.

   f.   Winding Thermostats. A number of generators incorporate a winding embedded thermostat, which opens in overload conditions when the winding overheats.

   g.   Transient Suppressors. These suppressors are installed to minimize the effects of intermittent spikes being impressed on the charging system. These could damage the rectifier and onboard electronics. The suppressor is usually a voltage dependent resistor (VDR).

4.24   Wind Generator Installation. Selection depends largely on available mounting locations; arrangements vary according to necessity. It is not easy siting a wind generator for optimum output on a yacht, as consideration of surface wind friction close to the sea surface creates turbulence. The higher the elevation, the better the wind flow. Some generators are mounted on the front of the mast; however, most are installed on stern posts.

a. **Stern posts.** The ideal mounting arrangement is on a stern post, which keeps the blades clear of crew and feeds them air coming off the mainsail. One of the major complaints is that wind generators create vibration under load. It is essential that the post section be as thick as possible and well supported. Usually, this extra support is on the stern pulpit (pushpit); some install stainless steel wire stays and others use stainless pipe supports.

b. **Masts and mizzens.** Wind generators are commonly mounted on masts and, in ketches, on the mizzens to improve wind capture with lower laminar flow effects. See the color image of a mizzen-mounted wind generator in the color insert.

c. **Stern arches.** With many installing quite strong and rigid stern-arch structures, wind generators have been installed on short posts. The stern arches are home to a couple of solar panels and a radar scanner along with GPS, AIS, VHF, and other aerials and antenna.

4.25 **Wind Generator Troubleshooting.** Always secure the turbine blades when installing, servicing, or troubleshooting a wind generator. Common failure modes include mechanical damage caused by blade damage from halyards and other causes. The following tests should be carried out.

a. **Safety.** Never approach a fast rotating or spinning generator. Never try to stop the spinning rotor blades using your hands; the blades are sharp edged. Consider the installation location, and be sure that it cannot create a hazard. Don't try to slow or stop the spinning rotor with any object, such as a pole. It may result in fracture or disintegration of the blade, which could create a hazard.

b. **Voltage check.** Check the open circuit voltage using a multimeter. Be aware that open circuit voltages can be as high as 50V in a 12V system and 210V in a 48V system, so a shock hazard exists.

c. **Current check.** If no ammeter is installed on the main switchboard, install an ammeter in line and check the charging current level. If there is no output from the regulator, check the system according to the manufacturer's instructions.

d. **Output check.** If there is no output and the generator has brushes, check that they are not stuck and are free to move. Instead of brushes and commutators, many generators have a set of slip rings installed with brushes to transfer power from the rotating generator down through the post to the battery circuit. They can jam and, on rare occasions, cause loss of power. Inspect the slip ring brushes and check that they are moving smoothly in the brush holders. Check that the O-rings are in good condition.

e.   **Winding thermostat.** Some generators have a winding embedded thermostat. Check with an ohmmeter that it is not permanently open circuited. If it is open circuited, the generator will not charge. The thermostat opens in high wind charging conditions. If the thermostat has not closed after these conditions and the generator case is cold, the thermostat is defective. Regrettably, it cannot be repaired unless a new winding is installed. To get the generator back into service, connect a bridge across the thermostat terminals. Remember that there will be no protection in high wind and heavy-charging conditions, so the winding may burn out.

f.   **Bearings.** Excess vibration may be caused by bearing wear. If the unit is a few years old, renew the bearings. Vibration can be caused by damage to one or more blades, and these should be carefully examined for damage that may cause imbalances. Consider carrying spare bearings.

g.   **Rectifier.** Check the rectifier to verify that it is not open or short circuited. If output is low, the rectifier diodes may have failed. Consider carrying a spare.

h.   **Regulator check.** If the generator output is correct, check for a malfunctioning regulator. The voltage input should be in the 14V to 18V range, and the output approximately 13V to 14V.

i.   **Connection check.** Ensure that all electrical connections are secure and in good condition.

j.   **Protection.** Check that the input circuit breaker or fuses to the battery are not open. Check that the stop switch, if installed, is not closed or damaged.

k.   **Vibration.** Excessive vibration is usually caused by rotor assembly imbalances. If vibration has just started, check the rotor blades for cracks, splitting, chips, edge degradation, dents, or accumulated material. Check all fastenings and ensure that bolts are properly torqued. Many vibration issues are installation related, and this includes improper cross section posts.

WATER GENERATORS

4.26   Prop Shaft Charging Systems. Prop shaft generator systems are either traditional alternators with prop shaft gearing to achieve rated output or alternators wound to generate outputs at low speed. These systems can be used as an extra energy source while under power—but this is not an economic proposition—or to take advantage of a freewheeling propeller under sail. The following points must be considered.

a.   **Cruising Patterns.** The viability of these units depends on your cruising pattern. Only about one-quarter to one-third of your time is spent passage making, so the shaft alternator is used for a limited period.

b.   **Drag.** Under any load, the alternator will brake the shaft by slowing shaft rotation, causing drag and a reduction in vessel speed. On a lightweight vessel, this can be as high as 0.5kt or more. On steel or other heavy-displacement vessels, the inertia of the vessel will generally minimize the drag effect. For such cruising yachts, prop systems are a possible proposition. With an increasing number of yachts opting for two- and three-bladed folding props and sail drives, prop shaft alternators are now relatively rare.

c.   **Output.** The maximum output will generally be in the region of 5A to 10A. The old Lucas unit had a maximum output of 12A, with an approximate output of 1A per knot. Cut-in speed is 600rpm and requires a shaft-pulley ratio of 5:1. A major concern has been gearbox damage due to improper lubrication while freewheeling, but many major gearbox manufacturers have dispelled this fear.

Table 4-4. Water Generator Output Table

Make and Model	Water Speed (kts)	Output (amps)
SailGen (Standard)	2	0.4 (5W)
	4	4.1 (50W)
	6	10.4 (125W)
	8	23.3 (280W)
Cruising 600	2.1	0.2/0.3
240mm/280mm Impeller	4.1	3.4/5.3
	6	12.4/17.6
	8	28.6/35
	10	44/48

4.27 **Hydrogenerators.** Also called submerged generators, or outboard leg type generators, they comprise a forward- or aft-facing, three-bladed propeller or rotor that drives a permanent magnet alternator. The propeller is mounted at the end of a tubular arm at a depth of
approximately 3ft (1m). As a water-driven power source, they are a good option, being easy
to lift and service. There has been a resurgence in their use, with some innovative technology
being used on various systems.

    a. **Drag.** The drag on a submerged generator is always a consideration. It varies
among makes and models and is usually rather overstated.

    b. **Physical Characteristics.** As the electrical alternator is underwater on most
models, the generator housing has double seals, as do the cable glands. The
alternator body is filled with hydraulic fluid to equalize external pressures when
fully immersed. A reservoir is fitted to allow for oil expansion and contraction.

4.28 **Cruising 600-Watt and Sea Hydrogenerators.** These have been used by some
entrants in the Vendee Globe and Volvo Ocean Races. The stern transom–mounted versions use
an aft-facing propeller, which is less prone to weed collection and impact damage. They also
have a permanent hull mounted Pod system not unlike a sail drive, in both Mini and Cruising
versions. The Mini 6.50 range comprises three transom installed models. The POD600 is an
under hull mounted system. They have a permanent hull-mounted pod system not unlike a sail
drive in both mini and cruising versions. Images of both variants are shown in the color insert.

    a. **Cruising 600 Generator.** The generator outputs a three-phase, 40VAC (volt-
age alternating current) output, which is then passed through an inverter to
derive the DC charging voltage. The unit is oil filled for bearing lubrication
and cooling and sealed against water ingress. The Cruising 600 model, with
a 240mm propeller, can deliver a very useful 10A at 5.6kts. The nominal
output power is 600W. The 5-knot output is 7A with the standard 240mm
propeller and 10A with a 280mm propeller. The propeller has been designed
for optimum power generation and is manufactured from polymer fiber.
Maximum speed with a 200mm propeller is 20kts.

    b. **Cruising 600 Regulator.** A maximum power point tracking (MPPT) charge
controller or regulator is used to regulate the battery charging. The solar
input to the MPPT regulator is limited to 50V and 14A, so you can use solar
and water through the same unit. The converter unit is Bluetooth-enabled to
monitor performance with a phone app.

    c. **Installation.** The maintenance is simple and includes checking and cleaning
the conical fitting and propeller. Power supply cable protection is recommended as a 50A fuse. The manufacturer recommends lifting the generator
when the boat batteries are fully charged, as the propeller will freewheel.
Alternatively, you can install a switch and relay to isolate the charge circuit.
There can be interaction when running the engine at the same time, as you
will get two charge sources going into the same battery bank; again, switch
and relay is required, or you can automate this. Check out wattandsea.com.

4.29   Eclectic Sail-Gen Water Generator. This is a variation and innovative reimagining of a towed turbine concept. The unit is easy to deploy and recover, which is an important feature. The system consists of a rigid welded aluminum frame, along with a carbon fiber drive shaft and a cast alloy impeller. Unlike other water generators, this is not a submerged turbine type. Reference images of a submerged and a retracted unit are in the color insert.

a.   **Sail-Gen Water Turbine.** The water turbine is supported by the frame, and the impeller is connected to the alternator on the boat via a glass and carbon fiber–reinforced composite drive shaft with a 1:1 alternator to impeller ratio. There is no gearbox or other mechanical interface, which makes it almost frictionless and increases reliability. The running depth is stabilized by a dive plane or hydrofoil that is mounted on the frame above the turbine. This ensures constant depth and no impact from boat pitching. The impeller running depth is adjusted by altering the dive plane angle minus 10 degrees to plus 20 degrees. Drag is less than 33lb (15kg) at 6kts.

b.   **Sail-Gen Performance.** At typical passage speeds of 5kts to 6kts, the Sail-Gen is capable of matching the power consumption on many cruising yachts. The cut-in boat speed for a standard impeller is just 1.7kts, 3.1kts with the high-speed impeller.

c.   **Sail-Gen Alternator.** This is a direct-drive unit with a three-phase AC permanent magnet generator. The output is rectified in the regulator; voltage outputs include standard 12V, 24V, and 48V, the latter becoming increasingly important.

d.   **Sail-Gen control and regulation.** Overcharging control is prevented by the use of a diversion or dump-type regulator, which is rated at a minimum of 40A at 12V. The excess current is switched from the battery charging circuit using either heavy-duty relays or an electronic pulse width modulated (PWM) circuit or chopper circuit. A chopper circuit converts a fixed DC voltage input into a variable or adjustable DC voltage output, so it is a DC-to-DC converter. It is called a DC chopper, as it chops or turns the DC supply on and off at high speed to control the output. The diverted charge is connected to a large wire-wound cylindrical ceramic resistor, where the energy is dumped and dissipated as heat; essentially, they are heating elements. Not all energy is diverted; some load remains on the generator and has a braking effect on the wind turbine. Installation considerations include a well-ventilated space where good air circulation is available. Interconnecting electrical cables between the regulator and the dump load should be as short as practicable and rated properly; it's best to oversize the cable size to reduce voltage drop. When a PWM regulator feeds a dump load, the pulsed current may create an audible buzz or tinkling sound from the dump resistors. MPPT-enabled dump regulators are not available for boats, although some exist for shore-based micro-wind systems.

e.   **Sail-Gen Duo-Gen Version.** Having a versatile charging system is an attractive option. The Sail-Gen is convertible to a wind generator. This offers some great options for maximizing alternative energy inputs. Changing over from water to wind is uncomplicated. In water mode, the performance is as stated above. The Duo-Gen uses the same alternator in wind mode. In water generation mode, the rated output is 100W at a speed of 6kts; at 7kts it is 150W; at 8 kts it is 200W. Typical daily production is around 200Ah. Typical drag when the system is operating is around 0.15kt. When in wind mode, the output at 10kts wind speed is 40W. At 15kts this is 90W; at 20kts it is 150W. (See image in the color insert.) Visit eclectic-energy.co.uk.

4.30   Towed Turbine Charging Systems. Towed turbines are still used by some and are typified by the Ampair 100 and Waterpower 200 from Hamilton Ferris. Water-based charging systems come in two configurations.

a.   **Towed Turbine Generator.** The towed turbine water generator is essentially a slow speed alternator with the drive shaft mechanically connected to a braided rope and turbine assembly. When streamed off the stern, the turbine turns and rotates the alternator: the same principle as the Walker towed logs for those who remember them. I haven't sighted one of these for several years now. Typical output is approximately 6A. The trail rope is typically around 100ft (30m) in length:

(1)   **Drag.** Typical drag speed reduction is around 0.5 kt. The trailing generator impeller, like the old-fashioned trailing log, is quite reliable; hungry ocean denizens rarely eat the turbine, although it happens. Paint the turbine black if you use one. You can get weed or fishing net and other debris tangled around it, which is often an issue.

(2)   **Turbine Skipping.** One problem is that the turbine or impeller tends to skip out of the water at speeds over 6kts. There are a variety of methods to reduce skipping, including adding sinker weights to the turbine, increasing the towline length, and increasing the towline diameter. The Ampair units have two turbine types: one for speeds up to 7kts and another coarse-pitch turbine for higher speeds.

4.31   Fuel Cells. Fuel cells are now available for sailing yachts and are an alternative power charging source of the future. The cell converts the chemical energy of a fuel such as hydrogen, natural gas, or methanol and an oxidant such as air or oxygen into water and outputs electricity. In principle, the fuel cell operates similarly to other electrochemical devices such as the battery. A fuel cell will produce electrical and heat output as long as the fuel and an oxidizer are available in the required quantities. The battery similarities are that both are electrochemical devices. Both have a positively charged anode and a negatively charged cathode. They both have an ion-conducting material that is termed an electrolyte. Fuel cells come in several varieties; each type utilizes different chemistry, and the classification is based on the electrolyte material. The proton exchange membrane fuel cell (PEMFC) device is the most common in boat charging systems.

4.32 **Fuel Cell Construction and Operation.** The basic construction consists of a fuel electrode, the anode, and an oxidant electrode, the cathode. The anode and cathode are separated by an ion-conducting membrane. Oxygen is continuously passed over one electrode, and hydrogen is continuously passed over the other. They chemically combine the molecules of a fuel and oxidizer without any combustion. The by-products of a fuel cell are a small quantity of carbon dioxide (CO_2), water, and minimal heat.

a. **The Anode.** The anode is the negative part of the fuel cell. The anode conducts electrons that are released from the hydrogen molecules; these electrons can then be used in an external electrical circuit. The anode has a series of channels etched into it. This evens dispersal of the hydrogen gas across the catalyst surface.

b. **The Cathode.** The cathode is the positive part of the fuel cell. It has a series of channels etched into it to aid in the even distribution of oxygen across the surface of the catalyst. The cathode conducts the electrons in the external circuit from the catalyst. They recombine with hydrogen ions and oxygen to release water.

c. **The Electrolyte.** The electrolyte is the proton exchange membrane (PEM). This material with special treatment resembles kitchen plastic wrap and will conduct the positively charged ions. The membrane will block the passage of electrons.

d. **The Catalyst.** The catalyst is made from a special material that triggers or is the catalyst for the reaction between the oxygen and hydrogen. The catalyst comprises a thin coating of platinum powder layered onto a substrate of either carbon paper or cloth. The catalyst is fairly rough and very porous. This is designed so that a maximum surface area of the platinum material is exposed to the hydrogen or oxygen, ensuring a maximum reaction. The catalyst is oriented toward the PEM.

e. **The Reformer.** Hydrogen used to be a major component of basic fuel cells; however, it is not readily available. To overcome this, a device called a reformer is used. The reformer converts hydrocarbon or alcohol fuels into hydrogen, which is then fed to the fuel cell. Reformers generate heat and produce other gases in addition to hydrogen, which tends to lower the efficiency of the fuel cell. Methanol (CH_3OH) is a farmed biofuel distilled from sugarcane.

4.33 **Fuel Cell Efficiency.** The ideal fuel cell is powered with pure hydrogen, and these cells can be up to 80% efficient. When a reformer is used to convert more readily available methanol to hydrogen, efficiency falls to approximately 30% to 40%. When the electrical energy is converted into mechanical work using either an electric motor or an inverter, the overall efficiency drops to about 24% to 32%. This makes the fuel cell considerably more efficient than solar or wind generators. The chemical reaction in a single fuel cell produces a fairly low 0.7V.

Alternator Charging Systems

The alternator is a robust and reliable piece of equipment and is the principal charging source on most boats. Automotive alternators, or derivatives of them, are used in the majority of boat charging systems. Most alternators, however, are incorrectly rated for the installed battery capacity and are therefore unable to properly restore the discharged current. The alternator is generally sized by the engine manufacturer and is designed to recharge the engine start batteries. Some have standard sizes, and some engine makers are finally offering much larger output units as well as additional alternators as an option. Alternator ratings have increased from 50A up to 160A in some cases. Manufacturers have managed to improve output while still maintaining the same physical footprint. They have improved fan cooling design and rectifier, voltage regulator, and stator.

5.1   Alternator Basics. The basic alternator consists of several components, described below. The typical automotive type alternator generates a three-phase AC alternating current, which is where it derives its name. This is then rectified to produce a DC output for charging via the full wave bridge rectifier. There are variations in design, and some twelve-pole units are rated to run at up to 20,000 rpm with modified cooling arrangements. The general principle is that the rotor, which comprises claw poles surrounding the windings, rotates inside the stator winding. The rotating magnetic field then induces a current into the stator stationary windings. The north and south poles of the rotating field intersect the stationary windings and generate an output current.

5.2   Alternator Components. The standard alternator is a claw-pole synchronous machine comprising the following components.

      a.   **Stator.** The stator is the fixed winding. It consists of a three-phase winding that is connected in a "star" or "delta" arrangement. The three winding coils are wrapped around an iron core to increase field strength. The windings are formed onto a solid laminated core and supply three phases of alternating current to the rectifier.

      b.   **Rotor.** The rotor is the rotating part of the alternator. The shaft has the pole or claw-shaped magnet poles attached; excitation or field winding, rotor shaft, cooling fan at one end; bearings; and the collector slip rings. The rotor has sealed and greased bearings at the drive and non-drive ends of the housing. Some alternators have internal cooling fans.

      c.   **Rectifier Assembly.** This consists of the diodes, heat sink or diode plate, and terminals. The rectifier consists of a network of six diodes, which are connected across the positive and negative plates. These plates function as heat sinks to dissipate the heat from power conversion. This rectifies the three generated AC phase voltages into the DC output for charging. Two diodes are used on each winding to provide full wave rectification. In some alternators, Zener diodes are used to limit voltage peaks that arise during sudden load changes.

d. **Exciter Diodes.** The exciter (D+) or preexcitation diodes consist of three low-power diodes that independently rectify each AC phase and provide a single DC output for the warning light or auxiliary control functions. They are required, as the residual magnetism (or remanence) in the iron core is insufficient at low speeds and starting to initiate the self-excitation required to build up the magnetic field. This only occurs when the alternator voltage is higher than the voltage drops across the two diodes. The warning lamp functions as a resistor and provides preexcitation current, which generates a field in the rotor. In this respect the power or watts rating of the lamp is important; 2W is typical.

e. **Brush Gear Assembly.** This comprises the brush housing and the brushes, which are made of copper graphite. The brushes are spring-loaded to maintain correct slip ring contact pressure, and the brush wires are soldered to the terminals. The brushes provide power to the field winding through the slip ring.

f. **Voltage Regulator.** The voltage regulator is usually combined with the brush gear; some have external regulators mounted adjacent to it. The field control output of the alternator is connected to one of the brush holders, which then supplies the rotor winding though the slip ring. Regulator sensing is normally connected to the D+ output circuit. The regulator maintains a constant voltage output over the entire operating range of the alternator and does so by modulating the field current to maintain a stable voltage. Earlier electromagnetic contact–type regulators are relatively uncommon now, with most being electronic types with no moving parts. The electronic regulator allows precise control, with short field switching periods. Regulators are varied in design and use a combination of transistors and Zener diodes. The latest electronic devices have various monolithic integrated circuit or voltage regulator chips; some have variable PWM output functions.

g. **Warning Light.** While not specifically part of the alternator, the warning light circuit is not simply for indicating failure; the lamp excites the alternator. In many cases, an alternator will not operate if the lamp has failed, because the remanent voltage or residual magnetism has dissipated. Ideally, a lamp should be in the range of 2W to 5W. Undersized lamps are often characterized by the need to "rev" the engine to get the alternator to "kick" in. This is often highly visible with alternator-driven tachometers. Many newer engine panels have a printed circuit board–type alarm panel.

5.3   Alternator Selection. Boat owners have a number of important factors to consider when selecting alternator output ratings. Along with regulators, the alternator is probably the most common item to fail on board; therefore, careful selection is required. The factors are summarized as follows.

a. **Engine Run Times.** The engines in a majority of cruising vessels are run excessively in an attempt to recharge batteries. The maximum run time goal is 1 hour in the morning and 1 hour in the evening, which coincides with mechanical refrigeration pull-down times.

b. **Engine Loading.** Diesel engines should not be run with light loads because unloaded engines suffer from cylinder glazing. Typical horsepower loads are approximately 1 hp for every 25A, so a 100A alternator will take 4hp when fully loaded and a 150A alternator around 6hp.

c. **Engine Speeds.** Ideally, the engine should be able to charge at maximum rates at relatively low speeds. The preferred speed is generally a few hundred revolutions per minute above idle speed. Alternator speed is dependent on the drive-pulley ratio and the alternator cut-in speed.

d. **Battery Capacity.** Nominal charging rates are specified by manufacturers, and they generally specify starting and finishing rates. A battery requires the replacement of 120% of the discharged current to restore it to full charge. This value is required to overcome internal resistances within the battery during charging.

e. **Charging Current.** As a battery is effectively self-limiting in terms of charge acceptance levels, we cannot simply push in the discharged value and hope the battery will recharge. During charging the battery is reversing the chemical reaction of discharge, and this can occur only at a finite rate. If possible, therefore, the alternator must recharge the battery at the optimum charge rate specified. Charging by necessity tapers off as full charge is reached, which is why start and finishing rates are specified. These ratings are largely impractical in marine installations. Alternator output current is the sum of electrical loads on the system during the charging period, plus the actual battery charging current.

f. **Charge Voltage.** The majority of alternators have a fixed output of 14V, with some makes having the option of regulator adjustment up to around 14.8V for isolation diode voltage drop compensation. Charge voltage is probably the single most important factor in charging, as all other factors are related to it.

g. **Alternator Output Current Selection.** From the power analysis table, we have calculated the boat's maximum current consumption. Added to this is a 20% margin for battery losses, giving a final charging value. One popular opinion is that alternator ratings should be approximately 30% of battery capacity. In practice, this is at best optimistic and difficult to achieve. I used to specify and install an 80A alternator, which is about the highest rating possible without going into high-priced or exotic alternators; however, that has changed with the availability of 160A alternators.

h. **Marine Alternators.** Marine alternators are essentially enclosed and ignition protected with an Underwriters Laboratories (UL) listing to prevent accidental ignition of hazardous vapors. Windings are protected to a higher standard by epoxy impregnation. Marine units have a corrosion-resistant paint finish and are designed for higher ambient operating temperatures.

i. **Marinized Alternators.** An alternator can be marinized to a reasonable degree. If you wish to marinize and improve your alternator, bearings should be totally enclosed; replace if they are not. Windings should be sprayed or encapsulated with a high-grade insulating spray. The back of the diode plate can also be sprayed with an insulating coating to prevent the ingress of moist, salt-laden air and dust, which can short out diodes and connections.

5.4 **High-Output Alternators.** It is a regrettable fact of life that many so-called marine electrical people push high-output alternators, typically 150A or more, as the first step toward solving battery charging problems. These alternators are expensive, require a lot of reengineering, and, in many cases, mask the more common problems of poor circuit design, poor installation, and inadequate regulation. Be warned! A high-output alternator will not necessarily solve your charging problems unless considered in a holistic approach to the battery bank requirements, the battery technology, and the charge regulation required. With new battery technologies with high charge acceptance rates, the sizes can be up around 160A. A considerably cheaper and more reliable solution may be to replace the regulator or use a smart regulator. If you choose to upgrade your alternator, install a quality unit such as those from Balmar. Balmar is probably the world's leading innovator of alternators and charging systems for boat applications; they are described below.

a. **Balmar 6-Series Alternators.** These are available in 70A, 100A, and 120A output models. They incorporate Balmar's patented Smart Ready technology. For optimum cooling, they have dual fan cooling and a high airflow frame. The maximum speed is 12,000 rpm. These alternators are built to comply with USCG Title 33, International Standards Organization (ISO), and Society of Automotive Engineers (SAE) and are CE compliant.

b. **Balmar XT-Series alternators.** These are available in 170A or 250A in a small frame package. They are able to output up to 180A at idle speeds and are suitable for lithium-ion batteries. For optimum cooling, they have dual fan cooling and a high airflow frame. They incorporate Balmar's patented Smart Ready internal regulator technology and are built to comply with USCG Title 33, ISO 8846, and SAE J1171 certification. What sets these alternators apart is that they are constructed using a braided stator wire design. These alternators have ninety-six slots, whereas traditional alternators have just thirty-six slots in an S-wound stator. This gives significantly improved electromagnetic efficiency. These alternators require dual-V or multigroove serpentine belts.

c. **Mastervolt Alpha Pro III.** This is a three-step—bulk, absorption, and float—charge regulator for both standard and high-performance alternators. Initial charging starts with bulk current output, and maximum output is limited by the alternator and engine speed. The battery charge voltage will increase until it reaches the absorption voltage level. This is then followed by the float stage, which maintains the maximum charge level.

5.5   48V High-Output Power Systems. There have been developments in this space, and the technology and applications will continue to grow. There are 48V deck winches, windlasses, thrusters, and air conditioners. Recharging 48V battery banks is also advancing rapidly to keep pace with these innovations.

a.   **Balmar 96-Series 48 Volt Alternators.** Given the rise in higher voltages in electric propulsion and other areas, Balmar has launched a 48V alternator that is combined with their new MC-620 regulator. They are available in 60A and 100A.

b.   **Integral Generator.** This innovation has a very high 9kW output. The system is essentially a very large alternator to leverage the power of your diesel engine. It is more than a high-output alternator with major improvements to the alternator. There is increased efficiency of the magnetic flux coupling between the stator and rotor. The winding design has been optimized for very high output at low engine revolutions. This is to enable outputs as high as 3.5kW at low engine idle speed, along with enhanced cooling airflow. The engine mounts have been upgraded to reduce engine side loads, reduce shock loads, and maximize belt life. A system comprises the following elements: the alternator, the integral intelligent controller, integral battery sensors, the integral power converter (48V to 12V or 24V), high-power automatic safety switch, and optional inverter to AC. The system power control innovatively uses pulse width modulation (PWM) to control and regulate battery charging, and the charging level is dependent on temperature, battery charge state, and other factors. I really like this concept; it leverages the engine power you have, along with the much more efficient 48V system and the lithium-ion battery bank. Visit integrelsolutions.com for details.

5.6   Overvoltage and Surge Protection. Some alternators are provided with separate surge protection units. Overvoltage protection comprises several methods.

a.   **Balmar APM (alternator protection module).** This newly released product addresses one of those serious issues. This unit is mounted on the rear of the alternator. It is designed to absorb voltage and current spikes that frequently occur on boats, which include faulty connections, overvoltage, and switching events. These are designed to accommodate lithium-ion battery issues, when a voltage surge is generated, and when a battery-initiated disconnect occurs. The result can wreck alternator diodes and regulators. The APM is designed to cope with both short- and long-duration transients. The APM-12 is able to clamp a 60V spike and overvoltage events that start at around 20V. The unit has both audible and visual alarms to indicate status. The device complies with ISO 16750-2 for Load Dump Protection and ISO 7637-2 for Surge Protection.

b. **Zener Diodes.** As described earlier, the rectifier diodes are Zener diodes that limit the high voltage spikes or peaks that arise below a safe value and can damage the regulator. The typical limiting voltages of Zener diodes in use are 25V to 30V for 14V alternators and 50V to 55V for 28V alternators.

c. **Surge-Proof Alternators.** Some alternators are equipped with high-specification components. The components are rated up to 200V for 14V systems and 350V for 28V systems. This is supplemented by installation of a capacitor across the alternator output and ground. The older Lucas/CAV alternators incorporate a surge protection avalanche diode within the alternator (ACR and A115/133 range). This protects the main output transistor in the regulator.

d. **Overvoltage Protection Devices.** Often these are installed only in 28V alternators. These electronic semiconductor devices are connected across the alternator output. They operate by short-circuiting the alternator though the excitation winding when peaks rise over a set value. Some alternators use what is called a freewheeling diode, anti-surge, or suppressor diode. This is connected in parallel with the excitation winding of the alternator.

e. **Additional Protection**. A metal oxide varistor (MOV) installed across the B+ and negative terminals will provide additional surge protection. Another good method is to solder a capacitor rated at 0.047µF/250V across each of the AC windings.

f. **Interference Suppression.** Alternator diode bridges create noise (RFI) that can be heard on communications or electronics equipment. Always install an interference suppression capacitor. As a standard, install a 1.0µF (1 microfarad) capacitor. In some cases, a suppression capacitor is required on the main output terminals.

g. **Alternator Filters.** The filter is rated for 150A and is designed for installation in the alternator output lead adjacent to the alternator. It will attenuate noise in the 70kHz to 100MHz range that commonly affects GPS and radios. Physically, they are relatively large and heavy, at 3lb (1.5kg), so will require careful fastening on the engine or an adjacent bulkhead.

5.7   Alternator Installation. Optimum service life and reliability can only be achieved by correctly installing the alternator. The following factors must be considered during installation.

a. **Alignment.** It is essential that the alternator-drive pulley and the engine-drive pulley be correctly aligned. Pulley alignment is critical to belt performance and must be correct prior to installing belts. The two types of misalignments are parallel and angular. The term "parallel misalignment" is given to pulleys that are located outside the plane of any other pulleys within the drive system, but with shafts that are parallel with the other components. Angular misalignment is where pulleys are installed within the drive system plane but are tilted, as the shafts are not parallel. The outcome is that belt tracking issues arise, along with excessive belt wear and belt instability, often characterized by a chirping sound from the belt in operation. Misalignment of just a few degrees can result in significant belt temperature increases and a belt life expectancy reduction of up to 50%. A general rule is misalignment should not exceed 0.1in per 10in of span (2.5mm per 250mm of span).

b.   **Drive Pulleys.** Drive pulleys between the alternator and the engine must be of the same cross section. Differences will cause belt overheating and premature failure. Ideally, the split, automotive-type pulleys on some alternators should be replaced by solid pulleys of the correct ratio as the one pulley on a yacht is absolutely critical to power generation. As the pulley rotates, centrifugal forces will act on it. If the pulley mass is unevenly distributed around the axis of rotation and is unbalanced, the centrifugal forces will consequentially be unbalanced; the result will be vibration. Uneven mass results from machining intolerances and material casting inconsistencies. These vibrations will be transmitted to bearings and any other machine parts, including mounts and bolts. Eventually you will get catastrophic failure—and who carries a spare pulley? Most pulleys are statically balanced to reduce this. There are international standards relating to V-belt drives.

c.   **Drive Belt Tension.** Belts must be correctly tensioned. Maximum deflection should not exceed 0.5in (12mm). When a new belt is fitted, the deflection should be readjusted after 1 hour of operation and again after 10 hours. If the drive belt does not have sufficient tension, it will begin to slip, causing heat to be induced into the belt and the system. This heat buildup, if left unchecked, will result in noise, a loss of system performance, and possibly system failure. The major causes of failures are improper low belt tensioning, a failure to run in and then re-tension with new belts, and no continuous monitoring and tension adjustment. Pulley groove wear contributes, along with belt sidewall wear and permanent belt elongation. If a belt is not tensioned correctly, the belt slips, the sidewalls wear smooth, and the belt eventually hardens as a result of glazing or heat aging. The greater the glazing of the belt surface, the more noise will be emitted, and the cause is belt stretch. Over-tensioned belts can reduce pulley bearing life due to excessive hub loads. Excessive tension results in increased belt wear, increased belt temperatures, and belt failure. Alternator bearing failure occurs as the bearing gets hot and melts out the grease. Be aware that belts can build up a static charge and, while not an ignition risk on diesels, can contribute to radio frequency interference (RFI).

d.   **Drive Belts.** Belts must be of the correct cross section to match the pulleys. Notched or castellated and serpentine belts are ideal in the engine area, as they dissipate heat easily. If multiple belts are used, always renew all belts at the same time to avoid varying tensions between them. For any alternator over 80A, a dual-belt or serpentine belt system should be considered because a single belt will not be able to cope with the mechanical loads applied at higher outputs, as they slip and overheat. See the section on pulleys below. When replacing belts, never attempt to force the belt onto a pulley using a lever or screwdriver. You can easily cause irreversible damage to the cord and fabric by tearing it. Loosen the alternator, and make it easier to install. The optimum tension is the lowest tension at which the belt will not slip under peak loads, such as fully loaded alternator output. When replacing a belt, check and re-tension after a 24- to 48-hour run period. Application of a belt dressing is not recommended. Serpentine belts are appearing on many vessels (and motor vehicles) to drive high output alternators, water pumps, refrigeration compressors, and other components. They are able to handle loads better with increased traction.

e. **Pulley Maintenance.** Even a pulley requires some regular inspection and maintenance. Pulley and belt problems usually arise because of improper alignment. Pulleys must be correctly aligned, and the belt must be in good mechanical condition, be clean, and have no contamination. Check the pulley for rust, oil, grease, dust, and dirt—contaminants accelerate wear and reduce belt life. Clean the pulley. Dirt and dust can cause belt slippage, while oil and grease will result in a reduction in belt traction and can destroy the belt surface. Inspect pulleys for wear; a worn pulley can significantly reduce belt life, and extreme wear leads to the belt bottoming in the groove. This causes belt slippage and excessive heat generation. You can utilize a pulley groove gauge to check the amount of wear. A V-belt should be riding at least flush with the top of the pulley. Check the pulley for burrs, nicks, gouges, and severe scratches, as these all lead to decreased belt life. It is good practice to use some fine emery cloth or similar annually to take the shine off and improve the contact surface for the belt.

f. **Belt Troubleshooting.** A squealing or chirping belt is indicative of belt slippage, usually from low belt tension. If the belt is correctly tensioned, the cause of the squealing could be oil, grease, dirt overload, or misalignment. The condition needs rectifying; otherwise, belt life will be shortened. Belt chirping is often due to misalignment and occurs when dust is present, or in very wet or dry conditions. Do not use belt dressing, which will degrade the belt surface and shorten its life. If a belt turns over, this is usually caused by severe misalignment, very badly worn pulleys, lateral vibration, foreign material, or pulsating loads. Cracked belts reduce the belt tensile strength and will degrade quite quickly, requiring renewal. Do not use belt dressings; they chemically attack the belt surface.

g. **Ventilation.** An alternator, similar to electrical cable, cannot achieve the rated output in high temperatures. Ideally, a cooling supply fan should be fitted to run when the engine is operating, and its airstream should be directed toward the alternator. Many alternator failures occur when boost charging systems are installed, because such systems run at near maximum output for a period in high ambient temperatures.

h. **Mountings.** There are variations in how engine manufacturers use single- and dual-foot alternator arrangements. Alternator mountings are a constant source of problems. When tensioning the alternator, always adjust both the adjustment bolt and the pivot bolt. Failure to tighten the pivot bolt is common and causes alternator twisting and vibration. Vibration fatigues the bracket or mounting and may cause it to fracture. Additionally, this can cause undercharging and radio interference. Ensure that the slide adjustment arm is robust. Most marine diesel engines have a vibration level that will fatigue the slide and eventually break it.

5.8   **Alternator Drive Pulley Selection.** Ideally, maximum alternator output is required at a minimum possible engine speed. This is typically a few hundred rpms above idle speed. Manufacturers install alternators and pulleys assuming that the engine is run only to propel the vessel, when in fact engines spend more time at low engine revolutions, functioning as battery chargers. Alternators have three speed levels that must be considered, and the aim is to get full output at lower speeds.

a. **Cut-in Speed.** This is the speed at which a voltage will be generated.

b. **Full Output Operating Speed.** This is the speed at which full rated output can be achieved.

c. **Maximum Output Speed.** This is the maximum speed allowed for the alternator; otherwise, destruction will occur.

d. **Pulley Selection.** An alternator is rated with a peak output at 2,300rpm. At a typical engine speed of 900rpm and a minimum required alternator speed of 2,300rpm, a pulley ratio of approximately 2.5:1 is required. The alternator's maximum output speed is 10,000rpm. Maximum engine speed is 2,300rpm, so 2,300 multiplied by 2.5 = 4,000rpm. This falls well within safe operating limits and is acceptable. A pulley with that ratio would suit the service required.

e. **Pulley Selection Table.** Table 5-1 gives varying pulley ratios with an alternator pulley diameter of 2.5in (63mm).

Table 5-1. Drive Pulley Selection Table

Engine Pulley Diameter	Pulley Ratio	Engine RPM	Alternator RPM
5in (127mm)	2:1	2,000	4,000
6in (152mm)	2.4:1	1,660	4,000
7in (178mm)	2.8:1	1,430	4,000
8in (203mm)	3.2:1	1,250	4,000

f.   **Alternator Characteristics.** The graphs below illustrate the typical relationship between output current, efficiency, torque, and horsepower against rotor
revolutions. The optimum speed can be selected from these characteristics.

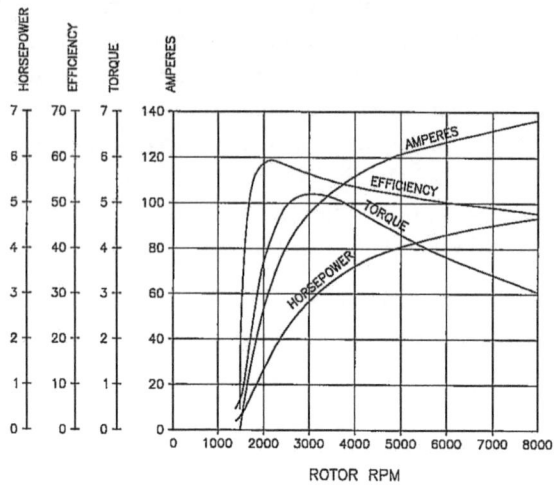

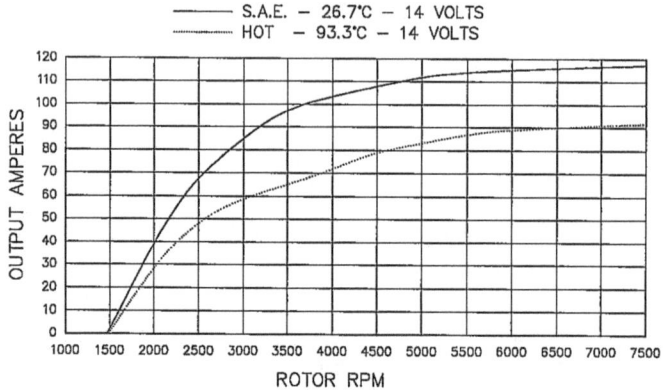

Figure 5-1. Alternator Output Characteristics

5.9   Alternator Maintenance. Many alternator failures can be avoided by performing
basic maintenance tasks. One statistic I have seen states that around 65% of alternator performance issues are directly attributable to battery issues. Around 15% of problems are directly
related to bad electrical connections, degraded ground connections, and battery cable and
wiring. Around 10% of actual alternator failures are caused by drive belts. They need to be
tensioned correctly, as noted earlier, and should not have any polished areas, cracks, or other
visible damage. The remaining 10% is due to various causes, such as overheating, bearing
failures, and spikes that blow rectifier diodes. The following tasks should be carried out.

a.   **Drive Belts.** Check the belt tension monthly and adjust. Deflection is 0.5in
(12mm) maximum. Examine for cuts, shiny areas, uneven wear, or fatigue
cracks. Ensure the belts are clean, with no oil or grease on the surface.

b.   **Connections.** Check monthly; clean and tighten all alternator terminals. Check cable and connectors for fatigue.

c.   **Vibration.** Check alternator monthly for vibration when running. Examine mounts for fatigue cracks.

d.   **Bearings.** Check every 1,500 operating hours. Remove the alternator and turn the rotor. Listen and feel for any bearing noises. Renew every 3,000 hours or at a major overhaul. See the section on bearings below.

e.   **Brushes.** Check every 1,500 operating hours. Check the brushes for excess or uneven wear. Check the slip rings for scoring.

f.   **Cleaning.** Clean yearly. Wash slip rings, diode plate, and brush gear with electrical solvent. Do not use any abrasives on slip rings. They must be cleaned only to preserve a film that is essential for brush contact. Ignore this step if you have a brushless alternator that does not have slip rings. Wash out windings and dry.

g.   **Pre-Cruise.** If you are about to embark on a long and extended cruise, I suggest removing the alternator and taking it to a quality marine or automotive electrical workshop. Take your starter motor along as well. Request a test of the alternator output for maximum current. Check the diodes. Clean the windings, slip rings, and brush gear. Renew worn brushes and renew the bearings with high-quality replacements.

5.10   Alternator Faults and Failures. Failures in alternators are primarily due to the following causes, many of which can be prevented with routine maintenance. An alternator with faulty diodes tends to make a whining or whistling sound. The greater the number of failed diodes in the rectifier bridge, the greater the sound level.

a.   **Diode Bridge Failures.** Diode failures are generally caused by the following:

(1)   **Reverse Polarity Connections.** Reversing the positive and negative leads destroys the diodes. This is a common occurrence.

(2)   **Short-Circuiting Positive and Negative.** A short circuit will cause excess current to be drawn through the diodes and the subsequent failure of one or more diodes. The most common cause is reversing the battery connections.

(3)   **Electrical Transients.** These comprise spikes and surges. A high-voltage surge is generated by the inductive effect of the field and stator windings. This occurs if the charge circuit is interrupted, most commonly when an electrical battery selector switch is accidentally opened. Short-duration, transient voltages several times greater than the nominal voltage can be caused by high inductive loads when starting up, such as a large pump, or a deck winch and windlasses.

Spikes are also caused by lightning strikes. Countermeasures are covered in the lightning protection chapter.

b. **Alternator Winding Failures.** Stator winding failures are usually due to the following:

(1) **Overheating.** Normally due to insufficient ventilation at sustained high outputs, causing insulation failure and inter-coil short circuits.

(2) **Short-circuiting.** Due to mechanical winding damage, overheating, or ingress of moisture.

(3) **Rotor Winding.** Short circuit or a ground fault due to overheating or overvoltage if the voltage regulator fails.

c. **Alternator Brush Gear.** Brush gear failures are not that common in a properly maintained alternator, but when they occur they are generally due to the following:

(1) **Brushes.** Brushes worn and sparking; characterized by fluctuating outputs and radio interference.

(2) **Slip Rings.** Scoring and sparking due to buildup of dust, causing radio interference.

d. **Alternator Bearing Failure.** The first bearing to fail is normally the front drive end pulley bearing as it is the load bearing end. Rotating it by hand will usually indicate grating or noise. It is important to check these regularly because a catastrophic bearing failure may cause bearing collapse; the rotor will be misaligned and tear up the stator. Causes are excessive vibration and misalignment of belts. Slipping belts and heat can raise bearing temperatures and melt the grease out, causing reduced lubrication and failure. Alternator bearings are designed to cope with both speed and temperature fluctuations. Cheap bearings will degrade quickly at relatively low temperatures. Top-quality bearings, such as those from SKF, Timken, NTN, FAG, and NSK, to name a few, incorporate high-quality polyamide ball cages and high-temperature contamination-resistant seals, usually double sealed with hybrid polyacrylic seals or similar. The bearings will be charged with a quality high-temperature grease rated to around 347°F (175°C).

5.11 Alternator Troubleshooting. Troubleshooting should be carried out in conjunction with charging system troubleshooting, as described in Table 5-4.

a. **Check Output.** This initially depends on the warning lamp and the regulator. Using a voltmeter, check that the output across the main B+ terminal and negative or case rises to approximately 14V. No output indicates total failure of either alternator or regulator. Partial output indicates that some alternator diodes may have failed or there is a regulator fault.

b.   **Regulator Check.** If there is no output either the alternator is faulty or the regulator is failing to excite the alternator. This is not difficult to address with external regulators, but if an internal regulator is fitted, the alternator will need to be removed, opened, and a wire attached to the brush holder. Switch off all electrical and electronic equipment at the switchboard circuit breaker before commencing test. If in doubt, don't try it. Check that the alternator gives full output by shorting the wire to negative in negative-type machines or positive in positive types. If the alternator gives full output voltage, the regulator is probably faulty.

c.   **Alternator Test.** The other components are tested after confirming the function of the regulator. Initially it is recommended that you remove the alternator and take it to any good automotive electrician with a test bench if in port. This saves a considerable amount of time and effort. If you don't carry spares, you can do little. To get you home with partial diode failure, the regulator may require disconnection and a full field voltage applied to get maximum output.

d.   **Auxiliary Diode and Warning Light Tests.** On some occasions, the auxiliary diodes may fail. Select the DC voltage setting on the multimeter and connect across 61/D+ and negative. If there is any reading, the diode may be faulty. Turn on the ignition key without starting. The reading should be around 1V to 2V. If lower, the wiring may be faulty; if higher, the diode may be faulty, or there is excessive rotor resistance or a bad connection. Check that the warning light is operating and on when the ignition switch is turned on. If the light isn't on, the lamp may be faulty, seated badly if a replaceable lamp, or there is a lamp connection fault. Check that the wire has not dislodged from the D+ terminal or the connection is loose.

e.   **Rotor Testing.** If a regulator has failed, particularly where it has failed due to an overcharge condition—and prior to replacing the regulator—the rotor should be checked for damage. The test is as follows:

   (1)   **Test Insulation Resistance.** Select the multimeter resistance setting, place one multimeter probe on a slip ring and the other on the rotor core. Resistance should be infinite or overrange.

   (2)   **Test Winding Resistance.** Place the multimeter probes on each slip ring. Resistance should be around 4Ω. If it is very high, an open circuit may exist; if very low, a coil short circuit may exist.

5.12   Alternator Terminal Designations. Alternators have a variety of different terminal markings, which are listed in Table 5-2.

Table 5-2. Alternator Terminal Markings

Make	Output	Negative	Field	Auxiliary	Tachometer
Bosch	B+	D–	DF	D+/61	W
Lucas	BAT	E	F	L	
Paris-Rhone	+	–	DF	61	W
Sev Marchal	B+	D–	DF	61	
Motorola	+	–	F	AUX	AC
CAV	D+	D–	F	IND	
AC Delco	BAT	GND	F		
Niehoff	BAT+	BAT–	F	D+	X
Valeo	B+	D–		D=	W
Mitsubishi	B+	E	F	L	
Nippon Denso	B+	B	F	L	
Prestolite	POS+	GND		IND LT	AC TAP

5.13   Alternator Regulators. The regulator is the key to all alternator charging systems. The function of the regulator is to control the output of the alternator and to prevent the output from rising above a nominal set level, typically 14V. Higher voltages would damage the battery, alternator, and equipment.

a.   **Principles.** An alternator produces electricity by the rotation of a coil through a magnetic field. The output is controlled by varying the level of the field current. This is achieved by applying the field current through one brush and slip ring to the rotor winding, and completing the circuit back through the other slip ring and brush. Essentially, the regulator is a closed-loop controller, constantly monitoring the alternator output voltage and varying the field current in response to output variations.

b.   **Regulator Operating Range.** A regulator does not control the charging process significantly until the battery's charge level reaches approximately 50%. When the voltage of the battery rises to this threshold, the regulator starts limiting the voltage level. The charge current levels off as the voltage level rises; this is called the regulation zone.

c.   **Standard Regulators.** The traditional automotive alternator is fitted with a regulator designed for vehicle service. This requires replacement of a relatively small amount of discharged power within a short time. The alternator then supplies the vehicle's electrical power as the engine runs. This arrangement is totally inadequate for marine applications. To recharge a battery properly, the charging system must overcome the battery's counter voltage, which increases as the charging level increases. The typical scenario is one of a high charge at initial start-up and then a rapidly decreasing current reading on the ammeter. As a result, few boat batteries are ever charged much above 70% of capacity. One of the many undesirable effects of standard regulators is that when a load is operating on the electrical system, charging current also decreases. As an example, I tested an alternator with a total output of 30A at 14V aboard a vessel with an electrical load of 24A. I found that only minimal amps were flowing into the battery, with a terminal voltage of only 13.2V. The more load you apply on the system during charging, the less goes to charging the battery. It is better to have as much load switched off as possible when topping up the battery.

5.14   Alternator Regulator Sensing. With any type of charging system, there is a voltage drop between the alternator output terminal and the battery. With a nominal alternator output of 14V, it is not uncommon to have a totally inadequate 13V reach the battery. The voltage drop will increase as the current increases. Regulator sensing consists of the following configurations.

   a.   **Machine Sensed.** The machine-sensed unit simply monitors voltage at the output terminal and adjusts alternator output voltage to the nominal value, typically 14V:

   (1)   **Charge Circuit Voltage Drops.** The machine-sensed regulator makes no compensation for charging circuit voltage drops. Voltage drops include inadequately rated terminals and cables and negative path back through the engine block.

   (2)   **Diode Isolators.** If a diode-isolator, charge-distribution system is installed, this contributes a further drop, typically 0.75V.

   b.   **Battery Sensed.** The battery-sensed unit monitors the voltage at the battery terminals and adjusts the alternator output voltage to the nominal voltage. Always install battery sensing if possible:

   (1)   **Charge Circuit Voltage Drops.** The battery-sensed regulator compensates for voltage drops across diodes and charge circuit cables. By sensing the battery terminal voltage, the regulator varies the output from the alternator until the correct voltage is monitored at the battery. Some alternator manufacturers, such as Bosch, Lucas, Prestolite, and Sev-Marchal, are introducing modifications so that regulators can be compensated with a separate sense connection that goes directly to the battery.

   (2)   **Caution.** In some cases, the voltage drop between alternator terminals and battery may be considerable; figures of 1.5V to 2V and above are not uncommon. Use a multimeter, and check the output and battery voltage to determine the voltage drop, ideally at the full

output current. An excessive voltage drop is a fire risk. Excessive current flow, along with high ambient engine space temperatures, can literally melt and ignite the cable insulation or, typically, first burn off the terminals. Check the output terminal to see if it is hot.

c.   **Temperature Compensation.** Very few alternator manufacturers incorporate temperature compensation, even though electrolyte is affected by temperature. In hot climates, charge voltage should be marginally decreased; in cold climates, it should be increased. Regulators with compensation usually have it sensed at the regulator, as the sensing element is part of the regulator circuit. In most vessels, however, batteries are not located near the engine, so the regulators reduce charging output when they sense high engine compartment temperatures. Compensation should be based on the ambient temperature of the batteries.

5.15   Alternator Regulator Types. It is extremely important to distinguish between a regulator and a controller.

a.   **Regulator Function.** A regulator is a fully automatic device that ensures a stable output from the alternator. The primary function of a regulator is to prevent overcharging the battery and damaging the alternator. This crucial function is frequently forgotten, with potentially disastrous results, when selecting a controller.

b.   **Alternator Control Devices.** There are now six main categories of alternator control devices, including the standard factory-fitted regulators that are standard on alternators. Then there is the latest generation of smart regulators. These devices use multistep charge profiles, and some can adapt or change charging to suit the requirements of the battery using an integral processor. Then we have cycle regulators, devices using a cyclic regulator control principle that is microprocessor controlled. Stepped cycle regulators use a timed cycle of voltage steps. Regulator controllers either parallel connect or override existing standard regulators. Finally, manual controllers are devices that have no regulator function and control alternator output manually by operator control.

5.16   Standard Voltage Regulators. Standard alternator regulators are simple and inexpensive voltage regulators with associated circuitry. They are an integral part of the alternator and integrated with the brush gear as a removable module or are located externally on the engine or an adjacent bulkhead. The best arrangement is to have a separate regulator mounted on an adjacent bulkhead to minimize engine heat and vibration damage. The electronic regulator comprises semiconductors that include transistors, operational amplifiers, and Zener diodes. Many now have voltage regulator integrated circuits. The closed-loop control opens and closes the alternator field circuit in response to the sensed output voltage.

5.17 **Regulator Polarity.** Regulators and field windings have two possible field polarities. It is important to know the difference when installing different regulators or testing regulator function. The two types are as follows.

  a. **Positive Polarity.** The positive regulator controls a positive excitation voltage. Inside the alternator, one end of the field is connected to the negative polarity. Alternators with this configuration include Bosch, Motorola, Sev-Marchal (older models), Silver Bullet, Lestek, and Balmar:

  (1) **Polarity Test.** To test, use a multimeter selected to the resistance or ohms (Ω) setting and connect across the field connection to an unpainted part of the alternator case or negative output terminal.

  (2) **Meter Reading.** The reading should be in the range of 3Ω to 8Ω.

  b. **Negative Polarity.** The negative regulator controls a negative excitation voltage. Inside the alternator, one end of the field is connected to the positive polarity. Alternators include Hitachi, Lucas A127, ACR 17-25, and AC5, CAV, Paris-Rhone, newer model Sev-Marchal, Valeo, AC Delco, and Mitsubishi:

  (1) **Polarity Test.** To test, use a multimeter on the resistance or ohms (Ω) setting and connect across the field connection to the alternator's positive terminal.

  (2) **Meter Reading.** The reading should be in the range of 3Ω to 8Ω.

5.18 **Alternator Field Circuits.** The field circuit is used to vary the output of the alternator. It can be simply defined as the alternator "controller" because all alternator output is controlled by the field current level. In general, the following measures are not common on modern boats, as alternators come with integral regulators and modification is a difficult task that requires a new external regulator. There are a number of variations in the connection of fields.

  a. **Advanced Field Switching.** This method is comparatively rare in modern integral regulator alternators. The field is taken through the battery selector switch auxiliary contacts so that the field circuit is broken, de-energizing the alternator immediately before the main output contact's break. This will prevent any accidental circuit interruption and subsequent diode destruction through generated surges.

  b. **Oil Pressure Switch Control.** This method has two configurations. The first senses battery voltage through an oil pressure switch on the engine. The alternator does not commence generating until after engine oil pressure has built up. The second takes the field directly through an oil pressure switch. The latter was common on older vessels.

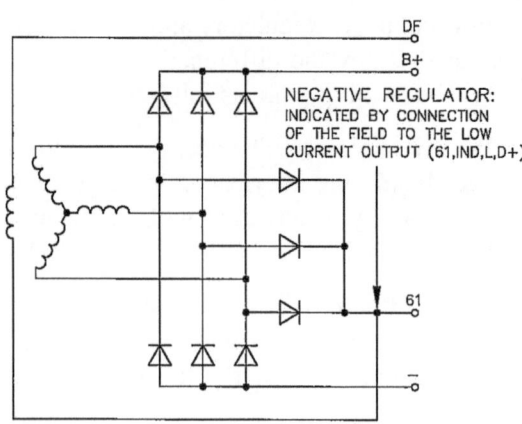

NEGATIVE REGULATOR:
INDICATED BY CONNECTION
OF THE FIELD TO THE LOW
CURRENT OUTPUT (61,IND,L,D+)

Figure 5-2.
Alternator Field
Polarity Circuits

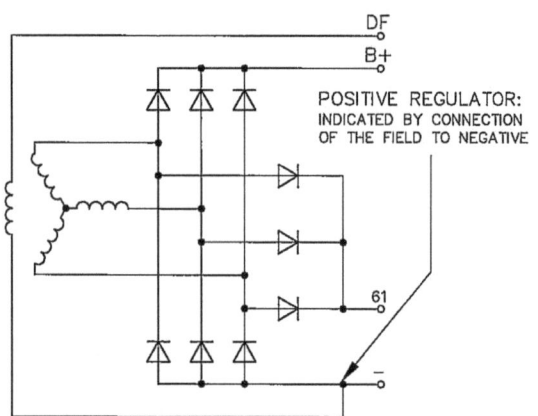

POSITIVE REGULATOR:
INDICATED BY CONNECTION
OF THE FIELD TO NEGATIVE

5.19   **Balmar Regulators.** Balmar has a range of very efficient regulators. Check them out at balmar.net.

a.   **MC-618 Regulator.** This new offering integrates the SG200 monitor. The unit has a color screen or smartphone app that facilitates programming. The system uses the same regulation as the MC-614 and, importantly, allows charging control on carbon foam batteries as well as lithium-ion batteries. The regulator has nine selectable programs for various battery technologies, including $LiFePO_4$. It has a 15A maximum field current, advanced programming modes, and alternator and battery temperature sensing and control. It incorporates an exclusive belt load manager function and integrates with the SG200 for both programming and data display.

b.   **Balmar Max Charge MC-618 Voltage Regulator.** This unit has nine selectable programs and is suitable for $LiFePO_4$ batteries. It has a 15A maximum field current and advanced programming modes. In addition, it has alternator and battery temperature sensing, control and belt load manager function, and integrates with the SG200 for programming and data display.

c.   **Max Charge MC-624 Voltage Regulator.** This unit has eight selectable programs that include LiFePO$_4$ and is designed for 24V installations. It has a 10A maximum field current and the same functions found on the MC-618 unit.

d.   **Max Charge MC-612 Dual Voltage Regulator.** This unit has eight selectable programs for batteries that include LiFePO$_4$. This is designed for the control of two alternators installed on a single engine, a common scenario, along with the same functions on the MC-618.

e.   **ARS-5 Voltage Regulator.** This unit has five selectable programs for batteries with a 9A maximum field current. Suitable for the 6-series alternators along with the same functions on the MC-618.

5.20   Wakespeed Alternator Regulator. The WS500 uses voltage, current, and temperature compensation and is configurable for 12V, 24V, and 48V systems. It is designed for all battery technologies, including LiFePO$_4$ lithium-ion batteries. See wakespeed.com.

5.21   Adverc Cycle Regulator. A cycle program is the basis of the charging system. The regulator is designed for parallel connection to the existing regulator, providing some redundancy should failure ever occur. Temperature compensation takes place, and the charging profile has a linear characteristic. The system has an indicator light warning system, with indication given for low and high voltage conditions or a loss of sense leads. Latest versions have circuit enhancements that attenuate RFI and have smoother control outputs. Check out adverc.co.uk.

5.22   Sterling Regulators. Sterling Power offers the Pro Reg DW advanced alternator regulator, and it is quite bulletproof, with an IP67 rating and ignition protected, along with two temperature sensors. It has fan cooling and a maximum positive field control of 6A. It has digital control with a memory chip coded with the charging profiles. The regulator uses a constant current four-step control program. The unit is programmable for different battery technologies. Along with different charge and temperature curves, the units have a comprehensive monitoring and protection program for both battery and alternator. Check them out at sterling-power.com.

5.23   Battery Charge Settings. Different battery types require different charge voltages, and regulator systems should be adjusted to suit the installed batteries.

Table 5-3. Battery Regulator Charge Levels

Temp	Flooded Hi/Float Volts	Gel Hi/Float Volts	AGM Hi/Float Volts
90°F (32°C)	14/13.1	14.0/13.6	14.4/13.9
80°F (27°C)	14.3/13.3	14.0/13.7	14.5/14.0
70°F (21°C)	14.4/13.5	14.1/13.8	14.6/14.1
60°F (15°C)	14.6/13.7	14.3/13.9	14.7/14.2
50°F (10°C)	14.8/13.9	14.2/14.0	14.8/14.3

5.24 **Alternator Manual Control Devices.** Manual devices are those that require total operator control of the alternator output without regulation. Some handbooks give information on how to make your own controllers. From personal experience, once these homegrown controllers and circuits are installed, it is a recipe for cremation of the charging system, batteries, and alternator not too far in the future. There is no such thing as a cheap solution, and if you really care about your power system, don't risk it. There is no sense in having and relying on electronics worth thousands of dollars only to balk at relatively small sums to improve charging. Use the following control methods at your own risk. The savings initially achieved with these methods are more than negated by one mishap, often shortened battery life through overcharging and battery plate damage.

   a.   **Field Switches.** A typical manual method is the switch directly connected to the field connection. This simply puts on a full field voltage, resulting in maximum alternator output. The results can be quite spectacular—and very damaging to both battery and alternator. While crossing a dangerous bar, a friend casually flicked a switch, which was followed by sparks and smoke curling out of the engine compartment. After investigation, I found this same setup, a potential disaster at the time.

   b.   **Field Rheostats.** The rheostat is the most common type of control. A rheostat is simply a variable resistance rated for the field current. Although the term "rheostat" is still in common usage, low-value variable resistances are generally termed "potentiometers." Operation is very reliant on operator control, with no safety cutouts or regulation. It is not recommended as a general alternator charging control, as both alternator and battery are easily and frequently damaged.

5.25 **Alternator Controllers.** Controllers are devices that require the yacht owner to manually select or partially override the existing regulator to fast charge. It is important to remember that the basic phases of charging a battery are bulk, absorption, float, and equalization, and that at no stage does battery voltage exceed gassing level. In most cases, controllers do not adhere to these basic charging principles.

   a.   **Operating Principles.** Controllers are either direct regulator replacement units or connected in parallel to the existing regulator. Some units have an ammeter to monitor output and require continual adjustment of field current to maintain the required charge current level. They do not monitor or take into account the high and damaging system voltages imposed while maintaining the initial high charging currents.

   b.   **Precautions.** All controllers have some beneficial outcome and can improve the charging process to varying degrees. There are, however, serious risks that must be considered to avoid damage:

   (1)   **Power System Disturbances.** If you apply excessive voltages or full alternator outputs, spikes and surges can arise in the system that will damage regulators and electronics equipment.

(2)　**Battery Damage.** Forcing current into batteries above their natural ability to accept charge will simply damage plates, heat the battery, and generate potentially explosive gases. Failure of automatic cutouts or forgetting about the regulator may cause all the noted problems.

c.　**Performance and Efficiency.** There are some important factors to consider before purchasing controllers:

(1)　**Efficiency.** At best, these types of units can offer a 10% to 15% improvement, bringing charge levels up to approximately 85% of nominal capacity.

(2)　**Performance.** It is interesting to note that virtually none of the controller manufacturers can offer verifiable proof or independent testing to support claims that they, in fact, improve charging.

5.26　**Regulator Troubleshooting.** There is a simple test to check whether your regulator or controller is working properly. This is not difficult with external regulators, but if an internal regulator is fitted, the alternator will need to be opened and a wire attached to a brush holder. Switch off all electrical and electronic equipment at the switchboard circuit breaker before starting this test. If in doubt, don't try it.

a.　**Alternator Test.** Check that the alternator gives full output. If the alternator operates after testing, the regulator is suspect.

b.　**Rotor Test.** If a regulator has failed, particularly in an overcharge situation, check the rotor for damage before replacing the regulator. The test is as follows:

(1)　**Test Insulation Resistance.** Set the multimeter on the resistance setting and place one multimeter probe on a slip ring and the other on the rotor core. Resistance should be infinite or over range.

(2)　**Test Winding Resistance.** Place the multimeter probes on each slip ring. Resistance should be around 4Ω. If it is very high, an open circuit may exist; if very low, a coil short circuit may exist.

c.　**Auxiliary Diode Test.** On some occasions, the auxiliary diodes may fail. Put your multimeter on the DC voltage setting and connect across 61/D+ and negative. If there is any reading, the diode may be faulty. Turn on the ignition key without starting. The reading should be around 1V to 2V. If less, the wiring may be faulty; if higher, the diode may be faulty and there is excessive rotor resistance or there are bad connections.

5.27　**Regulator Removal.** If a regulator must be removed or checked, certain procedures should be used to avoid damage. Mounting a separate regulator on the engine bulkhead makes replacement simple and inexpensive and facilitates testing. For a Bosch (K1/N1 series): Dismantle by unscrewing the two screws retaining the regulator. Carefully lift the regulator up and out. Be careful not to damage the brushes. Disconnect the (D+) lead from the back of the regulator. Hitachi alternators are typically installed to Yanmar engines.

The Valeo (includes Paris-Rhone and Sev-Marchal) alternators are usually a standard type fitted to Volvo engines; there are some differences in the design, although that is standardizing. Use the following procedure to disconnect and install a new external regulator system or replace the existing one: Unscrew and remove the four screws securing the regulator to the casing. There are four cables leading from the regulator (five on the new Valeo). If you are replacing the regulator with an external type, cut the cables at the regulator, as the regulator and housing acts as a spark arrestor cover for the brush gear. Remove the negative cable to the regulator entirely. The cable running internally under the plastic cover to terminal 61 should be soldered to one of the brush-holder connections. This cable was initially connected to the regulator until cut off. Solder a wire to the remaining brush holder and run it out through the cover for connection to the new regulator, creating the field control connection.

Table 5-4. Charging System Troubleshooting

Symptom	Probable Fault	Corrective Action
Reduced charging	Drive belt loose	Adjust to 10 mm
	Oil on belt	Clean belt
	Loose alternator connection	Repair connection
	Partial diode failure	Repair alternator
	Suppressor breakdown	Replace suppressor
	Regulator fault	Replace regulator
	Diode isolator fault	Replace diode
	Negative connection fault	Repair connection
	Solder connection fault	Resolder connection
	Underrated cables	Uprate cables
	In-line ammeter fault	Replace ammeter
	Ammeter shunt fault	Repair connections
Overcharging	Regulator fault	Sense wire off
	Sense wire off	Replace wire
No charging	Drive belt loose	Re-tension belt
	Warning lamp failure	Replace lamp
	Auxiliary diode failure	Repair alternator
	Regulator fault	Replace regulator
	Diode bridge failure	Repair alternator
	Jammed brushes	Clean brush gear
	Stator winding failure	Repair alternator
	Rotor winding failure	Repair alternator
	Output connection off	Repair connection
	Negative connection off	Repair connection
Fluctuating ammeter	Alternator brushes sticking	Repair alternator
	Regulator fault	Replace regulator
	Loose cable connection	Repair connections
High initial start current	Ammeter fault/overcurrent	Replace ammeter
Low charge current	Batteries sulfated	Replace batteries
	Battery cell failure	Replace batteries
	Battery charge very low	Recharge extended time

Battery Chargers

Battery chargers are generally used as the primary charging source in large vessels with AC generators in continual service. Many vessels have had batteries ruined by poor-quality chargers due to a marginal overcharge voltage level. In reality, battery chargers are not a principal charging source on a cruising yacht, and a relatively small-output automatic charger of approximately 10A to 15A will meet the normal requirements while in a marina. Given the rise of lithium-ion and carbon foam batteries, it is essential that chargers be able to charge these battery types correctly. ABYC A-31 provides recommendations and guidelines for installing battery charging devices.

6.1   **Charger Principles.** The basic principles of most common battery chargers are as follows:

     a.   **Transformation.** The AC voltage, either 220VAC to 230VAC or 110VAC, is applied to a transformer. The transformer steps down the voltage to a low level, typically around 15V to 30V, depending on the output level.

     b.   **Rectification.** The low-level AC voltage is then rectified by a full-wave bridge rectifier similar to that found in an alternator. The rectifier outputs a voltage of around 13.8V to 27.6V, which is the normal float voltage level.

     c.   **Regulation.** Many basic chargers do not have any output regulation. Chargers that do have regulation are normally those using control systems to control output voltage levels. These sensing circuits automatically limit charge voltages to nominal levels and reduce to float values when the predetermined full-charge condition is reached. Chargers are starting to get electronic control and monitoring systems.

     d.   **Protection.** Battery chargers have protective devices that range from a simple AC input fuse to the many features described as follows:

         (1)   **Thermal Overload.** This device is normally mounted on the transformer or rectifier. When a predetermined high temperature is reached, the device opens and prevents further charging until the components cool down.

         (2)   **Input Protection.** This is either a circuit breaker or a fuse that protects the AC input against overload and short circuit on the primary side of the transformer.

         (3)   **Reverse Polarity Fuse.** A fuse is incorporated to protect circuits against accidental polarity reversal of output leads.

         (4)   **Current Limiting.** Limiting circuits are used to prevent excessive current outputs or to maintain current at a specific level.

         (5)   **Short-Circuit Protection.** This is usually a fuse that protects output circuits against high current short-circuit damage.

e.   **Interference Suppression.** Most chargers have an output-voltage ripple superimposed on the DC. This is overcome by the use of chokes and capacitors across the output. This ripple can affect electronics and cause data corruption.

6.2   Battery Charger Types. There are a number of charger types and techniques in use.

a.   **Constant Voltage (Potential) Chargers.** Chargers operate at a fixed DC voltage. The charge current decreases as the battery voltage reaches the pre-set charging voltage. Unsupervised charging can damage batteries if electrolytes evaporate and gas forms. Additionally, such chargers are susceptible to input voltage variations. If left unattended, the voltage setting must be below 13.5V or batteries will be ruined through overcharging. They are essentially a step-down transformer with a rectifier.

b.   **Ferro-Resonant Chargers.** These chargers use a ferro-resonant transformer that has two secondary windings. One of the windings is connected to a capacitor, and they resonate at a specific frequency. Variations in the input voltage cause an imbalance, and the transformer corrects this to maintain a stable output. These chargers have a tapered charge characteristic. As the battery terminal voltage rises, the charge current decreases. Control of these chargers is usually through a sensing circuit that switches the charger off when the nominal voltage level is reached, typically around 15% to 20% of the charger's nominal rating.

c.   **Pulse Chargers.** These chargers output a series of short pulses of voltage and current with a time interval of about 20ms to 30ms between each pulse. This short rest period allows the battery chemical actions to stabilize, effectively equalizing the reaction before the next charge pulse. This reduces passivation and the formation of gas and crystal growth on the plates. These pulses have a unique waveform characterized by a precise rise time, which includes the width of the pulse, the frequency and amplitude, as well the time base. These instantaneous high voltages do not create battery heat effects and are effective at breaking down sulfate crystals, offering improved battery longevity. There is negative or reflex pulse charging, also known as "burp charging." This method employs a very short duration pulse, approximately 200% to 300% of the charging current for a period of 5 milliseconds (5ms). This is applied during the charge rest period and is for depolarization of the battery cell. The pulse dislodges gas bubbles that have evolved at electrodes, improving both stabilization times and charge times. The release of these gas bubbles is known as "burping."

d.   **Switch Mode Chargers.** Compact switch mode chargers are popular due to their compact size and low weights. These charger types convert the input line frequency from 50Hz to 150,000Hz, which reduces the size of transformers and chokes used in conventional chargers. An advantage of these chargers is that line input and output are effectively isolated, eliminating the effects of surges and spikes. These chargers are technically very advanced. The chargers are battery-sensed and temperature-compensated, have integral digital voltmeters and ammeters, and are lightweight and compact. These charger types have an LC output filter, which comprises an inductor and capacitor to smooth out the pulsed waveform.

e.   **Automatic Chargers.** This term covers a wide range of electronically con-trolled charging systems. These include chargers that have silicon controlled rectifier (SCR) control, a combination of current and voltage settings with appropriate sensing and control systems, as well as overvoltage and overcur-rent protection. The ideal charger characteristic is one that can deliver the boost charge required and then automatically drop to float charge levels so that overcharging does not occur.

f.   **Trickle Charger/Battery Maintainers.** This method of charging is specif-ically for battery self-discharge compensation and is called a maintenance charger. This type of charging is not suitable for lithium-ion batteries. Trickle charging is a continuous charge with constant current. Don't leave charging for more than a couple of days, as trickle chargers are unintelligent and don't know when the battery is charged.

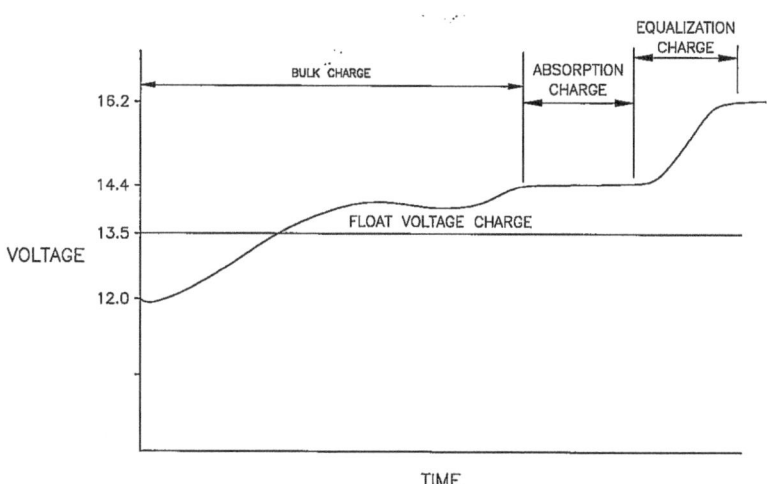

Figure 6-1. Battery Charger Phases

6.3   Battery Charger Output Ratings. Selection of a battery charger means assessing the battery bank size, along with the battery technology. A general rule is to select a battery charger based on a factor of around 20% of your house power battery capacity. As many battery chargers have dual and triple outputs, installing a battery charger that is capable of recharging a two-house battery bank or house and engine start battery is a good idea.

6.4 **Charger/Inverters.** Having the whole integrated package makes sense, and these utilize the inverter stage of the inverter to function as a battery charger. The Freedom XC PRO Marine inverter/charger is typical of the latest technology. It supports NMEA 2000 and is capable of charging dead batteries along with lithium-ion chemistry. Depending on the model, it has a 100A (2000W model) or 150A (3000W model) battery charger. Check them out at xantrex.com.

6.5 **Multiple Battery Charging.** Most marina-based boats have a battery charger connected permanently to charge a single house battery bank, although many boats have multiple house banks and twin engines with separate batteries. Either a separate battery charger or a method of splitting the charge to each battery is required. Gel cells, AGM, and lithium-ion batteries may have different charging requirements, and this should be verified prior to using any charger. When charging flooded cell batteries, be aware they can rapidly lose water when fully charged if charging voltages are imprecise.

     a. **Multiple Output Chargers.** Install a battery charger with multiple outputs where each battery bank has its own isolated charging outputs; this prevents any interaction and is an efficient way of having two or more separate chargers. The Xantrex Xplore battery charger is capable of charging multiple battery banks with varying battery chemistries at the same time.

     b. **Diodes.** A diode isolator can be used to split the charge between two or three battery banks. For three battery banks, use two diode isolators and link the diode isolator inputs. Problems of voltage drop across the diode have to be considered, and battery chargers with battery sensing are required to compensate for this. The typical voltage drop is around 0.7V, so the charger outputs without sensing will require adjustment of output voltage up an additional level equivalent to the drop. There are new diode units that do not have the voltage drop such as the Victron Argofet and Argodiode.

     c. **Relay/Solenoid.** A relay or solenoid can be used to direct the charge current to each battery bank. This is activated either with the monitored charging voltage or via a manually operated switch. The configuration effectively parallels all the batteries to form a single battery bank. Possible systems include the NewMar battery bank integrator (BBI). When a charge voltage is detected that exceeds 13.3V, the unit switches on. The unit consists of a low-contact resistance relay that closes to parallel the batteries for charging. When charging ceases and the voltage falls to 12.7V, the relay opens, isolating the batteries. The unit incorporates a voltage comparator and time-delay circuit, which prevents the unit from cycling in the event of a voltage transient or load droop on the circuit dropping voltage below the cutout level. These devices allow charging of two or three batteries from one alternator or battery charger. The units use a high-current switch rated at 800A and 1600A for alternator and charger ratings up to 250A.

6.6 **Battery Charger Installation.** Chargers should be mounted in a dry and well-ventilated area. The following precautions should be undertaken when installing battery chargers.

a. **Engine Starting.** Always switch off a battery charger during engine starting
if connected to the starting battery.

b. **Connection.** The AC connection should be an industrial-grade outlet in
engine areas or a normal outlet in dry areas.

c. **Grounding.** The metal case of any charger must be properly grounded to the
AC ground.

d. **Cables.** To prevent cables from moving, clips or permanent fasteners should
be used on cables if the charger is permanently installed.

e. **Power Isolation.** Switch off the charger before connecting or disconnecting
cables from the battery.

f. **Chargers and Inverters.** Where they are separate units, do not operate a
large inverter off a battery with a charger still operating. The large load will
overload the charger and may damage the circuits.

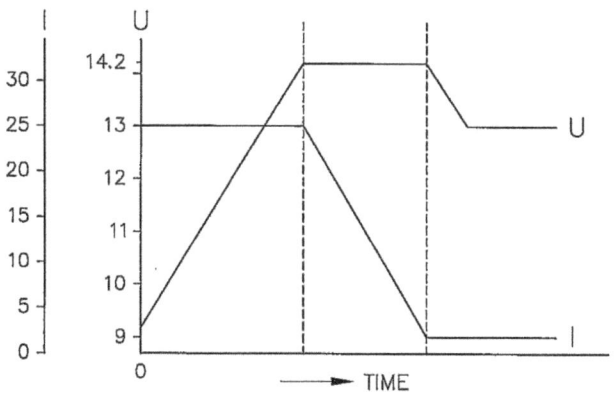

I = CHARGING CURRENT (AMPS)
U = CHARGING VOLTAGE (VOLTS)
13V = MAINTENANCE /FLOAT VOLTAGE

Figure 6-2. Battery Charger Characteristics

DC Systems Wiring

A significant proportion of boat electrical system failures and reliability issues are a direct result of incorrectly installed cable and wiring or substandard connections and terminations. Using accepted wiring practices can help eliminate these failures. Unfortunately, the common attitude is still to treat vessel low-voltage systems as similar to those in automotive installations, and the relatively high failure rates on boats reflect this attitude. The complexity of any boat installation varies by boat type and size and how you use it, whether for weekend sailing, coastal cruising, or ocean voyaging. A liveaboard has very different requirements than a weekend or vacation cruiser. Problems always arise when boat electrical systems are badly maintained, or you find yourself with an undocumented installation by the previous boat owner. The image in the color insert shows what a very well installed electrical panel looks like on a Hallberg-Rassy 34.

7.1   DC Wiring Failures. The principal causes of boat fires—based on a fire-related insurance claims study by BoatUS—are 55% AC and DC wiring and appliances, 24% engine and transmission overheating, 8% fuel leaks, 7% miscellaneous, and 5% unknown. Wiring and cable failures are attributable to several causes, and failure can cause fires. Wire and cable service life is a cause if improperly selected and installed. All cables age, including the insulation; the symptoms are cracking, embrittlement, and a failure of the sheathing and insulation. Once a conductor is exposed, the risk of fire and short circuiting rises substantially. The drivers for wiring longevity include excessive high or low temperature, mechanical resilience to vibration, stressing, abrasion, and ambient environmental conditions. Moisture ingress into the insulative sheath or conductor covering leads to failure. Improper installation and ingress of salt water into the copper conductors leads to oxidation. Conductor heating due to operation at close to maximum current ratings leads to impacts on the insulation. Exposure to UV radiation is a real insulation killer; anything external is at risk. *Note:* Make sure your boat is mice and rat free. Boats left unattended can have uninvited rodents aboard, and they just love to nibble on wiring insulation when no other food source is available.

7.2   Wiring Regulations and Standards. As a professional marine electrical and electronics person, even I find the various boat wiring rules, standards, and wiring recommendations challenging, and that includes within the commercial maritime space. They can often be complex, even unintelligible, if you are not a professional with the appropriate level of theoretical knowledge, understanding, and experience. What is often lacking is context. Boat electrical and electronic systems should be installed as far as practicable to comply with one of the principal sets of rules, standards, or recommendations listed below. The majority of standards are generally similar or overlap in many areas. Many boaters choose not to spend money purchasing copies of recommendations or rules, as they are relatively expensive. It is easy to be intimidated by the complexities and difficult to interpret them, as they are often couched in jargon rather than plain language. The generic and advisory recommendations are based on various standards. They encompass or exceed many of the provisions of the various standards listed below, and I've tried to infuse some context rather than just precise and dry technical requirements. Although they will not cover all of the provisions, they use best practices and will assist in getting your installation to a similar level. Whether you are required or choose to use official standards, a copy of the relevant standard is an absolute prerequisite. Select what is required for your local jurisdiction wherever you reside in the world, and be mindful that it may be mandatory for insurance purposes. It is always worth asking your boat builder, yacht manufacturer, or marine electrician which standard they are using.

7.3 **Boat Wiring Regulations.** The following are the most important and recognizable wiring rules and recommendations currently in use around the world. When a sailing or motor vessel must comply with classification society requirements, such as Lloyd's Register (LR), Det Norske Veritas (DNV), American Bureau of Shipping (ABS) survey, or other requirements and provisions, such as the USCG in the United States, the requisite wiring rules should be obtained and referenced for the particular boat installation. This may be a requirement for insurance or registration or a requirement by a national authority. The following regulations were current at time of printing; be sure to check for the latest versions.

7.4 **Boat Wiring Regulations—United States and Canada**

American Boat and Yacht Council (ABYC). Standards and Recommended Practices for Small Craft. E-11 AC and DC Electrical Systems on Boats. These voluntary standards and recommendations are widely used by many US sailing and powerboat builders and electrical professionals. They cover all areas of boat construction and systems. (Go to www.abycinc.org.) Note that the 2023 standard revision included E-13—lithium-ion batteries; A32—AC power conversion equipment and systems, and TE-12—three-phase AC electrical systems on boats.

The United States Coast Guard. Title 33, CFR 183. These contain mandatory requirements for electrical systems on boats. "Electrical Systems" are specifically 33 CFR 183.401-460; "Navigation Lights" are 33 CFR 183.401-460.

NFPA 302. Fire Protection Standard for Pleasure and Commercial Motor Craft, 2020 Edition. This standard is approved by the American National Standards Institute (ANSI) and is applicable to motorboat installations. The technical committee includes representatives from ABYC, USCG, Underwriters Laboratories (UL), and others, such as the National Association of Marine Surveyors (NAMS).

NMEA 0400 Installation Standard Version 5.0, 2021 Edition. This voluntary standard and set of wiring recommendations are stated as complementary to the ABYC recommendations, with an electronics installation bias. It defines competent installation best practices for the installation, servicing, or modification of marine electronics and electrical systems, along with associated peripherals. It is a very well developed and comprehensive set of recommendations. While it is a standalone document, it is designed to be used in conjunction with marine electronics manufacturer installation manuals.

Transport Canada TP1332 E—Construction Standards for Small Vessels. Section 8: Electrical Systems. This is a comprehensive set of standards for pleasure vessels under Canadian jurisdiction.

7.5   Boat Wiring Regulations—United Kingdom and Europe

European Recreational Craft Directive (RCD). A new RCD edition was released in 2022. These standards are now virtually mandatory on new construction power and sailing boats in European Union countries and include the following International Standards Organization (ISO) standards: ISO 13297: Small Craft—Electrical Systems—Alternating and Direct Current Installations, 2020. The standard specifies the requirements for the design, construction, and installation of the following types of DC and AC electrical systems, installed on small craft either individually or in combination.

a)   extra-low-voltage direct current (DC) electrical systems that operate at nominal potentials of 50VDC or less.

b)   single-phase alternating current (AC) systems that operate at a nominal voltage not exceeding 250VAC.

The standard excludes DC electric propulsion systems and so on, which are covered by ISO 16315. It excludes any conductor that is part of an outboard engine assembly. Also excluded are three-phase AC installations, which are covered by IEC 600092-507.

ISO 8846: 1990 Small Craft Electrical Devices—Protection Against Ignition of Surrounding Flammable Gas.

BMEA (British Marine Electronics Association) Code of Practice for Electrical and Electronic Installations in Small Craft, Fifth Edition (2013). These recommendations are used by members of the BMEA in the UK. They are essentially based on enhanced RCD and ISO requirements and are for sailing and power boats up to 24m in length.

Institution of Electrical Engineers Regulations for the Electrical and Electronic Equipment of Ships, BS7671:2001. These UK regulations generally apply to AC systems on boats.

Lloyd's Register. The Rules and Regulations for the Classification of Special Service Craft, July 2022. This supersedes the Rules and Regulations for the Classification of Yachts and Small Craft. These rules are applicable when a yacht exceeds 26m and is being classed by Lloyd's.

7.6   Boat Wiring Regulations—Australia and New Zealand

Australian/New Zealand Standard Electrical installations—Marinas and Recreational Boats Part 2: Recreational Boats Installations. AS/NZS 3004.2:2014. Reissued with Amendment No. 1, July 2015. This standard derived elements from (a) IEC 60092-507: Ed.1.0 (2000), Electrical Installations in Ships, Part 507: Pleasure Craft; (b) ISO 10133, Small Craft—Electrical Systems—Extra-low-voltage DC Installations; (c) ISO 13297, Small Craft—Electrical Systems—Alternating Current Installations; (d) American Boat and Yacht Council (ABYC), E-11—AC and DC Electrical Systems on Boats. One very important statement made in the preface of this standard holds true for all when working on boat wiring systems: "Designers are reminded that it is essential that the basic tenets of electrical and marine safety be addressed before any other equipment and installation design elements are considered."

National Standard for Commercial Vessels (NSCV). Subsection C5B Electrical (2020). This replaced the Uniform Shipping Law (USL) Code of Australia. NSCV Subsection C5B provides the standards for design, construction, installation, and repair of electrical systems on vessels less than 35m measured length and requires compliance with AS/NZS 3004.2:2014.

SYSTEM VOLTAGES

7.7   12-Volt Systems. The 12V DC system is the most common system used on boats. This is largely due to automotive influences, which have led to a large range of equipment being available. A common topic is 12V battery electrical systems and alternative system voltages. The most common electrical systems on boats of all types are configured around the 12V battery. But you know that already. There is a relatively small number of 24V battery systems to be found on both sail and motor boats as well as trawler yachts. Many vessels have hybrid systems with both 12V and 24V system voltages installed, and occasionally I run across an older vessel with 32V. There was some hype several years ago about the possible introduction of 42V within automotive technology. To really confuse things, 48V is being commonly used; this is primarily driven by the rapid advances in electric propulsion and electric vehicles.

    a.   **12V Basics.** The marine diesel engine 12V battery–based electrical system is primarily designed to start the engine and charge from a dedicated battery. This simple design factor actually influences and is the default design point for the entire boat electrical system. The system was never designed to charge a house 12V boat battery bank. In an automotive system, the battery starts the engine and recharges the start battery with the alternator; the alternator then supplies all electrical power when the engine is running. When the engine is off, the battery is not designed to withstand continual deep cycle drains on power; recreational vehicles (RVs) typically have a second battery bank installed for the same reason as boats.

    b.   **12V Engine Origins.** The same criteria apply to marine diesel engines; however, they have morphed into being a generator for battery charging when the propulsion system is not being used. This has spawned a multitude of design challenges, including how to reduce the battery charging time. This is the issue of charging a significantly higher capacity deep cycle battery bank that has different 12V start battery charge characteristics. To further complicate the process, the default 12V battery system with limited charging ability does not factor in the high electrical demands. This is not just charging but also running deck winches, anchor windlasses, refrigeration, water makers, and so on. The fact is, the 12V power system was never an ideal design configuration for boats.

c.   **Disadvantages.** The big disadvantage of 12V battery systems in a boating scenario is that of voltage drop within wiring. Any miscalculation in cable and wiring sizes or a faulty contact in either a cable joint or termination can have catastrophic effects on the circuit due to voltage drop. This can include seriously degraded equipment performance and subsequent failure of electric motors. The 12V power system also suffers from disturbances from DC motor–powered equipment, such as bow thrusters, anchor wind-lasses, deck winches, electric furling systems, hydraulics, and other systems. Every time a heavy electrical load is applied, this generates system transients that include voltage surges, spikes, and droops that create instability. These surges can and do interfere with other electrical and electronic equipment, causing what's known as "brownouts." See chapter 30 for information about electronics interference.

7.8   24-Volt Systems. The 24V electrical system is almost standard on most commercial vessels such as trawlers and work boats, as well as larger sailing boats, trawler yachts, and powerboats. The majority of large commercial vehicles, such as trucks and buses ashore, use this voltage. I have always been perplexed by the relatively low use and acceptance of 24V in smaller boats, given the significant advantages it offers. You can get deck winches and windlasses in 24V. Bow thrusters come in 24V, as do electric furlers and hydraulic power systems. If you use inverters to power up AC appliances, 24V is a better option. Toilet waste pumps, water pumps, and even lighting are available in 24V, which is simply derived by connecting two 12V or four 6V batteries in series. The usual reason quoted for not installing 24V in boat electrical systems is the difficulty in obtaining compatible appliances, lights, electronics, etc. This argument is somewhat flawed, as most equipment and lights are now available in 24V ratings, and the few items that are unavailable can be supplied by a DC-to-DC converter. Many of today's electronics are designed to function on either 12V or 24V. One drawback is that few marine engine manufacturers configure diesel engines in 24V, at least around 30 horsepower or lower; it is an option only as engines get larger. The big advantage of 24V boat electrical systems is significantly reduced cable sizes—by 50% as the current is halved. On long cable runs, this reduces voltage drop problems on large current consumers, such as a windlass.

7.9   32-Volt Systems. Occasionally an older vessel is found using a 32V battery system. They're often found on older boats and some vessels from the 1940s and 1950s, even into the 1960s. Golf carts and electric bikes are still using 32V battery systems. The 32V battery has been around for a long time and was used extensively in remote homestead power systems. It is derived from four 8V batteries to get 32V when connected in series. It may surprise you that 8V batteries are still available and being used in golf carts. For that reason, you can still buy new batteries, including lithium-ion types, and battery chargers to suit.

7.10   42-Volt Systems. The 42V power system just didn't happen. A new voltage standard was defined for automotive vehicle systems back in 2002, and the first vehicles appeared using the new standard system. The system used a 36V electrical system with 42V charging.

There was quite a bit of hype in various yachting and boating circles. Although it held great promise for a battery revolution afloat, it floundered and didn't proceed.

7.11   48-Volt Systems. The 48V battery voltage revolution has arrived. This system is becoming more prevalent, in particular for powering thrusters and with the emerging electric propulsion technologies. Many electric automotive vehicles now have 48V electrical systems; they provide power to the motors as well as electric turbochargers on hybrids. This can power up either mechanical or hydraulic power for everything including power steering, power braking, the water pump, radiator cooling fans, and of course air-conditioning. Unlike previous voltages, boat-friendly equipment is available. Some are asking if we are following the same route as the now-defunct 42V system changeover. There is a practical limit as to why these systems stay under 60V; after that, insulation value changes on wires are required and electric shock levels increase. This voltage system is the backbone of electric propulsion systems.

7.12   Why a 48-Volt Battery System? Many industries, such as telecommunications, have been using 48V battery systems for a long time. Data centers for computing have been using it as well; even some battery-powered hand and yard tools are using 48V. The trend is for electric vehicles standardizing on 48V. It should be noted that even shore-based solar power installations are using 48V as a primary power voltage. They have inverters, charge regulators, and other devices that offer a more efficient system. Bosch claims that by 2025, about 20% of all new vehicles will have a 48V battery installed. Some ask, "Why Stop at 48 Volts?" The Safety or Separated Extra Low Voltage (SELV) classification comes into play. The International Electrotechnical Commission (IEC) has defined this system as an electrical system in which the voltage cannot exceed ELV under normal conditions and under single-fault conditions. SELV also means the voltage is at a level that if you were to touch the live circuit either in normal operation or during a single-fault condition, you would not get an electric shock. Electrically separated means the extra low voltage circuit is electrically segregated from circuits carrying higher voltages. This means that voltages under 60V are classified as safety (separated) extra low voltage (SELV). There is an IEC standard for this, and 48V is under this threshold for practical purposes.

7.13   Equipment Availability. Commonly stated disadvantages of 48V battery systems are that you do not have a standard charging system and alternators are not available and therefore a more innovative charging arrangement is required. News flash! That is no longer true. Balmar, in an innovative first, now has a 48V alternator and, importantly, it is for boats, not vehicles. Unlike previous voltage developments, there are a number of 48V thruster systems now on the market, such as the Vetus Bow Pro units. Maxwell has come to the game with 48V windlasses, Maxwell P&S range (RC8 through RC12). A big driver is the emergence of many boat manufacturers with 48V battery–based electric propulsion systems. In the auto world, major companies including Valeo, Bosch, and Delphi are developing 48V systems. The big advantages of 48V power systems in the case of thrusters or hydraulic power pack motors are the consequential reductions in cable size and in voltage drop issues, which really impact high-load applications. The other major advantage is the reduction in size and weight of the electrical motors. This is combined with advances in electric motor technologies, with all being brushless and using permanent magnet technology.

7.14   **DC Voltage Conversion.** In many vessels, a mix of voltages requires the use of DC voltage converters to step down from 48V, 36V, or 24V to 12V. New-generation units are fully programmable. A number of technical points must be considered when selecting converters. Manufacturers such as NewMar, Victron Energy, and Mastervolt offer a range of converters. Check out victronenergy.com, mastervolt.com, and newmarpower.com.

   a.   **Power Input.** Converters may be either galvanically isolated or isolated only in the positive conversion circuit. Galvanically isolated units will totally isolate input and output, providing protection to connected loads, and these are preferable. Good quality converters have a stabilized output; stability is typically about 1% between line and load at rated output voltage. Typical power consumption of a converter without a load connected is approximately 40mA to 50mA, so there will always be a battery drain. The converter should ideally have an isolation switch on the input side. Most converters are installed with automatic thermal shutdown, short-circuit fuse protection, and current limiting and reverse polarity protection. Semiconductor heat is dissipated by large heat sink fins.

   b.   **Power Output.** Converters are able to withstand short surge current. Normally a 50% overcurrent can be applied for intermittent surges and approximately 70% for a very short duration of up to 30 seconds for peak loads. Some high-power units can withstand peak overloads of 200% for up to 30 seconds. Output ratings vary, but I usually install one rated at approximately 15A continuous. Duty cycle ratings are also applicable to converters. Intermittent overloads can only be sustained on a cycle of 20 minutes every hour, and peaks for 30 to 60 seconds per hour. Failure to observe these duty cycles will result in a burnt-out converter. Like most electrical equipment, converters are designed to provide an output at a specific temperature range, typically 32°F to 104°F (0°C to 40°C). At 122°F (50°C), converters should be derated by 50%. Some new generation units function as chargers.

   c.   **Protection.** Converters all have extensive protection that includes current limiting, short-circuit protection, reverse polarity protection, overvoltage protection, and automatic thermal shutdown. If a circuit breaker has tripped due to overload or a voltage transient, it should be investigated. Continued tripping is usually due to internal component failure.

   d.   **Installation.** Good ventilation is essential. Converters should be mounted vertically so that fins are vertical to facilitate convection cooling. Sufficient clearance must be allowed between top and bottom.

BOAT WIRING SYSTEMS

7.15 **How to Wire and Rewire Your Boat.** The average boat has many systems installed. Equipment is often purchased before the actual impact on the system is considered. Planning the installation requires a carefully considered systems approach. In the majority of cases, systems are overcomplicated, follow no accepted electrical practice, and have inherent system problems that are only overcome with costly total rewires. Use the KISS principle. As most already know, this is the acronym for "Keep it simple, stupid!" This design principle has its origins in the US Navy in the 1960s. The basic premise is that most systems are more reliable if they are kept as simple as possible.

7.16 **Boat Wiring Diagrams.** Each boat should have a complete wiring diagram showing all the wiring and systems installed. The diagram should include equipment identification, equipment current rating, cable sizes, circuit breaker and fuse ratings, and circuit identification. Perform the following wiring planning tasks.

a. **Planning.** Make a plan of your vessel and locate every item of equipment on it. Write down the equipment identification name. Assign a code to each item.

b. **Current Draw.** Check and write down the current draw for each item of equipment. Enter these into your battery load calculation table. This will allow calculations to be made on required battery capacity and charging requirements.

c. **Wiring Diagram.** Draw in the proposed cable route for each item of equipment, showing bulkheads, decks, or other obstructions. When the cable will be routed within bilge areas, or be exposed to mechanical damage, use an alternative route.

d. **Wire Sizes.** Determine the cable size by using the current draw and calculating the voltage drop within the circuit. It is best to standardize cables for most applications, as it is economical to buy cable by the roll. This eliminates the selection exercise based on lowest cable size to achieve specific volt drop values for most but not all circuits.

e. **Protection.** Enter the circuit breaker rating for the circuit and assign a circuit number.

f. **Wiring Diagrams.** The following illustration shows a typical wiring diagram. Diagrams can be devised in a boat geographical format, an electrical circuit and symbol-based format, or a combination. The main point is they are accurate. In many cases they can be broken down into a series of functional diagrams—one for main power, one for distribution, one for battery charging, another for electronics, and so on.

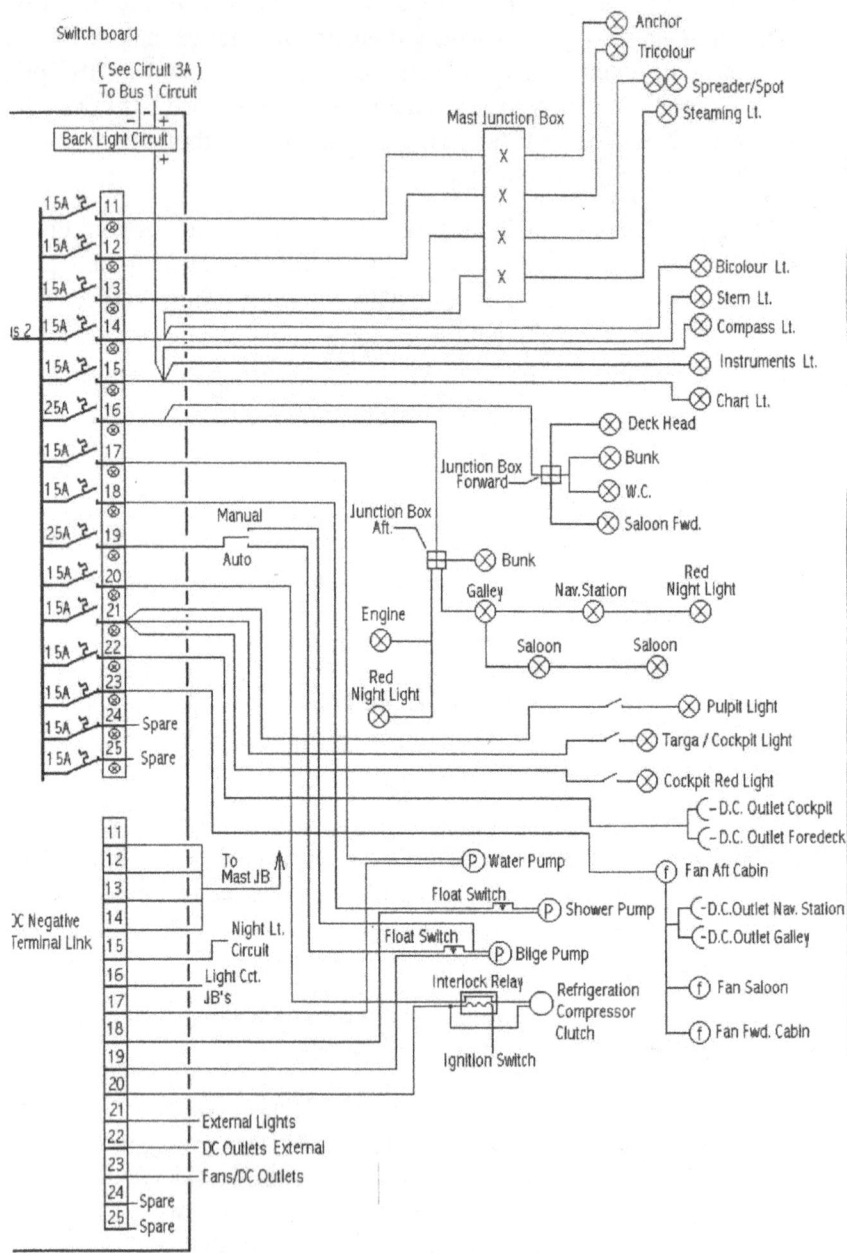

Figure 7-1. Typical Wiring Diagram

7.17   **Wiring Considerations.** There are a number of important considerations regarding wiring.

   a.   **Hull Material.** Hull material has important implications with respect to wiring systems as well as grounding and corrosion. This is important if building a new boat, as you can purchase an engine with a fully isolated electrical system for a steel or alloy boat. It's always wise to discuss this with your chosen boatbuilder and marine electrician.

   b.   **Boat Size.** The length of the boat affects the length of cable runs, with consequentially greater cable weights and voltage drop problems. This affects voltages, and large boats have a real case for selecting 24V, which reduces voltage drop problems. Boat size also impacts the battery weight; sizes for a given capacity are less, and the weight and size of equipment are generally reduced. Larger multihulls and yachts have a greater level of accommodation; therefore, more people are often aboard—and the parties are longer and better!—putting greater demands on batteries through lighting and electric refrigeration, with increased requirements on charging.

7.18   **Two-Wire Insulated System Wiring Configuration.** This is the preferred system of distribution for all steel and alloy vessels. In this configuration, no part of the circuit, in particular the negative, is connected to any ground or equipment. The system is totally isolated and is floating above hull; this includes engine sensors, starter motors, and alternators. In two-wire insulated systems, each outgoing positive circuit and negative supply circuit should have a double pole short-circuit protection and an isolation device installed. The isolator should be rated for the maximum current of the circuit. In this configuration, a short circuit between positive and ground will not cause a short circuit or systems failure. A short circuit between negative and ground will have no effect. A short between positive and negative will cause maximum short-circuit current to flow.

7.19   **Two-Wire with One Pole Grounded System.** Also called a polarized system, this system is preferred for Fiber-Reinforced Plastic/Glass-Reinforced Plastic (FRP/GRP) and wooden/timber vessels. It is the most common configuration and holds the negative at ground potential by connecting the battery negative to the mass of the engine block. The main negative cable is considered to be the grounded negative conductor in two-wire grounded circuit arrangements. In most installations, the main negative to the engine polarizes the system; the engine mass and connected parts, such as the shaft, provide the ground plane. There should be only one ground conductor. In a two-wire, one-pole grounded system, each outgoing circuit positive supply should have short-circuit protection and an isolation device installed. This may be incorporated within a single trip-free circuit breaker. The earthed pole should not have any protective device installed. In this configuration, a short circuit between positive and ground will cause maximum short-circuit current. A short circuit between negative and ground will have no effect. A short between positive and negative will cause maximum short-circuit current to flow. The single pole circuit breaker will break positive polarity only.

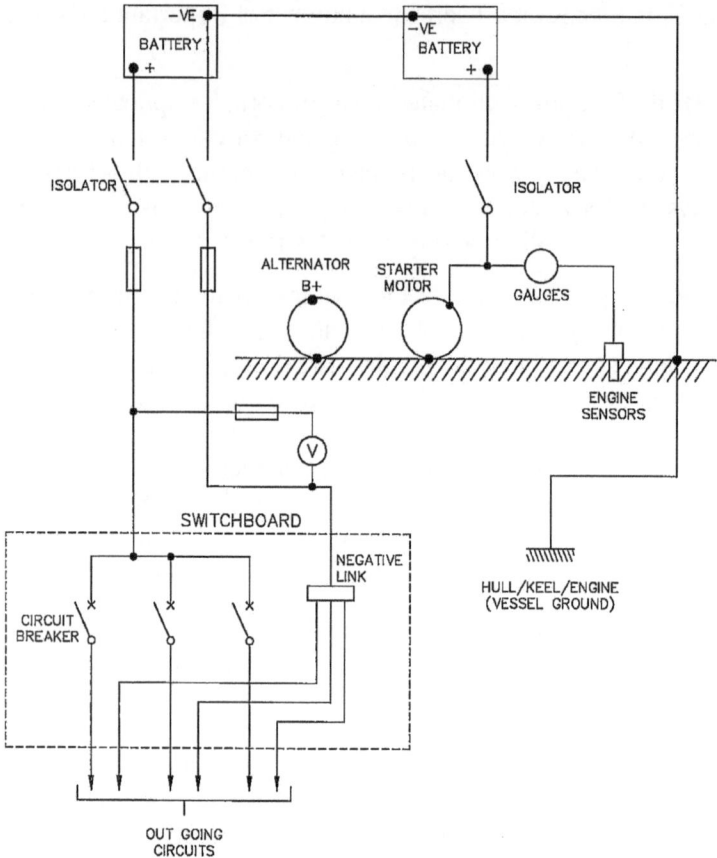

**Figure 7-2. Insulated Two-Wire
with One Pole Grounded System**

7.20   Distributed Systems. These systems are typically broken down into a system of subpanels, and they are becoming increasingly preferable on larger yachts and multihulls. There are a number of significant advantages over a centralized system, including separation of potentially interactive equipment such as pumps and electronics. Other options in this area include intelligent control systems that have remote control of circuits and systems with touchscreens to switch circuits.

a.   **Cable Separation.** Separation enables a reduction in the number of cables radiating throughout the vessel from the main panel to areas of equipment concentration. This is a cause of RFI and requires a greater quantity of cable. Most distributed systems run all the sub-circuits from the central panel, with each circuit protected by a circuit breaker.

b.   **Distributed System Topology.** The illustration below shows a possible methodology of sub-circuits and panels; it is based on the successful implementation on a number of vessels. In the case below, only essential circuits are installed, with metering on the main panel. Lighting panels can be located anywhere practicable, as circuits remain on with lights switched locally.

7.21   Voltage Segregation. DC systems should not be located on or installed adjacent to AC systems. Where DC and AC circuits share the same switchboard, they should be physically segregated and partitioned to prevent accidental contact with the AC section. The AC section must be clearly marked with "Danger" labels. This eliminates the chances of accidental contact with live circuits or confusion between wiring systems. Where systems are integrated, physical separation should be used to prevent contact, and separation should require tools to remove. The barriers should be well marked to warn of the danger.

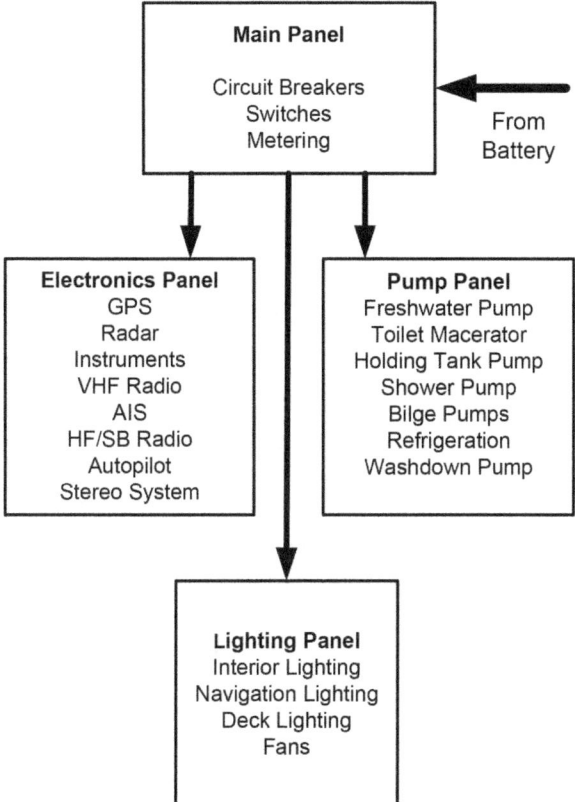

Figure 7-3. Distributed Power Systems

7.22   Modular System Controllers. These are centralized circuit control technologies. Carling Technologies has the MPower CLMD16 system, which comprises a sixteen-channel DC load controller module. This is designed to control DC systems using configurable keypads, switch modules, and multifunction displays. The Maretron CLMD16 has sixteen output channels implemented by DC electronic circuit breakers (ECBs). An ECB monitors output currents against nominal current, and if greater due to short circuit or overload, a semiconductor switch isolates the circuit. Tripping times are magnitude dependent, and all measurements, processing, and calculations are microprocessor controlled. Twelve breakers are able to switch up to 12A, and four breakers can switch up 25A, with a total current capacity of 125A. In addition to fast switching, low-loss solid state ON/OFF switches, it provides accurate current measurement for each load as well as short-circuit protection. The platform consists of twelve-channel and sixteen-channel DC load controller modules, Contura six-rocker digital switch modules, twelve-button customizable keypads, and a twelve-channel optional bypass module. All the MPower components connect directly to the NMEA 2000 network that facilitates circuit breaker control, monitoring, and reset from electronics, including the Maretron MBB300C Black Box, Maretron DSM color displays, and dedicated Maretron touchscreen.

Another recent innovation is the YachtSense modular digital control system from Raymarine. This system allows total control and monitoring of connected systems. Essentially, it is similar to an industrial-level PLC technology, only a bit better in some respects. There is a master module, controlled from a networked multifunction display (MFD), which is connected to a SeaTalk[ng] backbone. There is a power supply module (PSU). Remote modules comprise digital and analog input and output modules. Low-power modules have 4 x10A output channels, sized for operating water pumps and other equipment. The high power modules provide 2 × 20A output channels, which are sized for larger consumers like an electric toilet or refrigeration. There are signal modules that have 4 × digital or analog input or output channels and a reverse power module with a single 20A bidirectional channel. A single 100A power supply is taken to the module, and each output circuit has an integral fuse, so switchboards are redundant. Check out the details at raymarine.com/yachtsense. E-T-A has the PowerPlex digital switching system controlling and connecting everything via a CAN bus system that comprises DC and AV control modules. Visit the company at e-t-a.com.

7.23   Switchboard Circuit Control and Protection. The heart of all electrical systems is the switchboard or panel, which allows control, switching, and protection of circuits. The purpose of protection systems is to prevent overload currents arising in excess of the cable rating. They are designed to protect the cables and equipment from excessive currents that arise during short-circuit conditions. Circuit protection is not normally rated to the connected loads, although this is commonly done on loads that are considerably less than the cable rating, such as VHF radios or instrument systems. The two most common circuit protective devices are the fuse and the circuit breaker, which protect against the following.

       a.   **Short-Circuit Current.** A short circuit is where two points of different electrical potential are connected, that is, positive to negative.

       b.   **Overload Current.** An overload condition occurs when the circuit current carrying capacity is exceeded by the connection of excessive load. Excessive load can come from too many devices or equipment such as pumps with higher-than-normal load.

7.24   Switchboard Construction. The switchboard or panel should be constructed of nonhygroscopic and fireproof material. The panel should be rated to a minimum of IP44. Ideally, panels should be nonconductive; however, many are made of etched aluminum. They should be rated to meet either an IEC ingress protection (IP) code or National Electrical Manufacturers Association (NEMA) standard against the ingress of water. The switchboard interior should be fireproof or at least incapable of supporting combustion. Survey authorities specify that the internal part of the switchboard have a fire-resistant lining. Line all interior walls with appropriate sheeting; this will help contain any fire that may arise in severe fault conditions. Not many boats do this! The switchboard should be located in a position to minimize exposure to spray or water. Where occasional spray is possible, some protection is recommended, such as a clear PVC cover or similar measure.

7.25   Voltmeters. A voltmeter should be installed to monitor the voltage level of the start battery. A good quality voltmeter is essential for properly monitoring battery condition. A voltmeter will indicate that the battery is charging at the correct voltage level. As a battery has a range of approximately 1 volt from full charge to discharge condition, accuracy is crucial. Analog voltmeters are the most common; however, digital is taking over. The sense cable should go directly back to the battery, although on service battery connections, most connect directly to the switchboard bus bar. Direct connection gives greater accuracy and less influence from local loads. Voltmeters should be of the moving iron type and have a fuse installed on the positive input wire. A half-volt error is quite common. Remember to switch off the meter after checking. A voltmeter should be installed to monitor the voltage level of the service battery. A switch may be installed to enable monitoring of the service and start batteries from the same meter. The same provisions apply as for start battery voltage monitoring. In practice, more attention is given to house battery monitoring. Some switchboards use LED voltage-level indicators, and these devices are often used as a voltmeter substitute. They are not recommended, as they do not give the precise readings required. Digital voltmeters are relatively common and far more accurate. They are, however, susceptible to voltage spikes and damage, and many have maximum supply voltage ranges of 15V. There are a number of types, including liquid crystal displays (LCDs) and light emitting diodes (LEDs). LED voltmeters consume power; an LCD meter consumes much less power and is more practical. Where one voltmeter is used to monitor two or more batteries, switching between batteries to voltmeter is done through a double-pole, center off toggle switch or a multiple-battery rotary switch. Voltmeters should have fuses installed within the meter circuit to provide short-circuit protection. As a voltmeter is connected across the supply, that is positive and negative, protection against short circuit is recommended.

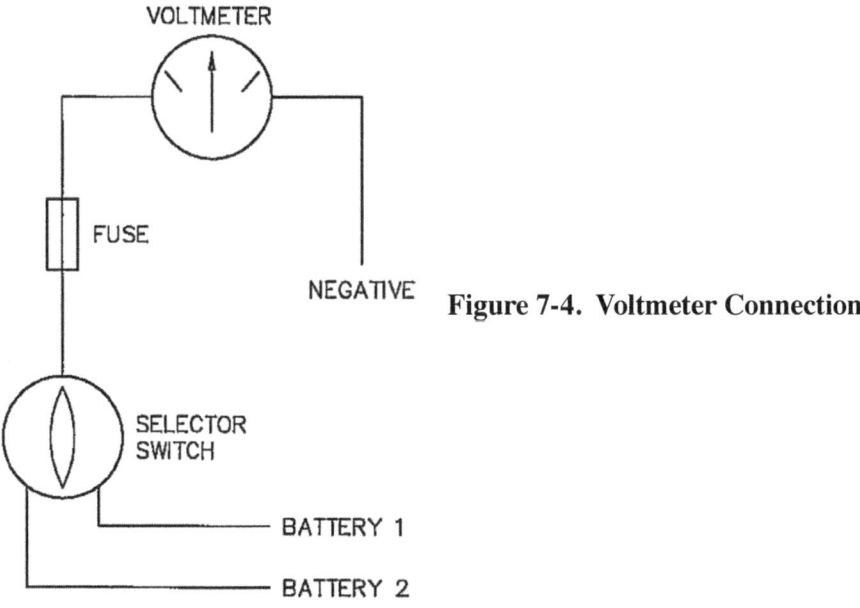

Figure 7-4. Voltmeter Connection

7.26   Ammeters. An ammeter should be installed to monitor the discharge current rate from the service battery. An ammeter is not required for the starting battery. The installation of an ammeter in the primary charging circuit to monitor the charging current is recommended. Ammeters are essential on the switchboard input positive to monitor service battery discharge levels. Analog ammeters should be selected on the calculated operating range. Shunt ammeters are used in these applications. Although useful, an ammeter on the charging system can indicate that current is flowing. You do not know what should be flowing into the battery, so peace of mind is the greatest benefit of this usage. Cheaper ammeters are of the series type, with the cable under measurement passing through the meter. The major failing of these is that very long cable runs are often required, with resultant voltage drops, and damage can occur if the meter malfunctions. Preference should be given to installing a shunt ammeter. An ammeter shunt is a resistance of defined low value that allows the main current to flow through it while monitoring and passing and displaying a small value in proportion to the current flowing. The advantage is that only two low-current cables are required to connect the ammeter to the shunt, reducing the risk of damage. Do not run the main charging cables to the meter and connect it; this can insert excessive voltage drop into the charging circuit. Install a shunt in the line wherever practical, and run sense wires back to the panel mounted meter. The digital ammeter often uses a different sensing system. Instead of a shunt, the digital ammeter has what is called a Hall effect sensor on the cable under measurement. The Hall effect transducer generates a voltage proportional to the intensity of the magnetic field it is exposed to. For vessel applications, a 0V to 10V transducer output corresponds to a 0A to 200A current flow. Sensitivity is increased and range is reduced by increasing the number of coils through the transducer core.

7.27   Integrated Battery Monitors. An integrated monitoring device that measures and displays all values is an acceptable alternative to installing separate ammeters and voltmeters. Unlike starting batteries, house battery charge levels cycle up and down, and power level information is critical to determining charging periods. Typical of early integrated monitors was the E-meter (Link 10), and I installed one of these on all my previous boats. Integrated or smart monitors are "intelligent" devices in that they monitor current consumption, charging current, and a range of monitoring functions that include voltage, high and low voltage alarms, amp-hours used and amp-hours remaining, and allowing the battery net charge deficit to be displayed. These systems maintain accuracy by taking charging efficiency into account, which is the Ah loss during charging. The charging efficiency factor (CEF) settings are nominally set at 95%. A falling CEF is indicative of battery degradation. Charge efficiencies are higher in Li-ion batteries at around 99%. In addition, monitors consider the "n" algorithm for calculating Peukert's coefficient, and this can be programmed in. For many they offer simpler diagnosis of battery power status without trying to "guesstimate" the actual level based on voltages.

7.28   Battery Monitors. Battery monitoring systems such as the Victron BMV-712 battery monitor incorporate Bluetooth technology, allowing monitoring from smartphones, tablets, and notebooks. I recently installed Victron BMV battery monitors on my own boat. A shunt is installed in the negative cable, and this is connected to the monitor via a standard RJ12 telephone cable. A quick-install menu and setup menu are available on the screen. Basic display parameters are available, including voltage, current, and ampere-hours consumption, and models display the battery state of charge and time to go, along with the power consumption in watts. Another model is the SmartShunt, rated at 500A. It is an integrated battery monitor without a display; your smartphone or tablet handles that via Bluetooth and the VictronConnect app. There are several others on the market, all with varying sophistication. These include the Veratron VL Flex gauge and the Xantrex Link Pro, which is designed for flooded-cell, gel, AGM, and lithium-ion batteries. This device has a history memory to record battery events, including both high and low voltage as well as a low-voltage alarm. The unit has an Ethernet cable port for interconnection to Xantrex chargers and inverters. Another unit is the Balmar Smartgauge 44-SG-12/24 battery monitor. This monitors battery condition and does not use a shunt or programming. It requires selecting the battery program that matches the installed battery technology.

a. **Shunt.** A meter shunt (500A/500mV) is usually installed in the negative load line. This is connected by twisted pair wires to prevent noise from induced voltages being picked up and carried into the meter, corrupting the data.

b. **Protection.** The battery sense lead and DC meter power supply have fuses installed, which should be checked if the meter fails to function.

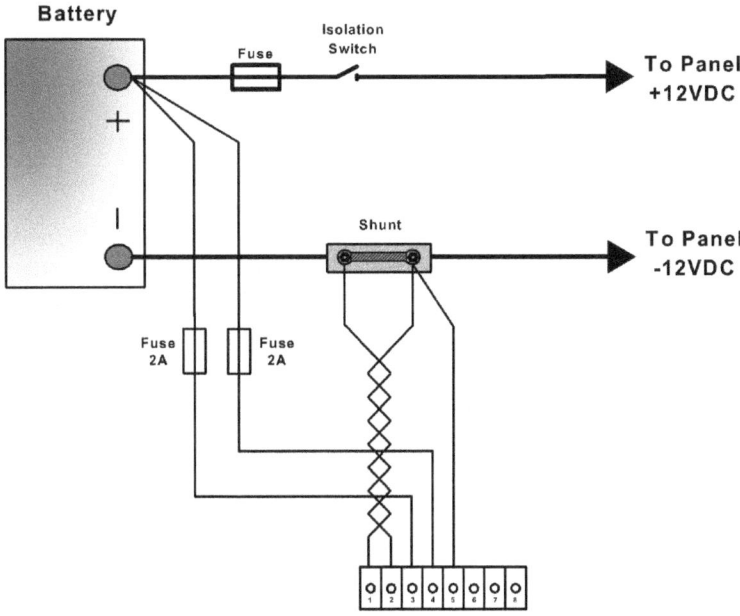

Figure 7-5. Shunt Ammeter Circuit

7.29 Marking and Status Indication. All protection and switchboard devices should be clearly marked to properly identify them. This should include the circuit number if practicable. *Note:* This should correspond with the boat electrical drawings. All circuit isolation and protection devices should have visual status indication. Circuit breaker status indicators normally consist of LED lights, filament lights, or backlit nameplates. Generally, green indicates ON and red is OFF, although many devices simply have a single red LED for ON. An LED requires a resistor in series, typically valued at 560Ω for 12V systems. Red filament lamps are disappearing. The one disadvantage of these is that they consume power, typically around

40mA. If 20 circuits are on, this adds up to a reasonable load on the system and a needless current drain. If you have a very large switchboard, allow for the status-light current drain. In many cases, people assume they have a current leakage problem when, in fact, the switchboard indicators are causing part of that drain. With the use of LEDs, this is now less of a problem.

7.30   Switchboard Cable and Wiring Installation. All cables, and cable looms to switchboard panels, should permit the panel to be opened without placing strain on connections or cables. In many installations, cable looms are too short to allow easy and full opening of panels for inspection or installation. It is common for connectors to be pulled off the rear of circuit breakers due to the strain on conductors or wiring looms. Looms should be neatly tied; in general, this may be in two or three separate looms. They should have sufficient length to allow complete opening of the switchboard, and the circuit cables should be secured to prevent undue stress on the connectors.

The image in the color insert shows badly installed switchboard wiring; you can spot many errors, from lack of labeling to cable support to improper terminations, among others. This is followed by a Hallberg-Rassy HR34 panel, installed to meet code. This is what one should aim for when installing.

7.31   System Interlock Wiring. There is always some interaction with engine systems and control functions within equipment wiring. Functions such as fans or solar panels are controlled through relays.

7.32   Circuit Protection. All protection and isolation devices should have an assigned DC fault rating and be approved by a relevant national or international standard. These standards may include NEMA, UL, Canadian Standards Association (CSA), now called CSA Group, and Lloyd's Register (LR), among others. Install only circuit breakers that are approved by one of the national approval bodies. Approvals for small vessel breakers categorize them as supplementary protectors; I normally use E-T-A, Ancor, and Carling circuit breakers. They must be approved for DC operation and marked with the DC rating.

7.33   Switchboard Feeder Protection. The power supply feeder to the switchboard should have short-circuit protection and circuit isolation installed as close as practicable to the battery in the positive conductor and be accessible. The isolator also should be accessible. This may be incorporated within a single trip-free circuit breaker. Fuses can be used; however, it is better to combine isolation and protection within one easily resettable device. They should be mounted as high as practicable above possible bilge and flooding levels. I have observed some recommendations for isolation of the negative pole as well, so a double pole breaker is a good option.

7.34   Battery-Connected Equipment. The power supply to auxiliary equipment connected directly to the battery should have short-circuit protection and circuit isolation installed as close as practicable to the battery, in both the positive and negative conductors, and be accessible. This may be incorporated within a single trip-free circuit breaker. These auxiliary supplies generally include high-current equipment such as thrusters, electric windlasses, winches, and toilets that are often connected directly to an engine battery or a separate battery bank. They should be mounted as high as practicable above possible bilge and flooding levels.

7.35   Navigation Light Protection. Each navigation light circuit should have a short-circuit protection device installed. A circuit breaker supplying all navigation lights creates the risk of a single fault tripping the breaker and all lights being unavailable until the fault is cleared. This may not be possible in adverse weather conditions. Where possible, separate circuit breakers should be installed. Alternatively, where a single breaker is used, each circuit should have a replaceable fuse installed. This may be a multicircuit fuse block or, at the rear of the switchboard, carrying fuses or circuit breakers for all circuits. I prefer protecting the anchor or tricolor light and, separately, the steaming light, stern light, and the forward bicolor light. Redundancy is the operative word here.

7.36   General Circuit Advice. Circuits for power and lighting should be separate. Circuits should not have mixed consumers, such as power to outlets or motors, also connected to lighting equipment. Lights should be just lights, and it is good practice to separate into zones, such as forward circuit, saloon circuit, and an aft cabin circuit. All fuses, distribution bus bars, and terminals should be covered. Covers should be fitted over all positive and negative bus bars, distribution bus bars, and fuse holders such as slow-blow fuses used with anchor windlasses. This is an ABYC requirement and protects against accidental contact and water.

7.37   Fuses. Fuses are still widely used, and although cheaper, they have many disadvantages. Control and DC circuit fuses are of the ceramic, blade, or glass type. There are either simple fuse holders or a combination fuse switch. Fuses are either fast-acting or dual-element time-delay types. Fast-blowing current-limiting fuses, also known as high rupturing capacity (HRC) fuses, are in many AC machinery installations. One of the advantages of fuses is a lower initial capital cost. The disadvantage is that they suffer from vibration and fatigue—and when you're in trouble, you can't find a spare fuse. Fuse parameters are as follows.

     a.   **Rating Variations.** The typical glass fuse is not always accurate and can rupture as much as 10% to 50% above or below the nominal current rating.

     b.   **Service Fatigue.** Fuse elements fatigue in service, with the fuse element properties altering; subsequently, the rated value may alter, causing "nuisance" failure. Vibration commonly causes failure.

     c.   **Voltage Drop.** There is added contact resistance in the fuse holder between each contact and the fuse ends, which commonly causes voltage drops, intermittent supply and heating, and increased corrosion.

     d.   **Troubleshooting.** Problems are amplified when a circuit has a fault and you go through a box of fuses in a trial-and-error troubleshooting exercise. A circuit breaker allows simple resetting.

7.38   Circuit Breakers. The circuit breaker is the most reliable and practical method of circuit protection. They are manufactured in pushbutton aircraft type, toggle type, or rocker switch. They are thermal, thermal magnetic, magnetic or magnetic hydraulic, tease, and trip free. Trip free means they cannot be closed with a fault current. Circuit breaker design considers many factors that include vibration, shock, humidity, degree of protection (IP rating), ambient temperature range, dielectric strength, and more. They are used for circuit isolation and protection, combining both functions, which saves switchboard space, costs, and installation time, as well as improving reliability. Single-pole circuit breakers are normally fitted to most boats; however, classification society rules only allow these in grounded-pole installations. This is because a fault arising on the circuit will provide a good ground loop, and the large current flow will ensure proper breaker interruption. Double-pole breakers are recommended for all circuits, as they will totally isolate equipment and circuits. This is a requirement of many classification societies or survey authorities.

a. **Circuit Breaker Selection.** Circuit breakers must be selected for the cable size they protect. The rating must not exceed the maximum rated current of the conductor. The cable sizes in the table below give recommended ratings for single cables installed in well-ventilated spaces. Bunching of cables and high ambient temperatures require derating factors. Ratings are given according to IEC Standard 157. Approvals and certification include UL 1077, TUV, CE, and CSA. Check out the Blue Sea range at www.bluesea.com, Carling at www.carlingtech.com, and E-T-A at www.e-t-a.com.

Table 7-1. Circuit Breaker Selection

Wire mm²	AWG	Circ Mils	Current (A)	CB Rating
1.5	15	3,260	7.9–15.9	8
2.5	14	5,184	15.9–22.0	16
4.0	11	8,226	22.0–30.0	20
6.0	9	13,087	30.0–39.0	30
10.0	7	20,822	39.0–54.0	40
16.0	5	33,088	54.0–72.0	60
25.0	3	52,624	72.0–93.0	80
35.0	2	66,358	93.0–117.0	100
50.0	0	105,625	117.0–147.0	120

b. **Tripping Characteristics.** The tripping characteristics are normally quoted by the circuit breaker's manufacturer. This is a curve of current against time. The curve is a graph of the relationship between overcurrent and the CB tripping time, and these graphs normally show the time-current operating zone. The tripping time is the time interval from the moment when the tripping current starts to flow and the moment when the current is interrupted:

(1) The greater the current value over the nominal tripping value, the quicker the circuit breaker will trip. In cases of short circuit, tripping is rapid due to the high current values.

(2) Slower tripping characteristics are seen where a small overload exists; tripping occurs some seconds or even minutes after switch on. This happens as the current levels gradually increase.

c. **Discrimination.** The principle of discrimination in both DC and AC circuits is extremely important and is not considered sufficient on boat electrical systems. A circuit normally should have two or more overcurrent protective devices, such as the main and auxiliary circuit breakers installed between the battery and the load. The devices must operate selectively so that the protective device closest to the fault operates first. If that device does not operate, the second device will trip, protecting the circuit against overcurrent damage and possibly fire:

(1) Use circuit breakers with different current ratings. This means that at a point on the time-delay curve, the first breaker will trip; if it does not trip and the current value increases, the next circuit breaker will trip. A point called the "limit of discrimination" is reached. At this point, curves intersect and both will trip simultaneously.

(2) Use circuit breakers with different time delay curves. This simply entails using circuit breakers with different time delay curves to achieve the same result.

(3) Use circuit breakers with different time delay curves, current ratings, and different circuit breaker types. This enables using the above to ensure discrimination.

7.39 Switchboard Troubleshooting. A number of faults routinely occur on switchboards and their protective devices. The following faults and probable causes should be checked first. It is assumed that power is on at the switchboard.

a. **Circuit Breaker Trips Immediately at Switch On.** This is characterized by the ammeter showing, in most cases, an off-the-meter full-scale deflection, indicating a high fault current:

(1) **Load Short Circuit.** Check the appropriate connected load; disconnect the faulty item before resetting.

(2) **Connection Short Circuit.** If the fault still exists after the load is disconnected, check any cable connections for a short circuit or, in some cases, cable insulation damage. Trip free circuit breakers may prevent closing of the circuit breaker if a short circuit exists.

b. **Circuit Breaker Trips Several Seconds after Switch On.** This is characterized by the ammeter showing a gradual increase in current to a high value before tripping off; it is typically an overload condition. In some cases, this may be an almost instantaneous trip:

(1) **Motor Seizure.** This fault may arise if the electric motor has seized or, more probably, the bearings have seized. In some cases, the pump has seized.

(2) **Load Stalling.** This fault is usually due to a nearly seized pump.

(3) **Insulation Leakage.** This fault is usually due to a gradual breakdown in insulation, such as a wet bilge area pump connection.

c. **There Is No Power after Circuit Breaker Switch Is On.** After checking that power is absent at the equipment connection terminals, check the following:

(1) **Circuit Connection.** Check that the circuit connection has not come off the back of the circuit breaker, a frequent occurrence. Check the cable connection to the crimp connection terminal.

(2) **Circuit Breaker Connection.** On many switchboards, the bus bar is soldered to one side of all distribution circuit breakers. Check that the solder joint has not come away. In some cases, breakers have a bus bar that is held under breaker screw terminals; check that the screws and connections are tight.

(3) **Circuit Breaker.** Operate the breaker several times. In some cases, the mechanism does not make proper electrical contact; several operations usually solve the problem by wiping the contacts.

(4) **Circuit Negative.** If all tests verify that the positive supply is present, check that the circuit negative wire is secure in the negative link.

d. **Circuit Power on but No Indication Light.** The LED, and in some cases the resistor, may have failed. Check the soldered connection to the circuit breaker terminal.

7.40 **Conductor Selection.** Selecting the right conductors is the foundation for a reliable yacht electrical system. Conductors should be selected based on the maximum current demand of the circuit. Ambient temperatures exceeding the rated temperature of the cable should be derated by a factor of 0.05 for each 9°F (5°C) above. All cables have nominal cross-sectional areas and current carrying capacities. The ISO 10133 standard specifies nominal capacities for a range of cross-sectional areas and temperature ranges. Temperature reference is typically 77°F (25°C). The table illustrates typical current ratings for equivalent cable sizes. It is recommended that standard cable sizes be used, which are cheaper and simpler to calculate. All cable current carrying capacities are subject to derating factors. In any installation where the temperature exceeds the nominal value, the continuous current carrying capacity of the cable is reduced. This is important in engine spaces. Where the temperature exceeds 122°C (50°C) the derated capacity is the nominal capacity multiplied by 0.75. Consult the actual cable manufacturer's ratings for accuracy.

Table 7-2. Typical DC Cable Nominal Ratings

Size AWG	Size mm²	PVC Insulation	Butyl Rubber (Lloyd's 100A1)	Resistance Ohms/100m
17	1.0	11A	12A	1.884
16	1.5	14A	16A	1.257
15	1.5	15A		1.050
14	2.0	20A	22A	0.754
13	2.5			
12	3.0	27A	30A	0.471
11	4.0			
10	6.0	35A	38A	0.314
9	7.0			
8	8.5	49A	53A	0.182
7	10.0			
6	16.0	64A	71A	0.1152
5	25.0			
4	20	86A	93A	0.0762
3	25.0			
2	35	105A	119A	0.0537
1	40	127A	140A	0.0381
0	50	150A	160A	0.0295
2/0	70	161A	183A	0.0252
3/0	95	200A		

The more common method of calculating cable current ratings or ampacity is the use of tables based on the formula CM = K × I × L/E, where CM = circular mil area of the conductors; K = 10.75 (a copper resistance constant per mil-foot); I = current in amps; L = conductor length in feet; and E = voltage drop at the load in volts.

Table 7-3. Battery Cable Ratings

Conductor Size AWG	Conductor Size B & S	Conductor Size Metric	Current Rating (60% Duty)
8	8	8	90A
6	6	15	150A
4	3	26	200A
2	2	32	245A
1	0	50	320A
00	00	66	390A

7.41   Voltage Drop. Conductor size should be selected with a maximum allowable voltage drop of 5% for all circuits. The voltage drop can be calculated using the formula in ISO Standard 10133, Annex A.2. Voltage drop must always be a consideration when installing electrical circuits. Unfortunately, the majority of voltage drop problems are created by the poor practice of trying to install the smallest cable and wiring sizes possible. The maximum acceptable voltage drop in 12V systems is 5%, or 0.6V. The voltage drop problem is prevalent in starting and charging systems, thrusters, windlasses, and long runs to equipment. The following formula is specified in ISO Standard 10133, Annex A.2. You can use one of the many charts available on the internet.

$$\text{Voltage Drop at Load (volts)} = \frac{0.0164 \times I \times L}{S}$$

S is the conductor cross-sectional area, in square millimeters

I is the load current in amperes

L is the cable length in meters, positive to load and back to negative

WIRING INSTALLATION

7.42   Wiring, Cables, and Conductors. All wiring conductors should be insulated and sheathed, called double-insulated cables. Additionally, insulation is temperature rated, which has important implications with respect to ratings. In most boats PVC-insulated and PVC-sheathed cables rated at 177°F (75°C) are used. For classification society rules, ship wiring cables that use butyl rubber, CSP, EPR, or other insulating materials are often specified; they have higher temperature ratings and, subsequently, higher current carrying capacities.

7.43   Wire Types. Conductors should be of stranded and tinned copper. When untinned copper is exposed to saltwater spray or moisture, it will oxidize very quickly, degrade, and fail. The argument against installation of tinned copper is cost. The price differential is typically 30% greater; however, the increased reliability and vessel resale value far outweigh the lower-priced plain copper conductor.

7.44   Minimum Wire Sizes. It is my personal contention that all conductors should have a minimum cross-sectional area of 16AWG ($1.0mm^2$). It is recommended that conductor sizes be standardized to 14AWG ($2.5mm^2$). Match the wire to the current carrying requirement, and then consider length of run and volt drop.

7.45   DC Wire Colors. In the UK, Europe, Australia, and other countries, DC conductors are nominally identified as red for the positive conductor and black for the negative conductor. Numbered cores are an acceptable alternative; the numeral "1" should be positive, the numeral "2" for negative, or any numbering system that works for you. This is standard in commercial marine systems installations. Some ship wiring cables often have white cores only, with numbering imprinted on the insulation. The United States primarily uses ABYC recommendations, which use color-coded insulated conductors. The ABYC nominates black and yellow as negative polarity colors. Yellow is also a primary AC phase, switching, or control circuit color, so caution should always be used. Make sure AC and DC are not in close proximity to avoid any possible confusion. Reference the circuit-color drawing in the color insert and the wiring color code table below.

Table 7-4. Wiring Color Codes

Color	Designation
Red	DC Positive Main Conductor
Black	DC Negative Main Conductor
Yellow	DC Negative Main Conductor
Green	Bonding Systems
Dark Blue	Cabin and Instrument Lights
Light Blue	Oil Pressure Sender to Gauge
Dark Gray	Navigation Lights
Dark Gray	Tachometer Transducer to Gauge
Brown	Pump Circuits
Brown	Alternator to External Regulator
Brown	Alternator Charge Light
Brown/Green	Freshwater Pump
Brown/Blue	Head Pump
Brown/Violet	Washdown Pump
Pink	Sender to Fuel Gauge
Orange	Feed to Accessory
Orange	Common Feed
Orange/Brown	Electric Head, MSD

(continued)

Table 7-4. *Continued*

Color	Designation
Orange/Blue	Communications Equipment
Purple	Ignition Switch to Electrical Instruments
Purple	Instrument Feed Distribution Instruments
Tan	Water Temperature Sender to Gauge
Brown w/Yellow Stripe	Bilge Blower Circuit

7.46   Cable and Wire Identification. Always mark cable ends to aid in reconnection and troubleshooting. Use a simple conductor sleeve marking system where PVC color-coded and numbered markings slide over the conductor. There are also numbered heat shrink sleeve markings, wrap around, and self-laminating systems. Some labeling systems use markers that are attached to wires with nylon cable ties. These simply require use of a marker pen. Just ensure that the marker is permanent. Reference the examples of poor wiring installations in the color insert section.

7.47   Conductor Installation. Good cable installation is essential if electrical problems are to be avoided. Follow these installation guidelines. Wiring and cable failures are attributable to several causes, and failure can cause fires. All cables age, including the insulation, and results include cracking, embrittlement, and failure of the sheathing and insulation. Once a conductor is exposed, the risk for fire and short circuit arises. The drivers for wiring longevity include excessive high or low temperature, mechanical resilience to vibration, stressing, abrasion, and ambient environmental conditions. Moisture ingress to the insulative sheath or conductor covering leads to failure. Improper installation and exposure to saltwater ingress to the copper conductors leads to corrosion and oxidation. Conductor heating due to operation at close to maximum current ratings impacts the insulation. Exposure to UV radiation is a real killer for insulation; anything external is at risk.

7.48   Cable and Wiring Runs. Cable runs should be installed as straight as practicable. Cable bend radii should be a minimum of 4 times the cable diameter. Tight bends should be avoided to reduce unnecessary strain on conductors and insulation. Minimum cable bend radii are applicable to all cables, but particular care should be taken with larger and more inflexible cables; times 6 is a better target radius.

7.49   Cable Bunching. Where cables are bunched, the cable ratings should be derated. When several (six to eight) cables are bunched in a large loom, the current capacity of the cable is reduced. The factor is typically around the nominal rating multiplied by 0.85. This may become an issue only in very large boats. I would recommend not bunching heavy-current cables with low-current ones. Where cables carry large currents for short time durations, they should be used subject to duty cycles. Heavy-current-carrying cables, such as those used on windlasses, winches, thrusters, and starter motors, are in fact used only for short durations. As there is a time factor in the heating of a cable, smaller cables can be used. Table 7-3 shows battery cable ratings that are rated at 60% duty.

7.50 **Cable Accessibility.** Cables should be accessible for inspection and maintenance. The emphasis must be on accessibility—for initial installation, maintenance, and addition of circuits. Under no circumstances should cables be encased in fiberglass. All cables, in particular those entering bulkhead transits, should be accessible for routine inspection.

7.51 **Cable Protection.** Cables should be protected from mechanical damage, either where exposed or within compartments. All cables should be installed to prevent accidental damage to the insulation, cutting of the conductors, or undue strain on the cable. Even though cables are routed through lockers, machinery spaces, and cupboards, they require protection. In many cases, faults are traced to what are considered safe areas, such as lockers. Objects and equipment are thrown into the space, and sharp edges deform or damage the cable or insulation. In machinery or engine spaces, cables are often damaged during engine repairs. Reference the image of cables in the color section. Although they are enclosed in black flexible conduit, they are at the bottom of a locker and prone to crushing when the locker is accessed and filled.

7.52 **Cable Transits.** Cables passing through bulkheads or decks should be protected from damage using a suitable noncorrosive gland or bushing. Cables transiting decks or watertight bulkheads should maintain their watertight integrity. Cable glands are designed to prevent cable damage and ensure a waterproof transit through a bulkhead or deck. A significant number of problems are experienced with the ingress of water through deck fittings, and I have seen a variety of methods used. In addition, running cables through FRP/GRP with some sealant invariably results in chafing and cable failure. Use circular multicore cables if it is possible to ensure proper gland sealing. The structural material of a deck must be considered before gland selection. A steel deck requires a different gland type than a foam sandwich boat. Reference the image of various cable gland options in the color insert.

7.53 **Cable Support.** Cables should be supported at maximum intervals of 8in (200mm). Supports and saddles are to be of noncorrosive material. When used in engine compartments or machinery spaces, these should be metallic and coated to prevent chafe to the cable insulation. Cable saddles should fit neatly, without excessive force on the cables or cable looms, and not deform the cable insulation. Cables can be neatly loomed together and secured with PVC conduit saddles or stainless saddles to prevent cable loom sagging and movement during service. While the recommendation is 18in (450mm) apart, closer support distances secure the cables more efficiently. Saddles should be placed no more than 6in (150mm) apart to prevent sagging and movement during service. I prefer standard electrical PVC conduit saddles, which come in a variety of sizes. It is important to have a neat fit only and not force saddles over cables or looms so that insulation is deformed. Metal saddles are often used in machinery spaces; however, a plastic sleeve should be placed on them to keep the sharp edges from chafing the cable insulation. The PVC cable tie, zip tie, or tie wrap is universal in application and should be used where looms must be kept together, or where any cable can be securely fastened to a suitable support. Do not use cable ties to suspend cables from isolated points; this invariably causes excessive stress and cable fatigue. Internal cable ties require only white cable ties; any external cable ties should be black UV-resistant type. PVC spiral wrapping is an extremely useful method for consolidating cables into a neat loom. If a number of cables are lying loose, consolidate them into some spiral wrap and then fasten the loom using cable ties. A hot-glue gun is often used to fasten small or single cables above shallow-depth headliners or in corners behind trim and carpet finishes. This is useful where there is no risk of cables coming loose. Do not use this method on exposed cable runs.

7.54   Cable Segregation. Cables should be, as far as practicable, separated into power, signal and data, and heavy-current carrying groups. Instrument and data cables should be installed as far as practicable from power cables and communications aerial cables. A minimum distance of 12in (300mm) is recommended. AC cables should not be run within DC system cable looms; they must be kept separate. Cables should be separated into signal or instrument cables, DC power supply cables, and, where space allows, heavy-current carrying cables such as windlasses or thrusters. This is to minimize induced interference between cables—in particular, on long, straight runs. All data and instrument cables should be routed as far as practicable away from power cables. Aerial cables should be routed well way from power cables.

7.55   Cable Heat and Lightning Clearances. As far as practicable, cables should be routed clear of mast steps and chain plates to increase separation distances for lightning protection. To reduce the chances of a side strike, try to install cables as far as possible from any lightning protection conductors or lightning system bonding connections. Where cables may be exposed to heat, they should be installed within conduits or otherwise protected from the heat source. Cables installed with machinery or within engine spaces should be rated for the maximum heat of the space. Cables should also be protected wherever they may be exposed to such heat sources as exhaust manifolds or piping.

7.56   Cables in Conduits. Where cables are installed within conduits, they should be supported within 3in (75mm) of both entry and exit points. Conduit ends should be treated or otherwise protected to remove sharp edges and prevent chafe to cable insulation. Conduits are often installed during the construction phase, which allows cables to be easily pulled in, replaced, or added. Conduits offer good mechanical protection to cables, and single-insulated cables are often run inside conduits back to the switchboard. As the cables are single insulated, they are exposed where they enter or exit the conduits and should be supported by saddle or clamp to prevent excessive movement. Try to avoid installing large bunches of cables in flexible conduits, as they tend to move around and chafe. PVC conduits should not be used in machinery spaces. Where cables exit conduits, the exit should be bushed to prevent chafing. When pulling in cables during installations, the insulation is frequently damaged as it rubs against sharp edges.

7.57   External Cable Installation. All externally installed cables should be protected against the effects of ultraviolet (UV) light. Continued exposure to UV light on external equipment cables will result in insulation degradation and failure. Small cracks in the insulation allow water to penetrate the conductor and subsequently degrade the copper. This is common on navigation lights, GPS aerial cables, radio aerial cables, and other equipment. All exposed cables should be covered in black UV-resistant spiral wrapping to prevent rapid degradation of insulation. Cable ties should be of the black UV type. Use tinned-copper conductors on all external wiring to navigation lights, spotlights, cockpit lights, etc. Check the navigation light image in the color insert, illustrating badly installed and exposed wiring, which will degrade quickly.

7.58   Cable Joints and Splices. Connections should be minimized within any circuit between the power supply and the equipment. Connections and joints in cables should be avoided. Any connection adds resistance to a circuit and introduces another potential failure point.

7.59   Equipment Grounds. Equipment grounds should be made to the nominated main ground point. (Refer to my notes on grounding.) Equipment grounds, such as pump casings or electronic device grounds, are usually connected to the boat ground. In many cases, a ground terminal block is installed close to instruments and a large ground conductor is taken to the main ground. Where you are running a ground, segregate the wires from main wiring looms to avoid picking up interference. You should never install a washer between the ground conductor terminal and the ground terminal. Inserting a washer under the connection inserts resistance into the circuit. You can install a star washer under the nut or bolt head.

7.60   Hazardous Areas. Electrical equipment, cables, and cable connection should never be installed within any compartment or space that may contain equipment or systems liable to have or emit explosive gases or fumes. This may include spark ignition engine fuel systems, LPG installations, or flooded-cell battery installations. Any installed equipment or fittings should be ignition protected in accordance with the appropriate national standards. Equipment or fittings must be classified as ignition-proof by an appropriate organization, such as UL.

7.61   Instrument and Data Cable Installation. Electrical and instrument cables are generally installed in close proximity to each other. If precautions are not taken, this may result in electrical interference on electronics systems. Instrument and data cables should be installed as far as practicable from power cables and communications aerial cables. Long parallel runs close to power cables should be avoided. Cables should be separated as far as practicable from power supply cables and heavy-current carrying cables, such as windlasses or thrusters, to minimize induced interference between cables, especially on long, straight runs. Cables should be routed well way from aerial and antenna feed cables. Where instrument and data cables cross power cables, this should be done as close to a 90-degree angle as practicable; induced interference is prevented with right-angle crossovers. Equipment screens should be grounded at one end only, or in accordance with specific manufacturer's recommendations. Conductor screens should be grounded as recommended by the equipment manufacturer. Equipment grounds are usually connected to the nominated boat ground. In many cases, a ground terminal block is installed close to instruments and a large ground conductor is taken to the same point as the battery negative connection point. This is not the battery but the actual termination point.

7.62   Equipment and Device Location. Navigation, autopilot, and position fixing equipment should be located as far as practicable from radar, satellite communications equipment, VHF, AIS, high frequency (HF) radio, amateur (ham) radio, and cellular telephone equipment, tuners and control units, cables, aerials, antennae, and related components. A minimum clearance distance of 39 inches (1m) is recommended. Where practicable, electronics equipment, control modules, processors, etc., should be located clear of cable looms and aerial cables to prevent interference. This should include all satellite communications and television system and cellular telephones. Autopilots are prone to interference, with the potential to initiate uncontrolled course alterations.

7.63   Grounding Systems. The term "ground" is a common descriptor covering a range of applications, but an explanation is required of what the various "grounds" are, as described in the chapters on lightning, corrosion, AC power systems, and radio systems. Grounding on shore is the mass of the Earth, and its resistivity and conductivity are known and fairly constant. Seawater has variable conductivity and resistivity; it is not actually at the same potential as the Earth. Instead it varies hourly with respect to tide and weather, and therein lies some of the challenges afloat along with interaction between shore and boat grounding systems. The various elements are described within ABYC E-11 and in UK/Europe RCD ISO 13297:2020.

    a.   **DC Negative Conductor.** The DC negative is not a ground; it is a current-carrying conductor that carries the same current that flows within the positive conductor.

    b.   **DC Polarizing Conductor.** In a single-circuit wiring configuration, the negative is bonded to a grounded point, usually the mass of the engine. The engine is connected to a seawater-immersed item, such as the propeller shaft. This is used to polarize the system; it doesn't actually carry current but holds the negative to nominal ground potential.

    c.   **Lightning Ground.** A lightning ground is a point at ground potential that is immersed in seawater. It carries current only in the event of a lightning strike, and its primary purpose is to ground the strike energy. It is not a functional part of any other electrical system, as explained in the lightning protection chapter.

    d.   **Cathodic Protection System Bonding Conductor.** The cathodic protection system "ground" is the equipotential bonding of protected underwater items to a sacrificial anode. It is not actually a ground, as explained in the corrosion chapter.

    e.   **AC Safety or Protective Ground (or Earth).** The AC safety ground or protective grounding conductor does not normally carry current. It is used to protect against electric shock. It is connected to AC equipment through the vessel nominal ground point or terminal and to the shore ground conductor through the shore power cable. In many wiring configurations it is interconnected to the AC neutral conductor. The nominal ground point is immersed in seawater. In a fault condition, the protective ground will hold all exposed conductive parts and extraneous conductive parts to ground potential and ensure operation of all short-circuit protective devices, such as fuses and ground fault circuit interrupter (GFCI) devices. The European RCD rules state that a ground connection can include any conductive part of the wetted surface of the hull in permanent contact with the water, depending on the overall system design. This of course outlines the challenges with grounding. A marine engineer and good friend of mine was killed due to an improperly grounded refrigeration vacuum pump, so this subject resonates very strongly.

f. **Radio Frequency (RF) Ground.** The radio frequency ground is not a ground; it is an integral part of the aerial system and is correctly termed the "counterpoise." The "ground" carries only RF energy and is not a current-carrying conductor. It is not, or should not be, connected to any other ground.

g. **Instrument Ground.** An instrument or device ground is connected to the nominal vessel ground. In many cases, a marshalling ground terminal link is installed behind the switchboard close to instruments. Cable drain wires, screens, and ground wires are connected to this, and a single ground wire is run to the instrument ground point. Noise immunity protection is negated as the grounding conductor is unshielded or incorrect. This link should not be connected to the DC negative link located behind DC switchboards.

h. **DC Ground Bus.** Frequently mentioned in ABYC recommendations, this is a copper bus installed to allow connection of all the various grounds and is connected to the main ground point.

i. **AC Ground Bus.** Frequently mentioned in ABYC recommendations, this is a copper bus. It is installed to allow connection of all the various grounds and is connected to the main ground point.

j. **Ground Point, Ground Terminal, Ground Plate.** These terms for ground are used interchangeably and can be confusing. The nominal vessel ground is the point where electrical current from a boat's conductive elements are conducted to water. This is variable depending on a vessel's configuration and varies considerably between boats. It can be the keel, the engine and immersed propulsion system, or a dedicated ground plate. Ground plate recommendations are varied. One is a copper plate of 1 square foot installed on the hull bottom. This is for internally ballasted boats, and given that modern yachts have an epoxy-coated cast-lead keel, they probably fall into this category. Dynaplates are used as an alternative to this type of ground plate, and many declare them to be ABYC compliant.

k. **Galvanic Isolator.** This device is installed in series with an AC protective or safety conductor of a shore cable. It is used to block low-voltage DC current while still permitting AC current flow within the safety or protective conductor. See the shore power and corrosion chapters.

7.64   Conductor Terminations. Conductor terminations are the single greatest cause of failure and should be made properly if you wish to maintain reliability. All conductors should be terminated, where practicable, using crimped connectors. The most practical and common method of cable connection is the tinned-copper crimp terminal or connector. These are color coded according to the cable capacity that can be accommodated. Terminals are usually designed and manufactured according to NEMA standards, which cover wire pull-out tension tests and voltage drop tests. Where possible, select double-crimp types, which should be used in high-vibration applications. When crimping, always use a quality ratchet-type crimping tool. Do not use a cheap pair of squeeze types, which do not adequately compress and capture the cable. Do not use pliers for crimping; this causes subsequent failure, and the cable pulls out of the connector sleeve. A good joint requires two crimps. Always crimp both the joint and the PVC insulation behind it. Ensure that no cable strands are hanging out. Poor crimping is a major cause of failure. A crimp joint can be improved by lightly soldering the wire end to the crimp connector. Avoid excessive heat. After crimping, give the connector a firm tug to ensure that the crimp is secure. It is actually very difficult to find manufacturers quoting maximum current ratings of crimp connectors; the table below is the closest I could come up with. It is often stated that the crimp is rated to the cable it is connected to, but that is not precisely accurate. You do need to check ratings, as ring terminals have different ratings to other terminal types.

a. **Ring-Crimp Terminals.** These are used on all equipment where screw, stud, bolt, and nut are used. They should be used on any equipment subject to vibration, or where accidental dislodgement can be critical, particularly switchboards. Always ensure that the ring is a close fit to the bolt or screw used on the connection to ensure good electrical contact. Consider using spring washers to maintain tension. One practical method used to prevent nut or screw creep is to dab on a spot of paint.

b. **Fork-Crimp Terminals.** These are an option where using ring-type terminals is difficult. It's simple to loosen a terminal screw, slide these under, and tighten. A yellow connector is rated at 36A, a blue is rated at 24A, and a red is rated at 18A. Check with your supplier for ratings.

c. **Push-On Blade Terminals.** These are commonly used, and it is important that no strain exists to pull them apart. The current rating of a blade terminal is less than the same ring connector. A yellow blade connector is rated at 20A, a blue is rated at 16A, and a red is rated at 12A. If you are using piggyback types to secure two cables on one terminal, the same ratings are applicable. Check with your supplier for ratings.

d. **Crimp Butt Splices.** Where cables require connection and a junction box is impracticable, use insulated in-line butt splices. This is more reliable than soldered connections, where a bad joint can cause high resistance and subsequent heating and voltage drop. Use heat-shrink insulation over the joint to ensure that waterproof integrity is maintained. Some connectors when heated form a watertight seal by fusing and melting the insulation sleeve. A yellow butt splice is nominally rated at 48A, a blue butt is rated at 27A, and a red is rated at 19A. Check with your supplier for ratings.

e. **Crimp Pin Bootlace Ferrule Terminals.** Bootlace ferrule, or pin, terminals can make a neat cable termination into connector blocks. However, from experience I have found these to be unreliable, simply because vibration and movement work them loose. In most cases, they do not precisely match the connector block terminal and make an inadequate electrical contact. A yellow connector is rated at 20A, a blue connector is rated at 16A, and a red is rated at 12A. Check with your supplier for ratings. It should be noted that some French- and German-manufactured ferrules are designed to either lightly solder or crimp. Twin wire ferrules are also available.

f.   **Crimp Bullet Terminals.** These quick-disconnect terminals are useful in cabin lighting fittings. I often use these on all cable ends—female on the supply and male on the light fitting tails. This makes it easy to disconnect and remove fittings. The current rating of a bullet is less than the same ring connector. A yellow bullet is rated at 20A, blue bullet is rated at 16A, and a red is rated at 12A. Check with your supplier for ratings.

g.   **Wire Terminations.** Where cables are terminated within terminal blocks, they should be secured to prevent contact with adjacent terminals. Cable ends should have the insulation removed from the end, without nicking the cable strands. The bare cable strands should simply be twisted and inserted in the terminal block or connector of a similar size. Ensure that there are no loose strands. If you are terminating into an oversize terminal block, twist and double over the cable end to ensure that the screw has something to bite on.

Table 7-5. Standard Crimp Ring Connector Table

Color	AWG	Cables Sizes	Current Rating
Yellow	12–10	2.63mm² to 6.64mm²	30A–48A
Blue	16–14	1.04mm² to 2.63mm²	15A–30A
Red	22–18	0.25mm² to 1.63mm²	10A–25A

7.65   Soldered Terminations. Conductor terminations should not be soldered. Do not solder the ends of wires prior to connection. In most cases, this is done to make a good low-resistance connection and prevent wire "corrosion." It is my experience that soldered connections cause many problems, with the solder traveling up the conductor and causing stiffness. This causes greater vibrational effects at the terminal, with resultant fatigue and failure. In most cases, the soldering is poorly done, with a high-resistance joint being made. A soldered cable end prevents the connector screw from spreading the strands and making a good electrical contact, causing high resistance and heating. The proviso is that you should use connectors of the correct size for the cable.

7.66   Termination Marking. Conductor terminations should be marked with a number. The negative and positive cables should be marked with the same number. Identification should be consistent with the wiring diagram. Always mark cable ends to aid in reconnection and troubleshooting. A simple, slide-on number system can be used. The stick-on adhesive types should be avoided, as they generally unravel and fall off as the adhesive fails. If wires are ABYC-compliant color coded, still use numbers, which are easier and much quicker to identify. The circuit positive should sequentially match the supply source, such as the circuit breaker. The circuit negative should match the positive and be placed in the same sequential order on the negative link. The numbering convention, if unmarked, is left to right.

7.67   Termination Location. Where connections are made within any area subject to water or moisture, such as bilges, the terminations should be made as far as practicable near the top of the bilge. Connections should be suitably protected against water ingress. Connections should be made above the maximum bilgewater level. Joints should be finished with self-amalgamating tapes or heat-shrink tubing. I have frequently seen connections permanently immersed and fail. In automatic bilge circuits, the live connection also contributes to corrosion problems in some boats.

7.68   Plug and Socket Location. When used to connect cables or equipment, plugs and sockets should incorporate screw-retaining rings and protective caps to prevent the ingress of water when not in use. They should be rated to a minimum of IP54. Deck plugs and sockets are often used instead of deck glands and junction boxes at a mast base or as outlets for hand spotlights. Many in use are of inferior quality and fail prematurely. Don't use the cheap chrome plugs and sockets; they aren't waterproof. When using deck plugs, ensure that the seal between deck and connector body is watertight. Leakage is very common on wet decks up forward, where they are usually located. Ensure that the cable seal into the plug is watertight. It is of little use having a good seal around the deck and plug to socket if water seeps in through the cable entry and shorts out the terminals internally, as is often the case. Most connectors have O-rings to ensure a watertight seal. Check that the rings are in good condition, are not deformed or compressed, and seat properly in the recess. A very light smear of silicon grease assists in the sealing process. Ensure that pins are dry before plugging in and that pins are not bent and do not show signs of corrosion or pitting. Do not fill around the pins with silicon grease, as this creates a poor contact. Keep plugs and sockets clean and dry.

7.69   Junction Boxes. Where connections are made, they should be protected within a suitable junction box and installed in a protected area. Junction boxes are the most practical way to terminate a number of cables, especially where access is required to disconnect circuits. To reduce the number of cables radiating back to the switchboard and minimize voltage drops, I use a junction box both forward and aft to power up lighting circuits. Terminal blocks are usually used, and in many cases the box is too small for the quantity or size of cables. In these cases, the box lid is forced on, applying pressure to the cables; this should be avoided, as it applies unnecessary stress to terminations. Cables terminated within a junction box should enter from the bottom and be looped to prevent water entering the box and connections. The upper surfaces of the junction box should have no openings that permit water to enter. Cables looped in at the bottom will allow water to drip off and prevent surface travel to the connections. Cables within junction boxes should be marked with numbers that correspond to circuit numbers used at the main switch panel. The numbers should match those on the wiring diagram. See junction boxes in the color insert.

7.70   Circuit Testing. Before you energize any circuit, you should verify that the circuit is ready. The insulation resistance between conductors, conductors and ground of all circuits, or the complete installation should be greater than $100,000\Omega$. All fuses, circuit breakers, and switches are to be closed. Tests should be made with all equipment, lamps, and electronic equipment disconnected. All circuits should be tested to verify that the levels of insulation are satisfactory on the whole system, as well as on each circuit. Supply circuit breakers should be switched on so that switchboard bus bars are included within the test. A multimeter set on the resistance range should be used between the positive and negative conductors. If readings are low, check that a load is not connected. All electronics equipment should be disconnected.

7.71   Mast Cabling. Mast cabling is a common source of failure. Many problems can be avoided if the cables are installed properly. Since masts are generally wired by mast manufacturers and riggers, vessel owners rarely take the opportunity to supervise or specify requirements. There are three major areas of concern in any mast installation.

    a.   **Mast Base Junction Boxes.** The most common area of failure is the junction box. If mounted inside the vessel, a good water-resistant box should be installed. If mounted externally—and this should be a last resort—a waterproof box is required. Always leave a loop when inserting cables in the box. If water does travel down the loom, it will drip off the bottom of the loop and not enter and corrode the junction box terminals or connections.

    b.   **Deck Cable Transits.** Cable glands are used to prevent cable damage and ensure a waterproof transit through a bulkhead or deck. A significant number of problems are experienced when water gets in through deck fittings, and I have seen some amazing systems utilizing pipes, hose, and so on. If figure 8–type cable is used or small, single insulated cables are installed, it is virtually impossible to adequately seal them in cable glands. To overcome this problem, use circular, multicore cables if possible. Or use the consolidation procedure described below to make a cable loom that can be put through a deck gland. You need to take deck material into account before selecting a gland. Steel decks require different glands than FRP/GRP and foam-sandwich decks.

    c.   **Cable Installation.** The following factors should be noted when installing cables:

        (1)   **Cable Types.** The major problem is the use of single insulated, untinned cables, generally of an underrated conductor size. Small conductor sizes cause many voltage drop problems, with unacceptable low light outputs as a result. Use 15A rated cable for each circuit.

        (2)   **Negative Conductors.** Masthead tricolors are normally connected to a dual-anchor light fitting. These use a three-wire, common negative arrangement. The same arrangement is used for combination masthead and foredeck spotlights. Never use the mast as a negative return, as I have found on some vessels. Always install a negative wire to each light fitting.

        (3)   **UV Protection.** All exposed cables should be covered in black, UV-resistant spiral wrapping to prevent rapid degradation of insulation. Small cracks in the insulation allow water to penetrate, which subsequently degrades the copper.

d. **Mast Cable Support.** Cabling must be properly secured within the mast. The weight of a cable hanging down inside a mast causes fatigue through stretching. If the cables are not fully enclosed in conduit, still a common practice, the internal halyards can whip against them. This will damage the insulation or cause severe damage to the conductors in multicore instrument cables. There are a number of methods for securing mast cables; a combination of all three is best:

(1) **Cable Glands.** Where a cable enters the mast base and exits at the masthead, it should pass through a cable gland. Once cables have been placed through the neoprene, the gland is tightened and compression around the cables takes the strain. The cables are protected from chafe against the mast entrance hole.

(2) **Messenger Line.** A small messenger line can be installed with the cables and supported at the masthead. The messenger should be tied or taped to the cable loom and then fastened to take the load off the cable ends. The messenger serves as a pull-through for adding or replacing cable. However, once the line is taped to the loom over its entire length, it is impossible to remove and replace single cables.

(3) **Cable Ties.** Where possible, use cable ties to fasten and support cables. The ideal place to do so is where cables come out of the mast to connect lights, radar, and so on, which usually provides three or four fastening points. There is generally sufficient space to insert a cable tie around the cables. A second hole large enough for a tie is required next to the main cable entry to enable the tie to be supported. Always use black, UV-resistant cable ties.

e. **Mast Cable Consolidation.** In many cases, the mast is wired with single insulated cables. To put these cables through deck cable glands, you need to consolidate them into a single loom. One method is as follows:

(1) Neatly make a cable loom and hold it in place with cable ties. Keep the loom as circular as possible.

(2) Apply silicone sealant to the loom and work it through all cables. This will ensure that a solid core is made. If done properly, it will prevent water from traveling down the cable loom.

(3) Apply a layer of black, UV-resistant spiral wrap to the loom. Again, spaces between the wrap should have silicone compound applied to fill any voids. The spiral wrap gives the cable loom a circular shape.

(4) Slide on a length of heat-shrink tubing and shrink it in place. This forms the outer sheath.

(5) Use a suitable deck gland, pass the cable through the deck, and connect into a suitable junction box.

f.   **Deck Plugs.** Instead of deck glands and junction boxes at a mast base, deck plugs are sometimes used. They can provide outlets for hand spotlights or other equipment commonly used. Many deck plugs are of inferior quality and fail prematurely, often when they are needed most. When using deck plugs, observe the following:

(1)   **Deck Seal.** Ensure that the seal between deck and connector body is watertight. Leakage is very common on wet decks up forward, where the plugs are usually located.

(2)   **Plug Cable Entrance.** Make sure the cable seal into the plug is watertight. It is of little use to have a good seal around the deck if the water seeps through the cable entry and shorts out terminals internally.

(3)   **Connector Seals.** Most connectors have O-rings to ensure a watertight seal. Check that the rings are in good connection, are not deformed or compressed, and seat properly in the recess. A very light smear of silicone grease assists in the sealing process.

(4)   **Connection Pins.** Ensure that the pins are dry before plugging in and that pins are not bent or showing signs of corrosion or pitting. Do not fill around the pins with silicone grease, which often creates a poor contact. Keep plugs and sockets clean and dry.

g.   **Mast Wiring System Maintenance.** Basic maintenance tasks will reduce mast wiring problems:

(1)   **Mast Base Cable Exits.** Regularly examine wiring where it exits the mast for signs of chafe. If the cable loom has not been protected with a UV-resistant sleeve, carefully examine insulation for cracks. Consider installing protection.

(2)   **Masthead Cables.** Regularly examine masthead cable exits for chafe. Ensure that coaxial, wind instrument, and power cables have a reasonable loom to allow for shortening and repair. Consider installing UV protection.

7.72   Mast Cabling Troubleshooting. Mast wiring faults are common because the mast will subject the wiring and cables to the worst damaging factors, such as excessive vibration, exposure to salt water, stretching, and mechanical damage. Fortunately, mast wiring is relatively easy to troubleshoot.

a.   **Tricolor/Anchor Lights.** If a light does not illuminate, lamp failure is the usual cause. Test from below before climbing the mast. LED navigation lights require a voltage check. If voltage is going to the light fitting and the light is not operating, the LED power supply driver circuit has failed. Perform the following tests:

(1) **Test Supply.** Open the mast connection box and locate the appropriate terminals. Using a multimeter set to the DC volt setting, check that voltage is present at the terminals with the power on. Many failures are due to poor contacts within terminal blocks or corrosion of the terminal and cable. With LED lights, if the power is on at the mast base connection and no light is visible, it is either a wire failure, an LED driver failure, or a faulty light connection.

(2) **Continuity Test.** Turn the power off, and with a multimeter set on the resistance setting, test between the positive and negative terminals. The reading should be approximately 2Ω to 5Ω with a good lamp installed. If the reading is above that range, the light fitting or connection has failed or the cable has been damaged. The mast cable entry and exit points should be examined first. Internal breaks occur only in masts without wiring conduits. Many tricolor/anchor lights have a plug-and-socket arrangement, which is an occasional source of trouble. LED lights are exempt from this test.

b. **Spreader and Mast Foredeck Spotlights.** The above tests are also valid for spreader lights. On many sailing boats, spreader lights are a sealed beam unit within a stainless steel housing, although these are being replaced with LED spotlights. It is very common to have short circuits to the mast, as cables chafe through on the sharp edges. This problem is notorious for causing circuit leakages and increased corrosion rates on steel vessels:

(1) **Mast Short Circuits.** With a multimeter set on the resistance ohms (Ω) range position, check between the mast and both positive and negative wires. The reading should be overrange. If you have any reading, you have either a short or a leakage from cable insulation breakdown.

(2) **Check Supply.** Open the mast connection box and locate the appropriate terminals. Using a multimeter on the DC volt setting, check that voltage is present at the terminals with the power on.

Corrosion

8.1   **Galvanic Corrosion.** For galvanic corrosion to occur, there has to be a potential difference between two dissimilar metals immersed in an electrolyte. This principle was discovered in the eighteenth century by Luigi Galvani, source of the term "galvanic." This was developed into the first practical battery cell by Alessandro Volta, hence the term "volt." Corrosion can be defined as the chemical deterioration or reduction of a metal or metal alloy due to interaction with the environment. In the corrosion process, metal atoms leave the metal to form compounds in the presence of water or gases. This is commonly called rusting. (Corrosion is often improperly called electrolysis.) Corrosion takes many forms, and corrosion with respect to boats and basic electrical systems falls into two main categories: galvanic corrosion and electrolytic or stray current corrosion. Stray current corrosion is damage resulting from current flow outside the intended circuit. Both corrosion processes are a result of DC electric current flow between the two metals in an electrolyte. The result is corroded hulls, propellers, shafts, rudders, stocks, and skin fittings. Other corrosion sources on boats include stress corrosion cracking, crevice corrosion, and pitting corrosion, among others. Have you checked your chain plates lately?

8.2   **Galvanic Protection.** The British Admiralty introduced anodic protection back in 1824 on copper-clad timber men-of-war. Galvanic corrosion, or electrochemical corrosion, is the process that occurs when galvanic cells or couples form between two pieces of metal with different electrochemical potential when they come into contact with each other. If the two metals have the same electrical charge or potential, they will not create a cell; no current will flow, and they are called compatible. Ideally, a vessel should be constructed so that most metallic items are compatible. If they are different, they must be either isolated or protected.

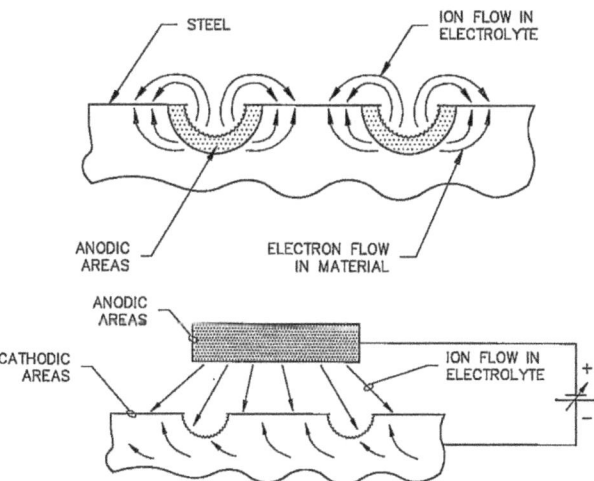

Figure 8-1. Galvanic Corrosion Process

8.3 **Anodes and Galvanic Protection.** Anodes are the normal protection method for achieving galvanic protection. Anodes are called sacrificial anodes because they are sacrificed instead of the hull or fittings. Because they are high on the nobility scale, they tend to corrode quicker than other items, such as mild steel and alloy. The zinc anode generates an electric current, and because the hull effectively has a higher potential, the anode allows current flow through it and bonded items, through the seawater, and back to the hull. The process corrodes the anode proportional to the level of current flow present, while preserving the hull and fittings. The rate of corrosion is affected by several factors, including the anodic corrosion current level, water temperature, and water salinity. A basic parameter is derived from Faraday's law: A known current acting for a known time will cause a predictable weight loss of metal. For example, 1A applied for 1 year will cause a loss of 22lb (10kg) of steel. The size of the exposed area of the cathodic metal relative to the anodic metal will affect the corrosion rate. The corrosion rate varies between metals, and the corrosion current within systems is typically rated in milliamperes (mA). Anodes have no association with stray current corrosion. For galvanic corrosion to occur, four basic requirements must be fulfilled.

a. **The Anode.** There has to be a positive or anodic area, which is called the anode. This will possess the lowest potential and be the metal that corrodes.

b. **The Cathode.** There has to be a negative or cathodic area, which is called the cathode. This will possess the highest potential.

c. **The Electrolyte.** There has to be a path for the current to flow. This is the electrolyte—in this case, seawater.

d. **The Circuit Path.** There has to be a circuit path for the current to flow; this is any interconnecting cathodic bonding connection.

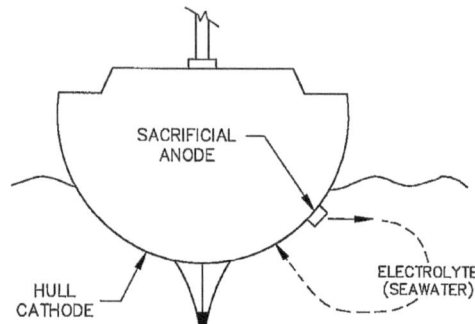

Figure 8-2. Galvanic Corrosion Protection

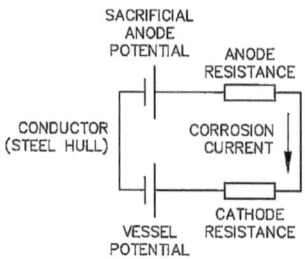

8.4 **Metal Nobility.** All metals can corrode, both ferrous and nonferrous. Base metals such as steel and aluminum corrode easier than the noble metals such as stainless steel and bronze. All metals can be classified according to molecular structure, and these electrochemical characteristics are listed in a metallic nobility table. The base metals at the top of scale conduct easily; the noble metals at the bottom do not. The materials with the greatest negative value will tend to corrode faster than those of a lesser potential. The closer in potential of metals close to each other on a chart, the less corrosion effect, since the voltage differential is less. The voltage difference between metals will drive current flow to accelerate corrosion of the anodic metal. Table 8-1 below shows the electrochemical potential of elements relative to a standard hydrogen electrode (SHE) in column 1 and is a standard electrode potential value. Some charts reference against an Ag/AgCl (silver/silver chloride) reference half-cell, which is the most common test unit. Other charts are derived from a saturated calomel (Hg_2Cl_2) electrode half-cell, as shown in column 2. It is important to note that values can alter when an electrolyte changes, such as variation in seawater salinity, pH, and temperatures.

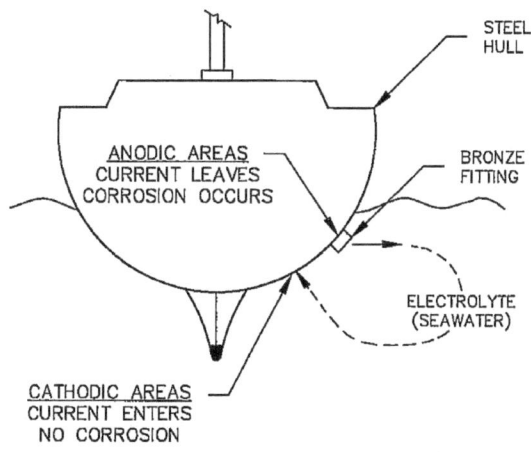

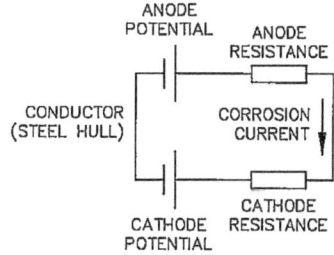

Figure 8-3. Steel Boat Galvanic Corrosion

Table 8-1. Metal Nobility and Galvanic Series Table

Metal	Volts (SHE)	Volts Hg_2Cl_2
Magnesium and Alloys	−2.340	−1.65
Zinc	−0.762	−1.10
Aluminum	−1.670	−0.86
Mild Steel/Iron/Cast Iron	−0.440	−0.71
Lead	−0.126	−0.55
Manganese Bronze	−0.270	−0.27
Copper, Brass, and Bronze	−0.340	−0.25
Silver	+0.779	−0.00
Gold	+1.420	+0.15

8.5   FRP/GRP and Wooden Boats. Vessels are generally categorized into specific groups; the following are the most relevant for yachts. The following are typical arrangements based on recommendations by leading corrosion specialists for FRP/GRP and wooden or timber vessels.

a.   **Type A Vessels:** Generally single-screw boats with a short propeller shaft length in contact with seawater. They generally have wooden or FRP/GRP rudders. Normally only one anode is required for propeller and shaft protection. The main anode should be located on the main hull below turn of the bilge at an equal distance between the gearbox and the inboard end of the stern tube.

b.   **Type B Vessels:** Single- or twin-screw boats with long exposed propeller shafts supported by a shaft bracket and in contact with seawater. One anode is required for each propeller and shaft assembly. Separate anodes are required for mild steel rudders. Bronze or stainless steel rudders with bronze or stainless steel rudder stocks must be bonded to the same anode.

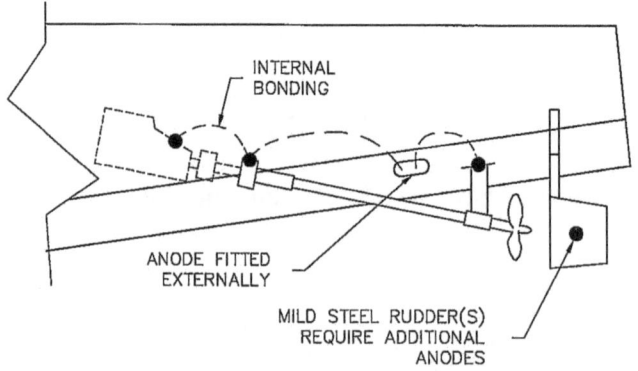

Figure 8-4. Class B Anode Arrangements

    c.   **Type C Vessels:** Single-screw boats with long exposed propeller shafts supported by a shaft bracket and in contact with seawater. They have FRP/GRP rudders and bronze or stainless steel rudder stocks. Normally only one anode is required for propeller and shaft protection. If there are mild steel bilge keels, they should have separate anodes affixed.

8.6   **Zinc Anode Purity Standards.** All cathodic protection systems generally use zinc anodes with approved purity standards. Zinc alloy anodes should conform to US Mil Spec MIL-18001K or, in Australia, AS2239-2003. If the anodes do not have a standard quoted, exercise caution using them. The average sailing vessel uses zinc sacrificial anodes for corrosion protection. They are called sacrificial because they are sacrificed instead of the item they are attached to and protect, such as the hull, propeller shaft, shaft bracket, or rudder. As they are high on the nobility scale and have a higher electrical potential, they will corrode faster than other items, such as mild steel. Pure zinc anodes are good conductors. The zinc anode generates an electric current and, as the hull effectively has a higher electrical potential, the anode allows current flow through it and bonded items to the seawater and back to the hull. The process corrodes the anode proportional to the level of current flow present while preserving the base metal, such as the steel hull or prop shaft. Although primarily used in commercial shipping rather than yachts, the anode consumption amp-hour per kilogram rate is calculated, given the impacts that hull corrosion can have. This term is used to denote the change in mass of a sacrificial anode over time and defines how fast an anode would be consumed within an electrochemical cell.

8.7   **Aluminum and Magnesium Anodes.** Vessels on freshwater rivers and lakes follow the same criteria as boats in salt water; the difference is that anodes are made from magnesium or aluminum. Magnesium anodes are effective in salt, brackish, and fresh waters. Fresh water has a much greater insulation value than salt water, so anodes such as magnesium and aluminum with a higher driving voltage than zinc anodes are required. When a boat moves into seawater or water of a higher salinity, anodes will become more active and should be inspected after just fourteen days. Given the rise in zinc prices, aluminum has become an economically viable alternative. Aluminum is actually a better material for anodes than zinc. The driving potential is higher, and it can be used in both salt and brackish water.

8.8   **Anode Number Calculations.** Calculations are normally based on the wetted surface area of the hull. The main vessel dimensions used are the waterline length, waterline beam, and the mean loaded draft. The area is calculated using the following formula:

$$\text{Waterline Length (LWL)} \times (\text{Waterline Beam} + \text{Draft})$$

This formula suits most heavy-displacement sailing vessels and motor cruisers. For medium-displacement vessels, multiply the calculated sum by 0.75. For light-displacement vessels, multiply by 0.5.

Table 8-2. One-Year Anode Selection

Wetted Area	Hull Anodes	Rudders
Up to 300ft² (28m²) SW	2 × 9lb (4.0kg) Zinc	2 × 2lb (1.0kg) Zinc
Up to 300ft² (28m²) FW	4 × 3lb (1.5kg) Mag	2 × 0.6lb (0.3kg) Mag
28.1–56 m² (>600ft²) SW	4 × 7.5lb (3.5kg) Zinc	2 × 2.2lb (1.0kg) Zinc
28.1–56 m² (>600ft²) FW	4 × 7.5lb (3.5kg) Mag	2 × 0.6lb (0.3kg) Mag
56.1–84 m² (>900ft²) SW	4 × 9lb (4.0kg) Zinc	2 × 2.2lb (1.0kg) Zinc
56.1–84 m² (>900ft²) FW	4 × 7.5lb (3.5kg) Mag	2 × 0.6lb (0.3kg) Mag
84.1–102 m² (>1100ft²) SW	4 × 14lb (6.5Kg) Zinc	2 × 5lb (2.2kg) Zinc
84.1–102 m² (>1100ft²) FW	6 × 10lb (4.5kg) Mag	2 × 1.5lb) 0.7kg Mag
102.1–148 m² (>1600ft²) SW	6 × 14lb (6.5kg) Zinc	2 × 5lb (2.2kg) Zinc

8.9   Anode Systems. It is essential for the anodes to be of the correct size, installed in the correct location, and of the correct number for the area being protected. It is quite possible to overprotect the hull and fittings; excessive numbers or larger anodes do not equal improved protection. If your vessel is in warm highly saline waters, you must make more frequent inspections of zinc anodes. Table 8-3 illustrates typical arrangements recommended by industry-leading corrosion specialists MG Duff Marine for steel, aluminum, fiberglass, and timber vessels. Note that metal and fiberglass hulls are treated separately. Anode position is not critical, but anodes must be able to "see" the parts to be protected. Anode fixing must be above the bilge line internally, and there must be a minimal internal bonding cable-run length. These anode sizes are based on propeller sizes and are approximate only. Check out mgduff.co.uk for everything corrosion protection.

Table 8-3. Anode Mass Table (Salt and Fresh Water)

Prop Size	Type A	Type B	Type C
10" SW Zinc	2lb (1.1kg)	2 × 2lb (1kg)	2lb (1.1kg)
FW Magnesium	10.5oz (0.3kg)	2 × 10.5oz (0.3kg)	10.5oz (0.3kg)
14" SW Zinc	2.2lb (1.1kg)	2 × 2.2lb (1kg)	2.2lb (1.1kg)
FW Magnesium	10.5oz (0.3kg)	2 × 10.5oz (0.3kg)	10.5oz (0.3kg)
19" SW Zinc	4.8lb (2.2kg)	2 × 2lb (1kg)	4.8lb (2.2kg)
FW Magnesium	14oz (0.4kg)	2 × 14oz (0.4kg)	14oz (0.4kg)
21" SW Zinc	4.8lb (2.2kg)	2 × 4.8lb (2.2kg)	4.8lb (2.2kg)
FW Magnesium	1.5lb (0.7kg)	2 × 1.5lb (0.7kg)	1.5lb (0.7kg)
26" SW Zinc	4.8lb (2.2kg)	2 × 4.8lb (2.2kg)	4.8lb (2.2kg)
FW Magnesium	1.5lb (0.7kg)	2 × 2.2lb (1.0kg)	1.5lb (0.7kg)

Prop Size	Type A	Type B	Type C
30" SW Zinc	4.8lb (2.2kg)	2 × 4.8lb (2.2kg)	4.8lb (2.2kg)
FW Magnesium	1.5lb (0.7kg)	2lb (1kg)	1.5lb (0.7kg)
36" SW Zinc	9.8lb (4.5kg)	2 × 9.8lb (4.5kg)	9.8lb (4.5kg)
FW Magnesium	2lb (1kg)	2lb (1kg)	2lb (1kg)
40" SW Zinc	9.8lb (4.5kg)	2 × 9.8lb (4.5kg)	9.8lb (4.5kg)
FW Magnesium	2lb (1kg)	2lb (1kg)	2lb (1kg)
48" SW Zinc	9.8lb (4.5kg)	2 × 9.8lb (4.5kg)	9.8lb (4.5kg)

8.10 Rudder, Skeg, and Bilge Keel Anodes. Anodes for mild steel rudders, skegs, and bilge keels are normally installed directly to steelwork. In most cases, the best solution is to bolt the anodes back-to-back; they may be welded on if required.

Table 8-4. Rudder, Skeg, and Keel Anodes Table

Protection Area	Anode Size
Up to 10ft² Steelwork	2 × 3lb (1.35kg) Strip Anodes
Up to 30ft² Steelwork	2 × 2lb (0.90kg) Anodes
Up to 70ft² Steelwork	2 × 4.8lb (2.22kg) Anodes

8.11 Anode Bonding. Anodes should be fixed in view of the parts and areas they protect and be bonded to the parts under protection. Anode positioning must be able to "see" the parts to be protected. When a zinc anode is 50% to 75% wasted, it must be replaced. White or green halos around zincs or other metals indicate that stray current may be affecting them. Bright zincs indicate excess current flow. A small amount of current can cause paint reactions. Rapid zinc wastage and degree of paint reaction indicate more-serious problems. Where a boat has moved into fresh water and back to salt, the anode will become encrusted with a white crust. This will stop it from functioning and must be cleaned off. Do not bond ferrous and nonferrous metals to the same anode; otherwise, you effectively create a cell or battery.

8.12 Grounding and Bonding. No bonding connections should be made to any skin fitting unless it is part of a cathodic bonding system. I do not support the practice of connecting every metal item, including through-hull skin fittings and bronze seacocks, and using stainless wire and hose clamps. Any other fittings that are isolated and are connected with rubber or PVC hoses need not be bonded. The current flow in a bonding circuit is very small, and any resistance introduced into the circuit from bad connections and cable resistances creates a difference in potential that will cancel any protective measures, and may actually create problems.

8.13 Equipotential Bonding. The equipotential bonding conductor is normally a non-current-carrying conductor used to connect exposed conductive parts of DC electrical devices and extraneous conductive parts at a substantially equal potential. Equipotential bonding conductors should be green in color or clearly indicate the function so as to avoid confusion with any AC ground. Many recommendations nominate green to identify bonding cables. Caution should be used, as this is a US AC safety ground color. Identifying each termination with the sleeved letters "EB" (equipotential bond) will reduce the chances of accidental disconnection. This possibility is real, as such bonds may be connected to the common ground point. Disconnection of an AC ground can create electrical shock risks, so this is for clear safety reasons—I speak from experience here. While some standards allow bare conductors, this can lead to early conductor deterioration and is therefore not recommended. Do not bond the lightning ground system or the down conductor to the anode bonding system. This is my personal view and at variance with the American Boat and Yacht Council (ABYC). (See my premise on that in the lightning chapter.) There have been several well-documented cases of skin fittings and thru-hulls being blown out in a lightning strike and the vessel subsequently sinking.

8.14   Bonding Cable Location. All cathodic bonding system cables should be installed clear of bilges or other wet areas. Any bonding cables should be installed well above the bilge line, or any other area that may be subject to water. They should interconnect only the items to be protected and should not indiscriminately bond all items. All interconnections must be of at least a 11AWG (4.0mm²) tinned-copper conductor and be bolted to the main bonding connection. Anode fixing studs should be connected to bonded parts by the shortest practical route to minimize resistance. It is critical that bonding be resistance free, so a heavy-gauge conductor is necessary. Use a minimum 11AWG (4mm²) cable or greater.

8.15   Bonding Circuit Resistance. The total resistance of any cathodic bonding circuit should not exceed 0.02Ω. The purpose of cathodic bonding is to equalize the electric potential of the underwater metals being protected and connected. It is not to dissipate any stray currents on 12V systems and spread the surface areas. It is critical that bonding cables be resistance free; therefore, the use of a heavy-gauge conductor is necessary. When the vessel is slipped, use a multimeter on the resistance range and check the resistance between anode and propeller; the maximum reading must be 0.02Ω. The current flow in a bonding circuit is very small; any resistance introduced into the circuit from bad connections and cable resistances creates a difference in potential that will degrade any protective measures and may actually create problems. Many recommendations call for a bonding loop or daisy chain. It is better to connect bonded items in a radial arrangement back to the anode bonding bolt to minimize resistance. If one bonding wire is accidentally broken, the majority of the bonding network will not be lost. It is important to check regularly that the main bonding points are secure and clean.

8.16   Shaft Collar Anodes. Most yachts have shaft anodes installed. When installing collar anodes to propeller shafts, make sure the shaft is clean and not covered with an antifoulant. I have frequently seen this done when around boatyards. The collars must be mounted as close as possible to the shaft strut (bracket) or cutlass bearing, typically a clearance of 0.25in to 0.5in (6mm to 12mm). These are standard and streamlined shaft collar anodes for the protection of the propeller shaft and come in zinc, aluminum, and magnesium. They come in two halves with locked-in stainless steel nuts and Allen screws. There are also shaft anodes called limited clearance collar zinc anodes. Before installation, make sure the shaft is perfectly clean and shiny. Install, ensure that the anode seats perfectly on the shaft, and then torque up the bolts correctly. Do not paint the anode!

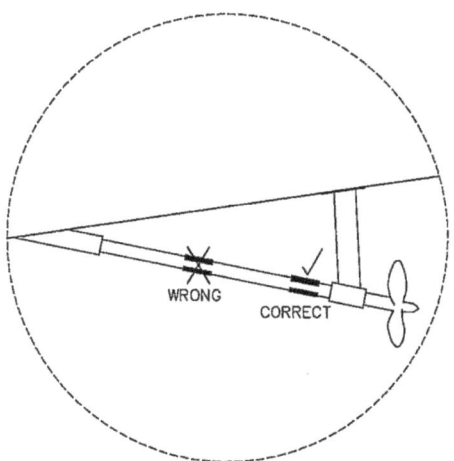

Figure 8-5. Shaft Anode Installation

8.17   **Anodes for FRP/GRP and Wooden Boats.** There are a few facts to remember with fiberglass or wooden or timber vessels. Anodes on fiberglass and timber hulls must have the internal bonding system connected to them. This is a common omission. An anode is working only when some corrosion is visible on it. I have frequently seen anodes mounted on a hull and not connected—and heard the owners of those vessels proudly proclaim that there was no corrosion problem.

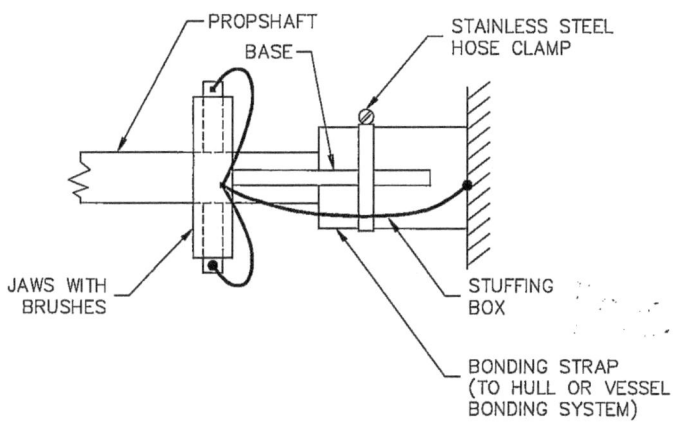

M.G. DUFF "ELIMINATOR"

Figure 8-6. Propeller Shaft Bonding

a.   **Propeller Shaft Bonding.** The usual method for bonding propeller shafts, both commercially and in small vessels, is to install a brush system. M.G. Duff's system is called the Electro Eliminator. Essentially, it is a brush system connected to the cathodic bonding system. If such a system is used, the shaft must be kept clean and free of oil, grease, and water. It is often better to bridge the coupling to the engine block and use a collar anode or separate anode bonded directly to the engine block.

    b.   **Seal Anode Bolt Holes.** Always seal the wood around the anode bolt holes; this can prevent wood electrolysis if an overprotection situation exists.

    c.   **Shaft Coupling Bonding.** Many engine installations incorporate flexible couplings to the propeller shaft. The coupling must be electrically bridged to ensure proper continuity of the system where the engine is not maintained electrically isolated above the bonding system.

8.18   **Anodes for Steel Boats.** On steel boats, the anodes are either welded to the hull or bolted on brackets. Many boats have studs welded to the hull and the anode is fastened onto the studs. It is important to ensure that the studs and connections are good and have a low resistance. It is common to see a stainless nut on a mild steel stud and the mild steel threads corrode. Use all 316-grade stainless steel. At each haul-out, take off the anode and ensure that threads and contact surfaces are clean. When the vessel is hauled out, use a multimeter set on the resistance range and check the resistance between the anode and propeller. The maximum reading should be 0.2Ω. The resistance reading between the anode and the hull should be zero. On ferro-cement yachts, forget the cement component of the vessels and treat them as steel vessels. Make sure bronze thru-hull fittings are isolated from the concrete reinforcing mesh. Personally, I would switch over to Marelon thru-hulls and did so on my previous steel boat. Some vessels use a zinc anode (also known as a guppy) on a wire over the side while in a marina or on a mooring. This must be properly grounded to the hull and have a low-resistance copper cable connected to the anode.

    a.   **Keel Anodes.** If a steel keel is fitted, an anode may be installed on each side. Follow the calculated wetted surface guidelines so that overprotection does not occur.

    b.   **Negative Cables.** Install isolated negative return starter motors, alternators, and monitoring gauge sensor units. These will ensure that no electrical items are connected to the block. Use double-pole engine starter isolators. If you do not have a fully isolated engine, make sure the engine is isolated above the hull, with insulated coupling and engine mounts. I have done this on a previous boat, and it is a viable method if rewiring the entire boat.

    c.   **AC Ground.** The AC ground can be bonded directly to the hull. This normally has no voltage flowing in it, except under fault conditions. It is essential that the main hull be part of the AC ground because in a fault condition, the hull could become partially live, up to rated voltage. This will not cause anode corrosion.

    e.   **Vessel Interaction.** Avoid mooring adjacent to any copper-sheathed (very rare indeed) or aluminum vessels. Do not use steel cables to tie up to shore—not common, but it does happen when people tie up for the winter.

    f.   **Paint Systems.** Although a vessel hull is protected by an extensive paint program, isolating metal from the seawater, the incorrect assumption is often made that corrosion cannot occur and cathodic protection is unnecessary. But it does occur, because air holes, or small areas of paint imperfection, occur along weld seams, and paint coatings are damaged due to abrasions from chains, piers, tenders, etc. Cuprous oxides used in antifouling paints can convert to copper sulfide and create a galvanic cell. I have encountered corrosion problems with steel vessels using copper-based antifouling paint although many have altered the coatings products being used. These antifoulants rapidly degrade the anode's performance, and in most cases, I have found corrosion and pitting where the paintwork has been chipped off the hull. Shaft struts (brackets) and propellers have suffered damage. Wherever possible, use copper-free antifoulants. When in doubt, you should always consult a corrosion specialist. Use the technical services of the paint manufacturer; all manufacturers will provide assistance.

8.19   Anodes for Aluminum Alloy Boats. Aluminum hulls have different requirements than steel vessels, and it is essential that they be correctly protected. Unlike steel vessels, an overprotected aluminum hull doesn't simply lose paint; it is eaten away by a caustic attack. The recommendations for steel hulls are also valid for the electrical systems on an aluminum vessel.

    a.   **Material Compatibility.** Insulate or use compatible through-hull fittings. Insulate any equipment made of metals above aluminum on the metal nobility scale. Avoid bronze fittings if at all possible. I have come across some who are using Lloyd's approved aluminum and plastic fittings, which resolve many problems.

    b.   **Vessel Interaction.** Avoid mooring next to steel- or copper-sheathed vessels for extended periods. The interaction can be very severe with aluminum, rapidly corroding a hull.

8.20   Crevice Corrosion. Crevice corrosion affects many other parts of the boat—typically the mast, standing rigging, and associated fittings. Everything from the backstay, forestay, cap shrouds along with shackles, chainplates, turnbuckles, and swage fittings are prone to this. Many a dismasting is directly attributable to this. If a Failure Mode and Effect Analysis (FMEA) was ever required, it should be performed on the rig, given the large number of failure points and modes. At the time of writing, I am in the process of replacing all my standing rigging, given several areas of weakness. One of the issues with crevice corrosion is that it is a hidden failure mode. Swage fittings are more prone to corrosion than compression ones, and compression fittings can be opened and inspected. Most rig failures are preventable, depending on inspection.

    a.   **Crevice Corrosion Principles.** Crevice corrosion is common and can initiate wherever there is a microscopic crack, weld seam, fissure, or any other metal flaw. Crevice corrosion is a localized attack on either the metal source or adjacent to any crevice or gap between two surfaces when they are exposed to a stagnant electrolyte. This occurs between any two metals or metal and nonmetal materials. Factors that impact crevice corrosion are depth and surface textures within the crevice or gap. Several variables dictate the establishment, speed, and aggressive nature of the corrosion. This includes the electrolyte or seawater temperature, oxygen levels within the crevice, pH levels, the level of chloride and halide ions, and other factors. This will result in electron sacrifice to the much greater cathodic surface area that surrounds it. When the electrolyte—salt water—infiltrates any crevice or gap, it eventually becomes stagnant and oxygen depleted, creating a microenvironment that undergoes chemistry alterations. These alterations result in a surplus of positive metal ions, and the crevice becomes anodic. The reaction results in more negative ions migrating into the crevice, usually in the form of chloride ions, and this accelerates acid formation and metal erosion.

b.   **Chain Plates and Keel Bolts.** These are common failure points. The factors impacting longevity are the design and installation. Material flaws are prime factors, along with age, fatigue, and corrosion levels. Keel bolts are another source of crevice corrosion that need to be monitored and checked; while total failures are relatively uncommon, severely corroded bolts are not. It is hard to define recommended sealants or other products that will reduce corrosion. See the images in the color insert of characteristic crevice corrosion rust weeping and bolts with bolt heads shearing off due to corrosion.

c.   **Propeller Shafts and Bearings.** A hidden area prone to crevice corrosion is at the end of the drive train. There are common cases of crevice corrosion on propeller shafts. Areas to monitor are the shaft seal and bearing areas, along with internal sources within the stern tube. Pitting and corrosion in the cutlass bearing or seal packing location require attention. Other areas include the gap between propeller and shaft if improperly installed.

d.   **Stress Corrosion.** This is caused by stress and heavy loading on the rigging components. When combined with crevice-corroded elements, it usually results in catastrophic failure. Swage fittings are a common location for this type of failure, where water is easily entrapped within the swage. Other failure points are mast attachment points and the fasteners that hold everything together with the ability to entrap water. The corrosion process can be very rapid once it establishes.

e.   **Passivation.** Stainless steel surfaces must be clean and have a lot of oxygen supply to maintain the protective passive surface layer. I have heard of some boat owners who regularly washdown fittings with fresh water and perform passivation with citric acid.

f.   **Troubleshooting.** When you sight rust stains and those light brown surface blooms also known as tea stains or, worse, both minor and major surface pitting, it indicates corrosion activity. Surface blemishes such as handrails are due to surface contamination from non-stainless materials. Grit from grinding and airborne particulates are usually at fault. This is often common immediately after a haul out, as boatyards are very dirty environments. There are various stainless cleaners that comprise acids, solvents, and stainless steel anaerobic thread sealants available; they are worth investing in to keep things in good shape.

g. **Mast Electrics.** Faulty mast-mounted electrical light fittings and cables can cause electrolytic corrosion when energized, contributing to corrosion. Check the integrity of systems regularly.

8.21   Corrosion Leakage Monitoring. Leakage to the hull on steel and alloy vessels can be monitored using suitable systems. Verifying the condition of a corrosion protection system requires testing and the measurement of the hull potential. This involves immersion of a reference electrode called a half-cell. This is made from silver/silver chloride, which is a silver wire coated with silver chloride. The reference cell is immersed approximately 6in (15cm) into the water away from the hull. The reference electrode is connected to a digital multimeter positive terminal. The negative terminal is connected to the boat ground, which is usually the battery negative ground point. The multimeter is then set to the DC voltage setting. This displays a value that is the actual hull potential. The readings must be interpreted with respect to water temperature and salinity.

8.22   Isolation Transformers. The isolation transformer is the most effective method for shore power supply circuit isolation. It is used to electrically or galvanically isolate all the normally energized conductors and the protective safety conductor from the shore power AC system. I installed one of these on my previous steel boat, and it is worth considering on a FRP/GRP power- or sailboat. An isolation transformer is similar to a step-down transformer; however, in the isolation transformer it is a 1:1 ratio. The input primary winding and the secondary output winding are electrically isolated. The secondary winding is not connected to ground. This results in complete galvanic isolation. Some transformers, such as those from Victron, incorporate a soft-start system, which ramps up the initial magnetization of the windings and reduces the inrush current magnitude, which can trip the shore supply circuit protection breaker. An isolation transformer called the Marine Puck, from Bridgeport Magnetics Group, is different from traditional transformers. It has a very low leakage current—under 1 milliampere (1mA)—which reduces GFCI nuisance trips, isolates the boat from the shore power, and eliminates the galvanic corrosion cell. Marine Puck transformers come in three power ratings: 3.6kVA, 6kVA, and 12kVA. This is a toroidal transformer; it is embedded in solid epoxy resin inside a cup-shaped nonmetallic enclosure. Marine Pucks are double insulated with no ground reference, and they have a lower inrush current than conventional transformers. Check out bridgeportmagnetics.com. Mastervolt offer the MASS GI transformers, which use high-frequency electronic switch conversion to input and output the AC the same as the traditional low-frequency transformer system, although they are significantly lighter and more compact and have integrated soft-start functions. Some units, such as Victron Energy transformers, have temperature-controlled fan cooling.

8.23   Shore Power Problems. When a boat is tied up in a marina and plugged into the pedestal, it connects to the shore ground. This effectively connects the boat to all other boats plugged into pedestals, and all grounds are then interconnected. Depending on how a boat is grounded on board—and if that ground is connected to the galvanic protection system. It is also connected through bonding to the DC system and immersed items such as the prop shaft, keel, propeller, and a large galvanic cell is created. This is further aggravated because marina piles are often steel and may or may not have anodes installed underwater. If DC leakage currents flow through the AC safety ground on board the boat, they not only can neutralize the boat corrosion protection system but also drive the system and rapidly affect the boat's underwater parts. The galvanic isolator is designed to galvanically isolate the AC shore ground from a DC bonding system when they are connected. The devices are for protection against galvanic currents, but if a "stray current" and "stray voltage" exceeding the diode threshold value were to flow, the diode could go into full conduction.

8.24   Impressed Current Cathodic Protection (ICCP). These are installed on larger steel and alloy vessels. Essentially protection is based on compensation of corrosion currents by the use of a countercurrent using an onboard power source. The reference anode senses the electrical potential of the seawater and sends a signal to the control unit, which then outputs an appropriate current to the active anode. The protective current is transmitted through the electrolyte to the areas under protection. The area under protection is converted into a cathode, preventing metal corrosion. Zinc anodes have very low and non-variable driving voltages, with reduced effectiveness. The varying combinations of water temperature, chemical composition, and exposed surface require monitoring and different current levels. The anode is made of a relatively inert material such as silicon iron, silver/lead alloys, tantalum, or platinum. The driving voltage and current outputs are adjusted at the power source to enable precise control. Corrosion will be inhibited as long as the protective potential is applied. In normal operation, mechanically damaged or porous hull areas will have an insulation layer form over them, caused by salts within the seawater due to the current flow.

      a.   **Cathodic Protection Problems.** It is important not to overprotect using too high a potential. The results are softening and blistering of traditional paint systems due to the formation of hydrogen bubbles under the paint surface. Chlorinated rubber paint systems are commonly used to counter this problem.

      b.   **Electrical Antifouling Systems.** Units such as those from Cathelco and Jotun are used to protect sea chests and seawater inlets against fouling by marine growth. These use a copper anode, which releases copper ions into the system, which acts as an antifoulant.

8.25   Galvanic Isolators. A galvanic isolator (also known as a zinc saver) is able to block low-voltage DC current flow within the safety ground but still allow AC fault currents to flow in the conductor. It functions by inserting a low value DC voltage drop into the circuit conductor. Galvanic voltages and currents are small and are measured in millivolts and milliamps. Galvanic isolators consist of internal diodes that are connected in an inverse parallel configuration. This means that diodes can pass current in both directions, but only if the level is greater than the forward threshold voltage of about 1.0VDC to 1.4VDC, where the diodes will start to conduct. The Dairyland units have a threshold voltage of around 1VDC and NewMar up to 1.4VDC; most are very similar. They incorporate an AC bypass capacitor that allows AC to flow and bypass the diodes. Isolators have surge current ratings, many as high as 500A, and units may have DC and AC LED status monitoring. The minimum voltage value where a diode will start to conduct is known as the forward threshold or cut-in voltage; the knee voltage is where the p-n junction will start conduction and increase rapidly. Prior to this, the diode is in the positive bias state, and the forward bias voltage is lower than the forward threshold voltage. The AC flows from the boat ashore in a fault condition and must be able to pass very low AC leakage currents. Latest designs are termed "fail-safe." A galvanic isolator does not prevent galvanic corrosion; it only prevents such action coming from a shore power source. Galvanic isolators should have a UL rating on the device label; if not, don't buy. These are certified as fail-safe; and even with diode failure, the AC ground is not affected. Check ABYC Standard A-28, and also visit galvanic-isolator.co.uk, and dairyland.com.

8.26   Galvanic Isolator Installation. You must install the correctly rated isolator. The isolator is installed in series with the shore power AC protective or safety ground green wire conductor. A 50A supply requires a 50A-rated isolator; a 30A supply requires a 30A-rated isolator. ABYC- and UL-approved units must be able to carry 135% of the rated current. In a short-circuit condition, the unit will suffer a very high current several times the rated current for a short period until fuse or breaker protection systems operate. Where a boat has two shore power inlets, each must have a separate isolator installed. An underrated isolator will possibly burn out under full-fault conditions. Isolators must be installed close to the shore power inlet socket, although it is most common near the switchboard. They must be installed in a well-ventilated location, as they can become very hot when in a fault current conducting mode. The American Boat and Yacht Council (ABYC) recommends that no part of the AC grounding system should bypass the galvanic isolator.

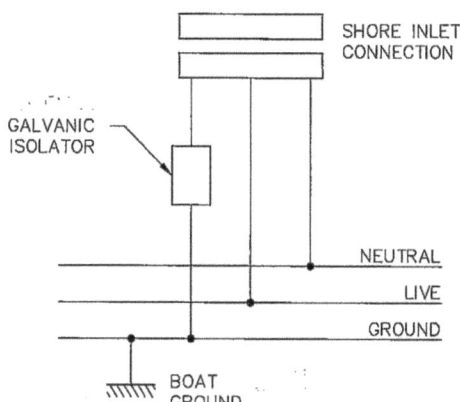

Figure 8-7. Galvanic Isolator Circuit

8.27   Galvanic Isolator Testing. Testing is to ensure that the galvanic isolator can safely conduct AC fault currents at all times and that the device has not failed. In accordance with ABYC recommendations, many isolators have a monitor with an LED indicator that displays the functioning of diodes and capacitors, along with AC safety ground continuity, reverse polarity, and other functions. It must be clearly stated that no shore electrical standards allow any device to be inserted within any grounding or earthing conductor; it is not allowed ashore anywhere in the world that I know of. Any person having an isolator installed may be in breach of local wiring rules and regulations and should check this out before installation. The common failure modes are short-circuited or open-circuited diodes.

a. **Shore Power Lead.** Disconnect the incoming shore power lead and select the "diode test" function on your multimeter (it will have a diode symbol next to it). Most modern multimeters have this function. Set the meter function to the diode test position, observe the open circuit reading, and touch the meter probes to the short-circuit position to verify. Short-circuit the isolator connection to remove any residual capacitor charge.

8.28 **Corrosion System Maintenance.** Perform yearly maintenance and inspection and at every haul-out as follows:

a. **In-Water Examinations.** Perform the following basic examinations:

(1) **Main Anodes.** Do an underwater check of anodes after 6 months. Check for increased corrosion rates if the vessel has moved into warmer or more saline conditions, as the corrosion rates increase. Rapid zinc loss and shiny zincs usually indicate a stray current problem.

(2) **Shaft Anodes.** Check that the shaft anode is still on the shaft. Check the anode corrosion rates.

b. **Slipping Examinations.** DC potential gradients may exist in the water around a vessel. This is caused by variations in salinity and temperature and will contribute to corrosion. If paint is chipped off under the bow, a circuit may be created with anodes or hull fittings in another part of the vessel with an area of water of differing potential. This can occur in small marinas with reduced tidal flows to flush out water heated during the day. Perform the following examinations:

(1) **Anode Replacement.** Replace anodes if more than 75% reduced, and check connections. Replace shaft anodes if necessary. Check that the mating surface of the shaft anode is clean. Check that the anode is correctly positioned.

(2) **Bonding Connections.** Inspect cathodic bonding system interconnections to make sure they are clean. Remove connections and clean so that the contact resistance is zero.

(3) **Check Bonding System Resistances.** Check bonding resistances between the anodes, the propeller, and the steel or alloy hull.

Table 8-5. Corrosion System Troubleshooting

Symptom	Probable Faults
Anode corroded 80%	Anode replacement required
Rapid anode corrosion	Hull electrical leakages Increased water salinity Increased water temperature Degraded bonding system Moored adjacent to vessel Marina electrical problems Galvanic isolator fault
Paint stripping off keel	Overprotected (too many anodes) Severe electrical leakage
Paint stripping off metal hull	Overprotected (too many anodes) Severe electrical hull leakage
Paint stripping around studs	Anode stud connections defective
No anode corrosion	Anode hull connections defective Bonding wires broken
Propeller and shaft pitted	Inadequate protection Degraded bonding system Shaft anode missing Shaft anode fitted over antifouled shaft Cavitation corrosion

8.29   Electrolytic (Stray Current) Corrosion. It must be understood that electrolytic, or stray current, corrosion has entirely different principles than galvanic corrosion; they should not be confused, although they often are. Protective measures for galvanic corrosion do not offer much protection against electrolytic corrosion. Stray current corrosion will, however, dramatically increase corrosion rates on underprotected hulls and anodes, degrading the galvanic protective system. If faults are undiagnosed, anodes will rapidly degrade; after depletion of anodes, paint stripping and antifouling stripping will occur and often require a complete repainting of the hull from primer upward. Automotive battery chargers are a common cause of corrosion in boats, particularly small boats without shore systems. Many automotive battery chargers often provide no isolation between the AC and DC windings and can energize the negative terminal, which energizes the boat's grounding systems. Stray current flows are variable with respect to magnitude and current paths. This results in differentiation between dynamic stray currents, which have an unsteady state, and static stray currents, which are steady state. There are direct stray currents and indirect stray currents.

8.30   Direct Stray Current Corrosion. This has its origins in direct current sources, such as the boat electrical system and cathodic protection systems. Principally such corrosion is caused by leakages across condensation or conductive salt deposits at DC connections or junction boxes, or tracking from diesel engine starter motor cable connections. Given the higher potential differences, 24V and 48V systems have higher risks than 12V systems. In some cases electrical leakages may be caused by damaged insulation. In a properly installed electrical system, there are relatively few opportunities for this situation to arise. The most common area is tracking across from the engine starter motor and solenoid connections, which are often never cleaned of grease, oil, and moisture. Ground faults on DC conductors occur when the cable insulation has been damaged and contact is made with the hull or connected metalwork. In many cases, the fault may not be sufficient to operate protective devices and as a result remains unnoticed for a considerable—and damaging—period. The most common areas causing faults are where cables enter grounded stainless steel stanchions, alloy masts, and engine charging and starter cables and terminations. In any area where a cable can contact grounded metal, leakage or fault currents can flow.

8.31   Indirect Stray Current Corrosion. This is caused by a direct current source that is outside the boat, such as another boat electrical system, usually from a shore power ground connection. This can transfer through the AC safety or protective ground wire. If it is not protected with a galvanic isolator for galvanic currents, it can impress itself onto the bonding system. If a galvanic isolator is installed and the stray current voltage exceeds the forward threshold voltage of the isolator, current can flow.

8.32   Corrective Measures. Install a DC leakage test unit so the hull can be monitored continuously, with any problems promptly identified and rectified. This is standard commercial ship practice.

  a.   **DC Leakage Currents.** All connections, and ideally there will be none in wet areas, should be in proper water-resistant junction boxes. They should be placed in dry locations away from any metalwork liable to conduct any leakages to the hull. Install a monitor on steel and alloy vessels.

  b.   **DC Ground Faults.** Ensure that all cables are double insulated. Check that all transits through metal bulkheads or stanchions have additional mechanical protection or grommets to prevent grounding.

8.33   Steel/Alloy Hull Leakage Inspections. It is always difficult to maintain a hull above electrical ground. Moisture and oil residues mixed with salt lower the isolation level. It is important to regularly examine isolation values to ensure that isolation is always maintained.

  a.   **Passive Insulation Test.** This test simply measures the level of resistance between the hull and both positive and negative. A multimeter set on the ohms scale is required. Perform the test as follows:

    (1)   Turn the main power switch off.

    (2)   Turn on all switches and circuit breakers to ensure that all electrical circuits are at equal potential or are connected in one grid.

(3)   Connect the positive meter probe to the positive conductor and the negative to the hull. Observe and record the reading.

(4)   Connect the positive meter probe to the negative conductor and the negative to the hull. Observe and record the reading.

b.   **Passive Test Results.** The test results can be interpreted as follows:

(1)   A reading of 10kΩ or above indicates that isolation above the hull is acceptable.

(2)   A reading in the range of 1kΩ to 10Ω indicates that there is leakage and that isolation is degraded. While not directly shorted to the hull, leakage can be through moisture or a similar cause. With the meter connected, systematically switch off each circuit to localize the fault area and rectify the problem. A common area is the starter motor connections.

(3)   A reading of less than 1kΩ indicates a serious leakage problem that must be promptly rectified to avoid serious hull damage.

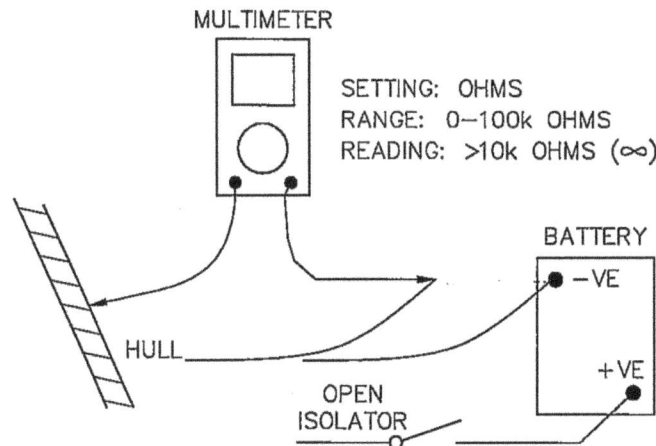

Figure 8-8. Passive Hull Testing

8.34   Voltage Insulation Test. While a passive meter test can show that all is satisfactory, the voltage of a system in use can break down resistances and cause leakage. To properly test the electrical isolation, a voltage test should be performed. With 220V to 115VAC power systems, this test must be performed using a 500VDC insulation tester. All results must exceed 1MΩ. This is not recommended for low-voltage installations, as the insulation values of cables are not rated this high. A low-voltage DC tester set at 100VDC should be used. Another easier test is as follows.

a.   Turn on all electrical circuits so that all are "alive."

b.   With a digital multimeter set on DC volts, place the positive probe on the supply negative. Place the negative probe on the hull.

c.   There should be no voltage at all. If there is even a small voltage, a leakage may exist on the negative.

d.   With a digital multimeter set on DC volts, place the negative probe on the supply negative. Place the positive probe on the hull.

e.   There should be no voltage at all. If there is even a small voltage, a leakage may exist on the positive.

f.   Systematically turn off electrical circuits to verify that there is a leakage, and that with all power off, the difference in potential is zero.

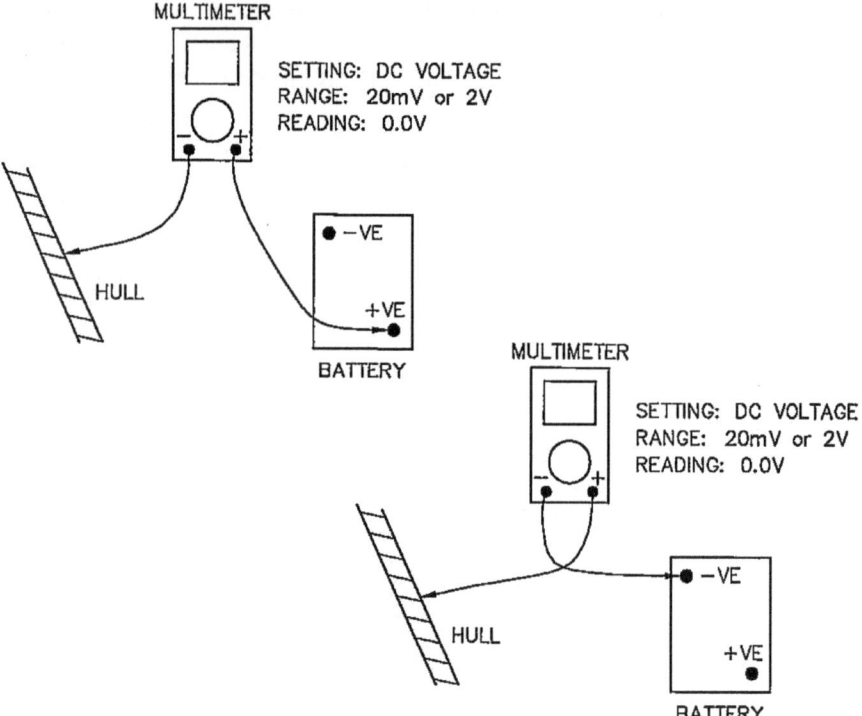

Figure 8-9. Insulation Testing

Lighting Systems

Boat lighting systems are often frustrating, and the selection of suitable lights is always challenging. Aesthetic considerations are of obvious importance, but there are other more important technical factors to consider, including function, technology, and budget. However, LEDs have made that decision easy. The planning of lighting is the same, regardless of the chosen technology. Lighting control is now possible from your MFD, and NMEA 2000 is incorporating lighting control into the standard.

a. **Illumination Area.** The size of the area to be illuminated is one important consideration. Where specific areas are to be illuminated, the factors to consider are as follows:

   (1) **Spot Lighting.** Spot lighting, such as berth reading lights and spreader lights or foredeck spotlights, requires special considerations. Projected light applications usually require a reflector or a special lamp. Factors such as beam power and beam angle are important.

   (2) **Large Area Illumination.** When lighting deck areas or the saloon, consider beam angles and light output power.

b. **Illumination Level.** The level of light reaching the working areas on the deck or the overall light level below must be sufficient to perform tasks safely. A number of factors must be considered:

   (1) **Background Lighting.** This is generally low output power lighting and uses interior surfaces and upholstery to reflect light for unobtrusive and shadowless illumination.

   (2) **Low-Level Lighting.** This is localized illumination that does not require levels sufficient to perform work. Typical are night lighting, courtesy lights, and general saloon lighting.

   (3) **High-Level Lighting.** This lighting is used in any application where safety or ease of work is important. It includes deck spotlights, handheld spots, saloon lights, bunk lights, engine space lights, and stern arch or transom lights, to name a few applications. Ideally, such lights should provide shadowless illumination without excessive glare.

c. **Light Colors.** The color-rendering properties of a light source play a major role in implementing effective lighting. Using lights with the right color-rendering properties can significantly alter the apparent richness of woods:

   (1) **Warm, or Soft, Colors.** Fluorescent tubes are generally warm-soft. Low-energy lights tend to have a softer light that strikes a balance

between good illumination levels and good color rendering. LED lights now offer warm white colors.

(2) **Cold, or Hard, Colors.** Halogen lamps and some fluorescent tubes generally have a cold and intense white light. LED lights are available with various color properties.

d. **Power Consumption.** Electrical power consumption is the very first factor to consider. Compare the main light types and make your decision based on the most satisfactory light for a given power consumption. Given the progress in LED light technology, they are the favored choice these days. A basic 2 × 10W LED, 120-degree beamwidth deck light has a 2,600 lumen output and consumes 1.6A.

9.1 Internal Lights. Lighting systems for cabins usually consist of a number of light types. Different lights may be used for different functions, and one of the main criteria is gaining maximum light output for a given power consumption. There are four main types available. Before deciding on lights for belowdecks, consider light reflectivity. Voice-activated lights have also arrived; the EOS-LUX Command Center with Alexa Integration (including voice control) from Lumishore allows voice command operation.

a. **Reflectivity.** The level of brightness and the contrast with background must be considered. In a teak-lined cabin, reflected light will be minimal, while a cabin with painted surfaces or light timbers will increase overall illumination levels. I have seen some beautiful wood-lined cabins with large numbers of lights fitted in the headliner and additional corner spots. Yet even with many lights on, they are still gloomy, with low light levels. Efficient interior schemes mean fewer lights with less cable, and much lower power consumption for a given light level. Reflectivity is usually expressed as a percentage:

(1) **Timber Surfaces**

Maple and birch: 60%

Light oak: 40%

Walnut and teak: 15%–20%

(2) **Painted Surfaces**

White and light cream: 70%–80%

Pale yellow: 55%–65%

Sky blue/pale gray: 40%–45%

Beige: 25%–35%

9.2 LED Lights. Light emitting diode (LED) lights have rapidly moved onto boats. The LED is a solid-state semiconductor device that converts electrical energy directly into light energy. These lights are low power consumption, low heat emission, and reliable, and they can be used virtually anywhere on board. Typical current draw is 0.54A at 12V for a tricolor light. LED lights are around ten times more power-efficient than an incandescent filament bulb. They have greater resilience to vibration and impacts.

a.   **LED Colors.** The basic diode is a very simple semiconductor device, also called a p-n junction. LED colors are a result of the semiconductor materials used. LED diodes come in three primary colors: green, red, and blue. There is no white LED; however, a combination of different LED colors or a yellow or orange phosphor coating converts into white light. White is derived by combining the different wave lengths of two or more diodes to mix colors. LED lights produce white light, which is classified depending on its warmth or coolness. Warm White = 2,700K–3,000K; Neutral White = 3,000K–4,000K; Pure White = 4,000K–5,000K; Day White = 5,000K–6,000K; Cool White = 7,000K–7,500K. The kelvin (K) is the Système International (SI) unit of thermodynamic temperature and is associated with the Kelvin scale. The term is used to describe the color temperature of a lamp. An incandescent lamp has a yellow hue and a color temperature of around 3,000K. This means the yellow-spectrum light resembles a hot object at 3,000K. A light with a color temperature of 5,000K to 5,500K, which contains more blue light, is denoted as daylight. Most LED lights fall within the 5,000K–5,800K range.

b.   **LED Construction.** LED clusters are usually encapsulated in resin to prevent any water ingress. In an LED, the conductor material is made from aluminum gallium arsenide (AlGaAs). The first component is gallium. Arsenic, aluminum, and phosphor are then added. Because they are able to produce photons, these contribute to the luminescence of the diode. Incandescent and fluorescent light bulbs are rated in watts; LED light output is measured in luminous flux, and the unit of measure is the lumen. In pure aluminum-gallium-arsenide materials, all atoms bond perfectly with adjacent atoms, with no free electrons (negatively charged particles) to conduct electric current. In doped semiconductor material, the additional atoms add free electrons or create holes where electrons can go. These changes make the material more conductive.

c.   **LED Drivers.** An LED or LED array requires a driver to prevent damage and enable stable light output. The forward voltage on an LED will change with an increase in temperature. When the temperature increases, the forward voltage decreases, resulting in an increase in current. If left uncontrolled, the LED will suffer from thermal runaway and ultimate destruction. The constant current driver prevents this by compensation for forward voltage changes. Each manufacturer has its own driver designs; for example, Hella has Advanced Multivolt electronics. Some manufacturers use PWM constant current drivers, which incorporate reverse polarity protection, thermal protection, and voltage stabilization under severe voltage fluctuations and low battery voltages. Earlier LED lights and indicators on switch panels used an LED with a series resistor. If you have RFI noise, these are often the cause.

d. **LED Installation.** LED bulbs are manufactured in LED clusters, and they are different from incandescent types because they are polarity sensitive. Correct polarity connection is essential. No, you don't need to buy new lights if you have incandescent cabin lights or navigation lights. You can convert to LED by obtaining LED clusters with matching lamp bases as the existing light fittings. **Caution:** Many of these LED clusters are not certified and are a common source of EMI on AIS and VHF.

e. **LED Noise.** Electromagnetic interference (EMI) can be an issue in some units with inferior design and lack of integrated suppression circuits. Transient suppression usually copes with voltages up to 60V. Always purchase LED lights that are certified for electromagnetic interference. LED lights are a functioning electronic circuit and radiate some EMI. In 2018 The USCG issued an alert about EMI between some LED lighting fixtures and equipment such as VHF radios and automatic identification systems (AISs). The issue seems to be most prevalent where an LED tricolor light and a VHF aerial are installed in close proximity at the masthead. If you are purchasing an LED light, check the label and accompanying certificate to make sure it is approved by the Federal Communications Commission (FCC). The FCC regulates the permissible frequency ranges and levels of electromagnetic emissions. To reduce the possibility of EMI, ensure that the VHF antenna and high-output LED lights have adequate separation. This can be mitigated by proper cable separation, so make sure AIS and VHF aerial cables are not close. Before you install a new LED light at the masthead, connect the light fitting down below, place a portable VHF near it, and go to a quiet channel. Adjust volume and squelch to minimal noise levels. Switch the LED light on and observe any increase in noise. If there is noise, you will have to consider purchasing a better light. Perform a check with the AIS and the lights, and observe if any changes. Check the light for the European CE label, which has more stringent requirements. Avoid cheap Chinese imports, which are usually noise producing.

f. **LED Light Troubleshooting.** Problems with LED lights are usually caused by a faulty driver, which may cause the light to flicker. Other issues can arise with a dimmer unit. Before you panic about the cost of replacement or dissatisfaction with longevity of the LED light, check whether the light starts flickering when you turn something else on. If it does, look for wires that are in relatively close proximity to lights or wiring. Check wiring connections in the light circuit; a loose wire termination can cause flickering. If all lights in the circuit are equally bright, wiring is probably not the issue. If the problem appears to be one specific light, remove that light and test it on the workbench. If a light driver is faulty, you won't be able to repair it.

9.3 Fluorescent Lights. Fluorescent lights have been the most common for a long time, but their prominence is coming to an end. They have drawbacks that must be considered if you choose to install or maintain them. DC tubes have a built-in inverter that raises the voltage to a higher AC value. Their elongated shape provides a good lumen/watt ratio with relatively low power consumption, 80% less than incandescent lights for the same light output. Typical output is 65 to 90 lumens. They withstand vibration and shock well, and their working life is five to eight times that of incandescent lights. The inverter in low-voltage DC fittings is generally the main cause of failure. In most of the cheaper light fittings, the quality of the electronics is poor. They fail in relatively small overvoltage conditions, such as when charging voltages rise to 14V. Always install fluorescent lights with a voltage input up to 15V. The fluorescent tubes for household use function quite satisfactorily with good-quality inverters. If the electronics are of poor quality, the tubes will show blackening in a short period. Tube output varies with temperature. Peak output is normally at 77°F (25°C). If hotter or colder, output is reduced. Fluorescent lights have a notorious reputation for radio frequency interference. This is due to the quality of the inverter electronics. High-quality inverters, such as those from Aquasignal, are suitably suppressed to international standards. Additionally, fluorescent units with high-quality inverters are far more reliable.

9.4   **Incandescent Lights.** Incandescent lights are the oldest and most common light types. The following factors should be considered, as many boat owners have yet to make the switch to LED. When switched on, power consumption can be fifteen times normal (hot) power consumption. The basis of the incandescent lamp is the heating of a filament; therefore, much of the energy is dissipated as heat. Incandescent lights are power hungry for the available light output, are subject to damage by vibration and overvoltage, and suffer rapid filament degradation. Overvoltage conditions significantly reduce incandescent lamp life expectancies. Operating at lower voltages extends service life but seriously reduces light output. For every 5% voltage drop, light output reduces by 20%. Many of you are familiar with that yellow glow as the battery voltage decreases. The secret to operating incandescent lights, especially navigation lamps, is to minimize voltage drop.

9.5   **Halogen/Xelogen/Xenon Lighting.** Halogen lights were the first high-output lights in common use; they have higher light outputs, typically around 20 lumens. Halogen lights have been supplemented and in some cases replaced by xenon and Xelogen lamps and, more recently, by LED lights. The halogen lamp base is designated as G4 and Xelogen G5. Halogen lights belong to the incandescent light category and are designed for use in commercial installations on a stable 12V and 24VAC power source. When used in DC installations, the life expectancy is significantly reduced; the higher voltages generated during battery charging also reduce life. Xelogen lamps have a service ten times that of a halogen lamp, and should last up to 20,000 hours. They have a lower operating temperature, and while their glass can be handled, it is better not to. Xelogen lamps are dimmable. Vibration resistance is relatively poor, and resistance to overvoltage situations is poor. Normally, a halogen lamp is operated in commercial applications with a very stable 12VAC supply, with maximum life being at around 11.8V. Operating on DC, and at charging voltages up to 14.5V, life can be seriously reduced. Halogen lamps degrade with salt air interaction with the pure silicon glass. Under no circumstances should the glass be handled; salts and impurities off the fingers will degrade the silicon glass and shorten life. Allowances must be made with halogen lights due to the high temperatures generated, which can reach 1,300°F (700°C) in normal operation. Good ventilation is required to prevent the lamp holder or wire reaching a maximum of 480°F (250°C). Most halogen fittings have high-temperature wiring.

9.6   Low-Energy Lighting. Low-energy lights, both AC and DC, are commonly installed on many vessels. They operate on a similar principle as fluorescent lights, in which an electrical arc occurs between two electrodes, located at each end of the tube. The arc is conducted by vaporized mercury and inert gases such as argon, neon, or krypton through a phosphor-coated tube. The emitted ultraviolet light makes the phosphor glow and emits visible light. These lights give a very high output for a relatively small power draw. They produce a light only marginally less than a 60W household bulb and have a power consumption of around 16W. Life expectancy is greater than standard fluorescent lights. Similar to halogen or fluorescent lights, they are intolerant of overvoltage conditions; most tolerate voltages up to 17V. In AC lighting, the compact plug-in fluorescent tubes have an average rated life of around 10,000 hours. These include the 2D square fluorescent, short twin-tube, quad-tube, and triple-tube. Some Edison screw (ES) types, including twin- and quad-tube types, do not require adapters. Quality light fittings have a printed circuit board (PCB) with an integrated circuit operating at 35kHz, suiting lamps in the 5W to 11W range.

9.7   Red Nightlights. Red lights are very useful in strategic locations, and it must be understood that it can take up to 45 minutes for normal night vision to return if the eye has been subject to a white light. Typical locations for night lighting are at the chart table and the steering station. Some fluorescent lights have dual tube fittings, and the use of LED cluster lights is effective. Some manufacturers make fittings with red diffusers. Make sure the steering position light faces down toward the deck and cannot be seen, or it might be construed as a port navigation light by anyone outside the vessel. If you have a coach house, a red-and-white unit can be mounted and switched locally.

9.8   Dimmers and Voltage Stabilizers. Many boats incorporate dimmers on lighting circuits; the following explains some of the factors.

      a.   **DC Dimmers.** Earlier types were variable resistance "rheostat" types; new-technology systems use electronic pulse width modulated (PWM) control. Features of PWM units are no heat, high outputs of around 100W at 12V, and high efficiency, with minimal losses of only about 2%. Control modules are overload and reverse-polarity protected. One feature on some models is a soft-start function that limits the initial inrush current, which can improve bulb life. Some modules have a no-load consumption, typically around 10mA, so light circuits must be isolated or switched off to prevent this when boats are unattended. When installing, the correct output rating must be used; overloading will lead to early failure. Ratings are 2 Amps = 12V/24W or 24V/48W; 5 Amps = 12V/60W or 24V/120W; 10 Amps = 12V/120W or 24V/240W; 20 Amps = 12V/240W or 24V/480W; 30 Amps = 12V/360W or 24V/720W. Dimmers often create electrical noise, and a filter capacitor should be installed as close as possible to the dimmer. Dimmers are reliable; when failures occur, always check the module power input and connections first. Where more than one push button is used, check all first; if all are out, the module is the cause. Push buttons and connection to them are the most common failure point. If you have an existing dimmer for halogen lights, the bad news is that you can't use them on LED lights. Halogen lights are dimmed by reducing the voltage; LED lights are dimmed by using PWM, which turns the diodes on and off very fast. You will not see this flickering, but control of the frequency controls the LED output brightness.

b.   **AC Dimmers.** Larger yachts having AC lighting systems use different dimmer types. One technique is using toroidal transformer units. Unlike normal transformers, there is virtually no magnetic hum or any noise due to mechanical vibration. The no-load power consumption is around 80% less than standard transformers, and overall efficiency is around 95%. The transformers have a primary voltage of 120V, with secondary voltage of 12V or 24V. Capacity ratings are quoted in volt-amps (VAs) and can range from 60VA up to 600VA. Units typically include overload, thermal, and short-circuit protection. If a unit suddenly stops, the thermal protection is the first possible cause. The transformers have AC input voltages of 120V 50/60Hz, with stable output voltages of 11.5V to supply low-voltage AC lighting such as halogens, an output frequency of typically 30kHz, and power outputs of 20VA to 75VA. Some units have a soft-start feature, which ramps up the voltage from zero to full output current, as well as the usual thermal and short-circuit protection. The units use toroidal transformers instead of ferro-magnetic ones to reduce heat and increase efficiency. Like all AC-powered equipment, caution must be used when installing and troubleshooting, and equipment must be switched off when working on it.

c.   **Lamp Voltage Stabilizers.** As stated, lamps are affected by higher voltages, which reduce life expectancy. Light manufacturer Cantalupi has developed a voltage stabilizer that can accept a variable input voltage in the range 12.5V–16V at 12V or 24.5V–29V at 24V. The output is stable at 12V or 24V over all voltage ranges up to a maximum power rating of 25W to maximize lamp life by least 200% to 400%, making it a good investment in vessels with large halogen light installations. Devices have short-circuit, thermal, overload, and reverse-polarity protection.

9.9   **Lamp Bases.** Lamp bases are extremely varied, and designations are often confusing. Following are many of the more common lamp bases and their designations.

a.   **LED Lamps.** All the standard lamp bases listed below are available in LED to allow easy installation. Caution is required here; check the certification of the LED lights, as they may cause radio and AIS noise. Many do not have anything to prevent noise. Caveat emptor (buyer beware)!

b.   **Halogen Lamps.** Lamp socket types include G1, G2, G4, GY4, GZ4, GX5.3, G6, G6.35, GX6.35, G8, GX9.5, GY9.5, and GU-10

c.   **Incandescent Lamps.** Lamp socket types include E14, E27, E40, B15D, B22D, P28, candle base E12, and medium base E26.

d.   **Fluorescent Lamps.** Lamp socket types include G5, G13, G23, G24, and G32.

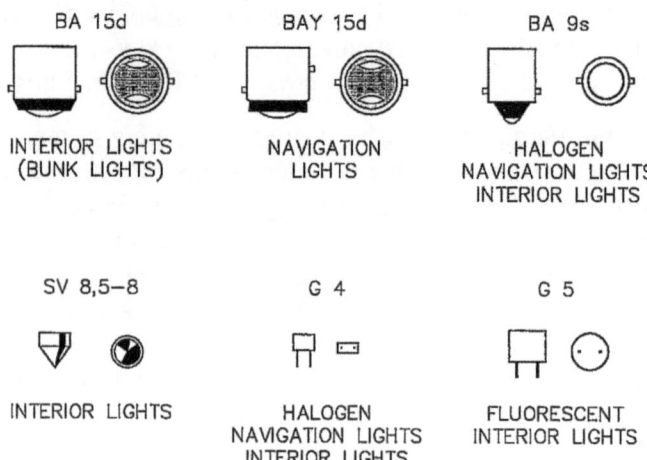

Figure 9-1. Lamp Base Types

9.10 **Deck Lights.** There are ranges of lights with different functions that must be installed properly and maintained. Some have safety implications.

a. **Courtesy Lights.** Courtesy lighting in the cockpit and transom areas is very useful, but there are points to consider. Many of the lights available are of very poor quality and quickly degrade, so always select quality fittings. A newer development is LED light fittings, which consist of high-output LED clusters. Units such as the Hella units have ten LEDs, which is equivalent to a 20W incandescent bulb using just 0.16A. Cheaper fittings that use festoon bulbs are a constant cause of problems, with poor bulb contacts being the primary issue. I recommend fitting a couple of LED stern navigation lights

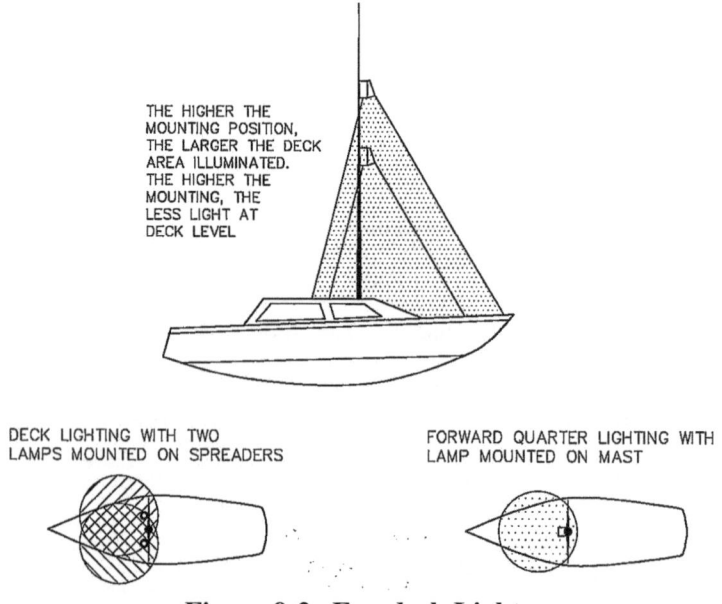

Figure 9-2. Foredeck Lights

168

facing downward off the stern pulpit, pushpit, or stern arch. They provide satisfactory low-level illumination, are weatherproof, and are a valuable safety feature when retrieving the dinghy or a crewmember.

b.   **Safety and Working Lights.** There are various options:

(1)   **Deck Lights.** Deck lights are essential for fishing, entertainment, security, and boarding at night, to name the important functions. Halogen and xenon, and now LED lights, are alternatives to the incandescent sealed-beam types, with big increases in efficiency and light output. Xenon lamps have working lives up to 2,500 hours and have internal xenon ballast modules. Foredeck spotlights are typically fixed on the mast at the spreader level. They also incorporate the steaming light. Install a rope guard when installing a foredeck spot light. For many years high quality, sealed beam halogen lights rated at 35W to 50W were the norm. Where stainless pulpits are installed, I have installed a small white navigation light at the pulpit, facing down to illuminate the anchor well; LED lights are a good option for this duty. The lights are switched from the steering station. The light is not too bright, but it is practical and required in the right area.

9.11   **Spotlights.** Spotlights are generally confined to spreader and foredeck spot lighting. The very severe environment they are subject to requires careful selection. A candela (cd) is the SI unit for luminous intensity. Candlepower (cp) is illuminating power expressed in candelas. Lux is the SI unit of illuminance and is equivalent to 1 lumen per square meter.

a.   **Spotlights.** Ratings are usually given as candlepower or lux, with lux being the amount of light at a nominated target distance, typically 300ft to 1,500ft (100m to 300m). Thus, a rating may be given as 330,000cp or 52 lux at 330ft (100m) and 2.1 lux at 1,500ft (500m). Some units use xenon lamp systems and have outputs of 1.5 million candelas; with a 24V supply, they consume 8A. The biggest failure area on spotlights is poor wiring and connections. Ensure that cables are rated for the power of the unit. As spotlights are exposed to weather, keep them covered when not in use. Operate them regularly through complete rotations, and pan and tilt to prevent seizing. If fuses are blowing, it is generally due to seizing and overloading. Deck spotlights have more concentrated beams—around 6 to 8 degrees compared to general floodlights of around 30 to 40 degrees. Spotlights should have a clearly defined beam pattern without scattering at the sides. I have changed over to the effective LED models available. Some LED searchlights, such as those from ACR, have a six-LED array spotlight that outputs a massive 220,000 candelas at peak luminosity with a 10-degree beam angle.

b.   **Hand Spotlights.** Spotlights should have a clearly defined beam pattern without scattering at the sides. If you want increased power for less-than-ideal conditions, choose a higher candlepower rating. Always select a light with a switch for emergency signaling

c.   **Spreader Lights.** Spreader lights are invaluable safety equipment. When severe problems are encountered on deck, good lighting allows for safer, faster work, which reduces deck exposure time. The following must be considered when installing spreader lights:

   (1)   **Illumination Levels.** Lights are designed to facilitate on-deck safety without excessive shadows. Do not use navigation lights under the spreaders; their deck-level illumination is generally very poor. For many years high quality, sealed beam halogen lights rated at 35W to 50W were the norm. These have been made redundant with new LED spreader lights. Typical of these are the Hella Sea Hawk-R LED floodlights, which I have just installed on my own boat.

   (2)   **Construction.** Use fittings made of plastic to avoid corrosion. Stainless units are prone to shake apart at the spot welds.

   (3)   **Installation.** Spreader lights are exposed to weather and subject to severe vibration. It is essential that lights be mounted securely in a location where these factors are minimized. Cables should be of sufficient length, and connections should be wrapped with self-amalgamating tape. Make sure you can change burned-out lamps.

9.12   **Navigation Lights.** Navigation lights are of the utmost importance, both for safety reasons and to comply with COLREGs 72 rules of the road. You should read and understand all of the content within the Convention on the International Regulations for Preventing Collisions at Sea, 1972. I am still surprised at the number of noncompliant sailing vessels showing incorrect navigation lights. My personal survey shows that only about 40% of vessels have the correct lights displayed. It is not sufficient to simply say you have lights installed and turned on; they must be mounted at the correct locations and must be the correct lights for your particular application, such as under power, at anchor, or under sail, etc. It is all very well to blame other vessels for running over pleasure craft, but if the correct lights are missing, nobody will be able to identify your vessel and status; even with AIS, you need to have your navigation lights on.

a.   **Navigation Light Legal Requirements.** All vessels are required by COLREGS 72 to display the correct lights. Failure to comply may void insurance policies in the event of a collision.

   (1)   **Navigation Lights.** Lights should be displayed in accordance with the provisions in Part C, Lights and Shapes.

   (2)   **Light Requirements.** Lights should be of an approved type and conform with the provisions of Annex I with respect to positioning and technical details of lights and shapes.

b.   **Tricolor Light (under Sail Only).** For yachts under 65ft (20m), the combination port, starboard, and stern light mounted at the masthead is the most energy-efficient solution. Because only one lamp is burning, it reduces battery power consumption; LED lights further drastically reduce current draw. This light must not be used under power, as is commonly done, or in conjunction with any other light.

c.   **Anchor Lights.** The anchor light is an all-round white light. It should not be masked at any point. See COLREGs Annex I, 9(b) regarding horizontal sectors. Vessels should install a combination tricolor anchor light for simplicity. Always use it if you are anchored where traffic is possible. If you do not and a vessel collides with or sinks your boat, it is probably your fault.

d.   **Port and Starboard Lights (Sidelights).** The port light (red) and the starboard light (green) must display an unbroken light over an arc of 112.5 degrees, from dead ahead to 22.5 degrees abaft the beam. On a vessel under 65ft (20m), the light can be combined into a bicolor fitting. On many yachts, these lights are installed on the bow pulpit, but a section of the pulpit often partially obscures the light. Ensure that the light is visible over the prescribed arc; otherwise, you are displaying a light that is technically illegal.

```
STARBOARD LAMP      112.5 DEGREES
PORT LAMP           112.5 DEGREES
MASTHEAD LAMP       225 DEGREES
STERN LAMP          135 DEGREES
ANCHOR LAMP         360 DEGREES
```

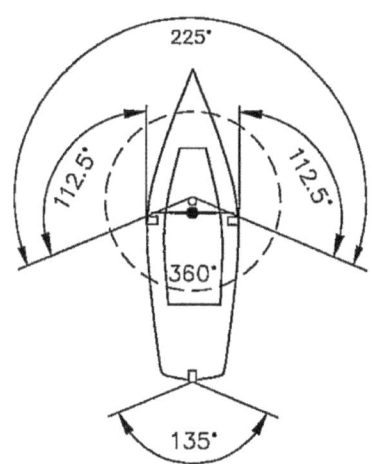

Figure 9-3. Navigation Lights

e.   **Stern Lights.** This is a white light placed at the stern on the centerline. The arc of visibility must total 135 degrees, from dead astern to 67.5 degrees each side:

(1)   **Display.** Stern lights must always be displayed when the vessel is under power, along with the port and starboard sidelights and the steaming or masthead light.

(2) **Mounting.** Do not mount stern lights on angled transoms without mounting plates that ensure they are vertical. In addition to not being legal, lights angled skyward are very difficult to see.

f. **Masthead Lights (Steaming Light).** This is a white light that must be visible over an unbroken arc of 225 degrees, from dead ahead to 22.5 degrees abaft the beam on each side. The light must be fixed on the centerline of the vessel, typically on the top of mast. There are vertical mounting requirements for the masthead light. See Annex I.2: "Vertical Positioning and Spacing of Lights":

(1) Vessels 40ft (12m) LOA or less: a minimum of 3ft (1m) above the sidelights.

(2) Vessels 40ft to 65ft (12m to 20m) LOA: a minimum of 8.2ft (2.5m) above the gunwale.

9.13 **Navigation Light Technical Requirements.** Chromaticity or color is covered in Annex I.7: "Color Specification of Lights" and Annex I.8: "Light Intensities."

a. **Navigation Light Color.** Color or chromaticity is defined by international collision regulations. By purchasing approved light fittings, you will know they meet the requirements.

b. **Lamp Ratings.** Lamp ratings are generally given by the manufacturer and are designed to give the required range and luminosity for which the light is granted approval. Do not increase the lamp rating to increase brightness or decrease it to save power. If you do alter the lamps and have an accident, your insurance may be invalidated. You also could be sued for damages because you were not displaying approved navigation lights:

(1) **Lamp Sockets.** Special sockets (typically BA15D, 1156, 1157) are used to ensure that filaments are correctly aligned to the lens and horizontal shade systems. New LED lamps have the same lamp bases.

(2) **Light Outputs.** Light output and wattage are designed for a high lumen-per-watt ratio.

(3) **Light Consistency.** The lights are designed to emit an even output through a 360-degree azimuth.

c. **Navigation Light Visibility.** The required minimum range of visibilities are as follows (note: 12m to 20m = 40ft to 65ft):

(1) Stern light: <12m = 2nm; 12m–20m = 2nm.

(2) Sidelights: <12m = 1nm; 12m–20m = 2nm.

(3) Masthead light: <12m = 2nm; 12m–20m = 3nm.

(4) Tricolor light: <12m = 2nm; 12m–20m = 2nm.

(5) Anchor light: <12m = 2nm; 12m–20m = 2nm.

d.   **Navigation Light Approvals.** Navigation lights should all be approved. Most manufacturers issue a certificate with each fitting; keep the certificate in your boat file. LED light fixtures are also type-approved. It is important to note that some fittings are only approved by a national port or marine authority and may technically be illegal in another country. Always keep the numbered approval certificate with your vessel files in case of litigation.

e.   **Navigation Light Maintenance.** Check lights regularly. Any defects may cause failure or even be illegal. Check and inspect the following:

   (1)   **Internal Moisture.** Check the lights' interior for moisture that can degrade lamp contacts or cause a short circuit.

   (2)   **Light Diffuser.** Check the light diffusers for cracks or crazing that will alter the light's characteristics, including chromaticity and color.

   (3)   **Electrical Connections.** Check and tighten all wire terminations.

   (4)   **Lamps.** Special incandescent lamp types are almost standard; however, LED-based lights are now common. The LED lamp has an average 50,000-hour life expectancy, with a light output increase of 20%; they consume 90% less power than traditional incandescent lights.

f.   **UK Light Approvals.** A navigation light in Europe carries the Conformité Européene (CE) mark. Additional to that is the IMO COL REG 72 mark. There also may be the German BSH mark along with US approvals American Boat and Yacht Council (ABYC A-16) and the United States Coast Guard (USCG). In the UK, prior to withdrawal (Brexit) from the European Union (EU), the regulatory regime was covered by the Marine Equipment Directive (MED). Following the withdrawal in January 2020 these have now been replaced by the UK Maritime and Coastguard Agency (MCA) regulations which cover conformity assessment and certification for marine equipment. This is called the UK Conformity Assessment (UKCA) mark, full implementation has now been delayed indefinitely. British boat owners should be aware of these changes when purchasing navigation lights and most other marine equipment. Type approvals also cover Lifesaving Appliances, Navigation Equipment, Radiocommunications Equipment and more. The UK Cruising Association (CA) is a good source of information on this. Go to theca.org.uk

Lightning Protection

10.1 **Lightning History.** Lightning has been a problem for mariners for centuries. As far back as the early 1800s on square rigged sailing ships, shipbuilders were devising and installing lightning protection systems to minimize the catastrophic effects of lightning strikes. These methods were essentially based on the grounding of spars and rigging. More than one vessel lost their mizzens and masts as a result, along with subsequent electromagnetic pulse–related compass problems. Lightning protection systems evolved as a response to the dissipation of lightning strike energy. There is some confusion as to what protection should be installed, but protection can never be achieved using a single method. A number of measures should be used to minimize and mitigate the risks, and the best protection involves a holistic, overall systems approach. The main objectives of any protection system are described, and while there never will be a 100% solution given the raw natural power of lightning, adoption of some measures will at least minimize the effects. It is a common narrative: radio aerials and masthead antenna blown off, wind instruments disintegrated, backstay insulators exploding, batteries exploding, electronics cooked, thru-hulls blown out, and much more.

10.2 **Lightning Hot Spots.** Lightning is one of nature's wonderful spectacles. I am forever in awe and can recall some spectacular shows when sailing. Locations such as the Caribbean's Mona Passage between the islands of Hispaniola and Puerto Rico spring to mind as one of my most memorable, entertaining, and somewhat intimidating experiences. Florida is considered the lightning capital of the world in a boat context. If you are sailing down to Key West or sailing along the Gulf Coast, your risks will be higher. The Mediterranean, Caribbean (Cuba and Colombia), Gulf of Mexico, Malaysia, Singapore, northern Australia, western Pacific Islands, West Africa, and Central America are all classed as high-activity areas. The Pacific coast of Mexico and Central and South America has the infamous chubasco, very violent and intense squalls accompanied by much lightning. The UK experiences "Spanish plumes," which result from layers of warm air migrating north from Spain and transiting France. The resulting temperature inversion entraps the warm and moist air, causing serious thunderstorms. These events are similar to those experienced in the US Gulf of Mexico, which create supercell storms and tornados. It is said that Java in Indonesia is the worst lightning spot on the planet. According to some estimates, there are more than 45,000 thunderstorms every day and 100 lightning discharge events per second. A leading marine insurance company reports processing around 200 lightning-related damage cases each year. Of course, claims have soared as many boats now have significantly greater levels of sophisticated and expensive electronics. If you want to investigate further, check out Convective Available Potential Energy (CAPE), a very helpful tool for monitoring and observing atmospheric instability, at windy.com. The odds of a lightning strike on your boat in any single year are around 1 in 1,000, which increases to 3.3 in 1,000 in lightning hot spots such as Florida. This website gives you real-time lightning strike activity around the world; check out map.blitzortung.org.

10.3 **Lightning Protection.** More than a 1,000 people are killed worldwide annually by lightning strikes, and in the United States transient lightning damage exceeds $1 billion.

The majority of lightning strikes take place between noon and 1800 hours. In Florida some 40% of deaths and injuries are related to water-based recreational activities. Of the many nonfatal shock injuries experienced yearly in the United States, between 50 and 300 are lightning related. The National Weather Service (NWS) provides a continuously updated weather forecast for Florida and the adjacent coastline on VHF channels WX1 (162.550MHz), WX2 (162.400MHz), and WX3 (162.475MHz). Given that lightning has geographic regions of increased activity, it is perhaps worthwhile to do a boat location–specific risk assessment.

10.4   Lightning Safety. There are some simple basic precautions when you are about to experience an electrical storm.

     a.   Stay belowdecks at all times.

     b.   Turn off all electronic gear and isolate circuit breakers if at all practicable.

     c.   Disconnect radio and other aerials and antennae if practicable.

     d.   Keep well clear of the mast or mast compression post until the storm activity passes.

     e.   Check your vessel position and plot it or write it down prior to shutting down, in case all electronic navigation equipment is destroyed.

     f.   Do not operate radios until after the storm passes, unless in an extreme emergency.

10.5   Post Lightning Strike. If you think you have been struck by lightning or had a very close strike near the boat, you do need to do some rapid inspections and assessments.

     a.   If you have bonded bronze thru-hull valves in an overall system, check all fittings for damage. If they are damaged or blown out, you will usually see water rising over the cabin sole. Start the bilge pumps! I carry wooden plugs for such eventualities attached to each thru-hull fitting.

     b.   Check your VHF and other radios, and verify that they are operational. A masthead strike often takes out the masthead radio antenna, along with masthead instruments.

     c.   Check all your electronics equipment such as GPS, AIS, and radar, and confirm operation. Mast-mounted radar scanners usually get destroyed in strike events.

     d.   Operate your engine; many new generation engine control panels are electronic, as are some engines with an electronic control unit (ECU).

     e.   Operate the engine and verify that the alternator is charging the battery, as diodes can be destroyed.

     f.   Check your compasses; in some cases, complete demagnetization may occur.

     g.   Check all running rigging and fittings; they may be damaged and seriously affect the vessel's capacity to sail. This includes mast, boom, and headsail furlers, especially electrically operated ones. Sheaves are often plastic, acetal, or aluminum. When strike currents circulate at the masthead, they do so through sheave pins and axles; if enough heat or arcing occurs, the sheaves start seizing up.

10.6   About Thunder. The scientific study of thunder is called brontology. Thunder is the sound generated during a lightning strike. Air in the lightning channel is superheated to temperatures as high as 50,000°F to 70,000°F (28,000°C to 38,000°C), and the rapid thermal expansion of the plasma (the air expands faster than the speed of sound) creates a shock wave. Thunder has a variety of audible qualities, including pitch, loudness, and duration. Different lightning events result in different sounds. Inversion thunder occurs when a cloud-to-ground strike occurs within a temperature inversion. The thunder generated in such events has far greater acoustic energy because the sound is being refracted back down. In close and direct strikes, you might hear an audible clicking sound that is associated with the leader, followed by the loud crack of the return stroke, and then a rumble of subsequent return strokes. It is worth noting that in the event your vessel gets struck by lightning and you are taking cover down below, the strike will be accompanied by an instantaneous thunderclap. The sound pressure level created is around 165dB to 180dB and can be much higher. The main frequency range of thunder is around 100Hz. The shock wave can also cause damage, including internal physical damage. This sound level can rupture eardrums and result in hearing impairment. If you are unlucky enough to experience this, you might experience temporary deafness. You can mitigate this effect by having packs of disposable ear plugs or covering your ears. Thunder travels at 1nm every 6 seconds, and typically the maximum range for hearing thunder is around 6nm. If you see lightning, count it out in seconds until you hear the thunder; divide by 6 and you will know the range. Some use the method of saying "one thousand, two thousand," and so on to time this out.

10.7   What Is Lightning? Is lightning alternating current (AC) or direct current (DC)? The answer is neither. Lightning is essentially a combination of both current types and has properties of both. It is termed a transient electrical pulse, transient impulse, or impulse signal. The AC component is that, given it is a high-frequency discharge, it can be unpacked into frequency components and radiates electromagnetic energy. It also has variable amplitude. But a lightning discharge is not a continuous wave and then becomes DC, since it has polarity and is unidirectional. Lightning is sometimes referred to as a capacitive discharge, as the current discharge is unipolar and flows from positive to negative, similar to a capacitor. The pulsed energy produced in a strike can result in currents of thousands of amps and many thousands of volts. This voltage pulse or wave presents as an overvoltage that is superimposed on the rated voltage of an electrical system. This is characterized by the rise time and the gradient of the pulse, along with the duration of that pulse. There is the phenomenon of wave reflection. This often results in a change in the down conductor system and impedance changes within it, such as an alloy mast, then a change to copper followed by a change to the ground plate and so on. This affects propagation, and reflection of the wave occurs; these are affected by the frequency and the steepness of rise time of the pulse. Depending on the amount of reflection, the voltage may increase substantially. This high-frequency pulse can be in the range of 1kHz up to 1GHz and over a very short time duration of a microsecond to a millisecond; this is the AC component. Much research has been done related to lightning protection and mitigation in shore installations. Approximately 50% of all lightning strikes have current levels exceeding 35kA; just 5% have a current that exceeds 100kA. The statistics for cumulative probability of attaining a peak current are 95% at 5kA, 50% at 35kA, 5% at 100kA, and 1% attaining 200kA. In a single lightning event, there can be a number of strokes, ranging from four to forty. The time interval between these strokes can range from 20ms up to 700ms. A lightning discharge event can generate billions of joules of energy, and both plasma and magnetic radiation are generated along with the characteristic visible light. The results of lightning are destructive because they have a thermal effect that creates high temperatures and material overheating, causing fires, and mechanical effects that cause structural deformation.

10.8   St. Elmo's Fire (Brush Discharge). When this phenomenon occurs, it usually precedes a strike, although the effect does not occur all the time. The vessel becomes a large ground mass. The discharge is characterized by ionized clouds and balls of white or green flashing light that polarize at the vessel extremities. The discharge of negative ions reduces the potential intensity of a strike. St. Elmo's fire is more common on steel vessels, with damage to electrical systems usually induced into the mast wiring, the boat steel hull acting as a large Faraday cage.

10.9   Faraday Cages. The Faraday cage or shield is simply a metallic enclosure or screen that is able to block electromagnetic and electrostatic fields or radiation, such as radio and microwaves, from entering or leaving. They can be constructed using any conductive material, including copper mesh. When an electromagnetic field, such as one created in a lightning strike, intersects a shield, the charge will remain on the exterior of the cage instead of entering. A cage constructed of copper mesh will prevent access by the field. This is applicable to electric fields, both static or constant. The Faraday shield or cage must be grounded to dissipate the accumulated charge on the cage.

10.10   Lightning Physics. Strong updrafts and downdrafts within cumulus and anvil-topped cumulonimbus thunderstorm cloud formations generate high electrical charges. The top of the formation develops a positive potential and the lower a negative potential. Lightning occurs when the difference between the positive and negative charges—the electrical potential—becomes great enough to overcome the resistance of the insulating air and create a conductive path between the positive and negative charges. Lightning is essentially a short circuit from clouds to earth. This is due to charge density and technically is not potential difference. Lightning is effectively a capacitive discharge. When the air insulation breaks down and ionizes, a DC current flows down that ionized path. Prior to and during the main lightning stroke, a large transient occurs.

      a.   **Negative Cloud to Ground.** These strikes occur when the ground is at positive polarity and the cloud's negative region attempts to equalize with the ground.

      b.   **Positive Cloud to Ground**. The positively charged cloud top equalizes with the negative ground.

c. **Positive Ground to Cloud.** The positively charged ground equalizes with the negatively charge cloud.

d. **Negative Ground to Cloud.** The negatively charged ground equalizes with the positively charged cloud top.

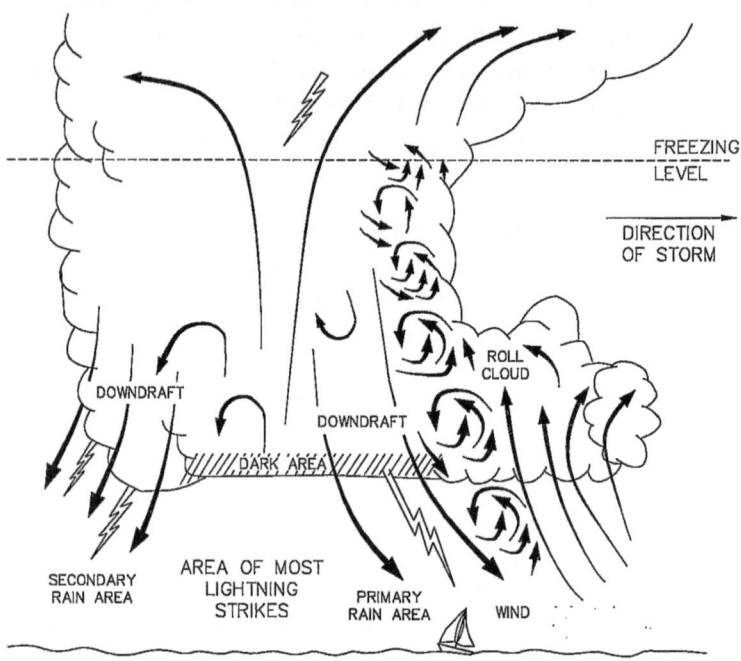

Figure 10-1. Cumulonimbus Storm System

10.11   Lightning Components. Lightning consists of a number of components that form a multidirectional flow of charges exceeding 100 million volts and 200,000 amperes (200kA) at over 54,000°F (30,000°C) for several milliseconds. The positive ions rise to the cloud top, and the negative ions migrate to the cloud base. Regions of positive ions also form at the cloud base. Eventually, the cloud charge levels have sufficient potential difference between ground and another cloud to discharge.

a. **Leader.** The leader, or step leader, consists of a negative stream of electrons comprising many small forks or fingers that follow and break down the air paths offering the least resistance. The charge follows the fork, finding the easiest path as each successive layer is broken down and charged to the same polarity as the cloud.

b. **Upward Positive Leader.** A positive charge rises some 150ft (50m) above the ground.

c. **Channel.** When leader and upward leader meet, a channel is formed. This is sometimes referred to as the attachment point and occurs when the breakdown distance or gap is exceeded.

178

d. **Return Stroke.** This path is generally much brighter and more powerful than the leader; travels upward to the cloud, partially equalizing the potential difference between ground and cloud.

e. **Dart Leader.** In a matter of milliseconds after the return stroke, another downward charge takes place along the same path as the stepped leader and return stroke. Sometimes it is followed by multiple return strokes. The movements happen so fast that it appears to be a single event. This sequence can continue until the differential between cloud and ground is equalized.

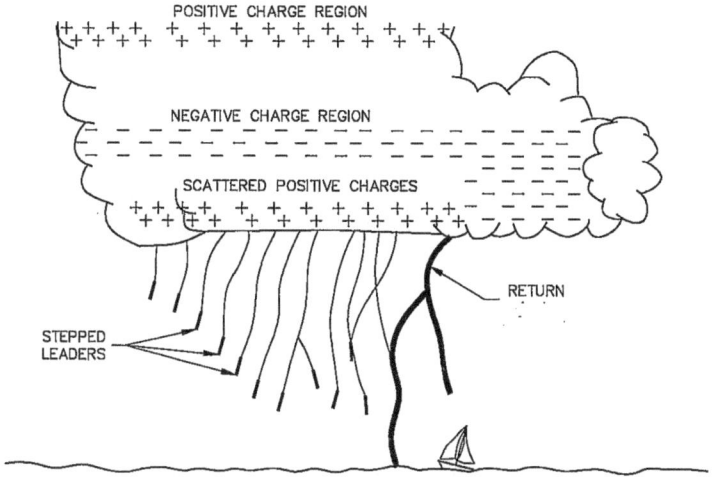

Figure 10-2. Lightning Process

10.12 **Lightning Protection Standards.** Marine classification societies and national marine authorities, including the ABYC (TE-4), lay down recommendations for lightning protection. The BS EN/IEC 62305 lightning protection standard is the international standard for lightning protection in shoreside applications. In addition, National Fire Protection Association (NFPA) Standard 780 is used in the United States. The NFPA standard categorizes lightning protection into five distinct groups from Class V to Class I, with the latter affording the greatest protection.

a. **Shore Facility Scenario.** A shore facility, solar PV array, or building installs a lightning rod air terminal and connects that to a down conductor. The down conductor is calculated on prospective current and is referenced to the lighting protection class outlined within EN 62305. These special cables are usually capable of carrying a 100kA surge current for a period of 1.2μs to 50μs (microseconds). They are flame-resistant and constructed with a twisted copper core surrounded by layers of insulation comprising meshed cross-linked polyethylene (XLPE) and a polyvinyl chloride (PVC) sheath. Some cables are available that can withstand 500kV, and some incorporate a semiconductive anti-corona layer. There are no side strikes or breakdowns that can degrade the down conductor. This conductor is installed vertically down through the building or facility to the earthing grid. That earthing grid is carefully calculated and installed based on the soil resistivity so that the grounding system is optimized with low and measurable impedance.

b. **Boat Scenario.** For the typical FRP/GRP-constructed boat, we will make the base assumption that an air terminal is installed at the masthead and that the base of the mast is connected by a copper cable to the keel. The alloy mast becomes a surrogate down conductor. At the masthead are wind instruments, a tricolor/anchor navigation light, and a VHF radio aerial. Further down the mast may be foredeck and spreader lights and a radar scanner. The supporting wiring and cables are running internally within the mast down conductor. In some cases, the mast has an in-mast furler, which may be electrically driven. Attached to this mast is a boom that may have an integral electric furler. We then consider that our down conductor, the mast, has a forestay on which there is a headsail furler with aluminum extrusion; sometimes a cutter rig with an inner forestay and furler is installed. There is also a backstay or running back stays. Then we have a mainsail halyard, sometimes made of stainless steel wire, that may be running internally and externally, creating a parallel pathway to the mast. We have the standing rigging, stays, and continuous and discontinuous shrouds running to the masthead or to intermediate locations. Added to this are the diagonals, diamonds, and spreaders, often two or more. Lastly we have the various tangs, chain plates, and so on. Why is this important? When lightning strikes the air terminal, all connected stays and shrouds will be "alive," and if they are not connected to the ground point, they will not carry current unless there is a flashover. The exception is the shrouds that run down and intersect the spreaders. This creates several parallel current paths as the strike current travels down the mast as well as down to the spreaders and back to the mast. This is often accompanied by arcing as the spreaders to shroud contact points do not have a low-resistance electrical contact. The next part of our circuit is the down conductor cable to the ground point, which is usually a battery cable, with no specific high-voltage insulation qualities. Finally, we connect to our nominated ground point; if, as usual, it is a lead keel, it is a poor ground, as described later. If it is a radio ground plate, it is a device with no verifiable electrical characteristics. The medium in which it acts as a ground interface is salt water, with variable conductivity. The question we ask is: How good is our protection when compared to a properly engineered solution? Therein lies the basis for so much inconsistent advice and recommendations.

10.13 Lightning Protection Zone. Lightning is classified as a high-frequency electrical phenomenon. It generates overvoltages on most if not all conductive items; that includes electrical wiring, antenna cables, and all connected devices. The most reliable protection system is one that grounds any strike directly to dissipate the strike energy. The main objective of a lightning protection system is to limit the overvoltages and the duration of a strike to levels that do not harm equipment and people. The objective is to limit or eliminate the collateral damage that may threaten the seaworthiness and safety of the vessel. The basic principles are as follows:

a. **Protective Area, or Cone of Protection.** Rules for determining this are found within BS:EN 62305, Part 3, and known as the protective angle method of protection. The air terminal or turned spike, mounted at least 6 inches (150mm) clear of all masthead equipment, gives an approximate 60-degree cone of protection below it. The cone base is approximately 1.5 to twice the diameter of the mast height. This protective cone prevents side strikes to adjacent areas and metalwork, including stays, rails, or other items lower than the masthead. It has a protection rate of 99% if the boat is within that cone and 99.9% if within 45 degrees. There are other methodologies called the rolling sphere, used within BS:EN 62305 to define protection classes; again this is for shore installations.

b. **Grounding.** The primary purpose of a grounding system is to divert the lightning strike directly to ground through a low-resistance, low-impedance circuit suitably rated to carry the very high instantaneous current values. This reduces the strike period to a minimum and also reduces or eliminates the problem of side strikes as the charge attempts to go to ground through alternative routes or paths.

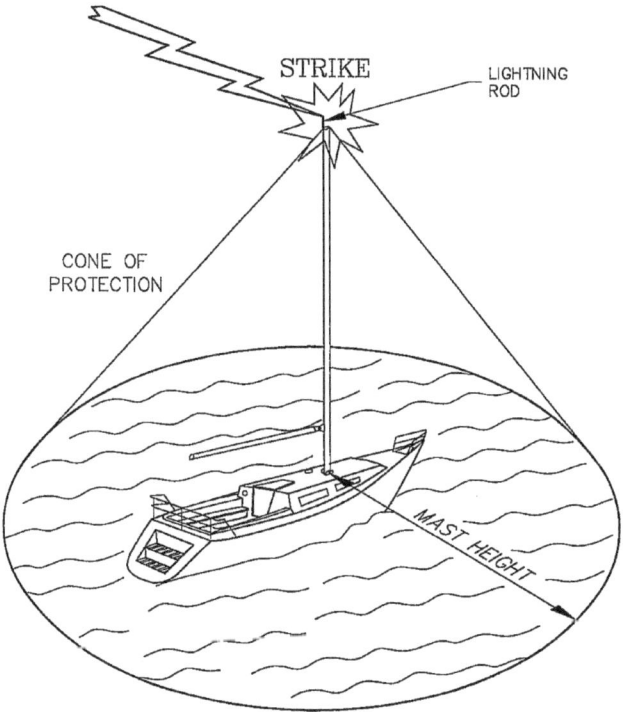

Figure 10-3. Cone of Protection

10.14   Catamarans and Lightning. In both the UK and United States, insurers insurance statistics show that a multihull is two to three times more likely to sustain a lightning strike. My own experiences in multihull wiring and installation activities show that in the right weather conditions, a static charge builds up on the large deck areas. It is my opinion that this plays a role in increasing the possibility of lightning strikes, although no research is available that I am aware of. The case often put forward is that it is too challenging to install lightning protection is simply incorrect. Several years ago, I installed the electrical and electronics systems on a large catamaran. It was on this same boat that I encountered the high static charge situation. After much discussion, I convinced the owner to install a lightning protection system that comprises the masthead air terminal and a down conductor cable from mast step to a radio frequency (RF) ground plate. I also postulated that it might help ground the static charges and bleed them off. Some weeks later the owner called and asked me to come down and check all the systems out. The story is that they sailed away for the weekend and, come nightfall, dropped the anchor in a nice little bay. In the middle of the night, they were awakened by authorities alongside. They had anchored inadvertently close to a very high voltage transmission line water crossing. When they swung on the anchor, the masthead hit the live transmission line, tripped the transmission grid protection, and blacked out half a city. Because the extra high voltage had gone directly to ground via the mast bonding conductor and ground plate, no one had noticed. There was no equipotential bonding, and the system did its job: keeping both people and boat safe. The only damage sustained was the mast-mounted radar scanner.

10.15   Electromagnetic Pulse (EMP). A lightning strike will generate and radiate a very large electromagnetic pulse, which is a strong magnetic field, along with associated high currents and overvoltages. This field is induced into and propagated through wiring and devices, as a high-voltage surge that can do nearly as much damage as a direct hit. Any electrical pulse—especially one with a fast rise time and decay rate, such as a lightning strike—will generate an electromagnetic pulse field. Given the magnitude of a typical lightning strike, it's easy to see that a substantial electromagnetic pulse will be generated as the electrical flow passes through the grounding system. This electromagnetic pulse will be transmitted into any electrical conductor inside the field as it expands and decays. The magnitude of these electromagnetic pulses will typically cause serious damage, even completely destroy, all electronics in the proximity. The fast rise time in an average lightning strike is around 1.2µs (microseconds); the decay time is around 50µs.

10.16   Lightning Dissipation Devices. Over the years, several dissipation devices have come on the market. These devices are typically brush or "bottle brush"–type arrangements. The principles are that all the spikes "bleed" off or dissipate electrons or ions, reducing the differential that may create a lightning strike. This bleed-off or dissipation rate is theoretically at the same rate as the buildup rate, creating a neutral condition. The latest generation of dissipators are dome-shaped devices. Much of this technology has come from shore-based installations where lightning protection is critical, such as airports and hospitals. One of these devices is Sertec's CMCE system. The claim is a 99% reduction in lightning strike probability within the protected area. The basic principle is that the device attracts and grounds excess negative charges, preventing or inhibiting streamer formation and so stopping strikes from forming.

10.17   Lightning Protection Systems. Most classification societies, the ABYC, and other advisory bodies generally recommend lightning protection in the form of a directly grounded mast and spike. There are a few basic elements in any protection system; however, they must be done correctly. Lightning will generally strike the highest point and take the path offering the least resistance to ground. The masthead air terminal is usually the strike point. The aim is to effectively capture the lightning strike current and then channel it safely to ground using the most direct route and avoiding sensitive equipment and associated cables and wiring. This is supplemented by the use of a surge protection device (SPD) and other devices to minimize the effects of induced and other indirect overvoltages through use of grounding or filters and SPDs. We not only aim for keeping the pulse as short as possible but also want to dissipate residual voltage.

10.18   Air Terminals. The first protection element on a sailing yacht is the air terminal. A lighting conductor or air terminal should be installed at the masthead. This should consist of a turned copper, aluminum, or stainless-steel spike of ⅝in to ¾in (14mm to 20mm) in diameter and project at least 6in (150mm) above the highest point, including VHF aerials and wind instrument transducer; this means a terminal 12in to 24in (300mm to 600mm) in height. The turned point should start 1in (25mm) from the tip and gradually come to a point, about a 30-degree angle; a simple short point is not sufficient. Some promote using a dome-shaped tip, but I am unable to determine the rationale for this. You can buy special mounting brackets and support bases to screw the rods into. The purpose of the point being sharp is that it facilitates what is known as point discharge. Ions dissipate from the ground and effectively cause a reduction in potential between the cloud and the sea. Note that a stainless-steel VHF whip aerial does not constitute any protection. The air terminal is mounted clear of all other equipment and gives a cone of protection below it. This protective cone underneath prevents side strikes to adjacent areas and metalwork, including rails or other items lower than the air terminal.

10.19   Down Conductors. On a sailing yacht, there is no conductor in the recognized shoreside sense. The mast by default is the down conductor.

     a.   **Down Conductor Purpose.** The purpose of the down conductor is to safely conduct the strike current through a low-impedance, low-resistance circuit suitably rated to carry the strike current to the ground point. It is meant to eliminate side flash dangers, minimize induction into other conductors, and maintain the strike energy time duration to the minimum possible. Much of the damage in a strike can result from heat, as the very large current flow into even a low-resistance down conductor cable will act as a large heating element. The overall impedance of the cable should not exceed 0.02Ω maximum, although various recommendations quote 1Ω.

     b.   **Down Conductor Sizes.** The normal shore-based system recommendations for down conductor sizes are not relevant. In sailing boats, this is the mast. It is difficult to get accurate data on the various aluminum mast extrusions, but the cross-sectional area (CSA) is probably greater than an equivalent to a copper conductor. Commonly quoted cable sizes are typically 1/0AWG to 3/0AWG (50mm² to 90mm²). In a shore installation, special purpose triaxial cables are used and the multiple screens reduce the large radiated fields; however, this is expensive and probably an option only on large yachts. There are recommendations on the resistance requirements of conductors, but in general it should not exceed that of a 4AWG (20mm^2) copper conductor.

c.   **Down Conductor Bonding Cable.** For all practical purposes, the mast bonding cable is as close to a down conductor as we get. This should go to the ground point, either keel or ground plate, and should be installed as straight as practicable, without sharp corners as side discharges occur; this is often called corona discharge. The minimum recommended radius of cable bends is 8 inches (200mm). It is useful to enclose the conductor internally with PVC flexible conduit, normally used in shore electrical systems to increase the insulation levels, as DC battery cable will break down under high-voltage conditions, given it is a low-voltage cable with very low dielectric properties. Minimum cable sizes are 3AWG (25mm^2) and greater.

d.   **Corona Discharge or Effect.** As mentioned, corona discharge occurs when the lightning strike's electric discharge ionizes the air surrounding the down conductor. This happens when electric field strength within the cable exceeds the dielectric strength of the surrounding air. This is more pronounced on cable bends; effectively it is leaking off into the surrounding air and it is often luminous and audible. This is directly in proportion to the surrounding air density. The insulation dielectric strength of down conductor bonding cables is very low unless you use purpose-made cable.

BONDING AND GROUNDING

10.20   Bonding Basics. In vessels with alloy masts, the base of the mast should be bonded to the mast step or the compression post. This then should be bonded to the keel or a dedicated ground plate. It is often easier to bond the base of an alloy mast to the mast step and then bond this to the compression post. The bottom of the compression post is then bonded to the ground plate or keel. Keel-stepped masts can be directly bonded to the ground plate or keel with a short and heavy gauge conductor. Timber masts ideally should have a conductor fastened externally to the mast. Some use a flat copper strip rather than a thick conductor, also bonding the external sail track. Having owned and sailed both wooden and steel boats, I had to use the mast track and ground that down to the keel. Some advocate that in yawls and ketches, the two masts should be bonded. I disagree with this and instead suggest that each mast has its own air terminal and down conductor or grounding circuit. This is not relevant in the case of steel and alloy ketches, but in wooden and FRP/GRP cases, I suggest running each to its own ground plate.

10.21   Grounding Circuit Resistance. The total resistance or impedance of the grounding circuit from the lightning conductor to the ground plate or hull grounding point should not exceed 0.02Ω. Lightning is sometimes referred to as a capacitive discharge, as the current discharge is unipolar and flows from positive to negative, similar to a capacitor. This needs to be unpacked to its component parts. Understanding resistivity and conductivity is important. Shore facilities have a lightning air terminal, a copper down conductor connected to a copper earth grid calculated for soil resistivity. In a boat you have down conductor elements of different metallurgy types and ground planes with different metallurgical properties. You have seawater of constantly changing composition, all of which affects the overall circuit resistance and impedance.

10.22   Mast Resistance. The mast is a central element in any boat lightning protection equation. A mast extrusion is manufactured from an alloy and heat-treated to achieve the required characteristics. In many cases, it is also painted. This all affects the ability to withstand and carry lightning strike energy to ground. It is not always about the overall cross-sectional area and current carrying capability, as there is a surface or skin effect when lightning travels down a mast. How does the paint affect that? There is a heating effect that varies with amount of strike current and duration of the strike, along with the impedance of the mast. Temperatures vary but can reach more than 2,730°F (1,500°C). One major spar manufacturer voids all warranty on masts if they are struck by lightning, as the heat can alter the metallurgical properties.

10.23   Carbon Fiber Spars. On boats with carbon fiber spars, the ABYC does not consider the spar as an acceptable down conductor. There is very little evidence regarding carbon fiber mast strikes and the resultant impacts. Unfortunately, the standing rigging does the down conductor duty if stainless steel; there is no practicable method to run a heavy copper cable up the mast to an air terminal, which voids the rationale behind having a lightweight spar. How much damage is sustained is hard to quantify, and masts usually are subject to ultrasound inspection to determine integrity. It is hard to get information on carbon fiber mast grounding recommendations. Strikes on carbon fiber spars tend to travel down the mainsail tracks. What about carbon, Vectran/Kevlar, and Dyneema rigging? Fortunately, they are good insulators.

10.24   Electrical Resistivity. This is represented by the Greek letter ϱ (rho). Resistivity is the measure of how much a material will oppose electrical current flow. The lower the resistivity, the easier a material will allow electron charge flow.

Table 10-1. Comparative Resistivity of Metals

Material	% Conductivity
Copper	100%
Aluminum	61%
Zinc	27%
Iron	17%
Phosphor Bronze	15%
Steel	3%–15%
Stainless Steel	3%–15%
Lead	7%

10.25   Electrical Conductivity. This is the measure of a material's capacity to conduct electric current. Conductivity is represented by the Greek letter σ (sigma), ϰ (kappa), or γ (gamma). Conductivity is the opposite, reciprocal, of electrical resistivity. Conductivity is affected by the cross-sectional area (CSA) of the down conductors, the length of the down conductor, and the temperature of the conductor. Temperature affects conductivity; when a conductor gets hot, molecular vibration increases and conductivity decreases. When a lightning strike occurs and conductors are relatively small, heating occurs and conductivity decreases.

Table 10-2. Conductivity and Resistivity of Metals

Material	Resistivity ϱ (Ωm) at 20°C	Conductivity σ (S/m) at 20°C
Copper	1.68×10^{-8}	5.96×10^7
Gold	2.44×10^{-8}	4.10×10^7
Aluminum	2.82×10^{-8}	3.5×10^7
Iron	1.0×10^{-7}	1.00×10^7
Lead	2.2×10^{-7}	4.55×10^6
Stainless Steel	6.9×10^{-7}	1.45×10^6
Seawater	2×10^{-1}	4.8
Hard Rubber	1×10^{13}	10^{-14}

a.   **Impedance (Z).** This is the measure of opposition to current flow in an AC circuit. Impedance has other factors, including capacitance and inductance, which form reactance. This is the measure of opposition to a change in current that is frequency dependent.

b.   **Resistance (R).** This applies to a DC circuit. Resistance reduces current flow and generates heat. A low-resistance and low-impedance grounding circuit is critical to the performance of the protection system. Any resistance will cause significantly greater heating effects, and strike energy will seek shorter and lower resistance ground paths. High-resistance circuits contribute to side strike activities.

c.   **Total Resistance (Ω).** The total resistance of the grounding circuit from the lightning conductor to the ground plate or hull grounding point shall not exceed 0.02Ω. A low-resistance grounding circuit is absolutely critical to the performance of the protection system. Any resistance will cause significantly greater heating effects, and strike energy will seek shorter and lower resistance ground paths. High-resistance circuits will contribute to a higher probability of side strike activities. The NFPA specification states that the resistance of the connection must not exceed that of a 2-foot length of the down conductor.

10.26 **Wooden Spars.** Having sailed a wooden boat for quite a period, I ran a cable from the timber mast mainsail track down through the deck, connected to the mast compression/ king post, and then connected the compression post step to a ground plate. A copper air ter- minal was bonded to the top of the mainsail track at the masthead. Several years ago, I was called to do a survey after a lightning strike on a wooden vessel with a wooden mast. It was moored in a very crowded marina full of alloy-sparred vessels. No one else got struck, which shows the vagaries of lightning events.

10.27 **Conductor Grounds.** The lightning conductor should be terminated at the hull, the keel, or an immersed ground plate with a minimum area of 2ft² (0.25m²), although recommen- dations vary from 0.1m² and greater. The larger the ground plate, the better the performance. In any boat, the ground plane is considered to be seawater, but that has serious limitations, as described elsewhere. It is important that strike energy is dissipated to ground with a minimal rise in ground potential through a low-impedance grounding system. Steel and alloy boats use the mass of the hull as the ground. In many FRP/GRP and wooden sailing boats, the lightning conductors are grounded to the keel bolts and, therefore, the keel acts as the ground plate; again, this is not always good practice. The latter has several limitations that should be considered. Connection of the rudder, the bronze thru-hull valves, propeller shaft, and propeller when interconnected are not a substitute ground terminal, as put forward by some. Running several thousand amps and volts through these along with your engine will cause serious damage.

10.28 **Conductor Terminations.** All terminations and connections should be crimped, and soldered joints should not be used. Under no circumstances use soldered joints alone— they will overheat and melt when carrying large strike currents, causing further havoc. It is very difficult to ensure a good low-resistance solder joint on large cables. Always crimp the connections and ensure that all bonded connections are clean and tight. All connections must be bolted to the ground point.

10.29 **Keels and Grounding.** Grounding to a keel sounds simple enough, but there are considerations before doing this. The majority of modern yachts have a bolted fin keel made of cast lead, often alloyed with antimony to enhance the strength. The keels are attached using 316 grade stainless steel bolts or K monel and some use silicon bronze. Keels are usu- ally faired and then given a couple coats of an epoxy barrier coat paint prior to the antifoulant. The problem is there are no exposed metal surfaces to function as a ground for anything. This is compounded by the fact that lead is a poor conductor of electricity when compared to copper. If you have an iron keel, you have different issues to consider.

10.30 **Keel Bolt Terminations.** A bridge or link should be installed between ground plate bolts, or at least two keel bolts, to evenly distribute strike current. It is recommended that you bridge out the two or four bolts with a galvanically compatible link to spread the contact area and therefore the current-carrying capacity. Links can be drilled and used to bolt the ground cable connector, as many ground shoes have relatively small bolts designed for RF grounds only.

10.31 **Ground Plates.** On many other boats, the choice is often the sintered bronze porous copper ground shoes, such as made of the Dynaplate from Marinco, Moonraker, Wonderbar, or Seaground. Similar to a sponge, the water permeates the bronze lattice and presents a very large surface area. If you are using these, preferably select the largest ones. It must be noted that NewMar clearly states that their radio ground shoes are not intended for lightning protection. As an alternative, up to three smaller ground shoes can be configured in what is called a radial or "crow's foot" principle. This radial system lowers the overall impedance to allow energy to diverge as each conductor and the ground shoe take a share of current. In a strike, the water permeating the sintered bronze ground shoe can literally boil, increasing the local resistance; any increase in surface areas will help reduce this effect. The high-voltage gradients around the shoe will be lower. Some quality ground shoes use gold-based grease under the shoe fastening bolt heads to ensure a good low-resistance connection. Do not use the HF/SSB radio RF ground plate as the lightning ground. The largest plate in the Dynaplate range measures 18 × 6 × ½ inches (460mm × 150mm × 12mm) and is recommended as a ground plane for an HF/SSB, or ham radio antenna counterpoise. Note the word "counterpoise"; the ground plate is not designed to conduct current, and that is the limitation. There are no current or voltage ratings quoted that I can find, and no test results showing performance as a lightning ground. This is not to say they are ineffective; I installed the same ground plate on my previous wooden sloop as part of my protection system.

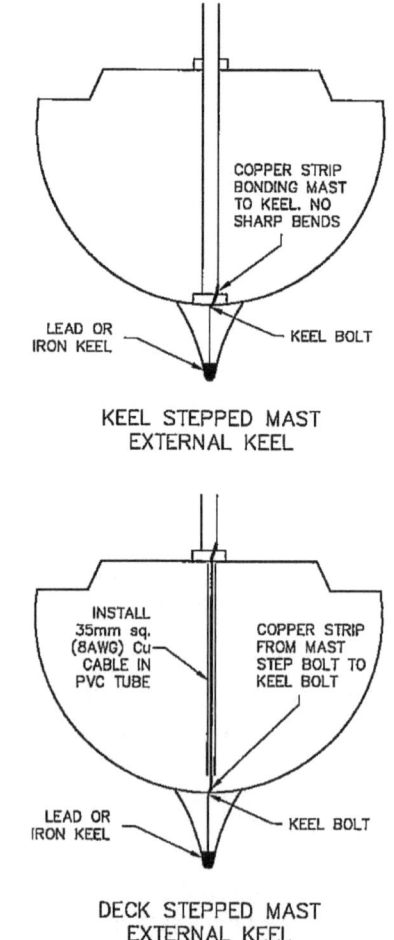

Figure 10-4. Mast Grounding Arrangements

10.32 **Seawater Properties.** The thermophysical properties of seawater impact how lightning energy can be dissipated. Seawater is an aqueous solution of salt and contains many dissolved solids, including sodium, calcium, potassium, magnesium, and chlorides. In coastal areas and on the open sea, seawater has a resistivity within the range of $0.2\Omega\cdot m$ to $30\Omega\cdot m$. Many factors contribute to seawater condition, including the temperature, pressure, and salinity — all of which impact electrical conductivity. Salinity is normally denoted in units ‰ (parts per thousand). When moored in estuaries, rivers, and up into brackish or freshwater areas, the salinity drops and the conductivity and efficiency of the ground plate drop dramatically. This changes continuously with tidal flows and rainwater flows into river estuaries.

10.33 **Grounding Stays and Chain Plates.** There are some recommendations to implement secondary bonding, which entails bonding the stays, shrouds, and chain plates to the lightning ground point. Besides spreading the strike energy a little further around the boat, this creates an additional high-impedance current path down the stays and chain plates that can result in crystallization of the stainless steel and possible loss of the rig under tension. This should be avoided, as it will fail while at sea and create another hazard. It is imperative that a single, low-impedance grounding system be installed. I have heard chain plates, shrouds, and stays referred to as secondary conductors. They are not; they are shrouds and stays and, in my opinion, should not be part of any grounding system. Basics time: If a voltage is applied across a resistance — for example, the alloy mast — which is grounded, a quantity of current will flow. If you add in another resistance — for example, the starboard stainless stay — in parallel and connect it to the same ground point as the mast, you create a second channel for current to flow. Add in a third resistance in parallel — the port side stainless stay — and you have effectively three resistances in parallel. Regardless of the resistance of the second and third paths, the total current flowing will be slightly higher than the single mast current value. If the total lightning current is higher, the equivalent resistance is lower. Large current values flowing through your stainless stays have the potential to alter the metallurgical properties of the wire and fittings, with a high chance of rig failure.

10.34 **Side Strikes.** There is anecdotal evidence that among closely moored vessels and in crowded marinas, lightning strike energy arcs from vessel to vessel as it attempts to find ground. Usually the strike exits from stays, shrouds, chain plates, and spreaders. In many cases, the strike goes to water from the chain plates, causing serious damage to hull and fittings. Even though you have an effective grounding system, side strikes may still occur.

10.35 **Boat Ground Review.** As outlined in the chapter on corrosion protection, a boat has several "ground" systems. They need to be understood in context to the many other device wiring configurations.

    a.   **HF/SSB Radio.** This is not a ground. It is correctly called the counterpoise and has no place in any ground system. See the SSB/HF radio chapter.

    b.   **AC System Safety Ground.** The ground or earth wire is to carry fault current; normally there is no voltage or current on this conductor.

    c.   **DC System Ground.** Technically it does not have a ground; however, it is used to polarize the negative in a typical system.

d. **Instrument Grounds.** Often referred to as the analog ground, signal ground, or signal common. The DC common for an instrument is the zero volt (0.0V); there is also a chassis ground and drain wires from shielded cables.

e. **Lightning Protection Ground.** The ground that carries lightning strike energy.

f. **Corrosion System Bonding.** This is not a ground. It is an equipotential bonding of items under protection by a sacrificial anode.

10.36   Equipotential Bonding. Equipotential bonding is the norm in shore-based installations, but the boat is a very different environment. In these installations this is done to minimize flashovers to equipment when differential potentials arise. A shore facility doesn't have corrosion protection systems, radio installations, instrument grounds, and battery polarization negatives. I am opposed to the commonly promoted method of a lightning protection system being bonded to the DC negative, radio ground, cathodic protection bonding system, seacocks or through-hull fittings, water tanks, and other components. This is a personal opinion, and there are sound reasons behind this position. There have been many documented incidents where bonding of the cathodic protection system, power supply negatives, grounds, and RF grounds have resulted in the vessel sinking because bronze seacocks and thru-hulls have been blown out and all communications, electrical, and electronics systems destroyed. If other systems are directly connected, there will be a higher voltage and actual current flow, which is probably more destructive. On steel and alloy vessels, the hull is the ground plane and the hull acts as a Faraday cage to minimize EMP effects, although you can still get them. On FRP/GRP and wooden boats, there are some recommendations for an internal equalization bus to bond systems and minimize side strike activity.

10.37   Lightning System Bonding. It is a common recommendation to interconnect all metallic items within 6ft (2m) of the mast base or ground point and bond to the ground plate. This is stated as a measure to prevent internal side strikes. Some recommendations call for the bonding of rails, stanchions, and all large metallic equipment, such as stainless water tanks, to the lightning ground. The bonding should be made at the point closest to the main conductor. Potential equalization bonding between ground plane systems is designed to eliminate earth loops and differentials, and reduce the level of potentially destructive transient currents and side strike activity that can occur when potential differences exist between unbonded grounding systems. I also don't fully subscribe to this practice. The reasoning behind bonding stays and chain plates is covered in the next chapter. Stainless-steel water tanks are often installed under bunks, and so on; I prefer to increase the dielectric strength levels. I do not like to assist the spreading of energy by bonding the tanks. The dielectric strength of FRP/GRP and timber used to surround a stainless water or fuel tank is relatively high.

10.38   Temporary Lightning Protection. A clamp is connected to the mast, and this is connected to a tinned-copper cable. The cable is terminated with a specially designed dissipation electrode that is then dropped into the water. This might be a possible option in a marina or anchorage if no permanent system is installed, but it has its drawbacks. Any temporary system carrying strike energy has the potential to seriously damage your standing rigging. In

a lightning event, given the single stay or shroud has a relatively high resistance and impedance, significant current will flow along with significant high-temperature heating effects. A
shroud or stay is a combination of several wires twisted together. The wires are terminated
with swaged fittings, bottle screws, and other hardware, all having resistance, and cumulatively they add up. Given that stainless steel has a relatively low conductivity compared to
copper, the outcome is not good.

10.39   Surge and Transient Protection. It is difficult to adequately suppress or control
the surges that arise during a lightning strike. Surge protection devices (SPDs) are high-impedance devices that are connected in parallel on a power supply input. Often termed
self-sacrificial, they shunt or bypass the overvoltage to ground. When the transient overvoltage from the lightning strike flows through the device, the SPD impedance decreases
and allows the current surge to flow and bypass the equipment it is protecting; it diverts the
current wave to ground and limits the overvoltage amplitude. There is an array of devices,
and they are capable of clamping or reducing a 20kV surge down to around 5kV, but that is
still not enough to protect equipment. In shore applications they use the cascade principle,
with several SPDs being used on critical systems or equipment to bring the overvoltage
down progressively. An SPD has one or more nonlinear components. The live part of an
SPD can have a varistor or a gas discharge tube (GDT). The SPD may incorporate a thermal
protection device to prevent thermal runaway.

10.40   Radio Antenna Protection. Radio antenna and aerials can draw a strike or have an
induced current flow through the coaxial conductor to the radio. All antennas ideally should
have arrestors fitted, although it is rare on small boats and yachts. Antenna cables can be fitted
with a two-way switch—one side to the radio, the other to ground. You can buy remote and
manual coax switches from NewMar. During a storm, or if the vessel is left unattended, place
the switch to the ground position. Ideally an arrester or spark gap device can be used. Coaxial
cable surge protectors can be used. Coaxial cable surge protectors via RF feeders are used
even in shielded cables and triaxial cables, which will confine most current. Some induction
can still occur due to magnetic and capacitive coupling.

10.41   DC Power Supply Protection. Power supplies should have double pole isolation
on both positive and negative supplies. Surge suppression units clamp voltage transients and
maintain them below specified threshold values to a safe level, which is usually the maximum
device voltage they protect. In general, protection is installed on the input power supply of a
device. The metal oxide varistor (MOV) forms the basis of most devices.

10.42   AC Power Supply Protection. Efficient clamping and filtering at the power supply point requires surge diverters. The purpose is to limit residual voltages to a level within
the immunity level range of the equipment. In 230VAC RMS systems, damage can occur
with peaks of just 700V. Typical tolerances of battery chargers are under 800V. Some shunt
devices can clamp the voltage at less than these voltages, but they do not limit the fast wave
front of the strike energy (dI/dt) before clamping action starts. In a lightning strike, the rate
of current rise can exceed 10kA/μsec, and this can be greater in multiple strikes and restrikes.
Low pass filter technology primary shunt diverters will reduce the peak residual voltage and
reduce the rate of current and voltage rise reaching equipment. Surge reduction filters (SRF)

will provide multistage surge attenuation by clamping and then filtering the transients on power input circuits, and these include MOVs.

10.43   Gas Discharge Tubes (GDTs). GDTs are commonly used in lightning protection. They are commonly connected in series or parallel configuration and used in protecting AC and DC power supplies. Basic surge suppression circuits incorporate a voltage dependent resistor (VDR) and GDT that are connected in series. GDTs are traditionally used in either the first or first two levels in a multilevel protection system. The surge arrestor principle of GDTs is one of arc or gas discharge, and they perform like a voltage-dependent switch. When in normal open-circuit state, the GDT has a high impedance. These devices incorporate an inert gas under pressure in a hermitically sealed ceramic chamber. Gases used include hydrogen, deuterium, neon, argon, and noble gases. The device has two or three special electrodes inserted and encapsulated in a ceramic tube, which contains one or more of these special gases. These devices have a high insulation resistance and low capacitance. The capacitance range is the gas discharge tube's ability to hold an electrical charge. GDTs have very high pulse ratings and can be used over a broad voltage spectrum. They are able to dissipate large energy values, which makes them suitable for lightning surge protection. When the lightning surge exceeds the device sparkover voltage, also called the impulse striking or firing voltage, the electric field strength between the GDT poles exceeds the gas breakdown strength. The electric arc is formed, effectively short-circuits, and all the excess current flows. The pulse breakdown voltage (Vsi) of most GDT devices ranges from 500V to 1,000V. This is instantaneous and eliminates the overvoltage as it shunts it to ground. When this contained plasma gas ionizes during a high-voltage transient event, they conduct. As soon as the discharge has dissipated, the GDT extinguishes and the internal resistance reverts to a high insulation value of several hundred megohms ($M\Omega$). There is a reaction time delay of only microseconds, but it does affect how GDTs are used. The reaction and switch on time is very dependent on transient slope; every lightning stroke is different. While GDTs have many advantages, they have a finite life cycle and a relatively high firing voltage, along with the slower response times. Basic selection criteria include that the DC discharge voltage of the GDT must exceed the maximum voltage during normal operation, typically around 15V. The pulse discharge voltage of the GDC must be less than the highest instantaneous value. The GDT holding voltage needs to be as high as possible. Once the overvoltage has gone, the GDT can be extinguished and revert to normal operation. The ground wire must be as short as possible and rated to pass high current transients. Depending on the event and the duration, a GDT can generate a lot of heat.

10.44   About Varistors. Varistors, the metal oxide varistor (MOV) and the zinc oxide varistor (ZOV), are often used on circuits for surge and spike protection. They are used as suppressors for transient voltages, including lightning surge and electrostatic discharge (ESD). When protecting against high surge currents (100A to 25kA), leaded disk varistors are used. For increased surge current (25kA and greater), block and strap varistors are used. An MOV has a maximum continuous operating voltage (MCOV), also called the "threshold voltage"; this is the voltage where the MOV resistance begins to fall.

10.45   Voltage Dependent Resistor (VDR). This has a nonlinear current and a voltage characteristic similar to a diode. The VDR is frequently used for excessive transient voltage protection and is used to shunt this to ground. They have a high clamping voltage and high capacitance and are used in a wide spectrum of voltage and current levels.

10.46   Zener Diodes. The Zener diode allows current flow in the forward and reverse direction. The reverse direction occurs when the nominal voltage is reached. There are specific Zener diodes designed for surge protection. Surge suppression diodes, or transient voltage suppression (TVS) diodes, are able to withstand a number of surges for pulses of less than 1ms.

10.47   VHF Coaxial Surge Protectors. DC blocked protectors are a possible solution on VHF cables from the mast. Relatively inexpensive, they are bulkhead mounted. They block all DC power and are able to pass the RF frequencies.

10.48   Lightning Strike Mitigation. The following are my own personal opinions on suggested mitigation measures, and they repeat earlier arguments. I have a dissenting opinion on the practice of equipotential bonding on boats. This controversial position is based on sound electrical reasons. The bonding of all items has its genesis in shore-based lightning protection configurations. A shore installation not only has a dedicated down conductor but also a dedicated and properly calculated earthing system or grid connected to that down conductor. Within that electrical system, all AC safety grounds and internal metalwork at risk are also bonded to a single ground point and earthing grid, so as to have an equipotential ground system. When lightning strike energy is directed down the conductor to that grid, there is a negligible rise in potential on the grounding network. When equipotential bonding is carried out on a boat, this is the bonding of everything from chain plates for all rigging, corrosion protection systems, bronze thru-hulls and seacocks, and, by default, battery negatives, water tanks, engine blocks, propeller shafts, and other components. I frequently read recommendations for the connection of mainsheet travelers, genoa tracks, stern rails, pulpits, and lifeline stanchions, etc. The argument put forward is to reduce the probability of side flashes internally to anything metallic and limit the rise in voltage potential on the boat wiring. If the boat had an effective grounding system that held all of these to earth potential, this might be a sound position, but it does not. In most cases, a keel does not give a good ground, as explained elsewhere. The RF ground plate is used and partially achieves this, although its effectivity or electrical performance is either a best guess or unknown. Seawater, as described elsewhere, is the surrogate for the soil ashore as the earthing medium. Given that we don't have any of the elements designed in accordance with international standards for shore systems, we have a system not really fit for purpose. The reality is that when a lightning event occurs, you have strike energy coming down the mast along with all the connected stays on a much higher impedance path than is optimal. When this energy tries to dissipate through one of the ground points, an extremely high potential is impressed on the equipotential ground network, and the results are usually catastrophic. The tales are endless; even the batteries have blown up along with everything else. Overlaid on this massive overvoltage event is the EMP component, which further adds to the carnage. It is my contention that, rather than adopt a bonding philosophy, one should use a segregation approach. The goal of lightning protection is to keep the overvoltages and EMP away from devices and systems. I want to channel all of that energy as efficiently as possible down to the ground point. I do not want to create several parallel paths down through stays and chain plates that will then carry full strike current and compromise the rig. That means trying to keep the overvoltages away from batteries, switchboards, electronics, and other devices. This entails a corrosion protection system that only bonds items that need to be protected. (Please revisit the corrosion chapter.)

10.49   Strike Mitigation Recommendations. The question remains as to what mitigation measures a boat owner can initiate to reduce the risk to self, crew, and the boat's seaworthiness. I will be controversial and initiate arguments, but these are suggested solutions if you live in a lightning hot spot. Do your own risk assessment; have conversations, listen to local anecdotal evidence, and decide on countermeasure requirements based on this information. Look at each measure and decide if it is for you, or consult a professional. You can simply cherry-pick items that are easy to implement and apply to your particular circumstances.

a.   **Ground Plate.** Consider installing an additional ground plate. Locate it aft, close to all the electrical and instrument devices and systems.

b.   **Rigging Measures.** Consider some standing and running rigging options:

(1)   **Main Halyard.** Replace the wire main halyard with Dyneema, removing the parallel grounding wires that carry as much current as the mast. This can reduce some of the flashover activity.

(2)   **Back Stay.** Replace the running back stay or the back stay with Dyneema or other similar synthetic rigging materials. This will remove the back stay overvoltage route down to the stern cockpit area.

c.   **Communications Options.** The following are some protection options:

(1)   **VHF Radio Aerials.** Consider whether you really need them at the masthead. Masthead mounting originated from the requirement for maximum range for contacting shore stations, and the need has decreased with the advent of mobile phones. For the average yacht, a stern-mounted VHF and AIS aerial work out well. Consider installing a coaxial lightning surge protector in-line. Consider replacing the aerial for a compact type if you have a stainless-steel whip at the masthead. If you have an in-line connection or deck plug, consider unplugging and disconnecting when not on board.

(2)   **HF/SSB Radio Aerials.** If you have an insulated backstay that comprises two insulators, insert another third one closer to the masthead to provide another layer of protection to reduce flashovers. Check on the electrical performance of any insulator you plan to buy. The Sta-lok backstay insulator has a published performance voltage proof test at 30kV for 2 hours with no breakdown. If you have a lead wire to the back stay, look at using a waterproof plug so you can easily disconnect the aerial when you leave the boat to isolate the ATU. If you do not have an HF/SSB radio, install back stay insulators. This reduces overvoltage conduction and flashover to the

stern. Alternatively see the KISS-SSB system, as described in the HF/SSB chapter, and take the aerial out of the equation.

d. **Grounding Electrical Protection Options.** There are several options to consider, some quite radical and others more achievable:

(1) **Wiring Configuration Change.** Consider adopting a metal boat system and converting the boat electrical system to a fully isolated above-hull configuration. The configuration is described in the wiring chapter.

(2) **Circuit Breakers.** The isolated system uses double pole circuit breakers for all primary and secondary circuits. This isolates both positive and negative.

(3) **Surge Protection.** Install surge protection devices such as a GDT on the main supply cables to shunt overvoltages.

(4) **Circuit Segregation.** Consider installing a separate distribution unit for all mast electrical circuits. Install double pole circuit breakers and an MOV on each circuit. This is the greatest overvoltage exposure point.

e. **Anchor Windlass Options.** The anchor windlass is located forward, making it highly exposed to overvoltages that travel down the forestay to the bow. There is a high flashover risk to the windlass and all the foredeck hardware. The mitigation measures are to limit or prevent the propagation of overvoltages back to the power supply and the electronics area:

(1) **Circuit Breakers.** Install a double pole circuit breaker near the supply point. Consider installing a double pole isolator forward in an accessible location close to the windlass, and isolate when alongside or at times of high lightning activity.

(2) **Surge Protection.** Install an MOV across the input terminals to the anchor control box.

(3) **Surge Protection.** Install an MOV across the anchor windlass supply dual pole circuit breaker.

(4) **Surge Protection.** Consider installing a GDT at the supply circuit breaker output and connect that to your new ground plate. If an overvoltage does impress itself on the supply cable, you can shunt this to ground.

(5) **Supply Cable Relocation.** Install or relocate anchor windlass cables so they run off the start battery if it isn't that way already. This is to eliminate surge from the forward area back to house batteries. If the cables run close to electrical or electronics wiring or close to other equipment, run them clear of everything.

(6) **Cable Shielding.** Install tinned-copper cable shielding mesh over the large anchor windlass and thruster cables that run forward. The cables run horizontally and the down conductor bonding cable is vertical, so it reduces induction. However, it can still get an induced overvoltage; improving the shielding can improve this.

f. **AC Grounding.** Consider taking the AC safety ground to a more reliable point:

(1) **AC Safety Grounds.** Connect the main AC safety ground wire from your AC switch panel to the additional new ground plate.

(2) **AC Circuits.** If you have an extensive AC power system, you can use surge protectors at your panel to protect downstream devices.

g. **Batteries.** Battery explosions in lightning strike events appear to be a common occurrence. They are primarily caused by the process of bonding and applying or channeling overvoltages and high currents into a battery. This creates excess heat and gas generation and can initiate an explosion. Lithium-ion batteries are more susceptible to overvoltage, and the integral BMS cell monitoring is easily damaged. Consider installing fine woven copper mesh around the inside of battery compartment to form a Faraday shield. Consider glassing mesh underneath the cover; ground this to the new ground plate.

h. **Water Tanks.** Many consider stainless-steel water tanks to be a flashover risk. Consider installing additional insulative material, such as rubber matting, around the water tank to increase the dielectric strength and reduce flashover risks. Alternatively, look at creating a Faraday mesh shield around the tank. Personally, I think this risk is a bit overstated.

Anchor Windlasses, Deck Winches, and Furlers

11.1   The Anchor Windlass. On my last boat, I didn't realize how much I would miss the electric anchor windlass until I didn't have it. It was really hard work and lots of serious manual exercise either using the double-action ratchet system or resorting to hand-over-hand pulling in chain when I was in a hurry. My latest boat has a manual windlass, and installing an electric windlass is a priority. For myself, the anchor windlass has a level of importance way up there with the sails, rig, and engine. It is crucial that the windlass be properly selected and installed. Unfortunately, they are rarely maintained properly and subsequently fail at critical periods. This chapter explains the process and the factors to consider when selecting and installing a windlass. All manufacturers offer very useful selection guides, so use them. Major manufacturers include Lewmar, Andersen, Hutton-Arco (in Australia), Maxwell, Quick, Lighthouse, Muir, and Lofrans.

a.   **Anchor Types.** It is prudent to select the correct anchor or two for your boat. Bower anchors come in many types. We can consider the Claw, Bruce, Kobra II, Wing, Lewmar Delta and Epsilon (a Delta replacement), Plow, Viper Pro Plow, CQR, Fortress, Ultra, Spade, Bügel, Danforth, Fluke, Grapnel, Lewmar LFX, Kedge, Rocna and Rocnar Vulcan, Mantus, Manson Ray, Britany, Sarca Excel, and several others including the original fisherman's anchor. Different seabeds favor different anchors and anchoring strategies. Sand, broken shell and sand, grass and weed, mud, and rock all have different requirements. Danforth, Bruce, CQR, Delta, and Spade are adaptable to most bottom types. Do some research and decide what works for you.

b.   **Windlass Selection.** Choose the windlass based on the weight of anchor chain and the vessel size, and also consider vessel displacement. Manufacturers have easy-to-follow charts.

c.   **Electrical Installation.** Install correctly rated cables and protective systems.

d.   **Electrical Control.** Install a reliable control system, from footswitches to wireless remotes.

e.   **Windlass Maintenance.** Do the recommended maintenance. There is much to consider, and I have covered the key factors below.

11.2   About Anchors. Much has been written about anchors and anchoring configurations—all of it excellent lived-experience advice. There are many great and informative articles about anchoring techniques, and I could probably write a few based on my own experiences. But to provide context, a quick run-through on anchor types is warranted. What do we want from an anchor? Holding power is the first consideration, and then we consider what type of bottom we want to hold in. When it comes to anchor windlass selection, the question of retrieval comes into play—how easy or hard will that be? Other factors that come into the equation are anchor strength and holding stability. Then we should consider anchor penetration—how easy will it be to set the anchor? "Scope" is the term used to indicate the ratio of rode length to water depth. Selecting the right anchor weight is important.

11.3   Anchor Parts. There are several parts to an anchor, and they can vary among the various anchor types. These terms apply to different anchor designs.

- **a.   Head.** The head is the part at the top of the shank.

- **b.   Shank.** The shank is the vertical stem at the top of the anchor. This is pulled to set or bury the anchor.

- **c.   Stock.** This is a either a fixed or removable crosspiece located at the top of the shank. This rotates the anchor into a position that facilitates the flukes to penetrate the seabed.

- **d.   Ring.** The anchor ring is located at the top of the shank at the crown for fastening the anchor; this might also be a D-shackle.

- **e.   Crown.** This connects the various parts of the anchor.

- **f.   Flukes.** Flukes are composed of the palm and the bill or pee. This is the part that gets buried in the seabed.

- **g.   Gravity or Balancing Band.** This is a movable band with rings on the shank at the balancing point or center of gravity of the anchor so that an anchor balances horizontally when lifted.

- **h.   Tripping Palm.** The tripping palm is used to tilt the flukes to the seabed.

- **i.   Arms.** These are attached to the shank at its base.

- **j.   Tripping or Hoisting Ring.** This is used to attach a tripping line for breaking out the anchor.

11.4   Anchor Setting. Setting the anchor is of crucial importance. Understanding what the bottom composition is, and understanding how your anchor will suit these bottom conditions, is part of successful anchoring. You have picked your anchor location; do you know what the bottom is composed of? I have a lead line, and I put some beeswax into the indent at the bottom of the lead to check bottom composition if in doubt. Sometimes old methods work well.

- **a.   The Approach.** Always approach your chosen anchoring spot at dead slow ahead. Head up the boat into the prevailing current or wind. Move up ahead of the location at the planned scope distance. Sometimes if it's a bit rough and choppy, you might choose to do a pass and determine the drift rate before dropping anchor. Plan on a 7:1 ratio to water depth to attain the anchor design holding power. This will depend on several factors, including the bottom material, how well you will hold, and weather and water conditions. The scope may need to be higher, up to and exceeding a 10:1 ratio. Things are rather different when you are in a crowded anchorage, when it can get down to 4:1 or less. Usually, I prefer to go somewhere less crowded if possible.

b. **The Drop.** Most choose to free-fall drop the anchor. Depending on the weather and water, once you have dropped the hook, you can allow the boat to drift back and feed chain out. Then it is time to set the brake and apply some tension on the rode to achieve bottom penetration. If the wind and current are absent, you can slowly motor astern until you get some resistance as the anchor buries itself. You will have to lay out more chain until you get to the planned scope and until a catenary curve is established. The catenary is that part of the rode that curves up to the bow in an arc. There may be a case for walking back the anchor chain, using the anchor windlass under power.

c. **Fouled Anchors.** Be alert for a fouled anchor event. You may get caught up with any number of underwater obstructions; that brings its own challenges and is where the use of trip line can be useful.

d. **Environmental Conditions.** You can have a tide rode where the bow is head to tide. You can also have a wind rode condition where the boat is riding head to wind. You can end up yawing based on the strongest influence of these factors. If either wind or tide veers, it becomes more problematic, and one has to be alert for the anchor breaking out and requiring resetting. Other influences, such as a lee tide, influence when both wind and tide are acting together. With weather tide, the stream is setting to windward and the wind and tide are in opposition. The takeaway is to set your depth sounder or devices to monitor position for anchor dragging and stay attuned to weather conditions.

e. **Coral.** If you are anchoring around coral reefs, exercise caution. Make sure you are not within restricted areas. Anchor in sand patches, not in living coral gardens. You should have enough chain to anchor in 100ft (30m) of water; there is generally no coral at that depth. Always arrive in daylight so as to observe the potential anchorage sites—ideally before 15:00 hours, when coral patches are easy to spot. Polaroid sunglasses are a big help. Be environmentally responsible.

11.5   Anchor Breakout, Heaving, or Retrieval. "Breaking out the anchor" is an older nautical term for the process of lifting an anchor and breaking it out from the seabed. The anchor chain is said to be up and down when the vessel is directly above the anchor and the chain rode is no longer lying on the seabed. The anchor is said to be aweigh when the anchor is clear of the seabed and all the weight is on the chain or rope as it is hauled up. The process of anchor retrieval is known as weighing, also called raising or lifting. The old nautical term of weighing dates back to around 1627. When an anchor is lifted clear of the seabed, it is said to be broken out. As it is raised or heaved away toward the vessel, the anchor is said to be coming home.

11.6   Anchor System Basics. There are many elements to what is termed ground tackle. There is a whole separate terminology that goes back through centuries of sail. The basic components are as follows; electrical and control systems such as footswitches are covered later.

a. **Horizontal Windlass.** The drive shaft and chain gypsy are horizontal to the deck.

b. **Vertical Windlass.** The drive shaft and chain gypsy are vertical to the deck.

c. **Anchor Rode.** This is the chain or rope, or combination of chain and rope, that connects an anchor with the boat.

d. **Rode Markers.** Markers are used to indicate how much anchor rode you have out. Some use a system of painted links. Others use small colored wire ties, but I don't recommend that; they can jam up the windlass chain gypsy. I do like the insertable colored plastic-block chain markers. Red, white, and blue are popular colors; I prefer white and yellow for better night visibility. Markings are typically every 20ft (6m) up to 50ft (15m); use what works best for you.

e. **Rode Counters.** These are commonly integrated into windlass control systems and are often retrofitted options. Lofrans, Muir, Maxwell, Quick, and Lewmar systems all have a magnet installed on the chainwheel or chain gypsy, and a reed-type magnetic sensor mounted on the windlass frame. As the chain gypsy turns, it counts each revolution; this equates to how much rode has passed around the gypsy circumference. This is indicated on a small electronic display, usually mounted in the cockpit. These systems now use microprocessor-based controls and detect direction of rotation. Some counters, such as the one from Quick, have smartphone connectivity.

f. **Chain Gypsy.** This is sometimes called the wildcat or chainwheel and is the vertical wheel, although some units have horizontal. They have pockets in the wheel for the chain and provide the mechanical advantage to haul in the chain. This is designed for the chain size to be used on the anchor rode. Rope and chain rodes have specially designed gypsies. Developments continue, with Lewmar having a hybrid polymer gypsy, which gives weight and strength gains.

g. **Free-Fall Release.** The clutch mechanism can be manually released to allow the anchor and rode to free-fall by force of gravity when letting go. Some anchor windlass control systems have auto free-fall functions.

h. **Manual Override.** Sometimes called the emergency crank, this allows the manual operation of the windlass to retrieve the anchor and rode in the event of motor, gearbox, or power failure.

i. **Bow Anchor Rollers.** The bow roller or stemhead fitting has the important task of controlling anchor movement to ensure heaving and weighing anchor goes smoothly. The rollers must provide friction-free operation when launching and retrieving anchors. Don't forget to inspect the bow roller; it should be rolling easily and not jamming. While there, check out your securing arrangement. If they are not rolling properly, disassemble, check for damage, lubricate, then reinstall and check again. Check that the spring-loaded pins for retention of rode and anchor are not bent or seized.

j. **Rode Swivel.** Installed at the anchor end between rode and anchor, rode swivels need to be in good working order. They are often neglected and either seize up or fail. There is much controversy about using or not using them. Their primary purpose is to prevent twisting of the rode when the anchor is set. Some use simple galvanized units; others, the more expensive stainless-steel units. These will be INOX 316 and will have the safe working load (SWL) marked on them. There are box swivels, multidirectional swivels, cup swivels, and more. If you are going to install one, consider a multidirectional swivel, as the swivel does the turning when swinging at anchor and not the rode. Many say that it aids in feeding chain into the windlass chain gypsy without twisting and allows more efficient chain drop through the chain pipe and locker. Others say it makes stowage into the bow roller easier. I have one on my boat, and it achieves all of those things. All of these advantages are true, depending on the boat and overall anchoring setup. Failure modes include fork bending and clevis pin failure under high loads; I have been through that myself. The obvious failure mode is that the bearing interface fails. Swivels also corrode or suffer metal fatigue. Buy the best quality units you are prepared to pay for, and ensure they are of the same load rating at the rode and not a weak link. Mantus makes a swivel and shackle.

k. **The Bitter End.** The bitter end, or bitt end, is the final part of the anchor rode; it is fastened to the boat, usually in the chain locker. Used for centuries, the final section of chain or rope traditionally has various color rags to indicate the end is near. When the marker rags were sighted it indicated there was no rope left to anchor with, usually because the water was too deep to set the anchor. Many use shackles at the bitter end that are corroded and inoperable. Change it to a lashing for the hopefully rare event that requires you to cut the lashing and let go the whole rode while under some anchoring pressure. Make sure a knife is installed ready for use, and make sure you corrosion-proof it with grease or something similar.

l. **Tripping Lines.** While not technically part of the overall setup, many use them, including myself. The trip line is attached to the anchor crown, tripping ring, or other designated point on the anchor. The line is then led up to a small marker buoy on the surface. The line is used to manually break out the anchor if the windlass is unable to do so.

11.7 Chain Stoppers and Snubbers. A windlass is not designed to take the surge and snatch loads when riding to anchor or when veering in large swells or heavy conditions. As a safety precaution, always transfer the load to a bollard or kingpost using a nylon rope snubber. There are cautions about the use of snubbers, and once I thought that the foredeck would tear out, such are the pressures applied to kingposts and other structures. Some folk deploy double snubbers. Chain stoppers do a similar task; however, they must be mounted on a reinforced deck area to absorb the shock loading that would be applied to the windlass. Some have stainless-steel chain hooks, grippers, and grabs with nylon tails attached. Chain hooks and grabs are purpose designed to secure a line to a length of chain under load. They are designed to fit over a link in one plane and rest on the following link.

11.8   Anchor Kellets. The kellet is also known as an anchor sentinel, chum, rider, and other names. I know some people who use kellets, and they are available store-bought or something more homemade. When an anchor is not setting properly or is holding badly and dragging, a kellet is an option. There are opposing views about their use, and it is something you can consider. The whole reason behind a kellet is to improve the holding power of an anchor. It is a movable weight, and many weigh around 30 pounds (15kg). A kellet is usually installed close to the anchor shank and up to halfway along the anchor rode. The weight then imposes a downforce on the shank to improve the attack angle of the flukes and has a dampening effect. This creates a steeper angle of the rode and levels out the rode on the seabed. Personally, I am not an advocate of them.

11.9   Lighthouse Windlasses. These have a long reputation as being precision engineered, which is a reason I include them here. The units have cast-bronze and chromed gypsies and stainless wildcats. The windlasses have bronze gearing and ball bearings, and hardened stainless-steel worm drives along with a complete stainless-steel gearbox. The clutches are manufactured from ethylene propylene diene monomer (EPDM) and urethane, which are then laminated to stainless steel. EPDM is a type of synthetic rubber; it is flexible, with a high heat capability. On the electrical motor side, these comprise a potted armature, and the brush holder plates are made from Micarta. Micarta is a composite thermoset laminate material that has very high physical strength and excellent insulative electrical properties, along with very high thermal and humidity capabilities. The brush holders are made from bronze with stainless-steel springs. The motor has sealed bearings using a lithium-based grease. Unlike the trend for neodymium-iron-boron ($Nd_2Fe_{14}B$) rare earth magnet motors, these utilize samarium-cobalt ($SmCo_5$ and Sm_2Co_{17}) fields. These have good magnetic qualities, are temperature stable, and have a high resistance to corrosion. Many magnets are anisotropic, which means they have a preferred direction of magnetization; $SmCo_5$ magnets do not. Another advantage is that $SmCo_5$ magnets can maintain magnetization in extreme temperatures, along with higher coercivity, which is a very high resistance to demagnetization. Performance of the 1501 is free running at 8A and 80A at rated pull. Continuous rated pull is 1,000lb (454kg) at 37fpm (0.19m/s).

11.10   Anchor Performance. A bower anchor's effectiveness is not all about the weight. The older designs have undergone some evolution since they were introduced. Quick bottom penetration is achieved through a process called dynamic setting. You want to achieve maximum weight to bear on the anchor tips when you apply force or tension to the chain. Newer generation anchors have improved tip loading designs and optimized angles. Shank and fluke shapes have also evolved. The latest anchors have concave flukes, which assist in faster and deeper penetration. Another important attribute for an anchor is how easily it resets. How many times have you broken out when the tide or wind changed? Some of the better anchors on the market reset easily, but not all. Having sailed with CQRs on my last few boats, and my current boat came with a stainless one, I am opting for a Delta or perhaps the Lewmar Epsilon this time around, along with my aluminum kedge and my fold-up grapnel in reserve. Getting the right anchor pull angle and how to optimize anchor setting are well debated, centering on scope increase and catenary. It is fundamental, but you need to know about whatever bottom you intend anchoring in; consult your chart. Mud and silt might suit some anchor types; sand or gravel bottoms have different results. Some suggest, quite correctly, that you should choose rode and windlass based not simply on boat displacement and length but also on the maximum water depth you might expect to deploy an anchor. Parameters to consider include windage and accurate vessel displacement, as numbers are based on average displacement.

11.11   Anchor Windlass Selection. The selection chart should be used, as correct weight selection is critical. Finding the right windlass chain is more difficult than might be imagined. The principal problem is that chain types do not always match the windlass chain lifter. Windlass selection is based on the weight of an anchor and the chain weight. Table 11-1 illustrates a selection of short link chain sizes for a variety of vessel lengths.

Table 11-1. Anchor Chain Weight Selection Table

Vessel Size	33ft (10m)	40ft (12m)	14m	16m	18m
Chain Size	8mm	10mm	10mm	13mm	13mm
All Chain	40m	50m	70m	80m	90m
Rope/Chain	40ft (12m)	14m	16m	18m	65ft (20m)
Chain Weight	1.42kg/m	2.22kg/m	2.22kg/m	3.75kg/m	3.75kg/m

    a.   **Winch pulling (working) power calculation.** Minimum windlass capacity is derived from the following formula, after working out the chain weight for your vessel size and this referencing Lewmar recommendations:

$$\text{Windlass Capacity} = (\text{Anchor Weight} + \text{Chain Weight}) \times 4$$

e.g.,   40ft (12m) vessel has a CQR of 35kg.

Chain Weight 111kg + 35kg = 146kg × 4 = 584kg.

    b.   **Rated Output Maximum Pull.** The windlass in this instance must have a rated pull of at least 584kg. Manufacturers have charts to assist in selection. Most windlass models have much higher maximum pull ratings of 1,235lb to 1,710lb (500kg to 775kg) for a windlass for this size boat.

    c.   **Maximum Line or Recovery Speeds.** Speeds are typically designed around a figure of 33 feet/minute (10 meters/minute) at a 220lb (100kg) load. The higher the load, the slower the anchor retrieval rate. Many have rates that vary from 65 to100fpm (19 to 30m/min) depending on the windlass type and rating.

11.12   Windlass Operation. When operating the windlass, observe the following.

    a.   **Engine Running.** Always operate the windlass with the engine running. The alternator supplies part of the motor load and keeps the motor from impressing a large voltage surge on the electrical system. More importantly, running the engine keeps the voltage from dropping too low.

b.   **Run Times.** In cases where the windlass is used without the engine running, the voltage drop is such that a severe drop in windlass power occurs after a few minutes. A further problem is that the motor may overheat due to the lower voltage, causing winding damage or burnout. Always pause for 20 to 30 seconds every few minutes and allow the voltage to recover. If you are having a problem with anchor retrieval, do not continue to load the anchor windlass until it stalls. Stop every 5 minutes and allow the motor to cool down.

11.13   Windlass Electrical Installation. Anchor windlass performance is frequently reduced by installation of incorrectly rated cables. Anchor windlass electrical supplies should run the most direct route to the engine starting battery, via the appropriate isolator and protective devices. At full-rated load, significant voltage drops can develop, with a corresponding decrease in rated lifting capacity. Underrated cables are probably the largest problem experienced. The following system components must be specified and installed correctly. Installing a separate battery, either at the machinery space or forward next to the windlass, is not recommended. Use the engine starting battery, as it has a high cranking amp rating and is more able to deliver the currents required by a windlass at maximum loads. A deep-cycle service battery cannot cope with these loads without being damaged.

a.   **Circuit Protection.** Current ratings vary depending on the manufacturer. Many windlasses have converted DC starter motors on the entire powered range. Typical current loadings are given as 55A at no load, 110A at half load, and 180A at full rated load. Protective devices are as follows:

(1)   **Circuit Breakers.** A circuit breaker should be installed on the power supply reasonably close to the battery and should be easily accessible. Typically, 100A and 125A circuit breakers are used. Use DC-rated circuit breakers, not AC-rated, as many commonly do.

(2)   **Automatic Thermal Cutouts.** I would caution against using automatic thermal circuit breakers. They trip automatically in overload conditions and reset. The problem is that you have to wait until they reset, which is usually when you desperately need the windlass.

(3)   **Slow Blow Fuses.** ABYC and USCG require a slow blow fuse to be installed on the system, and many manufacturers integrate this within the control box. The fuses are normally rated above the windlass's rated working current, typically 200A for 12V systems. Make sure you carry a spare.

b.   **Cabling.** Cabling must be able to cope with large currents over an extended distance. The ABYC specifies a maximum volt drop of 10%, but I think that is excessive. Instead I recommend a voltage drop that does not exceed 5%. Table 11-2 provides recommended cable sizes for length of cable run, not

Table 11-2. Windlass Cable Rating Table

Load Current (amps)	Cable Length (ft)	AWG	Metric (mm²)	Capacity (amps)
80	30	1	42	245
80	40	0 (1/0)	54	285
80	50	00 (2/0)	68	330
100	30	0 (1/0)	54	285
100	40	00 (2/0)	68	330
100	50	000 (3/0)	85	385
120	30	0 (1/0)	54	285
120	40	000 (3/0)	85	385
120	50	0000 (4/0)	107	445

for vessel length, and this is for positive cable to load and the negative back. Perform your own calculations for your specific installation. Ensure that you route cables well away from a mast step or lightning protection connection or cables. Table 11-2 is for a 3% volt drop, with a variable load of 10 minutes, which is typical for a windlass.

c. **Connections.** Connections are a common cause of failures.

　(1)　**Connector Types.** Always use heavy-duty crimp connectors. Do not solder connections; dry joints are commonplace, and solder can melt under maximum load. Soldered joints stiffen cables, causing fatigue.

　(2)　**Insulation.** Put a section of heat-shrink tubing over the entire crimp connector shank and cable to prevent the ingress of moisture.

　(3)　**Connections.** The lug terminal hole should always fit neatly to ensure maximum contact. Use a spring washer on the nuts to prevent loosening and subsequent heating under load. Coat terminals with a light layer of petroleum jelly. New windlass motor developments include Lewmar having a cover that goes over the connections to IP67, which will ensure dry terminations.

d. **Performance Curves.** The curves graphically illustrate the effect load has on power consumption and hauling speed. The higher the load, the higher the current, until a point is reached where the motor overloads and stalls. The higher the load when the windlass is operated, the shorter the operation time allowed on the motor. The higher the load, the slower the recovery speed. Hoisting the anchor is faster and causes less wear and tear if you motor up over the anchor and remove chain tension.

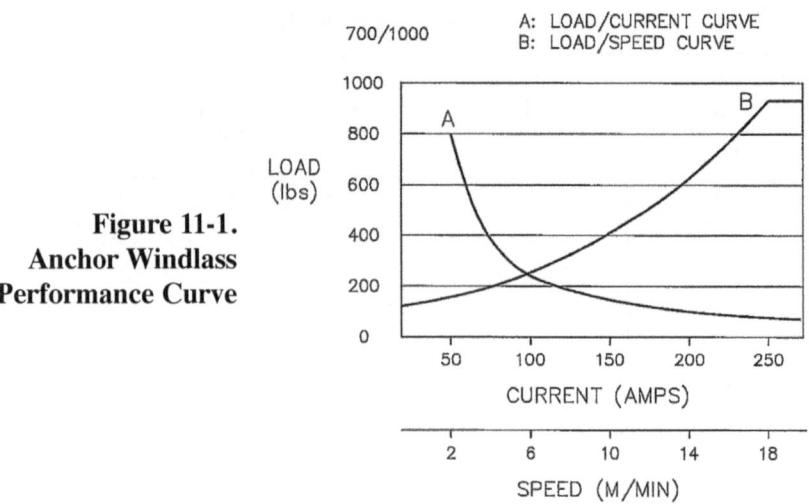

**Figure 11-1.
Anchor Windlass
Performance Curve**

11.14   Windlass Electrical Control. One of the most common failure points in an anchor windlass is the control system. Controls come in the following configurations:

a.   **Single-Direction Foot Switch.** A foot switch was often connected directly in the positive supply to the windlass motor. Thankfully that's no longer the case, and foot switches control the directional solenoids. Foot switches are notorious for filling with water, and usually in this type of control, a short circuit develops or the contacts and spring corrode. Shorting can result in brief, uncontrolled windlass operation and a burned-out switch. Foot switches from Lewmar are IP67 rated, as they all should be.

b.   **Single-Direction Solenoid/Foot Switch.** The foot switch is used to control a heavy-duty solenoid located belowdecks, which closes the main power supply to the motor.

c.   **Switch Operations.** Some install a waterproof toggle switch in the cockpit. Lewmar has a guarded rocker switch that can be mounted anywhere; these have a cover or guard to prevent accidental operation.

d.   **Pneumatic Deck Foot Switch.** These are no longer common. The units have a PVC tube connecting the switch to the control solenoid box. Air pressure from the switch operates a microswitch. There have been reports of spontaneous start-ups or shutoffs in extremely hot conditions, which in one case caused serious injuries. The problems were caused by pressure buildup in the air system. Evidently, earlier units are the most prone to trouble; newer units have a safety air bleed to correct the problem. Carefully follow the proper depressurizing procedures when installing switches.

e.   **Dual-Direction Solenoid Control.** A control box consisting of two or four solenoids is used for reversing the motor for both hoisting and lowering. Control is usually by a pair of foot switches and/or a remote panel.

(1) **Power Consumption.** Solenoids typically consume up to 1A.

(2) **Caution.** Never operate both foot switches together. In fact, many manufacturers specify that only the "up" foot control be fitted.

(3) **Protection.** Some control boxes incorporate fuse protection. Fuse failure is rare, but make sure you have a spare in the box for emergencies.

. **Remote Controls.** Remote control devices take a variety of forms:

(1) **Wired Remote Controls.** These are usually weatherproof wired handheld remotes that can be plugged into prewired socket-outlet stations. Ensure that the socket remains watertight. Many of these have an automatic lowering function and an up-alarm function. They have auto free fall and also anchor recovery functions if there is a sensor failure. Chain rode counters are included and chain speed is displayed. Some, such as those from Quick, have a CAN bus interface for data transfer. The handheld control units have LCD displays and usually have a range of programmable deployment functions and alarms.

(2) **Wireless Controls.** Several wireless control options are available. They have a base station and, for larger boats, an antenna for range extension. They use the 2.4GHz ISM band for communications; typical range is about 10ft (3m).

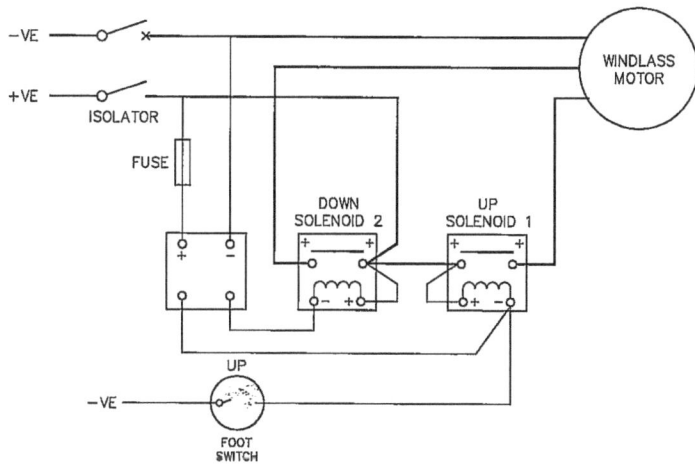

Figure 11-2. Basic Windlass Control Circuit

11.15   Windlass Maintenance. Like many, I have a customized canvas cover made for my windlass. I put this on and tie it down when passage making, as green water constantly over the windlass does nothing for its longevity. It may look pedantic, but it pays off and I recommend it. Following are some maintenance tips.

a. **Windlass and Anchor Cleaning.** To start your maintenance routine, give your windlass a good washdown with fresh water. An additional care step is to use warm soapy water; I use car wash liquid. Be careful about using pressure washers; the seals are not always rated for that. Washing down with fresh water after every outing to rinse off mud, silt, and other debris is a good practice. I have a canvas cover that is fitted over the windlass when not in use or when on passage to keep as much of the salty stuff out as possible. Some boats, including mine, have a deck wash-down pump installed, much like commercial shipping does. It is a smart idea and sure beats dropping a bucket over the side and trying to sluice away the mud. You can apply some automotive wax to the painted casings for added protection.

b. **Windlass Inspection.** Start your abovedeck checks by checking the entire windlass casing. Look for cracks and any mechanical damage, particularly around joints and seals. Check the chain gypsy for wear and tear. If your unit has a gear case oil-level glass, check it for level and oil quality. If the oil is milky, you have had water ingress and a probable seal failure. Strip the unit down, clean thoroughly, repair the leaking seal, and fill with new oil. Some windlasses have aluminum cast housings, and it is a good idea to remove stainless bolts and apply some Durolac, Tef-Gel, or equivalent to threads. Tef-Gel is a corrosion eliminator, anti-seize assembly lubricant. It really does inhibit galvanic corrosion between dissimilar metals. Follow the manufacturer's manual for windlass lubrication requirements, as they are all different. You can buy Lewmar multipurpose grease from West Marine, which I use; they have deck winch greases as well. Harken and Andersen have greases for their deck winches, and they should do okay. I suggest carrying a spare set of seals as part of your kit.

c. **Clutch Cone.** Annual disassembly of the chainwheel and clutch cone system and inspection, cleaning, and lubrication of all moving parts to prevent corrosion is highly recommended. Use the recommended lubricants, such as a lithium grease. Visual inspection for wear and lubrication will ensure good long-term performance. Check for scoring on all mating surfaces. Every windlass is different, so read your manual. Grease shaft threads and chain gypsy where the clutch cones rest.

d. **Deck Foot Switches.** Check the foot switches for water ingress. Check that the hinged covers are not damaged when closed and that they seal properly. Verify that rubber diaphragms are in good condition and that there are no tears or cracks. I have seen some recommend regular spraying of the covers with a UV shield protectant that includes 303 Marine Aerospace Protectant.

e. **Solenoid Box and Motor Terminations.** Check that all terminals are tight and not corroded. Remove, clean, and then retighten. Apply some protection such as petroleum jelly or other approved battery terminal protector or dielectric grease. Looking for corrosion is one thing; also look for signs of arcing, including charring, that indicate low-resistance tracking between the terminals. Wire brushing clean is important, especially treating any corrosion that is close to the terminals. Rust treat, prime, and paint any corroded surfaces. I have used Boeshield T-9 sprayed over everything belowdecks. It is effective on a range of other items as well. Install a desiccant bag inside the control box, or consider inserting a vapor phase corrosion inhibitor.

f. **Windlass Foundations.** Check that underdeck foundations and mounting studs are good. The hold-down bolts often degrade around the holes. If the reinforcing or deck is starting to soften, remove the windlass and repair. Look for signs of movement that indicate loose bolts or degrading foundations. Look for signs of leaks through the mounting bolt or stud holes. Leaks are generally a result of bedding seals. All of the above can be caused by overstressing the windlass. If you have a cleat or kingpost on deck for tying off a rope snubber, check the deck area and bolts as well.

g. **Motor Condition.** Underdeck units are often forgotten about, but they need checking. If a motor casing is starting to corrode, clean and apply rust converter, some primer, and a top coat. This is easily done with a spray can of quality paint.

h. **Anchor Locker.** Check that the locker drain is open and not blocked. I check the locker annually by letting the whole chain rode go and checking the bitter end fastening. Clean the anchor locker, clear any drain blockages, and you are ready to heave the rode back once the windlass has been maintained. A common problem area I have experienced is the chain not coming off the chain gypsy freely and falling down the spurling pipe. The chain heaps up, and when tension comes off the chain, the slack chain also jams the chain gypsy. The chain fall uses gravity, and if it does not fall freely, you get a cone-like pileup. The usual fix is to go below and spread that chain heap—not useful when you're shorthanded. A general rule is to allow at least 12 inches (300mm) below the windlass.

i. **Anchor Chain Rode.** When heaving in, check how the chain falls into the locker. Don't forget to rinse the erode; I use a pressure washer annually. Finally check the swivel and ensure that it rotates freely. If it is corroded or not rotating smoothly, lubricate or replace it, as it will affect anchor performance.

j. **Anchor Rope and Chain Rodes.** Check that the rode splice is in good condition in chain and rope arrangements. Nylon rodes have a tendency to stiffen and shrink at the splice and first chain link. Do a close check of the nylon rode. Examine for signs of fraying, or pigtails and hockles that arise from excessive abrasion.

k. **Windlass Routine Maintenance.** Mechanical equipment does not like to sit idle for long periods. Internal gears need lubrication, which helps combat corrosion from moist, salt-laden air inside. If you can find the time, take the chain off the chain gypsy and then run the windlass for a few minutes; don't forget to run the engine while doing so. Warming the motor helps dry out windings. Rotating bearings along with lubrication is good for your windlass.

11.16   Windlass Troubleshooting. Failures usually manifest at the worst possible times. Some problems can be avoided by simply testing your windlass before departure and arrival. Many causes are as simple as forgetting to switch on the isolator or circuit breaker—been there and done that, more than once. Mechanical problems are common; mainly, they are caused by the anchor chain jamming in the chain gypsy or an inherent poor fleet angle or offset causing the chain not to lead in properly. Bow rollers are often a problem and should be checked for jamming. More than once, a chain stopper has been left engaged and power applied, severely stressing the fitting.

Table 11-3. Anchor Windlass Troubleshooting

Symptom	Probable Fault
Windlass will not operate	Foot switch fault (most common cause) Circuit breaker switched off (next most common) Isolator switch off (very common) Foot switch connection loose Solenoid connection fault Solenoid fault Solenoid fuse blown (if fitted) Motor connection loose Motor fault (sticking brush is common) Motor fault (winding failure) Motor internal thermal cutout tripped Slow blow fuse ruptured
Windlass stalls under load	Excessive load Low battery voltage Motor connection loose Motor fault (brushes sticking)
Windlass operates slowly	Battery terminal loose Excessive load Low battery voltage Motor connection fault (hot) Motor fault (brushes sticking) Battery terminal loose
CB trips during operation	Motor fault Windlass seizing Windlass overloading
Control fuse ruptures	Fault in solenoid Fault in control circuit Fuse fatigue
Solenoid "chatters"	Low voltage Fault in control switch Control switch connection loose Solenoid connection loose

DECK WINCHES AND FURLERS

11.17   Electric Deck Winches and Furlers. Deck winches and furling gear are rapidly becoming electrically powered and are taking a lot of the muscle out of cruising for the shorthanded crew—and older husband-and-partner cruising teams like mine. They are generally treated in the same way as anchor windlass circuits, requiring good circuit protection and correctly sized cables. Electrical loads are considerable, and the following cable sizes are required for 12V systems. Electric winches generally consume far more power than windlasses, and careful power supply planning is required. Deck winches have high starting currents; this spike is instantaneous and has a duration of around 100 milliseconds.

Table 11-4. Deck Winch Cable Rating Table

Cable Length	Current Rating	AWG	Metric	B & S
up to 33ft (10m)	285A	0	54mm²	0
33ft–50ft (10m–15m)	330A	00	68mm²	00
50ft–65ft (15m–20m)	385A	000	85mm²	000

11.18   Electric Deck Winch Protection. Deck winches are becoming more sophisticated. Some factors to consider—and I quote Andersen winches here, as they appear to be leading the pack in technology. Electric motors require the usual electrical protective devices. Andersen deck winches have the following smart protections. In addition to integral protective functions, a deck winch should have a circuit breaker and a slow blow fuse for short-circuit protection, which is to allow for the high start currents.

a.   **Integrated Overload Protection.** The winch controller is preset to stop the deck winch if the maximum pull load is exceeded. The winch can resume operation as soon as the load returns below the preset limit.

b.   **Thermal Overload Protection.** The deck winch electric motor has an internal thermal cutout switch that will disable the winch motor if it overheats and then automatically reset when the temperature falls below the preset limit.

c.   **Reverse Polarity Protection.** The unit has an integrated 5A fuse for motor protection against incorrect cable termination.

d.   **Accidental Start Protection.** The winch is disabled when the push button is inadvertently depressed when the power supply is on.

e.   **Low-Voltage Detection.** The winch is disabled when low battery voltage is detected.

f.   **Continuous Run Time Limit.** The winch is disabled if the continuous run time exceeds 10 minutes.

11.19   Deck Winch Motors. Electric motors powering deck winches are usually brush-type series-wound motors that are sealed against water ingress and are usually mounted belowdecks. They provide high speed at low loads when rapid sheeting in is required and, conversely, low speed at high loads when sheet trimming. Innovative changes include illuminated push buttons to show power status along with hinged covers and water-resistant ratings to IP67. What is unique is the use of LED flashing codes in the push button to indicate performance issues, which assists in rapid troubleshooting. Other recent innovations include the Andersen Compact Motor electric deck winch. The deck winch's electric motor gearbox and controller are integrated into a single compact unit without the need for a control box. The Andersen Compact Motor is a brushless DC motor that is precisely matched to the low-profile planetary gearbox. The result is much lower current consumption than traditional configurations. The Compact Motor electric winches operate at variable speed in proportion to the pressure applied to the push button. The range of status codes appearing on the intelligent push button is considerable; if you invest in one of these deck winches, be sure to learn them. Selden also has electric deck winches. The Selden E40i deck winch is built around an electric motor that is completely integrated in the drum.

11.20   Battery Supply. Battery power is a consideration, and decisions need to be made on where to power a deck winch from. A single winch used for a total of 15 minutes every day will consume up to 40Ah. If you connect the winch to a normal house battery bank, the start-up surges will cause interference on electronics equipment; you should consider reevaluating your battery power arrangement. A dual house bank is possibly the best solution, and then running all your heavy consumers off that, such as deck winch, windlass, and electric toilet. Andersen advises caution when selecting battery types to be used with their Andersen Compact Motors. They advise that using non-approved battery types can result in motor and battery damage. Andersen Compact Motors that were manufactured before 2011 are compatible with flooded-cell and AGM batteries only. They should not be used with gel or lithium-ion batteries. Andersen Compact Motors manufactured after 2011 are compatible with flooded-cell, AGM, and gel batteries. The additional limitation is that they can be used only with lithium-ion batteries manufactured by Super B, Mastervolt, and Victron Energy (MG Energy Systems). These lithium-ion batteries have been compatibility tested. Andersen advises against the operation of a compact electric winch motor at high speed with no drum load, as spinning down can stress the motor and battery.

11.21   Capstans. The capstan is frequently called a drum, rope drum, or warping drum. The primary task of a capstan is for hauling rope, and every ship has one for handling mooring lines. It has similar electrical requirements as a windlass, and they are often integrated with anchor windlasses.

11.22   Electric Furlers. While once found only on bigger boats, these are now on boats as small as 33ft (10m). Manufacturers include Facnor, Bamar, Profurl, Reckmann, Seldén, and Harken. Similar requirements for deck winches and anchor windlasses apply. Furlers for boats down to about 33ft (10m) have power consumptions of 600W to 700W at 12V. They generally have a control box; some have wireless control or push button. Typical rotation speeds are around 40rpm, giving typical furling speeds of a big genoa at around 35 seconds. Seldén Mast has taken the technology a step further with the Furlex Electric. They have systems with a power supply unit (PSU) that uses a DC-to-DC converter in the motor control unit (MCU) to step up voltage from 12V/24V to 42V. This significantly reduces large power cables to the furler and results in a more compact unit. The worm gear drive motor is two-speed and has overload monitoring. Other features include a sleep mode when inoperative to conserve power.

The 42V brushless motor is connected to a gear box and a steel-bronze worm gear. The electric motor and primary gear box are installed in oil in a hermetically sealed inner compartment, and all units are pressure tested during assembly to keep the seawater out. The furler electric motor integrates a computerized controller that enables precise current consumption. When the preset level is reached, it cuts out to protect the system. The motor consumes 10A to 25A at normal load. To prevent overloading, the system has a built-in current limiter, and the torque setting is programmed into a memory chip. The Furlex Electric is used in tandem with the Seldén Power supply and SEL-bus system. The control system has an integral diagnostic system, with LED error codes on both the PSU and MCU. The PSU, MCU, and control buttons interconnect with a CAN bus, and they call this the SEL-Bus. The SEL-Bus system facilitates inter-unit communication, diagnostic, and other functions. In sleep mode, a small amount of current is consumed, so the system must be switched off when alongside. The SEL-Bus is a standalone system and cannot be connected to any other CAN bus system or NMEA 2000.

11.23   Powered Blocks. The Harken FlatWinder powered block is a self-contained, low-profile powered system developed for mainsheet traveler adjustment. This powerful block is easy to use and offers sailors huge benefits in mainsail control, allowing quick rig depowering and delayed reefing when the wind picks up. Like a compact captive winch for the traveler, the FlatWinder is completely self-contained. It operates in both directions, allowing the car to move anywhere on the track while keeping the traveler line off the cockpit floor. FlatWinders can be used for other applications, such as stern platform lifting or foil trimming. The FlatWinder is available with an electric or hydraulic motor. The compact horizontal motor is housed neatly belowdecks and has a maximum working load of 550lb to 1,100lb (250kg to 500kg). When used with 10mm line and a 4:1 purchase, this translates into around 2,205lb (1,000kg) of pull with the FlatWinder 250 or 4,410lb (2,000kg) with the FlatWinder 500. The FlatWinder 250 fits monohulls 50ft to 60ft (15m to 18m) and catamarans 45ft to 50ft (14m to 15m); the FlatWinder 500 fits monohulls 60ft to 80ft (18m to 24m) and catamarans 50ft to 70ft (15m to 20m). The FlatWinder is available in 12V, 24V, or 48V electric or hydraulic, depending on the boat's system. A Harken dual-function control box is included with the electric FlatWinders. This integrated load controller and control box conserves space and, with half as many wires as separate systems, is easier to install. Switches and circuit breakers are not included.

11.24   Electric Boom Mainsail Furlers. There are some on the market, although larger boats use hydraulics. The furler electric drives and associated electronic controls, including overload, are installed at the mast end of the boom. Similar rules regarding high-current wiring apply, along with battery power considerations and cable volt-drop issues. Seldén has similar electric options.

11.25   Keel Lifting and Canting Keel Systems. As some boats have lifting keels and canting keels, it is worth mentioning systems such as the Keel Servant. This is an innovative lifting system for telescopic keels. It is primarily an electromechanical lifting system, so similar electrical load–based calculations for winch power supplies and associated and protection apply. The lifting mechanism comprises a stem that is driven by an electromechanical screw jack. The system is patented and suits those who want both blue water performance and the ability to explore shallow areas. The system has a digital display to show precise keel position.

Bow and Stern Thrusters

12.1 **Thruster Basics.** Bow and stern thrusters are now common on many yachts. They offer increased maneuverability in confined areas such as marinas and are invaluable on boats during higher wind mooring conditions. I work a lot with large offshore vessel thruster systems, both fixed and variable pitch, tunnel and azimuth, and while these are somewhat more complex in terms of control systems, the average small-boat unit is relatively simple. The major suppliers include MaxPower, Lewmar, Sleipner, Vetus, Wesmar, Seamaster, and Quick Spa. There are many factors during both selection and installation that must be considered if thrusters are to be efficient. Thrusters don't really appear on boats much smaller than 45ft (13m). Basic operational factors to consider when using thrusters are as follows:

 a. **Vessel Behavior.** Learn and understand the behavior of the vessel in wind conditions.

 b. **Operation.** Avoid using the thruster unless stopped or moving very slowly.

 c. **Run Times.** Learn the maximum recommended run times and stay within the limits.

 d. **Operating Limits.** Do not use in short bursts. A single 5-second thruster period is more effective than five 1-second bursts; it causes less strain and overheating on the motor, as well as uses less battery power.

 e. **Isolation.** To avoid accidental operation, always switch off the control system when not in use.

12.2 **Thruster Ratings.** Thrusters are specified in terms of the thrust output capability: the thrust denoted in kilogram-force (kgF), not the output of the electric motor in kW or hp, determines effectiveness. Thrust is a result of the power of the electric motor, the propeller shape and dimensions, the speed in revolutions/min, and the tunnel efficiency losses. Typically thrust is in the range of 15kgF to 25kgF per kW of electric motor. With hydraulic systems, required flow rates are also quoted in lt/min or gal/min and pressure ranges, which vary between 2,393psi (165bar) and 4,061psi (279bar), depending on thruster models. Thruster output forces are selected so they equal or exceed the calculated or expected wind thrust forces, or to counter what is called the sail plane effect. The draft affects the drift rate, shallow draft vessels tend to have a greater wind effect than deeper draft ones. Factors that also affect performance include the hydrodynamic efficiency of the tunnel, propeller design and performance, gear leg design, motor design and control efficiency, and power supply capability when operating.

 a. **Wind Pressure.** The wind pressure on a vessel has a quadratic increase with wind speed. The equation often used is pressure P (N/m^2) = $\frac{1}{2}\varrho \times V^2$, where ϱ is the specific mass of air and V is the velocity of air in m/s. The applied force is determined by wind speed, the wind angle, and the lateral wind draft area.

Table 12-1. Wind Force Table

Wind Force Beaufort	Description	Speed m/s	Pressure n/m² (kgf/m²)
4	Moderate Breeze	5.5–7.9	20–40 (2.0–4.1)
5	Fresh Breeze	8.0–10.7	41–74 (4.2–7.5)
6	Strong Breeze	10.8–13.8	75–123 (7.6–12.5)
7	Near Gale	13.9–17.1	124–189 (12.6–19.2)
8	Gale	17.2–20.7	190–276 (19.3–28.2)

b. **Wind Draft.** The wind pressure must be multiplied by the wind draft area to calculate the actual wind force. This is determined by the boat surface area, wind speed, and wind angle. The lateral surface areas and effects vary with the superstructure size and shape, the hull freeboard and, therefore, the hull surface area. Worst case is where the wind is on the beam or at 90 degrees. A factor of 0.75 is used to account for streamlining and a less-than-flat and rectangular shape. The wind force can be substantial, as the infamous Suez Canal incident in 2021 showed. The mega container vessel *Ever Given*, with a very strong crosswind of some 40 knots against the massive stack of containers, veered into the canal bank and blocked the canal for 6 days. Never underestimate the power of wind.

c. **Turning Moment.** The turning moment can be calculated by the multiplication of the wind force by the distance between wind center of effort and the pivot point.

d. **Thrust Force.** Thrust force is an accurate measure of bow thruster effectiveness. It is not the electrical or hydraulic motor HP or kW rating that is important. The nominal thrust force of a thruster comprises parameters such as motor power, propeller shape, and tunnel efficiency losses. Sufficient thrust force to counter wind is calculated by the division of the turning moment by the distance between the bow thruster center and the pivot point.

e. **Torque.** Torque is calculated by multiplying wind force by the distance between the center of effort of the wind and the center of rotation of the boat. The center of effort is dependent on the shape of the deck structure; while nominally center, it may be either slightly forward or aft, depending on the actual location of the accommodation. The center of rotation tends to be more forward than center and is dependent on the underwater hull shape. The wind force torque is calculated by multiplying 50% of boat length by the wind force: Torque (T) = Wind Pressure (P) × Wind Draft (D) × 0.75 × Distance between the bow thruster center and the boat pivot point (one-half vessel length); nominally the pivot point is the stern. Thrusters should be installed as far forward as possible to maximize the leverage effect around the vessel's pivot point. Follow the manufacturer's recommendations and instructions.

12.3   Thruster Types. Thruster types are dependent on vessel types, sizes, and budgets. Many thrusters are single-propeller types, but there are also twin tandem propeller and twin counter-rotating types, some with a five-blade propeller design for optimum thrust outputs. The latter arrangements are to increase efficiency and reduce cavitation. Thrusters can be DC electric, AC electric, or hydraulic. Lewmar is using composites on gearbox housings, as well as anode-free designs. Some are producing skewed blades for noise reduction and reduced cavitation.

    a.   **Retractable Thrusters.** Swing retractable thrusters have the advantage of no drag; they are, however, more complicated, and are available in electric and hydraulic. Some used to use a rotating lead screw to activate the leg; others used an innovative folding system. Interlock limit switches are often installed in the control circuit for actuator position indication. These are prone to failure and must be maintained. Lewmar now has a patented low-pivot mechanism that minimizes the internal space requirements. Another feature of this is automatic retraction after 5 minutes of nonuse. Others have sensors that indicate if a door is partially open, auto-retract if the boat speed is too high, and lock out when partially retracted; all require maintenance and monitoring.

    b.   **Transverse Tunnel.** The tunnel thruster is the simplest and most common arrangement and is less complex in terms of mechanical components. Tunnels will cause some drag to the boat, although not enough to bother most boat owners.

12.4   Thruster Power Outputs. The table shows typical figures and is a good general guide to all fixed-pitch thrusters. Required thruster force (F) is calculated by dividing calculated torque (T) in Newton meters (Nm) by the distance between the bow thruster center and the boat pivot point in meters.

Table 12-2. Thruster Outputs Table

Thrust (kg/lb)	Boat Size ft (m)	Power (kW/hp)	Voltage (volts)	Battery (min CCA)
35/77	22–32 (6.7–9.75)	2.2/3	12	300
55/121	28–40 (8.5–12.1)	3.1/4	12/24	350/175
75/175	35–50 (10.7–15.2)	4.4/6	12/24	50/250
95/209	42–58 (12.8–17.7)	6/8	12/24	700/350
155/341	50–70 (15.2–21.3)	8/10.7	24	600
220/484	60–84 (18.3–25.6)	11.2/15	24	700
285/627	74–100 (22.5–30.5)	15/20	48	2 × 450 (24V)

12.5   **Thruster Power Supply.** The efficiency of thruster motors is directly related to battery supply voltage levels, the available battery capacity, and the voltage drop in the supply cables during operation. Slow blow fuses are used, as the fluctuating loads must not cause fuse failure in normal service and ratings should cope for up to 5 minutes. The table is a general guide only; all circuits must be measured and the appropriate cable size installed with the required protection.

a.   **Battery Power.** A common question is whether to use a separate battery located forward or not. This depends a great deal on boat size and practical space considerations. In smaller boats, it is better to have the supply run off an increased engine-start battery bank. This means an additional heavy cable is run forward to the bow thruster, but it is much simpler for charging purposes. The main engine is always operating when using thrusters, so the alternator provides additional power to the battery, with less battery drain; it is worth considering a high-output alternator. In larger vessels, a separate battery bank is preferable and additional battery charging must be considered. A major cause of reduced thrust is inadequate power availability from the batteries at full load; maintaining them in optimum condition is essential. If the battery suffers a major voltage droop, the thruster will suffer a reduction in thrust output. Where separate batteries are to be installed, my preference is for an AGM battery bank, which can deliver the required current, requires no maintenance, and has a high charge-acceptance rate. A separate high-output alternator is recommended, and the 24V option should be considered.

b.   **Supply Cables.** The power supply cables must be rated for the maximum current requirements of the motor and allow for voltage drop, which ideally should not exceed 5% at full load. Voltage drop is a major cause of reduced thrust.

Table 12-3. Bow Thruster Electrical Installations

Thrust (kg/lb)	Boat Size ft (m)	Voltage	Cable Size AWG/mm	Slow Blow Fuse Size
35/77	22–32 (10)	12	2/35mm²	125A
55/121	28–40 (11)	12/24	00/70mm²	250A
75/175	35–50 (12–13)	12/24	0/120mm²	355A
95/209	42–58 (12–13)	12/24	0/120mm²	355A
155/341	50–70 (13–19)	24	0/100mm²	355A
220/484	60–84 (18–30)	24	0/120mm²	425A
285/627	74–100 (25–35)	48	0/120mm²	425A

12.6 **Thruster Control.** In general, a joystick control lever is used on the latest thruster products. Electric motors generate heat, and thruster motors are limited by temperature rise. Generally, most thrusters are rated at between 3 to 5 minutes at full continuous output. Power supplies are critical, as high currents cause large voltage drops. A 3kW motor can draw 250A or more at initial starting, and less power means loss of thrust. Some thrusters offer AC motor power, which is a viable solution if you have a good generator. They are able to get good speed control, ramp time, and current control using a variable frequency drive (VFD) and joystick control, across the entire speed range. This results in an efficiency of 85% to 90%. Typical AC control is simple ON and OFF, and you get a start-up current of around seven to eight times nominal.

a. **Proportional Control.** This is another name for variable speed control. In DC electrical motors, new speed control uses high-efficiency permanent magnet brushless motors and PWM speed control. In hydraulic systems, the hydraulic pressure is controlled through a proportional valve. The big advantage is that you have greater battery power supply capacity when running less than full power; typically, run time is 3 minutes at that rate.

b. **Overheat Protection.** Once this limit is reached, thrusters are protected against damage by a winding embedded thermal cutout with heat levels rising to around 212°F (100°C). Depending on the amount of heat remaining within the winding, the time to reset the cutout varies from 1 to 5 minutes. To ensure availability, the proper use is about 10 to 30 seconds per maneuver to prevent cutout.

c. **Drive Reversal Protection.** Some thrusters incorporate electronic time-lapse protection against sudden drive reversal.

d. **Protection.** Motors and cables are provided with slow blow fuses for short circuit, as they must withstand the high instantaneous starting currents. Overload protection is provided by circuit breakers.

e. **Motor Testing.** Under no circumstances operate the motor out of water or with load off, coupling disconnected, or propeller off. The series-wound DC electric motor will accelerate very fast to a point where it will be seriously damaged.

12.7 **Thruster Maintenance.** Servicing and maintenance requirements are relatively simple. Some thrusters, such as the composite leg types, have sealed lubricated bearings and have no anodes or oil servicing. Check the manual for your installed thruster.

a. **Anodes.** Anodes should be inspected regularly, at least every 12 months, and new anodes installed on the propeller shaft should be secured using Loctite. Where installed on the leg, the same applies. Gear cases are usually bronze; however, the protection is for casing, propeller blades, and shafting.

b. **Oil Tanks.** The gear case oil header tank should be checked and topped up prior to each trip; Side-power specifies EP90 oil. Oil is used to overpressurize the leg and prevent water ingress. Oil consumption usually indicates a leaking seal. If installation is properly done, there are usually no problems; however, if the tank is installed at the wrong height, there will not be enough oil overpressure, which leads to water ingress. I have seen this mistake even in commercial installations. Check the oil tube to ensure there are no kinks or loops that can cause air locks or affect the oil pressure and flow. Oil should be changed every 2 years, or in accordance with the manufacturer's instructions, timed with slipping.

c.   **Antifoulant.** Coat only the gear case and propellers, not the seals, anodes, or propeller shafts. Use antifoulant designed for propellers such as Propspeed.

d.   **Electric Motor.** Check and tighten motor hold-down bolts every year. Vacuum out any carbon brush dust, and check the condition of the commutator and brush gear. Check and tighten electrical motor connections and directional solenoids.

12.8   Thruster Troubleshooting. The following are the most common faults that will be experienced. Read your thruster manual and always check the basics first.

a.   **Thruster Will Not Start.** Look at the following:

   (1)   **Power Supplies.** Check the obvious causes, such as isolator open, circuit breaker off, or slow blow fuse ruptured. Check that voltage levels at the thruster are correct; at no load it should be a minimum of 12.7V or 25.4V. If it is lower, check the battery voltage first; if that's acceptable, then check the terminations. Check voltage when trying to run the thruster; if voltage has dropped lower than 8.5V, the probable cause is the battery's condition and it cannot deliver the required power. Most newer generation thrusters will cut out when the battery is at 10.5V or below.

   (2)   **Control Systems.** If the solenoids do not operate, a control signal is probably absent; solenoids rarely fail. Check the voltage at the solenoid to confirm this. Check the power supply and fuse protection to the control panel; if it's good, check control cable connections and control panel outputs. Retractable thrusters may have interlocks on the retract systems; check that the thruster is down completely and interlock limit switches are operating. Feedback sensors are also on thrusters and may not permit a start.

   (3)   **Electric Motor.** If voltages are correct at the electric motor, the thermal cutout switch may be faulty or the motor brushes may be sticking. Isolate the power; open and manually check that brushes are moving freely in brush holders.

b.   **Reduced Thrust.** The most common cause is reduced voltages caused by battery failure or loose connections causing voltage drops. The brush gear can cause reduced thrust problems; the brushes and commutator should be checked. If electric items are good, the thruster may be fouled with marine growth or with plastic, fishing nets, or other marine debris.

c.   **Thruster Failure.** If the thruster stops in service, first check protection equipment such as fuses, thermal cutouts, and circuit breakers. If the motor is operating with no thrust, check that shear pins and flexible couplings have not separated due to the propeller jamming on debris.

12.9   Hydraulic Thrusters. The major advantages of hydraulic thrusters are that they have no restrictions on use and are continuously rated.

a.   **Hydraulic Power.** The hydraulic system is usually electrically powered, although some main engines have a belt-driven electromagnetic hydraulic pump. Oil is suctioned from the reservoir by the hydraulic pump through the oil filter and an oil cooler where installed. A typical nominal oil temperature value is 140°F (60°C). The cooler may be part of an existing seawater system on the engine or have a separate pump.

b.   **Hydraulic Basics.** The pressurized oil then passes through a directional control valve, which is activated from the thruster control station, either mechanically or electrically. The oil then goes to the hydraulic motor, which is coupled to the thruster. Speed control, where installed, of the hydraulic motor is by means of a proportional control valve, although some use variable displacement load sensing hydraulic pumps. Oil then returns to the oil reservoir.

c.   **Protection.** The system may have overload protection, which consists of a pressure safety valve (PSV) that vents back to the oil reservoir. In some installations, the system may power thrusters and generator, and each function will have an electrically activated main solenoid valve. In variable pitch control systems, both on propeller and thrusters, or azimuth thruster, hydraulic control is used; hydraulic system reliability is dependent on good maintenance practices. Solenoid electrical connections should be checked regularly and be dry and secure. See the chapter on hydraulics for maintenance and troubleshooting.

DC Motors and Equipment

13.1   DC Motor Introduction. Most installed pumps and machinery have DC motors installed, and most are now maintenance free. Where larger motors are in use, such as thrusters, windlasses, deck winches, refrigeration compressor drives, and starter motors, the question of proper maintenance becomes paramount. My commercial seagoing career started on 220VDC systems, and the first lesson was that DC motor performance and reliability is directly related to effective preventive maintenance. Many would think that DC motors have gone the way of the dinosaur, but they are still very much in use for many reasons. DC motors do have some advantages over AC motors. They have higher starting torque but need to be started under load. Never test run a DC motor without load, as it will run away and burn out. Another significant advantage of DC motors is that the speed-torque curve is more linear than for AC motors.

13.2   DC Motors. DC motors come in several types, including the series wound, the split series wound, shunt wound, and compound wound. DC series-wound motors are often used, as they develop a very large starting torque and can be run at low speeds. In this motor type the whole armature current flows through the field winding. In shunt DC motors, the shunt field winding is parallel to the armature winding and not the series field winding. Shunt DC motors have good speed regulation because the shunt field can be excited separately from the armature windings. Compound DC motors have both series and shunt field windings and have good starting torque. A DC motor comprises several main components, including the stator, rotor, poles, armature windings, field windings, yoke, and commutator.

13.3   Permanent Magnet DC Motors. Traditional brush-type DC motors in some applications are slowly being replaced by new developments in permanent magnet (PM) motor technologies. These are brushless, more compact, and maintenance free. In a DC motor, a voltage is applied through carbon brushes and the commutator to the armature windings; in permanent magnet motors, the stator is the electromagnet. The permanent magnet is used to supply the field flux. Advantages are very good starting torque and good speed regulation; however, they do have limitations on the load they can drive. In addition, motor torque is often limited to about 150% of rated torque to prevent possible demagnetization of the permanent magnets. These motors are being used by deck winch manufacturers such as Andersen with their Compact Motor electric winch. (See chapter 11 on windlasses and deck winches.) They are being used in 48V electric propulsion systems.

13.4   Pancake Motors. These increasingly common motors are called brushed pancake motors, printed circuit armature, flat armature motors, or disc armature motors. They are very compact and consist of a rotor disc that is made from a nonmagnetic and nonconductive material. The armature winding and the commutator are printed with copper on both sides of the disc. The disc armature is located between two sets of permanent magnets that are mounted on the ferromagnetic plates. Magnets are usually of two types: ferrite (AlNiCo) or high-power neodymium NeFeB (rare earth). Brushes are then placed around the inner periphery. The motor has low inertia, and the subsequent ratio of torque and inertia is very high. As the rotor does not have any iron, the inductance within the armature is low; the advantage is that there is a reduction in sparking and big increases in brush life. The low sparking results in much improved EMI. As there is no iron, thermal characteristics are much improved, with reduced cooling requirements. They do have some important operational features, and to get rapid acceleration and deceleration, the printed motor can cope with high overload currents and a peak current of around ten times the continuous rating. The nonmagnetic rotor also eliminates cogging torque. These motor types are the only type that do not have torque ripple. Torque ripple is defined as a periodic fluctuation, or increase or decrease, in output torque as the motor shaft rotates, creating torque pulsation.

13.5 DC Motor Inspection. Routine inspection of DC motors on thrusters, furlers, starters, windlasses, and propulsion systems consists of rigorous and thorough inspection. This should be done every 6 to 12 months.

    a.   **Brush Gear.** Inspect the carbon brushes for chipping, grooving, uneven wear, and loose or frayed wire connectors. Compare the length of brushes and replace the set of brushes if worn. Check that brushes move freely within the brush holders. Check that spring pressure is correct by simply pulling the brush back and snapping it against the commutator. Ensure that the wire brush connectors or pigtails are clear of any moving parts. Check that brush pigtail connections are tight. Inspect the commutator.

    b.   **Electrical Terminations.** Inspect the field winding connections. Inspect electrical connections and mechanical fasteners inside the motor and connection box. Inspect the motor interior for condensation, and access covers and gaskets. Most electric deck winches and anchor windlasses use a DC series-wound motor.

13.6 Commutators. The condition of a DC motor can be determined by observing the condition of the commutator surface.

    a.   **Good Commutator Surfaces.** Good commutator conditions can be found by observing the copper surface patina or surface markings. The skin that develops comprises oxide and graphite. A light tan film indicates a machine is performing correctly. A mottled surface condition, characterized by random film patterns on commutator segments, is normal. Slot bar marking has a film that is slightly darker and occurs in a definite pattern that relates to the number of conductors per slot. A heavy film condition is acceptable if uniformly found over the entire commutator.

    b.   **Commutator Deterioration Signs.** The following signs indicate degrading motor performance and require attention:

        (1)   **Streaking.** Commutator surface streaking that indicates the start of metal transfer from commutator to brush. Light brush pressures, a light electrical load, an abrasive or porous brush in use, or contamination can cause the condition by creating dust.

(2) **Threading.** Fine line threading on the commutator surface happens when an excessive quantity of copper transfers to the brushes. If severe, brush wear will be rapid and the commutator will require resurfacing. Caused by light electrical loads, light brush pressures, porous brushes, or contamination.

(3) **Grooving.** Grooves in the brush path are caused by abrasive brushes and contamination.

(4) **Copper Drag.** A buildup of copper material at the trailing edge of a commutator segment is caused by light brush pressure, vibration, abrasive brushes, and contamination.

(5) **Pitch Bar Marking.** Characterized by low or burn spots on the commutator surface. The number of marks equates to all or half the number of poles. The condition is caused by poor armature connections, unbalanced shunt fields, vibration, or abrasive brushes.

(6) **Heavy Slot Bar Marking.** Characterized by the etching of commutator segment trailing edges. The pattern relates to the number of conductors per slot. The condition is caused by poor electrical adjustment, electrical overloads, or contamination.

13.7 DC Motor Cleaning. Regular cleaning of DC motors is essential for reliability and long service life. Use a portable vacuum cleaner and soft-bristle brush to dislodge dust and other material. Don't use air pressure to clean. Disconnect the power before starting.

a. **Brush Cleaning.** Clean the brush boxes and brushes, and ensure that brushes move freely within the brush holders. Clean the accessible field and armature windings.

b. **Commutator Cleaning.** Clean the commutator and commutator risers. Use a small soft brush and clean any buildup of dust in between commutator segments. Buildups short out the insulation between the commutator segments. Do not polish or clean the commutator with emery cloth.

13.8 DC Motor Carbon Brushes. The carbon brush must have good commutating and contact characteristics, good mechanical strength and wear properties, a resistance to sparking, and a suitable contact voltage drop. Brushes are normally manufactured from hard carbon and either natural or electrical graphite. Metal graphite brushes are used for slip ring applications, as they have lower contact voltage drops. Abnormal brush wear may be caused by very low humidity, abrasive dust, intermittent loads, a commutator surface without a properly developed skin, incorrect brush grades, jammed brushes, or excess sparking. Sparking is caused by poor machine commutation, which may be due to wrong brush types and grades, incorrect brush pressures, badly undercut commutators, excessive vibration, or overloading. Typical brush pressure is in the range of 170–210g/cm². If the brush pressure is too low, the contact voltage drop will increase and brush wear will increase due to burning. If the pressure is too high, there will be increased friction and increased mechanical wear. Check the pressure on all brushes using a small spring balance. The wear surface of the brush can indicate performance. A very shiny surface indicates excessive friction or brush movement. A brush should always be semi-bright and have a surface covered with small pores. If brush replacement is required, it must be done correctly or risk damaging the commutator.

a.   **Brushes.** Ensure that you use the correct brush for the machine.

b.   **Brush Bedding.** Use a very fine grade strip of emery cloth slightly wider than the brush. Reverse it so the abrasive surface is under the brush. Move it back and forth around the commutator so that the carbon brush is shaped to that of the commutator.

c.   **Dust Extraction.** Use a vacuum cleaner to extract all the dust out of the machine to prevent accumulations of abrasive materials. Under no circumstances use emery cloth; this scratches the commutator surface and the conductive particles lodge in the commutator segments, causing shorts and arcing.

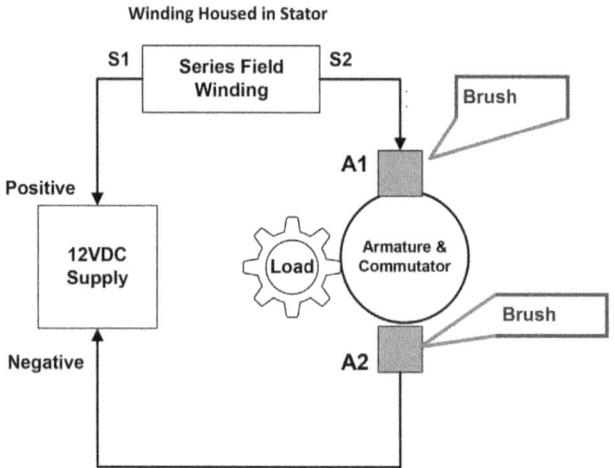

Figure 13-1. Series DC Motor Diagram

13.9   Motor and Circuit Control. Most motor circuits have small relays in control applications and contactors for supplying high power to equipment; the windlass is a good example. Both have a coil for operation and to close or open contacts. If you suspect that a relay or contactor is faulty, check the voltage at the coil; if voltage is applied and the coil does not pull in, the coil has probably failed. Generally, the outer coil covering will indicate heat damage, but not in all cases. In some cases, the mechanical armature part is the problem and may be jammed. In larger motor contactors, the current-carrying contacts may have deteriorated and pitted so that poor or no contact is made when the coil pulls in, or the contacts weld together. Most contactors have auxiliary relay contacts that are used to hold in the contactor. In some cases, the contacts on these get pitted or damaged, preventing the contactor from staying in.

13.10   DC Motor Troubleshooting. The DC motor generally gives many warning signs of decreasing performance. The following table gives common symptoms and causes.

Table 13-1. DC Motor Troubleshooting

Symptom	Probable Fault
Windings overheating	Motor overloading Run time excessive Ventilation insufficient High ambient temperature
Excessive commutator sparking	Motor overloading Oil on commutator Excessive brush dust buildup Brushes sticking Brush pressure too low Brushes worn Commutator dirty Commutator damaged
Motor overloading	Excessive mechanical load Bearings binding; lubrication failure
Excessive current draw	Excessive mechanical load or stalled load Bearings binding Valve closed (if a pump load) Electrical connection fault; high resistance
Excess noise and vibration	Bearing failure Motor hold-down bolts loose Motor load transmitting vibration Misaligned coupling; coupling will be hot Brushes bouncing on commutator Coupling damaged and out of balance

DC EQUIPMENT

13.11   Auxiliary DC Equipment. Many larger motor and sailing boats have several auxiliary electrical systems and equipment. The basics are covered briefly as follows.

   a.   **Horns.** Many power and trawler yachts have air horns, and larger vessels in survey must meet IMO COLREGs' standards. These are exposed to weather, so corrosion of parts and connection failures due to water ingress are common. Horns may have an integral motor driving a rotary air compressor. Like all rotating equipment, these can seize up. Many horns are air powered and electromagnetically operated, and the air solenoid valves often corrode and seize. Vessels larger than 75m must have horns with outputs in the 130Hz to 350Hz, 135dB range in single-, two-, or three-tone outputs. Units may have integral heating for cold climate use.

b. **Davits, Passerelles, and Gangways.** Many sailboats, particularly those in the Mediterranean, have powered systems using DC motors through gearboxes or hydraulics. Because they are exposed to the elements, motorized units often seize. Regular operation and wrapping with impregnated grease tape such as that from Denso or other sealing methods are good preventive measures for motors and control boxes. Some davit systems also have microswitches, which should be inspected regularly. If they jam and don't operate when recovering a dinghy, you may end up damaging or breaking the fall wires and damaging the boat. Outside control stations require regular inspection.

c. **Electric Line Pullers.** I like to drop a crab or lobster pot when opportunity permits. Major manufacturers include Brutus, EZ-Pull, Powerwinch, Scotty, Ace Trac, and Discovery Bay. These are 12V powered and can be stowed if required. Some are rated to haul around 110lb (50kg) and can draw up from 15A to 35A, dependent on load. Haul speeds can be as high as 140ft/min (40m/min).

d. **Electric Downriggers.** I like to use anything that improves my fishing outcomes and catching edible and legal-size fish. Although common on offshore fishing craft, I do run across sailing yachts and motorboats with electric motor–powered downriggers installed. Manual units come in quite powerful electrical versions that allow 10lb (5kg) leads. These units can have ion emission systems to attract fish and automatic bottom-tracking connected with the fish-finder. They allow you to drop a lead close to the bottom, along with your favorite bait or lure. The selection of downriggers is based on water depth. The deeper you are fishing, the greater the lead, and this requires larger downriggers. Smaller boats may use portable units and larger ones, electric. Manufacturers of electric downriggers include Penn, Big Jon, Cannon, Walker, and Scotty. Automatic units allow single-hand operation, which allows fast bait presentation and retrieval. Some have automation, with integral transducer and bottom-tracking mode. This allows bottom contour following and maintaining the bait at a preset distance above it. These units have a cycle function that will automatically jig the bait up and down. You need to consider the current draw of a full-sized electric downrigger and factor it into your power consumption calculations. The motor powers a gear drive, and some have belt drives. These have retrieval rates of around 20ft/min, with new ones having a capability of 100ft/min and 200ft/min. This depends on the weights and maintaining a stable voltage at the motor. If that droops or a voltage drop occurs, the power also falls. The motors do not run continuously. Retrieval may take several minutes, with average consumption being 3Ah to 5Ah of battery capacity. The typical power consumption draws around 6A to 7A with a 10lb (5kg) weight attached. On start-up this can momentarily jump to about 20A to 25A. If you get snagged on something, the downrigger can stall out and require around 30A. Cable ratings should be rated for 30A with a slow blow fuse or circuit breaker. Typical cable size is

12AWG or 16AWG, depending on distances from the battery and volt drop. Maintaining electric downriggers is all about keeping the units clean and dry. Operating problems are usually caused by loose switch connections, which cause high resistance and loss of power.

13.12   Window Systems. Many motor- and powerboats, along with trawler yachts and large yachts with pilot houses, have windshield wiper systems and washer systems; some have heating systems.

a.   **Windshield Wipers.** Wipers tend to have high failure rates due to corrosion and seizure of the pantograph arm mechanisms. Marine-grade systems, such as Exalto (imtra.com), have marinized components with sealed heavy-duty motor units. They come in various configurations—either fan or parallel type, single or dual wiper motors. Choose the inside motor versions if at all possible. Most have variable speed functions, and these are generally reliable. On DC units, this is a variable resistance. Some units have the option of a heater to prevent freezing. Power supplies are usually 12V or 24V. The typical power consumptions are in the range of 26W to 96W depending on the motor torque, which can range from 23 up to 80 Newton-meters (Nm). Some systems can cause RFI interference problems, and these should always be checked for. Heaters should be routinely operated; moisture ingress is the most common problem. The rubber blade inserts can suffer degradation due to ultraviolet exposure and should be checked for deterioration and replaced if necessary to maintain efficiency. Protection is typically a 3A to 4A fuse; spares should be carried.

b.   **Wiper Speed Control.** Wiper systems usually have a variety of control options, ranging from simple push-pull switches to touch pad control that uses relays for switching. The basic systems offer two-speed operation for single or dual motors, and self-parking of arms when switched off. More sophisticated systems also incorporate intermittent wipe action; electronic controlled systems offer more intermittent speed options and synchronization of multiple wipers. The function relays are generally reliable, and where microprocessor-controlled synchronization units are installed, they are generally resilient to voltage spikes; however, failure can be often attributed to such faults.

c.   **Clear-View Screens.** Rotating clear-view screens are popular on many boats, particularly oceangoing work boats prone to heavy-spray sea conditions, ice, and snow, which wipers cannot cope with. The optically ground armored glass screens are dynamically balanced and rotate at high speeds, in the range of 1,600rpm to 1,900rpm, and centrifuge the water off instantly. Nominal speeds take approximately 25 to 50 seconds to ramp up. Units may have a center-mounted drive motor or a belt-driven type. The rotating glass is relatively heavy, to provide inertia and maintain the high speeds. Typical power consumption is 2.5A to 3.5A for 12V systems. Some units may have an optional heating element and demisting; these typically consume 40W at 12V, 60W to 150W for 24VDC.

Marine Diesel Engines

The marine diesel engine is the main propulsion source for most cruising and sailing yachts, trawler yachts, and powerboats. In the marine diesel space, the most common are Volvo Penta, Yanmar, Nanni, Detroit, Westerbeke, Beta Marine, Steyr, Solé Diesel, Caterpillar, Perkins, Mercedes, BMW, Cummins, GM, Lombardini (Kohler), and Vetus Diesel. Most diesel engines on smaller yachts are marinized engines, often from proven industrial and agricultural diesel engines, including Kubota, Mitsubishi, Toyota, and Shibaura. Marine diesel engines are now required to comply with stringent emissions standards; these include the US EPA Domestic Marine Tier 2 and Tier 3 standards and the European Recreational Craft Directive (RCD). It is important to note that electric propulsion is now making inroads into this space, and, although small incremental steps at this time, will be significant in the decade to come.

14.1   Diesel Engine Basics. Diesel engines work on the principle of compression-ignition, where air is compressed to a point that fuel combustion will occur spontaneously when fuel is introduced. This chapter is about four-stroke compression-ignition engines. The four-stroke cycle comprises the air intake, compression, power, and exhaust cycles.

a.   **Air Intake (Suction) Stroke.** At top dead center (TDC) the inlet valves open and the air required for fuel combustion is drawn in via the air filter or the turbocharger, if installed, as the piston moves downward. At the bottom of the stroke, bottom dead center (BDC), the inlet valves close.

b.   **Compression Stroke.** In the compression stroke, the piston moves upward to compress the air and raises the temperature within the engine cylinder, typically to around 1,025°F (550°C). At just before TDC, fuel injection takes place and then, after an interval, ceases.

c.   **Power Stroke.** When the fuel is injected via the fuel injectors, it ignites spontaneously; once ignited, increased pressure is then generated in the cylinder, driving the piston down to BDC.

d.   **Exhaust Stroke.** At BDC the exhaust valves open to expel the exhaust gases; at the end of the stroke, the valves close at TDC.

14.2   Marine Diesel Subsystems. The principal subsystems of marine diesel engine are summarized below. Each should be looked at as a separate subsystem, and each requires different maintenance and troubleshooting tasks. Eventually the sum of all these subsystems provides the motive power for your boat, along with power generation.

a.   **Main Engine Components.** The engine comprises the block, the cylinder head, the head gasket, and the top cover. The block supports the crankshaft, camshaft, cam followers, cylinders, and pistons. Inlet manifolds and exhaust manifolds are attached. The top end comprises valve lifters, push rods, rocker arms, tappets, timing gears, valve springs, and valves. The piston assembly comprises components that include the piston pin, upper and lower compression rings, and oil rings. The bottom end of the assembly comprises the connecting rods, bolts and nuts, main or rod bearings, and rod cap.

b.   **Fuel System.** The newer engine technologies have electronic control with sensors and timing, and design innovations have improved pump, nozzles, and unit injectors. The diesel fuel injection system causes the most problems, often through a lack of maintenance or use of low quality fuel. This can result in poor fuel economy, excessive exhaust smoke, excessive carbon accumulations within the combustion chamber, and shortened engine life. Poor-quality fuels can cause starting issues and engine knocking. Engines have high injection pressures, and fuel pumps and injectors have very close tolerances for optimum fuel atomization and penetration. The diesel fuel must have lubrication capabilities to prevent excess wear and resultant damage.

c.   **Exhaust System.** A cause of major issues when improperly installed, particularly wet exhausts. It is a very reliable indicator of engine performance.

d.   **Cooling System.** From the seawater system to the closed freshwater system and heat exchanger, such systems require regular monitoring and maintenance. This includes strainers, water pumps, thermostats, and other components. The correct temperature-regulating function of the thermostat is important to efficient operation. A diesel engine running cold is almost as bad as an engine running too hot.

e.   **Lubrication System.** The engine lubrication system is crucial to the efficiency, cooling, and longevity of the engine. The oil pump supplies the oil from the sump through block passages to all the engine's working parts.

f.   **Air System.** The air system is crucial to good performance, and unlike their land-based cousins, clogged air cleaners and filters are not the prime cause of problems afloat. However, problems can arise due to lack of fresh air supply and other issues.

g.   **Starting System.** Probably the most common marine diesel engine system issue is the starting system. There are a number of factors to consider and these include the starter motor and solenoid performance.

h.   **Instrumentation System.** Normally a system of meters for monitoring the basics, from engine speed to oil pressure and water temperature, with associated alarms for out-of-tolerance conditions.

i.   **Charging System.** Most marine diesel engines rely on the engine charging system to recharge their own dedicated start battery and house power battery. Alternator charging is covered in chapter 5.

j.   **Engine Mounts.** Often forgotten, they do fatigue and need inspection. They absorb and dampen engine vibrations. Some shock-absorbing mounts, such as the VETUS hydro-dampener units, significantly reduce noise and vibration.

14.3   Combustion Efficiency. Combustion efficiency is a measure of compression ratio, and is highly dependent on the proper control of both ignition and fuel combustion. Factors controlling combustion are air quantity, fuel-air mixture, and compression temperature and pressure. The important factor is the delay period between the fuel injection and the ignition. Both engine design and fuel quality are crucial to this. It affects engine performance, cold-start characteristics, warm-up times, engine power output, engine noise, and the level of exhaust emissions. Short ignition delays do not generally cause problems; however, long ignition delays will allow fuel to accumulate in the cylinder prior to ignition. When this occurs, the cylinder pressure will rise rapidly, with incomplete and inefficient combustion. Ignition delay periods are mainly determined by fuel quality. Injection delay is the time required for the injection pump to build up pressure exceeding the opening pressure of the injector. This is dependent on fuel quality, compression temperature, compression pressure, and fuel droplet size. The ignition delay must be as short as possible. When ignition starts, combustion occurs quickly and pressure increases rapidly. When delay is excessive, pressure rise is also fast, causing engine "knocking." During the final part of the combustion process, the final fuel is burned off. When the temperatures and pressure are high, the fuel droplets ignite immediately. Good combustion efficiency ceases at this point.

14.4   Engine Efficiency. Efficiency and losses in the process relate to fuel energy, incomplete combustion, and air/fuel mixing less than 100% so that unburned fuel exits in the exhaust gases. Losses are typically around 65% and are higher at low-load than full-load operation. Sources of losses include leakage through piston rings, friction, heat through combustion chamber walls, incomplete expansion, thermal loss via exhaust gasses, and incomplete combustion. The air temperature, air humidity, and air pressure affect the power ratings of an engine. In many engine spaces, engine power is reduced due to lack of air. The higher the engine space temperature rises, the greater the efficiency loss, around 10% to 15%. Efficiency can be improved by installing an air inlet ducted to the engine air inlet, which provides cool external air. This is in addition to the normal machinery space ventilation inlets and outlets or exhaust fans.

14.5   Diesel Fuel Systems. Combustion is significantly influenced by the design of the fuel injection system. Fuel is drawn from the fuel tank, through pre-filters or separators and engine fuel filter, to the fuel pump. The pressurized fuel is then supplied to the injectors. Fuel is injected at high pressure. Although fuel is injected as a liquid, efficiency depends on optimal vaporization, which requires thorough and rapid mixing of both the hot air and the fuel vapor. There are several distinct fuel injection methods. These are the constant pressure system, common rail injection, individual pump system, multiple plunger system, in-line (jerk) system, accumulator system, pressure-time injection system, and the distributor pump system.

a.   **Constant-Pressure Common Rail System.** In the constant-pressure common rail system, the fuel is maintained at a constant pressure in the manifold. The manifold is connected to cam-actuated nozzles or a distributor, the timing valve, and pressure-activated injector nozzles. The pressure is maintained by compressing the diesel fuel using a pump and supplying fuel after each injection. Fuel is supplied from an accumulator and pressure regulating valve, which may be a governor or manually controlled. The quantity of fuel delivered at each injection is controlled by injection pressure, the injector nozzle orifice area, and the time at which the nozzle valve lifts.

b.   **Accumulator System.** The accumulator system uses upper and lower plungers in a common bore. The lower plunger is driven by an eccentric cam; the upper plunger is spring-loaded. As the bottom plunger is forced up, the fuel between the plungers is pressurized by the spring force applied to the top spring. Fuel continues to pressurize until a delivery groove in the lower plunger indexes with the outlet passage. This pressurized fuel is then injected and continues until the upper spring forces the plunger downward and closes the outlet passage.

c.   **Jerk (In-Line) Pump System.** The jerk pump system is the most common system in use for fuel pressurization, metering, and timing. These are constructed from separate pump and plunger units that are connected inline, with one per cylinder. A camshaft activates the plungers and controls injection. The jerk pump system is the basis for distributor pumps and unit injectors. Engine makers such as Caterpillar use hydraulically actuated, electronic controlled, unit injector (HEUIs), which have fuel injection pressures in the range 18,000psi to 24,000psi (1,200bar to 1,650bar).

d.   **Electronic Injection.** Each injector has a solenoid valve that controls the quantity of fuel to the injector. The gear-driven axial pump raises the fuel oil pressure for proper injection operation. The engine control module (ECM) transmits a signal to an injection pressure control valve, or dump valve, and a signal to each injector solenoid valve to inject the fuel. The control valve controls the injection pump outlet pressure by dumping oil back. Yanmar common rail technology utilizes a digitally controlled, high-pressure fuel injection and sensor system that reduces emissions. The system has multiple engine sensors that monitor operating parameters back to the engine control unit (ECU). This includes throttle position, various temperatures, oil, water, common rail and other pressures, and intake air pressures, as well as crank and cam positions. The ECU uses this data to regulate fuel from the pump into the high-pressure fuel rail and digitally-activated injectors. The ECU precisely controls fuel injection.

14.6   Injection Methods. There are two principal injection methods in use by marine diesel engine manufacturers, and these vary with engine size.

a.   **Direct Injection (DI).** This method is the most common. Air is drawn in through special-shaped inlet ports. This causes the air to swirl during compression, assisted by the piston crown. These engine types are more efficient, and deliver high power over small time periods. They have relatively small acceleration, as both the inlet and exhaust valves are restricted in diameter. This allows for a centrally located injector and inhibits the aspiration process. This method is very efficient, and the fuel must be atomized and combine with combustion air within a very short time duration. Very high fuel pressures are used, 25,000psi up to 30,000psi (1,725bar to 2,070bar).

b. **Indirect Injection (IDI).** These engine types have a greater fuel economy, and turbocharging can enhance this. An increased useful speed range is obtained as larger valves are used, and noise is reduced. The air is made to swirl by forcing air through a special auxiliary chamber connected to the cylinder head housing the fuel injector. Fuel, as a stream of fine fuel droplets, is injected close to completion of the compression stroke. When the fuel droplets mix with the heated air, combustion occurs. To increase the vaporization rate, the fuel must be injected into the cylinder as an atomized stream of fuel droplets. The pressure required to achieve this can exceed 7,000psi (480bar). The fuel injector must meter the correct quantity of fuel to match the engine power requirements, and the injection period must be determined precisely. The droplet size of the fuel is critical to achieving good combustion. Large fuel droplets will require a longer period to vaporize, delaying combustion. Small fuel droplets move relatively slowly, reducing the oxygen mixing times. Both conditions will lead to incomplete combustion, resulting in reduced efficiency, increased noise, and increased exhaust emissions.

14.7   Fuel Pumps. The fuel injection pump is often labeled the heart of the diesel engine. High-pressure fuel pumps are capable of producing anywhere from 2,000psi to 35,000psi (140bar to 2,400bar) or more, depending on the engine and system design. The different types include the continuous pump, distributor pump, and individual pump. Pumps might be an electric lift pump or a cam-operated lift pump. Common problems are leaks from seals and gasket failure (been there, done that!). Poor fuel quality or filtering also can gum things up; running fresh fuel through the system is important. I tend to use premium-grade diesel on my own boat.

14.8   Fuel System Cycle. There are effectively five functions within a diesel engine fuel system: metering, fuel injection, control timing, atomization, and pressure creation.

a. **Fuel Metering.** Fuel metering must be accurate so that, at the same fuel control setting, the same quantity of fuel is delivered to each cylinder for each power strike. This enables consistent speed and power output, ensuring even power distribution between each cylinder.

b. **Fuel Injection Control.** Rate of fuel injection is critical, as this determines the combustion rate. At the start, it is important that excess fuel does not accumulate within the cylinder during the initial time delay. Injection should be at a rate that ensures combustion pressure is not excessive and that injection is at a rate to obtain complete combustion.

c. **Timing.** The system must synchronize injection along with metering the correct quantity of fuel. This is to ensure efficient combustion and energy creation. If fuel is injected too early, ignition delay occurs. Excessive delay results in noisy and rough engine operation. Fuel is also wasted due to the wetting effect on the cylinder walls and the head of the piston. This can lead to fuel dilution of the lubrication oil.

d. **Fuel Atomization.** Atomization is the process of converting the pressurized fuel into small particles or droplets to form a mist. The mist or spray should be optimized for the combustion chamber. Different chamber designs have different atomization patterns. Each spray particle should be surrounded by oxygen particles to ensure proper and even combustion. When the high-pressure fuel passes into the injector and through the nozzle, it is forced through the injector nozzle holes at the spray tip, where friction then assists in breaking the fuel stream into particles, or micro-droplets.

e. **Pressure Creation.** The fuel injection process is required to increase the fuel pressure to a level greater than the compression pressure. This is required to ensure the correct level of fuel dispersion and ensure even combination of air and fuel for efficient combustion. If fuel particles are too small, they will not penetrate the chamber for even and complete combustion.

14.9   Diesel Fuel Quality. The quality of the diesel fuel is critical to good engine combustion. The US standard for diesel is the American Society of Testing and Materials (ASTM) D 975.

a. **Contamination.** The deterioration of fuel is almost inevitable, as contaminants are introduced into the fuel system through mixing, transferring, and storage. During transportation and storage, diesel fuel comes in contact with air and water vapor and can undergo change as a result, sometimes forming sludge. This may block the fuel filters or create gums that can damage the fuel injection equipment or leave deposits within the engine.

b. **Injector Fouling.** Low quality fuels cause injector fouling. When fouled, simple hole injectors in a directly injected (DI) engine can change the spray patterns; the needles and seats become sticky, causing leakage, poor combustion, and excess smoke. The pintle injectors within an indirectly injected (IDI) engine can also become fouled. If the needles and injector bodies are dirty, the initial small fuel burst does not occur, and all the fuel is delivered at once.

c. **Fuel Quality.** Diesel fuel has variable quality and purity, ranging from very good to extremely poor. API specifications allow for acceptable levels of impurities such as sulfur, wax, and other contaminants, including water, dirt, and ash. Fuel contaminates are classified as either precipitates or particulates. Precipitates are commonly noncombustible materials that form when fuel oxidizes; being heavier than fuel, they normally fall to the bottom of fuel tanks. Particulates are also known as asphaltenes, black tar-like substances that plug fuel filters.

d. **Cold Temperatures.** In subzero temperature conditions, diesel fuel forms a suspension of paraffin crystals; this is known as waxing. When the concentration of crystals reaches a high level, they clog the fuel filters. Wax is added to fuels; however, if not suitable for the climate, this wax may thicken and gel, causing fuel blockages.

e. **Water Contamination.** Water is the worst and most common contaminant in fuel, destroying the fuel's lubrication qualities and damaging fuel pumps and injectors. The most common method of water removal is stripping, which uses a silicon-treated medium to inhibit water flow. The coalescing filter uses gravity to remove water droplets from the fuel, and absorption filters use a filter medium to absorb water out of the fuel. The more commonly installed filter/separators use the stripping method.

14.10 **Fuel System Troubleshooting.** The basic fuel system problems are as follows:

a. **Connections.** A loose fuel line connection on the suction side can cause ingestion of air into the system and cause a low fuel pressure problem.

b. **Mechanical Damage.** A fuel pump can be damaged internally by a seized plunger or some other fault, such as a ruptured diaphragm.

c. **Fuel Filters.** The fuel filter can be clogged and cause a restriction in fuel supply, causing low pressure, so check the filter first. Note that after changing the filter, you frequently have to bleed air out of the system by cranking over or running the engine. You can get debris within a fuel regulator valve if installed; again, cleaning is the answer.

d. **Fuel Lines.** A kinked or crushed flexible fuel supply line can restrict fuel flow and cause low fuel oil pressure.

14.11 **Engine Space Fuel Safety.** This is a repeat of some content, but it deserves being here on safety grounds. Standards include the EI 15 Energy Industry requirement for oil lines above 1-bar pressure, and this is focused on oil mist prevention. The International Convention for Safety of Life at Sea (SOLAS) is a marine industry regulation that includes requirement for shielding and screening of oil spray within vessel engine rooms.

a. **Fire Causes.** The vast majority of engine room or compartment fires are initiated by an oil leak and subsequent ignition of the oil mist. Oil leaks are caused by small perforations of fuel lines; failure of a flange, connection, or valve; or a fracture or leak of pressurized oil, which can be diesel, lubricating oil, or hydraulic oil. When an oil leak occurs, oil mist is projected into the space. Leaks have sources in injectors or engine high-pressure fuel systems and oil lines that atomize the fluid when it escapes. Initially such escapes are invisible. Oil mist can be very small droplets, as small as 1 to 10 microns, and disperse within the surrounding air. When that mist attains the lower explosion limit and comes into contact with a heat source of 390°F (200°C), an explosion and fierce fire are often the result.

b.   **Ignition Sources.** Ignition sources have diverse origins that include bearings, turbochargers, exhaust systems, and electrical sources such as electric contacts, faulty wiring, motors, and static electricity. There are things you can do to mitigate the risk; while this is an issue in larger sailing yachts, do a basic risk assessment regardless of your boat's size.

c.   **Protection Systems.** The commercial maritime and offshore oil industry install what are called pipe and flange safety shields. Pipe safety spray shields are installed on boats to protect people from injuries and damage, fire, and explosions arising from high-pressure leaks and atomized oil mist. This hazard is created by spray-outs of fuel oil and hydraulic oil at failing pipe connections, including flanges, valves, and expansion joints. These spray shields are commonly called by the brand name FlangeGuard and come in a variety of materials. The original bag-type safety shield is a very cost-effective way to protect flange joints. The shield is simply wrapped around the outside of the flange and the protective fabric drawn down around the bolts to the pipe. The bag has drawstrings to hold the shield tightly in place. This design type is able to control the oil leak and prevent both oil spray-outs and the formation of oil mist. The escaping fluid coalesces inside the shield and then drains vertically in a drip or stream. The basic design has been improved by incorporating an outer band with specially formulated, multilayered internal mesh. This mesh is held tight against the flange and is the basis of what is termed pressure diffusion technology (PDT). The pressure is diffused at a controlled rate within the mesh. This is self-draining and prevents spray-outs and mist or vapor formation. There are valve protection options that fit butterfly, globe, check, gate, and bull valves. Check klinger-thermoseal.com for FlangeGuard details. Another product is Spray Stop Anti-Splashing Tape. It meets SOLAS requirements and is approved by all ship classification societies. Check them out at t-iss.com. Do a survey and a risk assessment on your own boat.

14.12   Diesel Fuel and Water. Water in your fuel is the great enemy. Fuel tanks should always be topped up to reduce condensation, and, where possible, tanks with a drain valve should have the water drained off. If you don't have a drain valve at the lowest point of your fuel tank, consider installing one. After filling fuel tanks, the pre-filters should be monitored for excess water and drained if necessary. During transportation and storage, fuel is in contact with air and water vapor and can undergo a quality change as a result. Sometimes sludge is formed, which may block fuel filters, or gums are formed that can damage fuel injection equipment or leave deposits within the engine. Another often forgotten source of water is fuel deck fills. Either the caps are not tightened or the O-ring seals are degraded. Be sure to check your deck fill cap. The single greatest investment that you can make with the fuel system is to install a filter/separator. Popular units include those by Parker Racor, Baldwin Dahl, Keenan, Vetus, and Delphi. The main filters are described below.

a. **Dahl Filter Operation.** Fuel from the tank enters the filter inlet port and is directed down through the center tube. The de-pressurizer cone then spreads the fuel. As fuel is discharged from the de-pressurizer cone, 80% of the contaminants are separated from the fuel. The fuel rises upward, and most of the solid contaminants and water settles into the bowl quiet zone. The system includes a reverse-flow valve to hold prime in the fuel system and prevent fuel from flowing back to the tank during shutdown. There is a removable primer plug at the top, for use when complete priming is required. As the fuel rises upward, remaining small water droplets collect on the cone, baffle, and bowl surfaces. The size and weight of the water droplets gradually increase, causing downward flow into the sump. Fuel is then filtered completely by the 2-micron paper element. The clean fuel continues up through the outlet port to the pump and injection system. The transparent bowl holds up to 24fl oz (710ml) of water capacity to reduce draining intervals; this is done via the drain cock.

b. **Racor Filter Operation.** Units have optional heaters and electrical water continuity probe alarms. Some have a vacuum gauge to monitor pressure drop across the fuel filter elements. In the separation stage, a turbine separates large solids and free water using centrifugal force. In the coalescing stage, the smaller water droplets and solids coalesce on a conical baffle and drop into the collection bowl. The filtration stage uses a fine-micron Aqua-Bloc water repelling paper element.

c. **Keenan FilterBOSS.** These are either a single or dual filter and polishing system. The single-element MK60SP has a remote warning panel to alert system problems and an integrated fuel pump for bleeding, priming, and backup of the main engine fuel pump. It also uses Racor filters, which are easy to obtain. The unit has a vacuum gauge that's color coded to determine if operation is normal. The integral pump allows a return line to the tank to polish stored fuel. The system has optional smart phone connectivity. Check out shop.keenanfilters.com.

d. **Vetus Filters.** These are spin-on type filters based on a patented full-fuel flow system that separates water and particles prior to fuel flow through the filter element. They have a clear bowl that's approved by CE (ISO10088) and ABYC. For offshore blue water sailing, a dual filter system is recommended, as rough weather motions generally result in tank agitation and increased input of particles. Dual systems have a vacuum gauge to indicate filter clogging.

14.13 Microbe Growth and Fuel Additives. The chemical makeup of fuel will designate the performance characteristic. There is no perfect fuel, so it is a compromise. Additives are often added to diesel fuel to improve ignition-delay periods, providing improved combustion efficiency, reduced noise, and reduction in smoke emissions. Detergents are added to keep the injectors clean and allow correct fuel metering. Anticorrosive additives are used to protect the injection system, and antifoam compounds are used to limit frothing. Uncontaminated fuel is essential to good combustion and efficient operation of the engine. There are regular accounts of water-contaminated fuels being supplied to unsuspecting boats, and I have experienced this myself. Moisture and water within the fuel system can encourage microbiotic growth within the fuel, including fungi, bacteria, algae, molds, and yeast. Aerobic bacteria require oxygen, anaerobic bacteria do not require oxygen, and facultative bacteria can thrive with or without oxygen. Microbial growth is highly dependent on temperature, thriving in the 50°F to 115°F (10°C to 45°C) range. Once the system is "infected," considerable flushing is required if the infection is ever to be eliminated. The solution is to add chemical biocides to kill the contaminant or maintain quality.

14.14   Fuel System Bleeding. If any part of a fuel system is disconnected or air has entered the system, the system must have all air bled or purged. The injector pump and injector operate by having hydraulic pressure high enough to open needle valves so that fuel can enter under sufficient pressure to atomize. Air absorbs the pump force, and the injector will not open. Do not attempt to start the engine until the injection pump is filled and primed or damage can occur, as fuel is also a lubricant for the pump. Paint the bleed screws with bright yellow paint to make them visible in low light. Have the correct tools ready by the engine for use. Use caution with high-pressure fluids such as fuel—escaping pressure can penetrate the skin, causing serious injury that can result in gangrene. Fluid injected into skin must be surgically removed within hours. Tighten all connections before pressure. Check your engine manual for the correct bleeding system procedure, since every engine is different. Before you start bleeding, check that the fuel tank has sufficient fuel. Visually check that all pipes and connections are tight. Check all hoses, fuel line fittings, and steel lines where they may have chafed. Step 1 is to loosen the bleed screw on top of the engine fuel pre-filter or separator. Step 2 is to operate the thumb or hand-priming lever on the lift pump until a stream of bubble-free fuel exits. Tighten the bleed screw. Step 3 is to loosen the bleed screw on top of the engine fuel filter after the lift pump. Operate the thumb or hand-priming lever on the lift pump, and pump it until a continuous stream of bubble-free fuel exits. Tighten the bleed screw. Step 4 is to repeat at the injection pump inlet connection, and so on.

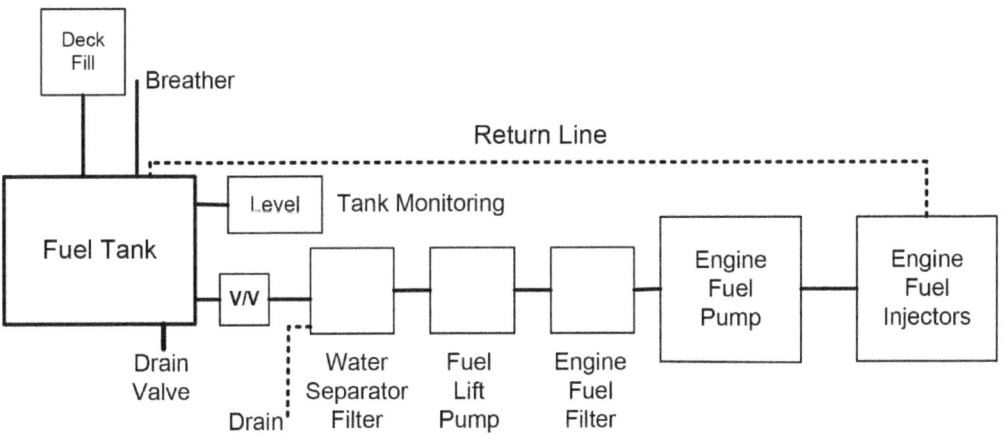

Figure 14-1. Fuel System Schematic

14.15   Engine Freshwater Cooling System. Diesels now universally have a heat exchanger with a closed-circuit freshwater (FW) cooling system. The FW system has a pump that circulates water from the expansion tank, through the various engine water galleries, to carry engine heat away, through the cooler and then back to the expansion tank. The coolant system controls the overall operating temperature of the engine in operation; proper heat transfer is essential. The freshwater cooling system must remain saltwater contamination free to prevent corrosion or formation of sludge and scale that may impede coolant flow or block coolers. This reduces heat transfer rates by coating engine block water passages with insulating scale buildup. This will result in gradual overheating, with all the damage that comes from it. Inhibitors must be maintained at the correct concentrations if performance is to be maintained and damage avoided.

    a.   **Coolant Additives.** A number of additives are available to improve the performance of engine coolants. Coolant water may contain sulfates, chlorides, dissolved solids, and calcium. Coolant should have an antifreeze additive to prevent freezing and engine damage in cold climates. Most ethylene glycol–based antifreeze solutions contain the required inhibitors for normal operation. Check your engine manual.

    b.   **Corrosion Inhibitors.** These are generally water-soluble chemical compounds that protect the metallic surfaces within the system against corrosion. Compounds can include borates, chromates, and nitrites. Soluble oil should not be used as a corrosion inhibitor.

    c.   **System Cleaners.** One well-known cleaner is Sea Flush, which is a patented formulation for cleaning and flushing engine, generator, and air-conditioning heat exchangers, oil coolers, and other system components. Descalers are usually caustic, so care must be taken. I once had a high engine temperature issue and discovered that the cooler was caked in fine silt from the very brown, silted river we were transiting in Europe.

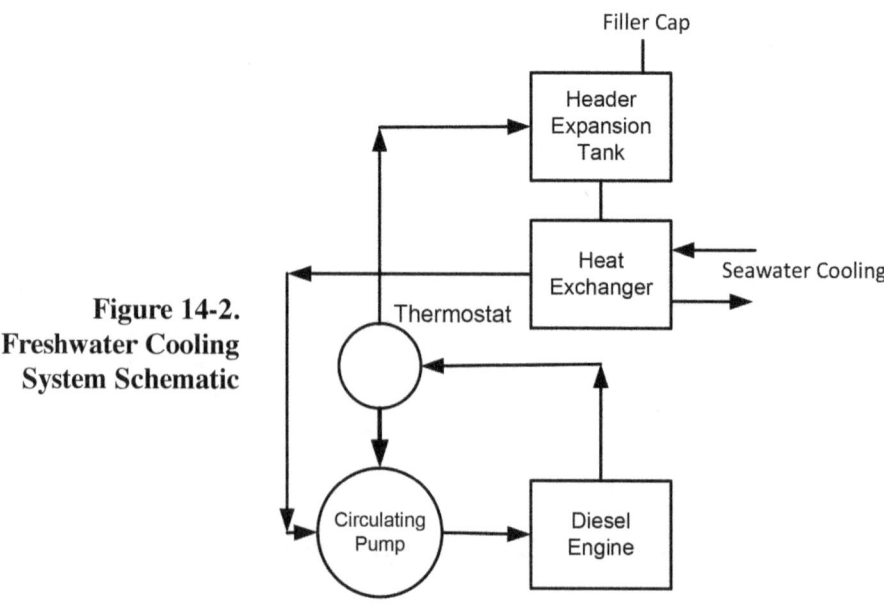

**Figure 14-2.
Freshwater Cooling
System Schematic**

14.16   Engine Seawater Cooling System. The heat exchanger provides the medium for transferring engine heat from fresh water to the primary seawater coolant. The seawater is drawn through the seawater inlet valve, the strainer, and to the suction side of the seawater pump. The flexible impeller blades create close to a perfect vacuum. As the impeller rotates, each successive impeller blade draws in seawater and transfers it from the inlet to the outlet port. The pressurized water is then pumped to the heat exchanger or cooler, where it passes through the cooler tubes and transfers heat from the fresh water. The seawater then may be injected into the exhaust line and discharged overboard through the exhaust outlet. Seawater cooling circuits are dependent on regular maintenance, and coolers require regular inspection and cleaning. Do not neglect pump impeller inspections. Change them every season; carry a spare plus a removal tool if required.

a.   **Water Pump Impeller.** Many impeller pumps are of the Jabsco type and are powered using a drive belt driven off the crankshaft pulley. When temperatures are high, the impeller may be fatigued. Follow this simple change procedure: Remove the pump cover, making sure not to damage the gasket. Carefully pry out the impeller with a screwdriver. Check the impeller for damage, cracks, and flexibility. Replace or renew, and coat with water pump grease only; do not use petroleum jelly, which will degrade the impeller. Refit the gasket and pump cover; do not overtighten the screws. Make sure you have the pump details recorded, including model and serial number, so you have the right spare on hand.

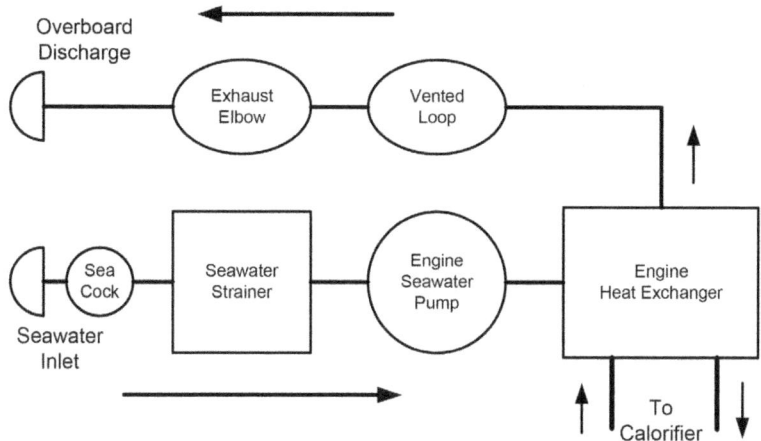

Figure 14-3. Seawater Cooling System Schematic

b.   **Water Pump Impeller Troubleshooting.** If the impeller is damaged, the following are possible causes:

(1)   **Impeller Condition.** Check if there are pieces missing out of the blade tips at the center of the impeller, pitting at the ends, or the edges have a hollowed-out appearance. This is caused by cavitation due to low pressures at the pump inlet. It can be rectified

by reducing inlet pipe restrictions and lengths and increasing inlet pipe diameters.

(2)   Impeller Wear. Check if the impeller blade tips and end faces are worn or if the impeller drive is worn. This is caused by cavitation, which can be caused by low pressures at the pump inlet. The same corrective measures apply.

(3)   Impeller Fatigue. Check if the end faces of the impeller have a hard and polished appearance, or if some or all blades are missing. This is caused by running the pump without water. The pump should not be run longer than 30 seconds without fluids and should be stopped as soon as the fluids are gone.

(4)   Impeller Degradation. Check if the impeller blades have excessive or permanent distortion or curving. This is caused by chemical action, excessive pump storage periods, or the end of normal service life. Chemical actions are caused by pumping incorrect fluids. If an engine is stored for long periods or over winter, remove the impeller to avoid damage.

(5)   Impeller Binding. Check if the impeller binds inside the pump housing or if the blades appear longer than the hub. Check if the impeller rubber is sticky and soft; this is caused by chemical actions, high fluid temperatures, or long immersion. Pumps should be flushed clean after use and drained if being stored. High-fluid temperatures should be avoided.

(6)   Impeller Cracking. Check if the impeller blades are cracked by 50% or more and if parts of the blade are missing. This is caused by the impeller reaching the end of normal operational life, perhaps due to high output pressures. High- or low-temperature fluids and running the pump dry can cause similar damage. Check and reduce the pump pressures and outlet pipe restrictions, such as long pipe runs or blockages.

(7)   Impeller Access. Most water pumps have an impeller access plate. Years ago I had a Speedseal Safety Cover, a quick-release cover plate to enable rapid impeller access. Always ensure that gaskets are in good condition; suction of air lowers pump efficiency.

14.17   Seacocks and Thru-Hulls. The seacock and its attached hoses are all that is separating you from being a boat or a reef. The seacock is innocuous, but when you unpack the elements of watertight integrity, things change. Most boats have several seacocks to provide seawater to propulsion engines and generators; for toilet seawater flushing, air-conditioning, refrigeration cooling, and other functions. We also have an overboard discharge for all of these. Refer to ABYC Standard H-27. There are other recommendations in addition to ABYC. ISO 9093:2020 requires that seacock and through-hull fitting components formed of a metallic material show no degradation to the point that their operation is impaired. This ISO specifies requirements for through-hull fittings, seacocks, hose connections, their fittings, and their installation in small craft.

14.18   Scoop Strainers. Scoop strainers are often installed on the seawater inlet and function as a coarse first-stage filter that is normally cleaned by the water flow across it. They are designed to deflect larger debris away from the seawater inlet when under way; this includes seaweed and plastics. There are the traditional oval slotted strainers, and there are round ones for stationary applications. They are normally manufactured from bronze in a variety of designs. TruDesign has glass-reinforced nylon composite scoop strainers, which eliminate the potential corrosion issues. Scoop strainers have strainer area numbers and also area ratios of strainer to skin fitting, typically around 1.8:1. Make sure to install them correctly; they should be facing aft to avoid pressurization of the water intake system. This position is controversial, as installing them facing forward increases water flow into the system. Some engine manufacturers, such as Volvo and Beta, caution against doing this, as water may be forced through the pump under dynamic pressure into the exhaust system when under sail and back into the engine.

14.19   Seacock Types. Selection of seacock materials is often controversial, ranging from traditional to the latest technologies.

   a.   Seacock Material. The standard for a very long time has been either bronze or what is known as DZR (dezincification-resistant brass) seacocks. The metallurgical properties comprise copper, zinc, tin, and other metals to increase the physical strength. Bronze is an alloy of copper, tin, nickel, and various other metals. The Blakes & Taylor (B&T) seacock has a tapered core plug assembly that facilitates valve adjustment without removal; it is easy to disassemble and service. I had a B&T seacock on a previous old wooden boat. You do need to annually disassemble it, clean thoroughly, lubricate it correctly, and lap it in if necessary. The seacock was over 50 years old and still leak-free and suitable for use. Similar bronze fittings are also manufactured by both Perko and Groco; however, they are not the conical type but instead utilize ball valves made from stainless steel or a polymer.

   b.   Composite Materials. This has been joined by the Forespar Marelon valves. These are not PVC or plastic, as they are often disparagingly described. They are a formulation of polymer composite compounds that use either composite-reinforced DuPont Zytel polymer and additives or glass-reinforced DuPont Zytel; they are ABYC recognized, UL approved, and ISO certified. Other composite seacocks are those from TruDesign, which have a Teflon-impregnated ball. There has been a trend toward monitoring the status of valves, either open or closed, and TruDesign has position-monitored ball valves via a display panel.

14.20   Seacock Maintenance. The prime cause of seacock failures of all types is lack of basic maintenance. If you don't do the maintenance, you threaten the seaworthiness of your boat and the safety of everyone on board. Always have wooden bungs tied close to the valve for emergencies.

    a.   **Winterization.** Seacocks are very prone to freeze damage in some locations. They should not be left full of water. Prepping for winter entails closure, and draining water through the drain plug. Once drained, reinsert the drain plug.

    b.   **Lubrication.** Marine growth can occur within the valve, but you can reduce this. Many seacocks have a grease nipple or zerk fitting. You should lubricate according to the manufacturer's recommendations. Groco recommend their own lubricant, Groco U-LUBE, which is injected into the space between ball and valve body. This is a specially formulated grease for ball valves and is marine growth inhibiting. Groco recommends opening the valves when greasing and removing the drain plug. Then charge the valve with grease; when grease appears at the drain plug, stop and reinsert the plug. In most cases you will have to remove the hose and hose clamps to access the valve to lubricate it. You can use winch grease; a product called LanoCote grease is very effective. Some use water pump lubricants, which are okay. Then there are polytetrafluoroethylene (PTFE), or Teflon, lubricants, which do not attract and hold dust or grease. PTFE lubricant is a low-molecular lubricant formulated for applications that require a low coefficient of friction, good surface adhesion, and corrosion resistance, which is exactly what a seacock requires.

    c.   **Valve Exercise.** Regular maintenance requires monthly operation of the valve; repeat five times each time you do this. It should close without the requirement for excessive force or offer much resistance. If it does, valve servicing or investigation is required.

    d.   **Corrosion.** Visually inspect for signs of corrosion or evidence of dezincification on both body and hose tail and any other attached fittings. Do not ever mix various fittings of different metal composition. I have come across a bronze valve with brass adapters and fittings; no doubt, the easiest way to source something was from a hardware outlet. Different metals are a source of galvanic corrosion, and the valve with mismatched metal attachments will not last long. Check your valves every 6 months for signs of dezincification. Gently scrape the surface; if it looks pinkish or is acquiring a dark red color, it is time to replace it, as the zinc part of the alloy is leaching out. A common failure point is the barbed hose tail. I have experienced this myself; the barb literally breaks off when removing a hose due to it being very brittle. UL-1121 states: "The components of a through-hull fitting or sea valve shall be formed of galvanically compatible materials having the strength and resistance to corrosion necessary to withstand intended and abnormal use to which they are likely to be subjected." I think this is very sound advice.

    e.   **Hose Tails.** The hose tail, spigot, tail pipe, or pipe to hose adapter is a male fitting that connects the hose to the body of the valve. The hose tail has "barbs" that hold the hose in place by friction. Hose clamps are used to secure the hose in place. Installing and removing hoses to hose tails is always a challenge. You can put the end in hot water or gently use a hair dryer to heat and soften the hose end.

f.   **Hoses.** Use the right hoses, and spend the money to get the best. For cooling fluids, hoses manufactured from ethylene propylene diene monomer (EPDM) rubber that incorporate synthetic fabrics along with spiral steel reinforcement are ideal. These are suitable for cooling water and have a wide range of temperature capabilities. These hoses will not fold shut or kink, which often happens on seawater inlets and creates many overheating incidents.

g.   **Hose Ends.** Hoses are another variable to this equation. PVC hoses are often used for this, as well as reinforced hoses. Some are heavy walled, others wire-reinforced. The hose diameter should be the same size as the hose pipe. I have frequently come across people trying to clamp on an oversized hose or trying to push an undersized pipe onto the hose pipe. Before installing Jubilee clips or worm-drive clamps, the hose ends should be clean and fresh. By that I mean, do not install any hose end that is aged, brittle, or fatigued. Hose ends deteriorate because they are partially stretched over the barbed hose tail. The metal barb and the valve body suffer from cycles of hot and cold, which contributes to hardening and degradation. When installing a hose, always allow some slack for trimming after each hose removal and reinstallation. As many know, removing hose ends is very difficult; it is quicker and easier to simply cut it off and remove from the barbed hose end.

h.   **Hose Clamp Installation.** Use clamps that are the correct size, not oversized. When installing the hose clamps, do not overtighten and deform the hose end. Installing two hose clips or clamps increases the surface area being clamped and reduces failure. I install the two clamps with opposite-facing worm drive in order to exert force more evenly. See the section below on hose clamps. If you are reusing the same clips and observe signs of rust and corrosion, chances are they are 304 grade and not 316 grade stainless steel.

i.   **Failure Modes.** The most common failure mode is the distortion or breaking off of valve stems or handles, or shearing the stem completely. People try to force them the wrong way or force them against the stop, rendering the valves inoperable. The valves were often seized or people simply do not know how to operate a valve. When operating a valve, if it is hard or stiff to move and requires force, you need to service the valve. If a valve is installed correctly, you should be able to operate the valve handle through 90 degrees, allowing easy identification of status, whether open or closed; better still, label the "open" and "close" positions for less-aware crew members. Another key failure mode is stepping on or accidentally applying force to the attached hoses. Often seacocks are installed in areas prone to damage. Stepping on a hose creates what is known as the lever-arm effect. The stiffer and larger the hose, the more prone it is to damage. Another failure mode is where people are trying to pull hoses from barbed hose tails and end up damaging the valve through excessive force.

14.21   Hose Clamps. The humble hose clamp, Jubilee clip, or worm drive clip has a lot of responsibility. These clamps are also called pipe clips, hose band clamps, or hose locks. It is an essential part of the thru-hull and seacock arrangement, along with the hoses that connect to it. Many install the wrong grade of clamps, such as plated steel or 304 grade stainless steel. Always select 316 grade stainless steel and check the clip for markings. They are used on seawater intakes and discharges and on freshwater, sewage, and bilge pump overboard lines, with many systems being under pressure. If they fail, you can end up with a very big mess or, the worst case, sink—all riding on the integrity of a hose clamp.

    a.   **Clamp Selection.** Use the correct clamp for the hose; if it is too large, you will be unable to apply pressure evenly and the tail may be damaged. Long tails are a danger, and I have frequently been lacerated from sharp stainless edges, once requiring an emergency room (ER) visit. You can buy small plastic hose clamp end guards or protectors from Clamp-Aid; given my past painful experiences, I absolutely recommend them. You can use color-coded ones to mark your seawater and freshwater systems and so on, which is a bonus. The traditional Jubilee worm-drive clip is identifiable by the short and thick worm drive, which has an integral grub screw or set screw mechanism located on the band top. Go to www.clamp-aid.com and www.force4.co.uk in UK.

    b.   **Clamp Tightening.** When you tighten clamps, they pull in and grip the band circumference on one side for maximum band tension and clamping force. Most clamps are tightened using a large screwdriver blade, a small socket spanner, or a hex key. Personally, I use a ratchet drive and socket, which is more effective than a screwdriver or hex key to torque properly in confined work areas. Do not overtighten, as you end up damaging the clamp threads and the hose. Another alternative is the Super Clamp from Jubilee. These allow torquing to 18ft-lb (25Nm) and incorporate 8.8 high-tensile bolts and are specifically designed for suction and delivery hoses where a tight seal is required. Bolt drive or T-bolt versions have a thinner profile and are also known as T-bolt clamps or head bolt clips. They are distinguished by the small bolt located across the band top. When the bolt is tightened with a hex drive, it pulls the hose clip in from both directions and evenly applies the torque.

14.22   Seawater (Raw Water) Strainers. The raw water strainer is crucial to efficient engine cooling. It is frequently fouled and partially blocked with everything from jellyfish, seaweed and grasses, the remains of small fish, and crabs to an ever-increasing amount of plastic and other assorted detritus. Strainers have to cope with a variety of water conditions, from clear offshore water to highly silted estuarine water. Many strainers are made from plastic; there are also stainless steel and bronze units, such as the one on my boat. Note the following.

a. **Flow Rates.** It is important to select the appropriate strainer based on the required engine flow rate. A typical strainer with a 1in (25.4mm) hose connection has a 55gal/min (210l/min) capacity. Check that your strainer is correctly rated for the engine.

b. **Strainer Baskets.** Most strainer baskets are polypropylene or stainless-steel mesh. Mesh size is important, as water flow should not be impeded or allow debris into the pump suction. Factors such as pressure drop and flow reduction become important when they start to clog; surface area is a design factor.

c. **Manufacturers.** Groco is the leading strainer manufacturer in the United States. Their strainers are different in design in that while most strainers are top-entry for basket removal, these units are accessed from below and have a clear sight glass that unscrews. Strainers can be Monel, plastic, or 304 grade stainless steel. Check out the range at groco.net. In Australia, the Hallmark Marine bronze strainer is the best. Check them out at hallmarkmarine.com.au. Various makes are on the market in the UK, including Vetus, Guidi, and Maestrini.

d. **Automatic Cleaning Strainers.** Larger yachts, trawler yachts, and commercial vessels have automatic strainer cleaners. Typical are the Groco Hydromatic Self-Cleaning Raw Water Strainers. The Hydromatic unit performs a 30-second cleaning cycle. In this cycle, all the accumulated debris and material is macerated and discharged overboard; the timing is user definable. Cycle periods often require extension when the area has a lot of jellyfish or seagrass. Another innovation is the ElectroStrainer, an integrated sea strainer with integral biofouling prevention system along with flow monitoring. The biofouling system uses electro-chlorination technology to inject low-level chlorine to prevent marine growth. The strainer basket is optimized for maximum chlorine diffusion and effective straining of debris with minimal pressure drop. The control unit display alerts when the strainer requires checking and cleaning and has a real-time flow sensor using CLEARVIS Flow Sensing Technology. The technology employs ultrasonic technology for seawater velocity measurements, which transmits and receives sound waves between two transducers. Check it out at electrosea.com. Another new innovation is the TITAN T-Strainer, which has refined the normal design to allow a high-water flow path and unrestricted up to 80g/min (300l/min). Check it out at forespar.com.

e. **Strainer Maintenance.** Strainers need to be cleaned weekly when out on the water, and very muddy conditions can quickly result in strainer basket clogging when combined with other material. Strainers are required to continue allowing water flow as they become clogged, and the pressure drop between 0.1bar and 1bar can be as much as a 60% water flow decrease. Wear gloves when cleaning strainers—you don't know what is in there. I have been stung by various jellyfish when putting bare hands into strainers. Strainers have a transparent polycarbonate top or sight glass to allow inspection. A common problem is overtightening the fastening nuts, so use care when reinstalling. If the strainer has a screw-on plastic top, don't overtighten and distort the seals. Check the seals for deterioration. Groco recommends lubricating cap threads and O-rings yearly with a Teflon or silicon grease.

14.23   Engine Lubricating Oil System. Lubricating oil separates the various engine working surfaces to prevent metal-to-metal contact. The goal of lubrication is to reduce friction by substituting fluid friction for sliding friction. Oils assist in bearing cooling and prevent corrosion. The pistons must receive adequate oil supplies to prevent expansion and seizure that is caused by excess heat due to increased friction on a dry cylinder wall. Lubrication has the dual function of lubricating the engine's moving parts and removing heat generated during the combustion process and friction, which reduces wear. Oil removes heat from the outside of cylinder liners, removes heat from the pistons and inner walls of liners, and cools the main bearings. Friction can create enough heat to melt bearing metals. The following outlines the various aspects of lubrication in diesel engines.

a.   **Oil Circulation.** Oil is taken to the oil pump from a submerged suction in the oil sump or pan. Oil pumps are usually a two-stage, positive displacement, gear type pump. Oil pumps are driven off the timing gear train to maintain constant oil flow. It pumps a constant volume of oil, although flow will decrease slightly with increasing pressure. A nonadjustable pressure regulator valve is installed on the pump outlet side. A simple bypass valve is used to limit maximum oil pressure by diverting excess oil back to the sump. The pressurized oil is passed though the oil filter then distributed around the various oil galleries and points within the engine. As oil reaches the crankcase bearings, flow is restricted, and the pump forces oil into the clearances between the main bearings and crankshaft. Oil is carried to the crank, connecting rod bearings, main journal, and connecting rod journals. Oil may be used in turbo bearings and sometimes under piston crowns. Liner, piston, and rings are lubricated by oil thrown by the camshaft and crankshaft, and the oil manifold supplies oil to the governor and rocker gear. Oil usually flows off the end of the rocker arms to splash-lubricate valve springs and stems. Oil also lubricates the timing gear bearings. The oil then returns to the sump via gravity. In many engines the heat is dissipated through the oil cooler and through oil sump surfaces.

b.   **Oil Filtration.** All oils contain contaminants, and the content levels must be minimized. These may be metals, fibers, or microbial growth, and most damage is caused by hard particles that are slightly larger than the clearance of contact surfaces. Critical sizes are 4.0 to 15.0 microns and usually in the range of 3 to 5 microns. The oil filters remove these abrasive materials and contaminants. Oil protects against corrosion and assists piston rings by sealing cylinders and reducing compression loss through the piston rings. Oil provides a strong film between all the surfaces of working engine parts so that wear is reduced. Oil films may be typically around 30thou (0.08mm) thick between the shaft and bearings, and they must be strong enough to prevent surface contact.

c. **Oil Contamination.** In a four-stroke diesel engine, lubricating oil reduces engine wear and absorbs contaminants that enter the engine. Most contaminants are expelled from the engine with exhaust gasses, but some remain in the cylinders and crankcase to corrode metal parts and form sludge and lacquer deposits. Most events that occur in the engine have an effect on oil quality. When injection and combustion pressures are raised, the loads increase on pistons and bearings. Excess oil will bypass pistons and rings into the combustion chamber, where it burns, forming carbon deposits. Lubricating oil is affected by chemical composition and the presence of nitrogen, phosphor, potassium, sulfur, and external growth factors such as temperature, pH values, water, and oxygen. Oil neutralizes the acids that form and breaks up the deposits caused by combustion blowby. Oil cleans by dissolving sulfur that is converted to acid when the fuel is burned. High oil temperatures cause chemical breakdown of oil, so maintaining the correct engine temperature is vital. Reduce the effect of acid formation by operating the engine at proper working temperatures. Humidity in combustion air also assists in acid formation. When sulfur is low, around 0.02%, there is little effect, but at 1% humidity will affect acid formation. Lubricating oil will at some point lose its effective viscosity as it transfers heat during combustion and absorbs combustion by-products. When oil completely oxidizes and forms black sludge, your engine is being damaged. Regular oil changes are essential for your engine's well-being.

d. **Lubricating Oil Standards.** It is essential that the correct grades of oil be used for the prevailing temperature conditions, and that the filter be changed regularly along with the oil. The nominal rating of oil viscosity must be maintained if correct lubrication is to be achieved. This is dependent on the engine remaining within the proper operating temperature ranges. Engine oil should comply with API CC/CD specifications and may be synthetic, mineral, or blended. Mineral oils have various additives and are suitable for infrequent use or under less difficult service, such as boats. Oils in generators are changed at 500 hours; synthetic oils usually last longer by a factor of 10. This results in a cleaner engine, rings, and liners; increases engine life by 40%; and decreases lube oil consumption by 75%. SAE standards match oil viscosity to operating temperatures. The "W" notation is the winter service rating defined at 0°F (−18°C).

Table 14-1. Engine Oil Viscosity

Viscosity	Temperature °F	Temperature °C
SAE 10W	−10°F to 70°F	−23°C to 21°C
SAE 30	+20°F to 100°F	−7°C to 38°C
SAE 40	+45°F to 120°F	7°C to 49°C
SAE 10W-30	−10°F to +100°F	−23°C to 38°C
SAE 15W-40	+5°F to 120°F	−15°C to 49°C

14.24   **Engine Lubricating Oil Problems.** The following problems can occur.

a.   **Fuel in Oil.** Fuel in oil can create a crankcase explosion risk and is characterized by low lube oil viscosity. Fuel mixing with lube oil reduces viscosity and reduces lube oil effectiveness, which increases wear due to a thinner film of oil on moving parts, including main crankshaft bearings. Causes are faulty injectors, incomplete combustion, wear on engine parts such as valve guides and injectors, and cold starting.

b.   **Water in Oil.** Water in the oil can cause emulsification and degrade the oil's lubrication properties. Water causes internal corrosion, and dissolved water creates oxidation, resulting in accelerated wear, increased friction, and high operating temperatures. Water should not exceed 500 parts per million (ppm). The system must be completely flushed out after a leak repair so that no moisture remains. Common causes are external contamination, such breathers and seals, and internal leaks, such as heat exchangers and condensation.

c.   **Microbial Growth.** Where water and oxygen are present, microbial growth can occur within the oil and system. Anaerobic microorganisms can live in oxygen-free areas on partly mineralized hydrocarbons, and oil additives can stimulate growth of organisms. As the oil degrades, oil characteristics also change; this is caused not by the degradation but by the organisms that produce extracellular biopolymers (slime), usually bacteria and fungi, and yeast contamination. Organisms start to produce surfactants and biopolymers when in contact with a hydrophobic culture medium (oil) that they can break down. This leads to increased exposure and oil breakdown. Optimal temperatures range is around 50°F to 122°F (10°C to 50°C). Once the system has microbial infection, considerable flushing is required to sanitize the system. There is a range of additives, but nothing beats good maintenance, diligent monitoring, and constant vigilance.

14.25   **Lubrication Oil Testing.** With larger boats and larger engines, it makes sense to take oil samples at regular intervals for analysis. All leading oil companies offer this important service. Analysis of the various trace metals found in the sample is a good indicator of wear and engine condition. Once an oil sample is submitted, the results can usually be obtained from most testing companies via the internet. Following is some sampling advice.

a.   **Oil Sampling.** The oil sample must be taken uniformly, as wear metals are heavy and sink to the bottom of the oil sump if the oil is settled and not circulating. Take the oil sample within 20 minutes of engine shutdown so that the oil material is equally dispersed in suspension. Do not sample from the sump bottom but from a point before the oil filter or suction, or out of the dipstick tube. Drop the suction hose to the sump bottom, then raise an inch for sampling. New engines will show higher wear metal numbers than old engines, which is why many manufacturers call for an oil change within a specific period, often around 50 hours. The baselines are best established after the engine is broken in.

b. **Sample Reports.** On the oil analysis reports, depending on the reporting format, "normal" status indicates that the physical properties of the lubricant are within acceptable limits and no signs of excessive contamination or wear are present. "Monitor" status indicates that specific test results are outside acceptable ranges but not serious enough to confirm abnormal conditions. Initial abnormalities often indicate the same result patterns as temporary overloading or extended operations. "Abnormal" status indicates that lubricant physical properties, contamination, or component wear are unsatisfactory but not critical. "Critical" status is serious enough to warrant immediate diagnostic and corrective action to prevent major long-term performance loss or in-service engine component failure.

14.26   Lubricating Oil Viscosity. Understanding the viscosity of your engine oil is an important maintenance parameter, so investing in a simple test kit is worthwhile and is outlined below.

a. **Oil Viscosity Testing.** Viscosity testing is a simple test method using a viscosity test stick. Both the new and used oil must be allowed to stand for an hour and stabilize at room temperature. The test stick is angled so that each oil sample runs down a channel. When the new oil reaches a mid-scale point on the graduated scale, the position of the used oil is read off. If the used oil has not reached the same point as the new oil, it has a high viscosity, usually due to oxidation or high levels of insolubles. If the old oil has run past the scale point, the additives may be failing or the fuel oil may be contaminated. Simple and easy to use, these test kits should be part of any testing regime. Viscosity tests measure a lubricant's ability to flow at a specific temperature. If the viscosity is reduced, an oil film cannot be properly established at the friction point. Improper viscosity leads to overheating, accelerated wear, and failure. This can occur due to heat and contamination not being removed at the appropriate rates.

b. **Oil Viscosity Grades.** Lubrication oils are classified by the ISO viscosity grade (VG). The ISO VG refers to a lubricating oil's kinematic viscosity at 104°F (40°C). If an oil's viscosity is within plus or minus 10% of its ISO grade, it is normal. When oil viscosity exceeds plus or minus 10% and less than plus or minus 20% percent, it is marginal. Viscosity that exceeds plus or minus 20% from grade is classified as critical. Low viscosity creates many engine problems; good filtration and regular oil changes result in improved engine life and performance.

c. **Low Oil Viscosity Impacts.** Loss of oil film quality results in excessive wear due to increased friction on surfaces. The increased friction results in greater energy consumption and increased heat levels, as well as increased exposure to particle contamination due to reduced oil film levels. Thin oil films lead to high temperature at higher loads, or during engine start-up. High viscosity leads to engine problems and is why you should gently warm up a very cold engine. It can cause excess heat generation that results in oxidation of oil along with sludge formation. A condition called gaseous cavitation can occur, which is caused by inadequate oil flow to both pumps and bearings. High viscosity causes oil lubrication starvation due to inadequate flow and surface film creation. Poor air dissipation or demulsibility also can occur.

14.27   Engine Wear Metals. There are typically more than forty different elements present in lubricating oil. Oils are analyzed using a spectrograph, or elemental spectroscopy. Elemental spectroscopy is used to determine the concentration of wear metals, contaminant metals, and additive metals in a lubricant. Spectroscopy cannot measure particles larger than approximately 7 microns. As each element is burned in an electric arc, the sample emits at a unique light frequency. The test results show amounts of each metal present in parts per million (ppm). The spectrochemical analysis is given in parts per million by weight. The analyst looks for wear metals, contaminants, and oil additives. Oil analysis should be done at regular intervals, as trend analysis is more important than a single test. Wear metals have differing thresholds, usually indicating ranges of increased probability that problems are developing. Sharp increases in wear metals or a major shift in physical properties signals impending problems. Oil analysis results indicate expected and unexpected elements. Some contaminates from within the engine are collected as the oil circulates, as different parts of an engine have different metals, all washed out by the lube oil. Other contaminates enter the machine from external sources, such as faulty seals and breathers, and during service and maintenance. Contamination usually presents in the form of insoluble materials. These include water, metals, dust particles, sand, and rubber. Even the very smallest particles, below 2 microns, are able cause significant damage. Typically, these are silt, resin, or oxidation deposits. Elements that are indicators of contamination include silicon, that is, dust and dirt or de-foamant additives; boron from coolant corrosion inhibitors; potassium from coolant additives; and sodium from detergent and coolant additives. Tests include the following:

a.   **Soot Test.** This determines the amount of fuel soot—carbon suspended in the lubricant—and higher values indicate reduced combustion efficiency. This may be due to air intake or exhaust restrictions, injector malfunctions, or excess idling.

b.   **Oxidation Test.** This looks at chemical incorporation of oxygen into the oil, which causes a loss of lubrication performance and degradation due to aging, adverse or abnormal operating conditions, or interval overheating. The main driver behind oxidation is heat, and rate of reaction will double with every 18°F (10°C) increase in temperature. This may contribute to increases in viscosity, acidity, and sludge formation in lubricants. A period of prolonged or elevated oxidation can result in deposits characterized by very dark color with a granular texture.

c. **Nitration Test.** This is a measure of organic nitrates formed when combustion by-products enter the engine oil during normal service or as a result of blowby past the compression rings. This may be caused by excessive aeration, leading to compression combustion of air molecules called micro-dieseling, resulting in thermal breakdown of the oil molecules, and will result in increased nitration values.

Table 14-2. Engine Oil Analysis

Element	Indicators
Iron	High levels indicate wear from rings, shafts, gears, valve trains, cylinder walls, pistons, or liners.
Chromium	May indicate excessive wear from chromed parts, such as rings, liners, and some additives.
Nickel	Secondary indicator of wear from some bearings, shafts, valves, and valve guides.
Aluminum	Wear from pistons, rod bearings, and certain shaft types.
Lead	An overlay on main rods and bearings. Lead is an expected wear metal in any machine using plain bearings, as lead and tin are the most predominant metals used in Babbitt overlay, with lesser amounts of copper, antimony, and/or arsenic. Typically, increasing levels of lead from this layer are not considered actionable until metals like copper or nickel from a lower layer start appearing.
Copper	Wear from bearings, rocker arm bushings, pin bushings, thrust washers, and other brass-bronze parts.
Tin	Wear from bearings and pistons in some engines.
Silver	Wear from bearings. A secondary indicator of oil cooler problems when coolant is detected.
Titanium	Used as an alloy in steel for gears and bearings.
Silicon	Airborne dust/dirt contamination indicates poor air cleaner servicing and can accelerate wear.
Boron	A coolant additive and possible coolant leak. A detergent additive in some oils. A problem only if boron level deviates greater than 25% from the new oil or reference value.
Sodium	A coolant additive in some oils. Indicative of a coolant leak. Often seen in conjunction with potassium and/or boron. Sodium is also an indicator of contamination with salt.
Potassium	Indicative of a coolant leak; present due to the additives used in coolant formulations.
Molybdenum	Wear from rings; an additive in some oils.

(continued)

Table 14-2. *Continued*

Element	Indicators
Phosphorus	Anti-rust agent and combustion chamber deposit reducer.
Zinc	Antioxidant, corrosion inhibitor, anti-wear additive; detergent and extreme pressure additive.
Calcium	Detergent, dispersant, and acid neutralizer.
Barium	Corrosion inhibitor, detergent, and rust inhibitor.
Magnesium	Dispersant and detergent additive; alloying metal.
Antimony	Bearing overlay alloy or oil additive.
Vanadium	Heavy fuel contaminant.
Lithium	Grease thickener. Trace levels may migrate into oil but are not harmful.

14.28   Air Systems. The air for fuel combustion is drawn in through an air filter and compressed. The amount of fuel that can be burned, and therefore the power of the engine, is limited by the air mass within the cylinder. The options are to either pre-cool the air to increase the air density or use turbocharging. This raises the air density by increasing the pressure at which the cylinder is filled with air during the air intake stroke. This increases the engine power for the same cylinder size. The turbocharger is essentially a small air compressor driven by a turbine placed in the exhaust line. As the engine load increases, the exhaust gas output velocity also increases. This increases the turbine speed to drive the air compressor faster, which raises the air pressure into the cylinders. As the air is compressed into the engine cylinders, the air temperature increases, reducing available oxygen. Some engines may have intercoolers installed. These cool the compressed air, which improves combustion. Air filter cleanliness is absolutely essential, and it is surprising how boat engine air filters can get so dirty on the water. It doesn't take much filter blockage to cause power reductions. Make sure you carry a spare filter. VETUS has an air filter housing designed to attenuate airflow noise. Another enhancement to the combustion air system is the addition of a supply fan or blower. ABYC Standard H32, Ventilation of Boats Using Diesel Fuel, has a formula. The normal heat load needs to be considered and compensated for with increased air supply. Not only does the engine receive a greater amount of air but the whole space runs cooler, including rubber drive belts, alternators, and other parts. The Rule 4-inch turbo bilge blower has an airflow rate of 235 cubic feet per minute (cfm) and consumes 4.3A; it is usually interlocked with the engine control system.

14.29   Engine Exhausts. Stringent emissions legislation is now being introduced and is becoming mandatory in many parts of the world. Maintaining a clean exhaust is important as an indicator of good engine health. It is inevitable that some fuel will remain unburned after combustion and come out in the exhaust, with all those visible particulates. Maintenance is essential to reduce or maintain this emission to a minimum. The explosion of combustion is the source of an exhaust sound. The shock waves resonate through the cylinder liner and engine and then reverberate through the exhaust outlet. Wet exhausts use engine cooling water that is injected into the exhaust line to reduce the exhaust gas temperature to around 104°F to 122°F (40°C to 50°C). This reduces diesel exhaust fumes. Systems vary, but a basic system comprises the exhaust hose, the waterlock or muffler, the gooseneck, and overboard fittings. While some waterlocks are made from FRP/GRP, VETUS makes heavy-duty units made from Navidurin, which has a temperature resistance up to 500°F (260°C). It is important that water be prevented from running back to the diesel engine. The location of the water injection point becomes critical. When the water injection point is 6in (15cm) or more above the waterline, the cooling water can be injected directly into the exhaust system. When it is less than 6in (15cm) above or below the waterline, the cooling system is capable of siphoning water through the intake when the engine is off. Water is then able to fill up the exhaust system and backflow into the engine through the exhaust valves. This can be prevented by using a breather hose in the cooling water system or an air vent. Most engine makers have clear guidelines on correct installation. It is always advisable to install an exhaust temperature alarm, which is used to monitor the waterlock and sound the alarm if seawater suction is reduced, usually because of plastic obstructing the strainer or thru-hull and seacock. The indication is much faster than freshwater cooling or low lube oil pressure. Use the best high-temperature hose you can. They are made from high-quality silicon rubber incorporating woven synthetic material and encapsulated steel spiral reinforcement. Typical temperature ranges are –65°F to 350°F (–54°C to 177°C).

14.30   Anti-Siphon Valve. This is also known as a vented loop, siphon break, or anti-siphon breaker. In addition to the anti-siphon note above, it deserves a separate section. This is a subject that requires careful attention, given that these valves are critical to the safety of the boat but are often never inspected or maintained and are often almost inaccessible. Whether installed on the engine or generator exhaust system or on the toilet system, they need to be monitored and maintained. A valve malfunction can lead to your boat sinking. The vent is a single-direction, or one-way, breather valve that is installed at the top of the loop. The loop is installed as high as practicable above the dynamic or heeled waterline; recommendations vary between 1ft and 2ft (300mm and 600mm). The loop acts as a hill for water to flow up and over. The vent valve will allow air to enter the line when no water flow is following, which equalizes pressure and prevents the siphoning effect from initiating. When the engine is running and water is flowing up through the loop, the valve will close under water pressure and seal. The valve types vary from diaphragm types to duckbill or joker valves. Consider the following.

　　a.   **Vent Inspections.** Inspection will require the removal of the vent installed at the top of the loop arch. Some disassembly is required, which is often challenging because they are in hard-to-access locations. Remove vents with care to avoid damage. Check the rubber, silicon, or neoprene membranes for deformities; if degraded, fatigued, or encrusted with salt and debris, it should be replaced. Other valve types incorporate spring-loaded flaps or balls, which should be cleaned and checked; in some cases, they may require lubrication.

b. **Vent Tubes.** I inherited this arrangement on my current boat. It consists of a tube from the loop vent to overboard. It is important to check that it is not obstructed or kinked, because a blocked tube is the same as no valve and facilitates siphoning. The water outflow is a good indicator of water flow when operating the engine.

c. **Engine Exhaust Leakage.** Hose joints to the vented loop are often improperly installed, and hose clamps are not tight. There may be evidence of salt crystals around the hose joint. Check the hose end condition and tighten the hose clamps.

14.31 **Diesel Exhaust Smoke.** In normal operation, well-maintained diesel engines do not create smoke, and exhaust gases are clear. Monitor exhaust at all times; when the color changes, investigation and action are required. The mantra for a properly performing diesel engine is given as "clean fuel, clean air, and clean lubrication oil." While it is good practice to allow a diesel engine to warm up, when you have finished a run, allow a few minutes for the engine to cool down slowly.

a. **White or Gray Exhaust Smoke.** If you observe white exhaust smoke, it is possible you have fuel injection problems and incomplete fuel combustion. This may happen when an engine is started at very low temperatures but soon disappears as the engine warms through. Causes include diesel passing unburnt through the engine, coolant is entering the combustion chamber, or the engine temperature is too low. A thick white cloud is serious, and the engine should be stopped. Diesel engines require precise injector pump timing and fuel delivery at the correct pressure. A decrease in either results in white smoke from incomplete combustion:

1. Low cylinder compression: broken or leaking valves

2. Low cylinder compression: worn piston, piston rings, and cylinder

3. Low cylinder compression: stuck piston ring

4. Low cylinder compression: leaking or blown head gasket

5. Low cylinder compression: liner glazing

6. Low cylinder compression: cylinder head or block cracked

7. Injector fault: stuck open; timing gear worn

8. Injection timing problem

9. Fuel pump low pressure: worn

10. Fuel pump timing problem

11. Clogged fuel filter

12. Glow plug problems

13. Coolant water mixed with diesel fuel: cracked head

14. Cylinder head gasket failure: coolant leak

15. Cylinder head or block cracked: coolant leak

16. Cylinder head damage: coolant leak

b. **Black Exhaust Smoke.** Black exhaust smoke indicates combustion issues. It can be an indicator of too much fuel or not enough; excess air or starvation of air. Bad combustion is economically poor, environmentally damaging, and could herald major and expensive damage if not dealt with promptly:

1. Damaged turbocharger or intercooler

2. Dirty or clogged air cleaner: air restriction

3. Excessive engine sludge buildup

4. Low compression and worn piston rings

5. Dirty, worn, or malfunctioning injectors

6. Incorrect injector timing

7. Fuel pump damage

8. Cylinder valves cracking or clogging

9. Incorrect valve clearance

10. Cracked or clogged valves in cylinder head

11. Excessive carbon building up in the combustion chamber

12. Cold operating temperatures

c. **Blue Exhaust Smoke.** Blue exhaust smoke is an indicator that oil is being burnt. It often can be observed in a cold engine until it comes up to temperature and the rings expand and reseat back to normal. No quantity of blue smoke is considered normal. Always allow the engine to warm through before applying load:

1. Piston ring wear and blowby

2. Cylinder wear and glazing

3. Worn valve guides, stems, or seals

4. Engine overfilled with oil

5. Incorrect oil grade for your climate

6. Oil fuel dilution: injector stuck open

7. Lift pump worn or damaged

8. Cylinder glaze burning

9. Injector ring damaged

10. Injector pump damaged

11. Turbocharger seal leaks

12. Cracked cylinder head

13. Head oil drain line obstructed

14. Cylinder head gasket failure: breather clogged

15. Crankcase over pressure

14.32   Turbocharger Troubleshooting. Turbocharger faults cause reductions in power output, generate black exhaust smoke, and increase oil consumption. The turbocharger shaft assembly should be inspected where possible to determine the fault. This requires removal of the inlet and exhaust trunk. The turbine should be rotated by hand, and the housing should be examined for signs of contact or rubbing. The oil drain should be checked and cleaned if fouled, and oil leaks should be investigated. Engines subject to low speeds or extended idle periods tend to leak; this usually disappears when the engine is loaded up. The most common cause of turbocharger failure is hot shutdown. When an engine is shutdown suddenly, the turbocharger continues to rotate without oil. Turbo service life is reduced due to bearing wear from inadequate lubrication. Eventually, bearing wear allows turbocharger casing–to-turbine contact and out-of-balance conditions. This can cause serious damage, even destruction. When stopping an engine, it is good practice to operate at slow speeds for a few minutes. This allows the turbocharger to spool down and cooling to take place. The engine should never be revved prior to shutdown, as the turbocharger will continue rotating without lubrication, which may damage the bearings. Turbochargers rotate at high speeds and generally use engine oil for lubrication, so maintaining clean oil is essential. Signs of problems are as follows:

a.   **Turbocharger Noise.** Turbochargers tend to make serious noises when something is wrong or degraded. The most common causes are restricted or clogged air inlet filters. The rotating turbine assembly may be binding or touching the turbocharger housing. Flanges on the manifolds are possibly loose or initiating leaks. There is an object inside the compressor housing, inlet ducting, or manifold.

b.   **Turbine Assembly Binding.** This may be due to ingress of material causing turbine or compressor damage. The turbine or compressor wheel is contacting the turbocharger housing due to bearing wear or failure. There is an accumulation of carbon deposits in the turbine housing or on the turbine blades.

c.   **Seal Leakage.** This may be due to restricted or clogged air inlet filters. The oil drain lines are clogged or the crankcase breather is clogged. Other causes may be worn or failing bearings, or piston ring leakage, or high crankcase pressures. Worst case is the compressor wheel is damaged.

14.33 **Engine Operation Advice.** If you want long-term reliability and maximum life, operating a diesel engine properly will help. Note that some engine makers are installing covers over pulleys and belts to comply with the EC Machinery Directive for safety. Operate as follows:

a. **Warming Up.** Operate at 1,200rpm or less for 5 to 10 minutes; extend this to 15 minutes in very low temperatures. Running a very cold diesel engine stresses everything from the crankshaft and rods to the camshaft. Proper lubrication and combustion require a warmed-up engine. Slowly bring the engine up to speed and allow the engine to come to normal operating temperature; it saves fuel and reduces wear. Run your engine at optimum cruising speed. This varies among marine diesel engines, but 200rpm to 500rpm below peak speed is the general rule.

b. **Glow Plugs.** If the diesel engine combustion chamber is not heated correctly when starting in cold conditions, cold fuel on the semi-heated glow plugs can result in the diesel fuel gelling. This can stick to various internal surfaces and result in damage. When you activate the glow plugs, wait until they heat properly. If you suspect gelling once warmed up, replace your oil filter; it may be partially plugged.

c. **Battery Power.** As described in the battery sections, battery capacity availability is severely reduced at low temperatures. Make sure the battery is warm, if practicable, and fully charged.

d. **Lubricating Oil.** A diesel engine is two to three times harder to turn over at freezing temperatures. Cold and heavy-viscosity oil creates resistance and requires more power to turn over. At the end of every season, preempt cold-start problems with an oil change with the correct temperature-rated oil along with filters. An engine run to ensure that all internal parts are coated is recommended before shutdown. When starting the engine at any time, stop the engine if the oil pressure does not rise to nominal pressure within 10 seconds.

e. **Engine Idling.** Avoid excess idle periods. Prolonged idling causes the coolant temperature to fall below the normal range. This causes crankcase oil dilution due to incomplete fuel combustion and permits formation of gummy deposits on valves, pistons, and piston rings. It promotes rapid accumulation of engine sludge and unburned fuel in the exhaust system. If practicable, do not run at idle for longer than the normal warm-up period.

f. **Cooling Water.** If the cooling water is not coming out of the exhaust, or out of the overboard discharge, stop the engine and investigate immediately. Don't allow the pump to run dry or allow the engine to overheat. This should be the first task after starting your engine.

g. **Overheating.** If the temperature rise is gradual, the seawater pump impeller may be failing, the heat exchanger may be clogged, or the seawater inlet strainer may be clogged. A rapid rise in temperature usually means total seawater strainer blockage, pump impeller failure, or, the worst case, water pump rubber drive belt failure. Seriously consider installing an engine or exhaust alarm to provide very rapid warning of high-temperature issues.

h. **Vibration.** Although an engine vibrates in operation, it is not the only source, and you should consider other sources if you are experiencing vibration. Cavitation is one possible cause, along with worn stern bearings. An incorrect propeller pitch or an unbalanced propeller causes vibration; investigate bent propeller shafts and misalignment of propeller shaft couplings. In some cases, the propeller clearance relative to the hull and stern post can create vibration. A rotating propeller can result in impulses as the blades pass the stern gear components. The rudder, stern post, or supports can also create turbulence if improperly faired and can create vibration. Identifying the source is a process of elimination.

i. **Shutting Down.** Do not just shut your engine down. Let it cool down at idle speed for several minutes before shutting down. If you have a turbocharger, this allows the bearing temperature to reduce. If you have a turbocharger, do not give the throttle a surge before shutting down. This will spool up the turbocharger; when you shut down, the bearings will become oil starved, ruining the bearings.

14.34 **Diesel Engine Maintenance.** Maintenance should always be carried out, as a minimum, in accordance with the engine manufacturer's recommendations. I have been extensively involved in preparing and implementing planned maintenance programs on offshore vessels, as well as commissioning large marine diesels. Good maintenance will significantly reduce machinery downtime and reduce overall operating costs. The price of filters and oil is a small price to pay in comparison to a breakdown at sea and major engine damage. A good maintenance schedule, properly documented, may have significant benefits on the boat's resale value. The following maintenance tasks are guidelines only and are valid for both propulsion engines and generators. Table 14-3 outlines a typical maintenance schedule. In reality the average cruising yacht diesel engine only gets an average of 40–50 hours run time per year. Liveaboards and those on extended voyages will have longer run times, although the goal is always to run only as necessary. This means greater scrutiny of lubrication and other fluids as they can absorb moisture. 150 hours between tasks equates to around 3 years, so a more frequent inspection is required, preferably at the end or prior to each sailing season.

Table 14-3. Diesel Engine Maintenance Schedule

Task	Pre-Start	150	300	500
Check engine lube oil levels.	X	X		
Change engine lube oil.		X		
Replace engine lube oil filters.		X		
Perform engine lube oil analysis.			X	
Replace fuel pre-filters (Racor/Dahl/Vetus).		X		
Drain water from fuel tank.		X		
Replace engine fuel filters.		X		
Check and clean crankcase breathers.		X		
Replace air filter elements.			X	
Check coolant levels and expansion tank (if fitted).	X	X		
Check antifreeze additive concentrations.		X		
Check water system anodes.		X		
Check seawater pump impeller.		X		
Check freshwater pump impeller (if fitted).			X	
Check that cooler tubes are clean.			X	
Check and tighten alternator rubber drive belts.		X		
Check alternator brackets for fatigue and tightness.				
Check engine cable harness for chafe.		X		
Check that engine mounts are tight and clean.		X		
Check that flexible couplings are secure.			X	
Tighten starter and alternator connections.			X	
Check and tighten engine negative connection.			X	
Check transmission oil levels.	X	X		
Change transmission oil.		X		
Check and tighten all hose clamps.		X		
Check for oil and water leaks.	X			
Check cooling system zinc anodes.		X		
Adjust valve clearances.				X
Check for oil leaks on sump and vents.	X			
Check exhaust system for leaks.		X		
Check that all bolts, nuts, and fasteners are tight.		X		
Perform compression test.				X

Table 14-4. Diesel Engine Troubleshooting

Symptom	Probable Fault
Engine will not start	Incorrect start procedure Fuel supply valve closed/stop solenoid Fuel lift pump fault Fuel filter clogged
Low cranking speed	Air in fuel system Exhaust restriction Fuel pressure low; fuel pump fault Injector fault Low battery voltage
Engine hard to start	Gearbox engaged Air in fuel line; bleed the fuel line Cold weather; use starting aids Slow starter speed; battery voltage Water, dirt, air in fuel system Clogged air filter; replace element
Engine starts and then stops	Cold start; low engine temp Fuel filter clogged Air in fuel system Injector nozzles
Lack of power	Intake air restriction; replace element Clogged fuel filter; replace elements Overheated engine Low temp; check thermostat Valve clearances out Dirty or faulty injection nozzles Injection pump problem Turbocharger malfunction if installed Leaking exhaust manifold gasket Restricted fuel line; replace hose
High temperature	Water pump impeller fault Water pump drive belt loose Low coolant level; hose leaks Seawater pump fault Seawater strainer clogged Low lubrication oil level Air cleaner clogged Oil filter clogged Heat exchanger clogged Thermostat fault Injector fault Engine overloaded Head gasket failure Air in the cooling system
High temperature	Thermostat fault

Table 14-5. Diesel Engine Troubleshooting

Symptom	Probable Fault
Low oil pressure	Oil filter clogged Oil cooler clogged Low oil level Oil pump fault Bearing overheating Wrong oil viscosity
High oil consumption	Oil leaks, lines and gaskets Crankcase vent clogged Head gasket damage Worn piston rings High load operations
High fuel consumption	Clogged air filter Engine overload Valve clearances out Injection nozzles dirty Injection timing fault Reduced compression
Engine misfiring or irregular	Low engine oil level One injector clogged or faulty Injection pump timing Low coolant temp, thermostat Engine overheating Fuel quality; water-contaminated Air in fuel system
Engine knocks	Dirty or faulty injection nozzles Water, dirt, air in fuel system Clogged fuel filter; replace element Low coolant temp; check thermostat Exhaust or valve leak Valve damage
Engine will not start	Isolator switched off Battery voltage low Control power failure, blown fuse/breaker Stop solenoid seized Start circuit relay fault Starter cable connection fault
Starter will not crank (Power on solenoid operates)	Starter brush jammed Starter bearings seized Starter mechanical failure or seized Starter windings failed Valve damage
Low cranking speed	Low battery voltage Battery terminal loose Starter motor fault

Engine Electrical and Instrumentation

15.1   Diesel Engine Starting Systems. The systems that make up a typical marine diesel engine electrical system include the battery, engine control panel, wiring loom, preheating system, starter motor and solenoid relay, shutdown solenoids, instrument sensors and transducers, and alternator. A basic sequence of electrical functions takes place when starting the engine. When the key switch is turned to "ON"; this closes the circuit to supply voltage to the control circuit, and generally initiates alarms. When no audible or visual alarms occur, it indicates that no power is on. When the key switch is turned to the "Preheat" position, this manually or automatically energizes the heating glow plugs or heating elements. When the key switch is turned to "START" or the engine "START" button is pressed, voltage is applied to the starter motor solenoid coil. The solenoid then pulls in to supply main starting circuit current through a set of contacts. When closed the contacts supply current to the starter motor positive terminal. This then turns the starter motor to start the engine.

15.2   Diesel Engine Controls. These controls are becoming all electronic in newer engines. Throttles are still generally morse push/pull cables. Electronic controls use a CAN bus protocol to transfer commands and communications with the engine and gearbox. Digital CAN bus systems have multifunction digital engine displays and eliminate the gauges and alarms. (See the later chapter on engine controls.)

15.3   Electric Engine Starters. Essentially, the electric starter consists of a DC motor, a solenoid, and a pinion-engaging drive. The DC motor is typically series wound, as it provides the high initial torque required to exceed friction and inertia, such as oil viscosity, and cylinder compression and accelerate the engine to a point where self-ignition temperatures and combustion starts—typically in the range 60rpm to 200rpm, depending on whether glow plugs are used. The starter motor torque is transmitted by the pinion and ring gear on the flywheel. The drive gear pinion has a reduction gear of around 15:1.

    a.   **Starter Solenoid Relay.** The starter solenoid relay is essentially a large heavy-current relay that consists of a coil and armature, along with moving and fixed contacts. The solenoid is mounted directly to the starting motor housing, which reduces cables and interconnections to a minimum. When the solenoid coil is energized by the starting circuit, the solenoid plunger is drawn into the energized core; this closes the main contacts to supply current to the starter motor. On some starters the solenoid has a mechanical function. The solenoid activates a shift or engaging lever to slide the overrunning clutch along the shaft to mesh the pinion gear with the flywheel; when engaged, the starter motor then turns the engine, so meshing occurs before starting.

    b.   **Starter Motors.** The motor consists of four pole shoes or magnets; some use permanent magnets. The poles are fitted with an excitation winding, which creates the magnetic field when current is applied. The rotating part, called the armature, also incorporates the commutator. The four carbon brushes provide the positive and negative power supply. There are four basic DC motor types in use, and they are based on connection of the field windings. The field windings are connected either in series or parallel with the armature windings:

(1) **Shunt (Parallel)-Wound Motors.** The motor operates at a constant speed regardless of loads applied to it. It is the most common motor used in industrial applications and is suited to applications where starting torque conditions are not excessive.

(2) **Permanent Magnet–Excited Motors.** The permanent magnet starter offers the advantages of reduced weight, physical size, and less heat generation than normal field type starters. Current is supplied through the brushes and commutator directly to the armature. Another feature is a reduction gear, which allows faster speeds and increased torque.

(3) **Series-Wound Motors.** The speed of this type of DC motor varies according to the load applied; speed increases with load decrease.

(4) **Compound Motors (Series/Shunt-Wound).** This configuration is often used on large starter motors. It combines the advantages of both shunt and series motors and is used where high starting torques and constant speeds are required.

c.   **Pinion-Engaging Drives.** The pinion-engaging drive is located within the end shield assembly of the starter and consists of the pinion-engaging drive and pinion, the overrunning clutch, and the engagement lever or linkage and spring. When the motor operates, the drive gear meshes with the ring gear or flywheel teeth to turn the engine and then disengages after starting. The overrunning clutch has two important functions. The first is to transmit the power from the motor to the pinion; the second is to stop the starter motor armature from overspeeding and being damaged when the engine starts. Preengaged starters generally use a roller type clutch; larger multi-plate types are used in sliding gear starters.

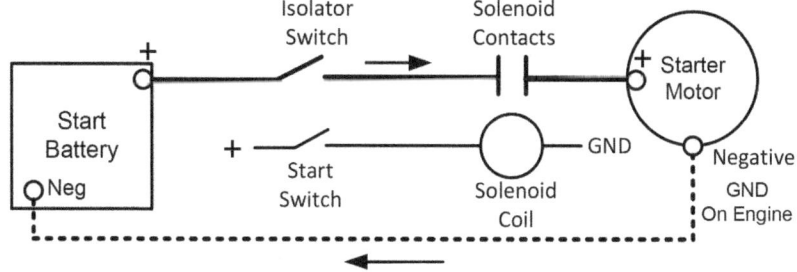

Figure 15-1. Basic Engine Start Circuit

15.4   Starter Types. There are several types of starters in use, the most common being the overrunning clutch starter; the inertia-engagement Bendix drive is now less common. There are four basic groups of starter motors.

a.   **Preengaged (Direct) Drive Starters.** The most common type of starter motor is the solenoid-operated direct drive unit; the operating principles are the same for all solenoid-shifted starter motors. When the ignition switch is placed in the "Start" position, the control circuit energizes the pull-in and hold-in windings of the solenoid. The solenoid plunger moves and pivots the shift lever. This moves the pinion along the shaft to mesh or engage with the flywheel toothed ring gear. When the solenoid plunger is moved all the way, the contact disc closes the circuit from the battery to the starter motor. Current now flows through the field coils and the armature. This develops the magnetic fields that cause the armature to rotate and turn the engine.

b.   **Gear Reduction Starters.** Some manufacturers use a gear reduction starter to provide increased torque. The gear reduction starter differs from most other designs in that the armature does not drive the pinion directly. In this design, the armature drives a small gear that is in constant mesh with a larger gear. Depending on the application, the ratio between these two gears is between 2:1 and 3.5:1. The additional reduction allows a small motor to turn at higher speeds and greater torque with less current draw. The solenoid operation is similar to the solenoid-shifted direct drive starter in that the solenoid moves the plunger, which engages the starter drive.

c.   **Sliding Gear Drive Starters.** These two-stage starters have either mechanical or electrical pinion rotation. The electrical units have a two-stage electrical pinion-engaging drive. The first stage allows meshing of the starter pinion without cranking the engine over. The second stage starts when the pinion fully travels and meshes, and then allows full excitation and current flow to the starter motor. The first stage of mechanical units is a solenoid switch, which pushes forward the pinion-engaging drive via a lever. When pinion meshing occurs, current is applied to the starter through the solenoid switch.

d.   **Bendix Drive Inertia Starter.** The Bendix friction-clutch mechanism drive was developed in the early twentieth century. These starters use a drive friction clutch, which has a drive pinion mounted on a spiral-threaded sleeve. The sleeve rotates within the pinion, moving the pinion outward to mesh with the flywheel ring gear; the impact of this meshing action is absorbed by the friction clutch. Once started, the engine turns at a higher speed and drives the Bendix gear at a higher speed than the starter motor. The pinion then rotates in the opposite direction to the spiral shaft before disengaging. The drive pinion being thrown out of mesh and then stopping is a common fault. Always wait several seconds before attempting to restart or the drive mechanism may be damaged. Another fault is when the pinion does not engage after the starting motor is energized. If the starter emits a high-pitched whine, turn off the ignition immediately or the unloaded DC starter motor will overspeed and be seriously damaged. Problems can be minimized by ensuring that the sleeve and pinion threads are clean and lubricated so that the pinion engages and disengages freely. The Bendix gear, shaft, bearings, and end plates can be cleaned of dried grease with a suitable cleaner such as CRC Lectra Clean. You can lubricate with WD-40 multipurpose penetrating oil or a fine 3-in-1 sewing machine oil. There are starter motor specific greases with cold temperature capability. The PolySi G-MAN Lubricant, PST-433 extreme low temperature grease is a well-known product. If you lay up your boat every winter in a very cold climate, consider this or an equivalent product.

15.5   Starter Installation, Maintenance, and Troubleshooting. Starter installation is generally limited to two factors. The first is being mechanically secure; the second is that the attached electrical cables are of the correct rating and that terminal nuts are properly torqued so they do not work loose. In addition, the negative cable should be attached as close as practicable to the starter. Starter motor design is generally robust, as it must withstand the shocks of meshing, engine vibration, salt- and moisture-laden air, water, oil, temperature extremes, high levels of overload, and other factors. Preventive maintenance is essential to ensuring reliability.

a.   **Starter Shaft Corrosion.** A common problem, especially on idle vessels, is the buildup of surface corrosion or accumulated dirt on the shaft and pinion gear assembly; lack of lubrication causes seizure or failure to engage. It is good practice to remove the starter every 12 months and clean with CRC Lectra Clean and lightly oil the components according to the manufacturer's recommendations or use a starter motor specific lubricant.

b.   **Starter Motor Maintenance.** Problems often occur with seized brushes, and this is primarily caused by lack of use. Always manually check that brushes are moving freely in the brush holders and that the commutator is clean. Remove all dust and particles using a vacuum cleaner. If badly soiled, wash out with a quality spray electrical cleaner such as CRC Lectra Clean. Follow the DC motor maintenance procedures. Never clean or polish the commutator with any abrasive materials.

c.   **Starter Troubleshooting.** Many are familiar with the silence and then a loud click when the start solenoid operates but the starter fails to turn over. The main causes are a poor negative cable connection, a poor positive cable connection caused by loose or dirty connections, or a solenoid plunger sticking and not closing fully, preventing the main contacts from closing. See the image in the color insert of degrading starter motor termination.

d.   **Negative Conductor Connections.** To prevent the connection becoming loose from vibration, the main negative cable should be secured on the engine using a spring washer. The main engine negative cable connection is prone to vibration from the engine operation. Cables frequently come loose, causing starting problems, with high resistance and intermittent equipment operation, interference, and, in some instances, alternator failures. In most cases they are simply fastened to a convenient bolt, which is not acceptable if you value reliability. The mating surface must be cleaned to ensure a high-quality, low-resistance electrical contact. Install a spring washer to maintain tension.

15.6   Preheating Systems. Many marine diesel engines will not start without preheating, and extended engine starting turnover times may overheat and damage the starter.

a.   **Glow Plugs and Pins.** Direct injected (DI) engines commonly have a glow plug or pin heater installed within each cylinder. They preheat the air in each cylinder to facilitate starting. In cold weather, this will dramatically decrease the electrical power required to start the engine.

(1)   **Activation.** Prior to engine starting, the plugs or pins are activated for an operator-selected time period or interlocked to a timer. Beta Marine recommends 6 seconds maximum when activating; otherwise you can burn out the plug or pin.

(2)   **Power Consumption.** The glow plugs and pins can draw relatively large current levels for a short time. If your battery voltage is low, allow a few seconds after preheating before starting; this enables the battery voltage to recover from the heater load.

b.   **Air Intake Heaters.** These grid-resistor heaters are installed in the main air intake of IDI engines, and there is normally only one heating element.

c.   **Preheater Control.** Many preheating circuits have relays, either timed or untimed. Timed relays are a common cause of failure. It is advisable to have a straight relay with a separate switch, and simply preheat manually for 5 seconds and then start.

15.7   Preheater Maintenance. The following maintenance tasks should be carried out to ensure system reliability.

a.   **Electrical Connections.** Preheater glow plug connections must be regularly checked if they are to function properly. The connections must be cleaned and tightened every 6 months.

b.   **Cleaning.** The insulation around the glow plug connections must be cleaned. It is a common fault to have tracking across oil and sediment on the engine block, with a serious loss of preheating power.

c.   **Glow Plug Cleaning.** The plugs should be removed and cleaned yearly. Take care not to damage the heating element.

15.8 **Preheater Glow Plug Troubleshooting.** The following table lists the most common faults on preheating systems.

Table 15-1. Preheater Glow Plug Troubleshooting

Symptom	Probable Fault
No preheating	Loss of power (fuse failure) Connection fault on engine Relay failure Connection on ignition switch disconnected Circuit short-circuiting to engine block
Partial preheating	One or more glow plug failed Glow plug interconnection failure Dirt around glow plug causing tracking

15.9 **Engine Starting Recommendations.** The power supply to the engine starting system should have an isolator installed as close as practicable to the battery in both the positive and negative conductor. The isolator should be accessible. Short circuit protection is not required. The isolator should be rated for the maximum current of the starting circuit. It should be rated for the maximum current starting circuit. The main starting circuit positive and negative conductors should be rated so as not to exceed a 5% voltage drop at full rated current. Cables should be kept as short and as large as possible to minimize losses and maximize power availability. Voltage drops mean less volts at the starter and slower turning speeds, with harder starting.

15.10 **Engine Starting Systems.** A starting system must be viewed not simply as a collection of series-connected components but as a system. The typical starting system comprises the following elements.

    a.   **DC Positive Circuit.** This includes connections at the battery, the isolator or changeover switch, the solenoid connection, solenoid contacts, and the starter motor (which includes several components, such as brushes, brush gear, commutator, bearings, and windings).

    b.   **DC Negative Circuit.** This includes cable connections at the battery, engine block, the cable back to the battery, the engine block, and the meter shunt (if fitted).

    c.   **Engine Control System.** This is from the panel and includes the key switch, stop and start buttons, wiring harness, connectors, and fuses.

    d.   **Preheating System.** This includes heating elements, called glow plugs, and interconnections, relays, and connectors.

    e.   **Electrical Power Supply.** This is the start battery and is covered elsewhere.

15.11   Starting System Configurations. There are several engine starting configurations and arrangements.

a.   **Remote Battery Isolators.** Many boats have simple mechanical isolation switches to isolate the engine starter motor power supply. In many cases, remote isolation using relay-type isolators is used. The control relay may be operated from a separate switch or interlocked to the main key switch so that power is applied when the switch is turned.

b.   **Two-Pole Engine Systems.** In many engines that have dual-pole isolated systems, two battery isolation relays are installed—one on positive and one on negative. The relay coil is connected to the alternator D+ terminal, and this energizes the coil when the alternator is operating. In remote isolation relay systems, one relay can be used to energize both switches.

c.   **Parallel Battery Starting Systems.** Some vessels have a 12V power system and a 24V engine system. The batteries are configured through the relay so that the parallel-connected batteries are series connected to 24V when the engine start switch is operated.

d.   **Parallel Connected Starters.** Larger engines use two starter motors, which keeps the size down. The system uses a large-capacity double-acting relay to supply current to both starter motors simultaneously.

15.12   Starting System Failure Mode Analysis. The math of any simple analysis shows a total of about fourteen connection points plus the solenoid coil, the starter motor, the battery, and the key switch that can impact the starting system. Each point represents a single point of failure, with subsequent total system failure with no apparent redundancy. If a person persists with turning over an engine that will not start, the starter motor may burn out. Other less-common scenarios include:

a.   **Starting Circuit.** Relocate and connect the main negative cable to, or as close as possible to, the starter motor. This maintains two connections but takes the engine block out of the circuit and generally reduces voltage drop in the circuit.

b.   **Maintenance Factors.** Perform the recommended maintenance on all critical equipment and systems.

c.   **Starter Motors.** Starter motors have low failure rates, as actual operating hours are relatively low. Failures are dependent on operational frequency, with seized bearings or stuck brush-gear being the major causes of failure. Regular operation reduces failures. In addition, this generates heat, which assists in displacing moisture within windings. Starter motors should be cleaned or overhauled on a regular basis, ideally not exceeding 2 years.

15.13   Starting System Diagrams. In the previous three editions of this book, I included simplified wiring diagrams for a variety of engines. Over the years, there have been so many engine variants and technology changes that this is no longer practical. Always check the diagrams supplied in the operator's manual for your specific engine model; you can download a copy from the manufacturer's website to your tablet or notebook. Make sure you have the correct circuit diagram for the installed engine; I suggest printing and laminating a copy. Wiring varies considerably, even between older and newer engine models. The following table gives equivalent color codes for various manufacturers. The illustration is a typical Volvo engine starting circuit; although newer engines have different arrangements, it does illustrate the various elements.

Table 15-2. Engine Wiring Color Codes

Purpose	US Codes	Yanmar	Bukh	Volvo	Perkins	Nanni
Ignition Start	yellow/red	white	blue	red/ yellow	white/red	brown
Ignition Stop	black/ yellow	red/black	black	purple	black/blue	white
Diesel Preheat	orange/ yellow	blue	yellow	orange	brown/red	orange
Negatives	black	black	black	black	black	black
Alternator Light	brown	red/black	green	brown	brown/ yellow	green
Tachometer	gray	orange	yellow	green	black/ brown	blue
Oil Press. Gauge	light blue	yellow/black	green	light blue	green/ yellow	gray
Oil Press Alarm.	white/blue	yellow/white	brown	blue/ white	black/ yellow	gray/red
Water Temp Gauge	tan	white/black	brown	light brown	green/blue	yellow
Water Temp Alarm	white/ brown	white/blue	yellow/ gray	brown/ white	black/ light green	yellow/ red

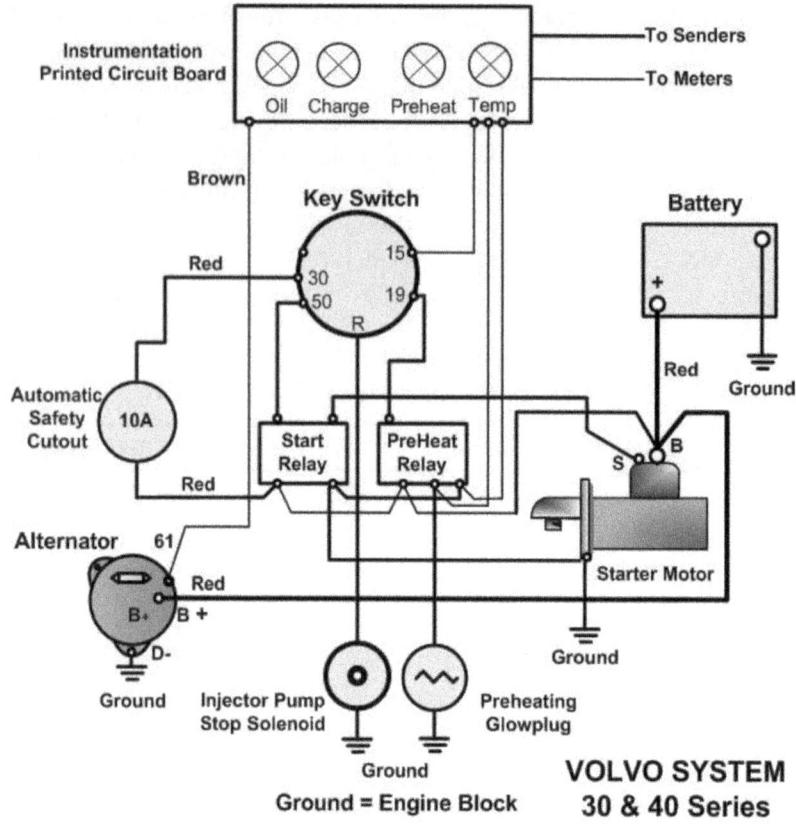

Figure 15-2. Typical Volvo Starting Circuit

15.14   Engine Instrumentation. Engine instrumentation is crucial to ensuring that engines operate correctly within the designed parameters. Instruments may consist of a bank of discrete analog meters or an integrated system with digital and visual screen displays, and most manufacturers have such systems. The latter is becoming more prevalent and consists of trend analysis, alarm set-point management, alarm logging, and other advanced features. Check all sender-unit terminals and connections regularly along with a test of all alarm functions, preferably before you start your voyage. Typical of the new-generation digital screens is the Yanmar multifunction display. The processor is able to display virtual gauges as well as alarms and diagnostic troubleshooting codes. Parameters include engine speed, engine coolant temperature, engine hours, oil pressure, and engine load, along with wind, speed, depth, and AIS data. The ultrawide full-color display has a 170-degree viewing angle. The NMEA 2000 connectivity allows data transfer to all compliant devices; an example is Raymarine i70 control heads, which can display engine operating data. The keys on many control stations are made from waterproof silicon. Alarm functions are as follows:

a.   **Oil Pressure Alarms.** A pressure alarm either is incorporated into a gauge sender unit or is a separate device. It consists of a pressure-sensitive mechanism that activates a contact when the factory-set pressure is reached. They are grounded to the engine block on one side, and operation grounds the circuit, setting off the panel alarm. To test the alarm circuit, simply lift off the connection and touch it to the engine.

b.   **Water Temperature Alarms.** These standalone alarm devices consist of a bimetallic element that closes when the factory-set temperature is reached. To test, simply remove the connection from the sender terminal and touch on the engine block to activate the alarm. The sensor has two terminals; "G" is used for the meter, and "W" is used for the alarm contact. In many boats, damage is often done because the alarm either did not function or was not noticed. The first reaction is often "What's wrong with the alarm?" not "What's wrong with the engine?" It is good practice to add a very loud audible alarm, as some of the engine panel units are difficult to hear at times over ambient engine noise.

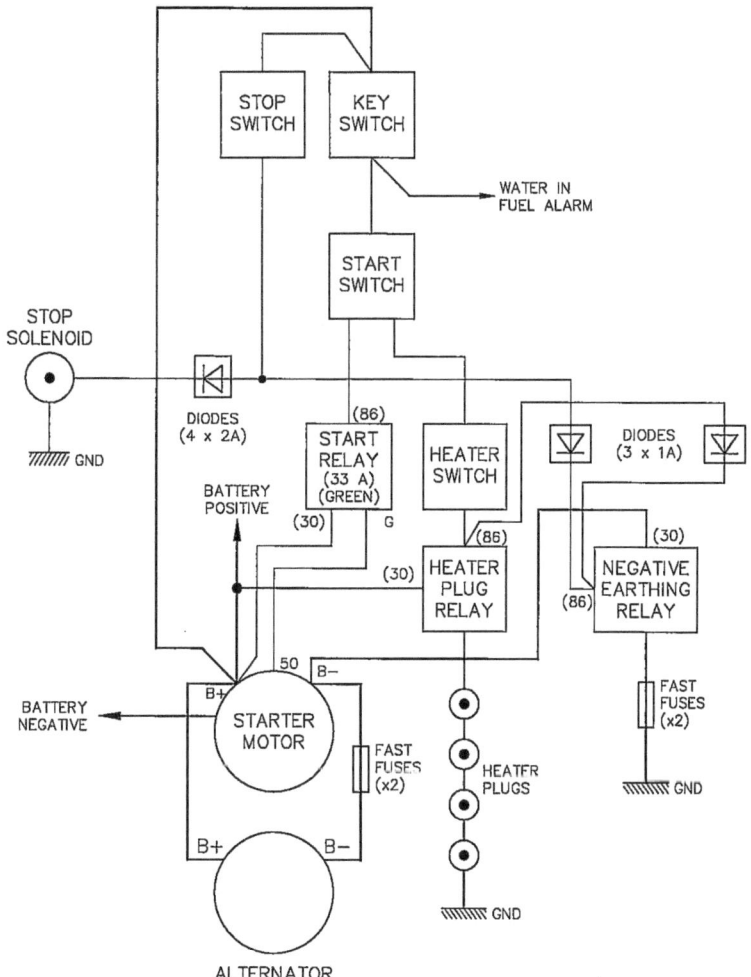

Figure 15-3. Typical Perkins Starting Circuit

15.15   Pressure Monitoring. Monitoring pressures is fundamental to proper operation of any engine. This includes lubricating oil and filter differential pressures; fuel and filter differential pressures; coolants, both seawater and fresh water; turbocharger charging air pressure and air inlet pressures; gearbox and transmission oil pressures; and engine crankcase pressures.

      a.   **Oil Pressure Monitoring.** The oil pressure sender unit is a variable resistance that alters proportional to pressure. It is very common in alarm situations to assume the meter or alarm is wrong. Low oil pressure readings are caused by low lube oil level or a clogged oil filter causing a lowering in oil pressure. A faulty oil pump can cause a lowering in pressure or a rise in oil temperature caused by an increase in engine temperature or an oil cooler problem.

      b.   **Sensor Checks.** Oil pressure sender units should be removed every year and any oil sludge cleaned out of the fitting, as this can commonly clog, causing an inaccurate or no reading. Sender units are poorly grounded or Teflon tape is improperly applied to threads to make a high-resistance contact.

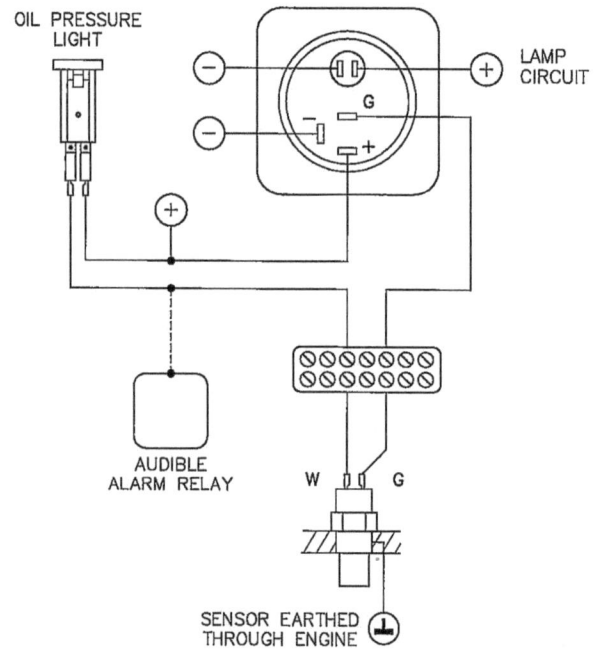

Figure 15-4. Oil Pressure Monitoring

15.16   Temperature Monitoring. The main temperature monitoring points utilizing the same sensor types include lubricating oil; transmission oil; coolants, both seawater and fresh water; fuel temperature; aftercooler; and turbocharger inlet air.

      a.   **Water and Oil Temperature Gauges.** Monitoring water temperature is essential to safe operation of the engine. Temperature extremes can cause serious engine damage or failure. Sender units are resistive and give a resistance proportional to temperature in a nonlinear curve.

b.   **Sensor Checks.** If the gauge readings are not correct and a gauge test shows it to be good, check the sensor. Before you check the sender unit, the main causes of high temperatures are loss of freshwater cooling caused by a faulty water pump impeller, loose rubber drive belt, low water levels, fouled coolers, and increases in combustion temperatures. Loss of saltwater cooling is caused by a blocked intake or strainer, faulty water pump impeller, clogged cooler, or aeration caused by a leak in the suction side of the pump. Increased engine loadings caused by adverse tidal and current flows or overloading. Sender units are poorly grounded or Teflon tape is improperly applied to threads to make a high-resistance contact.

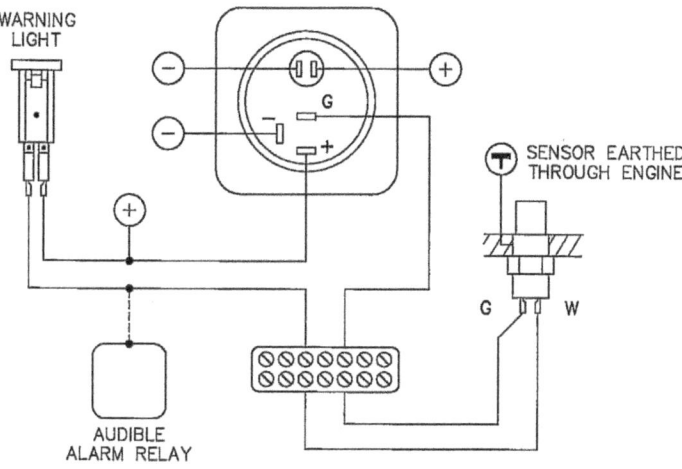

Figure 15-5. Water Temperature Monitoring

15.17   Exhaust Gas Temperature Monitoring. Exhaust gas temperature monitoring is used in commercial ships and is recommended on all yachts. Engine problems are easier and faster to identify than water temperature and oil pressure monitoring. These can be problems within the cooling water system; increased engine loads that may be caused by adverse tidal and current flows; air intake obstructions caused by clogged air filters or, where installed, blocked air coolers; combustion chamber problems caused by defective injectors, valves, and other components. Larger-engine boats will have cylinder monitoring; this allows identification of problems specific to cylinders to be identified and monitored. Smaller engines may have a sensor installed on the main exhaust manifold. Pyrometer compensating leads and wiring should be routed clear of other cables to avoid induction and inaccurate readings.

a.   **Operating Principle.** Exhaust temperature sensors are called thermocouples or pyrometers. These sensors consist of two dissimilar metals—iron/constantan, copper/nickel, platinum/rhodium, nicrosil/nisil, nickel/aluminum—which at the junction will generate a small voltage proportional to the heat applied to the sensor. The voltage is measured in millivolts (mV). The typical thermocouple consists of a sensing junction and a reference junction. The open circuit voltage is measured with a high impedance voltmeter and is the temperature difference between the sensing junction and the reference junction. The thermocouple junction is also called the "hot junction." The compensating cables between the junction and the measurement meter are electrically matched to maintain accuracy; they are polarity sensitive, so must be connected positive to positive.

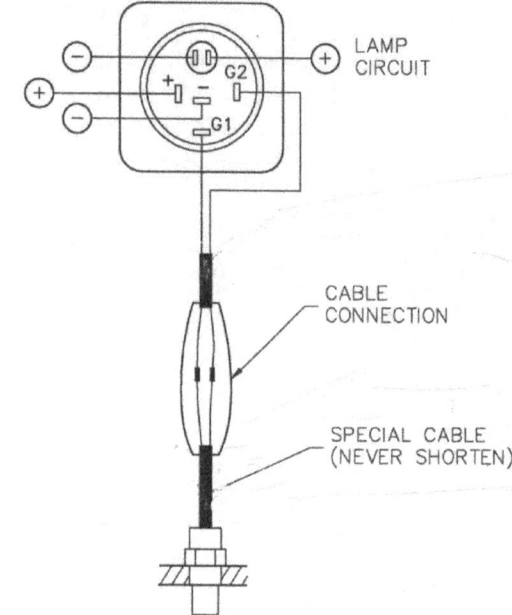

Figure 15-6. Exhaust Gas Temperature Monitoring

15.18   Engine Tachometers. The tachometer is used to monitor engine speed, differential or synchronization, shaft revolutions, and turbocharger speed.

a.   **Generator Tachometer.** These tachometers take a signal from a mechanically driven generator unit. The generator outputs an AC voltage proportional in amplitude to the speed, and this is decoded by the tachometer. Variations in speed give a proportional change in output voltage and a change in meter reading. The most common fault on these units is damage to the drive shaft mechanism.

b.   **Inductive Tachometer.** These tachometers have an inductive magnetic sensor that detects changes in magnetic flux as the teeth on a flywheel move past. This sends a series of on/off pulses to the meter, where they are counted and displayed on the tachometer. Ensure that the sender unit is properly fastened. A common failure is damage to the sensor head by striking the flywheel when adjusted too close.

15.19   Engine Alternator Tachometers. These tachometers derive a pulse from the alternator AC winding, typically marked "W." The alternator output signal is a frequency directly proportional to engine speed. The pickup is taken from the star point or one of the unrectified phases. Typical connections for VDO tachometers are illustrated. If the alternator is faulty, there is no reading. There are a number of different alternator terminal designations used by various manufacturers; the main ones are W, STA, AC, STY, and SINUS. If there is no output terminal, a connection will have to be made, as shown in the illustration, and an alternator tachometer installed.

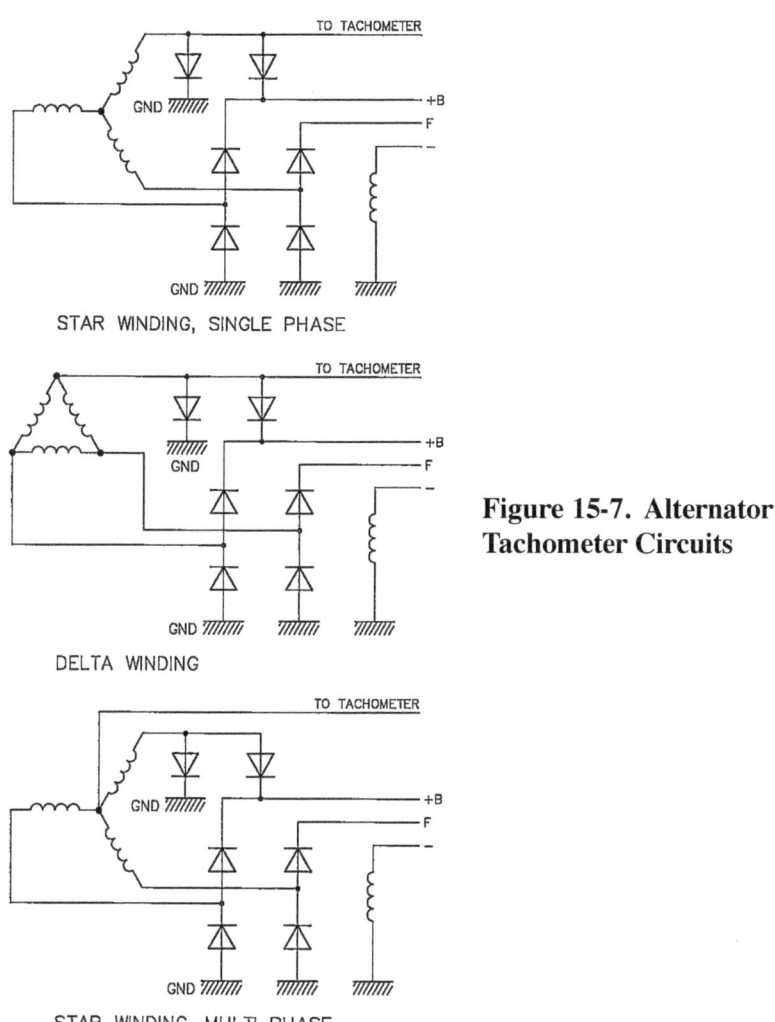

STAR WINDING, SINGLE PHASE

DELTA WINDING

Figure 15-7. Alternator Tachometer Circuits

STAR WINDING, MULTI PHASE

15.20   Electrical System Monitoring. There are a number of electrical monitoring parameters and methods of instrument installation. These are linked to alternator battery charging and include the following:

   a.   **Charging Voltmeters.** Many instrument panels incorporate a voltmeter to indicate the state of the charging. As they have a coarse scale, they are only partially useful in precisely assessing battery voltage states; however, they are a useful indicator on the charging system. Many voltmeters have a colored scale to enable rapid recognition of condition—red for under- or overcharge and green for proper range. Digital voltmeters are quite common now,

275

usually as part of a battery monitoring system. Many switch panels also have digital voltmeters installed.

b. **Charging Ammeters.** Charging ammeters are reasonably popular and, again, are an easy guide to the level of charge current from the alternator.

(1) **Series Ammeter.** This ammeter type has the main charge alternator output cable running through it. In many cases, the long run to a meter causes unacceptable charging system voltage drops and undercharging. An additional problem with installing such ammeters on switch panels is that the charge cables are invariably run with other cables, causing radio interference. If you are going to install this type of ammeter, ensure that the meter is mounted as close as possible to the alternator. If these ammeters start fluctuating at maximum alternator and rated outputs, this is generally due to voltage drops within the meter and cable. The underrating of connectors is a major cause of problems.

(2) **Shunt Ammeter.** The external DC ammeter shunt, or shunt resistor, is a precision calibrated resistance connected in heavy load paths such as an alternator output. It provides a low resistance path, typically 0.001 ohms, to the current flow and has a very small voltage drop across it. Shunts can be installed in either the positive or the negative current carrying conductor. Shunts are rated specifically

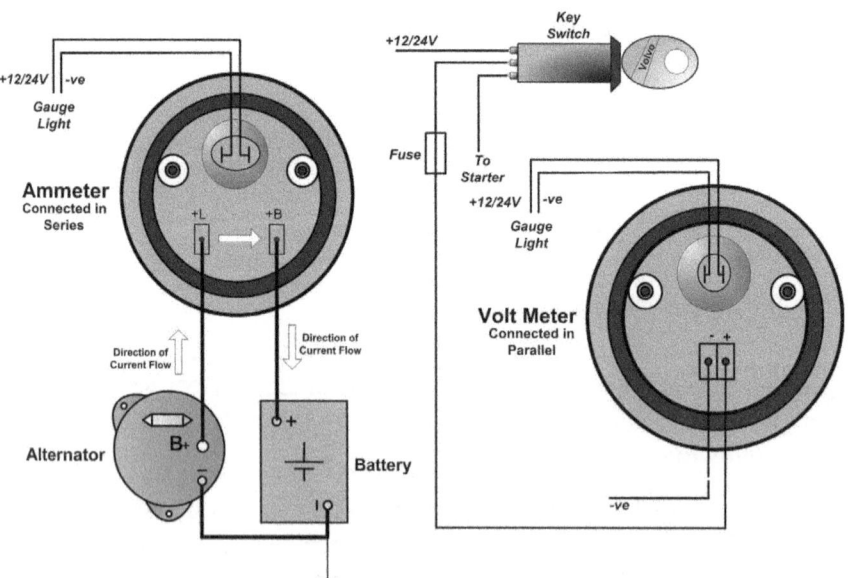

Figure 15-8. Voltmeter and Ammeter Connection

for the current, such as 500A to 50mV, and ratios vary. Wires connected across the shunt are led to the ammeter to shunt or carry a millivolt value that is proportional to the current passing through the shunt. Some meters require a shielded and twisted pair cable 16AWG (1.3mm²) to reduce interference in the millivolt signal to the meter. The ammeter, which is a millivolt meter, then displays the current flowing through the shunt. It is important that cable terminations to a shunt are very tight and low resistance as this will affect reading accuracy. The meter interconnecting wires are polarity sensitive and that must be observed. A shunt should always be rated for the alternator output.

15.21   Hour Counters and Clocks. Operating hour counters are essential to keeping a record of maintenance intervals. Essentially, an hour counter is a clock that is activated only when the engine is operating. There are a number of methods of activating hour counters.

a. **Ignition Switch.** This is the easiest and most practical method. The meter is simply connected across the ignition positive and a negative so that it operates when the engine is running.

b. **Oil Pressure Switch.** This is not common, although some installations activate through the switch so that operation is only when the engine is operating.

c. **Alternator.** In many installations the counter is activated from alternator auxiliary terminal D+ or 61.

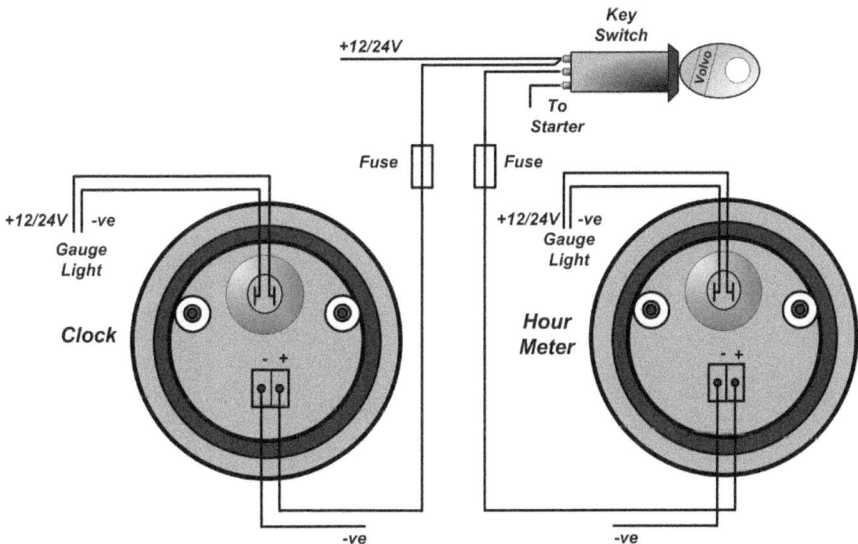

Figure 15-9. Hour Meter and Clock Connection

15.22 **Acoustic Alarm Systems.** Acoustic alarms are generally interconnected to warning light circuits, and the buzzer is activated by a relay. Acoustic warnings are activated along with lamp from sensor contact "W." The acoustic alarm should be activated through a relay, not through the sensor contact, which is not rated for such loads.

a. **Buzzer Test.** Using a lead, connect a positive supply to the buzzer positive terminal and ensure that a negative one is connected. If the buzzer operates, remove the bridges. Ideally, a test function should be inserted into the circuit so that alarm function can be verified.

b. **Operating Test.** With alarm lights on, put a jumper from negative to the buzzer negative, as sometimes a "lost" negative is the problem. Connect a positive supply to the relay positive, typically numbered "86." If the relay does not operate and the buzzer is working, the relay is suspect. Verify this after removal using the same procedure. Note that sometimes a relay may sound like it is operating but the contacts may be damaged and open-circuited. If a buzzer is not operating along with lights, either a cable or connection is faulty or the operating relay is defective.

c. **Mute Function.** On many home-built engine panels, it is essential to silence the alarm, which entails placing a switch in line with the buzzer; the lamp will remain illuminated to indicate the alarm status.

d. **Time Delays.** During engine start-up, a time delay is necessary to prevent alarm activation until oil pressure has reached normal operating level. Time delays are typically in the range 15 to 30 seconds.

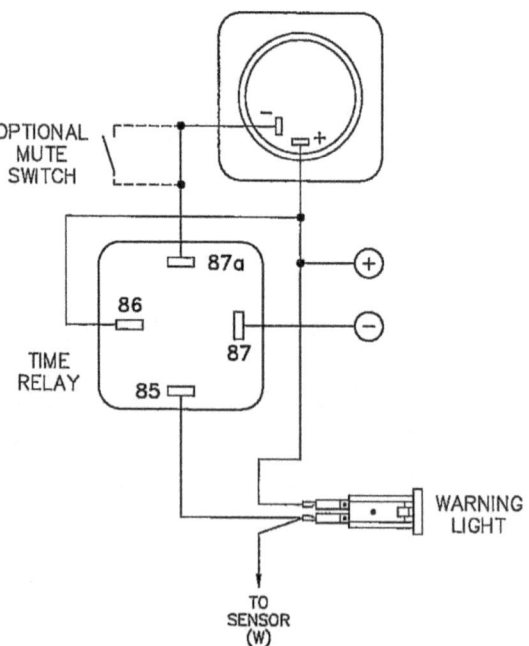

Figure 15-10. Acoustic Alarm

15.23   **Instrumentation Maintenance.** Maintaining instruments is relatively simple.

    a.   **Electrical Connections.** A regular check of all sensor unit terminals and connections, along with a test of alarm functions, is all that is required.

    b.   **Oil Pressure Sensors.** Oil pressure sensors should be removed every year and any oil sludge cleaned out of the fitting. Sludge-clogged sensors may be inaccurate or show no reading.

15.24   **Gauge Testing.** Use the following procedure on suspected gauge faults.

    a.   **Open Sensor Test.** Remove the sensor lead marked "G" from the back of the gauge. Switch on meter supply voltage. The gauge needle should now be in the following positions:

        (1)   **Temperature gauge:** Left-hand hard-over position

        (2)   **Pressure gauge:** Right-hand hard-over position

        (3)   **Tank gauge:** Right-hand hard-over position

    b.   **Sensor Ground Test.** This test involves bridging sensor input terminal "G" to negative. The sensor lead must be removed and the meter supply on. The gauge needle should now be as follows:

        (1)   **Temperature gauge:** Right-hand hard-over position

        (2)   **Pressure gauge:** Right-hand hard-over position

        (3)   **Tank gauge:** Left-hand hard-over position

15.25   **Engine Alarm Set Points.** It is important to know when maximum values are being approached. Use the table to check your own, and enter the specific values from your engine manual for reference. These figures are ranges for alarms and automatic shutdowns for either main engines or generators (14.5psi = 1bar; 1psi = 0.07bar).

Table 15-3. Engine Alarm Set Points

Parameter	Typical Value	Your Boat Value
Lube Oil Temperature	194°F–203°F (90°C–95°C)	
Lube Oil Pressure	14psi–28psi (1bar–2bar)	
Lube Oil Filter Differential Press	14psi–20psi (1bar–1bar)	
Pre-lube Oil Pressure	14psi–28psi (1bar–2bar)	
High Crankcase Press.	0.2psi (0.01bar)	
Fresh Water Cooling Outlet Temp	212°F–220°F (100°C–105°C)	
Fresh Water Cooling Maximum Temp	220°F–230°F (105°C–110°C)	
Fresh Water Pressure	3psi–6psi (0.2bar–0.4bar)	

(continued)

Table 15-3. *Continued*

Parameter	Typical Value	Your Boat Value
Fuel Pressure	25psi–40psi (2bar–3bar)	
Fuel Filter Differential Press	7psi–10psi (0.5bar–0.8bar)	
Fuel Temperature Max Temp	149°F (65°C)	
Air Inlet Manifold Temp	187°F–202°F (85°C–95°C)	
Air Inlet Manifold Pressure	35psi–50psi (2.5bar–3.5bar)	
Exhaust Temp Cylinder	933°F–1,023°F (500°C–550°C)	
Exhaust Temp Manifold	1,113°F–1,203°F (600°C–650°C)	
Exhaust Temp Deviation	122°F (+/– 50°C)	
Exhaust Temp to Turbo	933°F–1,023°F (500°C–550°C)	

15.26   Sensor Testing. These tests are good for most makes of sensors. Disconnect the cables and using a multimeter—digital or analog—set on the resistance (Ω) range. Place the positive (red) meter probe on the terminal marked "G" on the sensor; if a dual alarm and sensor output, the alarm output is marked "W." Place the negative (black) meter probe on the sensor thread.

    a.   **Temperature Sensors.** Readings should be as follows:

        (1)   104°F (40°C) = 200Ω to 300Ω.

        (2)   248°F (120°C) = 20Ω to 40Ω.

    b.   **Pressure Sensors.** Readings should be as follows:

        (1)   High pressure (engine off) = 10Ω.

        (2)   Low pressure (engine running) = 40psi:105Ω; 60psi:152Ω.

    c.   **Fuel Tank Sensors.** Reading should be as follows:

        (1)   Tank empty = 10Ω; tank full = 180Ω.

15.27   Crankcase Oil Mist Detection. Large motorboats, sailing yachts, and superyachts have larger output medium-speed diesels where crankcase explosions can occur, causing fatalities and serious engine damage. Oil mist detectors use an air supply at 7psi (0.5bar) to take atmosphere samples from each crankcase compartment and measure the turbidity or opacity. The sample is passed between an infrared emitter and detector, where the oil mist absorbs some light, reducing the light level at the detector. This opacity is measured as a percentage, with a typical operating range of 0% to 7% and maximum alarm levels around 2.4%. The friction of nonlubricated metals causes very rapid overheating, and oil mist can rise to critical levels in 20 to 30 seconds. Oil mist is explosive at a quantity of 50mg of atomized oil per liter of air, with ignition at 930°F (500°C).

15.28   Engine Space Oil Mist Detectors. The vast majority of engine room or compartment fires are initiated by a fuel leak and subsequent ignition of the oil mist. When a fuel leak occurs, oil mist is projected into the space. Leaks are caused by small perforations of fuel lines or a fracture or leak of pressurized fuel, which can be diesel, lubricating oil, or hydraulic oil. They have sources in injectors or in engine high-pressure fuel systems and oil lines that atomize the fluid when it escapes. Initially, such escapes are invisible. Oil mist can be very small droplets, down to 1 to 10 microns, and disperse within the surrounding air. When that mist attains the lower explosion limit and comes into contact with a heat source of 400°F (200°C), an explosion and fierce fire is the result. Ignition sources have diverse origins that include bearings, turbochargers, exhaust systems, and electrical sources such as electric contacts, faulty wiring, motors, and static electricity. Oil mist detection systems are what are termed "optical opacity meters." Older systems used infrared light, but technology has advanced and lasers are now employed. Laser light is transmitted from the transceiver to a reflector and back, and as the optical qualities of the laser are precisely known, any oil mist present will be detected by opacity of the laser light. These detectors are programmable, and alarms and warnings can be calibrated to any opacity parameter. They are also applicable to other machinery spaces with oil mist potential, including generator and hydraulic spaces such as stabilizer and thruster compartments.

15.29   Programmable Logic Controllers (PLCs). PLCs can be found in almost everything ashore these days. Trawler yacht, superyacht, and larger sailing yacht systems now have more sophisticated PLC automated control systems. PLCs are now found in engine control systems, water makers, hydraulic control systems, alarm systems, air compressors, and other systems. A PLC is a microprocessor-controlled computer management system. The relay, timer, and interlock functions are all programmed and can have hundreds of operations. The program makes the electrical connections between inputs, outputs, internal relays, timers, and counters. Manufacturers include Siemens, Omron, GE-Fanuc, Allen Bradley, and ABB. The main component parts are the power supply unit (PSU), which powers the system, and the central processor unit (CPU), which contains the program in memory, uses ladder logic, and initiates and controls all the control, monitoring, and output commands. The input/output (I/O) modules input sensor and status data, such as temperature, pressure, and level information, or positional status and interlock data from limit switches, microswitches, and proximity detectors. Limit switches are common on all boat systems, including thrusters. They work by having the object travel until it strikes the actuator; this activates a set of either normally open or normally closed contacts. These contacts may be part of a control or indication circuit. It is important to check switches regularly and hand-actuate or exercise them when the system is off, as this is a very common failure mode. After processing, there are output commands to field devices such as actuators or alarm signals, relays or contactors, solenoids, and other devices. Where actuators are hydraulic rams or other mechanical devices, check them regularly, as they tend to cause the most problems.

15.30 **Control Circuit Devices.** Motor starters, hydraulic control systems, and many other systems and equipment operate on sensor information.

 a. **Pressure Switches.** Pressure switches are used for alarm activation, shutdown and interlock functions, control functions, and inputs to PLCs. Pressure switches in general use are the industrial brass and nitrile rubber diaphragm types on low pressure—20psi to 200psi (1.4bar to 13.8bar)—with high accuracy; piston types when high-pressure, high-cycling use, such as air compressors, is required; and Bourdon Tube for high-pressure and high-accuracy applications. Switches are selected based on the pressure range, with the usually adjustable set point being at midrange. Some switches require opening to adjust; on others the screw is top mounted. The set point is when the switch will activate. All switches have a dead-band, which is the pressure range between the set point and the point at which the switch resets on falling pressure. Switching mechanisms may be a simple three-terminal changeover or sets of normally open or closed contacts.

 b. **Temperature Sensors.** Temperature gauges or meters are generally of the bimetal or glass bulb type. PT100 platinum resistance sensors offer high accuracy and are used on many engine monitoring and automation systems. Thermistors have resistance changes with temperature changes. The negative temperature coefficient (NTC) units are used to drive alarms and temperature controllers. The positive temperature coefficient (PTC) units are used in over-temperature and overcurrent applications on motors and transformers and in warning and trip circuits. Thermocouples are described elsewhere. Resistance temperature devices (RTDs) like thermistors have a changing resistance with temperature. Bimetal devices are seen frequently in snap action devices on fridge compressors or surface mounted on equipment. Infrared devices are typically on handheld units, which are very useful tools on boats. These can be used to monitor bearings, motors, or any object subject to heat. As they are noncontact, measurements are easy to take.

 c. **Level Switches.** The traditional level switch is the float type, which is used in many applications. Other types in use are optical sensor switches, which use an infrared (IR) sensor. Some industrial switches that use probes in conductive liquids should be used with caution on boats, as they have been known to cause corrosion problems.

 d. **Flow Switches.** Some systems use flow switches to monitor water flow. Flow switches generally have a normally open (NO) or normally closed (NC) reed switch. Reed switches can fail, so the continuity of the switch must be checked. The reed switch is usually connected to a small control relay, which is then connected into the main control circuit. Flow switch paddles can jam or seize and may require checking.

e. **Limit Switches.** These switches can use a plunger pin, roller plunger, roller lever, and adjustable rod. They work by having the object travel until it strikes the actuator, and this activates a set of either normally open or normally closed contacts. These contacts may be part of a control or indication circuit. It is important to check switches regularly and hand-actuate or exercise them when the system is off. In addition, application of lubrication to prevent corrosion of the mechanism is necessary. Problems are caused by mechanical damage; the switches are designed to have an object strike the operating mechanism, and damage is often sustained.

f. **Proximity Switches.** Switches are either capacitive or inductive devices that detect objects at close ranges of around 0.6in (15mm). Inductive sensors detect metal objects, and sensing distances vary for different metals. Capacitive sensors detect both metal and nonmetal objects, including liquids. Sensors have either NPN or PNP solid-state outputs. NPN outputs are often referred to as "sinking type devices," as the load is wired between the output and the positive terminal. PNP outputs are often called "sourcing types," as the load is wired between the output and the negative terminal. Sensors are very robust and the circuits are encased in epoxy. They are resilient to shock, vibration, heat, and water. Most problems are from mechanical damage, as they are set up too close to sense surfaces. Regularly clean and check the sensor faces for damage.

g. **Photoelectric Sensors.** These sensors use light to detect the proximity of objects. They have a greater sensing range than inductive or capacitive sensor types. The two types are light energized, which have an output when light is detected, and dark energized, which have an output when no light is visible. The light sources are generally a red LED. The through-beam device uses a light source and a detector, which are aligned opposite each other; the sensor operates when the beam is interrupted. Retroreflective detectors have a light source and detector integrated in one fitting. The source emits light, and reflected light comes back to the detector from a reflector. Diffuse reflective sensors have source and detector within one fitting and are able to distinguish different reflector target types and colors. Photo sensors are either NPN- or PNP-type semiconductor outputs rated at 100mA to 200mA to switch relays or provide inputs to control systems. The typical response times for detection to output switching is around 0.5 to 2.5 milliseconds on DC detectors.

h. **Solenoid Valves.** Solenoid valves are integral to many systems, including hydraulic, refrigeration, air-conditioning, air compressor, engine starting, and water systems. Applying a voltage to a coil activates the solenoid. The most common use on boat systems is the direct-acting type, which does not require differential pressure to operate. The energized solenoid coil directly activates the valve as the coil magnetizes the plunger, pulling it upward along with the valve spindle. These valves are used on low- to high-pressure systems. AC-powered solenoids tend to be more powerful and faster than DC, and they have a higher inrush current. Solenoids generally have a duty cycle, so repeated energization can overheat and burn the coils out. Continuous rating means the solenoid can stay on all the time—duty cycle plus the "ON" time divided by the "ON/OFF" time. Many solenoids have plug-on electrical connections, and these should have the retaining screw tightened. Regular inspection of connection pins is recommended for signs of corrosion and tracking between terminals. Solenoids that are not used regularly should be activated several times every month, if possible, to prevent seizure of the valve or solenoid mechanism. The solenoid has an audible metallic "click" when operating. Coils do burn out, and if you are testing, place a multimeter across the coil terminals or in the plug connector to check whether voltage is present. If voltage is present, the valve is jammed or the coil has failed.

i. **Remote Valves.** Larger vessels may have remote-controlled gate and ball valves on seawater and freshwater systems. They should be routinely operated over the full range to exercise the mechanical systems, and to check the operation of limit switches. Operation time is typically in the range of 15 to 40 seconds. The control boxes should be inspected every year, cleaned, and all connections tightened. Lubrication should be applied in accordance with the manufacturer's instructions.

j. **Relays and Contactors.** Most circuits have small relays in control applications and contactors for supplying power to equipment. Both have a coil for operation and close or open contacts. If a relay or contactor is suspected to be faulty, check the voltage at the coil; if voltage is applied and the coil does not pull in, the coil has failed. Generally, the outer covering shows heat damage, but not in all cases. In some cases the mechanical armature part is the problem and may be jammed. In contactors, the current carrying contacts may have deteriorated and pitted so that poor or no contact is made when the coil pulls in, or the contacts may have welded together. A humming contactor is a coil issue.

Table 15-4. Engine Instrument Troubleshooting

Symptom	Probable Fault
Gauge does not operate	Power off
Temperature gauge needle hard over	Gauge supply cable off Sensor fault Cable fault
Pressure/tank gauge needle hard over	Sensor fault Cable fault
Alternator tachometer no reading	Alternator fault Lead off alternator terminal Alternator not "kicked" in Meter fault
Generator tachometer no reading	Broken drive mechanism Meter fault Generator fault Cable fault
Inductive tachometer no reading	Sensor clearance excessive Sensor mechanically damaged Meter fault Sensor fault
Low gauge readings	Negative connections to engine block
Oil pressure alarm activated	Low oil pressure (oil pump fault) Low oil level High oil temperature (cooling fault) Blocked sender unit Sender fault Cable fault
Water temperature alarm	High water temperature Low cooling water level Saltwater cooling inlet blocked Cooling water pump fault Loose drive belt Sensor fault (pump not working) Relay fault
No audible alarm	Audible alarm fault Connection fault Lamp failure Lamp connection fault Alarm circuit board fault

Electric Propulsion Systems

In the previous diesel engine chapter, I opened with the statement that the marine diesel engine is the main propulsion source for most cruising and sailing yachts; and it is, for now. This is a very rapidly evolving technology space. Professionally I have been working with electric propulsion for decades on large maritime and offshore systems. It has evolved with technology, with the use of high voltage (HV) power systems, variable frequency drives (VFDs), and other advances. Many cruise liners, deepwater drill ships, and ferries have electric propulsion. However, that said, electric propulsion for sailing yachts and other pleasure vessels are a different proposition with different challenges. Yacht manufacturers, such as X-Yachts in Denmark, have teamed up with Finland's Oceanvolt and launched an electric propulsion yacht. The 49-foot X-Yacht yacht is installed with two 10kW Oceanvolt sail drives. Other early adopters and major players include Sweden's Ancona and Electric Yacht in California. It is worth noting that the ABYC has an electrical propulsion standard, "E-30 Electric Propulsion Systems," revised in 2021.

16.1   Electric Propulsion Basics. I started this conversation in the battery chapter about 48V and the new lithium-ion battery technologies, and these elements are crucial to performance. The prime consideration with electric propulsion systems is how to maintain enough power to operate them. Sufficient power is always a challenge without electric propulsion; with electric propulsion, the challenge is amplified. The range of electric propulsion systems has many factors, including boat size and displacement, battery capacity, boat speed, and sea and wind states. The average ranges vary from 20nm to 50nm in calm conditions, depending on how fast you are going; the lower the speed, the greater the range. Many scenarios demonstrate the effectiveness of electric propulsions systems. Day sailing and limited coastal cruising have entirely different demands. Similarly, a long-distance coastal cruise and a Caribbean cruise have different demands than ocean cruising and crossings. I have read wildly overoptimistic suggestions that solar panels can replenish house loads. If you operate a fridge, as we all do these days, no solar panel will keep up with that unless, perhaps, you're in a catamaran with a small solar farm on board. Given that deck winches and windlasses are available in 48VDC, a refrigerator operating on 24VDC through a converter is a viable option.

16.2   Electric Propulsion Configuration. The Holy Grail for yacht electrical propulsion systems is sustainability, power on board without the need for fossil fuel support of a diesel engine or generator, and a zero-emission environment. That involves several factors, including the battery bank sizes and technology to be used, but also, more importantly, how to keep the battery bank charged. Many plans fail on this important point. There are a number of configurations for electric propulsion systems; these are summarized below.

> **a.   Parallel Hybrid System.** The parallel hybrid system configuration uses an electric motor that is added to the engine-driven shaft. This is via a belt drive and an electric clutch system. When operating, the motor generates power back to the batteries but so does the engine alternator.

b. **Generator Support System.** This system uses an electric sail drive, pod, or direct electric motor–driven shaft. This system comprises a generator as the backup power source to charge the battery power that supports the electric propulsion and house power system. In cruising scenarios, this is supported by the use of renewables to restore depleted battery capacity. The aim in this configuration is to use the generator as a backup, only used when power levels cannot be restored sufficiently with renewables. The generator option can comprise one of these very compact generators, such as those from Fisher Panda and Whisper, and charge through a high-output battery charger. You can consider installing a compact DC generator; one approach is to use a small 10hp (7.5kW) engine and install a couple of the new Balmar 48VDC high-output alternators and regulators so that maximum charge is achieved in the shortest possible time. With electric propulsion and the generator running, your system effectively becomes a diesel electric propulsion system, so range is limited by the diesel tank.

c. **Electric Propulsion System.** This system uses an electric sail drive, pod, or direct electric motor–driven shaft. This system has no other battery charging system except the primary regeneration capability of an electric propulsion system. This system is relatively precarious and possibly unsustainable. It doesn't account for periods of no sailing, which is the norm for most cruisers, or periods of low power regeneration. It doesn't properly account for all the other power demands placed on the battery system. You can end up with a very depleted battery bank and not much to power the propulsion.

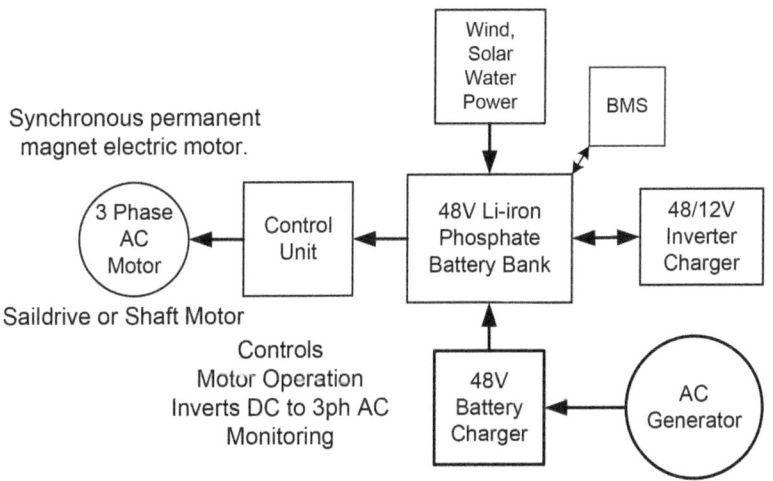

Figure 16-1. Basic Electrical Propulsion System

d. **Electric Propulsion with Renewables.** This is where electric propulsion battery and house power are replenished with the full suite of renewables, which include solar, wind, and hydro-generator, all of which are covered in other chapters. To achieve this, you have to do very detailed calculations, work out worst-case scenarios, and have contingency plans. If you are not under way, the hydro-generator is not in play. If the wind isn't blowing in your secluded anchorage, it is all up to the solar system. For monohulls, solar is often minimal given the lack of space for a large output system; multihulls can install a larger system. It is very easy to slowly discharge the battery bank when overall charging input does not keep up with power consumption.

16.3**Propulsion System Batteries.** Battery technology has enabled many developments, and the lithium-ion battery has transformed the space. Developments include significantly increased cycle capability, ability to provide constant voltage across the majority of its range, and the ability to accept charge at a very high rate. Read about lithium-ion batteries in the battery chapters.

16.4**Battery Charging.** The Oceanvolt ServoProp system is innovative in that it uses the variable pitch sail drive propeller to generate power when sailing. The propeller is variable pitch so that it can be adjusted to get maximum charging or feathered to reduce drag. The greater the boat speed, the greater the charging. Given that most cruising yachts are not like high-performance hybrid racing yachts, the results are not as effective when you are sailing at 5 to 6 knots. It is important to separate the two design factors; the primary one is electric propulsion. The regeneration process is an innovative battery charging function. However, it is prudent not to depend on this as the sole charging source.

16.5**Propulsion Electric Motors.** One of the catalysts for change is the innovations in electric motors. The new generation of permanent magnet, brushless motors and rim-driven motors that deliver greater torque with lower energy consumption make electric propulsion viable. Electric vehicles are drivers in this space, and the same motor technology is already being implemented in thruster motors, deck winches, and anchor windlasses. They use the many advantages of a higher 48V supply. There have been many breakthroughs in motor design, and Siemens leads in this space. One of these developments is a spinoff from aviation, with a motor weighing just 110lb (50kg) being able to deliver a continuous output of 348hp (260kW), paving the way for electric aircraft; hopefully this trickles down to yacht electric propulsion systems. Some electric motor systems still require water cooling to maintain efficiency levels, although pod and rim drives are surrounded by water and have some advantages.

16.6**Electric Boats.** One of the factors in boat electrical design is how to handle the overall electrical system. There are similar challenges to using hybrid mixed-voltage systems such as 24V and 12V systems. These are not that common, but if you are going to install a 48V propulsion system, your core power bank is nominally 48V. Retrofitting a propulsion system will create some changes to system architecture. I have noted the importance of redundancy in other chapters; creating two parallel battery banks assists in this. But the refrain is usually about other equipment power supplies. Some manufacturers are installing a separate 12V house bank to handle electronics, lighting, and other functions, but I think that is overcomplicating things. It is, in my opinion, poor economics and use of an expensive battery bank solely for propulsion. A DC-to-DC converter will handle all the LED lighting and the electronics power supplies; units such as the Victron converter work very well.

16.7 **Electric Dinghy Outboards.** These are now becoming common. Typical of these is the ePropulsion Spirit 1.0 Plus. They have a rating of 3hp (1kW) and can push your dinghy along at 3 to 4 knots with a range of up to around 22 miles at half speed. The motor is a brushless type and has a weight of 10.6kg for the short shaft model, with a battery weight of 8.7kg. The integrated battery is a lithium type rated at 1276Wh. The standard charger unit recharge time is 8.5hrs and the fast charger time is 3.5hrs.

AC Power Systems

The increasing use of AC appliances such as microwaves, air-conditioning, washing machines, stoves, hot water, TV, power tools, and other appliances and devices has increased the need for power both at and away from the marina. The AC power systems on many sailing yachts now consist of several elements that include shore power, AC inverters, generators, AC motors and starters, and AC installations that include grounding and circuit protection.

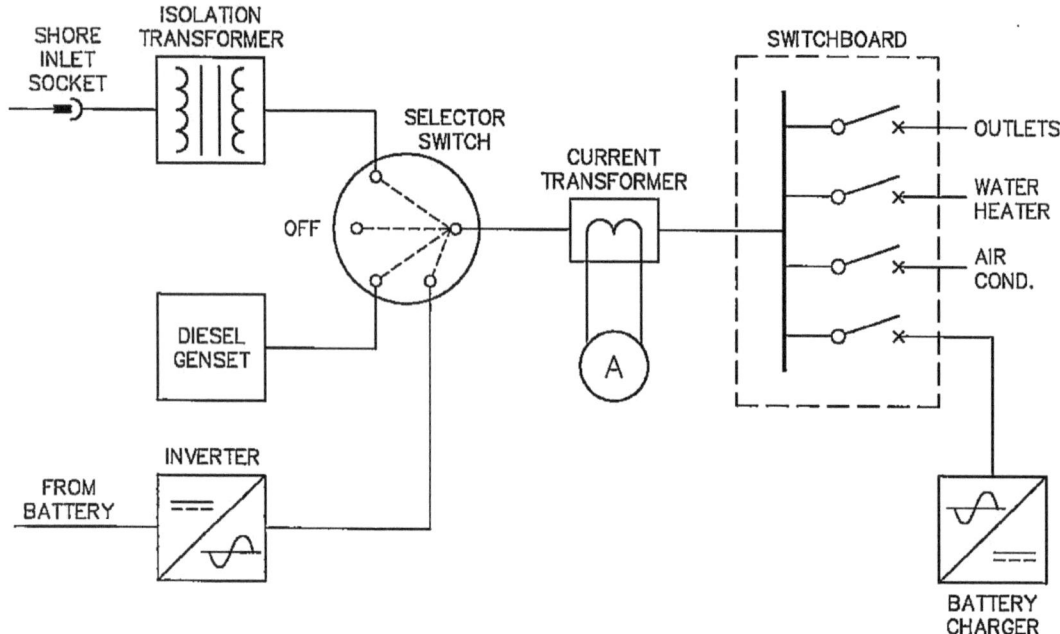

Figure 17-1. AC Power Systems

17.1   AC Power Safety. AC is potentially lethal, and AC systems and equipment must be correctly selected, installed, and maintained. The following safety precautions must be undertaken at all times when working with AC. Fifty-five percent of all shoreside house fires have an electrical origin, and there are 500 to 1,000 deaths from electrical injuries every year in the United States. The US Bureau of Labor Statistics reports that the majority of electrical accidents are preventable. Around 5% of all ER burns unit cases are due to an electrical injury. Death rates in the UK and Europe are higher, as they use 220VAC and not 110VAC. The takeaway here is that you absolutely need to be very careful on your boat when doing anything with AC power.

Disclaimer. This chapter is for informational purposes only, and readers should observe the safety recommendations given. Readers should also refer to the disclaimer at the front of this book.

WARNING

a. **Always use a qualified and licensed AC electrician to perform your AC system and equipment repairs and maintenance. Check the license!**

b. **Never work on energized "live" equipment. Always isolate and lock out the equipment energy source before opening any system or equipment. Attach a "Danger" or isolation tag at the power source. Always isolate your onboard AC circuit breaker and prove the system is de-energized.**

c. **Always remove the shore power plug and isolate the local main switch. Remove the shore power cable completely to prevent another person from plugging it back in.**

d. **If there is an inverter, or inverter/charger unit, always isolate or disconnect the DC input.**

e. **If there is an AC generator, always isolate any auto-start function.**

f. **Before starting work, always use a voltage tester to verify and prove that the circuit is de-energized before starting work.**

g. **Never work on AC equipment or circuits alone! Always have someone ready to assist if you accidentally receive an electric shock.**

h. **Learn how to perform CPR. Take an approved CPR training course.**

17.2   115VAC 60Hz Single-Phase System. This voltage (also often stated as 110VAC, 117VAC, or 120VAC and 127VAC) is used primarily in the United States and Canada. It is also used in Guam, Puerto Rico, Brazil, and most South and Central American countries including Colombia, Costa Rica, Cuba, Ecuador, Honduras, Mexico, Peru, and Venezuela. In the Caribbean it is used in Anguilla, Antigua and Barbuda, Aruba, Bahamas, Bermuda, British Virgin Islands, US Virgin Islands, Cayman Islands, Dominican Republic, Sint Maarten, Turks and Caicos, and Trinidad and Tobago. In the Pacific it is used in American Samoa, the Philippines, Micronesia, and Palau. It is normally a 30A supply requiring a supply cable of 12AWG (4.0mm²). The following circuit illustrates a typical shore power system. In this configuration, only one wire in the power inlet is "hot," or energized.

    a.   **Black cable.** The black cable is an active, or "hot," conductor.

    b.   **White cable.** The white cable is the neutral conductor.

    c.   **Green cable.** The green cable is the safety ground, or earthing conductor.

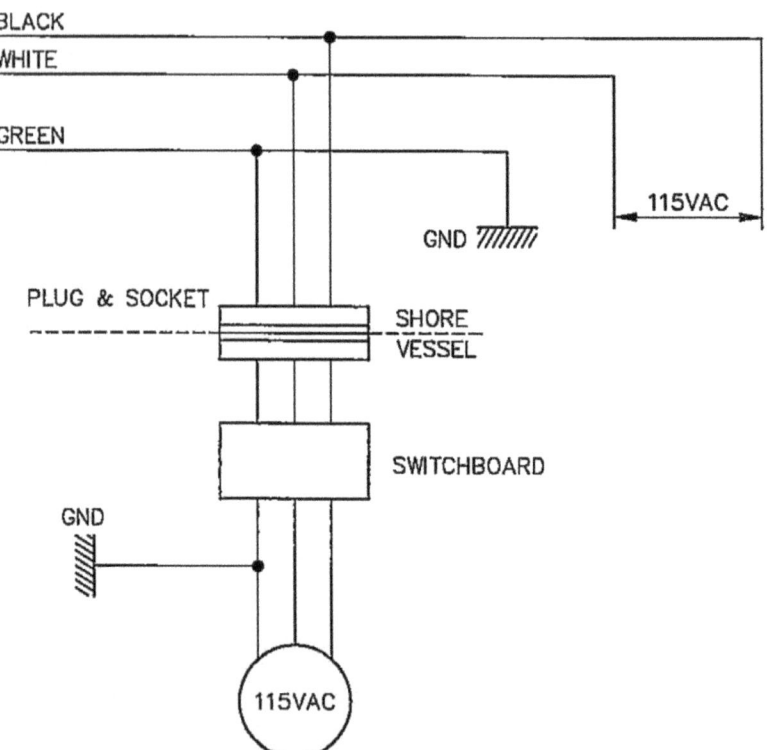

Figure 17-2. 115VAC 60Hz Single-Phase System

17.3 **115/230VAC 50Hz Single-Phase Systems.** The following circuit diagrams and color codes are for typical American dual-voltage shore supply systems.

      a.   **Red cable.** The red cable is an active, or "hot," conductor.

      b.   **Black cable.** The black cable is an active, or "hot," conductor.

      c.   **White cable.** The white cable is the neutral conductor.

      d.   **Green cable.** The green cable is the safety ground, or earthing conductor.

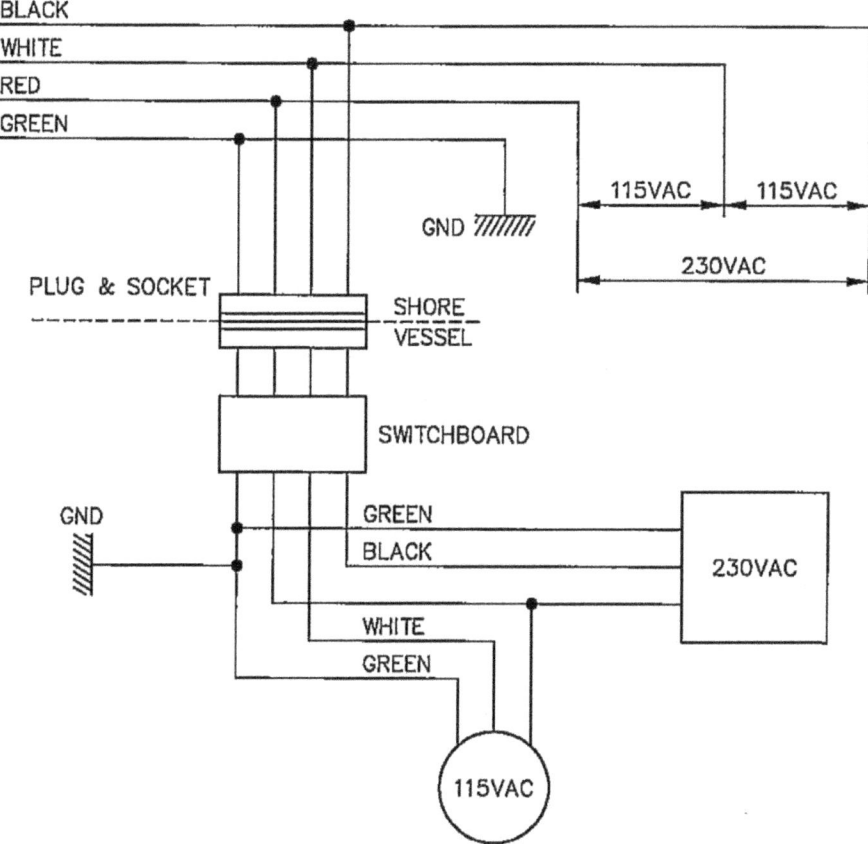

Figure 17-3. 115/230VAC Shore Power Systems

17.4   220/240VAC 50Hz Single-Phase Systems. This voltage is the IEC standard and is used in the UK, Europe, Australia, New Zealand, South Africa, most Pacific Islands including Fiji and New Caledonia, and most other countries. The IEC is standardized in most countries at 230VAC. The following circuit diagrams are for typical systems using IEC standard color codes and incorporating an isolation transformer—normally a minimum 15A supply rating requiring supply cable of 14AWG (2.5mm²). Many marina power pedestal supplies only have a 10A supply. The shore power cable should be a maximum of approximately 45ft (13m) long; anything over this will start to introduce volt drop problems at rated load.

a.   **Brown cable.** The brown cable is the active, or "hot," conductor (this used to be red).

b.   **Blue cable.** The blue cable is the neutral conductor (this used to be black).

c.   **Green/yellow stripe cable.** This is the safety ground, or earthing conductor (this used to be green).

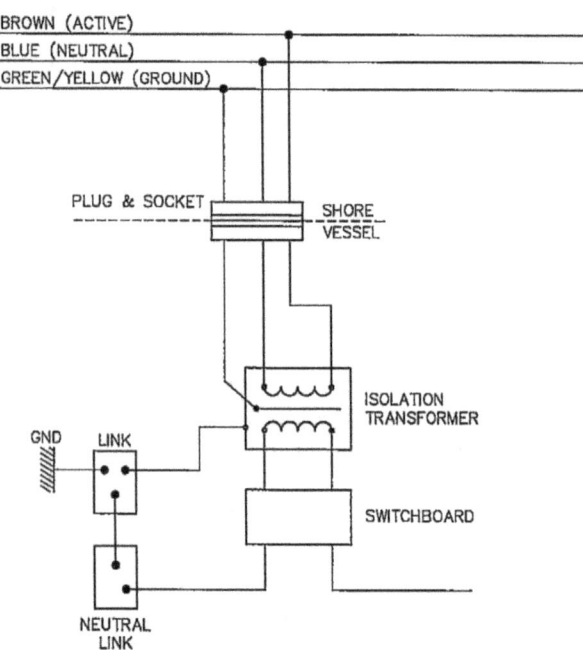

Figure 17-4. 220/240VAC
Shore Power Isolation Systems

17.5   440/460VAC 60Hz Three-Phase System. This arrangement is more commonly used on superyachts and larger sailing yachts. In larger yachts having three-phase power supplies, connection to shore power is less common or unavailable. Few normal marinas are able to offer power supplies. A three-phase offers equipment 20% lighter and smaller than single-phase and with 150% greater efficiency. In a three-phase vessel, power is usually generated in a three-phase, star, or wye configuration. This means the center point is grounded at the neutral. In a 208VAC three-phase wye system, you get 208V between phases and

120VAC phase to ground. Connected loads are usually symmetrical with electric motors; however, where loads are connected to each phase, unacceptable imbalances can occur. Single-phase loads must be evenly distributed across phases. In some instances, 115/220VAC transformers are used to power lower voltage systems.

a.   **Phase Marking.** Circuit designations are R, S, and T for each phase; it also may be R, Y, and B to represent the phase cable colors. In the United States this may be L1, L2, and L3. Secondary motor circuits may be also called U, V, and W or U1, V1, and W1.

b.   **Phase Colors.** The US phase colors for LV (120V/208V) are black, brown, blue, and white neutral. For HV (277V/480V) designations are now L1, brown; L2, orange; L3, yellow; neutral, gray. In most other parts of the world, the primary colors of red, yellow, and blue were used and are still used on older systems. In the UK, Europe, and Australia designations are L1, brown; L2, black; L3, gray; neutral, blue. In Canada designations are red, black, blue, and white neutral.

c.   **Phase Sequencing.** Phases must be connected in the correct sequence for a motor to run in the correct direction. If a motor is running in reverse rotation, reverse any two-phase cables to reverse the direction.

d.   **Phase Current Loads.** In any three-phase motor, each phase should have the same current. If there is an imbalance, this indicates a possible high resistance in the connections or, in some cases, a failing motor winding.

17.6   AC Domestic Plug Types. The various power plug types required when traveling can be confusing. It is recommended that you acquire an international adapter or several of them, as described here:

a.   **United States.** Two parallel pins (United States, Canada, South America, Philippines).

b.   **UK.** Three square pins (UK, South Africa, Portugal, Singapore).

c.   **European.** Two round pins (all of Europe, South Africa).

d.   **Australia.** Two and three pins (Australia, Fiji, New Zealand).

17.7   AC Transformers. Transformers are used to raise or lower voltages in electronics equipment power supplies, starters for control voltages, and isolation transformers for shore power supplies. Larger boats transform three-phase from 480VAC to 208/120VAC for power and lighting circuits. It consists of a primary winding and a secondary winding on an iron core. Power applied to the primary magnetizes the core and induces a voltage into the secondary winding. The most common type is the "isolation" transformer, which electrically separates input and output. Autotransformers are lighter and cheaper, with connected primary and secondary windings. Transformer maintenance is a simple cleaning of dust and an insulation test every year. Visually check windings and wire insulation for heat damage. The frequency cycle "hum" can cause vibration, which can loosen screws, so all connections should be tightened.

At energization there are high inrush currents, up to 25 times the rated current, and these must be considered when selecting protection devices. There are booster transformers available that also act as isolation devices and are used to boost lower dockside supply voltages.

17.8   Isolation Transformers. Isolation transformers electrically isolate the vessel from the shore power system, eliminating the ground path ashore. This galvanic separation eliminates electrolytic (stray current) corrosion. Isolation transformers should be rated for the maximum current rating of the shore supply inlet. Most marina outlets rarely exceed a maximum of 15A/240-220VAC or 30A/115VAC. Some transformers are dual input with inputs of 120/240VAC or 480/240VAC, and this provides some flexibility. Transformers do not alter frequency, which may be different in some countries. This will affect motor speeds (50Hz is slower than 60Hz), but for battery charging, domestic appliances, and resistive loads, there is little problem. Victron has incorporated a "soft-start" unit, which manages high start currents that would otherwise trip the marina supply breaker and short-circuit protection. Soft starters should be considered for all AC electric motors.

Table 17-1. Isolation Transformer Rating Table (0.8pf)

Output (kW)	Output (kVA)	Current 120V	Current 240V
3kW	3.74kVA	34A	17.0A
4kW	5.00kVA	45A	22.7A
5kW	6.25kVA	57A	28.4A
7kW	8.75kVA	80A	39.8A

17.9   AC Switchboards and Panels. These basic guidelines should be followed:

a.   **Construction.** The AC switchboard should be of the dead-front type and constructed of nonhygroscopic and fireproof material. If protective and isolation devices are not to be integrated into the main electrical panel, consider an industrial or domestic consumer distribution panel or module as an alternative. These panels are made of plastic with a splash-proof cover, and have all earthing and neutral conductors, main switch, or residual current device (RCD) and miniature circuit breaker (MCB) within one compact unit.

b.   **Location.** The switchboard should be located in a position to minimize exposure to spray or water. Locate the switch panel or distribution unit in a dry place, such as a cupboard or other suitable and safe area. Many locate them at the navigation station; while it looks good, it does expose the board to damage.

c.   **Mixed Voltages.** AC systems should not be located or installed adjacent to DC systems. Where DC and AC circuits share the same switchboard, they should be physically segregated and partitioned to prevent any accidental contact with the AC section. The AC section must be clearly marked with "Danger" labels. Where DC and AC are installed on the same panel, which is common on many yachts, cover exposed AC connections at the rear of the panel or, preferably, enclose them in a separate compartment. I think it is preferable to move the AC panel elsewhere.

17.10   AC Circuit Protection Principles. In any AC system, there are several control and protection devices. It is normal to have these coordinated; this is called discrimination or selectivity. The principle is that a fault appearing in any part of the system is cleared or disconnected by the protection device immediately upstream of the fault by that device only. Ideally discrimination is full, but in most cases it is only partial. Partial discrimination may be achieved up to a specific current level; any fault current levels above that cause all breakers to trip. This is achieved by having different current settings, with a motor thermal overload, switchboard distribution breaker, and generator supply breaker all set at a different level. Tripping times are dependent on the rate of current increase. If a fault such as an overload rises up to a level of 0.75 above normal current, the thermal relay will trip. If the current continues to rise, such as in an impedant short circuit, the magnetic circuit breaker supplying the circuit should trip. With a full short-circuit current, the supply breaker will operate very fast. If there is no discrimination, all circuit breakers up to the generator supply may trip.

17.11   AC Short Circuits. Short circuits are relatively rare in yacht electrical systems. A short circuit happens when two points of different electrical potential are connected—that is, live to neutral, live to ground, phase to phase, and phase to ground. A short circuit causes a very rapid rise in current; this can reach several hundred times the nominal value in milliseconds and cause high thermal and mechanical stresses, which can destroy the cables and the bus bars that feed the fault. There are two thresholds in short circuits. The first is where electrical arcs commence, insulating materials start to break down, and parts start to deform. The second level is where contacts melt and weld together, electrical arcs continue, and insulators start to carbonize. Short circuits can be either a lower-level condition, called an impedant short circuit, or an intermediate short circuit, usually caused by deterioration in insulation.

17.12   AC Short Circuit Causes. Typically, short circuits are caused in order of probability by loose connections, insulation damage or failure, and metallic bodies or conductive deposits on terminals within junction boxes and motor terminals. The next grouping is broken conductors within a cable, dust and moisture gradually tracking and burning across terminals to create a short circuit, and crossover errors after new installations or repairs.

17.13   Short Circuit Calculations. Prospective short-circuit current (ISC) can be calculated for circuits, although this is often done only in larger yachts with large power systems and built to survey or classification society rules. This is defined as the calculated RMS (root-mean-square) value of short-circuit fault current at any point in the system should a conductor of negligible impedance replace the protection device. Values depend on the supply voltage and line impedance.

17.14   Short-Circuit Protective Devices (SCPDs). Typically, an SCPD is either a fuse or a circuit breaker. Fuses are devices whose conductors melt to break the fault current. Circuit breakers detect short circuit current and open the poles to clear the fault. An SCPD must detect and break high fault current levels quickly before reaching peak current values. Given the high-rise times and magnitude of currents, this must be fast to be effective. Speed of the SCPD is determined by the peak current and breaking capacity of the contactor thermal device to withstand overload. A direct online (DOL) starter has a start curve of approximately 7.2 × current for 10 seconds. Thermals must protect against low-level faults but allow for machine start-up times and current. A miniature circuit breaker (MCB) combines the action of an isolation switch, overload protection, and short-circuit protection.

a.   **Overload Protection.** This function is thermally operated. Tripping values are normally to hold 110% of rated value and trip at 137% of rated value at 77°F (25°C).

b.   **Short Circuit Protection.** This function is magnetic. A solenoid coil within the breaker trips when the factory-set short circuit current value is reached. Under short circuit fault conditions, a large arc can be generated; breakers use the generated magnetic field to direct and quench the arc in a chute.

17.15   Protection Equipment Selection. Equipment must be selected properly if it is to perform properly under fault conditions. The main switch or protection circuit breaker should be rated for the maximum current capacity of the circuit. The MCB must be selected to protect the cable, not the equipment. The MCB must be rated to hold at the maximum demand, such as motor starting loads, which can be four to six times the rated load. The device voltage must suit the system voltage. The interrupting capacity must be able to cope with any prospective fault current levels. The MCB current rating must hold at 100% of operating current and trip at 125% at 104°F (40°C).

17.16   AC Cable Installation. The cabling installation requirements for AC systems are virtually the same as those defined for DC systems. These are briefly outlined as follows:

a.   **Cable Ratings.** Conductors are to be selected based on the maximum current demand of the circuit. Ambient temperatures exceeding the rated temperature of the cable should be derated. Where cables are to be installed in hot machinery spaces, consideration should be given to derating if the ambient temperature exceeds the cable rating temperature.

b.   **Conductor Volt Drop.** Conductor size should be selected with a maximum allowable voltage drop of 5% for all circuits. The voltage drop problem can be prevalent in high starting current equipment. Where equipment may be running at maximum load, a larger cable size may be a good option.

c.   **Conductor Construction.** Conductors should be stranded, insulated, and sheathed (double insulated) and, where possible, of ship wiring standard. All conductors should have a minimum cross-sectional area of 18AWG (1.0mm^2). Cable ratings and insulation materials should conform to recognized national standards. Where possible, cables classed as "Shipwiring Cables"—three core (two core and ground)—should be used. Ship wiring standard cable is expensive. If you do not want to pay the cost of such cable, and do not have to meet classification society rule requirements, install 15/30A heavy-duty outdoor-rated, orange or yellow-sheathed extension cable. Never install domestic triplex single strand–type cables.

d.   **Conductor Color Codes.** Conductor color codes should conform to either IEC or US standard codes:

    (1)   **IEC Code.** The brown wire is active, "live" or "hot"; blue wire is neutral; green/yellow stripe wire is ground, or earth. Three phase systems used to be red, yellow, and blue for each phase; in some cases, white is used for the yellow phase. As stated elsewhere, in the UK, Europe, and Australia this is now L1, brown; L2, black; L3, gray; neutral, blue. Switching and control circuit wires may be any other color that cannot be confused or mistaken.

    (2)   **US Code.** The red wire is active, or "hot"; the black wire is "hot" (active); the white wire is neutral; the green wire is ground.

e.   **Cable Installation.** Cable runs should be installed as straight as practicable. Cable bend radii should be a minimum of six times the cable diameter. Tight bends should be avoided to reduce unnecessary strain on conductors and insulation. Cables should be run as far as practicable from DC power, network, data, and signal cables and should never be within the same loom or cable bundle. They should be run clear of any data, network, or signal cables.

f.   **Neutral Conductors.** All neutral conductors terminated in a neutral link should be identified to correspond to the marked circuit number. When connecting AC circuits, the neutral should be marked and connected to the same numbered terminal in the neutral terminal block. Similarly, if an earthing or ground terminal block is used to collect the grounds, the ground for the circuit should have the same terminal number. For example, circuit No. 1 has neutral No. 1 and ground No. 1.

g.   **Cable Access.** All cables must be accessible for maintenance and inspection; in particular, those entering transits should be capable of access for routine inspection and adding other circuits.

h.   **Cable Protection.** Cables should be protected from mechanical damage, either where exposed or installed within compartments. All cables should be installed so as to prevent accidental damage to the insulation or application of excessive stress on the cable.

i.   **Cable Transits.** Conductor cables passing through bulkheads or decks should be protected from damage using a suitable noncorrosive gland, bushing, or cable transit. Cables transiting decks or watertight bulkheads should preserve or maintain the watertight integrity.

j.   **Cable Support.** Conductor cables should be supported at maximum intervals of 8in (200mm). Supports and saddles are to be of a noncorrosive material. When used in engine compartments or machinery spaces, these should be metallic and coated to prevent chafe to the cable insulation. Cable saddles should fit neatly, without excessive force, onto the cables or cable looms and not deform the insulation. Cables should be neatly loomed together and secured with PVC or stainless saddles to prevent cable loom sagging and movement during service.

17.17 AC Cable Grounding. The ground conductor should not have any switch, fuse, or other device installed or anything connected in line to it that may cause opening of the circuit. (See the notes on galvanic isolators.) All ground connections should be mechanically secure and be protected against mechanical damage or accidental disconnection. The connection should be clear of water or moisture so that corrosion cannot occur and should be coated for protection. All AC equipment should be grounded, and the maximum resistance between any ground point and the boat ground should be 1Ω. The maximum resistance of the grounding (earthing) system must be 1Ω between the main ground terminal block and the boat ground. Any fault arising on an ungrounded or inadequately grounded item of equipment may cause exposed metal to be "alive" up to rated voltage. Accidental touching and grounding by a person may cause serious electric shock, injury, even death. Grounding provides a low-impedance, low-resistance path for any fault arising on exposed and bonded metal. AC grounding is always challenging, and Moonraker manufactures copper alloy ground plates that are rated for AC use. This meets Australian rules when two of these are connected and provide the minimum surface area of 2.5ft² (2,500cm²) and an impedance of less than 1Ω at 5MHz in seawater. During fault conditions, extremely large current levels of several hundred to thousands of amps may flow. This high current usually ruptures fuses or trips circuit breakers. Improper or degraded grounding or a high-resistance ground may cause circuit conductor heating and fire. While corrosion protection is an important consideration, it is not part of the electrical considerations. In multiple grounded systems ashore, any current flowing in the grounded neutral conductor will cause a voltage rise on that conductor above the mains supply voltage. This will also be impressed on the grounding system, which is typically a few volts above ground potential. In a fault on the mains, the supply neutral voltage on the installation grounding system will rise in proportion to the load, and the value will be dependent on the impedance of the main ground system. At all times, it is recommended that an AC-licensed electrician be used to ensure that installations are done correctly. Check the license!

17.18 Ground (Earth) Leakage Protection. The most reliable and accepted way to protect people is by installing earth leakage-protection devices. The RCD provides protection against indirect contact, supplementary protection against direct contact, and protection against fire and thermal effects. These units are not primary circuit protection; they are in addition to the circuit breakers and fuses that provide the overload and short-circuit protection. Many combine both to provide total protection. Virtually all marinas now install these on each circuit. RCDs are unable to limit voltage or current; they provide protection by limiting the actual time a specific maximum current can flow to ground. Devices are now considerably more advanced. Residual current devices (RCDs), ground fault circuit interrupters (GFCIs), and residual current circuit breakers (RCCBs) are combination circuit breakers. Called arc fault circuit interrupters (AFCIs), they are available in combined GFCI/AFCI units or RCCB units and should protect all power circuits on a boat. They were not previously required on water heaters or electric cookers, as element leakages could cause nuisance tripping. However, it is now recommended that they be installed on all circuits, including lighting.

17.19   Ground Fault Current Interrupter Selection. GFCI/RCDs are defined by three main characteristics: the rating in amperes, the rated residual operating current of the protective device in amperes, and whether the device operates instantaneously or has an intentional time delay to permit discrimination. Other important factors are the current waveform and the possible waveform of any fault current to ground or earth.

a.   **Residual or Fault Current Basics.** Some devices and equipment have DC leakage current due to the technology being used and the design. Some equipment can generate DC currents under fault conditions, and these are often those with semiconductors. The waveform is affected on circuits under fault and load conditions.

b.   **About GFCI/RCD Types.** Consider the type of electrical equipment to be connected. The final factor is the type of device to be used, which in the UK, Europe, and Australia are Type A, Type AC, Type F, and Type B. Regulations state that the correct RCD should be selected in accordance with the equipment connected to the circuit. The different GFCI/RCDs are based on the behavior of the device when various DC components and frequencies are present. Similar designations are also in the United States. A GFCI for residential use are Class A devices under UL943 and limited to circuits of 240VAC and lower. Class A devices are found in commercial and industrial settings as well as boats, and under the National Electrical Code (NEC) they are required on all 15A and 20A 125VAC utility outlets. In 2009 Underwriters Laboratories (UL) published UL943C, which created three new classes of special-purpose GFCIs that are intended for use with higher voltages; these are denoted as Classes C, D, and E. These will trip at 20mA instead of the mandated 6mA current for Class A devices. The trip level is raised based on the UL assumption that there is a reliable parallel ground with the body. When a fault occurs, the safety ground conductor will shunt the fault current through the body and cause the device to trip. This provides let-go protection, and the 20mA threshold will provide fibrillation protection. It is important to note that the GFCI/RCD types should never be confused with the different types of circuit breakers, which are manufactured based on their time/current characteristic. The types are as follows:

(1)   **Type AC.** This is the minimum-level device and will trip on detection of alternating sinusoidal residual current that is either suddenly applied or increases smoothly. This has been the most commonly used type to date. They are often ineffective as a result of the residual DC fault current. The regulations in the UK, Europe, Australia, and elsewhere have now changed to prohibit or ban the use of Type AC devices, and many suppliers have ceased their manufacture. Type AC RCDs have been the standard for many years because they were effective for equipment and circuits that are inductive, resistive, or capacitive, with minimal electronic components. These RCD types do not have a time delay and operate instantaneously on detection of

an imbalance. Circuits where these are used include electric shower units, ovens, cooker hobs, and hot water immersion heaters. Type AC units were banned in early 2023, so check what you have installed. If you have Type AC, invest in a Type A device.

(2) **Type A.** This device will trip on alternating sinusoidal residual current and on residual pulsating direct current that is either applied suddenly or increases smoothly. Tripping occurs for residual pulsating direct currents that are superimposed on smooth direct current up to a level of 6mA. Type A devices are used for circuits with inverters, Class 1 IT equipment, power supplies for Class II, lighting equipment with dimmers, and LED drivers. The recommendation is that Type A devices should be the minimum device installed to allow the connection of electronics based equipment.

(3) **Type F.** This type has the same trip criteria as Type A plus composite residual currents and is intended for a circuit supplied between line and neutral or line and earthed center conductor. Tripping occurs for residual pulsating direct currents that are superimposed on smooth direct current up to a level of 10mA. These are used with appliances with variable-speed drive (VSD), such as air-conditioning, some power tools, and those with synchronous motors.

c. **Equipment DC Fault Currents.** Much new technology equipment is able to create DC residual fault current. This can include VSD, LED lighting, washing machines, clothes dryers, and dishwashers. Solar photovoltaic systems that have inverters for DC to AC conversion also fall into this category. Modern technology has resulted in many devices and equipment incorporating electronics microprocessor control circuits. This includes VSD motor frequency speed control, which has an inherent DC residual fault current and also uninterruptible power supplies (UPSs).

d. **Selectivity (Discrimination).** The term "discrimination" has now been redefined as "selectivity." When installing multiple GFCI/RCD devices in series, which is common, it is important to achieve selectivity. The common misconception is that an GFCI/RCD with a higher mA sensitivity will provide the required selectivity. However, due to the device's instantaneous operation, selectivity requires installation of upstream timed-delayed devices.

e. **Tripping Requirements.** The selection and installation of an GFCI/RCD is based on the tripping values and, therefore, the level of protection. The values are as follows (tripping times are 30ms to 50ms):

(1) **30mA Setting.** This is the recommended setting and is becoming mandatory protection against personal shock. A 30mA GFCI/RCD is required to trip when a current is in the range of 18mA to 28mA.

(2) **100mA Setting.** This level is designed to provide fire protection and is common on the upstream primary circuit protection.

f. **GFCI (RCD) Operation.** The GFCI/RCD units work on an electromagnetic principle as illustrated below:

(1) A toroidal transformer is used to detect magnetic fields created by current flow in the active and neutral conductor of the protected circuit.

(2) Under normal conditions, the vector sum of the currents, known as residual current, is effectively zero and the magnetic fields cancel.

(3) If a condition arises where current flows from active or neutral to ground, the residual current will not be zero and the magnetic field will establish a tripping signal to the protected circuit.

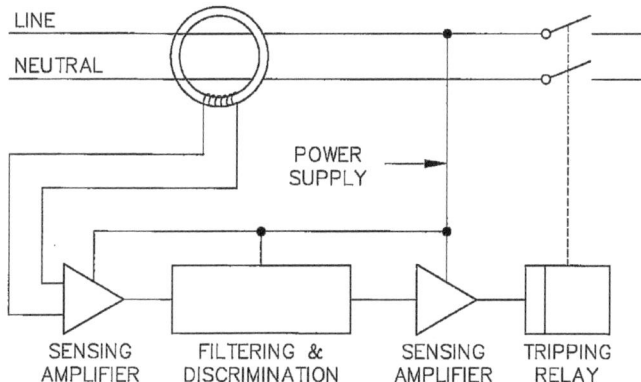

Figure 17-5. Residual Current Protection Devices

g. **GFCI Tripping.** Due to the marine environment, earth leakages are commonplace and nuisance tripping is very frequent at marina slips. Not all trips are "nuisance"; they indicate a problem that must be investigated and rectified. Most nuisance trips occur either during or following rain or in periods of high humidity. The principal causes of trips are as follows (again, call a licensed AC electrician for assistance):

(1) Connection of a neutral and ground (earthing) connection downstream of a GFCI.

(2) A crossed neutral between protected and unprotected circuits.

(3) Deterioration in cable insulation.

(4) Water and moisture in terminal boxes; cumulative leakages from various small leakage paths from a number of sources.

(5)   Absorption of moisture into heating elements, including steam irons, fridge defrost elements, stove and hot water elements, and electric kettles. This problem disappears if an element operates for half an hour or more.

(6)   Tracking across dirty surfaces to ground.

(7)   Intermittent internal arcing in appliances.

(8)   High voltage impulses caused by switching off inductive motor loads.

(9)   High current impulses caused by capacitor start motors.

h.   **Installation Checks.** Installation should be performed by an AC-licensed electrician and tested using purpose-made test equipment. The following tests must be performed using a 500VDC insulation resistance test meter (Megger/Mega):

(1)   Disconnect supply, neutral, and earth. Test between active and earth. On new installations, readings must exceed 1MΩ, a minimum of 250kΩ on existing systems.

(2)   Test between neutral and earth. Readings must be a minimum of 40kΩ.

i.   **Testing.** Where earth leakage protection devices such as RCDs and GFCIs are used, they should be tested monthly to verify operation using the integral test facility. The GFCI has an integral test button that simulates a ground fault to test the tripping function and a reset button. In reality, an GFCI/RCD may trip well before the nominal setting. Devices require checking with a special test unit, as the self-test button is not always a reliable function test.

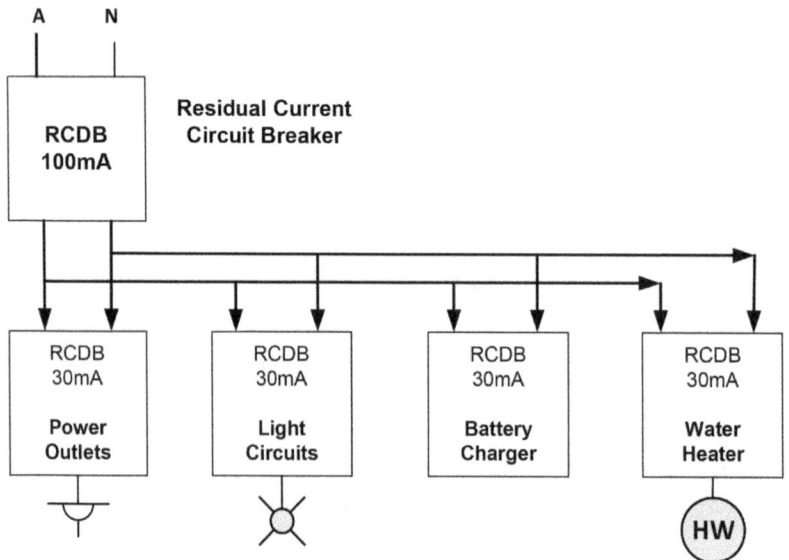

Figure 17-6. RCCB Circuit Protection

j.   **Corrosion Systems Affects.** There is much conversation on how these DC leakages from AC with DC elements impact the corrosion protection system. If you have a 6mA DC leakage onto your bonding system, there will be impacts. Review and upgrade your devices. Check what your marina is doing and what is installed. Invest in a galvanic isolator.

17.20   AC Circuit Testing. The following MUST be tested before putting any circuit into service and as part of routine inspection and testing:

a.   **Insulation Resistance (IR) Test to Ground.** The insulation resistance between all circuit insulated poles and ground, and between poles and ground, must be greater than $1M\Omega$. A 500VDC insulation (Megger) insulation resistance (IR) test meter should be used. Disconnect all electronics and appliances, turn the power off, and disconnect the main grounding conductor. All switches should be in the "ON" position. Insulation resistance between ground and live conductors must be a minimum of $1M\Omega$. Water heater elements must be at least $10k\Omega$. An ohmmeter should be used to check between all active and neutral poles on each circuit to ensure that only load resistances are present and that with all switches on there is no short circuit through either cable damage or incorrect equipment connection.

b.   **IR Test Busbars.** The insulation resistance between all switchboard panel bus bars, and between bus bar and ground must be greater than $1M\Omega$. A 500VDC insulation (Megger) tester should be used. Disconnect all electronics and appliances, turn the power off, and disconnect the main grounding conductor. All switches should be in the "ON" position. Insulation resistance bus bar to ground must be a minimum of $1M\Omega$.

c.   **IR Test Cables.** The insulation resistance of all cables to and from generators and motors, as well as windings, control gear, and grounding must be greater than $1M\Omega$. A 500VDC insulation (Megger) tester should be used. Insulation resistance between all parts and ground must be a minimum of $1M\Omega$. Tests should be made on hot machines.

d.   **Transposition Checks.** There must be no transposition of active and neutral conductors. All switches, circuit breakers, outlet live pins, and equipment terminals must be checked and be of the same polarity. No transposition of neutral and actives (crossed connections) is allowable.

e.   **Transposition Neutral and Ground.** There must be no transposition of ground and neutral conductors. All equipment and outlets must be checked to ensure that there are no crossed connections.

17.21   AC Rotating Machinery. The most common type of AC rotating machinery on board is the electric motor. Electric motors are generally very robust and provide years of trouble-free service. The fixed-speed and constant torque squirrel-cage, single-phase motor is the most common in use. Larger boats have three-phase motors, with higher starting torques and efficiency with lower power consumption. Motors should be selected with the right rating for the load and the right degree of National Electrical Manufacturers Association (NEMA) and IEC Ingress Protection (IP) ratings. Insulation ratings should be as high as possible, with Class B the most common. On larger boats and ships, Class F at the top is the most common. Totally enclosed fan cooled (TEFC) motors should be used in marine environments. Motor failures are primarily caused by bearing failures, often due to overgreasing and undergreasing. Sealed bearings offer a good maintenance-free solution. It is important for connected loads such as compressors and pumps to be properly aligned so that vibrations do not transmit through to the bearings and cause early failure. Onboard repairs are generally limited to bearing replacement. Rewinding and similar repairs will need to be undertaken by a shore repair facility. On many occasions, an AC motor may have a very low winding insulation value due to moist air or being flooded. To undertake repairs, the following should be done to get up and running again:

a. **Motor Winding Drying.** If you have submerged your motor, you will need to disassemble the motor completely. If the motor has been totally immersed for an extended period, replace the bearings. Wash out the motor stator with fresh water. Place the stator in the oven at approximately 158°F (70°C) for at least 4 hours. Check insulation value to case with an 500VDC (Megger) insulation resistance (IR) test meter. The insulation resistance reading should be at least 1MΩ. Recheck the insulation reading after 4 hours to ensure that the reading remains high. Don't forget to clean and dry out the cable termination box. In some cases, you can dry out moisture with heat lamps or a hair dryer if needed.

b. **AC Motor Maintenance.** The maintenance requirements of AC motors are minimal and consist of insulation testing every 6 months; the winding-to-ground insulation resistance should be tested using a 500VDC (Megger/Mega) insulation tester. Readings should be a minimum of 1MΩ. IR testing is all about looking at trends; if the insulation value is decreasing every time you check, investigation is required. Terminals and connections should be checked and tightened. If bearings are not sealed, they should be repacked every 2 years, depending on run times. If a motor is stationary for much of the time, maintain the bearings by manually turning the shaft at least once a month.

17.22   Motor Start Currents. The main problem with AC motors is the starting current, which may be two to five times rated load at start-up. Thermal overloads are usually adjustable within a range. The thermals will withstand the short motor start overloads. Starter ammeters are supplied via a current transformer (CT). The ratios vary but typically may be 30:5, which means that 30A through the CT will give a 5A output. The meter is scaled to read actual current. These are black cylindrical devices, and either the active or a phase wire is passed through the center of the CT. The output is connected to the ammeter.

17.23 **Motor Troubleshooting.** The main troubleshooting checks are outlined in the table below. In general, tripping circuit breakers due to overload is the main problem.

Table 17-2. AC Motor Troubleshooting

Symptom	Probable Fault
Ground fault	Insulation resistance broken down (moisture, overheating, aging, winding damage)
Low insulation reading	If below 1MΩ, check termination box
Motor overheating	Mechanical overload (check amps) Bearings seizing (check amps/bearing temp) Ground fault
Circuit breaker tripping	Mechanical overload Ground fault Winding intercoil short circuit Terminal box cable fault
Motor overload	Stalled load Seized bearings Terminal connection loose
Bearings hot	Bearing lubrication failure Bearing collapsed or worn Drive belts over-tensioned
Vibration	Coupling misaligned Bearing failure Loose frame foundation holding bolt

AC Generators

Generators are a viable source of AC power on many yachts above 40ft (12m). Generators have shrunk in size, and smaller units are possible on yachts down to around 34ft (11m). The majority of generators are single phase, with three-phase machines being used on larger vessels, and new generation inverter units have made lightweight and compact units possible. Most units come complete with sound shields and only require external connection of cooling water, fuel, electrical, and exhaust systems. Most manufacturers have significantly reduced weight, physical size, noise emission, and vibration levels. Diesel generators generally have the same principles and requirements as those for main propulsion diesels, and in practice they often operate for longer hours than main engines. There is also a hydraulic-powered generator option, with these units hydraulically powered from an engine power takeoff (PTO). They vary in output from 2kW through to 15kW and more, and they offer a way to utilize your engine. Yacht generators come in many types; leading suppliers include Fischer Panda, Mase, Northern Lights, Onan, WhisperPower, Kohler, Nanni, Sole Diesel, and Next Generation Power. There are many innovations within this space. WhisperPower is one of them and has developed the very compact lightweight Genverter range. They have designed a very compact permanent magnet alternator that is located just behind the engine rather than the flywheel. The output is then fed to an inverter, which outputs into a pure sine wave with fixed frequency output. The generator rotation speed is programmable from 2,000rpm to 2,800rpm. There is an optional power cube for DC battery charging. Visit whisperpower.com for details.

18.1   Generator Selection. The generator selection process should include an analysis of projected loads that will be powered. Typical loads are shown below; check the nameplates on all your equipment for precise values. Consider how many you are likely to run either alone or simultaneously. Add your own specific equipment and ratings.

Table 18-1. AC Appliance Loadings

Appliance	Typical Power in Watts
Air-conditioning	1,000–3,600
Electric Water Kettle	1,800–2,400
Hot Water System	1,200–1,500
Battery Charger	250–500
Television	60–150
Refrigeration	1,000–1,500
Coffee Maker	800–1,200
Corded Power Tools	500–800
Power Tool Chargers	70–150
Microwave Oven	800–3,000

Appliance	Typical Power in Watts
Hair Dryer	1,000–1,500
Bread Toaster/Pop-up Toaster	900–1,500
Air Fryer	1,500–2,500
Electric Wok or Fry Pan	1,000–1,200
Device Charger	25–40
Sandwich Maker	650–800
DVD Player	25–45
Dehumidifier	200–300
Food Blender	300–400

.2   **Generator Installation.** The basic installation factors are as follows (reference the vice outlined in the diesel engine and diesel engine electrics sections):

a.   **Fuel Systems.** Clean fuel is essential, as is the installation of a separator system such as a Racor, Dahl, Keenan, Delphi, or Vetus filter. Check them daily when operating and drain any accumulated water. If you are getting a lot of water and sediment, it is time to drain and clean the fuel tank.

b.   **Exhaust System.** Improperly installed exhaust systems are a major cause of problems. The exhaust outlet should be installed above the loaded waterline, and the transom is the best location. Ensure that the vented loop (siphon break) is installed where required, and regularly check that it operates. Ensure that exhaust lines are properly installed with respect to loops and slope, and ensure there are no points where water can be trapped. Like all engines, use caution when cranking over the generator, as water can fill the muffler and back up into the engine.

c.   **Seawater Systems.** The raw seawater suction should not be fitted with a scoop-type inlet. They can pressurize the water and subsequently force water past the pump impeller, which can cause the muffler to fill and, in worst cases, into the exhaust manifold and engine cylinders. Ideally, a separate water system supply and strainer should be used. Ensure that the strainer is cleaned regularly; strainers are often forgotten until a high temperature occurs. Where the generator has anodes, check these regularly.

d.   **Control Systems.** Generators are generally remote start and stop systems with automatic engine protection systems. In most generators, all shutdown functions are suppressed for 10 to 20 seconds at start-up; this may include low water pressure and low oil pressure. Some larger generators on large yachts and superyachts will have generator units that have a pre-lubrication pump operating when the engine is stopped; a failure in pre-lube pressure will inhibit the start. On vessels with multiple generators and automatic st and start functions, all systems must be selected to "Auto." In any system is important to know what interlocks are in the start ready chain. Most ge erators will trip automatically on overspeed, high crankcase pressure, hi temperature, low oil pressure, and low oil levels.

e. **Governor.** The mechanical governor controls the fuel rack and maintai constant speed. On starting, the engine runs up to nominal speed; then t governor maintains it. Many engines now have electronic governors.

18.3 Generator Fuel Consumption. The table below gives approximate fuel consum tion rates at full rated load for a number of engine output ratings. These will act as a gene guide in working out similar onboard consumption values.

Table 18-2. Generator Fuel Consumption

Cylinders	Capacity (liters)	Speed (rpm)	Output (kW)	Fuel Rate (liters/hour)
2	0.5	3,000	7.5	1.5
3	1	1,500	7.7	2.5
4	1.3	3,000	9.8	3.3
4	1.3	1,500	19.4	5.9
4	1.5	1,500	11.9	2.6
4	1.8	1,500	13	4.5
3	2.5	1,500	21	6.1
4	3.9	1,500	34	10.2

18.4 AC Alternators. Alternators are generally robust and constructed to marine sta dards. The alternator consists of the main stator winding, exciter stator winding, main wou rotor, rotor exciter winding, cooling fan, terminal box, bearings, and rotating rectifier.

a. **Single-Phase Alternators.** The single-phase two- or four-pole alternator the most common configuration and may be either brushless and self-exci with permanent magnet or brush with slip rings and externally excited. Alt nators may be single bearing and directly coupled to the engine or be d bearing machine.

(1) **Operation.** At initial start-up, there is sufficient residual (remane voltage left in the machine to enable establishment of the main fie Once rated speed and output is reached, the automatic voltage reg lator (AVR) controls the output voltage in response to system va ations. Frequency is a function of speed, and the engine govern maintains speed.

(2)   **AVR Operation.** The AVR controls excitation voltage level. The control voltage is applied through the brushes and slip rings to the rotor-mounted excitation winding and diode rectifier. The rectifier DC output goes to the main rotor excitation winding rotating field and controls field strength.

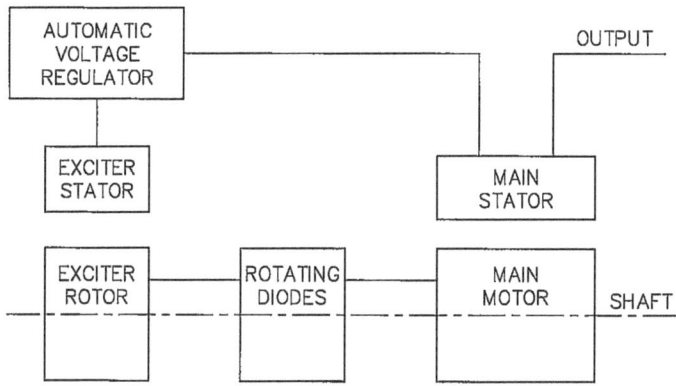

Figure 18-1. Single-Phase Alternator Diagram

.5   **AC Alternator Parameters.** The alternator has the following parameters:

a.   **Voltage.** Typical rated output voltages are 115/230VAC or 220/240VAC for single-phase machines and 440/480VAC for three-phase machines. The AVR maintains nominal output. The AVR is an electronic regulator that senses output terminal voltage and varies the field strength to maintain correct value. Regulation is typically within 2% of nominal rating. The AVR must be able to control the output rapidly in response to large load fluctuations; recovery in good machines is typically 3% of rated output within 0.25 second when full load is applied. Voltage is not a function of speed or frequency when an alternator is running at or near rated speed.

b.   **Frequency.** Frequency is specified in hertz (Hz) and is the number of alternating cycles per second. Output frequency is a direct function of speed and varies in response to speed fluctuations. Stability is dependent on the ability of the machine to maintain nominal frequency over the complete power output range, typically within 1%. The engine governor controls engine speed. When a large load is applied, such as a motor starting, the generator loads the engine, causing it to momentarily slow. The governor reacts by increasing fuel flow and speeding up the engine. When the load is removed, the reverse occurs. Stability depends on response time, and governors are factory set. A small time lag is inherent in the system and helps minimize hunting, which is caused by continual alterations based on small load fluctuations. Frequency and specified engine speed is dependent on the number of poles within the alternator. Two pole machines generate one cycle per revolution and require an engine speed of 3,000rpm for 50Hz. Four-pole machines generate two cycles per revolution and require only 1,500rpm for 50Hz. The Fischer Pan iSeries generators use variable speed technology, which allows engine spe regulation corresponding to the electrical loading. Inverter generators outp AC, which is then rectified before being inverted back to pure sine wave A This allows an electronic governor to control engine speed and maintain stable output frequency.

c.   **Power.** Power output is stated in either kVA or kW ratings:

(1)   **kVA Rating.** The kilovolt-amp (kVA) rating is simply the pow output, which is the current multiplied by voltage to give volt am and divided by 1,000 to give a kVA rating.

(2)   **kW Rating**. Kilowatt rating is the kVA rating, which is then m tiplied by the power factor, typically 0.8. This is the actual pow output.

18.6   AC Alternator Rating Selection. Rating selection must consider a number of fa tors. A total expected load analysis must also be undertaken to calculate the peak loads th might be encountered.

a.   **Starting Currents.** Starting current values may be as high as five to ni times that of actual normal running current. The inrush current at starti causes these high currents, and the energy required to overcome bearing a load inertia. Duration of the peaks is typically less than 1 second, and mc alternators can withstand 250% overloads for up to 10 seconds.

b.   **Power Factor.** In simplified terms, power factor (PF) is the ratio of us ful power in watts to the apparent power (volt amps) of the circuit. Pow (watts) = Volts × Amps × Power Factor. In a purely resistive circuit such a: heater, the alternating current and alternating voltage sinusoidal wavefor are said to be in phase with no phase shift between them. The average pow over a complete cycle is the product of the voltage and current in volt am When reactance is introduced into the circuit, the voltage and current becor out of phase so that during any cycle, the current is negative and the vo age positive. The resultant value is less than the volt amp value. Inducti reactance causes current to lag the voltage; this will be an electrical ang between 0 and 90 degrees. Resistive loads are said to be in phase, with angle of difference, and these are termed "unity power factor" or "one." electrical circuits, capacitive reactance cancels out inductive reactance. T use of capacitors can improve low power factors; this is generally limit to fluorescent lighting systems. Most machinery nameplates specify pow factor ratings. Available alternator output power decreases with any decrea in system power factor values, so the higher the better.

18.7   Generator Rating Calculation. From a load analysis, the following calculatic can be performed to estimate required minimum alternator size.

Installed Batteries
C. Payne)

Proper Battery Terminals Installation
(John C. Payne)

mini Flexible Solar Panel Installation *(Courtesy Marlec Engineering Co Ltd.)*

Stern Davit Mounted Solar and Wind Installation *(John C. Payne)*

Stern Mounted Solar and Wind Installation *(John C. Payne)*

Rutland 914i Wind Generator
(Courtesy Marlec Engineering Co Ltd.)

Solar MPPT Regulator *(Courtesy Marlec Engineering Co Ltd.)*

Eclectic D400 Wind Generator on Aventura IV
(Courtesy Eclectic Engineering)

Marine Kinetix Mk4 Wind Generator
(Courtesy MarineKinetix)

Superwind 350-II Wind Generator
(Courtesy superwind GmbH)

Superwind 350-II Regulator *(Courtesy superwind GmbH)*

Superwind 350-II Hub Assembly
(Courtesy superwind GmbH)

Mizzen Mast Wind Generator Installat
(John C. Payne)

**Watt & Sea Water
Generator System**
(Courtesy watt&sea)

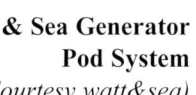**& Sea Generator
Pod System**
(ourtesy watt&sea)

Sail-Gen System Retracted
(Courtesy Eclectic Energy)

Sail-Gen System Subm
(Courtesy Eclectic Energ

Duo-Gen System—Wind Mode
(Courtesy Eclectic Energy)

Alternator Connection Issues
(John C. Payne)

Substandard Main Panel Wiring Installation *(John C. Payne)*

Substandard Main Panel Wiring Installation *(John C. Payne)*

Hallberg-Rassy HR34 Electrical Panel
(Photo R. Tomlinson, Courtesy Hallberg-Rassy)

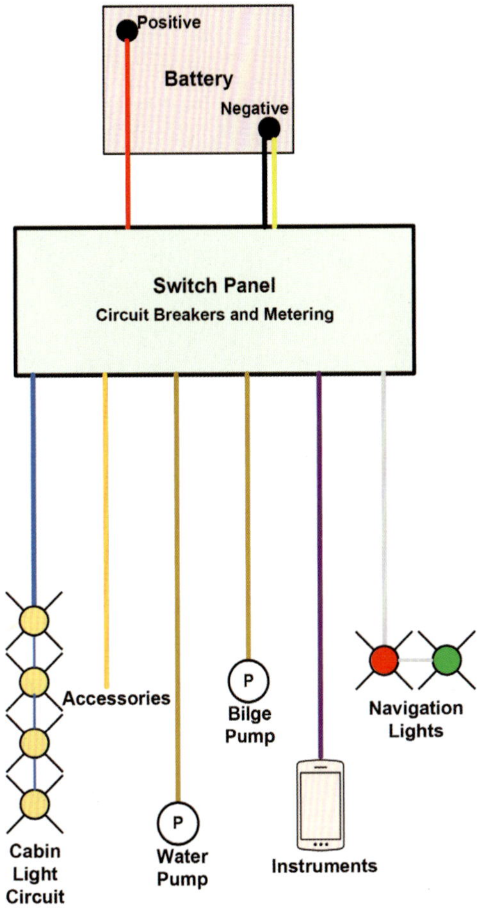

DC Wiring Circuit Color Codes
(John C. Payne)

Cable Glands and Junction Boxes *(John C. Payne)*

Poor Navigation Light Wiring
(John C. Payne)

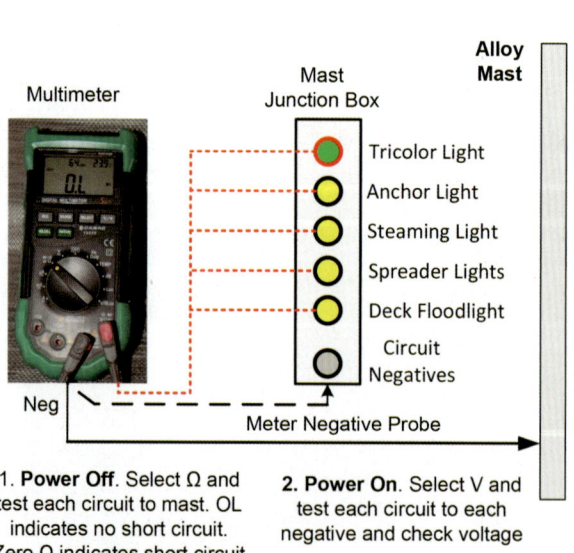

Mast Circuit Testing *(John C. Payne)*

Chainplate Bolt Crevice Corrosion
(John C. Payne)

**Chainplate Bolt Head Crevice
Corrosion Failure** *(John C. Payne)*

Poor Starter Motor Connection *(John C. Payne)*

Properly Terminated Starter Motor Connection *(John C. Payne)*

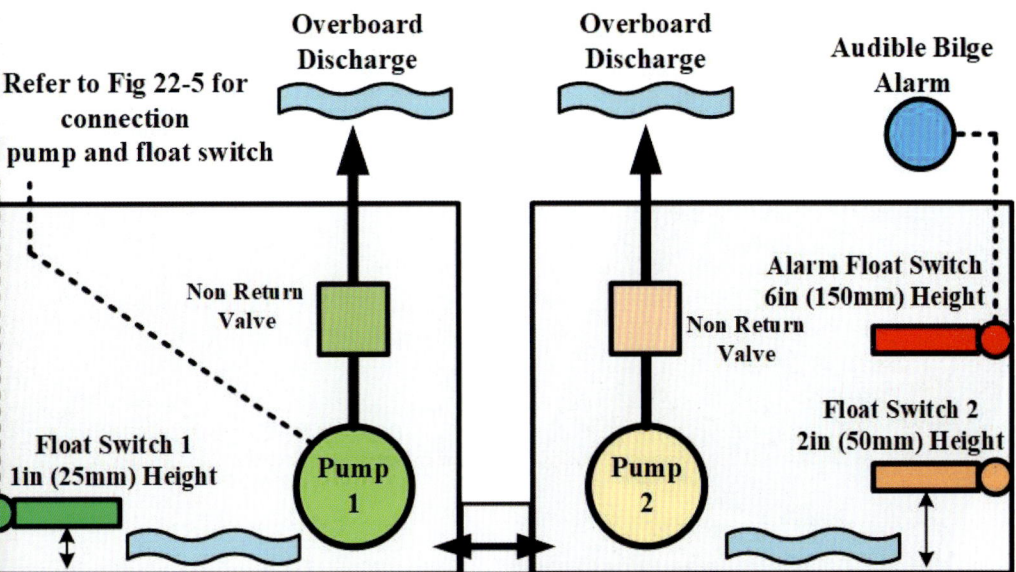

Refer to Fig 22-5 for connection pump and float switch

Overboard Discharge

Overboard Discharge

Audible Bilge Alarm

Non Return Valve

Alarm Float Switch 6in (150mm) Height

Non Return Valve

Float Switch 2 2in (50mm) Height

Float Switch 1 1in (25mm) Height

Pump 1

Pump 2

Dual Bilge and Pumps with Alarm Float Switch *(John C. Payne)*

Bird Wattmeter Forward Power VHF Aerial Test *(John C. Payne)*

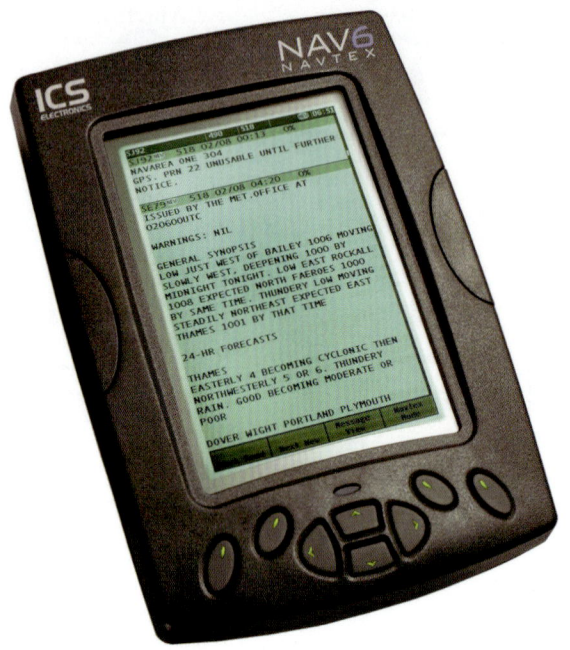

ICS NAV 6 NAVTEX Terminal
(Courtesy ICS Electronics Ltd.)

Container Ship Passing Singapore *(Photo Piet Sinke, Maasmond Maritime)*

a. Generator to run at 80% maximum load with 20% reserve:

Largest Single Load Value = 2,400 watts

$$\text{Max. Start Current (Ir)} = \frac{\text{Power (watts)}}{\text{Volts} \times \text{PF}} = \frac{2,400}{240 \times 0.8} = 12.50\text{A}$$

Max. Start Current (Is) = Ir $\times$ 4 = 50A

b. If alternator can withstand overloads of 250% for 10 seconds. Starting current (Is) must be divided by 2.5 = 20A.

c. Rating is therefore $20 \times 240 = 4.8$ kVA or $0.8 \times 22.8 \times 240 = 3.8$kW. Adding a 20% minimum margin to operate at 80% = 4.5kW.

d. In selecting the margin to the calculated required output rating, an estimate must be made of the maximum load likely to be applied. Some loads, such as kettles and heaters, are resistive and do not have large starting currents. If the device having the start-up current is to operate simultaneously with other equipment, the other loads need to be added. So if an air conditioner with 3.8kW rating is to be run at the same time as the hot water system, then add 2.4kW, and a water kettle 1.8kW then a figure of 8kW, plus a 20% margin for other small loads such as lights and TV, you require 9.6kW. The higher the rating, the higher the initial capital cost and the greater the weight and space required. A decision must be made as to what equipment usage regime will be followed in order to reduce the generator to the lowest suitable size. A generator should be loaded to at least 35% and ideally operate at 75% to 80%. A major factor increasing maintenance is the running of gensets diesels on light loads, causing cylinder glazing.

18.8   Generator Systems. There are several players in this space, each with its own technologies. Things to look for are size and weight, sound shield options, noise and low vibration levels, and, importantly, serviceability with access to key components such as oil and fuel filters. Given it is so easy to go online and download data sheets as well as service and operating manuals, checking out the detail has never been easier. In previous book editions, I outlined some of the key features and operating notes, but the range of generators currently available makes repeating this information far beyond the scope of this book.

18.9   Generator Protection. Diesel generators have several protection systems.

a.   Low Oil Pressure Shutdown. Most generators have automatic shutdowns on low oil pressure to protect the engine; this is usually a pressure switch–activated function. Do not ignore it; check out possible causes immediately.

b.   Low Oil Level. This activates when the oil level in the sump falls; it is a level switch. It means that oil requires topping up and possible oil leakage.

c.   High Water Temperature Shutdown. Generators will shut down on detection of a high jacket cooling water temperature. Causes are outlined in the diesel engine chapter, but raw water intake blockage is the most common.

d.   **Low Water Level Shutdown.** This is activated when the expansion tank level falls.

e.   **Low Seawater Flow Shutdown.** Loss of seawater cooling, due either to a plugged strainer or impeller failure, will initiate an automatic shutdown.

f.   **High Exhaust Temperature.** A high exhaust temperature will initiate an automatic shutdown.

g.   **Overspeed Shutdown.** If the generator overspeeds, an automatic shutdown is initiated. This can be caused by governor failure.

h.   **Overvoltage Shutdown.** If the AVR malfunctions and the voltage goes high, the generator will be shut down.

18.10   **Alternator Protection.** The alternator system requires the following:

a.   **Protection.** Protection of generator electrical circuits consists of one or more of the following, which may be integral to the main circuit breakers, for larger vessels, or be part of a separate protection system:

(1)   **Overload.** Some gensets have an overload circuit breaker fitted at the genset control box. Reset if tripped; if repeatedly tripping, fault-find the system and remove the cause of the overload. In many cases the problem may simply be too many appliances operating and overloading. Overloads are characterized by time delays between reset and tripping. If a major motor is installed and is seized or has a locked rotor, tripping may be immediate due to high current.

(2)   **Short Circuit.** This generally is a circuit breaker, which is mounted at the genset control box. Again, if it immediately trips after initially resetting, fault-find and correct the fault.

(3)   **Reverse Current.** This protection is generally seen only on larger installations or where two units are paralleled. If the load-sharing function fails or, when manually taking a generator off the board, the load is taken to zero, this will trip.

(4)   **Low Frequency.** Not all generators have this protection; adjust only according to manufacturer's instructions.

(5)   **Undervoltage.** Undervoltage trip relays are used in larger installations and are normally interlocked with main circuit breakers. It usually indicates an AVR fault.

b.   **Grounding.** All exposed metal capable of carrying a voltage under operating or fault conditions must be grounded to an equipotential point. The generator frame should be securely connected to the boat ground system. In most generators the starter negative is bonded locally to the AC ground. This will require bonding to the main boat ground. In single-phase installations, the neutral is connected to the distribution system neutral at the main switchboard, not the frame. Connection should be in accordance with the manufacturer's installation instructions.

c. **Alternator Troubleshooting Safety.**

Disclaimer. This chapter is for informational purposes only and readers should observe the safety recommendations given. Readers should refer to the disclaimer at the front of this book. The same safety warnings apply for troubleshooting live AC generator systems. Call a qualified service technician or qualified marine electrician. There are serious risks of electric shock and death when performing voltage checks and testing.

18.11   Alternator Maintenance. Alternator maintenance is simple and easily carried out. Isolate and disconnect the start battery supply before working on a generator to eliminate accidental starting.

a. **Alternator Inspection.** Carry out the following tasks:

(1)   Remove the alternator access covers and check the gaskets and seals.

(2)   Check that connection boxes are clean and dry.

(3)   Inspect the interior for dirt, dust, oil, and water. Use a vacuum cleaner to clean all accessible surfaces and windings as required.

(4)   If the generator has a brush excitation system, check the excitation brushes for breaks and chips and that they move freely in the holders. Check brush springs and shunts.

(5)   Check that the slip rings are smooth, have no scoring, and have a shiny surface patina. Do not polish or use emery paper.

(6)   Check for loose electrical and mechanical connections.

b. **Insulation Test.** Use a 500VDC insulation tester (Megger) to test all active conductors to ground. Measure and record the temperature of the alternator stator circuit insulation resistance in the boat maintenance log. Isolate the exciter and measure the rotor insulation resistance and record. Any reading less than $1M\Omega$ should be rectified.

18.12   Generator Mechanical Systems. The alternator prime mover gives the most in-service troubles. The operating principles are identical to those described in the main propulsion diesel engine section. If the lubrication, cooling, fuel, and air quality are maintained properly, long-term, trouble-free operation is assured.

a. **Coolants.** Genset diesels may be cooled by seawater or have a heat exchanger with a closed-circuit cooling system. The coolant provides the medium for transferring engine heat to the primary seawater coolant and for controlling the engine's overall operating temperature. It is essential that adequate heat transfer is maintained. The coolant must remain free of saltwater contamination to prevent corrosion or the formation of sludge and scale that may impede coolant flow or block coolers. A coolant that has no inhibitors, incorrect inhibitors, or improper concentrations of inhibitors will ultimately cause problems with rust, sludge, fatigued water-pump seals, and reduced heat transfer rates as engine block water passages become coated with an insulating layer of scale. This will gradually result in overheating and all the damage that goes with it.

(1) **Additives.** A number of additives are available to improve the performance of coolants, including sulfates, chlorides, dissolved solids, and calcium. Coolant should have an antifreeze additive to prevent freezing and engine damage in cold climates. Most ethylene glycol–based antifreeze solutions contain the inhibitors required for normal operation. Coolant additives do not remain effective forever; consider replacing every year or in accordance with your generator manual.

(2) **Corrosion Inhibitors.** These are generally water-soluble chemical compounds that protect the metal surfaces within the system against corrosion. Compounds can include borates, chromates, and nitrites. Inhibitors with soluble oils should never be used as a corrosion inhibitor.

b. **Lubricating Oil.** Lubrication has the dual function of reducing friction between moving parts and taking away some of the heat generated during combustion. It is essential that the correct grades of oil be used for the prevailing temperature conditions, and that the filter is changed regularly along with oil. Oil viscosity must be maintained if correct lubrication is to be achieved, and this depends on the engine remaining within proper operating temperature range. Lubricating problems include:

(1) **Fuel in Oil.** Fuel in the oil creates the risk of a crankcase explosion and is characterized by low lube oil viscosity.

(2) **Water in Oil.** Water in the oil causes emulsification, which destroys the oil's lubricating properties. After repairing a leak, completely flush out the system. No moisture must remain.

(3) **Microbe Growth.** Moisture in the system can encourage microbial growth in the oil. Once the system is "infected," considerable flushing is required to eliminate it. Check the diesel engine chapter for more detail.

c. **Fuel.** Clean, uncontaminated fuel is essential to good combustion and efficient engine operation. There is a history of water-contaminated fuels being supplied to unsuspecting yachts from bunker station and fuel barges. Observe the following installation precautions:

(1) **Filters.** Install a filter and water separator (Racor or Dhal), preferably with a water-in-fuel alarm. In most cases, the generator takes fuel from the same tanks as the main engine, which should have similar protection.

(2) **Purge Fuel Tanks.** Microbial growth can occur in water-contaminated fuels. Tanks must be regularly purged of water and kept topped up to avoid condensation. The tank is usually the same as the main propulsion engine and should be drained, inspected, and cleaned annually for optimum cleanliness.

d. **Exhaust System.** The same rules and requirements required for diesel engines apply. In particular, follow the manufacturer's recommendations on installation of antisiphon loops.

e. **Electrical System.** The electrical and monitoring system is similar to the main engine system. It consists of the following:

(1) **Starting System.** Always install a separate start battery for the generator. Starting batteries for a typical 4kVA to 6kVA genset are recommended at 70AH/325CCA. This provides backup power for the main engine in an emergency or, alternatively, ensures that the genset can be started if the main engine is out of service. Some generators have an interlock in the starting system that prevents starting if the generator is running.

(2) **Charging System.** The alternator charging system on most gensets is very small, typically around 15A to 25A maximum. Normally, it is only used to charge the start battery, although alterations can be made to charge the main engine or house batteries. In some cases the alternator can be uprated to 55A or even 80A, which, when connected to a cycle regulator, provides the main battery charging source, limiting the run time requirements of the main engine. Care must be taken, as output shafts cannot always cope with large mechanical side loads.

(3) **Monitoring System.** Most generator units have basic control panels with alarms only for high water temperature or low oil pressure. Most current generation units have electronics in the control and monitoring system. They all incorporate an hour meter, pilot light, overload alarm, and water warning alarm.

(4) **Preheat Systems.** Many generators have a glow plug preheating system, which uses a traditional cylinder glow plug in the combustion chamber. Some engines have preheaters in the air intakes, referred to as a "cold start aid." Use them properly in very cold conditions.

317

18.13   Generator Maintenance. The maintenance tasks for the main engine are also valid for generators, and these are typical for most generators. Check your manufacturer's manual for recommended service routines and durations.

a. **Fuel System.** Renew and clean filters every 200 operating hours. Check the filter and drain water weekly.

b. **Lube Oil System.** Replace oil and filters every 200 operating hours. Always check oil viscosity and for signs of water, fuel, or microbial growth that may affect viscosity and quality.

c. **Air System.** Replace the filter element every 500 operating hours.

d. **Coolant Systems.** Check seawater strainer weekly or more frequently. Check coolant levels monthly and the expansion tank weekly or more often. Renew antifreeze and inhibitors annually. Replace the seawater pump impeller every 500 hours. Check and replace anodes yearly or more frequently, depending on condition. Check all coolant hoses every 100 hours; check and tighten hose clamps.

e. **Cleaning.** Keep the engine clean of oil and dirt every 200 hours.

f. **V-Belts.** Check and tighten alternator and water/fuel pump drive belts every 100 hours.

g. **Charging System.** Using a voltmeter, check that the charge voltage is approximately 14.4V.

h. **Mountings.** Check rubber mountings for cracks and fatigue.

i. **Electrical.** Check that battery and starter connections are clean and tight.

j. **Anodes.** Sacrificial zinc anodes in the cooling system should be checked every 6 months and replaced if corroded.

18.14   Generator Operating Notes. Consider the following when operating the genset.

a. **Starting.** If the generator doesn't start after 5 seconds, stop cranking. Cranking for too long will overheat and burn out the starter motor. Leave a few minutes between each start attempt to allow the motor to cool down. Check that the preheat glow plugs are operating.

b. **After Start.** After starting, always check that seawater coolant is discharging overboard to ensure that coolant is passing through the engine. Don't wait for a high-temperature alarm to warn you of possible engine damage.

c. **Operating Temperatures.** Run the genset for 3 to 5 minutes before putting a load on the generator. This gives the engine a chance to increase to a normal operating temperature.

d. **Generator Loading.** Do not let the generator run on light or on no-load for extended periods. This will cause cylinder glazing and deposits within the engine, which will increase maintenance costs. If you have an electric water heater or other suitable appliance, turn it on to increase load and make the most of the available energy. Typically aim for 50% or greater load.

e. **Generator Exercising.** It is good practice to run the generator weekly at approximately 50% load for about an hour. This warms up and dries out the alternator windings. It runs fresh diesel fuel through the system and uses up stale fuel, which helps lubricate and flush through the fuel system. The run also circulates and lubricates all internal engine and fuel system components.

f. **Generator Stopping.** When stopping, always run the generator for 3 to 5 minutes and allow it to cool down.

Table 18-3. Generator Troubleshooting

Symptom	Probable Fault
Generator will not hold load	Fuel filter clogged Air filter clogged Air in fuel system Governor fault Voltage regulator fault
Generator will not stop	Check stop solenoid Check governor Check local stop function and remote stop
Generator will not remote start	Check local panel start function Control circuit power tripped
Generator will not start from local	Control circuit power breaker tripped Low battery voltage Start solenoid fault (check terminals) Starter motor fault (solenoid is closing) Check battery negative terminations
Engine cranks and will not start	Fuel starvation; check filters, supply valves Air in fuel system; prime system Glow plug not operating correctly Fuel pump problem Low battery voltage; check battery Battery low; check V-belt and alternator
High exhaust temperature	Seawater strainer clogged Seawater inlet valve not open fully Seawater pump impeller worn Heat exchanger clogged

(continued)

Table 18-3. *Continued*

Symptom	Probable Fault
Low oil pressure	Oil level low; top up and monitor Oil pump fault
High engine temperature	Seawater strainer clogged Seawater inlet valve not open fully Seawater pump impeller worn Heat exchanger clogged V-belt tension loose Overloading generator
Engine instability	Fuel filter clogged; change filter Fuel system air leak; check connections Faulty governor or seizing up Injector issues; check smoke color
No output voltage	Circuit breaker tripped; if it repeats, check Alternator connection failure Alternator failure
Voltage instability (high or low)	Fault in automatic voltage regulator Engine governor fault
Diesel engine problems	See diesel engine chapter

AC Inverters

Inverter technology has advanced considerably in the last few years. They range from small portable units of just 150W up to large fixed systems of 5kW with some units that can be paralleled with an automatic synchronization module. The combination charger/inverter is very common. The latest units use insulated gate bipolar transistors (IGBTs), which provide precise output control and waveforms, over older MOSFET-type systems. Transformers have been reduced in size and weight as conversion is done at high frequencies. Reliability on most systems is greatly increased over earlier-generation units that unfairly gave inverters a questionable reputation. IGBTs open and close multiple times per cycle; this is known as pulse width modulation. If a cycle is broken up into many small segments, the controller instructs the IGBT on how long to close during each segment. An inverter is essentially a DC-to-AC transformer that has a voltage inversion process. The 12VDC input undergoes inversion into high-frequency and high-voltage AC. The core of the inverter is an integrated PWM controller. The PWM controller provides several functions, such as internal reference voltage, an error amplifier, oscillator and PWM overvoltage protection, undervoltage protection, short circuit protection, and an output transistor. ABYC A-25 has guidelines for the installation of power inverters. New Xantex units, such as the 3kW/150A Freedom XC Pro inverter/charger, have NMEA 2000 connectivity.

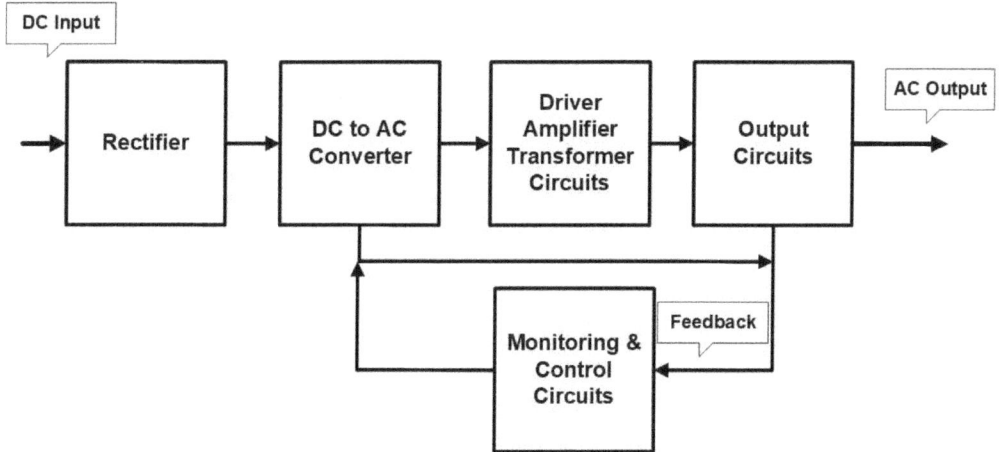

Figure 19-1. Basic Inverter Diagram

19.1   Inverter DC Loads. The typical inverter is capable of drawing large DC current loads, and it is essential that battery capacity is capable of supporting these loads without affecting the existing electrical system and connected loads. A simple method of calculation is to divide the power in watts by 10 for 12V systems or 20 for 24V systems. Some manufacturers offer inverters for 12V, 24V, and 48V inputs, depending on system voltage. The minimum battery capacity required for an inverter is 20% of the inverter capacity—i.e., 2,000W, 400Ah capacity.

Table 19-1. Inverter 12VDC/220VAC Current Loads

AC Load	DC Current Draw	Peak Overload
200W (0.8A)	18A	42A
400W (1.6A)	37A	100A
600W (2.4A)	56A	145A
900W (3.6A)	84A	180A
1,200W (4.8A)	120A	290A
2,400W (9.6A)	240A	580A
3,000W (12A)	300A	750A

19.2   Inverter Output Waveforms. Output waveforms are an important consideration when looking at proposed applications.

a.   **Trapezoidal.** The majority of inverters have a trapezoidal waveform. This is suitable for most equipment, but microwaves and some inductive loads do not operate at full output, dropping efficiency by some 20% or more in some cases. Fluorescent lights may be less efficient at starting; a 3μF capacitor across input will improve starting characteristics. Interference is possible on these waveforms.

b.   **Modified Sine Wave (MSW).** Some units have a modified or quasi sine wave output, which closely resembles a pure sine wave. It is not and offers slightly less performance. Interference is possible on these waveforms. Appliances and equipment such as microwaves and VCR units with clocks often run either slow or fast. Battery chargers for cordless portable drills are susceptible to early failure. If a charger gets excessively warm or hot, turn it off, as some incompatibility may exist.

c.   **Pure or True Sine Wave (TSW).** Sine wave output units are the ideal and offer quality better than shore mains power but are more expensive. If you are using sensitive equipment, then sine wave should be used. They are available in most ratings up to 1,800W in 12V systems, with outputs at 230V RMS +/– 5%. Frequency stability is typically +/– 0.05%. Quality units have low harmonic distortion, typically less than 3%, and low EMI (electromagnetic interference) levels. These types of inverters are high frequency and drive switches at around 50kHz. This makes inverters smaller and lighter.

d. **Ratings.** Units generally have an output rating based on a resistive load and for a nominal period, typically 30 minutes. The continuous rating is the normal continuous operation rating. The peak or maximum rating is the maximum short-duration load the inverter can withstand. Most units are capable of withstanding the short duration and intermittent overloads that are required, especially with motor starts. The surge rating enables them to withstand short time overloads of up to approximately 200% over the continuous rating for 5 seconds.

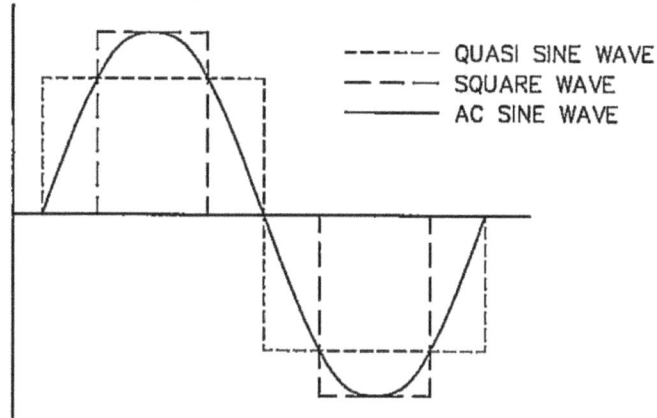

QUASI SINE WAVE
SQUARE WAVE
AC SINE WAVE

Figure 19-2. Inverter Waveforms

19.3   AC Transfer Systems. If another AC power source is connected across the inverter output, the inverter electronics will be seriously damaged. Shore power supplies, generator, and inverter outputs must not be paralleled at any time.

a. **Rotary Switches.** The normal selection system is a rotary cam switch. The switch should be a center-off type, with the inverter to the side opposite the generator and shore power.

b. **Automatic Transfer Systems.** Many units are able to automatically switch the supply from shore power or generator to the inverter on supply loss, giving the same features as a UPS. Freedom series inverter/chargers have an automatic "load sharing" facility that offers protection to connected AC loads from variations in shore or generator power. When the shore or generator power is connected, the unit transfers the power to any connected AC loads and some power is used to charge the batteries. When in the charge/transfer mode, the incoming AC power is monitored; when a voltage drops (or increases) longer than 8 milliseconds, charging ceases, load transfer ceases, and inverter mode then starts.

323

c. **Auto Start.** Most units have an auto start capability. This means they remain in a standby or idle mode until a load is switched on.

   (1) **Idle Mode.** Input current in idle mode is typically around 10mA to 50mA (1.5W). This load value must always be included in DC load calculations.

   (2) **Activation Load.** The load required to activate most inverters is approximately 6VA to 10VA or greater. If the vessel is to be left unattended for an extended period, the DC supply should be switched off. In some cases, loads such as fluorescent lights or electronics equipment may not activate; another load should be momentarily switched on to cut the inverter in.

19.4 Combination (Combi) Charger Units. Most of these units have a modified sine wave output. The chargers are settable for various battery types, such as flooded cell, gel, or AGM batteries; those from Heart use a three-stage microprocessor charging process. An automatic transfer switch and relay transfers between charger mode and inverter mode when AC power is off. *Caution:* Always remember that combi units may automatically supply AC. Proper isolation from the battery source is essential before opening or working with the unit.

19.5 Inverter Installation. Follow these guidelines when installing an inverter. For general installation recommendations, also refer to ABYC A-31 and E-11.

   a. **Input Cables.** Always install the maximum size cables possible, as considerable volt drop problems are possible at peak loads. Ensure that cable connections are tight.

   b. **Inverter Grounding.** Inverters must be grounded. There are documented cases of fatalities where this has not been done. Grounding introduces a number of factors that require consideration. Most standards or recommendations (ISO, ABYC, USCG) specify that the AC ground be connected to the DC negative. This requirement has raised considerable controversy in shore power installations as well as inverters. Many do not do this to reduce the perceived risks of galvanic corrosion. The ABYC recommends that the inverter chassis be bonded to the DC negative, although this often doubles up the grounding, as the AC ground and DC are often already bonded. Check your installation manual or consult an AC-qualified electrician.

   c. **Ground Fault Protection.** It is normal to install a ground fault circuit interrupter (GFCI) on the AC output of the inverter. As indicated, the GFCI must be tested regularly (refer to the chapter on GFCI). The GFCI should be function tested when inverting or when transferring generator or shore power supplies. Do not test when the inverter is in idle mode; the unit may not trip, and the GFCI electronics circuits could be damaged by the inverter idle mode sense pulses. Nuisance tripping does occur on inverters; this is generally caused by neutral-to-ground leakages, usually from surge suppression circuits, which contain capacitors connected across active and ground or neutral and ground. This is attributed to waveform harmonics on modified sine wave outputs. Another cause of GFCI tripping is due to the improper connection of the inverter AC output neutral to the main neutral bus. Check your installation manual or consult an AC-qualified electrician.

d.   **Protection.** Most inverters have an undervoltage cutout that is typically set at around 10.5V. Units have overload protection and a high voltage cutout, thermal overload protection that will shut down the inverter if an over-temperature condition is reached. The unit should have reverse polarity indication and protection and short circuit protection of the output and input if the voltage ripple is too high; some units have AC back-feed protection and a GFCI installed on the output.

e.   **Ventilation.** Good ventilation is essential to reliable operation and the availability of full rated outputs. The unit should be installed in a dry, well-ventilated area. Sufficient vertical clearance should be allowed for natural convection of heat from the unit. Derating factors are illustrated in the efficiency.

Table 19-2. Temperature Derating Factors

Temperature	Temperature	Output Rating
104°F to 122°F	+40°C to +50°C	80% Rated Output
14°F to 104°F	−10°C to +40°C	100% Rated Output
14°F to −04°F	−10°C to −20°C	140% Rated Output

19.6   Inverter Efficiency. Inverters have become very efficient. This is directly related to the load applied and also the heat buildup. All electronics under load generate heat, and this reduces overall efficiency. As the performance curve shows, the time an inverter can withstand overload is directly proportional to the load. The typical inverter is now approximately in the range of 85% to 95% efficient at rated output. This efficiency level has evolved by new electronic switching technologies such as metal oxide semiconductor field effect transistor (MOSFET) and insulated gate bipolar transistor (IGBT).

19.7   Inverter Interference. Inverter control electronics, such as logic circuits and memory circuits in inverters, can be corrupted. This may be from lightning strike surges and onboard electrical power system surges, voltage dips, and spikes. In many cases the unit may simply require resetting if no signs of catastrophic failure have been detected, such as smoke or burning smells. Try the following procedure first before calling in technical assistance.

a.   Switch off power at remote panels if fitted.

b.   Switch off the inverter main power switch.

c.   Disconnect the AC input power source.

d.   Disconnect the DC negative cable for at least 5 minutes.

e.   Reconnect the DC negative cable (sometimes a small spark will be seen as the filter capacitors start charging).

f.   Switch on the inverter power switch and switch on an AC load to ensure the inverter has a load. Check that the inverter supplies AC power.

g.   Reconnect the AC input power source and check that the automatic transfer circuit functions.

h.   If this does not restore operation, check that fuses have not blown or circuit breakers tripped. Check that all connections are secure.

19.8   Inverter Troubleshooting. There is little troubleshooting to do with an inverter. The principal issue is low battery voltage. The next most common issue is overheating, and note the derating factors. Installation issues comprise undersized DC cables and voltage drop. Read the manual and understand the error messages. **Disclaimer:** This chapter is for informational purposes only and readers should observe the safety recommendations given. Readers should refer to the disclaimer at the front of this book.

Inverter Troubleshooting Safety. AC is potentially lethal, and inverters must be correctly selected, installed, and maintained. The same safety precautions described in AC Power Safety must be observed at all times. Always use an AC-qualified and licensed marine electrician or a shore-based licensed industrial electrician.

Table 19-3. Inverter Troubleshooting

Symptom	Probable Fault
Noise from inside case	Cooling fan fault if installed Blocked air inlet filter if fitted
Fault display error codes messages	Check the manual to understand
Low output	Low battery voltage Undersize DC supply cables Battery termination high resistance
Tripping off on over-temperature	Cooling fan issue Overloading Ventilation obstructed
Does not start up	Power supply off Control fuse blown (power surge) Control system fault Reset button tripped

Shore Power Systems

Shore power requirements are ever-increasing as many choose to weekend alongside or spend days tied up and chilling out. The average yacht has accumulated many appliances to maintain all the home comforts. Air-conditioning tops the list, along with TV and a hot water system. The following basic recommendations should be used in conjunction with the relevant national or other standard.

20.1   **Shore Power AC Power Safety. Disclaimer:** This chapter is for informational purposes only and readers should observe the safety recommendations given. Readers should refer to the disclaimer at the front of this book.

WARNING

a. **Always engage an AC-qualified or -licensed electrician to do your shore power AC system repairs and maintenance. Check the license!**

b. **Never work on "live" equipment. Always isolate and lock out equipment before opening.**

c. **Always remove the shore power plug and lead, and isolate the local main supply switch or circuit breaker.**

d. **Always prove that the circuit is de-energized before starting work.**

e. **Never work on AC equipment alone; always have someone ready to assist if you accidentally receive a shock.**

20.2   **Rules and Regulations.** Connection of the vessel to a marina power system imposes certain obligations on the boat owner. Any boat that is connected to a marina or any other shore power circuit is generally required to comply with the relevant provisions of local and national electrical codes. Recommendations contained within this section are advisory only and do not override the legal responsibilities of boat owners to meet specific requirements. Major standards include the following:

a. **ABYC E-11: AC & DC Electrical Systems on Boats.** These recommendations are the de facto standard in the United States. There also may be shore-based electrical codes at a marina, such as the National Electrical Code (NEC) NFPA 70.

b. **The United States Coast Guard.** Mandatory requirements for electrical systems in Title 33, CFR 183, Subpart I, Section 183.

c. **NFPA 302: Fire Protection Standard for Pleasure and Commercial Motor Craft, 2020 Edition.** This standard is approved by the American National Standards Institute. The technical committee includes representatives from ABYC, USCG, UL, and others, such as the National Association of Marine Surveyors.

d. **European Recreational Craft Directive (RCD).** These RCD standards are now mandatory on new construction sailing and power boats in European Union countries and include the following ISO (International Standards Organization) standards: ISO 13297: Small Craft—Electrical systems—Alternating and Direct Current Installations (2020). The standard specifies requirements for the design, construction, and installation of the following types of DC and AC electrical systems, installed on small craft either individually or in combination.

e. **BMEA (British Marine Electronics Association) Code of Practice for Electrical and Electronic Installations in Small Craft, Fifth Edition.** These recommendations are used by various members of the BMEA. They are essentially based on enhanced RCD and ISO requirements and suit both sailing and power boats.

f. **Australian/New Zealand Standard Electrical Installations—Marinas and Recreational Boats, Part 2: Recreational Boats Installations** (AS/NZS 3004.2:2014. Reissued with Amendment No. 1, July 2015). This standard derived elements from (a) IEC 60092-507: Ed.1.0 (2000), Electrical Installations in Ships, Part 507: Pleasure Craft; (b) ISO 10133, Small Craft—Electrical Systems—Extra-Low-Voltage DC installations; (c) ISO 13297, Small Craft—Electrical Systems—Alternating Current Installations; (d) American Boat and Yacht Council (ABYC) E-11: AC and DC Electrical Systems on Boats. One very important statement was made in the preface to this standard that holds true for all when working on boat wiring systems: "Designers are reminded that it is essential that the basic tenets of electrical and marine safety be addressed before any other equipment and installation design elements are considered."

20.3 **Shore Power Components.** The following items or components make up a deceptively simple boat shore power system:

a. **Dock Shore Power Receptacle.** This is the dock or marina boat shore power pedestal and includes the socket and local circuit isolation.

b. **Boat Shore Power Cord Holder.** This is what you use to support your boat shore power lead or cord so it doesn't fall in the water or present a trip hazard.

c. **Boat Shore Power Cord Adapter.** This is used for adapting various boat shore power plugs to different ones so you can plug into a shore power pedestal. There is also the marine shore power Y adapter, which is used to power a single boat shore power lead through two supplies to increase power.

d. **Boat Shore Power Lead.** This is the lead, cable, or cord connecting the shore power pedestal with your boat. Shore power leads for boats come in a range of types and ratings: 15A shore power cord, 20A shore power lead, 30A shore power lead, 50A shore power splitter, and 50ft and 75ft shore power cords.

e. **Boat Shore Power Plug.** This is the power plug on the boat end of the shore power lead. There are many types of shore power plugs. Many have asked me about the Smart Plug shore power connection. Both Marinco and Hubbell manufacture high-quality shore power equipment, so invest in safe and reliable equipment.

f. **Dock Shore Power Receptacle.** This is the boat shore power receptacle installed on your boat for plugging in the shore power lead.

g. **Shore Power Isolation Transformer.** These transformers are sometimes installed on steel- and aluminum-hull vessels.

h. **Shore Power Circuit Breaker.** This is part of the onboard protection; it includes primary overload and short-circuit protection, including the GFCI installed on your boat.

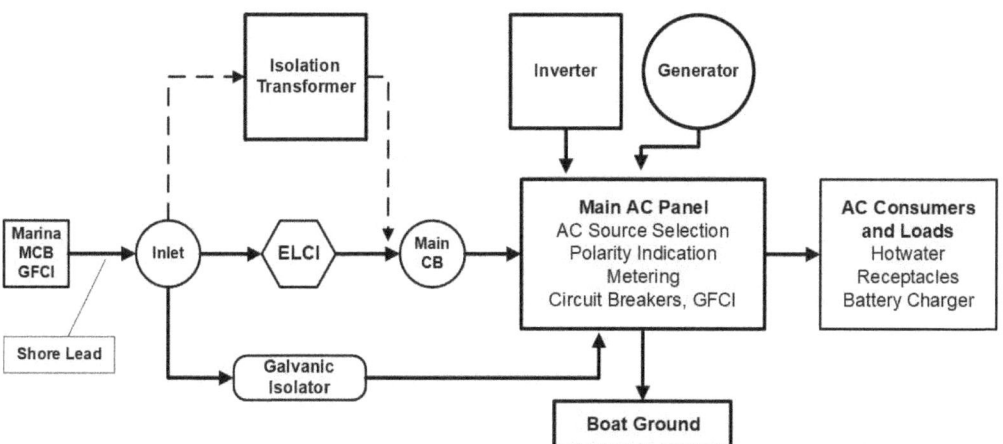

Figure 20-1. Shore Power Systems

20.4   Shore Power Inlet Sockets. Shore power inlet sockets should be of the self-closing type and rated to IP55. All inlet sockets should have a means for locking in the plug when inserted. Inlet sockets should be of a weatherproof standard in accordance with International Protection Standard, or equivalent NEMA rating, which requires protection against heavy seas. Inlet sockets must have spring-loaded, self-closing, and locking covers. The inlet sockets must be of a male type connector only. Do not use inlet sockets that are designed for caravans, trailers, or recreational vehicles—they are not of the standard required. The plug should have a screw locking ring that prevents the plug from being pulled out with boat movement. Suppliers such as Marinco (marinco.com) and Hubbell (hubbell.com) can supply high-quality equipment. Ratings of inlet sockets and plugs is 15A for 240VAC systems; where domestic extension cables are used, these are often 10A only. The nominal rating for 115VAC systems is 30A. Check that the O-ring seals are in good condition and have not deteriorated.

20.5   Shore Power Inlet Socket Installation. Shore power inlet sockets should be mounted at a position and height that prevents immersion and mechanical damage. The sockets should be in an accessible location and as high as possible above the deck line. They must be located so that there is no risk of mechanical damage. They should be shielded from precipitation so that a driving rain at a 45-degree angle will not enter the plug and socket.

20.6   Marina Shore Power Pedestal Outlets. The yacht should not be connected to any marina power supply outlet that is damaged. Inspect the condition of marina pedestals before you connect your boat shore power cable. Look for damage to outlets and tracking on pins, wires, damaged conduits, and other elements, which are a major cause of problems. If the marina systems are in a degraded condition, this represents a risk to you, your crew, and the boat, as well as possible nuisance tripping of GFCI units. Check the dock supply voltages with a multimeter. If they are low (greater than 5%), this is not good for onboard electric motors. Some battery chargers will have lower outputs. Start your largest appliance, plus operate all proposed systems, to see whether the circuit will withstand the load; in many cases, it will not. It is incumbent on you to check and report deficiencies; if you use the outlets in a degraded condition, you may become liable for any consequential damages to your boat and the marina pedestal. If defects and faults are not reported, they will not get repaired.

20.7   Shore Power Leads. Shore power leads should be weatherproof and have a high-visibility outer sheath. Shore cables should be suitable for outdoor use. The cable must be heavy-duty rated, which is typical of most outdoor-rated extension cables and special shore power lead suppliers. High-visibility outer sheaths are typically yellow or orange. These outer sheaths are also UV stabilized, unlike standard-duty domestic leads, which should not be used. It is good practice to have a strain-relief grip over the cable to prevent unnecessary strain on the cable, particularly where the plug into the supply pedestal has a screw-locking ring fitting. The flexible cable should be arranged so as to permit normal movement of the vessel without stress in the full predicted tidal range. Cable arrangement must prevent water traveling along the cable to the inlet receptacle and be secured so that immersion in the water is not possible. Additionally, provision must be made to prevent the plug falling into the water if it is accidentally disconnected. Plugs should be double insulated and be made of impact-resistant material. If a plug has accidentally been immersed in water, do not use it. Disconnect and dismantle for cleaning and drying first. Cables should be regularly inspected and the outer insulation checked for cuts or other damage. Do not repair or join a damaged cable; it must be replaced. Cables should always be run from the pedestal so as to minimize exposure to damage and not present tripping hazards to other people in the marina.

20.8   Shore Power Grounds. Never disconnect a shore power ground conductor as a method for reducing corrosion. Some boat owners disconnect the shore power ground connection either in the plug or at the socket inlet to open the ground path ashore. This should never be done. In many cases it removes the equipment's safety ground, creating a serious electric shock risk and entering the area of criminal negligence.

20.9   Polarity Indication. The AC power panel should have a polarity indicator and a polarity reversal switch. Switchboards should incorporate a reverse polarity indicator and have a changeover switch. It is quite common to find marina supplies with the neutral and active (hot or live) conductors reversed. This condition is indicated, and the switch simply reverses to the correct polarity. Most marina supply outlets are protected by a circuit breaker or a GFCI, which may trip on connection to the boat if ground and active are wrongly connected.

20.10   Input Isolation. The AC power panel should have power input isolation, power source selection, and short circuit protection. The input isolation can be a trip free circuit breaker rated at the maximum power input of the panel. The selection switch is for switching each input source, such as shore power, inverter, or generator, without paralleling. Parallel connection of an inverter will destroy it.

20.11   Circuit Protection. AC power circuits should all have ground fault monitoring and protection. Due to the safety issue of electrical current leakage, the ABYC has developed specific requirements. ABYC E-11.11.1 requires that an equipment leakage circuit interrupter (ELCI) be installed either with or in addition to the main shore power-disconnect circuit breaker. Circuit breakers with integral ELCI are available and should be installed. In addition, a ground fault circuit interrupter (GFCI) should be installed either on all circuits or on outlets located within the galley, weather decks, head and shower spaces, and other spaces. I install one on every circuit on my boats.

20.12   New Shore Power Technology. VoltSafe from Canada has developed a safe shore power plug system. The innovative system replaces the traditional pronged plugs with flat mating pads that connect magnetically to boat and shore supply. The big breakthrough is that arcing is eliminated and, along with it, the corrosion pathway ashore. The system comprises internal intelligent circuits that restrict current flow until it recognizes a shore power signature and then switches on within 5 milliseconds. Of course, wireless networking is part of the capability. The system has integrated arc and ground fault circuit interruption (AFCI & GFCI) for safely eliminating arc, fire, and electric shock. Visit voltsafe.com for details. Smart Plug is a product that features a receptacle locking system for taking cord tension and strain. Its manufacturer claims that the contact area is around 20 times greater than the standard plug. What is innovative is the integral thermal overload sensor with automatic power cutout when the connector reaches 200°F (93°C). While it will reconnect at 120°F (49°C), you really need to be investigating the heating issue by then.

20.13   Microwave Ovens. Yacht-installed microwave ovens have increased rapidly over recent years. Coupled with good freezer capacity, they make meal preparation much easier, especially in bad weather conditions. They also enable conservation of limited liquefied petroleum gas (LPG) reserves. Essentially, microwaves convert the AC input voltage into very high frequency energy using a magnetron. Typically, they are around 50% efficient, so the quoted rating often seen, such as 650W, is the microwave power. Actual power consumption is in fact around 1300W, so an inverter would have to be able to supply that value if on maximum output. It should be remembered that if operating off a square-wave or modified square-wave output inverter, efficiency could drop by up to 30% due to the power supply waveform. As they are essentially domestic appliances, microwaves should be permanently installed to prevent moist, salt air, and condensation getting into the magnetron, which may cause corrosion and premature failure. It is a good idea to put a bag of silica gel inside the case. The microwave converts the input voltage into a high voltage and then supplies the magnetron, which emits radio waves. These vibrate food molecules to heat and cook the food. The best maintenance is to keep the internal parts clean and the interior dry.

331

Refrigeration and HVAC

21.1   Refrigeration Systems. A well-found galley is important on any yacht, and refrigeration is central to this. I served on some of the most automated and advanced computer-controlled refrigerated cargo ships afloat as electrical officer and engineer. We primarily loaded out bananas in Honduras, Costa Rica, Ecuador, and Colombia for the United States and Europe; grapes and apples out of Chile; and tangerines, satsumas, and tomatoes out of Morocco; as well as voyages out of Brazil and Australia with frozen lamb, beef, and chicken. It was maintenance intensive then and still is, even on small-scale yacht systems, as are all HVAC systems.

    a.   Gas Compression. The fundamental principle is that when a high-pressure liquid or gas expands, temperature reduces. In a refrigeration system, a compressor is used to pump the refrigerant fluid around the system. The compressor pumps the refrigerant vapor; this then increases the refrigerant gas pressure, which becomes hot. The high-pressure hot gas then passes through to the condenser.

    b.   Condensation. The hot gas passes through a condenser, which acts as a heat exchanger and releases heat. The condenser is either air cooled by natural convection or a fan or cooled by water passing through coils. Coils are typically made from cupronickel/copper. The gas condenses into a hot liquid back to the receiver. A receiver is essentially a pressure vessel that maintains the refrigerant in a liquid state before passing it through an expansion valve. It then passes through to the expansion valve as a high-pressure liquid.

    c.   Driers. There will always be a small amount of water vapor remaining in a system regardless of purging and evacuation. Water causes ice formation at the expansion valve, causing either total blockage or bad operation. The drier is installed in the liquid line between the receiver and expansion valve, serving as both a filter and removing water. The drier desiccant materials are made of silica gel or activated alumina. Flared units enable easier change-out than soldered ones when saturated; the removable cartridge type are even better. Internal corrosion begins at above 15 parts per million (ppm), causing oil breakdown, making it acidic and contributing to motor burnouts in hermetic systems.

    d.   Moisture and Liquid Indicator. The indicator sight glass allows easy visual inspection of the liquid. All indicators incorporate a moisture indicator. The sight glass has a porous filter paper. Indicators are chemical and change color. Dark green means the refrigerant is dry; yellow means it is wet. The sight glass will show bubbles if refrigerant is low, which is often seen at start-up and stop. It will show restrictions or blocked filter driers ahead of the sight glass. The color reverses when the refrigerant dries. Sight glasses are suitable for all refrigerants. This is important, as moisture within any refrigerant system can chemically affect the hydrolysis of the lubricants. In addition, it can cause corrosion, facilitate ice formation, and cause degradation of a hermetic compressor motor's winding insulation.

e.   **Refrigerant Control.** The thermostatic expansion (TX) valve regulates the rate of refrigerant liquid flow from the liquid receiver high side into the evaporator low side. This is to maintain the pressure difference; therefore, expansion into the evaporator is in exact proportion to the rate of liquid leaving the evaporator. Flow is regulated in response to both pressure and temperature within the evaporator. The thermal sensing element is placed on the outlet end of the evaporator and, where installed, motor thermal sensing element; it consists of a bulb and capillary. Valves can normally be adjusted for optimum temperature. This pressure reduction causes a fall in the temperature of the liquid. The cold liquid then passes to the evaporator. The TX valve is controlled by two conditions: the temperature of the control element and the evaporator pressure. Automatic TX valves allow refrigerant flow only if evaporator pressure falls when the compressor operates. The single greatest cause of TX valve failure is dirt, acids, moisture, and sludge in the system. All of these will freeze up or jam the valve.

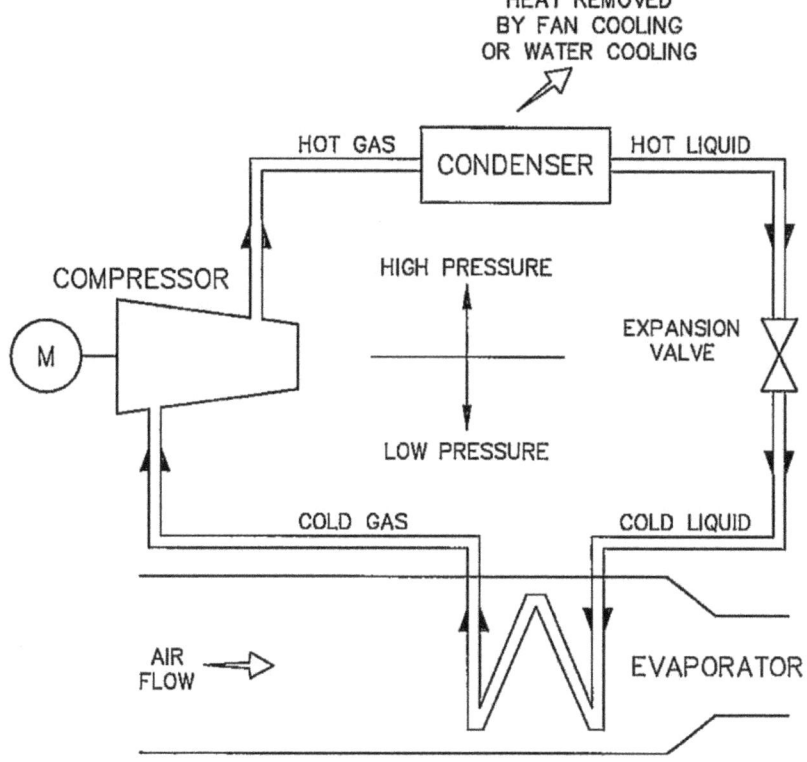

Figure 21-1. Basic Refrigeration Cooling Cycle

f.   **Evaporator.** The cold liquid passes through to the evaporator cooling surfaces (or eutectic tanks). Heat within the refrigerator space is absorbed by the cold refrigerant, causing the air to cool. The absorption of the heat causes the refrigerant liquid to evaporate into a gas. Heat is removed by either conduction on the evaporator, radiation, or convection. Isotherm developed an innovative new system called the self-pumping (SP) cooling system. A special integrated condenser and through-hull fitting have been developed that replace the galley sink fitting. The movement of the vessel causes water to "pump" in and out and remove waste heat.

g.   **Recycle.** The cold gas is suctioned back into the compressor to repeat the cycle.

h.   **Refrigerants.** For many years the most commonly used refrigerant was Freon 12 (R12), which is harmful to the ozone layer. Like all CFC gases, it was replaced by Freon R22. Now HFC-134a is the standard refrigeration system gas. This gas is an obvious choice, since automotive air-conditioning systems use the same gas, having already made the conversion. You cannot use HFC-134a in an existing Freon 12 system—virtually all the system components are incompatible and will require renewal. There is now a move in automotive systems to the gas R-1234yf over the next decade.

21.2   Eutectic Refrigeration Systems. This is the most common and efficient method of vessel refrigeration. The evaporator is replaced by a eutectic plate or tank. Operation is as follows:

a.   **Eutectic Principles.** A eutectic system uses brine or a fluid mixture that freezes at what is called eutectic temperature. Originally aqueous brine solutions were used, but systems now have an ethylene glycol–water or similar mixture of special mineral salts. The mixture has a much lower freezing point than water, which is 32°F (0°C). Once the mixture is frozen completely (eutectic point) and refrigeration is removed, the tank will cool the refrigeration space, gradually thawing out as it absorbs heat. Eutectic fluids are selected based on the temperature required.

b.   **Holdover Period.** The period of time that the space will remain within required temperature ranges before refrigeration is required is called the holdover period. When specifying a system, the holdover time and the temperature required are critical to the size of the plates or tanks and of the type of eutectic solution required.

21.3   Compressors and Motors. There are five compressor types in common use: reciprocating, rotary, scroll, screw, and centrifugal. Many systems use an engine-driven reciprocating compressor or either DC or AC motor powered compressors with a belt drive. Engine-powered units have a belt drive off the engine and drive pulley and are installed with an electromagnetic clutch for operating the compressor. The most common compressors in use are the reciprocating and swash plate types.

a. **Swash Plate Compressors.** These are typified by automotive air-conditioning compressors and are satisfactory where temperatures are required down to approximately –5°F (–15°C). These are suitable for most average applications. These compressors are not really designed for eutectic refrigeration systems; although they work well, failure rates are higher than reciprocating units.

b. **Reciprocating Compressors.** The reciprocating compressor consists of cylinders, piston intake and exhaust valves, and connecting rods to the crankshaft similar to an engine. The compressors are driven by rubber V-belts from the drive motors. Compressor maintenance is as follows:

(1) **Daily.** Check operating pressure gauges, temperatures, compressor oil levels; check for abnormal noise or vibration.

(2) **Weekly.** Check the evaporator and defrost if necessary. Check that all valve covers are on and tight.

(3) **Every 6 months.** Check operation of both the high-pressure and low-pressure cutout switches. Perform an oil sample test. Inspect and clean condenser. Check V-belts and adjust if necessary.

c. **Lubrication.** Lubricants in use within refrigeration systems are miscible, wax-free oils. These do not degrade under low temperatures or high pressures. Lubricating oils are carried around the system with the refrigerant and eventually return to the compressor sump. It should be noted that only reciprocating compressors have an oil sump; swash plate units do not. On reciprocating compressors, check the oil. Oil should be shiny and clear, have no visible particles, and feel smooth and greasy when rubbed between the fingers. If in doubt, renew the oil. Samples will help determine internal component condition and wear. Wear particles should not exceed the following (in parts per million): lead (10), copper (10), silicon (25), iron (100), chrome (5), nickel (5), aluminum (10), tin (10).

d. **Reciprocating Compressor Servicing.** Perform the following:

(1) **Condenser Pressure and Temperature.** High pressures indicate reduced cooling or air in the condenser. Low pressures indicate that refrigerant may be restricted to the evaporator.

(2) **Filters.** Liquid line, oil return, suction line, and TX valve require cleaning; clogged filters will cause restrictions in evaporator supply.

(3) **Moisture Indicators.** If these alter from green to yellow, moisture is in the system and the filter dryer requires replacement.

(4) **Leak Detection.** Regular checks should be made on a new installation every month until joints and flanges settle and are retightened. Refrigerant should be recharged.

(5) **Pressure Switches.** These should be checked and adjusted.

(6) **Condensers.** Open and clean tubes. Check and replace anodes.

(7) **Belts.** Rubber V-belts should be checked and retensioned.

21.4 **Auxiliary Refrigeration Controls.** A few different control devices are essential for safe and efficient operation.

a. **Clutch Engine Interlocks.** Many electromagnetic clutches are operated from a dedicated circuit breaker on the main switch panel, giving protection on the clutch coil and cabling. It is common to see the switch inadvertently left on and subsequently flatten the batteries, as typical current draw is 3A to 4A. On some occasions the operating coil can burn out. To prevent this, an interlock should be installed into the ignition system so that the clutch is de-energized when the engine is shut down.

b. **High-Pressure Cutout.** The high-pressure cutout is to protect against high pressures caused by loss of cooling water, plugged condenser, or, in the worst-case scenario, serious contamination of the refrigeration system with water and air. The cutout is usually wired in series with the compressor contactor or clutch. Typically, this is above 75psi (5bar) in the condenser. To test operation, close off the cooling water and wait until head pressure builds up and activates the cutout.

c. **Low-Pressure Cutout.** The low-pressure switch monitors suction line pressure. The cutout operates when gas discharge from the evaporator is too low. Operation of the cutout is indicative of a low refrigerant charge, typically below 30psi (2bar). To test the switch, slowly close the suction valve to activate the switch. Low suction pressures increase compression ratios and can cause compressor damage.

d. **Thermostatic Control.** To test, vary settings and observe cut-in and cut-out. Many are now microprocessor controlled, with the temperature sensor located at the holding plate, although some use the general box cooling space.

e. **Reduced Holdover Times.** A common complaint is that holdover times have been reduced; the usual causes are as follows:

(1) **Warm Foodstuffs.** A fridge or freezer system is often pulled down to the required temperature and then has a full load of unfrozen food or warm drinks, such as a case of beer, dumped in it with the expectation that the system will rapidly cool them. It will not.

(2) **Climate Change.** More often than not, the system worked well in a temperate climate, but the first extended cruise into tropical waters results in a dramatic reduction in apparent efficiency. A seasoned cruiser opens the fridge sparingly, but people new to the lifestyle are probably opening it far more than necessary, and far more than they did on an average weekend cruise. Keep access to a minimum.

(3) **Mechanical Causes.** Engine drive belts are not retensioned, and resultant belt slip under load causes decreased refrigeration.

(4) **Seawater Temperatures.** A voyage to warmer waters can also cause changes in condenser cooling efficiency. In many cases the eutectic plate takes longer to pull down, so refrigeration operation times need to be extended.

21.5 Electric Refrigeration Systems. These systems are typically self-contained. Electric refrigeration requires either AC or DC battery power for operation. The average power consumption is approximately 50Ah for a refrigerator, and approximately 100Ah for freezers operating on a 50% duty cycle. Restoring battery capacity requires a far greater run time than an equivalent engine-driven system, and installation of either a higher output alternator or a fast-charging regulator should be considered. I installed an Isotherm Magnum water-cooled unit on my previous two boats.

a. **Self-Contained Electric.** Self-contained refrigerators are DC powered and have eutectic holdover plates. Insulation on the units is reasonable but, where installed, the surrounding area should be insulated further. Power consumption of these units averages around 35Ah per day, depending on ambient temperatures and frequency of opening. As many units are built-in, good ventilation must be provided to carry heat away from the compressor unit. Many units do not function properly as a result of this omission; a fan and ducting make a difference. Webasto Isotherm, Sea Frost, Technautics, Frigoboat, Dometic, Frigonautic, and Norcold are well-known systems manufacturers. Evaporator cooling is more efficient with water-cooled systems. Isotherm has a self-pumping system; others, such as Frigoboat, have keel cooling, which looks like a ground plate and has to be installed through the hull; they are very efficient.

b. **Energy Utilization.** Some manufacturers have introduced circuitry that enables the overriding of thermostats during engine run periods. If temperatures are down, this enables the alternator to supply loads for an additional pull-down period that reduces electrical consumption later. A similar function is used on Webasto Isotherm ASU (automatic start-up) systems. The controller senses the raised system voltage from the alternator and operates at double the speed to pull down temperatures and maximize the energy available. Another feature of Isotherm systems is the control of compressor speed with respect to refrigeration requirements. This utilizes the SECOP (formerly Danfoss) hermetically sealed compressors and uses electronic control on the three-phase motor supply. I opted for a freezer model with spillover plate and stainless butterfly vent to adjust cooling in the fridge space. Note that the seawater cooling system anode must be removed and cleaned every 6 months, or it will shed enough material to clog the pump suction lines. When shutting down for any period, simply flush the system with fresh water.

c. **Peltier Effect System.** Still used by manufacturers such as Koolatron, this system of thermoelectric refrigeration uses the principle of heat transfer via electrons instead of a refrigerant. The principle uses the property of semiconductors. These are either P type, which conduct by flow of positive charged particles or "electron holes," or N Type, with negatively charged particles or electrons. If a current is forced to flow from a P type to an N type material, the junction where they connect absorbs heat; this is the Peltier effect. The opposite ends become hot and radiate heat. When a series of N-P junctions are combined, a significant cooling effect is possible. A normal thermostat is used for control, and thermal efficiency is low. In practice, a power supply is transformed if it flows from AC and is rectified to DC. This is then fed to a thermoelectric module. The junction inside the fridge cools and the external junction is heated. There are usually a number of hot and cold junctions series connected to make up required capacity. The positive DC is fed to the N end of a junction and the P is series connected to the next and exits the cooling space on an P, although these are part of a module with simple terminals. Module exteriors have a heat sink to facilitate heat transfer away from the junction module.

21.6   Hermetic Compressors. The majority of electric fridge systems use hermetically sealed SECOP (formerly Danfoss) type compressors. The most common compressor models in use are the BD35, BD50, and BD80. The electric motor is sealed within a domed housing along with the compressor. The motor and compressor assembly are supported on a spring suspension system to absorb vibration. These are used because the starting torque characteristics are very good. The motor has two windings, one for start and one for running. The motor has a capacitor, which is the black cylinder on the unit, wired in series with the motor start winding. A starting relay is used, which may be a current or potential relay switch, and this is located on the outside of the compressor unit. When power is switched on, current passes through the start and run windings. The capacitor alters the phase angle and effectively converts the motor to a two-phase motor; when the motor speed increases and the start current decreases at approximately 60% to 75%, the switch opens and the start winding and capacitor are disconnected. Motors have a thermal overload.

a. **Compressor.** Hermetic units are typically twin-piston reciprocating compressors with a valve plate assembly. Units with fan-driven condenser cooling should be cleaned every 3 months. Tube flare nuts and service valve caps should be tight to prevent leaks.

b. **Troubleshooting.** Motor failures due to internal faults are rare, and failures are usually due to external causes. Continuity and resistance checks will indicate the status of windings. Start winding C to S resistance is approximately 5Ω. Run winding C to R resistance is approximately 2Ω. R to S resistance for both windings is approximately 7Ω. Megger R or S to case is at least $1M\Omega$. Testing of both open-circuit and on-load voltage will indicate problems. A difference exceeding 10V indicates an overload or motor winding fault. Use a clamp ammeter to check operating current. If the compressor current draw exceeds 7A, it is close to failure. Check all external control devices first before assuming compressor failure. Discharge the capacitor first; use a multimeter and test across the capacitor terminals. A short-circuited capacitor will indicate zero ohms; a high reading indicates an open circuit. If the capacitor is good, the reading will initially go to zero then slowly rise. Regular failures in capacitors are usually caused by slow starts (typical maximum is 3 to 4 seconds), too many starts (typical rate is 3 to 4 starts per hours), low voltages, or faulty starting switches.

21.7   Refrigeration System Installation. There are kits for fridge installations; however, the best practice is to get a good refrigeration mechanic to install the system.

a.   **Insulation.** If a fridge system is to be effective and reliable, it must be of sufficient size to meet the expected needs and be well insulated. Insulation thicknesses should be at least 4 inches or more. Inadequate insulation levels cause the majority of inefficiencies of vessel fridge systems, so install as much as you can. Every insulating material has varying degrees of thermal conductivity. The ideal insulating material is urethane foam, followed closely by glass fiber wool and then polystyrene foam. In many installations this is done by foaming in place using a two-part mix, but great care must be taken; failure to have the mix correct will produce inadequate results, without a good closed-cell finish that is required for good insulation. Ideally, the use of preformed slabs is much more reliable; fill any outstanding voids with foam mix. The whole insulation block should be surrounded with plastic to prevent the ingress of moisture, with a layer of reflective foil such as that used in domestic house construction to minimize heat radiation. A two-layer system of foam slabs and foil is the ideal combination. Two common insulation brand names are DuPont Styrofoam Brand square edge SM XPS (extruded polystyrene) foam insulation, often known as blueboard, and Owens Corning FOAMULAR INSULPINK-Z XPS Rigid Foam Insulation board. Vacuum insulation panels are also an option.

b.   **Refrigeration Size.** Do not build refrigeration spaces greater than the actual requirements. It is common to construct oversized boxes that remain half empty, which is a subsequent waste of energy and greater installation costs for a larger, more powerful system.

c.   **Access Hatches.** It is very important that your hatch gaskets do not leak air. Many top hatches are designed with a center hinge in the hatch lid so that only one side needs to be opened. These are prone to constant leakage of air and heat, which is conducted by the metal hinge. The ideal thermal solution is to replace this with a one-piece design. The next option would be to block the airflow with insulating tape placed between the two sections under the hinge. A cubic foot of air at 100°F (38°C) contains approximately 18 BTUs of heat. If all the cold air in your box were to spill out when you open the hatch, it would take only a relatively short compressor run time to remove the heat that entered.

d.　**Compressor Brackets.** Engine compressor mountings and brackets must be extremely robust to prevent vibration. Overengineer them to ensure that vibration will not fracture any part.

e.　**Compressor Drive Belts.** Alignment of the compressor and engine drive pulleys is essential to ensure proper transfer of mechanical loads. Belts are usually dual pulley arrangements. Ensure that both belts are tensioned correctly.

f.　**Energy Saving Measures.** Energy conservation and efficiency improvement measures can be implemented. Fill any empty spaces in the fridge compartment with either blocks of foam or inflated empty wine cask bladders. This will decrease the fridge space when partially full and reduce energy requirements. If all frozen goods are placed at the bottom of the compartment, purchase a mat. Place the mat over the food so that cold air is retained within the food below the mat.

g.　**Battery Voltages.** Ensure that battery voltage levels are maintained. Low battery levels will cause inefficient compressor operation. Do not let the battery level sink to the normal minimum level of 10.5V. It takes far more energy and engine run time to charge a nearly flat battery than a half-charged one.

h.　**Ventilation.** Ensure that the compressor unit is well ventilated. Install an additional fan and ducting to ensure positive ventilation on non–water cooled units.

i.　**Condensation.** Condensation issues can arise even with the best materials and construction techniques. One common complaint is condensation behind a settee cushion adjacent to the cold box. This happens independently of the quality of the box insulation. The extra insulation provided by the cushion makes the back side of the cushion colder than the ambient air. If the air temperature reaches the dew point, condensation will occur. The air cools more with the condensation process, causing a low-pressure area that pulls in more moist air. Remarkable quantities of water can be generated this way. The solution is to provide ventilation between the cushion and the cold box wall. Another condensation trouble spot can be the hatch frames, especially with a front-opening freezer door. Two ways to minimize the amount of heat conducted, and therefore the chances for condensation or even freezing of the gaskets, are to aim to minimize the mass and maximize the length of the heat path through the hatch frame.

21.8　Refrigeration System Troubleshooting. Few boats carry vacuum pumps, bottles of refrigerant, gauge sets, and spare parts; in fact, carrying out such work may be a breach of environmental laws if not certified. Knowingly releasing Class I (CFC/chlorofluorocarbon) and Class II (HCFC/hydrochlorofluorocarbon) substances into the atmosphere can result in severe penalties, even imprisonment. Besides working on refrigerated cargo ships, known as reefer vessels, I used to work between seagoing assignments part-time as a refrigeration mechanic repairing refrigerated shipping container systems. All repairs were done in filtered clean areas, and it is highly unlikely that conditions will be suitable on your boat for proper overhaul and repair of compressors. There are few exceptions, and these are largely limited to professional refrigeration mechanics. The first—and best—way to avoid problems is to have the system properly installed in the first place. This chapter does not include procedures for disassembling and checking compressors, purging, and recharging because you are more likely to do further damage. If, after checking the control systems, you are unable to rectify problems, call in a licensed refrigeration technician. It is illegal to do anything else. It is important to determine what is going on in the system. Pressure gauges are used to check system pressures; thermometers are used to measure evaporator, line, and condenser temperatures. There are EPA regulations relating to licensing of technicians under the Clean Air Act; this includes system testing and HVAC. In the UK and Australia, you must have refrigerant handling qualifications for HCFCs and HFCs.

a.   **Condensers.** If a condenser is undersized or dirty, internal and external, the head pressure and condensing temperature rise. The higher temperature will make the compressor pump to this higher pressure and temperature. It is important to check and clean condensers regularly.

b.   **Refrigerant Loss.** Refrigerant loss is a common fault and will cause a gradual reduction in cooling efficiency, eventually tripping the low-pressure cutout, if fitted, and system failure. Low refrigerant levels can be observed in the sight glass, and bubbles will be seen. An empty sight glass indicates no refrigerant at all. If all the gas has escaped, and after the leak has been located, the system must be purged of air and moisture before gas recharging. You will have to get a qualified refrigeration mechanic to do this. A frost-covered evaporator generally signals that gas levels are okay. If the system is undercharged, refrigerant does not properly liquefy before passing through the TX valve; effective latent heat is reduced, so refrigeration is poor. Some vapor will pass through the TX valve, reducing fridge control capacity, and the vapor passing at high velocity will increase the wear on the TX valve needle and seat. Air in the system will increase total head pressure. Total head pressure equals fridge condensing pressure plus air pressure in the condenser. The refrigerant will then have to condense to a higher temperature and pressure, so the cylinder head and exhaust on compressor and top tube of the condenser will all be at higher temperatures. This will then affect the oil quality. It is important to ensure that caps on service valves are replaced and tight to reduce leaks. Some buy a small R134a bottle and slowly top up the refrigerant via the Schrader fitting. My advice is call in the licensed experts and have it leak tested and charged properly.

c.   **Leak Detection.** Perform leak detection by pressurizing the system and then checking all possible leakage points at connections and fittings. ***Caution:*** Do not use a torch with HFC-134a refrigerants. Refrigerant leaks have an environmental impact and increase power consumption. In Europe there are the European Fluorinated Gas (F Gas) Regulations. This applies to systems with a refrigerant charge between 3kg to 30kg. It is illegal to top up a system refrigerant without leak testing. Even though most boat systems are below 3kg, there is an implied responsibility:

(1) **Halide Torch.** The most common test will require use of a propane halide torch. Air is drawn to the flame through a sampling tube. Small gas leaks will give the flame a faint green discoloration; large leaks will show as bright green.

(2) **Soapy Water or Leak Detection Spray.** A simple check is using soapy water, generally dishwashing liquid, and applying it to all piping joints with the system running. If there is a pressurized leak in the joint, a bubble will form.

(3) **Electronic Leak Detectors**. These are used by refrigeration professionals and must be calibrated and maintained.

Table 21-1. Refrigeration Troubleshooting

Symptom	Probable Fault
Slow temp pull-down times	Drive belt slipping Low refrigerant level Compressor fault High cooling water temperature Plugged condenser Low battery voltage—motor running slow Fridge space seals damaged High ambient temperature Insulation failure Thermostat faulty
Clutch circuit breaker tripping	Clutch coil failure Clutch cable shorting out Compressor bearing failure, seizing up
Expansion valve icing up	Drier requiring replacement Low refrigerant charge Faulty TX valve
High discharge pipe temp	Discharge valves leaking
Moisture in system	Condenser leaking Compressor bearing failure Compressor gasket failure Liquid line filter clogged Pipe compression fitting
Refrigerant gas leak	Condenser leak Isolation valve and gauge connection leaks Pipe compression fitting and flare nuts Damaged piping Valve caps off Compressor seal failure

Symptom	Probable Fault
Abnormal compressor noises	Low oil pressure Oil foaming Liquid in suction line Coupling misalignment Faulty oil pump Piston rings or cylinder wear Faulty discharge valves Faulty solenoid valve oil return Clogged oil filter Compressor mounting loose Low cooling water flow Bearing failure
High condenser pressure	High cooling water temperature High-pressure cutout activated Refrigerant overcharged if just after a service Cooling water loss (check strainer or inlet filter) Clogged condenser
Low condenser pressure	Inlet cooling water valve closed Low refrigerant charge Excess cooling to condenser
Low oil pressure	Piston rings or cylinder wear Low oil level Oil pressure switch has activated Oil pressure too low at regulator Oil foaming in crankcase Liquid in suction line Defective oil pump defective Clogged oil filter Oil temperature is too high
Poor or reduced cooling	Clogged oil filter Refrigerant leak and low level Engine clutch failure, wire fault Engine clutch coil failure High-pressure cutout activated Low-pressure cutout activated Clutch drive belt slipping Faulty thermostat

HVAC

21.9   Air-conditioning. Air-conditioning is possible even on small boats and is virtually standard on larger vessels. Like refrigeration, air-conditioning cools a cabin by transferring heat out. In most marine installations, seawater is used generally for condenser cooling; although fan-cooled systems are available, they are less effective. There are two types of marine air-conditioning systems: the single-stage direct expansion type and the tempered (chilled) water two-stage type. Manufacturers such as MarineAir, Pompanette, Webasto, Veco, Climma, Dometic-Cruisair, Mermaid-air, Thermowell, Frigomar, and Clion-Marine offer extensive ranges with options. Inverter technology is becoming as common on boats as ashore, having 12V chiller systems with inverters for compressor drives, which is much more energy efficient, with lower starting currents. Connectivity is also common, the Webasto BlueCool series has the Blue Cool Connect app for monitoring and control, with NMEA 2000 interface and connectivity with MFD from Raymarine, Simrad, Garmin, Lowrance, and B&G. Veco systems have the Veco Hub for remote control and monitoring.

21.10   Air-conditioning Basics. Under the Montreal Protocol, Freon-12 refrigerants have been replaced by Freon-22, and most systems now use 134a. Systems are generally rated in British thermal units (BTUs), which is the energy required to heat or cool an area. In metric this is kilocalories (Kcal). The conversion is approximately 4BTU = 1Kcal. Gas charging and testing must be performed by certified technicians and using Environmental Protection Agency (EPA) approved equipment. The same applies in Canada, the UK, Europe, Australia, and New Zealand.

a.   **Self-Contained and Remote Condensing Systems.** These single-stage direct expansion units may be either a self-contained unit (SCU) or have a remote condensing unit installed within the machinery space. The self-contained reverse cycle system is normally a relatively compact module that can be installed under a bunk or locker. These are rated between 5,000 and 24,000BTU/hr. They are precharged with refrigerant at the factory, are seawater cooled from a remote pump, and have an integral reciprocating, rotary, or scroll compressor, depending on the model. Units have integral condenser, evaporator, blower, and safety switches. Many systems have a remotely installed condensing unit within the machinery space, with seawater supply also adjacent. Refrigerant is carried to the air-cooling unit. These units have cooling capabilities in the range of 5,000 to 60,000BTU/hr. Dometic Marine has the innovative Voyager Series TX variable capacity SCU. The system controls the compressor speed based on cooling or heating demands, and then the compressor operates at reduced speed. The variable speed results in significant power savings. Other advanced features include a titanium condenser and an electronic expansion valve for very precise refrigerant control. Power consumptions vary between 18A starting current to 7A to 10A running.

b.   **Drop-In (Carry-On) Hatch Mounted Units**. These portable air-condition-ing units are relatively light and easy to install. They sit over the open hatch and have an integral fabric hood to seal the opening. The Pompanette unit has a 6,000BTU/hr rating, uses R-410A refrigerant, and has start current of 14A and a running current of 5A at 115VAC. West Marine's carry-on is rated at 7,000BTU and uses 6.9A.

c.   **48V Air-Conditioning.** There are developments in this space, with higher voltages, lighter weights, and more compact designs. The Ergon unit uses a variable frequency drive (VFD) for compressor speed control, the VFD varies speed in the 20Hz to 100Hz range. In ECO mode the power draw is only 150W to 200W, with a peak start current of 2A. This technology is very efficient.

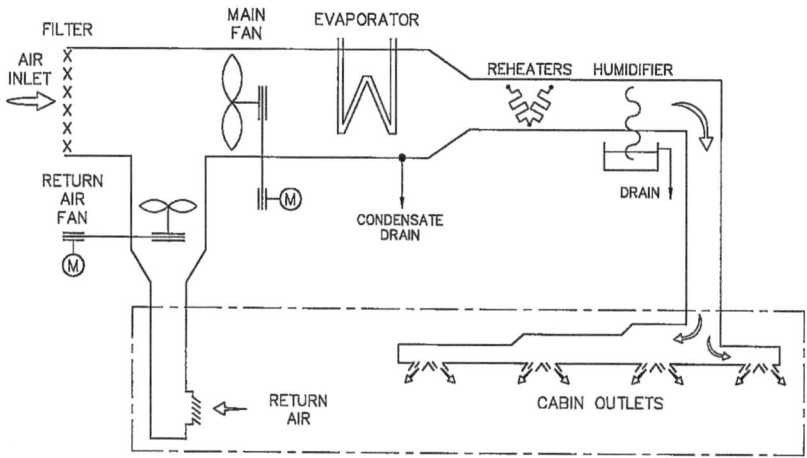

Figure 21-2. Typical Air-Conditioning Schematic

21.11   Electrical Power Requirements. Air-conditioning system power requirements are as follows:

a.   **AC Systems.** A system normally requires a constant AC power source to operate, so the generator must run continuously. If an air-conditioning sys-tem is to be installed on the vessel, the generator must consider the maxi-mum loads. As systems use AC induction motors on the compressor, there is a significant start-up current surge that must be allowed for in generator load calculations, typically three to four times full-load amps. Hermetically sealed compressors have high starting currents that are reduced by capacitors to around three to five times running current.

b.   **Seawater Pumps.** Electrical load calculations should factor in the seawa-ter pump. Pumps are generally not self-priming and must be positioned at or below waterline. Like all seawater pumps, they are prone to corrosion. Self-priming impeller pumps should be installed where possible. (Follow notes on AC motor maintenance.) The seawater pump is generally controlled via a relay box, and this should be mounted in a dry location. Pump capacities are typically 100g/h (380l/h) for a 5,000BTU unit up to 250g/h (950l/h) for a 12,000BTU system. Seawater is usually pumped through a strainer; ensure that it is cleaned.

c.   **Control Systems.** Controls range from simple on and off switches, speed control, and thermostat to programmable controllers. These offer timing functions, high and low temperature settings, systems monitoring, fan speed controls, and compressor restart time delays, including fault condition automatic shutdown and even automatic dehumidification. Protection and control systems are similar to those in refrigeration systems. There are variable speed fan controllers that use small variable frequency drives.

21.12   Capacity Calculations. The capacity of the system must be calculated by determining the volume to be cooled; the following are guidelines used by HFL. For ambient temperatures exceeding 85°F (30°C) add 20%; for water temperatures exceeding 75°F (25°C) add an additional 20%. This is to maintain 60°F to 72°F (16°C to 22°C). The estimated seawater cooling is 3.5g/m (19l/m) for a self-contained unit.

a.   For below decks cu ft × 14 = BTU (m³ × 504 = BTU).

b.   For above decks cu ft × 17 = BTU (m³ × 612 = BTU).

Table 21-2. Air-Conditioning Capacity Table

Capacity (BTU/hr)	Below Deck ft²	Mid-Deck ft²	Above Deck ft²
6,000	90	60	45
7,000	115	75	55
9,000	165	110	85
12,000	200	150	100
16,000	267	178	135
20,000	335	250	167
24,000	405	300	200

21.13   Air-conditioning Maintenance. A number of maintenance tasks are required if you want reliability and optimum performance.

a.   **Weekly.** Check the seawater inlet strainer and clean. Blocked and clogged strainers are the most common cause of high-pressure cutouts.

b.   **Monthly.** Filters on air-cooling units should be checked and cleaned.

c.   **Every 6 months.** Seawater cooling condensers should be cleaned where this is possible. Where systems have anodes in the cooler, they should be checked and replaced.

21.14   Air-conditioning Troubleshooting. Air-conditioning systems have many similar faults to refrigeration systems. The most common are as follows:

>   **a.**   **Controls.** Check external control equipment, such as thermostats.

>   **b.**   **Cutouts.** Check system high-pressure and low-pressure cutouts where installed. A high-pressure cutout usually indicates a cooling issue, which in water-cooled systems is typified by a blocked seawater strainer.

>   **c.**   **Evaporator Cooling.** Check that evaporator cooling systems are clean and functioning. Check cooling water flows; check and clean the seawater strainer.

21.15   Ventilation Fans. Good ventilation is essential in any boat, especially the galley, the main saloon, engine space, and cabins. There are a number of ventilation fan options, and all have uses in particular applications. I use a wind scoop, made by my local sailmaker, on my forward hatch to get airflow through the cabins; however, when the wind drops it is very stuffy, something most experience. The principal aim in any ventilation system is to achieve a satisfactory number of air changes. Depending on whom you ask, the required air exchange in cubic feet per hour (cfh) is anything from 1,000 to 1,500. I have solar powered vent fans that operate 24/7 that are very effective; I just added one to the head compartment. Besides maintaining fresh air below, the airflow is essential for combating excess moisture and dampness and inhibiting mildew and mold growth, along with odor removal. Fans can be classified either as extraction fans or as blowers.

>   **a.**   **Extraction Fans.** Extraction fans take air out of a space, either to increase natural ventilation flow rates and air changes or to remove heat or fume concentrations.

>   >   **(1)**   **Solar Fans.** Solar fans have a small solar cell powering the fan motor. Newer Marinco models have a small, solar-charged battery so the fan can operate at night, the period when it is most required. I have three of these on my own boat. They extract 1,000cfh (28m³) or 24,000cfh (672m³) every 24 hours. The three fans, along with passive measures such as my wind scoop, are helping a lot. The vents have a rechargeable 2500mAh, 4.8V, nickel–metal hydride (NiMH) battery that when fully charged will operate the fan for 24 hours.

>   >   **(2)**   **Engine Extraction Fans.** These are used to extract heat from engine spaces. In warmer climates, it is preferable to leave the fan operating for 30 minutes after the engine stops to reduce heat buildup and stop the increase in lower deck temperatures from radiated heat.

>   >   **(3)**   **Powered Ventilators.** These units have two speeds and are reversible, allowing them to be adapted to the conditions inside. At 25cfm (cubic feet per minute), air displacement is very good, which suits normal cabin environments. Power consumption is relatively low, at only 1.7A on the fast setting. Air extraction rates are a reasonable 36cfm.

(4)   **Cabin Fans.** Reliability and longevity are the issue with these. I recently scrapped an old oscillating fan and installed one of the new Caframo types, which has been very successful. The Canadian-made Scirocco has three speed settings, four timer settings, and has a very quiet, long-life motor with a power consumption of 0.12A on low speed, 0.22A on medium speed, and 0.35A on maximum speed. It has a maximum flow rate of 85cfm. The Caframo 757 Ultimate is a grill-free unit with very large surface area blades. It has only two speeds and draws 0.5A at top speed and 0.2A on low.

(5)   **Hatch Fans.** These units can be mounted under a hatch and are airflow reversible, with a typical flow rate of 500cfm. Typical power consumption is 0.2A at low setting and up to 4.5A at full speed.

b.   **Blowers.** In-line blowers push air into a space and are used either to displace existing air such as in bilge blower applications or, in most cases, to direct air in large volumes over specific areas, such as in alternator cooling or engine combustion applications. Airflow rates are typically around 100cfm to 250cfm and have a power consumption of 4A to 6A. Blowers used in areas where hazardous vapors are concentrated must be ignition proof. In most cases, they are interlocked to run with the operating engine. It is good practice to interlock the fan to the engine start with a relay to ensure that it always operates and switches off at engine shutdown. Check the certification: ISO 9097 Marine, Ignition Protected; USCG 183.410; NMMA Type Accepted; and ABYC compliant.

DIESEL HEATING

21.16   Diesel Heater Systems. Boat heating design and efficiency have improved over the years. Power consumption figures, heat outputs, and fuel consumption rates for typical Webasto and Eberspacher models are illustrated in Table 21-3. The new Airtronic units from Eberspacher have brushless fan motors. Heaters have the following operational cycles:

a.   **Starting.** Cold air is drawn in by an electric fan to the exchanger/burner.

b.   **Ignition.** Fuel is drawn at the same time by the fuel pump, mixed with the air, and ignited in a combustion chamber by an electric glow plug.

c.   **Combustion.** Combustion takes place within a sealed exchanger; gases are exhausted directly to the atmosphere.

d.   **Heating.** Heat is transferred as the main airflow passes over a heat exchanger to warm the air to the cabin. A thermostat in the cabin shuts the system down and operates it to maintain the set temperature.

Table 21-3. Diesel Heater Data Table

BTU Output	Fuel (liters/hour)	Fuel (gallon/hour)	Power Draw
6,100	0.21	0.06	40 watts
11,000	0.38	0.10	45 watts
15,000	0.57	0.15	70 watts
28,000	1.05	0.28	115 watts
41,000	1.40	0.37	190 watts

e.   **Power Consumption.** Typical power consumption is 40W (3.33A) during running. At start-up, the draw can be up to 20A for a period of 20 seconds during the glow plug ignition cycle.

f.   **Heater Maintenance.** The following maintenance tasks should be carried out to ensure optimum operation:

(1)   Check that all electrical connections are tight and corrosion free.

(2)   Check exhaust connections and fittings for leaks. Leakages can cause dangerous carbon monoxide gases to vent below deck.

(3)   Remove and clean the glow plugs. Take care not to damage the glow plug spiral and element. Use a brush and emery cloth, and make sure all particles are blown out afterward.

(4)   At 2,000 hours, take the unit to a dealer to decoke the heat exchanger and replace the fuel filter.

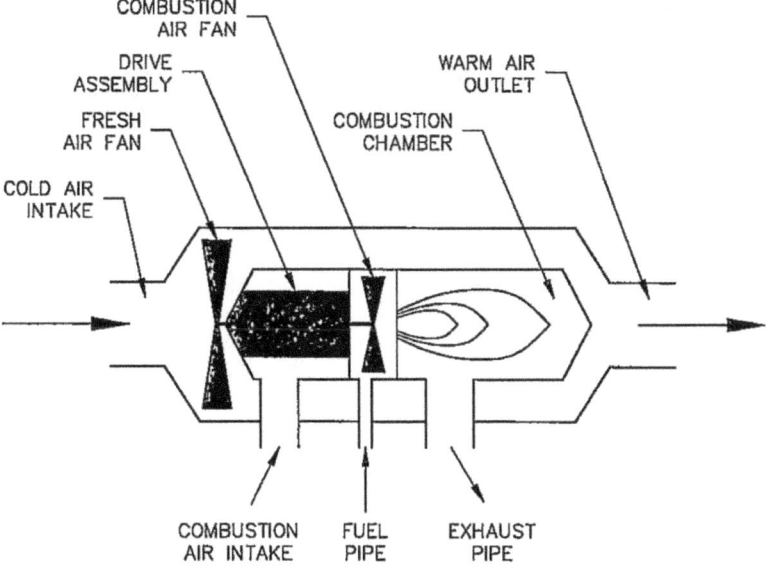

Figure 21-3. Diesel Heater System

Table 21-4. Diesel Heater Troubleshooting

Symptom	Probable Fault
Heater will not switch off	Temperature switch fault
Heater smokes and soots	Combustion pipe clogged Fuel metering pump fault Blower speed too low
Heating level too low	Hot air ducts clogged Fuel metering pump fault Blower speed too low Temperature switch fault
Heater will not start	Supply fuse blown Low battery voltage Blower not operating Fuel-metering pump fault Thermal cutout tripped Fuel filter clogged No fuel supply Glow plug fault Control unit fault
Heater goes off	Fuel-metering pump fault Thermal cutout tripped Fuel filter clogged Fuel supply problem Control unit fault

21.17   Diesel Galley Stoves. This is a good place to mention the options of diesel galley cooking. We have diesel heaters and hot water systems, and occasionally I run across diesel powered galleys, which eliminate liquefied petroleum gas (LPG) from the boat. These are typified by the 87D and 89D product range from Finnish company Wallas-Marin. The stove fuel pump supplies fuel; the electronic control unit manages combustion air and fuel metering. At switch-on, a glow plug is energized and ignites the diesel fuel. A sensor monitors heat levels and activates an ignition LED. The heat is transferred to a ceramic stovetop. The fuel consumption is 0.1 to 0.22 l/hr. Power consumption is 0.5A; when the light is on, this will increase to 1.5A. During ignition it draws 8A for about 7 minutes. Another manufacturer is Canadian company Dickenson Marine, which offers a wide range of diesel stoves that are worth considering. Check them out at dickinsonmarine.com.

Water and Sewage Systems

22.1   **Water Systems.** The modern sailing boat, trawler yacht, and motorboat have several water systems that must be properly planned, installed, and maintained. A failure in the freshwater or sewage system makes life uncomfortable. A failure in the gray water systems, such as the shower drain and sink units, is inconvenient. A failure in bilge pump systems is dangerous and renders the boat unseaworthy. The various onboard systems are summarized as follows.

a.   **Pressurized Water Systems.** This is the potable freshwater system and includes pressurized systems and safe water storage.

b.   **Hot Water Systems.** These systems heat water either via your engine cooling system, with AC and DC powered heating elements, or with a diesel water heater.

c.   **Desalination (Water Maker) Systems.** These systems desalinate seawater for onboard potable water.

d.   **Bilge Pump Systems.** Both submersible and diaphragm pumps, these are essential safety systems for removing water from within the hull.

e.   **Sewage (Black) Water Systems.** These include the sewage system, toilet to holding tank, and electric and vacuum flush toilets.

f.   **Gray Water Systems.** These are the shower drain and sink drain systems.

g.   **Fluid Transfer Pumps.** These are used for diesel and oil transfer duties.

h.   **General Seawater Pumps.** These are used for washdown duties and for cooling applications in air-conditioning, refrigeration, and similar systems.

i.   **Shore Water Systems.** Some marinas allow hookup to watermains, but there are considerations.

22.2   **Water Tanks.** It is good practice to have two separate tanks for water storage. Before filling a tank, transfer any remaining water to one tank. The new water can be put in the tank without contaminating water you know to be good. If the water is of poor quality and you have to dump it, you won't lose the whole tank.

a.   **Water Tank Cleaning.** Toxic by-products from bacteria are characterized by unpleasant smells. Cleaning regimes should be undertaken at least twice a year to ensure the integrity of your water.

b.   **Water Tank Disinfection.** New water and the tank must be disinfected to prevent bacterial growth. Water chlorination is easily accomplished by adding a solution of household bleach or sodium hypochlorite in a ratio of 5:100 of tank contents, or a level of 50ppm concentration. Let some amount run though all outlets to disinfect all parts of the system. Then top off the tank and allow the contents to stand for 4 hours (I do so for 12 hours). Reflush the system another three times. If you want to be super clean, now add vinegar in the ratio of 1 pint to 13 gallons (1 liter to 50 liters) of system capacity and allow to stand for 2 days. Refill with fresh water and flush three times again. The tank is then ready for use and will maintain potable water quality for several months. An easier and quicker method is to use Puriclean or Aquatab Marine effervescent tablets, which will sanitize the water tank. After filling the tank and adding the cleaning solution, always let it stand a few hours before flushing.

c.   **Water System Disinfection.** This is often called shocking the system. Before you empty and flush your water tank, flow the treated water through to each outlet in the galley and bathroom faucets. Once it is flowing, close and hold the treated water for 12 hours to kill all bacteria in the system. Do this for every outlet. Once done, start flushing the system through. Fill and flush out the tank at least three times.

22.3   Pressurized Water Systems. Water is the one essential on board required for drinking and washing. Of course, it is also a scarce resource and has to be carefully managed. To quote from that famous poem by Samuel Taylor Coleridge, *The Rime of the Ancient Mariner*: "Water, water, every where, And all the boards did shrink; Water, water, every where, Nor any drop to drink." This epic poem recounts an adventure at sea on an icebound ship near the South Pole, the infamous slaughter of an albatross, and the curse that befalls the mariner and his shipmates. However, I digress; that being said, we have very efficient water makers now. A typical pressurized water system arrangement is illustrated below.

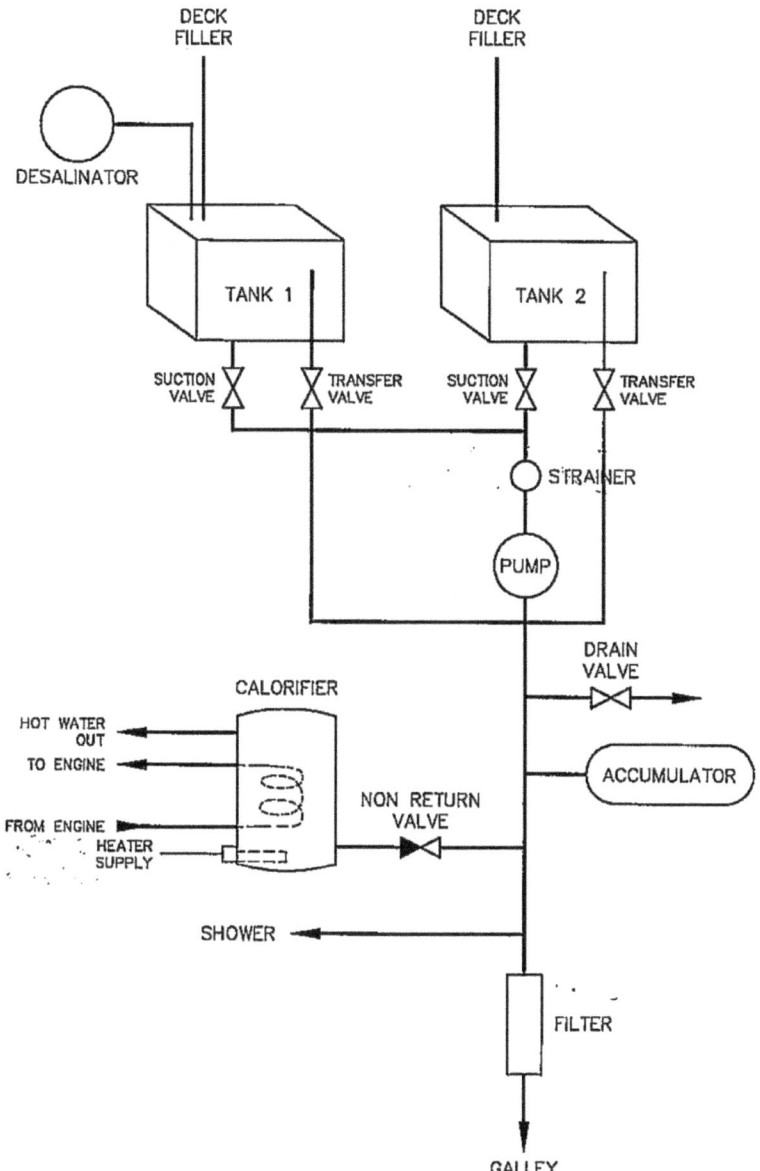

Figure 22-1. Pressurized Water System

22.4 **Water Pressure Pumps.** The primary purpose of the pressure pump is to supply and pressurize the water from the tank. A pump is selected based on the number of outlets to be supplied, the required pressure (psi), the liquid flow rate in gallons per minute (g/min) or liters per minute (l/min) required, and the maximum head, which is the difference between the pump and discharge level. In most boats, that can be 6ft to 10ft (2m to 3m). Other considerations include how much pressure is lost due to friction when flowing through the pipe, as long pipe runs will create losses. Pumps are self-priming, which means they expel air from the system and achieve suction. Good suction means very tight water inlet connections. If the pump is incorrectly rated for the system, the flow will drop off when another outlet is opened.

a.   **Diaphragm Pump.** These units are the most robust and are designed for multioutlet systems. They are self-priming, more tolerant to dry running conditions, relatively quiet in operation, and have built-in hydraulic pulsation dampening. The design makes them inherently liquid tight, so they are seal-less. Typical flow rates are 2g/m (15l/min) to 4g/m (30l/m). Maximum power consumption is around 3.5A to 10A.

b.   **Impeller Pump.** These units normally have a pump with bronze casing and nitrile or neoprene impeller. They are self-priming but less tolerant to running dry.

c.   **Variable Speed Pumps.** These are very efficient and reduce electric load and water flow surges. The Johnson Aqua Jet Flow Master WPS 5.0 is a typical model. They use electronic speed control, using an inverter for pump speed control.

d.   **Multiple-Chamber Pumps.** Pumps have different chamber numbers, usually three or four and can have up to five, as in the Johnson Aqua Jet. Multiple chambers reduce pump water pulsation.

22.5 **Shore Water Systems.** There has been a rapid increase and improvement in marina facilities, enabling connection of most utilities, including water. Water connections pose some problems in that pressure of a watermain is significantly higher than with onboard systems. The mains water system is filtered and regulated down to boat system pressure. Some are simply pressure reducers, such as those from Pentair Shurflo. Some systems incorporate an electronic bilge sensor that will automatically close an inlet solenoid and activate an alarm should the bilgewater level rise, as well as activate the bilge pump should there be a system leak so there is minimal flooding.

22.6 **Water Pump Wiring.** Observe the following when installing water pumps.

a.   **Cable Sizes.** Ensure that the cable is rated for maximum current draw and voltage drop, as voltage drop problems are very common.

b.   **Connections.** Electrical connections directly onto water pumps are common; ensure the crimps are done properly. Apply some petroleum jelly over the connection to stop moisture from getting to the connection lugs. Where wire tails require a butt splice, ensure they are crimped properly and covered in heat shrink tubing. Some pumps with integral pressure switches require connection directly to one side, so ensure that this is properly done.

    c.   **Protection.** The supply cable must be protected with the appropriate size fuse or circuit breaker. Typically, this is around 10A. Many pump motors have integral thermal protection.

22.7   **Water System Strainer.** The strainer is installed in the water suction line to protect the pump from damaging sediment and particles from the storage tank.

    a.   **Strainer Element Cleaning.** It is essential that the stainless-steel element is regularly cleaned; blockages are most frequent at new vessel commissioning, or after refilling an empty tank. I have seen a number of vessels where the element has been removed because the owners were tired of cleaning blockages. A 40-mesh screen is the typical value. The result will be early pump failure, so it is good practice to check and clean the system regularly.

    b.   **Strainer Bowl Seals.** After cleaning the element, ensure that a good seal is made with the transparent inspection cover, as this is a common cause of air being drawn into the system. Ensure that the seal is in good condition and apply a smear of petroleum jelly. Do not overtighten the cover.

22.8   **Water System Accumulators.** Most pumps require an accumulator. The basic principle is that air will compress under pressure and the water will not. The accumulator is a tank filled with air that fills approximately 50% with water when the pump operates. Typical accumulator sizes are around 0.5gal (2l). After the pump stops running, the compressed air provides pressurized water stored within the accumulator. It serves two functions, the first being a pressure buffer, or cushion, that absorbs fluctuations in pressure. The effect is to operate quietly, and the pump pressure switch is able to reach the cutoff pressure, which increases the life of the pump, motor, and pressure switch. The life of the pump is extended, as the accumulator will prevent the pump from operating as soon as the water outlet is opened. There are two system types.

    a.   **Non-pressurized.** These units are typically plastic cylinders, which are installed upright within the system. They have a cock at the top to vent off air within the water system. With tanks that do not have bladders to separate the air and water, the tank must be drained every few months; as the air gradually disappears and the tank no longer functions as an accumulator, there must be air inside.

    b.   **Pressurized.** These accumulator types have an internal butyl rubber membrane or bladder that can be externally air pressurized with a bicycle pump; some are factory pre-pressurized with nitrogen. A pre-pressurized accumulator is at 12psi (82kPa). At installation, the following procedure must be performed: Turn the pump off. Open outlets, and release system pressure. Using a car tire pressure gauge, release nitrogen until pressure falls to 5psi (34kPa) below pump cut-in pressure. If too much pressure is relieved, use a bicycle pump to increase to the correct pressure. Typical accumulator pre-charge pressures are 10psi (0.7bar), and setting range up to around 116psi (8bar).

22.9   Water Filters. Clean water is essential on any boat, and with the variability in water quality, especially when cruising foreign, it is cheap health insurance. Water filters should be fitted to all drinking water outlets; in most cases, this is the galley outlet.

a.   **About Filters.** A filter will remove small particles, off-tastes and smells caused by tank water purification chemicals, and most bacteria. Bacteria will form in pipes and tanks during extended periods of inactivity. You can sterilize the system, using the tank sterilization procedure, and flush the system. Always install a filter with easily replaceable filter elements and replace them promptly at the manufacturer's recommended service life completion. Always cleanse the water system before installing a new filter.

b.   **Filter Types.** Filters are generally manufactured from granular activated carbon (GAC), activated carbon block (ACB), or carbon-wrapped paper. Bacteriostatic kinetic degradation fluxion (KDF) carbon filters are higher-standard filters with long service lives and have efficient heavy metal removal capability. KDF elements comprise high-purity copper-zinc granules that reduce water contaminants through a redox reaction, or oxidation-reduction. The redox reaction results in an electromechanical reaction that exchanges electrons with contaminants, converting them into a harmless substance. Sediment filters are made from spun polypropylene.

c.   **Ceramic Filters.** Higher water quality can be had from filters that use porous ceramic media. They are capable of removing bacteria, dirt, sediment, cysts, turbidity, and other contaminants. If a ceramic filter is combined with an activated carbon filter, additional contaminant removal, including volatile organic compounds (VOCs) and chlorine, is possible. Chlorine-based disinfectants are used in water tanks, and these filters remove that taste.

d.   **Seagull IV System.** One of the most well-known water filters is the Seagull IV unit from General Ecology. The Seagull IV X-1B Drinking Water Purification System uses the company's proprietary Structured Matrix purification technology. These systems are independently certified against the EPA's "Guide Standard and Protocol for Testing Microbiological Water Purifiers against Viruses, Bacteria and Water-borne Cysts." They remove all microbiological material, including bacteria, cysts and viruses. They also remove pesticides, herbicides, chlorine, foul and off-tastes, discoloration, and smells or odors. A good filter should always provide a test report issued by an appropriate authority. It should be rated for the expected flow rate and renewed at the recommended due date.

22.10   Water Pipes and Fittings. Water pipes should be of a high-quality material that is suited to both hot and cold water. ABYC Standard H-23, "Installation of Potable Water Systems for Use on Boats," is worth observing. Piping and fittings for potable water systems are required to comply with National Sanitation Foundation (NSF) and American National Standards Institute (ANSI), NSF/ANSI 61 02022 Drinking Water System Components – Health Effects and the US Food and Drug Administration (FDA) standards. Consider the following when selecting and installing piping.

a.   **Pipe Standard.** The piping should be nontoxic, suitable for potable water systems, and must not be able to support microbiological growth. There are two types:

(1)   **Semirigid Piping.** There are color-coded water tubing systems — blue for cold and red for hot. These are made from high-quality acetal, polypropylene (PP), polyethylene (PE), and polysulfone (PSU) tubing. Acetal has relatively low impact resistant and temperature qualities. Polysulfone possesses the highest temperature and impact-resistant properties. Be aware that PSU is not resistant to certain volatile organic compounds or caustic chemicals. When using PSU piping, do not use liquid thread sealants or anything (almost everything) on board that includes hydrocarbons, thinners, fuels, cleaners, paints, piping cement, and similar compounds. To meet ABYC recommendations, always install above normal bilge level. This piping is easy to install using compression fittings. Ensure that the pipe is not kinked and that you assemble the connectors correctly. Where tight bends are required, install a bend, following the minimum bend radii guidelines. Use a tubing cutter, not a knife or hacksaw; the tubing must be square and clean. Manufacturers include PEXtite, Sea Tech, Flair-It, Whale, and John Guest (Speedfit). Use the same principles as for electrical systems and support the piping properly. I have installed my new water system using the Whale Quick Connect system using color-coded piping.

(2)   **Flexible Hose.** Hose is the most common piping. Ensure that it meets the required standards. Hose is prone to kinking, so installations should be done with care. Do not use translucent tubing, which encourages algae growth.

b.   **System Pressures.** Piping must be able to withstand water system pressures. Whale piping is rated at 60psi (401kPa) and 194°F (90°C). When installing piping over longer runs, larger pipe diameters are required to reduce friction losses.

c.   **Fittings.** Fittings must be able to withstand system pressures. Nuisance leakage can be avoided. Plastic hoses generally have PVC T-joints and are installed with clips.

d.   **Outlets.** There are many different taps, valves, and shower heads on the market. Always choose good-quality items and choose only those that are compatible with the entire plumbing system. This makes finding spare parts easier. Reputable names include Whale and Jabsco. If you are using a non-flexible, permanent shower head, opt for one of the domestic, low-water-consumption fixtures.

e. **Connections.** Ensure that all piping or hose connections are double clamped. Quick-connect fittings should be assembled correctly and tightened properly.

Table 22-1. Water System Troubleshooting

Symptom	Probable Fault
Water pump does not run	Circuit breaker tripped off Pump seized Pressure switch fault
Pump runs but no water	Blocked water inlet filter or strainer Suction inlet air leak Ruptured pump diaphragm No water in tank Blocked one-way check valve Inlet hose kinked Inlet valve closes Debris under flapper valves Outlet restriction Low voltage to pump Discharge head too high
Pulsating water flow	Restricted pump delivery Pressure switch fault
Pump cycling excessively	System pressure leak Water outlet leaking Accumulator problem
Pump will not switch off	Empty water tank Pump diaphragm ruptured Leaking discharge line Pressure switch fault Debris under valves
Low water flow and pressure	Pump inlet air leak Clogged strainer Pump impeller worn Pump diaphragm ruptured Pump motor fault High discharge head Incorrectly rated pump

HOT WATER SYSTEMS

22.11   Hot Water Calorifier. The calorifier or hot water system is becoming one of those hard-to-do-without luxuries. It is not difficult to install or incorporate into a water system; it even functions as an additional water reserve. The term "calorifier" is given because most marine hot water systems heat from inbuilt coils using calorific transfer, supplied from heated engine cooling water or, on the old tramp ships I once served on, steam. It makes sense to utilize all the available energy consumed by the engines. Manufacturers include Isotemp, Force 10, Raritan, Kuuma, Whale, SureCal, Sigmar, and Quick.

a.   Heating Coils. The majority of units are fitted with a single copper heating coil. Beware of the cheaper imported units, as the coils are very small and have only a turn or two. Good calorifiers will have several turns installed to ensure good heat transfer rates. Some units have twin coils to improve heat transfer.

b.   Electric Elements. Calorifiers should incorporate an auxiliary electric heating element for mains AC heating or DC powered capability.

(1)   AC Elements. AC element ratings should not exceed 1,200W to 1,800W due to electrical supply limitations of marina shore power and small generators, unless you have a reasonably high-output generator set.

(2)   Thermostats. A thermostat is essential for controlling temperature and preventing overheating and, therefore, overpressure conditions.

(3)   Additional DC Elements. SureCal units' water heaters have a 12V, 24V, or 48V element that allows water heating off the engine when charging or by using surplus wind and solar. The AC element is rated at 700W; the DC element is 300W.

(4)   DC Element Only. The Whale 12V water heater is an option if you don't want AC or engine calorifier water circuit installation. Given the increase in wind and solar, the heater has ample power. These are small capacity at 3gal (11l) and have a rapid heat-up time of less than 1 hour and a heat-retention period of up to 10 hours. It suits a dual alternator system where extra power is available and draws around 30A. Interlocking with an engine start system is recommended; this is not an issue with wind- and solar battery–charged battery banks, although caution about battery drain is important.

c.   Pressure Relief Valves. All calorifiers should have a pressure relief valve (PRV). The valve should be regularly operated manually to ensure that it is not seized and to eject any insects or debris from the overflow pipe.

d. **Valves.** The inlet of a calorifier should always have a nonreturn valve fitted to prevent the heated and expanding water in the tank from backflowing into the cold-water system and pressurizing it.

e. **Insulation.** Ensure that the calorifier has a good insulation layer or cover to avoid heat wastage. If the engine is run every alternate day, good insulation will keep it warm over the extended period.

f. **Mounting.** The calorifier must be mounted with the coil on the same level as, or below, the engine cooling water source. This is because the engine pump must circulate water through a longer system, which introduces resistance and could overload the pump.

g. **Air Locks.** There must be no air locks in the system, as these go through the engine cooling system and affect cooling. The calorifier must always be installed lower than the engine water filling point.

h. **Hose Connections.** Use heat resistant rubber hoses to connect the heating circuit. Ensure that all hose connections have double hose clamps and that air locks cannot form in the hoses.

i. **Thermostatic Mixing Valves.** As water can rise to 194°F (90°C) or above, a risk of high-temperature scalding exists when washing or showering. Always use a mixer tap or valve to maintain temperature to around 86°F to 158°F (30° to 70°C); this also saves water.

22.12 Diesel Hot Water Heaters. The diesel hot water system is becoming commonplace on vessels. This unit can also be part of a central heating system. Eberspacher, Hydronic, and Webasto have very efficient systems. The Webasto unit is illustrated below. The typical operational cycle is as follows.

a. **Starting.** Cold air is drawn in by an electric fan to the heat exchanger/burner. This is normally from the engine area.

b. **Ignition.** Fuel is drawn in at the same time by the fuel pump from the main tank and mixed with the air. The fuel is ignited by an electric glow plug in a combustion chamber.

c. **Combustion.** Combustion takes place within a sealed exchanger; the exhaust gases are expelled to the atmosphere.

d. **Heating.** An integrated water pump circulates the water through the heat exchanger and subsequently to the calorifier and heating radiators. A thermostat in the cabin shuts the system down and operates to maintain set temperature. Eberspacher has developed an automatic quarter heat control to reduce unnecessary cycling, thereby improving fuel economy.

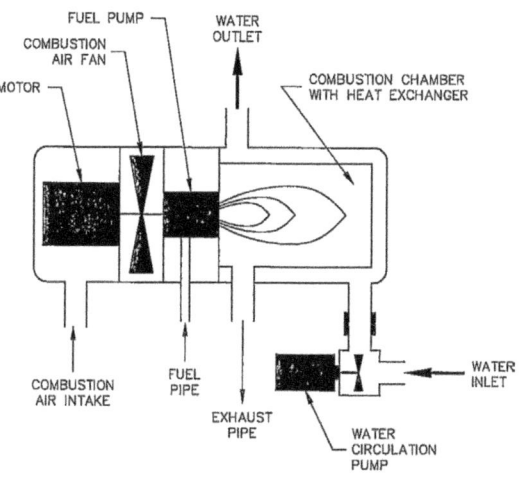

Figure 22-2. Diesel Hot Water Systems

22.13   Diesel Hot Water Boilers. Some larger boats have a more substantial oil-fired boiler system, and steam boilers have similar principles. I had a very reliable Kabola diesel heater on an old Dutch barge I cruised around Europe. Kabola makes a good range of diesel and hybrid diesel and electric heaters. Kabola's KB Ecoline Hybrid series of heaters are automated, carbon-free, oil-fired boilers of 8kW to 38kW, with no soot emissions. Boilers provide hot water for air-conditioning, heating, hot water, and central heating. The Canadian Hurricane diesel water heater can burn No. 1 and No. 2 diesel, stove and furnace oil. Installation is critical, and the exhaust system must be properly specified and installed. Good combustion depends on good drafting, and the length, diameter, and limited bends all affect this. The Hurricane SCH25 heater is a hydronic heating system that uses circulating water for heating. A hydronic heating system utilizes a solution of water and antifreeze (inhibited propylene glycol) for heating of spaces using fans and water heating. There is a facility for engine preheating using a circulation pump and waste heat recovery using a heat exchanger from an operating engine. It is a very versatile system for cold-weather boating. Typical power consumption is around 5A, with a fuel consumption of only 0.84lph. Check out itrheat.com in Canada and dieselheating.com in the UK. Visit Kabola at kabolaheaters.nl.

22.14   Diesel Hot Water Heater Troubleshooting. Burners require regular cleaning, as they can clog or soot up, and atomization affects the efficiency. Flame failure devices require cleaning. Solenoid valves can seize and should be checked. The usual electrical faults of poor connections also occur, as maintenance tends to be a lot lower because people avoid the boiler. A fuel filter should be installed and checked regularly. High temperatures or failure to ignite during the start cycle are often flame failure sensor problems.

Table 22-2. Diesel Water Heater Troubleshooting

Symptom	Probable Fault
Heater will not start	Supply fuse blown or breaker tripped Low battery voltage Fuel-metering pump fault Thermal cutout tripped Fuel filter clogged No fuel supply Glow plug fault Control unit fault Check error codes
Heater smokes and soots	Combustion pipe clogged Fuel metering pump fault
Heater goes off	Thermal cutout tripped Fuel filter clogged Fuel-metering pump fault Water pump fault Control unit fault Flame failure detection fault Check error codes
Low water temperature	Thermostat control fault Heat exchanger fouled Overheat sensor fault

WATER MAKERS

22.15   Desalination or Water Maker Systems. Cruising to foreign places is fun, but unfortunately, once you get there, the water is often scarce or not fit to drink. Water makers are becoming more popular on many vessels because they give you a greater degree of freedom. Onboard water resources are limited, which affects maximum cruising ranges. The most practical system is the reverse-osmosis desalinator, as evaporative systems require long-term engine use for reasonable economy. It must be stressed that water should not be made within 10 miles of a coastline or within inhabited atolls in the Pacific. These waters are generally polluted to levels well above World Health Organization (WHO) recommendations, and this pollution can be carried into the tanks with product water if the system does not have high quality filtering and sterilization systems. Principles are as follows.

    a.   **Reverse Osmosis.** In natural osmosis, when fresh and salt water are separated by a semipermeable membrane, fresh water flows through to the saltwater side. To reverse this process, salt water is pressurized to force the fresh water out through the membrane. Seawater is pressurized by a priming pump and filtered to remove particles. Then pressure is increased with a high-pressure pump, which forces fresh water through the membranes. The membranes are housed in a high-pressure casing.

b.   **Efficiency.** Desalination membranes remove salt, bacteria, viruses, protozoa, and cysts from seawater. This includes virus families such as hepatitis A, Norwalk virus, rotavirus, poliovirus; bacteria such as *E. coli* (*Escherichia coli*), salmonella (*Salmonella typhimurium*), cholera (*Vibrio cholerae*); and protozoans that cause amoebiasis (*Entamoeba histolytica*), giardiasis (*Giardia lamblia/Giardia intestinalis*), and cryptosporidiosis (*Cryptosporidum parvum*).

c.   **System Components.** The heart of any system is the osmotic membranes. Membrane quality is the key to good unit performance, and cheaper units with poor-quality membranes usually end up being quite costly because of the high maintenance and replacement costs. Pumps can be either engine driven or AC and DC power driven. Power consumption can be up to 2kW and generally requires a minimum generator capacity of 3kW for starting currents and approximately 1.5kW for running currents. Well-designed systems incorporate prefilters for the salt water. Prefilters typically have a rating of 50 microns, followed by a secondary filter of 5 microns. Dometic's new SeaXchange XTCII Series water maker is a single-touch unit with programmable logic control. The unit has a touchscreen for status display and control. It features automatic freshwater membrane flushing for scaling and fouling inhibition. NMEA 2000 connectivity is possible. Manufacturers include SeaWater Pro, Sea Recovery, Katadyne, Spectra, Rainman, Zen, Sea Maker, SpotZero, Schenker, and Pur.

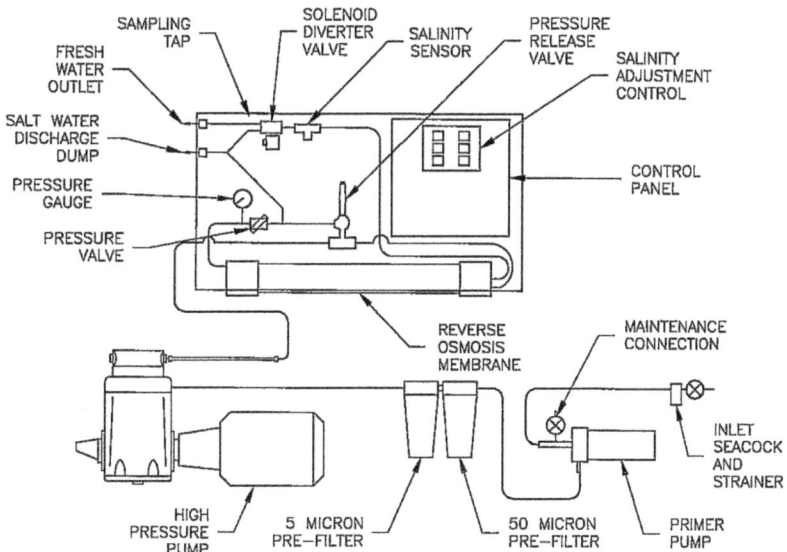

Figure 22-3. Water Maker System

d. **Installation.** Space considerations are always of critical importance, and many systems, such as Sea Recovery, Spectra, and Pur, have resolved this problem. Thru-hull fittings are required for raw seawater intake and the overboard brine discharge. It is not good practice to take the input from auxiliary engine or generator water inlets, as this may starve the system of water. Install a separate seawater intake with strainer.

e. **Monitoring and Control.** A typical system and operation are as follows:

(1) Raw seawater is supplied through the seawater inlet valve and sea strainer to the booster pump suction.

(2) The seawater is then pressurized to 20psi (138kPa) by the booster pump and supplied to the media filter and the 5-micron prefilter and oil water filter, where sediments, suspended solids, silt, and oil are removed. This water is pressure monitored with a gauge and low-pressure switch, which stops the system when low pressure is detected.

(3) Water is pressurized to around 900psi (62bar) by the high-pressure pump and regulated by a valve. This is controlled by a high-pressure switch.

(4) The pressurized water enters the reverse osmosis membrane, which forces out the salt and minerals. A salinity probe monitors the product water quality, which adjust for water temperature. The brine flows through a monitor and is then dumped through a discharge valve.

(5) The product water is monitored, passed through a charcoal filter and UV sterilizer, and then sent to the potable water tanks.

(6) The system has an automatic freshwater flushing system, which flushes the system to reduce fouling.

f. **Outputs and Membrane Correction Factors.** In a good system, salt rejection rates are typically 99% in the pH ranges 4 to 11 and operating pressure range of 700psi to 900psi (48bar to 62bar). In these conditions, output is unaffected by pressure and temperature. Where temperatures and pressures change, correction factors must be applied to ensure improved production rates.

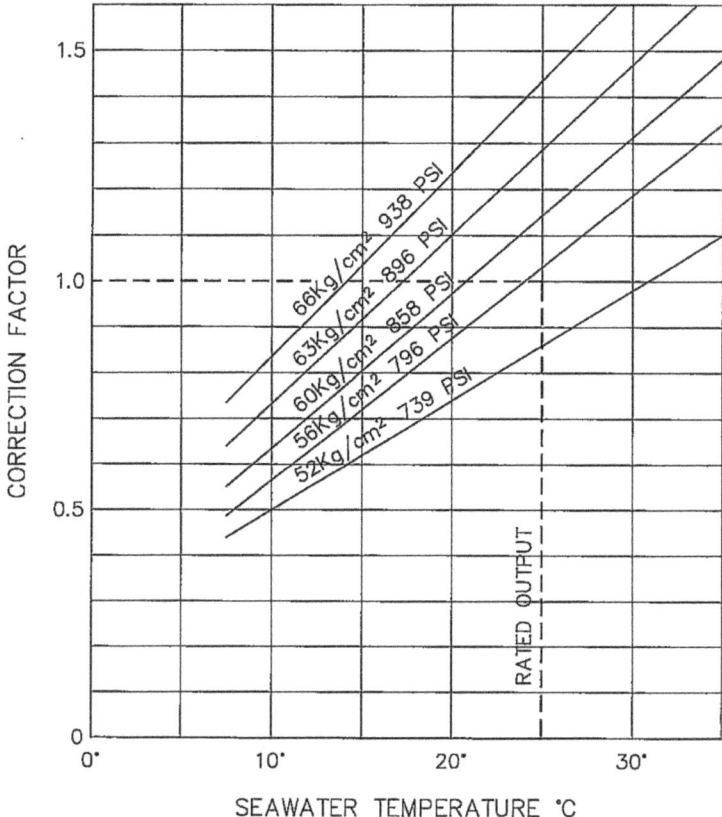

Figure 22-4. Desalination Temperature Correction Factors

g. **Water Treatment.** Membrane fouling is a common problem and is caused by organic molecules, suspended solids, bacteria, and algae, as well as minerals. Prefilters take out some of the material, but not all. Cleaning is time-consuming and requires the use of chemicals. One system that significantly increases intervals between cleaning, increases the permeation rates, and lowers feedwater pressure, and transmembrane pressure drops is the Zeta Rod. This entails using an electrode in the feedwater line. The rod effectively forms a capacitor with the water and the impurities, the piping, and vessel walls. The system charges particles that ultimately form sludge and scale to the same polarity. This causes repulsion and prevents bonding and formation of deposits in piping and membranes. The units use a high voltage to generate the electrostatic field, and the electrical current consumption is low.

h. **Maintenance.** These are typical maintenance procedures:

(1)   Clean the inlet strainer at the same time as the engine strainer.

(2)   Prefilters can be washed five to six times before replacement. This equates to approximately 80 hours' operation in clean waters.

365

(3)   Use recommended biocides to disinfect membranes to prevent bio-
        logical fouling for any shutdown period exceeding 14 days. Failure
        to do this will significantly reduce output and damage membranes.
        Never allow membranes to dry out.

(4)   Check pressure pump oil levels and renew every 500 hours.

(5)   Check and retension rubber drive belts every 6 months.

(6)   Clean membranes when output drops below 15% of rated output or
        when product salinity increases. This is due to the buildup of grime,
        biological material, and mineral scale. Do not open the pressure vessel
        to do this. Clean according to the manufacturer's recommendations.
        This usually entails alkaline and detergent cleaning to remove organic
        material and acidic cleaning to remove mineral scale.

Table 22-3. Desalinator Troubleshooting

Symptom	Probable Fault
Low water flow	Blocked strainer; check and clean Blocked prefilter; clean or replace Fouled membranes; clean Pump belts loose; re-tension
No product water flow	Pump stopped; check circuit breaker Drive belt broken Pump clutch coil failed
Circuit breaker tripping	Clutch wire grounding out Pump seizing, motor fault, connection fault
Low working pressure	Leaking relief valve Pump fault High-pressure leak
Product water salty	Brine dump valve fault Fouled membranes Working pressure incorrect

22.16   Bilge Pump Systems. Bilge pumps play a crucial safety role in any vessel, yet many
owners tend to get the cheapest units they can, along with float switches, and install them
incorrectly. Bilge pumps should be of the highest quality available, and they should be properly installed and regularly maintained. In the troubleshooting and maintenance chapter I talk
about Failure Mode and Effects Analysis (FMEA); this concept, along with risk assessment,
should be applied to bilge systems. Consider that a pump and associated float switch are all
that stands between you and oblivion. Emotive, I know, but when water starts flooding in,
having reliable bilge pumping and alarm systems is crucial to survival. They either keep you
afloat or buy you valuable time to assess and act, even if that act is abandonment. The FMEA

is about risk assessment and redundancy, and having a dual system is a logical response. Although all boats have a hand-operated bilge pump, if you have ever tried pumping for any length of time, it gets rather tiring. Examine and assess your system, determine whether it can cope with worst-case scenarios, and look at upgrading. I installed two bilge pumps on separate circuits with integral alarms and a separate high bilge warning alarm on my own boat.

22.17   Bilge Pump Selection. There are basically two types of electric bilge pumps: submersible and centrifugal. The following factors should be considered when installing electric bilge pumps:

a.   **Head Pressure.** Head pressure is related to the height that water must be lifted to. All pumps have maximum head figures for a particular model.

b.   **Pressure Cutoff.** This is the pressure at which the pump will cut off when the faucets or output is closed. Typically this is around 41psi (2.8bar).

c.   **Flow Rate.** Most bilge pumps are listed with flow rates, which are designated as gallons (or liters) per minute. Electric pumps with bronze housings are rated up to a maximum of 11g/m (50l/m). Always select a pump for the highest expected flow rate.

d.   **Pump Impellers.** Pump impellers come in a number of different compounds. Choose the correct type for optimum life and efficiency. Centrifugal pumps should never be operated dry for more than 30 seconds; they are designed to be lubricated by the pumped liquid. Operating without liquid generally means a ruined impeller. Impeller types are as follows:

(1)   **Neoprene.** These are typically found in bronze pumps (Jabsco) and are suitable for bilge pumping in temperatures ranging from 40°F to 178°F (4°C to 80°C). Use at the outer temperature limits reduces performance and service life. They must not be used to pump oil-based fluids, as the neoprene impeller can absorb oil compounds and expand. On the next start-up, the binding impeller is destroyed. Always flush out a line if oily fluids are used.

(2)   **Nitrile.** These are designed for pumping fuel but are also suited to pumping oil- and fuel-contaminated engine bilges in temperatures from 50°F to 195°F (10°C to 90°C). Use at the outer temperature limits reduces performance and service life. Nitrile impellers have a flow rate 30% lower than neoprene impellers, so they should not be used in any high-temperature applications.

e.   **Submersible Pumps.** These pumps are by far the most common. It is important to always buy and install the very best quality you can. Cheaper pumps are unreliable and tend not to last very long. Pumps such as the Orca Auto 1300 from Whale have an integral level sensor. The unit incorporates an electronic circuit for sense and control. Gulper IC pumps are remote mounting, receive signals from the Strainer IC for automatic pump control, and have a soft start control. The Gulper 320 has a remote strainer with a nonclog electric valve. Submersible pumps have the following general characteristics:

(1) **Motor Rating.** Motors are rated continuously, but the bilgewater normally assists motor cooling while pumping.

(2) **Motor Type.** Motors generally use a permanent magnet motor, which means no brushes.

(3) **Dry Running.** Pump impellers are not damaged by dry running, though motors require water to cool them.

f. **Diaphragm Pumps.** The Whale Gulper series is very reliable and effective, and I have installed these on all my boats. They remain my pump of choice.

22.18 Bilge Alarms. The main bilge should have a separate visual and audible alarm to indicate levels above the normal operating range of automatic systems. Install the loudest unit you can so it is audible under all conditions; around 100dB is ideal. Most boats have automatic bilge pumping systems. If the pump is running and cannot keep up with the water inflow, an additional alarm will indicate the high bilge level. If the float does not operate and bilge starts to fill, a separate alarm will indicate the condition. I have experienced both conditions, and additional protection is prudent. Where bilge pump control circuits incorporate automatic operation (e.g., float switch), caution should be paid to the risks of pollution by uncontrolled or unmonitored discharge of oily bilgewater. Most bilge pumps incorporate a float switch to enable automatic unattended operation. When pumps are running in this mode, you should be aware that uncontrolled discharges of oily bilgewater into the water might render you liable for stringent penalties and fines. Consideration should be given to installing suitable bilgewater filtering equipment.

a. **Pollution Prevention.** There are very heavy penalties for the willful or accidental discharge of oily water and waste into harbor and coastal waters. It is the environmental responsibility of all boat owners not to discharge any waste into the sea. No bilge, which is capable of having oil in it, should ever be fitted with an automatic pumping system. All vessels have obligations under the International Convention for the Prevention of Pollution from Ships (MARPOL).

b. **Bilge Pump Controls.** Automatic switches are notoriously unreliable. If the float or device stays on, the bilge pump may burn out and probably ruin a set of batteries by totally flattening them. There are a number of control devices, which are explained below.

c. **Bilgewater Filters and Separators.** If you suffer from oily bilges, either control the oil leaks or introduce a new bilge cleaning regimen. The Vetus filter is a typical unit. It comprises a replaceable filter element capable of removing around 95% of oil within bilgewater. They recommend bilge pumps with a max capacity of 6.6g/min (25l/min). Even the slightest oil sheen can get you in major trouble, so consider installing a good filter. The Wavestream System 1 uses XOil bonding technology and is a filter medium that bonds with oil. Each filter cartridge will absorb three times its own weight in oil. Some filters such as the MyCelx BK-2 BilgeKleen will remove 99.9% of hydrocarbons, and for compliance must not exceed 5ppm.

22.19   Bilge Pump Float Switches. Float or level switching devices are used to operate pumps when fluid levels reach a specific level and are also used as alarm activation sensors. They may use several different operational principles. Suppliers include SPXFlow Johnson, Rule, Jabsco, Whale, Aqualarm, Marine Products International, Attwood Corp, Seewater, and Water Witch, along with other specialist manufacturers. (See the color insert image of a typical dual bilge well arrangement with float switch and bilge alarm float switch.) Common technologies are as follows:

a.   **Pivot Mechanical Float Switches.** The pivoted float type switch is the most common device and uses a variety of principles:

   (1)   **Mercury Switches.** These have been around for many years and are still available. They incorporate a vial of mercury, although you will see some makers describing their switches as mercury free. The mercury, being conductive, is used to close the circuit.

   (2)   **Metal Ball Switches.** In this type of float switch a metal ball rolls down inside the float and activates a microswitch. West Marine automatic float switches use this method.

   (3)   **Reed Switches.** Many float switches employ magnetic-operated reed switches. They operate by the movement of a magnet along a hermetically sealed stem as water rises or falls. When the magnet passes over the reed switch, the reeds are pulled together to make contact.

b.   **Reed Switch Float Switch Failures.** Failure modes are generally caused by overcurrent. When bilge pumps are run through the switch, there is an arc condition when the switch makes and breaks the inductive pump load. Switch ratings are steady-state current values and do not consider the normal operating condition. The power spike is several times greater than the rated current. It doesn't take long for arcing to weld or seriously degrade contact surfaces. There are UL resistive power ratings for reed switches, along with maximum switching voltages and currents.

c.   **Vertical Float Switches.** These comprise a vertical tube that has a rig-type float with an integral magnet. When the bilgewater increases, either a pod or a ring floats up the guide and reaches the activation point, activating a magnetic reed switch. Some units have a stainless-steel vertical float shaft. The Johnson Electronic Switch (part #36152) and the West Marine Magnum (part #3685443) use this principle.

d.   **Field Effect Sensors.** These use the Mirus field effect for operation. The switches incorporate a field effect detector cell composed of an active device

369

and an electrode surrounded by a second electrode. The applied power creates a micro-electric field around the Mirus cell. When a dielectric material, the bilgewater, interrupts this field, the device gives a signal output. This is output to an alarm or pump relay.

e. **Capacitance Sensors.** A capacitance sensor uses changes in capacitance when water is present. The capacitance change detection activates the switching element. The Water Witch 230 has an exposed capacitance sensor element embedded on its matchbox-sized case.

f. **Optical Devices.** These devices resolve many of the problems normally encountered. The pump units are controlled by an optical fluid switch that emits a light pulse every 30 seconds. If the lens is immersed in water, the light beam refracts and the beam's change in direction is sensed by a coating inside the lens. This triggers the pump. Time-delay circuits can be adjusted for periods of 20 to 140 seconds so that the pump will continue draining the bilge after water clears the sensor. I have tried some of these devices, and I found them effective.

g. **Air Pressure Devices.** These are relatively old but simple devices that work reliably. They depend on sensing the air pressure of water in a tube to activate a switch via a diaphragm. If a detection tube becomes blocked or fouled, it will not work correctly. These devices take the electrical factor out of the water area.

h. **Conductivity Probe Devices**. These devices have probes and sense the electrical conductivity of the bilgewater. Dirty bilges with oil can reduce effectivity.

i. **Installation.** Pivoting arm float switches should be installed on the bilge centerline, with the pivot arm oriented aft in a fore-and-aft plane. This is to cope with any surging bilgewater. The switches should be mounted as close to the bilge pump as practicable. Bear in mind that when a vessel is heeled, the bilgewater will surge to the leeward side. While many float switches use long leads to allow connection above bilgewater levels, it is important to either terminate in a waterproof junction box or install a butt splice and protect with heat-shrink tubing. Cable fouling of floats is common, and wires must be installed to allow movement of a float without restriction. There is much said about floats and pumps being obstructed with debris. If you manage your boat in that condition, perhaps it is time to change and keep bilges clean. It is not that difficult. This minimizes the common and serious risk of corrosion problems should a leakage occur. Float switches must be rated for the relay or pump current rating they control, typically around 10A to 15A maximum. Alternatively, consider switching a relay for the pump. I have two switches installed at different levels for redundancy. This circuit diagram, provided for steel vessels, isolates the positive supply to the float switch.

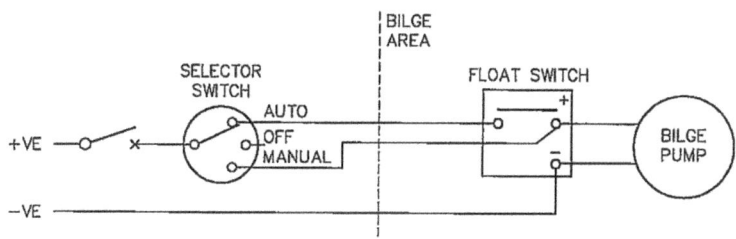

STANDARD BILGE CONTROL

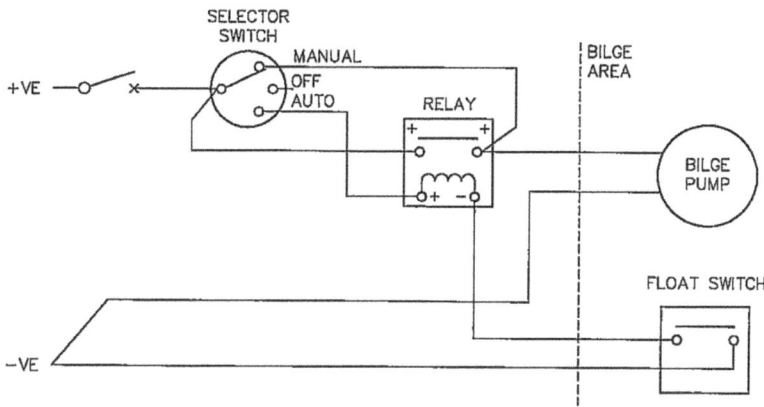

ISOLATED SYSTEM FOR STEEL/ALLOY VESSELS

Figure 22-5. Bilge Pump Control Schematics

22.20   Tank Level Monitoring. Monitoring onboard fuel, water, waste tanks, and bilge levels is an essential task. A simple electrical gauge can be installed that provides the necessary information; they have become more sophisticated and "smarter."

    a.   **Resistance Sensor.** The majority of tank sensors currently in use operate by varying a resistance proportional to tank level. The two basic sensor types are as follows:

        (1)   **Immersion Pipe Sensor.** This sensor type consists of a damping tube with an internal float that moves up and down along two wires. These units are suitable only for fuel tanks. Their big advantage is that they are well damped; therefore, fluctuating readings are virtually eliminated.

        (2)   **Lever Sensor.** The lever type system consists of a sensor head located on the end of an adjustable leg. The sensor head comprises a variable resistance and float arm pivot. As the float and arm move relative to fluid levels, the resistance alters and the meter reading

changes; typical resistance readings are 10Ω to 180Ω. Lever type units should be installed longitudinally, as athwartships orientation can cause serious problems with vessel rolling. Where these units are used for fuel or water, the primary difference is that for water sensor units, the variable resistance is located outside the tank to avoid water problems, while the fuel unit has a resistance unit.

b. **Capacitive Sensors.** This type of transducer operates on the principle that the value of a capacitor is dependent on the properties of the dielectric layer between the plates. The sender unit measures the capacitance difference between air and the liquid. The sensing circuit outputs a voltage proportional to the level in the typical range of 0V to 5V. The most common fault in these systems is water damage to the circuit board from tank condensation.

c. **Pressure Sensors.** These sensor types are considerably more expensive, but are very accurate and less prone to damage. The transducers are either placed at the bottom of the tank or on a pipe that is taken out of the side at the tank bottom. The sensors output either a 4mA to 20mA or 0.6V to 2.6V proportional to the pressure of the fluid in the tank. The pressure value is proportional to the tank volume. If the sensor is located on a small pipe, it may become clogged.

d. **Air Sensors.** Air-operated bilge switches, such as the Jabsco Hydro Air, use an air column to pressure-activate a remote mounted switch. They can switch 20A, are ignition protected, and are less prone to jamming than float switches.

e. **Ultrasonic Sensors.** These sensors can be affected by temperature, pressure, and vapors. Ultrasonic sensors emit sound waves that return to the sensor when the waves meet and reflect off a fluid such as water. They are suitable for diesel fuel, black and gray water, and salt and fresh water. Sensors are generally easy to calibrate and can suit almost any tank shape. They do not suit stainless or metal tanks.

f. **Micro-Radar.** These innovative new sensors are from Swedish company Gobius. They can be installed on the outside of the tank, and can measure through plastic and fiberglass but not metal. They have fast measurement times, with a 1-second update rate. They have a new tank monitor system, GobiusC, that is able to give precise levels and volumes, irrespective of tank geometry, with an integral tank calculator; they also incorporate wave motion reduction. This unit mounts at the tank top for a stainless steel tank. Data can be output via Bluetooth to a smart phone, with two digital outputs for alarms, or activation and stop of a pump or alarm. There is a sensor LED to show operating status. Head to gobiusc.com for details.

g. **Nonelectric.** The Tank Tender is a pneumatic system that has been well proven over the years. It can monitor up to ten water and fuel tanks. Visit tanktender.com for more information.

22.21   Bilge Pump Installation. Bilge pumps must be installed as follows to operate correctly and reliably.

a.   **Location.** Mount the pump or suction line in the lowest part of the bilge. It is best to keep this a short distance from the bottom to avoid drawing in bilge sediments.

b.   **Strainer.** Always install a strainer on the suction side of centrifugal pumps. Submersible pumps have a strainer as an integral part of the base, but these are rather coarse. It is quite common for bilge debris to jam the impeller.

c.   **Discharge Piping.** Select flexible hose that will not kink. Many pumps are rendered ineffective due to kinks or constrictions in the discharge line. As a safety precaution, always use two hose clamps on every hose connection. The discharge should be as far above the waterline as possible so it will be clear even when heeled.

d.   **Bilge Pump Antisiphon Loops.** When a boat is under sail and the bilge discharge is below the dynamic or heeled water line, it is possible to have siphoning. It is worth installing antisiphon loops on each bilge discharge line. The main point is that the valve is above the waterline under all sailing conditions.

e.   **Electrical Connections.** If the cable is long enough, make connections above the maximum bilgewater level. I recommend soldering each connection, covering the joint with heat shrink insulation, and covering the entire cable with heat-shrink insulation or wrap it in self-amalgamating tape. This will generally prevent the joint from interacting with salt water and ultimately failing. The circuit must be fused on a circuit breaker rated for the cable size, typically 15A. Always run the pump after installation to ensure that pump rotation is correct.

22.22   Bilge Pump Maintenance. Regular maintenance is essential for reliable pump operation. Regularly clean bilges of sediment and debris. Run pumps every month with water in the bilge. Many bilge pumps can seize after years or even months of disuse.

Table 22-4. Bilge Pump Troubleshooting

Symptom	Probable Fault
Low water flow	Strainer blocked with debris Pump impeller fouled Suction hose kinked Suction hose blocked with debris Suction line has airlock
Pump will not operate	Circuit breaker tripped Float switch fouled (usually with debris) Float switch connections corroded

(continued)

373

Table 22-4. *Continued*

Symptom	Probable Fault
Pump will not switch off	Float switch jammed Float switch fouled by debris Float switch mounted too low Float switch connection short-circuited
Circuit breaker tripping	Pump impeller seized Bilge area connections short-circuited Pump motor winding fault

SEWAGE (BLACK WATER) SYSTEMS

22.23   Marine Sanitation Devices (MSDs). Many sewage systems are being altered from hand-pump toilets to electrics, and the stringent requirements for holding tanks and pump-out systems require careful consideration in systems planning. In the United States the Clean Vessel Act of 1992 is the primary relevant legislation. In many countries, similar legislation has already been enacted. It is important to comply with laws pertaining to illegal discharges. An MSD is any equipment for installation on board the vessel that is designed to receive, retain, treat, or discharge sewage, as well as any process to treat the sewage. In many locations there is considerable confusion relating to the use of devices; boaters should make sure they understand the requirements.

22.24   MSD Certification. The USCG certifies MSDs within the USA, which fall into three categories.

a.   **Type I.** This device treats sewage using chemicals to disinfect it. The discharge must be free of visible solids and meet standards for bacterial content. To do this, the sewage must be macerated to break up solids. The fecal coliform bacteria output must be 1,000 colonies per 100 milliliters or less. The Raritan Electroscan and ThermoPure 2 units meet this standard. Units typically operate by macerating the waste in the holding tank then transferring it to treatment chamber, where it is heated at a low level for bacterial elimination. Type I MSDs operate using both salinity and electric current that consists of electrodes to break down the seawater to form chlorine (hypochlorous acid), a chlorinating agent that kills bacteria and disinfects the sewage. After treatment, the acid recombines to reform as salt water. The Raritan PuraSan MSD uses a solid chlorine tablet that produces a halogen solution to treat waste. Both systems draw around 45A for 3 minutes on each flush, which has considerable impact on electrical system power requirements. These units cannot discharge into "No-Discharge" zones. The SeaLand SanX can be used in all waters, but installation requirements are considerably more than other systems and may suit larger vessels. The SanX injects a chemical disinfectant agent into the treatment tank to mix with the macerated waste. Check raritaneng.com.

b. **Type II.** This is a device similar to Type I devices but with a higher level of treatment and higher quality discharge. The fecal coliform bacteria output must be no more than 200 colonies per 100 milliliters, with suspended solids of less than 1,000 parts per 100 milliliters. The best-known devices are those from Galleymaid.

c. **Type III.** This is a holding tank, meaning no discharges within nominal limits. Emptying is usually via a deck fitting, and tanks have a vent line overboard. In the United States, 33 CFR 159.7 clearly and unambiguously lays down requirements regarding the discharge of untreated sewage and the locking of toilet waste outlets. When using Type III devices, the valves must be clearly labeled and must be secured so as to prevent any discharge. This entails either closing the discharge valves and removing handles or padlocking valves in the closed position.

22.25 **MSD Devices Elsewhere.** In the UK, for most vessels you need to use a holding tank and discharge at a pump-out station or sail 3 nautical miles offshore to discharge. Holding tanks must be a secure container and constructed, maintained, and inspected to local guidelines. The guidelines are currently quite loose, with advisories on using toilets in estuaries and anchorages and other directives. Australia has varying regulations, with those of the most popular cruising area of Queensland and the Great Barrier Reef being the most stringent. There are many non-discharge areas, including marinas, and you should have a sewage management plan. Either imported MSDs or a locally made unit can be used; the latter is from Sani-Loo: sani-loo.com.au.

22.26 **MSD System Maintenance.** MSD systems have several process components, as well as various sensors and electrical elements. Some MSD units are PLC controlled, so correct operation of all input sensors and output devices is critical for proper operation. The first-stage treatment tanks contain level sensors. The second-stage sedimentation tanks do not have any components. The macerator, sludge, and discharge pumps all require routine maintenance. Backwash systems have both a water pump and solenoid valves. The disinfection system has a flow control system with chemical feed pump, and chlorination units have a power supply. Ensure that no toilet chemicals incompatible with sodium hypochlorite are being used on the boat. Normal troubleshooting principles apply: Check the power supply; if auto and manual selection are wrong, the PLC may require reset. Then check inputs and outputs to level switches or solenoids. Operational checks include the following:

a. **Daily Checks.** Check macerator, discharge, backwash, and sludge pumps and electric motors for unusual noises and vibration. Check all hose and pipe connections for leaks. Check that the chlorination liquid (bleach) reservoir is full. Check that pump pressures are normal where they are installed.

b. **Six-Month Checks.** Check control panel for moisture ingress and corrosion, and ensure the cover is closed tightly. Replace desiccant crystals if installed.

c. **Annual Checks.** Check treatment and sediment tank anodes if installed. Check that solenoid valves are operating correctly, and remove to check and clean. Grease bearings only in accordance with operating hour requirements, typically 10,000 hours. Check and tighten all electrical connections.

d.   **Hose Maintenance.** Hoses tend to have calcium carbonate buildups along with uric acid scale, and eventually they clog up. Raritan Engineering C.H. (Cleans Hoses) is an effective cleaner for holding tanks and hoses. The formula is inserted at a ratio of 1 part C.H. to 5 parts water.

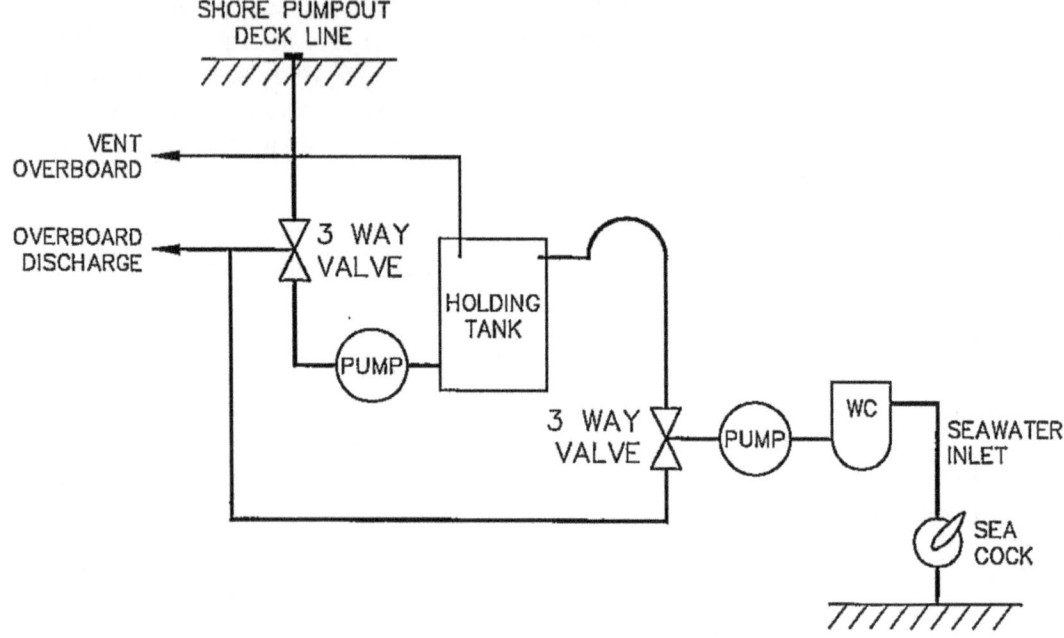

Figure 22-6. Typical Sewage System

22.27   Electric Toilet Systems. Toilets come as either a manual system or an electric unit with integral macerating function. Manual systems use a piston rod–type pump to pump out the combined sewage and seawater. Electric pumps replace the piston rod pump with an impeller type pump and macerator that effectively liquefies waste. These systems use and require a lot more water to ensure that all materials are flushed out properly. Many power vessels have vacuum flush systems, such as the VacuFlush and Environvac; they use a vacuum pump to operate and require small water quantities. These units have vacuum ejector pumps or generators and discharge pumps that require routine inspection and maintenance, along with automatic valves, vacuum and level sensors, and the control system with alarms. The 12V vacuum pumps have typical power consumptions of 4A to 6A, and large boat systems have pumps up to 0.5 hp. Discharge pumps consume around 6A. Xylem (which owns Jabsco, Flojet, Rule, and others) has the Quiet Flush E2 toilet. Simple to maintain, it has a waterproof membrane digital control panel with normal, eco flush, and dry bowl options. The discharge macerator pump motor is a permanent magnet type, so brushless. The toilet has an integrated rinse pump and solenoid valve for use with existing water systems. External water suction requires a water pump. It is rated at 20A intermittent operation.

a. **Electric Toilet Power.** One of the biggest problems with toilets is the failure to install adequately sized cables to the units or allow for voltage drop for macerator pumps. The PAR unit consumes 18A and requires a heavy-duty supply cable. The Pentair Shurflo macerator has a 13gpm (49lpm) flow rate and draws 17A maximum. As toilets are always located in a wet shower area, ensure that all electrical connections are taped up with waterproof self-amalgamating tape. Always allow sufficient cable length to pull the toilet out, as motor disconnection is generally difficult. Check the motor connections monthly to ensure no corrosion is occurring. Lightly coat the terminals with silicon or petroleum jelly. I have started to respray motors with an additional paint layer to seal and prevent water seeping into the motor housing flanges, as corrosion occurs here easily.

b. **Electric Toilet Controls.** Electric toilets have become more sophisticated. The Raritan units have Smart Toilet Control (STC). It has automatic flushing start if "Normal" flush or "Water" flush is selected. There is optional Bluetooth control from your phone with the appropriate app.

c. **About Waste.** It is essential that only normal waste be put through the toilet. To quote that readily available plaque for marine heads: ***"Don't put anything in the bowl that you haven't eaten and already digested."*** Macerator cutter plates are easily jammed or damaged by putting cigarette and cigar butts, rags, disposable wipes, and other materials down the bowl. Cleaning macerators is my number-one most unpleasant task on a vessel, so it is well worth making the effort to avoid potential problems. Always flush with enough water to properly clear piping of waste.

d. **Macerators.** Macerator pumps are usually connected to the holding tank discharge and are used to pump out waste to shore facility tanks or overboard. Units grind waste to 3mm size and are self-priming. It should be remembered that pumps are not rated continuously and that run times should not exceed 10 minutes. Heavy-duty models are available for larger systems and greater pump-out capabilities. After pumping out holding tanks, flush out the macerator pump with clean water to expel any debris that may cause bacterial buildups. Raritan Engineering has reengineered the macerator pump; their Model 53101 has a wastewater gate valve assembly that facilitates pump removal without plumbing disconnection. Their Smart Macerator Control (SMC) has automatic pump stop if the pump senses dry running and will do a brief pump start every 7 days to prevent the impeller from binding. They have improved the nitrile rubber pump impeller and seals to prevent bearing failures on the motor. The motor is a permanent magnet type. Typical power draw is around 11.5A.

e. **Toilet Antisiphon Valves.** These are also known as vented loop, siphon break, or antisiphon breaker valves. Although they are critical to the safety of the boat, they are often neglected or poorly maintained and are often difficult to access. A valve malfunction can lead to the sinking of your boat. The vent is a single-direction breather valve installed at the top of the loop. The loop should be installed as high as practicable above the dynamic or heeled waterline; recommendations vary between 1 and 2 feet. Read your head installation manual. The bowls of most marine toilets are much lower than the water intake opening and so are significantly exposed to siphoning after flushing. Regular inspection is important to verify valve condition. Hose joints to the vented loop are often improperly installed, and hose clamps are not tight. There may be evidence of salt crystals around the hose joint. Check the hose end condition and tighten the hose clamps. Improper hose joints are often the source of odor. Sometimes it is easier to remove the loop and check everything. Some advocate regular flushing with vinegar through the head to reduce any deposits.

f.   **Solenoid Vent Systems.** TruDesign has a system that utilizes a solenoid valve at the vent. This allows the vent to be activated to fully closed, eliminating any air ingress. The solenoid valve is normally open and is powered to close the valve. The power consumption is 710mA. The vent in toilet applications is interlocked with the pump circuit.

GRAY WATER SYSTEMS

22.28   Gray Water, Shower Drain, and Sink Systems. Gray water systems comprise shower drain systems and galleys or sinks and require specific pumps. The following pumps are commonly used.

a.   **Diaphragm Pumps.** Jabsco and Whale have a range of purpose-designed shower diaphragm pumps that do away with a sump pump and float switch. The pumps are self-priming and have four chambers; they are connected directly to the drain outlet and simply have a strainer installed in-line on the suction side. The pumps have power consumption rates of around 5A to 8A at full load. The Whale Gulper series is very reliable and effective, and I have installed one myself. The pumps can run dry, and strainers are recommended to prevent blockages. Pumps are generally repairable with the appropriate spares kit.

b.   **Submersible Pumps.** These plastic units are common, and most makes have a fully integrated sump, suction filter, pump, and float switch. The sump pump units usually have a check valve to prevent back-siphoning and a clear cover for inspection. The Gully IC senses the gray water level and inputs that to the Gulper IC; it has a soft start and low current switching and an inbuilt time delay. The electronic control circuit is potted to prevent water ingress. Submersible pumps come in ranges from small flow rates of 380g/h (1,438l/h) up to 2,000g/h (7,569l/h). Typical power consumptions are 1.7A for a 380g/h unit up to 10A for 2,000g/h pumps.

c.   **Centrifugal Pump.** Some manufacturers recommend a centrifugal self-priming pump. These have a flow rate of around 3.5g/m (12l/m) and use 3.6A at full load. This offers the chance to match all the pumps so that bilge, wash-down, and shower pump motors are all interchangeable.

d. **Pump Electrical Protection.** Pump motors have integral thermal overload trip protection. If they trip on internal overload, determine the cause of the trip; typically the overload is due to pump bearings seizing or the pump being jammed with debris, stalling and overheating the electric motor.

e. **Pump Maintenance.** Shower and sink drain pumps are prone to rapid filter clogging due to hair, soap residue, and other debris. In automatic float switch units, hair and solidified soap often cause the float to stick. The filters should be checked and cleaned weekly. This also reduces bacteria formation.

22.29   Shower Water Conservation. Water conservation is always an important issue. This water-saving shower idea is not for everyone, but it has worked for me on two boats where water conservation was everything. It allows long hot showers with low water consumption.

a. **Option 1.** If you have hot water from a calorifier, start showering and then start recirculating the water via the shower drain sump pump. When finished with the soap end of the exercise, run a rinse water through. The sump suction has a filter to strain suds and hair and then recycles water back through the system. When finished, water is diverted overboard or into your gray water holding tank; you rinse off using the same procedure.

b. **Option 2.** Heat some water in a solar bag, empty the contents into the shower sump, and start recycling water through the shower head. After getting wet and cold, the solar bag still keeps heating and makes for a quick recovery from the elements.

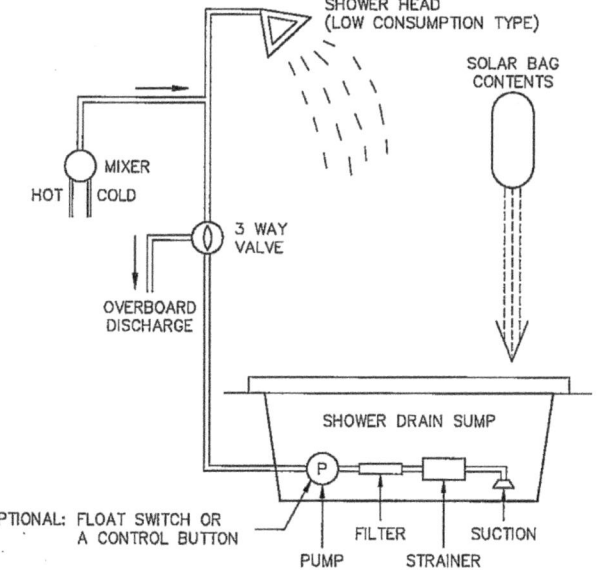

Figure 22-7. Shower System

UTILITY PUMPS

22.30   Transfer Pumps. These are used for diesel and oil duties. One recent innovative solution is a reversible electronic pump with flow regulation. This is a heavy-duty gear pump with an output of 2.5g/min (10l/min) at 50psi (3.5bar). One pump on the market is a self-priming unit that has helical polyether ether ketone (PEEK) gears for transfer of low-viscosity liquids. They have an integral check valve and an electronic pressure sensor. This allows proportional speed control. The PEEK gear configuration gives constant pressure, and the gears have optional remote-control capability. The keypad allows flow reversal and flow rate adjustments with automatic flow rate fine-tuning to compensate for fluid viscosity to prevent pump cavitation. Power consumption is around 4A. Go to www.marco.it.

22.31   Washdown and Seawater Service Pumps. Seawater pumps are used for many functions, from cooling water to air-conditioning to deck washdown services. Many are traditional types; however, pump technology is also changing. The SeaStrong Pump from Electrosea has a seal-free magnetic drive that couples the pump to the motor; pump bodies are constructed to be corrosion free, which is an issue on seawater pumps. Johnson, Shurflo, Jabsco, and others all have a range of quality washdown pumps. They vary in range from 4g/min (15l/min) to 5g/min (19l/min) and higher. While I am unable to say which is the better, many magazines have run comparison tests, so get online and research. Some metal-cased pumps, such as washdown pumps, must be grounded. Typical power consumption is 10A to 15A.

22.32   Bait Tank (Livewell) Pumps. More for the fishing boat and motorboat people, but I have come across yachts with livewell bait tanks and pumps whose owners are active anglers who like to actively supplement their diet with fresh fish. A baitwell pump supplies fresh, cool, oxygenated and purified water to keep baitfish alive. Ammonia and waste build-ups quickly cause fish to become sluggish. Some pumps simply aerate the water. Plan on a pump that replaces all the tank water every 6 to 8 minutes. Some have oxygenator units that connect with the pump body and oxygenator spray heads. Motors are designed as cartridges for easy removal and replacement. Options range from variable flow control valves to strainer and filters. Like bilge pumps, they must have the proper sized cables, installed to the pump motor with watertight connections. Flow rates vary according to models and are typically 4g/min (15l/min) for the Pentair Shurflo Pro Baitmaster II livewell pump, which can double as a washdown pump with some pipe and valves. Maximum power consumption is around 8A. Other higher throughput livewell pump models, such as the Pentair Shurflo Bait Sentry, have a magnetic drive with a 500g/h (1,893l/h) capacity and 1.7A maximum current draw. Some of the high-flow models have a patent-pending electric motor cooling system. Aerator pumps have power draws of between 2A and 5A.

22.33   Ballast Pumps. These are not common on sailing yachts but are used on wake boats and powerboats. Companies such as Pentair Shurflo have pumps designed for high-capacity ballast tank applications. They provide both rapid fill and drain rates. The bronze self-priming pumps have a reversible, high modulus neoprene compound impeller. They have a high torque electric motor. Current consumption is 15A to 20A. Similar pumps are used in yachts with water ballast systems.

Hydraulic Power Systems

There are several hydraulic power units on the market, and these are an economical solution for many larger boats. They are suitable for yachts from 40ft (12m) in size and upward, as the weight and physical envelope of these units has shrunk considerably. Hydraulic steering is commonplace on many yachts. The hydraulic power unit (HPU) is very versatile and is able to power an impressive list of equipment and systems, including bow and stern thrusters, windlasses, capstans, deck winches, captive winches, davits, lifting and swing keel systems, canting keel systems, mast and boom furling systems, transom garages, vangs, headsail furlers, passarelles (folding gangways), stabilizers, deck cranes, transom doors, and rig controls such as backstay adjusters. It is not just large superyachts or megayachts that use hydraulics; they are also common on sailing boats and motor and trawler yachts. Refer to ABYC H-30, "Hydraulic Systems," which addresses the design, construction, installation, operation, and control of hydraulic components and systems used to transmit force. This standard applies to all boats equipped with hydraulic systems.

23.1   Hydraulic Power Basics. The basic principle is that an incompressible oil is suctioned from a reservoir, pressurized, and then transfers energy by way of oil pressure through piping to an actuator, cylinder, or hydraulic motor. Many factors need to be considered within hydraulic system fluid dynamics, including friction losses, laminar flow, and turbulent flow. Flow and flow velocity are the two key factors.

23.2   Hydraulic Power Safety. Many are not aware how dangerous a hydraulic system can be; it may look innocuous but is far from it. Hydraulic systems operate under high pressure, with many systems above 2,000psi (140bar). Incorrect procedures can be near fatal, and damaged hoses, hose terminations, and system component damage can result in severe injuries. This can arise from several hazards, including hot, high-pressure fluid spray and bruises, cuts, or abrasions from flailing hydraulic hose ends, burst hoses, and hose terminations. High-pressure fluid injection into the skin is a real risk, and the most common injury is associated with pinhole leaks. Often the signs are oily or dirty spots on a hose or line. If you run your finger or hand over one of these leaks, you may end up with high-pressure oil injected into the skin. If this happens, go immediately to your local ER, as you are at risk of losing a finger, hand, or arm—even your life. Before working on any hydraulic system, you must de-energize the system. Always assume that stored energy remains, even with the pump off; check and prove that it is de-energized. Know how to completely depressurize the system. Some systems may retain pockets of pressure and so must be made safe. Never crack open a connection point to vent or relieve pressure. Never attempt to tighten or loosen a connection when the system is under pressure and operating. Never adjust a pressure relief valve when it is under pressure. Before working on hydraulics systems, don safety clothing, safety shoes, safety glasses, and other safety gear. Always wear high-quality work gloves. If you see oil spraying out under pressure and forming a mist or fog, switch the system off immediately. If ignited, it could cause a fire or explosion. **Repeat:** Don't crack open connections under pressure for the same reasons.

23.3   Hydraulic Power System Elements. In general, hydraulic power is supplied from an integrated hydraulic pump unit (HPU). The HPU comprises several operational elements.

a.   **Hydraulic Tank or Reservoir.** The reservoir contains the hydraulic oil required for the system. It allows oil to settle and cool, and for entrapped air to dissipate, ready for suction by the pump back into the system. It generally contains enough oil for system operation and allows for thermal expansion. The tank will have isolation valves for pump suction and return lines, and a sight glass for checking oil level. There may also be a breather on the tank top. Tanks can be very heavy with oil and quite large. Lewmar has been in this space for a long time, since the first edition of this book. They have introduced a patent-pending vortex hydraulic reservoir. This innovation actively removes air from the hydraulic system during operation due to a unique internal flow pattern, where oil returning to the reservoir removes free and dissolved air from the oil so that little air remains entrapped. The efficiency of this results in a small amount of air remaining entrapped in the oil, with a much smaller quantity of reserve oil required, stated as a 95% reduction. This results in a 20% increase in hydraulic motor performance. The system is designed so that once bled, no air can be introduced into it. This leads to a much smaller and lighter HPU, making it viable for much smaller boats.

b.   **Hydraulic Pumps.** Pumps convert mechanical into hydraulic energy. Pumps are categorized into vane pumps, gear pumps, and axial piston pumps. Piston pumps have the greatest performance in high-pressure applications. Wesmar utilizes load-sensing pumps.

c.   **Hydraulic Motors.** There are several types of hydraulic motors, including axial piston, radial piston, hydraulic gear, and hydraulic vane motors. The hydraulic motor uses hydraulic pressure to rotate and is similar to a pump. The amount of oil being fed to a pump determines the speed, and torque depends on oil pressure.

d.   **Hydraulic Pump Electric Motors.** Motors for DC-powered systems are typically rated at 3kW, up to 14kW, and require considerable battery power, with maximum outputs of 2,000psi (140bar). To prevent derating, voltage drop must be minimal. Pumps can be powered by single- and three-phase AC motors, with soft start options, and on larger boats are often configured to run off a generator. Pumps can also be installed via a PTO on a generator. These are generally through an electromagnetic clutch.

e.   **Hydraulic Valves.** Valves control the hydraulic flow and isolation. The directional solenoid valves are mounted on manifold blocks and feed each equipment hydraulic circuit. Hydraulic control may include directional control valves, pressure flow valves, and check valves to control flow and pressure. Check valves are used to prevent back pressure to the pump. Cushion valves reduce or dampen surges and overloads. Systems incorporate pressure relief valves (PRVs) to vent overpressure back to the reservoir.

f.   **Oil Filtration System.** This usually consists of duplex oil filters placed in the oil reservoir return line; they are typically 2 to 3 microns.

g.   **Heat Exchangers and Coolers.** Heat is generated within the oil, and hydraulic oil must be maintained within a specific temperature range to avoid oil quality breakdown and equipment damage. They are generally seawater cooled.

h.   **Accumulators.** Accumulators store energy and absorb oil pulsations and shocks. They come in several varieties, including bladder, diaphragm, piston, spring, and weight-loaded types.

i.   **Hydraulic Oil.** Hydraulic oil, both mineral and synthetic, is required for energizing as well as heat removal and lubrication. There are two primary standards for oil cleanliness: National Aerospace Standard (NAS) 1638 and International Standards Organization (ISO) 4406. NAS 1638 is extensively used within the commercial maritime and offshore oil industries. It defines five categories of particle size in oil: 5 to 15 microns, 15 to 25 microns, 25 to 50 microns, 50 to 100 microns, and >100 microns. NAS 1638 further classifies contamination levels by a number ranging from 00, the cleanest level, to 12, the dirtiest; this is based on the number of particles per 100ml of fluid for each of the particle size brackets. ISO 4406 only rates particles within three overlapping size classes. Most hydraulic systems should meet NAS 5 to 6 as a minimum.

j.   **Control System.** On some systems, this can be an integral electronic printed circuit board (PCB), which controls the directional solenoid valves. More-complex systems can be controlled by a programable logic controller (PLC).

k.   **Motors, Rams, and Actuators.** Hydraulic cylinders convert hydraulic energy into motion and linear force. Motors convert this into rotary motion.

23.4   **Hydraulic Maintenance.** Hydraulic system reliability and performance absolutely depend on regular inspection and maintenance, as follows:

a.   **Hydraulic Oil System.** Hydraulic oil must be kept clean, so filters must be changed and oil levels checked and topped up; many gravity tanks are allowed to empty before refilling. Filters remove metal particles that result from equipment wear. Oil filters should be cleaned or changed regularly. Oil coolers should be checked and cleaned. Cleanliness is the key to good performance, so the oil condition should be checked regularly. Having an oil analysis done every year is a good system check:

(1)   Oil is dark: indicates oil oxidation or overheating.

(2)   Oil is milky or emulsified: indicates water in oil.

(3)   Bubbles: indicate air in oil, low oil levels, or a suction air leak.

(4)   Contaminants: indicate wear, or dirt in oil.

(5)   Black oil with a burnt smell: indicates oil aging, overheating, or seriously contaminated.

b.   **Hydraulic Cooling System.** Ensure that oil coolers are maintained so that maximum heat transfer takes place. Anodes, where installed, must be checked and renewed. Where a separate seawater pump and inlet are used, ensure that strainers are checked and cleaned regularly.

c.   **Hydraulic Hoses and Lines.** Always check hoses for bends and kinks; check hose covers for abrasion, cracking, or exposed wires. Older hoses are subject to damage to hose reinforcing or rusting. Check for evidence of bubbling on hoses and seepage of oil around couplings and terminations. In many installations, stainless steel tubing is used. This then interconnects equipment through flexible hoses. Bends, pipe friction, and restrictions can create turbulence and flow issues. In commercial applications and in the maritime and offshore oil industry, a grease impregnated tape generally known by the trade name Denso Tape is wrapped around every joint and termination to protect them.

23.5   Hydraulic Power Troubleshooting. Water and air contamination are the usual cause of problems in more than 80% of hydraulic system failures. This contamination is caused by faulty pumps, over-temperatures, and ingress of moisture and air into the oil. Consider the following:

a.   **Whining Noise.** The pump is able to extract dissolved air out of the oil and create cavitation, which can damage the pump.

b.   **Knocking Sound.** This indicates probable oil aeration or air being pulled into the pump cavity. Check for loose connections or inlet leaks. Oil aeration can also contribute, and this is due to air ingress through failed O-rings and seal failures. If a pressure relief valve is faulty, it makes a noise as it lifts and diverts oil back to the reservoir.

c.   **Blocked Filters.** Check filters and, if installed, blockage or differential pressure indictors. Check filters for metallic contamination that may indicate a failing actuator or motor. Pump cavitation is caused by oil inlet restrictions, clogged oil filters, hoses collapsing, or high oil viscosity. A pressure drop is created across the restriction and can result in serious pump damage. Oil aeration can contribute and is due to air ingress through failed O-rings and seal failures.

d. **Oil Over-Temperature.** The main cause is a plugged cooler and blocked seawater strainers. High oil temperatures rapidly degrade oil quality.

e. **Low Oil Pressure.** The primary causes are clogged filters, low oil levels, and system leaks. Pumps powered by DC pumps also contribute if voltage levels are low. Failing hydraulic pumps are a problem, along with frequent PRV operation back to the reservoir.

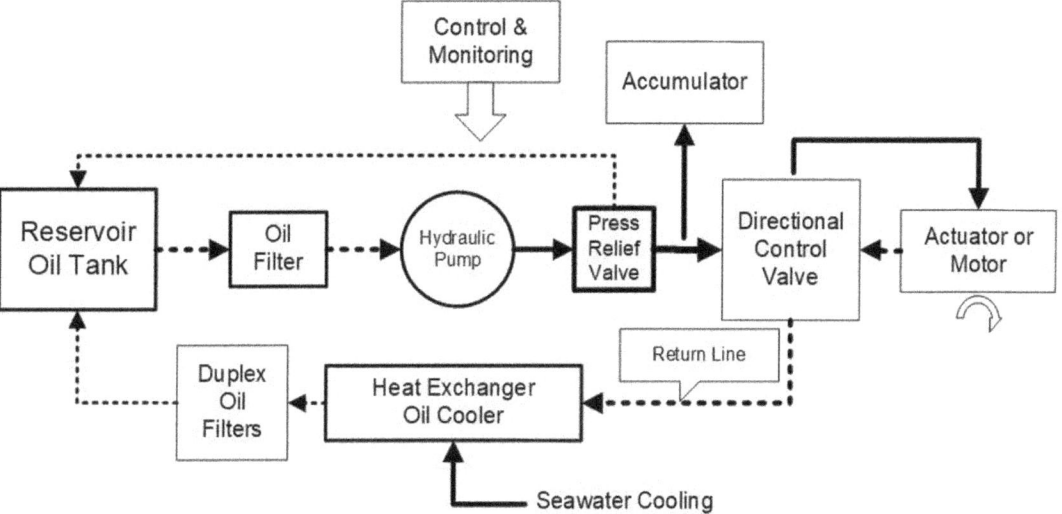

Figure 23-1. Basic Hydraulic System

SECTION TWO

MARINE ELECTRONIC SYSTEMS

Marine Electronics

Marine electronics has evolved and developed at a breathtaking pace. When the first edition of the *Marine Electrical and Electronics Bible* was published, many now-obsolete systems were standard. Radio direction finding (RDF) was the norm. The Transit Satellite Navigation system was relatively new and LORAN-C was almost standard before GPS appeared and made them both redundant. Carrying a sextant as a backup was also normal for ocean voyagers, but I doubt I could manage to take an accurate sight now. I am currently in the process of relearning that skill. Radio communications have also evolved. Morse code transmissions ceased on July 12, 1999, and signed off with Samuel Morse's first message of 1844 and the prosign "SK," which means "end of contact." Morse is fortunately kept alive by radio amateurs. Sometimes you will hear Morse on high frequency (HF) radio bands as well as on some of the ham radio bands. Satellite communications and GMDSS have transformed safety and communications in the entire maritime world. For coastal cruisers, the popularity of smartphones has changed everything, with myriad apps for almost every purpose. Discrete instrumentation systems were the norm and networking was a relatively new concept. Now we have integrated systems incorporating electronic charts, radar display, fish-finder screen, navigation data, boat performance data, and a lot more, all conveniently located at the helm station on a single multifunction display (MFD). Electronics processing power and software have assisted this change, which continues at a rapid pace. Computer processing power doubles every 2 years, and that is by a factor of about 1,000 in 20 years. There is a view that marine electronics have an average 7-year life cycle. After 7 years, the electronics are superseded, made redundant, with spares and support unavailable. I know many people afloat who have 20-year old electronics systems that are still going strong—like my own, which I am currently upgrading.

24.1   Technology Advances. The penetration of the internet around the globe was about 60% in 2021, along with nearly 5 billion phone users. It is forecast that there will be 75 billion devices connected to the internet by 2025; such is the growth of the Internet of Things (IoT). Quantum computing is the big game changer and will transform computing in the years ahead. It will allow solving of very complex problems simultaneously. Artificial intelligence (AI) is already transforming the world, from manufacturing to professions like law, finance, and economics and anything else data driven. AI and robotics make a very powerful force, and COVID-19 has helped accelerate this as skilled professionals become scarce. When this is used in conjunction with cloud-based technology, the outcome is intelligent data processing. Given the IoT revolution and the explosion in satellite low-latency internet access such as Starlink, big changes are coming and, in fact, are already here resolving connectivity issues.

24.2   Artificial Intelligence (AI). You hear this term used frequently. Essentially it means the simulation of human intelligence processes by machines and computer systems. The gamut of AI is substantial and includes natural language programming, speech recognition, and machine vision. Programming AI centers on the three principal cognitive skills: reasoning, learning, and self-correction. AI incorporates other human traits, including linguistics, psychology, philosophy, and computer science. The outcome is a system or machine that is able to perform some of the intelligent and imaginative functions of humans, and be able to do so independently or autonomously. This facilitates problem-solving and decision-making.

24.3   The Internet of Things (IoT). You hear a lot about this, but what does it mean? IoT is generally defined as the interconnection to the internet of "things" such as computer devices, appliances, and other "things." These "things" are embedded with software and sensors along with various technologies that enable them to receive, exchange, and send data to other systems and devices via the internet. This Internet of Things comprises a large network of physical devices, automated equipment and machinery, and any other smart objects that are able to collect data and then transmit it via the internet. This is already entering the boating and shipping space.

24.4   Maritime Autonomous Ships. The maritime sector is advancing rapidly with the use of AI and machine learning. In 2022 a large, fully laden liquefied natural gas (LNG) carrier, *Prism Courage*, successfully departed Texas, transited the Panama Canal, crossed the Pacific, and arrived safely in South Korea, totally autonomously. This was done under American Bureau of Shipping (ABS) supervision. Big container ship companies such as Evergreen Marine had a new 12,000-TEU vessel gaining a digital safe security certification from Lloyd's Register. Ships are now designed with smart sensors and actuators, along with digital telecommunications and computing; the drivers are economics, safety, and efficiency. When ships founder, run aground, or collide, the resulting ecological effects are huge. It is assessed that 75% to 96% of maritime accidents and incidents result from human error. This has ramifications for yacht visibility, including AIS and radar reflection in the relatively near future.

Figure 24-1. Sailing with Axiom + 7 (*Courtesy Raymarine*)

Marine Electronics Installation

25.1   Marine Electronics. There are standards for the installation of marine electronics that should be followed, including NMEA 0400 "Installation Standard Version 5.0, Edition 2021." This voluntary standard is a set of wiring recommendations that are stated as complementary to the ABYC recommendations, with an electronics installation bias. This very well-developed and comprehensive set of recommendations defines competent best practices for the installation, servicing, or modification of marine electronics and electrical systems along with associated peripherals. Although it is a standalone document, it is also designed to be used in conjunction with marine electronics equipment manufacturer installation manuals.

25.2   Navigation Station Design. Someone recently emailed me and asked whether I thought that the traditional navigation station was now somewhat redundant. The answer is no, it has probably been simplified a lot. Before you start installing navigation equipment—especially if you are fitting out a new vessel—consider the following:

> **a.   Aesthetic Considerations.** There is a certain amount of satisfaction in having an impressive navigation station. It is nice to show off and will attest to a seamanlike attitude that is not lost on your guests. The trap is that a nicely presented navigation station is worthless if the equipment malfunctions or is unreliable because of a lack of planning or a failure to consider the technical requirements of the equipment. Make it look good, but, above all, make sure it all works.

Figure 25-1. HR55 Navigation Station Photo by P. Szamer
(Courtesy of Hallberg-Rassy)

b. **Location.** The navigation station is invariably located at the bottom of the companionway steps, where it is easily accessible. In many cases the electronics are exposed to spray or even solid water if the washboards are carelessly left out in the event of a knockdown or a wet seaway. Many problems are associated with this exposure. Precautions should include the following:

(1) **Equipment Selection.** Select equipment rated as splash-proof so it can withstand exposed positions and intermittent spray.

(2) **Waterproofing.** Mount instruments in a panel that prevents water from getting behind to connectors and power connections. In exposed positions, look at protection options.

25.3   Marine Electronics Installation. Consider the effects the instruments can or may have on one another. Interference is the enemy of all marine electronics.

a. **Cable Routing.** Route all radio transmission cables well clear of signal cables. Where cable crossovers are required, make sure they are at 90 degrees. Properly space out and secure cables with the required separation distances. Position electronic equipment so that aerial cables and inputs exit the navigation station directly. Don't route cables behind other instruments or close to other cables.

b. **Electrical Equipment Location.** Where possible, do not locate the main electrical switchboard next to the electronic equipment. In many cases, however, this is nearly impossible. The growing trend is to install a smaller sub-board containing circuit breakers for the electronic equipment only. This removes a great deal of interference caused by electrical equipment.

c. **Interference Protection.** Interference sensitive equipment such as AIS, chart plotters, and GPS, along with autopilot control units, should be located in a block. Construct an aluminum housing around the section and ground it if this is a problem, although new equipment is much more resilient.

d. **Accessibility.** Make sure you have easy access to rear connections, plugs, sockets, and rear fuse holders. Do not install tight wiring looms.

25.4   Marine Electronics Ergonomics. Instruments should be positioned so they are easy to operate and monitor. Instruments should be grouped into functional blocks where possible. Keep communications equipment in one block, position fixing equipment in another. Equipment must be fitted so that access is unobstructed. Many navigation stations are a jumble of systems crammed into any available space. With the increased use of MFDs, many stations now have one of these along with one at the helm station. Important considerations are as follows:

a. **Display Visibility.** Position displays at an angle that is normal to observation. Many instruments are mounted vertically for observation when sitting down, but when at sea in normal operation, they are generally monitored when standing up. I really like new-generation MFDs—one mounted at the helm and one at the nav station, all with wide viewing angles.

b.   **Ergonomics.** Make sure you can easily reach and operate controls. On some badly designed stations, you either have to stretch awkwardly or a knob is placed in such a tight corner you can't get to it.

c.   **Lighting.** Make sure there is adequate lighting, with a good deckhead light above or at the chart table. An effective red LED night light is essential for those still using paper charts, consulting a pilot book, or anything else at night.

Electronic Charting

The UK Hydrographic Office (UKHO) announced in 2022 that they would end the production of paper charts by late 2026; this was extended in early 2023 until 2030. Paper charts have been used for centuries to plan voyages and plot navigational positions, so the change is a major step. In the age of electronic charting, updating electronic charts automatically is the norm since the SOLAS mandate for the Electronic Chart Display and Information System (ECDIS) was implemented. Given the widespread use of electronic charts, the demand for paper charts has fallen to uneconomic levels. The UKHO has over 18,000 electronic charts and digital products and will focus on new and innovative next-generation digital charting technologies, including enhanced satellite connectivity and data transmission. It should be noted that the UKHO portfolio of Admiralty standard nautical charts and thematic charts will be gradually withdrawn up to late 2026. In the United States, the National Oceanic and Atmospheric Administration (NOAA) has announced nautical paper chart phase-out plans for January 2025. US seafarers will still be able to create paper and PDF charts from current NOAA electronic navigational chart (ENC) data.

26.1   Electronic Charting Basics. Electronic charting plotters now universally incorporate GPS, or are included within GPS units, and are effectively position-fixing devices for many. The chart plotter is essentially a display with processor that decodes the data on the chart cartridges for display on the screen. The information is often layered so that chart areas can be expanded and the lights, buoyage, and contours can be called up as required. Many functions and stored data are available, including tidal predictions, sun and moon rise and set, navaids such as lights and buoys (10,000-plus items is typical), waypoints and routes, and other information. Digital selective calling (DSC) radio interfacing is also possible showing the location of a vessel in distress. Technology convergence is complete with navigation data, GPS, AIS, chart plotters, radar, and fish-finders enabled for multifunction displays (MFDs).

26.2   Chart Plotter Features. Current dedicated chart plotter displays are generally touchscreen with optional sizes from 7-, 9-, and 12-inch displays. They are also brighter, with 1,200 nits and wide display-viewing angles of 80 degrees from top to bottom and port to starboard, and have micro-SD (secure digital) slots. Many, such as those from B&G, support Insight, Navionics (Gold, NAV+, Platinum+), C-MAP (MAX N, MAX N+), and NV Digital (Raster US Charts). Free software upgrades are available for these. Garmin GPSMap chart plotters have voice command similar to Alexa or Siri; it is called "OK Garmin," as each voice command is prefaced by that phrase. The unit then responds with a ping as the cue to voice your command or question. You can request information on water depth, tidal phase, fuel levels, and so on. The unit requires the Garmin USB Voice Control Bundle to enable pairing with a compatible USB headset and also supports AirPods.

26.3   Chart Cartography Systems. Software developments for electronic charting have been rapid, with packages offering very powerful navigation.

   a.   Lighthouse Charts. These raster and vector charts from Raymarine are specifically designed for use with Axiom, Axiom+, Axiom Pro, Axiom XL, and

Raymarine Element displays. The coverage area includes coastlines of the contiguous United States, Alaska, Hawaii, thousands of inland freshwater bodies, and the Bahamas. Coverage also includes the Great Lakes and related waterways, British Columbia, the St. Lawrence River, and the coastlines of Newfoundland, Nova Scotia, Prince Edward Island, and New Brunswick.

b.   **Navionics Charts.** Navionics has several subscription offerings: Gold, NAV+, and Platinum+. They offer high-quality cartography and HD bathymetry but are not compatible with Garmin chart plotters.

c.   **C-MAP Charts.** Their range of high-quality charts include C-MAP Essentials and C-MAP 4D MAX+ C-MAP (MAX N, MAX N+ Reveal). Check out their products at www.c-map.com.

d.   **Lowrance Nautic Insight Charts.** These use NOAA ENCs and have quality bathymetry data. They are compatible with all Lowrance HDS, Mark, Elite-HDI, and Elite Chirp fish-finder and chart plotter devices.

e.   **Garmin.** They have their own BlueChart g3 and BlueChart g3 vision charts

f.   **AusENC.** These official vector Australian ENCs published by the Australian Hydrographic Society (AHS) provide coverage of Australian and Papua New Guinea waters. They are authorized for use in IMO-compliant electronic chart display and information systems (ECDISs) and can be used in compatible electronic chart systems (ECSs).

g.   **UKHO AVCS** (United Kingdom Hydrographic Office, Admiralty Vector Chart Service). They offer around 18,000 official Electronic Navigational Charts (ENCs) from hydrographic offices worldwide. See amnautical.info/avcs/.

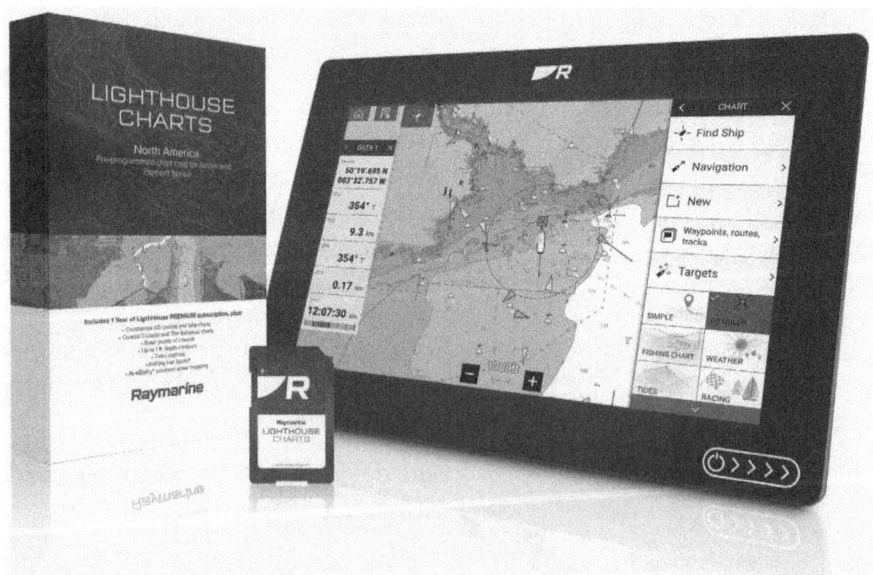

Figure 26-1. Raymarine Lighthouse Charting (*Courtesy of Raymarine*)

GPS Systems

The advances in electronic position fixing systems have been nothing short of spectacular. Equally, the fall in global position system (GPS) prices, making them affordable to everyone, has ensured that nearly everyone has one on board—now more as a data input to chart plotters and autopilots than a discrete instrument. Many are curious about the status of the various position fixing systems, given the rapid integration of GPS-based systems. Satnav, the Transit Satellite Navigation System, was switched off in 1997. Decca was shut down in 2000. Radio Direction Finding (RDF) is no longer used on small boats, although it is still used by rescue authorities in locating distressed vessels via your VHF transmission signal. LORAN-C was shut down by the USCG in 2010. The umbrella term is now "global navigation satellite systems" (GNSS), which covers all the various systems. The US GPS system went live in 1993. The Russians have GLONASS (Globalnaya Navigazionnaya Sputnikovaya Sistema), and the Chinese have the BeiDou Navigation Satellite System. India has a system for their local area, Indian Regional Navigation Satellite System (IRNSS), or NavIC. Japan has the QZSS system with four satellites in quasi-zenith orbits. Galileo, the European alternative to GPS, was created by the European Space Agency (ESA) and went live in 2016. Unlike the others, which are military controlled, this is a civilian service.

27.1 **Basic GPS Principles.** It is useful to understand what is accuracy when it comes to any positioning system.

        a. **Repeatable Accuracy.** This is defined simply as the ability to sail back to a position or waypoint previously fixed by the receiver and is vitally important with man overboard (MOB) functions. If any system is placed in a static situation and the positions are plotted at intervals, there will be a wandering of position. It is important to remember that all displayed positions must be used with the understanding that errors exist or can occur. After hitting the rocks, claiming the position fixing system was at fault is not a valid defense.

        b. **Predictable Accuracy.** This is less concise than repeatable accuracy. Essentially this the difference between the position from your position fixing system and that indicated on your chart, if you still use them. These errors are often induced by the vagaries of electronic fixing systems, such as signal propagation problems. These errors can be attributed to datum variations, inaccuracies in the electronically-derived position.

        c. **Chart Datum Variations.** Plotting a position on a chart has inherent errors. These errors can be caused by the GPS fix error or the transformation between GPS datum and chart datum. There may be a discrepancy that requires correction, and many charts carry appropriate notes. A wide variety of datums are used around the globe, and new charts are generally being compiled on WGS84 datum, the same datum used by GPS. At one stage of the 3,337 British Admiralty charts, sixty-five different datums were used, and it was reported that a typical error was a 140m offset in the Dover Strait. An official warning was issued many years ago not to rely on any position within 3nm of land in the Caribbean. Note that Datum NAS83 on US charts is virtually identical to WGS84 (GPS) datum on UK charts. All chart types, either paper or electronic, are derived from a mathematical model of the Earth. The World Geodetic Spheroid 1984 (WGS84), which is used by GPS and the several global navigation satellite systems (GNSSs), is a mathematical model, a compromise that is generally representative of our planet for practical navigation purposes, and is the default datum used in satellite positioning receivers. Most charts are not yet referred to this datum, so some incompatibility exists. Be aware of the danger this presents.

27.2   GPS Operation. The US Department of Defense (DOD) operates the GPS NAVSTAR system. The constellation consists of twenty-four satellites in six polar orbits so that at least four will always be visible above the horizon at any time. Twenty-one will be in operation, with three used as spares. GPS position fixing involves trilateration (often incorrectly called triangulation) of position from a number of satellites, satellite ranging to measure the distance from the satellites, accurate time measurement, the location of all satellites, and correction factors for ionosphere conditions. Normal GPS is called standalone, single point positioning (SPP) or standard GNSS. Standalone is the standard GNSS practice, and there are no error corrections made. The GNSS satellites provide the best standard signals available. Typical operation of a GPS receiver is as follows at power-up.

a.   **Initialization.** Turning the power on initializes with the closest visible satellite and ephemeris or almanac data (relating to the orbital parameters of the satellites), the approximate Doppler shifts of each visible satellite, receiver position, and other stored receiver data. This is downloaded into the receiver memory. A period of up to 20 minutes is sometimes required to stabilize a position and verify the status of satellites, availability, and other factors. After a GPS is switched off, the last position is retained in memory. If your position remains within 50nm, a position will generally be available within approximately a few minutes the next time the power is turned on.

b.   **Acquisition.** The receiver collects data from other satellites in view. Based on the data, it locks on to a satellite to commence the ranging process. Once the initialization process is completed, the receiver then commences a search for each visible satellite.

c.   **Position Fix.** Based on the data on position and time, the receiver triangulates the position with respect to the positions of satellites. Normally this will be displayed in two decimal places. Some units give three decimal places, but such accuracy is highly suspect and should be treated with caution; also reference the comments about chart datum errors.

27.3   **GPS Accuracy.** GPS accuracy has been the subject of widespread debate and controversy. GPS satellites broadcast their signals with a defined accuracy. Reception at your GPS receiver is subject to various factors, including satellite geometry, the effects of atmospheric conditions, and the quality of your receiver design. A smartphone GPS-enabled unit has an accuracy of 16ft (4.9m); when near trees and buildings, this accuracy further degrades. Users such as military and specialized commercial are able to improve the standard accuracy with the use of dual frequency receivers and augmentation systems that can give accuracy down to a few centimeters. The US government has a commitment to broadcasting GPS signals with a global average user range error (URE) of ≤ 6.6ft (2.0m), with a 95% probability, across all healthy satellites in constellation slots. The actual performance is generally much better than this. In April 2021 the global average URE across all satellites was ≤ 2.1ft (0.643m) 95% of the time, which is impressive.

    a.   **Precise Positioning Service (PPS).** This service is primarily for military use and is derived from the Precise (P) code. The P code is transmitted on the L1 (1575.42MHz) and L2 (1227.60MHz) frequencies using an encrypted navigation data message. Accuracy can be 3ft (1m) or less but can be up to 120ft (40m).

    b.   **Standard Positioning Service (SPS).** This service is for civilian use and is derived from the course and acquisition (C/A) code. SPS accuracy levels are typically around 10ft to 35ft (3m to 10m) but can be up to 130ft (40m).

    c.   **Selective Availability (SA).** The use of SA was officially abandoned in May 2000. The United States has indicated that it will not be reintroduced. The following is included for historical informational purposes. SA was the process of intentionally degrading positional accuracy by altering or introducing errors in the clock data and satellite ephemeris data. SA is characterized by a wandering position, and often a course and speed over the ground of up to 1.5 knots while actually stationary. When this was enabled by the US DOD, it added 150ft (50m) of error horizontally and 300ft (100m) vertically to GPS signals.

    d.   **Position Dilution of Precision (PDOP).** Complex mathematical algorithms are used to derive a DOP number. The four classifications of DOP are geometric or position (3D) dilution of precision (GDOP or PDOP–3D); horizontal dilution of precision (HDOP); vertical dilution of precision (VDOP); and time dilution of precision (TDOP).

    e.   **Horizontal Dilution of Position (HDOP).** Accuracy is determined by (geometric) horizontal dilution of precision (HDOP), which indicates the dilution of precision in a horizontal direction. The cause is poor satellite geometry, which is due to poor satellite distribution. It is generally measured on a scale of 1 to 10. The higher the number, the poorer the position confidence level; the lower the number, the higher the position confidence.

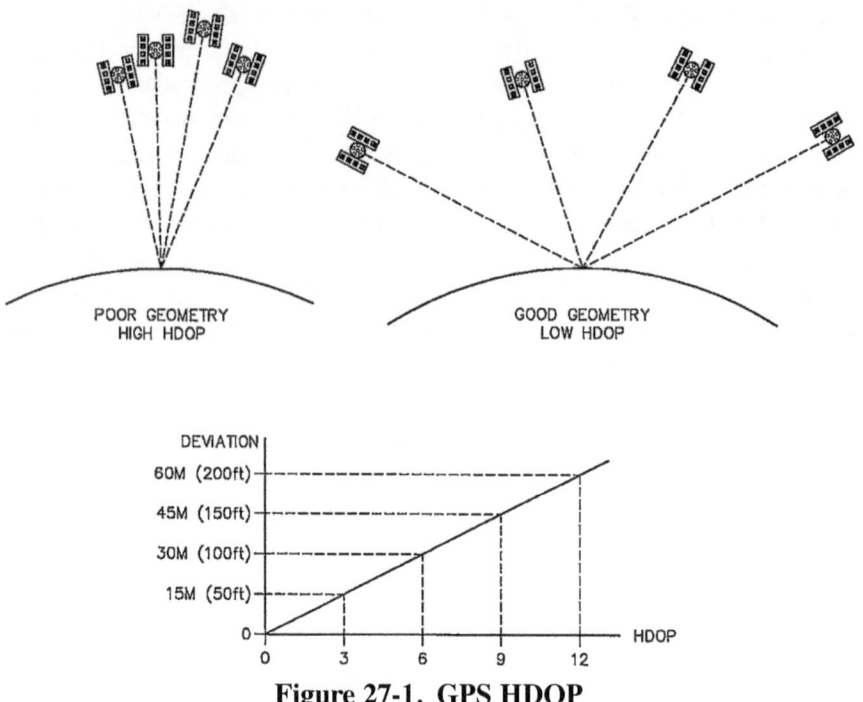

Figure 27-1. GPS HDOP

27.4   GPS Error Sources. The GPS has inherent errors that can decrease its accuracy as a navigation tool. These errors are in addition to the HDOP errors as described, and it is important to understand them. These errors can be caused by major solar storms and flares, interference or deliberate jamming or spoofing, satellite maintenance, and other factors.

    a.   **GPS Clock Errors.** Each GPS satellite has two rubidium and two cesium atomic clocks. These clocks are monitored against terrestrial atomic clocks. Based on this information, the entire GPS system is continually calibrated against Coordinated Universal Time (UTC).

    b.   **Ionosphere Effects.** Like radio signals, both ionospheric and tropospheric conditions can affect GPS accuracy. Errors occur in signal transmission times that can impose signal propagation delays. This signal refraction introduces timing errors that cause positional inaccuracies. Like radio propagation, this alters due to changes in atmospheric conditions, solar activity, etc. Errors can be as great as 60ft to 100ft (20m to 30m) during the day and 15ft (5m) at night.

    c.   **Multipath Effects.** This occurs when signals from a satellite traveling to a receiver arrive at slightly different times due to reflection or alteration. The effect is that positions may be derived from the "bad" signal, resulting in an inaccuracy.

    d.   **Satellite Integrity.** If the signal being transmitted from a satellite is corrupt due to a malfunction, it will have subsequent effects on position computations.

e. **GPS Week Number Rollover.** The GPS rollover occurs every 1,024 weeks, or 19.6 years, and is caused by programming limitations for storing dates. The first GPS rollover took place at midnight August 21–22, 1999; the second occurred midnight April 6-7, 2019, and may have affected older equipment and systems. Most manufacturers have fixes for these problems, although there are residual issues. Some older Furuno GPS products had a rollover on January 2, 2022, with resultant errors. A GPS rollover on many INMARSAT satellite communications TT mini C units happened on March 25, 2023, resulting in noncompliance for long range identification and tracking (LRIT), as these have an imbedded GPS module. Readers should be aware of the potential for errors on older equipment and systems, with generally unpublicized problems continuing to arise. The next GPS rollover is forecast for November 21–22, 2038.

27.5   Differential GPS (DGPS). DGPS ceased operation in 2020. The following is included for historical informational purposes. This system was designed to overcome the position errors with respect to selective availability, and it substantially reduced errors in position fixes. DGPS used a shore-based reference station located in an accurately surveyed location. The position was compared with the GPS derived position to produce an error or position offset. These errors may be due to SA or other previously covered issues. A correction signal to satellite range data (pseudo range differential) was then broadcast by radio beacon (285kHz to 325kHz in standard format RTCM SC104), which was then received by a radio beacon receiver, then incorporated into the vessel GPS receiver position computation to derive a final and more accurate position. The accuracy came down to around 6ft (2m) in some instances. DGPS services used UHF radio correction signals as well as spot beam, Inmarsat, and Marine Radio Beacon and International Association of Lighthouse Authorities (IALA) VHF transmitted data.

27.6   About GDGPS. Global Differential GPS (GDGPS) is a high-accuracy GPS augmentation system. This was a development of the NASA Jet Propulsion Laboratory (JPL) and was designed to support the real-time positioning, timing, and determination requirements of NASA science missions. The system has a comprehensive ground network of real-time reference receivers and real-time data processing software. GDGPS is able to provide sub-decimeter (<10 cm) positioning accuracy and subnanosecond time transfer accuracy at any location around the world independent of local infrastructure. Don't expect it to be available for yachts anytime soon, though.

27.7   Satellite Based Augmentation System (SBAS). This is a wide-area augmentation system that employs geostationary satellites to broadcast primary GNSS data. The SBAS satellites are provided with ranging, integrity, and correction information by a network of SBAS ground stations. The primary purpose of SBAS is to provide integrity assurance, and it increases accuracy and reduces position errors. Designated ground stations in various geographical locations calculate the differential corrections, not unlike the defunct DGPS system. These corrections are then uploaded to the satellite and subsequently broadcast to any receiver subscribed to receive it. There is much research being done in this sector, and one such project involves Australia and New Zealand collaborating to implement an SBAS.

The technology aims to improve unaugmented GPS accuracy and other positioning services from the current 16ft (5m) to 33ft (10m), to as little as 4in (10cm).

27.8   Wide Area Augmentation System (WAAS). This was discontinued in 2020 and the following is for informational and historical purposes. WAAS was primarily developed for aviation users. For marine users it was known as Nationwide Differential GPS (NDGPS), which was discontinued in 2020 pending the rollout of the new-generation GPS III satellites, and with SA shut down it wasn't deemed necessary. This also ceased in Australia in 2020, with the UK following suit in 2022. For aviation people WAAS comprises 25 ground reference stations across the United States. These stations monitor GPS satellite data. Two master stations on each side of the country collected data and derived a GPS correction message. The message included orbit and clock drift as well as signal delays. This ground base calculated ionosphere differential correction signal was then uploaded to one of two geostationary satellites and rebroadcast back to WAAS-enabled GPS receivers. Accuracy was typically around 10ft (3m) against 30ft (10m). The WAAS service was interoperable with other satellite-based augmentation system (SBAS) services, such as the European Geostationary Navigation Overlay Service (EGNOS) in Europe, Multifunction Augmentation System (MSAS) in Japan, and GPS Aided GEO Augmented Navigation (GAGAN) in India.

27.9   Galileo EGNOS System. This system is the European alternative to GPS. Created by the European Space Agency (ESA), it went live in 2016. The formal name is the European Geostationary Navigation Overlay Service (EGNOS). The EGNOS infrastructure consists of a ground network of 40 ranging and integrity monitoring stations (RIMS), 6 navigation land earth stations (NLES), 2 mission control centers and signal transponders deployed on 3 geostationary satellites (Inmarsat III satellite and SES ASTRA GEO satellites SES-5 and ASTRA-5B). The system transmits signals from two satellite transponders parked in geostationary orbit. The system enhances GPS signal accuracy with an average precision of 5ft (1.5m) over European territories. The system is intended to provide horizontal and vertical position measurements within 3ft (1m) precision. The system has four atomic clocks for timing, which are synchronized with the ground control station network. The system comprises twenty-four satellites plus spares located in medium earth orbit (MEO) and eight active satellites inserted in each of the three orbital planes that are inclined at a 56-degree angle in reference to the equator. There are developments for the introduction of a maritime service to EGNOS. Implementation is forecast for 2023. If vessels and seafarers want to utilize SBAS correction data and information, they will be required to upgrade their GPS receiver to a type approved for SBAS use. Similar SBAS systems under commissioning or deployment include Wide Area Augmentation System (WAAS) in the US, Southern Positioning Augmentation Network (SouthPAN) in Australia and New Zealand, Michibiki Satellite Augmentation System (MSAS) in Japan, as well as development programs in India, China, South Korea, Russia, and Africa.

27.10   GLONASS Positioning System. The Russian system operates using a thirty-one-satellite system (24 active and 3 spares). The constellation has three orbital planes with eight evenly spaced satellites in each plane. The claims are that the system is more accurate than GPS, and this has been proven in higher latitude locations such as the UK and

Europe. It is claimed to be more reliable because it is not subject to experimental shutdowns or position degradation, which is no longer true for GPS.

27.11   Space Weather and GPS Effects. The ionosphere is well known for the effects it has on HF and ham radio, as described in the communications chapters. A lot less known are its effects on GPS. It is an important source of range and range rate errors for users of GPS satellites where high accuracy is required. Ionosphere range error can vary from a few meters to tens of meters, with troposphere range error at a peak of 6ft to 10ft (2m to 3m). The ionosphere has a dispersive effect; it can alter rapidly in value, changing significantly over 1 day. In practice the troposphere range error does not alter more than plus or minus 10% over long periods. GPS signals pass through the ionosphere but suffer propagation delays. Ranging errors of tens of meters can occur in extreme ionosphere conditions and typically are 15ft to 30ft (5m to 10m). These generally equatorial events are often associated with plasma bubbles that characterize the unstable state of the equatorial ionosphere at night.

     a.   **Plasma Bubbles.** Ionosphere plasma bubbles are a natural phenomenon consisting of large regions within the atmosphere where there are large depletions of the ionosphere plasma. Subject to considerable research, they were first detected in Brazil in 1976 and continue to be a major problem within the offshore oil industry. Plasma bubbles are known to interfere with satellite communications in the frequency range VHF to 6GHz, and are known to interfere with GPS, causing position errors. The plasma bubbles are closely aligned with the Earth's geomagnetic field lines, along which they may extend thousands of miles (kilometers); across geomagnetic field lines, they measure 60 miles to 250 miles (100km to 400km). They occur after sunset and exist at night only, and they are generally more active during periods of maximum solar activity. The next predicted solar activity maximum is Solar Cycle 25 in July 2025.

     b.   **Scintillation.** Irregularities in the ionosphere produce diffraction and refraction effects, causing short-term signal fading, which can severely stress the tracking capabilities of the GPS receiver. Signal enhancements also occur, but the GPS user cannot get any benefit from brief periods of strong signal. Fading can be so severe that the signal level will drop completely below the receiver lock threshold and must be continually reacquired. The effects are called ionospheric scintillations, and the region can cover up to 50% of the Earth in varying degrees. Strong scintillation effects in near-equatorial regions are observed generally 1 hour after local sunset to local midnight. If precise measurement using GPS can possibly be avoided during 19:00 to 24:00 hours local time during periods of high solar activity and during months of normal high scintillation activity, the chances of being significantly affected are small. There are seasonal and solar cycle effects that reduce chances of encountering scintillation in near-equatorial regions. April to August, the chances are small of significant scintillation in the American, African, and Indian regions. In the Pacific region, however, scintillation effects maximize during these months. In September to March the situation reverses. The regions where the strongest scintillation effects are observed are Kwajalein Island in the Pacific and Ascension Island in the South Atlantic. The occurrence of strong amplitude scintillation is closely correlated with the sunspot number; in years with near minimum solar activity, there are little if any strong scintillation effects on GPS. The caution is where GPS is used for autopilot waypoint steering in these regions. Skippers should be alert for course changes, the GPS beeping as you lose lock, or simple unexplainable and short-term periods of inaccuracy.

c.   **Signal Spoofing and Jamming.** Deliberate jamming or interference with GPS signals by masking the signal with noise has been widespread by Russia, both before and during the invasion of the Ukraine. While common in Russian cities in the maritime space, it has been prevalent in the Black Sea and the Russian Far East. Such activities remain a risk near any conflict zone.

27.12   GPS Installation and Troubleshooting. The reliability and accuracy of your GPS position data system depends on proper installation. Now that almost all sailing yachts have GPS as their primary navigation source and, in many cases, it is input directly to charting and autopilot systems, it is essential that the antenna be properly installed.

a.   **Antenna Installation.** The GPS antenna should be sited so that it is clear of spars, deck equipment, and other radio aerials or antenna. Where possible the antenna should have as wide a field of view as practicable, without shadowing while being located as low as possible. In installations that utilize a stern arch or stern post with mounted radar, ensure that the GPS antenna is not within the beam spread of the radar antenna. Ensure that the location is not prone to fouling by ropes and other equipment that may damage the antenna.

b.   **Cabling.** Many GPS problems are a result of cabling problems. Power supply cables should be routed as far as practicable from equipment cables carrying high currents. Antenna cables should be routed well clear. It is extremely important for the antenna cable not to be kinked, bent, or placed in any tight bend radius. This has the effect of narrowing the dielectric gap within the coaxial cable, which may cause signal problems. Ensure that all through-deck glands are high quality in that they properly protect the cable and prevent water from going below. Glands include the Thrudex/Index range of cable glands, which enable the plug to be passed through along with the cable. Do not shorten or lengthen an aerial cable unless your manufacturer approves it.

c.   **Connectors.** Ensure that all connectors are properly inserted into the GPS receiver. Ensure that screw-retaining rings are tight, because plugs can work loose and cause intermittent contact. The coaxial connector from the antenna into the receiver should be rotated properly so that it is locked in. External antenna connections should be made water resistant where possible. Self-amalgamating tape is useful for doing this. If you have to remove and refit an antenna connector, ensure that you use considerable care and assemble the connector in accordance with the manufacturer's instructions. Use a multimeter on the resistance range, and check the center pin to shield resistance. Low resistance generally means a shorted shield strand. Resistance is typically 50Ω to 150Ω.

d. **Grounding.** The ground connection provided with some GPS units must be connected to the instrument ground system in accordance with the manufacturer's recommendations.

e. **Power Supplies.** A clean power supply is essential to proper operation. Either use an in-line filter or install suppressors across "noisy" motors and the alternator. The power supply should not come from a battery used for engine starting or used with any high-current equipment, such as an anchor windlass or electric toilet. Note that many cheaper, unsuppressed fluorescent lights and low quality LED lights also cause interference that may cause data corruption.

f. **GPS Maintenance.** Many problems can be identified and rectified before the system fails. Perform the following routine maintenance checks:

(1) Check the antenna to make sure the connections are tight and the plugs are in good condition. Ensure that the antenna is mounted vertically and has not been pushed over, which is a common problem.

(2) Ensure that all connectors are properly inserted. In particular, examine the external aerial connector for signs of corrosion, especially the outer shield braiding.

(3) Many GPS units have internal lithium batteries with a life span of only around 3 years, so ensure that the battery is renewed prior to any voyage.

(4) Make a hard-copy list of all your waypoints for reference and reprogramming if required.

g. **GPS Troubleshooting.** You should attempt some basic troubleshooting before you call a technician or remove a GPS unit for repair by the service agent. Many problems are related to peripheral equipment rather than the unit; simple checks of the following may save considerable sums of money:

(1) **Large Fix Error.** The GPS system may be down, or a satellite may be shut down. Check your NAVTEX MSI transmissions or other navigation information source for news of outages. The HDOP may simply be excessive due to poor satellite geometry in your location. With older multiplex receivers, loss of signal may be a problem in heavy sea states.

(2) **Small Fix Error.** Errors that are not significantly large but consistently outside normal accuracy levels are attributable to a number of sources. The signal may be subject to an excessive number of atmospheric disturbances, such as periods of extensive solar flare activity. This may be confirmed by similar HF radio reception difficulties, which suffer propagation problems. The antenna connections and part of the installation may have degraded, so check the entire system. Make sure antenna orientation is vertical and not partially pushed over. Check that some antenna shadowing has not been introduced.

(3)   **No Fix.** This is often caused in receivers by loss of a satellite view or when a satellite goes out of service. Another common cause is the antenna being pushed over to horizontal, so check that it is vertical. Antenna damage after being struck by other objects is another major cause of a sudden fix loss. Check all cables and connections. If these show no defects, a check of all initialization parameters may be necessary; then reboot and see what happens. If those check out, the receiver and antenna may require shore servicing.

(4)   **Data Corruption.** This error is often caused by power supply problems. Check whether the incident coincides with engine or machinery run periods. Radiated interference is also a possibility, often from radio equipment. A lightning strike with resultant electromagnetic pulse can cause similar problems. Another quite common cause of data corruption is "fingers." Has another person unfamiliar with operating the GPS altered configuration parameters, such as time settings or altitude? Poor system grounds have the capacity to do this along with cable installation and crosstalk.

Table 27-1. GPS Troubleshooting

Symptom	Probable Fault
Large fix error	GPS satellite system down High HDOP Severe atmospheric problem Satellite acquisition loss (heavy weather) Antenna pushed over or damaged
Small fix error	Atmospheric propagation problem Antenna misalignment
No fix	GPS satellite system down Network fault, connector Antenna fault Antenna cable or connector fault Antenna pushed over to horizontal Antenna "view" obstructed
Data corruption	Power supply interference Radiated interference

AIS Systems

28.1   About AIS. The Automatic Identification System (AIS) is probably one of the most significant maritime safety developments since GMDSS was implemented. It is important to note that AIS is not part of the GMDSS environment. AIS is part of the International Convention for Safety of Life at Sea (SOLAS), and while not technically part of GMDSS, it is effectively part of that system now due to the introduction of the AIS search and rescue transponder (AIS-SART). These are allowable as an alternative to the search and rescue radar transponder (SART) under GMDSS. When used with a display, AIS significantly improves situational awareness and collision avoidance for all vessels, and AIS gives visibility in all weather conditions, including fog and heavy precipitation. Although not all vessels are equipped with AIS, there are an estimated 500,000 ships or more with AIS installed. It is considered an aid to navigation and is not an automated collision avoidance system, so the Convention on the International Regulations for Preventing Collisions at Sea, 1972 (COLREGS 72) still applies.

28.2   AIS Functions. An AIS unit automatically transmits data at defined time intervals, and this is called dynamic data. This includes course, heading, speed, and the vessel name, dimensions, cargo information, and navigational status such as whether it is at anchor or under way. The AIS is a very high frequency (VHF) radio broadcasting system that transfers data packets over a VHF data link (VDL). Vessels and shore-based stations equipped with AIS are able to transmit and receive this data for display on a chart plotter, computer, or other display device. In commercial vessel installations, AIS units are interfaced with Electronic Chart Display and Information System (ECDIS) and radar displays. When interfaced with a radar, it provides information on targets and also to an automatic radar plotting aid (ARPA). An AIS system uses compass inputs for heading, rate of turn (ROT), and other information. AIS includes information on vessel speed, destination and estimated time of arrival (ETA), and Course over the ground (COG), and this information is available on various internet tracking sites, such as shipsnow.com and www.marinetrafic.com.

Figure 28-1. i70 Instrument Head AIS
(*Courtesy of Raymarine*)

28.3   AIS Principles. Each ship-based AIS station comprises a single VHF transmitter; two VHF receivers, AIS 1 and AIS 2; and a single VHF DSC receiver (Channel 70). These are connected using a marine electronic communications link and related sensor systems. System positional data and timing are derived from a global navigation satellite system (GNSS) receiver. AIS employs what is called a time-division multiple access (TDMA) system to share the VHF frequency; this is called the VHF data link (VDL). Two frequencies dedicated to AIS are used: AIS 1 (161.975MHz), denoted as Channel 87B, and AIS 2 (162.025MHz), denoted as Channel 88B. Each frequency within the VDL is segregated into 2,250 time slots. These are repeated at intervals of 60 seconds. An AIS unit transmits packets of data in these slots, and every vessel that is listening to these slots within the AIS range can receive and display the information. There is an International Maritime Organization (IMO) performance standard for AIS with specific requirements, including ship-to-ship mode for collision avoidance. This facilitates countries having data about any vessels and the cargo being carried and acts as a vessel traffic system (VTS) management tool for ship-to-shore traffic management.

28.4   AIS Types. There are two classes of AIS installed on vessels: Class A and Class B. Other AIS types are used for shore AIS base stations and AIS navigation aids, search and rescue (SAR) aircraft AIS, and AIS-SARTs. Class A AIS is mandated by the IMO for all vessels of 300 gross tonnage and above that engage in international voyages; commercial cargo vessels of 500 gross tonnage and above that do not voyage internationally; and all passenger vessels capable of carrying twelve or more passengers. Class B AIS is for non-SOLAS vessels and is not mandated by the IMO. These AIS units have less functionality than the Class A units. They are the types used by yachts and other pleasure vessels. The AIS-SART allows SAR transmitters with AIS to help determine the location of a vessel in distress as part of the Global Maritime Distress and Safety System (GMDSS). For the purposes of this chapter, we will look at only Class B AIS systems, as these are more relevant, as well as being lower cost and lower power systems. Although Class B units have less functionality that Class A units, they can communicate with Class A and other AIS units.

28.5   AIS Advantages and Limitations. In addition to aiding in collision avoidance and improving situational awareness and safe navigation of vessels, AIS provides several important advantages. This includes the ability to look around the bend, which is very useful in trafficked rivers and estuaries, as AIS uses VHF and this is not line of sight. The ability to know the course and actions of other vessels is important, and AIS allows you to know the heading of other vessels, allowing quick assessment in the case of rapid course alterations. AIS allows identification of other vessels, with the vessel name, call sign, and maritime mobile service identity (MMSI) number.

Negatives include that not every ship has the correct data inserted or programmed. Also, as AIS uses GNSS positional data and if that is dysfunctional, or if the AIS data is corrupted. Not all vessels keep the AIS switched on, and these ships often do so in high piracy areas; known as silent mode, this is a viable strategy off Somalia, in the southern Caribbean, and in other potentially dangerous waters. (Read more about pirates in my book *Piracy Today*.) This method also is employed by black, or dark, vessels engaged in illegal activities. Going dark is a method used by illegal fishing boats, sanction-busting ships, and other illicit vessels. While switching AIS off and going dark might be the default position for those engaged in illegal activities, those vessels are not free from surveillance. Satellite imaging is one method to reveal these vessels, but maritime authorities are also adopting military-level technology of radio frequency data acquisition. Specialist companies use satellite receivers to monitor large amounts of RF spectrum for all types of radio emissions. The satellites are used to geolocate the radio emission source from radio communications emissions. Each ship has a unique RF signature or fingerprint, and you are visible no matter the weather. When synthetic aperture radar and imaging are used, identification is easy. If you even momentarily switch on your AIS, the data and all the information are sampled. When AIS was implemented in 2002, the concepts and new technologies now installed on modern vessels, such as cybersecurity, Internet of Things (IoT), artificial intelligence (AI), large-scale commercial satellite internet communications, and smartphones, were not in existence. The downside is that AIS does not have data security or the ability to interact with satellites; the biggest deficiency is its inability for data encryption.

28.6   AIS Installation and Operation. If an external AIS antenna is installed at 50ft (15m) above sea level, the AIS receiving station will get information and data from all other AIS vessels within a 15nm to 20nm range, which equates to a Class A transponder output. Range is all about the antenna type, the altitude, and weather conditions, such as high humidity. The higher the antenna, the better the reception. Class B AIS units have three RF output power levels: 2W, 5W, and 12W. Setting your unit at its lowest setting will still give you about a 10nm range. If the RF propagation conditions are not optimal, higher settings may need to be used. The AIS reception range is limited to that of a VHF radio, which is 10nm to 40nm. VHF is effectively line of sight, although reception distances can be greater than this, as radio waves tend to bend a little around the planet's surface. Most normal VHF aerials will give decent reception, but the better the aerial, the better the performance. Many aerials come with a preinstalled down-lead, usually with a PL-259 connector installed. Some have a socket that requires a cable and connector to be installed, and AIS receivers often have BNC connectors, which require adapters. The required coaxial cable is RF 50Ω coax; RG-213 or LMR195 coax is preferable to the smaller RG-50. Don't use TV coaxial cable. The lower the wiring losses, the better the performance. Many AIS units now have inbuilt aerial splitters so that a single VHF aerial can be used for both devices, although on my own boat I have opted for two Glomex "Aloud" series VHF aerials. Course, trajectory, and closest point of approach (CPA) are calculated by your AIS unit, which processes each packet of data that is received. With each segment of data, the CPA also alters. If a Class B–equipped vessel is under way with a speed of less than 2 knots, the nominal reporting rate decreases to 3 minutes, so a reliable CPA calculation is difficult. It is always worth monitoring the rate of turn (ROT) data from Class A–equipped vessels, as this indicates whether that vessel is altering course. If you have radar, you can cross-check using the "old-fashioned" method of Mark I eyeballs and hand-bearing compass to get a relative bearing.

28.7   AIS Developments. Specifications for Class B AIS transmitters were modified in 2006. The Class B AIS broadcasts on the same two channels at 2W power level, with a static reporting rate of once every 30 seconds. This employs the carrier-sense time-division multiple access (CSTDMA) scheme and is denoted as Class B/CS. Class B/SO units arrived on the scene in 2013. These units use the self-organized time-division multiple access (SOTDMA) system. These units transmit AIS data every 5 to 30 seconds at a power of 2W or 5W. Class B/SO transmitters were devised for non-SOLAS-class vessels, with improved output power and increased range. The SOTDMA channel-sharing system ensures that more transmissions are heard by other AIS stations, especially in congested waterways. That brings us to 2019, when Vesper Marine, now owned by Garmin, launched their Cortex safety and communications platform. They have converged AIS, digital selective calling (DSC), VHF, wireless, and cellular into a software-based platform that incorporates a Class B/SO transmitter. This electronic black box is installed and connected to one or more handsets along with two smartphone apps. The system uses a touchscreen and appropriate buttons. These innovative solutions are the sign of developments to come in the next decade. Given that the two VHF channels currently used, 87B and 88B, have limited bandwidth, solutions are being worked on to address the expanding requirements for cybersecurity, increased functionality, and growing number of installations and current system limitations.

28.8   VDES System. The latest development is the VHF data exchange system (VDES). When it is finally operational, this bidirectional communications system will enable satellites, vessels, and shore stations to receive and transmit high-speed data. When the system is launched in the next few years, you will be able to purchase a VDES module that will bundle various communication frequencies. This will lead to AIS becoming simply a collision avoidance advisory system without all the extra data. VDES comprises four components: AIS, VDE satellite, VDE terrestrial, and ASM channels. The system will be based on eighteen frequencies, with two being for long-range satellite AIS, two for the standard AIS system, two for application specific messages (ASMs), six satellite uplink channels, and six satellite downlink channels. When two vessels are close, there will be an automatic data exchange on future routes as well as position. Shore stations will be able to broadcast digital updates that include safety text messages and other relevant information. Cybersecurity will be enhanced and VDES will have the ability to transmit encrypted positions that restrict spoofing by those undertaking illegal activities.

28.9   AIS Troubleshooting. Troubleshooting AIS problems comprises two elements: first, the aerial or antenna, coaxial and power cable plugs, sockets, and connections; and second, software, setup, configuration, and operation. Most AIS units have LED status and fault indication, and the manual should be consulted for descriptions.

Table 28-1. AIS Troubleshooting

Symptom	Probable Fault
No AIS targets displayed	Data or network problem VHF antenna fault GNSS (GPS) antenna fault GNSS (GPS) antenna view obstructed AIS unit setup incorrect No AIS targets in range
VSWR alarms	Antenna incorrect for AIS unit Antenna open or short circuited Software version is out of date

Symptom	Probable Fault
AIS unit does not power up	Supply circuit breaker open Low battery voltage Input power plug not inserted correctly Software issue, upload latest version
Erratic or missing data	Configuration or software issue Hardware interconnection faults

28.10 AIS-SART. An AIS search and rescue transponder (SART) is a homing beacon designed for emergency location use. When activated, the AIS-SART will transmit a GPS position using AIS within a dedicated SART message. This message is recognized by the AIS unit display on other vessels as an emergency message and triggers an alarm. Units are available in vessel versions or as personal locator beacons (PLBs). An AIS-SART at water-line level with a 2W output rating would have a range up to around 5nm; a PLB would have a 2nm to 3nm range.

Figure 28-2. AIS 7000 Axiom 9 Target Summary (*Courtesy of Raymarine*)

Instrument Systems

Integrated instrument systems have led the marine electronics revolution due to the rapid advances in microprocessor computing power, networking, and associated software developments. Networking and the introduction of the high-resolution multifunction display (MFD) have all contributed to systems that can calculate and display increasing amounts of parameters and data. Radar images can be overlaid with electronic charts, and helm-mounted MFDs can have split screens displaying almost everything from instrument and engine data to radar and electronic charts to a fish-finder. The principal players in this space are Raymarine, Navico (Simrad), Brookes & Gatehouse (B&G), Tacktick, NASA Marine, Ockum Instruments, CruzPro, Autonnic Research, NKE Marine Electronics, and Garmin. Many of these brands are now under the same corporate umbrella.

29.1   Systems Architecture. The basic architecture of an instrument system varies among manufacturers, and the BoatSmart system from Airmar is a big step forward. It is a sensor management system and a multinetwork protocol gateway. You can plug multiple analog sensors into the module, which then coverts all data to NMEA 2000 and outputs to the network and MFDs. It has wireless Bluetooth access for a browser-based interface for sensor control and configuration, and wireless communication with MFDs. Visit Airmar at www.airmar.com.

a.   **Discrete Instrument Systems.** These systems have a transducer serving each dedicated instrument head and powered separately. The head processes and displays the information. Data can be exchanged between each instrument on a dedicated network using NMEA or manufacturer communications protocols for computing related data.

b.   **Networked Systems.** These systems interface single or multifunction instrumentation displays, all transducers, radar, GPS, chart plotter, autopilot, and fish-finder to a single network. All data is transferred over the network. Transducers are connected to local pods, which are then connected to the network. These systems have an intuitive menu structure and split-screen capabilities and use a simple keypad with trackball and point-and-click control.

c.   **Active Transducer Systems.** Each transducer has a microprocessor embedded in it where all raw sensor data is processed. The transducers are all connected by a single cable network, and all data is available through user-definable instrument displays. These multifunction displays can be configured with simple keystrokes.

d.   **Wireless Systems.** Each transducer transmits data back to the processor. Raymarine has this system and now owns Tacktick, which pioneered it. Airmar, the leading innovator and manufacturer of transducers for the marine market, has recently launched a smart transducer, incorporating an innovative new paddlewheel design, along with a depth and temperature transducer. While it connects to the NMEA 2000 network, it has wireless access for calibration, setup, and so on with a smartphone app.

29.2 **Device Interfacing.** Interfacing is the interconnection of various electronic equipment and systems so that digitally encoded information can be transferred between them and used for processing tasks or display. There are a number of considerations for manufacturers, including the type of physical equipment involved, such as connectors and cable; the voltages, impedances, and current values; and signal timing. At a more technical level are the data structure and transfer rate and the protocol, which determines the information to communicate, the time to communicate, the frequency, and error correction. The data messages must also have compatible structures and content. The US National Marine Electronics Association (NMEA) devised the first general digital standard in 1980, denoted as NMEA 0180. Then followed NMEA 0182 and NMEA 0183; now we have NMEA 2000. A good resource is NMEA Boater at nmeaboater.com.

29.3 **NMEA 2000.** Sometimes referred to as NMEA2K or N2K, the NMEA 2000 interface standard has been developed in conjunction with the International Electrotechnical Commission (IEC). Version 3.000 of NMEA 2000 was released in April 2022. It is a low-cost, bidirectional serial data protocol permitting multiple talkers and listeners to share data. NMEA 2000 is based on the controller area network (CAN) protocol originally developed for the auto industry. The NMEA 2000 protocol is organized around a bus topology, which enables the building of a network of connected instruments and devices. NMEA 2000–compliant instruments are simply connected to a backbone that comprises one central cable with T-connectors and drop cables. This enables sharing of data between all elements of the system. If one unit fails, the network continues to operate and communicate data. This type of system has a requirement for a single 120Ω termination resistor located at both ends of the bus. As these resistors are parallel connected, a properly terminated bus has a total resistance of 60Ω. The electrical power supply cables are connected close to the backbone center for proper operation of the network. Compatible devices include GPS, depth-sounders, navigation instruments, wind instruments, chart plotters, and autopilots.

    a. **NMEA 2000 Networking Basics.** There are basic rules and requirements to adhere to when setting up your NMEA 2000 network:

        (1) The backbone cable length must not exceed 328ft (100m).

        (2) The maximum single drop cable length must not exceed 19ft (6m).

        (3) The total length of all the drop cables must not exceed 55ft (76m).

        (4) The maximum number of devices connected to the network must not exceed fifty.

        (5) The network must be properly terminated. There should be one terminator installed at each end of the backbone.

        (6) The supply voltage must be in the 9V to 16V range, unless it is a 24V system.

(7)   The voltage drop from each end of the network must not exceed 1.5V.

(8)   The supply current for the network must not exceed 3A or 60 load equivalency number (LEN).

b.   **What is the load equivalency number (LEN)?** This is always listed with NMEA 2000 devices. It is denoted as a whole number that specifies the current value of the device that it draws from the network. One LEN equates to 50mA, which means a device that consumes 200mA from the network has a LEN of 4. When planning your system, consider the maximum power requirement.

c.   **Network Voltage Drop.** Like all electrical wiring, there is a voltage drop when current flows. When devices are connected to the backbone and the device drop cables, there is a voltage drop. This can cause voltages at each extremity of the network to be different. Overall voltage drop must not exceed 1.5V or data errors can result. The general formula for calculation of voltage drop is given as Volt Drop = LEN × Backbone Length × 0.0006. If maximum voltage drop is exceeded, an additional power cable may be required.

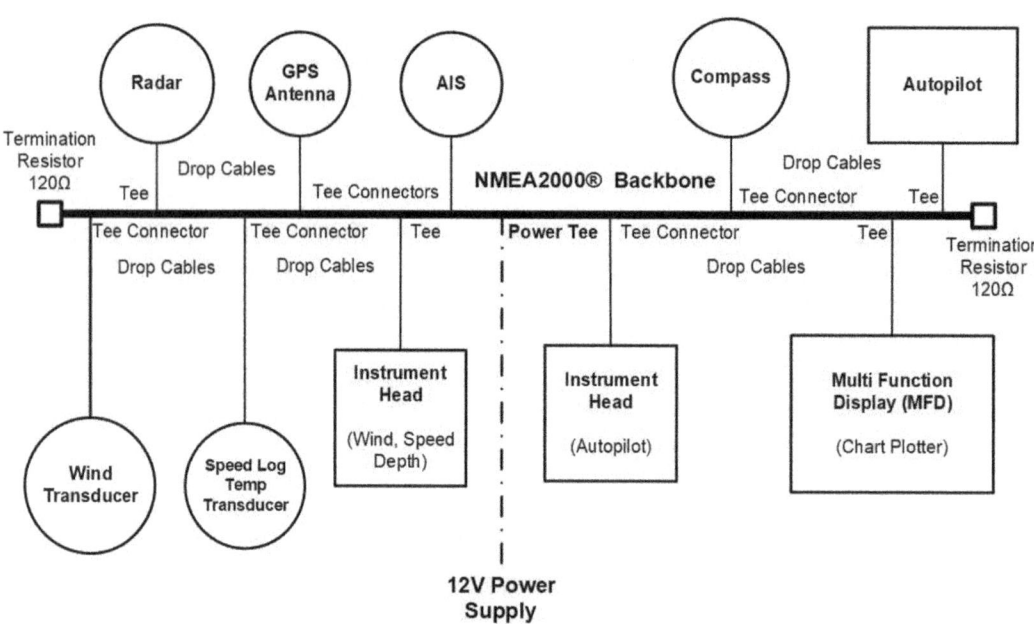

Figure 29-1. NMEA 2000 Backbone

29.4   NMEA OneNet Ethernet Standard. The National Marine Electronics Association introduced the NMEA OneNet Standard, Version 1.000, protocol in 2020. This is based on Internet Protocol Version 6 (IPv6) and the IEEE 802.3 Ethernet local area network (LAN). The protocol is specifically designed to complement NMEA 2000 networks, as it facilitates high-capacity data transfer and is a standard method for sharing NMEA 2000 data over a LAN. OneNet provides a common network infrastructure for marine devices and services on IPv6. Marine data network standards currently have bandwidth capacities that are below 1Mbps. OneNet has the capacity for hundreds or thousands of megabits per second and much greater bandwidth. The rationale for this is that additional bandwidth is required for the transfer of unprocessed sensor data from sonars and radars. It allows the connection of up sixty devices. Speed is a big advantage, and it is claimed that OneNet is 40,000 times faster than NMEA 2000. A power over Ethernet switch allows powering of all devices where NMEA 2000 required device-specific power supplies are specified by the device manufacturer.

29.5   NMEA 0183. This standard was designed to enable transfer of a variety of information between position-fixing systems, radar, compass, plotters, and autopilots, as well as any other systems either sending or requiring data. NMEA 0183 is what is called a single talker, multiple-listener architecture. Compliance with these standards is voluntary, and there are cases where implementation of the standards has been technically flawed and communication poor or impossible. The NMEA 0183 comprises standard message sentences. They may be divided into input and transmit sentences, where many are simply transmitted as inputs to processors; other information is transmitted to appropriate systems or displays. Message sentences have the following formats: HDM = Compass Heading, Magnetic; WPL = Waypoint Location; XTE = Cross Track Error. There are many parameters, and listing them all does not serve any practical purpose. One important recommendation was the use of opto-isolation on circuits. The opto-isolator is commonly used in many high-noise environments, and an LED and phototransistor are used to provide total electrical or galvanic isolation. This prevents any transfer of electrical noise into equipment circuits. Version 4.11 was released in 2018.

29.6   Other Communications Protocols. The trend has been for implementation of in-house communications protocols. The main reason is that fast broadband data transfer is a requirement for video and graphics images such as radar, plotter, and fish-finder screens. Some of the major company in-house protocols being used are as follows:

a.   **Furuno NavNet.** This has evolved, and the VX series is now current. This uses an Ethernet 10BaseT (twisted pair) system, which is common in many shore data systems. Systems have a star topology, with each device having a separate set of wires radiating out from the hub. When a fault arises, it is contained to that one device or cable. Ethernets have high data rates, and cables must be unshielded twisted pair (UTP) standard to ensure data integrity. Make sure cables are routed well clear of fluorescent lights, transformers, and so on to avoid interference.

b.   **RayNet.** From Raymarine, this is an address-based data packet network capable of handling high speeds of 10/100/1,000Mbps. Ethernet networks are designed with a star topology. These systems having three or more nodes that require components such as a hub, router, or switch to enable all point-to-point connections.

c.   **SeaTalk.** This is a Raymarine protocol that allows data exchange between devices such as radar, chart plotter, and navigation instruments. When using between three and eight devices, a network switch will be required. This incorporates five ports with four RayNet ports that will support 10 or 100Mbps network speeds.

d.   **Controller Area Network (CAN).** This is a fast serial bus designed as an efficient and reliable link between sensors and actuators. The CAN uses a twisted pair cable for communications at speeds up to 1Mbps with up to forty devices connected. Originally, Bosch developed the electronics standard for automobiles. The system requires an interface for NMEA based communications. Features include any node access to the bus when the bus is quiet, and use of 100% of the bandwidth without loss of data and automatic error detection, signaling, and retries.

29.7   Interface Installation. Virtually all problems with interfacing occur at installation, and most systems are simple plug-and-play, meaning errors are hard to make. The majority of faults are related to the following, although Raymarine and others are using waterproof connectors. These replace the standard RJ45 connector with a vibration resistant and waterproof twist pin design that is more resilient to boat conditions.

a.   **Cable Connections.** Unless an equipment manufacturer supplies the interface cable and connector, ensure that the correct pins are used on the output port connector. These vary between equipment and manufacturers. Check with the supplier, or get them to make up the cable and connector. All connections should observe the correct polarity with respect to ground references. Incorrect connections mean no signal transfer. If the system is fiber-optic, ensure that the connection is properly inserted, rotate to lock, but do not force the connector on.

b.   **System Setup.** At commissioning, ensure that the appropriate interface output ports are selected with the correct NMEA or output format selected. In many cases, problems are directly attributable to this, so carefully go through the setup procedures in your manual.

c.   **Electrical Cables.** All cables should be shielded, twisted pair unless stated otherwise. Using other cables may lead to data corruption due to induced noise from adjacent electrical cables and radio transmissions. Flat cables are generally untwisted; round ones are generally twisted. Use only CAT 3, CAT 4, or CAT 5 cables within data networks. This has a 100Ω impedance and wire size of 22AWG to 26AWG ($0.34mm^2$ to $0.14mm^2$).

d.   **Electrical Grounding.** Ensure that screens and reference grounds are properly terminated and connected where terminations are required. In many cases, data corruption occurs, or the system simply does not work.

e. **Cable Drain Wires.** Many instrument and data cables include drain wires. They are found on many devices, including radar scanners and other devices. Drain wires are generally tinned copper to prevent galvanic corrosion reactions with the dissimilar aluminum foil screen and copper conductor. Cables are screened for electromagnetic protection. The drain wire provides a low-resistance and continuous connection with the cable foil shielding and effective grounding of electrical noise.

29.8   Interface Cable Designations. There are a number of variations in designating interface cable connections. The standard NMEA terminology is "signal" (positive) and "return" (negative). NMEA output port variations are as follows. Equipment NMEA ports are configured in what is termed a "balanced pair," with both wires carrying the signal. The signal level, the difference in voltage between the pair, is also known as a differential data signal. The connection of wires is simple: The transmitting device has the transmit connected to the receive port of the other. The receive port is similarly connected to the transmit port of the other.

a. **Data Signal Output:** Data O/P; Tx; Tx hot; A Line; Positive data; Signal O/P; NMEA O/P; NMEA Sig Out; O/P Sig; Data Out; Tx-ve; Tx Data O/P.

b. **Data Return Output:** Gnd; Tx Cold; Ground; Signal Rtn; Return Out; O/P Return; NMEA Rtn; Data Rtn; I/P Gnd; Ref; Negative.

c. **Data Signal Input:** Signal I/P; NMEA Sig In; I/P Sig; NMEA I/P; Rx Data I/P.

d. **Data Return Input:** Signal Rtn In; Signal Rtn; I/P Rtn; NMEA Rtn; Gnd; Negative; Reference; Ref.

29.9   Instrument Selection Criteria. Selecting a system requires consideration of a number of factors. Consider location mounting and the use of purpose-made pods and housings from specialist companies such as NavPod (https://navpod.com). It is not uncommon to confront a confusing array of digital displays, and the aviation and vehicle industries have invested heavily in researching easier assimilation of data from a situational awareness and safety perspective. Instrument display types are noted below; many are combination LCD and analog and are now made to Federal Communications Commission (FCC) and CE electromagnetic compatibility (EMC) standards.

a. **Multifunction Network Displays.** These have changed over the years as technology advances. Video graphics array (VGA), super video graphics array (SVGA), and thin film transistor (TFT) liquid crystal display (LCD) screens are being superseded by high-density in-plane switching (IPS) displays. IPS displays are used, as they have both color accuracy and consistency along with wide-angle viewing capability. These are wide super VGA (WSVGA) and wide extended graphics (WXVGA) displays. The display resolution is the width and height dimension measured in pixels. The new generation Raymarine MFD touchscreens have a hydrophobic coating called HydroTough for protection. Brightness levels are between 1,500 and

1,800 nits. The nit is derived from the candela per square meter and is the unit for illuminance. The term is derived from the Latin verb "to shine" (*nitere*). The Raymarine Axiom+ epitomizes the development innovations in multifunction fluid-touch and hybrid-touch displays. They are optimized for high-speed data processing and incorporate powerful quad-core processors to handle all chart and other powerful image-handling data requirements. They have very wide viewing angles, which is important on any boat. The award-winning Garmin GPSMAP 8616xsv is the MFD that is all things for all types of boating. The system is available in three sizes, with wide viewing angles, sunlight readability, and high definition. While the fishing community gets full fish-finder input, it utilizes the Garmin SAILASSIST features, including laylines, race start line guidance, enhanced wind rose, heading and course-over-ground lines, true wind data fields and tide-current-time slider, wind angle, set and drift, wind speed, polars, and much more.

b. **Digital Liquid Crystal Display (LCD).** Many displays used to have a seven-segment display with those characteristic chunky numerals. This has changed with multisegment displays and a dot matrix field for graphics. Some displays are difficult to read at wide angles or in bright sunlight, and you should consider this, although technology is improving, with higher contrasts and wider viewing angles. All units generally have a multilevel backlit illumination system. Displays average around 25mA power consumption without illumination and up to 400mA with backlight on.

c. **Analog Display.** The analog display is still used on some instruments, and it is practical on ergonomic grounds, as it can make overall instrumentation displays easier to monitor. It is often a needle position or change rather than a value that is monitored. I personally have a preference for analog displays; they are easier to see, particularly on depth displays going into coral reefs with the sun behind. Analog is now digital on some Simrad, Raymarine, and other manufacturers' instruments—not a mechanical needle but a digital or virtual one with virtual needles.

**Figure 29-2.
i70 Instrument Heads**
(*Courtesy of Raymarine*)

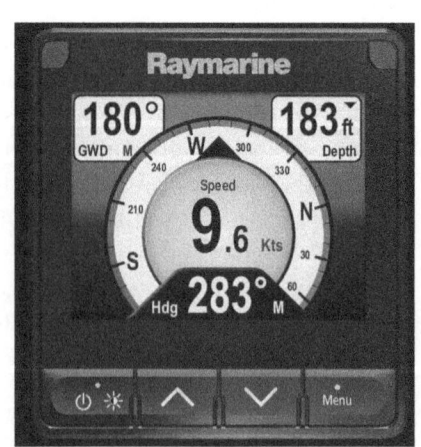

ELECTRONIC COMPASSES

29.10   Fluxgate Compasses. A fluxgate sensor detects the Earth's magnetic field electronically, sampling it hundreds of times per second. The sensing part of the compass consists of coils mounted at right angles in a horizontal plane. Each coil is fed with precisely controlled current that is subsequently modified by the Earth's magnetic field. The processor compares the signals within each coil, automatically correcting for variation. The resulting analog output is subsequently converted to digital signals for processing. Leading manufacturer KVH has the C100 Compass Engine, a standalone sensor that outputs heading data in six user-selectable digital or analog formats. The C100 is a microprocessor-controlled fluxgate compass that incorporates a toroidal fluxgate sensor element with a small electronics processor board.

29.11   Electronic and Solid-State Compasses. These compasses, made by KVH, are not fluxgates and are entirely solid state. The purely electronic sensing overcomes the problems of analog-to-digital conversion by the output and processing of a digital signal. This combines a digital magnetic compass and a three-axis gyro sensor. The KVH 1000 measures the boat magnetic deviation and performs automatic compensation to an accuracy of 0.5 degrees. It has an LCD with backlight and rotating compass rose and interfaces with other NMEA 0183–compatible devices. Current consumption is 63mA without backlight and 125mA with maximum lighting. Simrad has the Precision-9 compass, which provides very accurate magnetic heading, rate of turn, pitch, roll, and heave (heel and trim) data. It has an array of solid-state, rate-stabilized sensors that measure motion and orientation on nine different axes. It has NMEA 2000 connectivity. Raymarine has the innovative EV Sensor Core, which comprises a nine-axis gyro sensor that monitors all boat motions of pitch, roll, and yaw and the boat heading. This consists of a three-axis digital accelerometer, three-axis digital compass, and three-axis gyro digital angular rate sensor. The accelerometer measures boat speed acceleration forces, and the digital angular rate sensor senses the roll, yaw, and pitch angular rate of rotation. It is incorporated into their Evolution series autopilot systems; it utilizes the Automagic system, and no calibration is required. In the commercial and offshore maritime space, they are called motion reference units (MRUs) and measure pitch, roll, heave, surge, sway, heave, accelerations, and angular rates of velocity. Recent launches are the NMEA 2000 sensor from Furuno, the SCX20 Satellite Compass. This multi-GNSS compass for direct NMEA 2000 data input utilizes four separate GNSS antennae and, unlike single baseline calculations, uses any one of six baselines drawn between the antennae. This quad antenna design facilitates very accurate calculation of heading, pitch, roll, and heave data and precise data for input to the radar, fish-finder, and autopilot.

29.12   Sensor Location. The sensor must be mounted in the area of least magnetic disturbance so that no interference is induced into it that results in errors and degraded accuracy. It must be positioned close to the center of vessel motion, as errors are caused by vessel heeling and pitching. Steel vessels pose problems, and the sensor must be at least 5ft (2.5m) above the deck, although new electronic units are reducing this challenge. Accuracy is dependent on proper location, clear of interference. Accuracy is typically plus or minus 1 degree, although some self-compensate to 0.5 degrees.

29.13   Compass Compensation. Many have automatic deviation compensation and some will require steering in a 390-degree turn circle at a steady turn rate of around 2 to 3 degrees per second once or more times at commissioning calibration, most require just a 360-degree turn. The compensation takes place with respect to current boat magnetic deviation and creates its own curve in the process. This may vary if you have no electrical devices operating, but accuracy could alter with everything running. This will involve another recompensation that is simple and quick. Like speed log calibration, it is best done on a slack tide with no wind or current so you can do the turns smoothly. As a quality check, it is a good practice to check the compass against a magnetic compass on board, fixed or hand-bearing.

29.14   Compass Damping. Typically, damping can be from five to ten levels for some models. The rougher the sea state, the more damping required. A low damping level will result in erratic or rapidly altering headings.

Figure 29-3. i70 Instrument Suite (*Courtesy of Raymarine*)

29.15   Speed Logs. The speed log has the obvious function of indicating speed through the water and distance traveled. The speed log is a critical part of the integrated instrument system and is normally interfaced with other instruments. For the sake of nostalgia, not so long ago, the first merchant vessels I served on had a towed Walker log. As soon as we were clear of port and full away on passage, the turbine was streamed by the third mate and the mechanical counter was mounted on the poop rail and set to zero. I left this subject in this edition, despite the technological advances, as some yachtsmen still use them, although it has been a long time since I encountered someone with a trailing log. Unlike earlier versions, such as the reliable unit from Stowe, these trailing logs do not have a rotating line but instead have a sensor at the end of a 30ft (10m) cable that sends a signal to the freestanding control box. There are a few basics to remember when using these logs. Prior to streaming the log, make sure the line is hooked onto the log. This is a common error! When streaming, pay out the line quickly and at a constant speed before launching the turbine. Do not pay out the turbine first and allow the line to follow, as its rotation will cause tangling. When recovering the log, the challenge is to retrieve both line and turbine without tangles. Ideally, you should slow the vessel to reduce drag on the line and then initiate a small turn to put some slack into the line. As soon as this is done, disconnect the line from the log and pay it under or over the stern pulpit. This will take out the turns put in the line by the turbine before the turbine is recovered. Dry the rope before stowing.

29.16   Paddle Wheel Logs. The older and once common paddle wheel had magnets embedded in the wheel blades and a detector giving a pulse that can be counted and processed. Earlier units had a glass reed switch that was prone to impact-induced mechanical failure; new-generation units have a Hall effect sensor. The signal pulses are normally seen as a voltage change, such as 0 and 5 volts, to give a stepped characteristic that can be counted; the result is directly proportional to the speed and distance traveled. Counting may be either the pulses per second or based on pulse length proportional to distance. It is pretty standard these days to have multisensor transducer units that have speed, depth, and temperature functions.

29.17   Smart Transducers. Airmar has a smart sensor, the DX900+, which has embedded microelectronics, with the transducer element and the signal processor located adjacent to each other. The signal processing takes place and is then supplied to any NMEA 0183–compliant display head or into any NMEA 2000 network. This eliminates noise and interference and simplifies wiring. The smart sensor can transmit data to your phone app, as it has integrated Bluetooth technology built right in. The speed input employs an electromagnetic speed sensor to provide dual-axis 360-degree water speed measurement and calculate and provide leeway data. The DX900+ leeway value differs from the calculated value, which, when used as part of the calculation algorithm, provides a more refined and accurate true wind calculation. The latest innovation in smart transducers was launched in mid-2022 by Airmar. Called the DST810 New Gen2, it is not overstating to say this is a game-changer. They have reinvented the paddle wheel and come up with a highly accurate speed sensor that is part of the DST810 Smart Multisensor. The paddle wheel redesign is very sensitive; it starts transmitting data at just 0.3 knot and becomes linear at just 0.6 knot, with accurate and stable speed data through a large speed range. The transducer has a 5.7Hz (5X per second) speed through water output with higher resolution. While the DST810 delivers water depth (frequency of 235kHz), water temperature, and boat attitude data to your NMEA 2000 network, it can be accessed wirelessly from your mobile device using the Cast app. This powerful app allows easy calibration and corrections. This includes speed through water and angle heel calibration. The sensor incorporates an attitude sensor for heel and trim data and updates 10X/second. This facilitates heel-compensated speed calibration across multiple heel angles and speed ranges. The wireless protocol uses Bluetooth Low Energy (BLE) and has a range of 30ft (10m). The transducer has a LEN of 3.

29.18   Ultrasonic Logs. The transducer consists of two 2MHz piezo electric crystals; these transmit short pulse acoustic signals simultaneously and reflect the signals off water particles approximately 6in (15cm) away, which is clear of the turbulent boundary layer. The water particles pass through the forward and then the aft beam; the transmission time of the acoustic sound signal between the two crystals is then measured. The time delay is used to determine precise speed based on the known distance between the two transducers. Airmar has patented the UDST800, which provides an ultrasonic speed input along with depth and temperature. These units have both advanced filtering and sampling rates, resulting in high accuracy for the whole boat speed range. The unit has automatic boat speed, water depth, and water clarity adaptation. The depth-sounder operates on a frequency of 235kHz, which eliminates interference from other depth-sounders. The speed range is 0.1 to 50 knots, with an operating frequency of 4.5MHz.

29.19   Electromagnetic Logs. On this system, there is no impeller to foul up. One manufacturer is NASA Marine. These systems measure changes to the magnetic field in the water, which alters with boat speed. The transducer generates an alternating magnetic field in the water passing across the transducer head. As the sensor passes through the water, it will generate an alternating electric field; this field is proportional to the boat speed. The field is detected by small stainless-steel electrodes on the sensor. The interface box will then amplify the signal and convert into both boat speed and distance. This unit has an LED indicator that flashes when NMEA 0183 data is transmitted and output sentences are VHW and VLW.

29.20   Dual Transducer Systems. Catamarans require a transducer in both hulls, and some monohulls use a dual system to compensate for heeling. Gravity switches have been commonly used in racing monohulls to turn on the appropriate transducer for port or starboard tacks; but in multihulls, where heel angles are less, a switch that activates when the mast rotates is often used. Raymarine used their Micronet technology, which employs wind direction to automatically select which hull transducer is supplying the data, eliminating the switch requirement.

29.21   Doppler Logs. Unlike other logs, which give speed through the water, these logs report actual speed over the ground by transmitting acoustic pulses that reflect off the bottom. They are commonly used on commercial ships.

29.22   Log Installation. Correct installation is essential if the log is to be accurate and reliable. Observe the following notes:

   a.   **Location.** The log transducer is normally mounted in the forward third of the hull and must be in an area of minimal turbulence, called the boundary layer. Dead rise angle for best performance is under 7 degrees.

   b.   **Cabling.** Do not run depth-sounder and log cables together, as interference may result. This doesn't apply to triducers that have temperature, depth, and speed logs in the one unit.

29.23   Speed Log Calibration. Every instrument system has varying instructions in the manual to calibrate the log. Read the manual! Calibrating a log normally requires using a measured mile. These are always clearly marked on charts. Many new logs are self-calibrating or have an optional manual calibration. The calibration run should be carried out at slack water in calm, current- and wind-free conditions to minimize inaccuracies. Some specify doing a run and checking it against the GPS speed; but, again, perfect conditions are required. Before making a run, check that the vessel is on the correct magnetic course; this means making appropriate corrections for variation and compass deviation. Make the runs under power at a constant throttle setting; it's best get up to speed before you start the run and reach your start position. If you struggle to steer a straight course to a visible marker, you can always put the autopilot on. Ensure that your transits are accurately observed at the start and finish of each run. Tidal area runs require multiple runs, as shown below. Speed under sail may be different, as heeling errors and leeway come into play. The formula for determining log error is shown below. The resulting figure will show either underreading or overreading, which is used either to calibrate the log or correct the log readings:

$$\frac{\text{Runs 1 + 2 (ground measurement)}}{\text{Runs 2 + 2 (through water)}} = \text{Correction K}$$

29.24   Log Transducer Maintenance. Logs in general require little maintenance, although paddle wheels require more than most. The paddle wheel log transducer has a typical life of around 5 years, a lot longer if you pull them out when not in use. Perform the following checks:

a.   **Paddle Wheel.** Regularly remove the paddle wheel to ensure that it is rotating smoothly and freely. Apply some light oil to the spindle.

b.   **Seals.** Check to see if the O-ring seals are in good condition to prevent leakage into the bilge.

Table 29-1. Speed Log Troubleshooting

Symptom	Probable Fault
No display	No power Cable connection fault Network plug connection Instrument transducer fault Instrument head fault
Partial display	Processor fault Display fault
Erratic readings	Transducer fault Connection degradation Interference from radios, electrical systems
No or low boat speed	Transducer fault Transducer not installed (blank still installed) Transducer not connected Fouled transducer Paddle wheel seizing

29.25   Wind Instruments. The wind transducer experiences a very harsh and unforgiving life at the masthead—constant vibration, movement, extreme weather, and more. The typical wind system comprises an integral anemometer for measuring windspeed and a wind vane for measuring true wind direction. The data is fed either to dedicated instrument heads and displays or into a network for use in linked displays or multifunction displays. Masthead wind transducers are now very lightweight and made of carbon fiber or aluminum. They have low friction sealed bearings and very low start-up wind speeds. Most important sailing data is derived from just five values. The first two are relevant to this chapter: apparent wind speed and apparent wind angle. Combined with boat speed, compass heading, and boat heel angle, they allow many performance parameters to be calculated.

a. **Wind Speed.** The term "anemometer" is derived from the Greek word *ane-mos*, which means "wind." The anemometer is essentially a rotating pulse counter similar to the speed log. The pulses are counted and processed to give speed, although technology is overtaking this with wireless ultrasonic transducers. While most use a cup system for rotation, there are propeller versions. Mechanical wind sensors have limitations. They are not able to handle turbulence from rapid changes in wind direction and wind speed, such as gusts. Factors that influence and limit accuracy and performance include the start-up torque of the anemometer cup and the wind vane. This creates a small time delay, so accuracy is impacted. There are other emerging technologies in wind measurement, and these are generally used with large commercial and scientific weather station applications and include light detecting and ranging (LIDAR) and hot wire anemometers. In the average anemometer, a sensor is used to count the wind cup rotations over a set time base. Each wind instrument manufacturer employs different methodologies to process this data. They vary the sampling rate for averaging the data, so short intervals produce high-resolution data and low rates give the opposite. It comes down to the software algorithms used to process the data.

b. **Wind Direction.** This part of the masthead unit consists of a simple wind vane. The wind vane sensor measures the vectoral change in wind angle through 360 degrees. A number of methods can be used to measure the angle and transmit the signals to the instrument head or processor. Some units use an electromagnetic sensing system. Others use an optical sensing system to identify coded markings that relate to the wind vane position and direction. Some units used to employ a potentiometer, which generated a variable resistance output that could be measured and displayed. Others use a Hall effect device.

(1) **Apparent Wind Direction.** The measured wind direction is apparent wind. The display often indicates the close-hauled angles and gybe points.

(2) **True Wind Direction.** True wind data is a result of the instrument processing vessel course and speed and the apparent wind direction and speed.

29.26 Wireless Wind Systems. Pioneered by TackTick, wireless wind systems have advanced with technology. Eliminating mast wiring in a racing yacht reduces the spar weight and also provides a marginal decrease in the center of gravity. Raymarine wireless displays are solar powered and have integrated lithium batteries that power the unit for life. The power consumption is so low that unit operating time is up to 300 hours. The Raymarine T220 vertical wind transmitter is the face of new technology. The Raymarine wireless remote display controls, configures, and repeats all the required performance, including wind trends, wind speed trends, velocity made good (VMG) to wind and waypoint, speed over ground (SOG), course over ground (COG), and performance graphing. It gives highly accurate data that includes true wind speed (TWS) and true wind angle (TWA). This is possible, as errors that arise from disturbed wind flow are minimized by elevating the wind sensor by 4ft (1.35m) into cleaner air. Garmin has the gWind Wireless 2 Transducer. Accurate wind data is transmitted via an advanced and adaptive network technology (ANT) connection directly to a GNX wind instrument or to a compatible Garmin GPSMAP chart plotter. The transducer has the unique Nexus twin-fin technology, which has a three-bladed propeller.

29.27   Wired Wind Systems. NKE Marine Electronics have the HR series wind instrument sensor with an integral processor composed of a wind calculator that integrates true wind tables, along with performance polar diagrams and noise reduction. It has very accurate sensing and is within 1 degree for wind angle and less than 1% for wind speed linearity; sensitivity can measure wind speeds below 2 knots. This sensor is available with a long vertical arm called Carbowind HR. This meter-long rigid carbon fiber pole locates the wind sensor above the masthead and clear of disrupted wind. This unit incorporates a temperature sensor. It is different, as it integrates an adjustable counterweight for fine-tuning the balance of the wind vane, as does one of the B&G units.

29.28   Ultrasonic Wind Instruments. Sound travels by causing air molecules to move back and forth. Wind speed also affects the speed at which sounds travels. The propagation of sound waves in the air is directly affected by the wind speed component parallel to the direction of propagation. Ultrasonic wind speed units use high-frequency sound. The ultrasonic anemometer comprises either two or three pairs of sound transmitters and receivers that are located at right angles to each another. The sonic anemometer transducer transmits this high-frequency pulse ultrasonic sound from the north-facing side of the sensor. The microprocessor then calculates the time the pulse takes to travel to the south-facing transducer. The windspeed is then calculated on the pulse travel time between two transducers. The Airmar WX series is quite unique due to its ability to calculate both theoretical and apparent wind speed and direction. This unit incorporates an integrated compass and a GPS. It calculates the theoretical wind based upon the apparent wind, speed of the boat, and compass heading. The ultrasonic sensors have no moving parts. Unlike their mechanical cousins, ultrasonic sensors are unaffected by mechanical inertia, so changes in wind speed and direction are detected in real time. Recently Maretron released its ultrasonic weather station, which is able to measure wind speed and direction, air temperature, barometric pressure, relative humidity, wind chill factor, heat index, and dew point. The wind measurement utilizes six ultrasonic sensors in a delta configuration, no moving parts at all, and is NMEA 2000 compatible as well. It's not available for masthead installation yet, but I hope it will be soon.

29.29   Masthead Unit Installation. The masthead unit is always mounted on the end of a boom in front of the mast to reduce turbulence. The position is not perfect—the masthead unit is subject to updrafts and turbulence from the sails—but it is still the best alternative. In recent years these booms have extended significantly to give transducers clear air without influences from turbulence as wind passes across the masthead.

a. **Fastening.** It is important that the unit be properly fastened down, especially as masthead units are often installed in a simple bracket assembly and are removable. Check fore-and-aft alignment to reduce inaccuracies in angle readings. Besides seagulls and lightning, the main cause of masthead damage is vibration.

b. **Alignment.** As it is very difficult to get perfect mechanical alignment of a wind sensor unit on the boat centerline, many units require an offset adjustment. Precise wind values require exact alignment.

c. **Unit Inspection.** Check that the anemometer cups and the vane are not damaged. Ultra violet (UV) slowly degrades the plastic, and the cups split. Check that the unit is securely mounted to the main mounting block. Check electrical connections and make sure the cable connector is securely fastened. It is good practice to put a few wraps of self-amalgamating tape around it to prevent water entry. If you must apply petroleum jelly or silicone grease, do not smother the socket. It simply gets pushed into the masthead unit and contributes to poor electrical contact. It is better to keep electrical connections dry with tape. You can apply grease on the screw threads to minimize seizing.

d. **Mast Base Installation.** Install cable connections in a water-resistant instrument connection box and check that all connections are tight. Recheck every 6 months:

   (1) Check securing bolts and frame, and tighten as required.

   (2) Check cable connector for moisture and water, as well as for signs of corrosion on the pins. Smear a small amount of petroleum jelly or silicone grease around the threads when replacing the connector, and then rewrap with self-amalgamating tape. Examine cable insulation for signs of chafing at any mast access point.

   (3) Check that the anemometer rotates freely without binding or making any noise, which may indicate bearing seizure or failure. Check the cups for splitting or damage, which frequently is caused by those pesky seagulls.

   (4) Apply a few drops of the manufacturer's light oil into the lubrication hole and rotate to ensure that it penetrates the bearing.

   (5) Check the terminations in the connection box located at the mast base. They should be tight and show no signs of corrosion. If they do, replace them so you have a good quality joint without any circuit resistance.

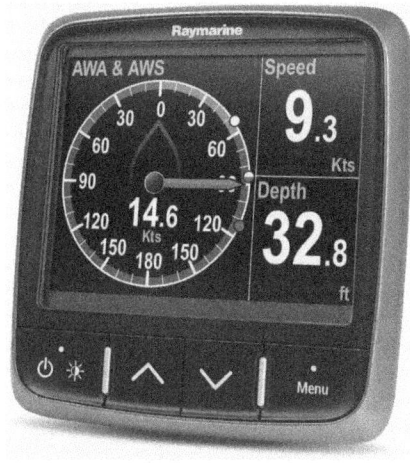

Figure 29-4.
i70 Head with Wind
(*Courtesy of Raymarine*)

29.30   Velocity Made Good (VMG). The VMG instrument head or display readout is among my favorites, as I still like to sail to windward as fast as possible. VMG is the speed or progress of the boat either upwind directly toward or downwind away from the true wind direction. A sailboat's VMG to a mark or waypoint is an important piece of data with respect to steering and sail trim. VMG is derived from calculation of true wind, course, and speed. VMG enables the helmsman to sail the optimum course so that maximum speed is made toward the destination. Having had more than my share of uncomfortable slogging to windward and making that windward beat a few hours faster is well worth the effort. The following are used to achieve optimum VMG, which is indicated with a higher reading.

    a.   **Sail Trim.** Adjusting sail trim will increase or decrease speed and alter the VMG.

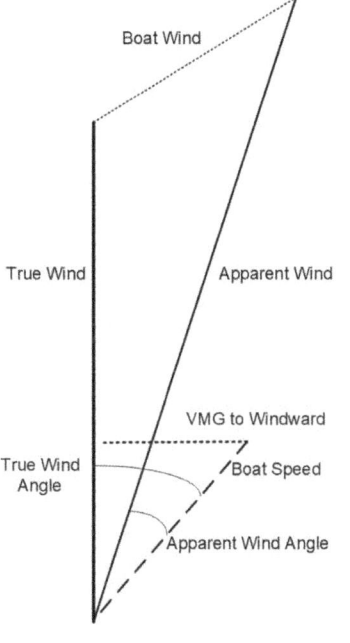

Figure 29-5. Wind Vectors

b. **Course Adjustment.** Changing course into or off the wind will change VMG reading.

c. **Target Speeds.** The maximum boat speed at windward and downward VMG is called the target speed. Instruments now display VMG in real time, and many chase the maximum numbers. Downwind sailing benefits immensely from VMG information.

Table 29-2. Wind Transducer Troubleshooting

Symptom	Probable Fault
No wind speed	Electrical interference Mast base connection fault Masthead unit plug fault Anemometer seized Anemometer fault Processor fault
Erratic wind angle	Electrical interference Corroded masthead unit plug Masthead unit plug fault Vane seized or seizing Loose connections Water in masthead unit plug Masthead unit fault
Intermittent wind speed or angle	Water in masthead unit plug Loose connections Processor fault

29.31 Rig Tension Monitoring. Cyclops recently launched a wireless load pin for accurate monitoring of rig stay tension called smarttune. It is calibrated to transmit data to either the Cyclops app or with a gateway NMEA 2000 connection and display to your instrument system. Data can be viewed on the Cyclops Marine Smart Fittings Manager app. The systems uses the Smartlink Nano wireless load sensor. The sensor is powered by a CR2032 battery with a life of around 200 hours. It is easily installed to forestays, kickers/vangs, or sheets. Go to https://www.cyclopsmarine.com for details. Rig tension monitoring is relatively common. Spinlock has the Rig Sense tension gauge with a phone app. There are several rig tension gauges on the market, including the Loos, which I have on my boat, SureCheck, and Rig-Sense. Generally, backstay tension monitoring uses load cells, which are essentially strain gauges that output a small 0–5V in proportion to the applied load. Some devices use a 4–20mA output. A strain gauge is able to detect changes in the geometry of an item by measuring changes in the resistivity and conductivity. When the gauge is stretched under tension, the length increases and cross-sectional area decreases, causing an increase in resistance. When compressed, the converse occurs. As measurement values are very small, circuits incorporate a Wheatstone bridge.

29.32 Depth-Sounders. The echo or depth-sounder is an indispensable item of marine electronics. I have spent considerable time working with underwater acoustics systems in the offshore oil industry and a couple of years working on a submarine sonar program, and this technology is very complex. Equipment performance depends on the output power of the transmitter, the efficiency of the transducer, and the sensitivity of the receiver, along with the processing software that filters out the spurious noise. Many fish-finders have user-selectable noise filters to enhance noise rejection processing. The price of equipment reflects all these elements, with the most expensive systems having the highest performance specifications on all parameters. The word "sonar" is derived from "SOund, NAvigation, and Ranging" and has its origins in World War II anti-submarine warfare. The depth-sounder normally projects the acoustic signal directly downward at a set beam angle so that a cone of coverage is made with respect to the bottom or contours being passed over. Many depth-sounders operate at a frequency of 200kHz; lower transmission frequencies of around 50kHz give greater depth capability.

a. **Digital.** The most common depth instrument is a vertical unit with a digital or analog display incorporating depth alarms, anchor watch alarm facilities, and other functions. The information displayed is generally several seconds old due to signal processing times. That has changed; depth transducers are now very sophisticated and accurate, with improved transducer technology and processing software, and can be directly input into a NMEA 2000 network.

b. **Accuracy.** Acoustic signals suffer from propagation delays and attenuation as water and various bottom formations cause absorption, scattering, refraction, and reflection. This can be caused by biological matter such as algae and plankton, as well as suspended particulate matter and turbidity containing silt, dissolved minerals, and salts. Water density and salinity levels as well as water temperatures all affect signal propagation. While cold layers of water called thermoclines can affect signal, this is more relevant to deep water. Bottom formations consisting of sand and mud, or large quantities of weed beds, will absorb or scatter signal; hard bottoms that comprise shale, sand, and rock will reflect signal, with strong returns. The power output of a device is also important with respect to range and resolution; the higher the power, the greater the depth range and signal return.

29.33   Basic Sonar Theory. The basic principle is that a high-voltage electrical pulse at a set frequency is converted to mechanical energy, sound pressure, or acoustic signal via a piezoelectric element; this is then transmitted toward the seabed. This sound wave travels in accordance with the transducer radiation pattern. When the transducer transmits the acoustic energy, it expands to form a cone-shaped characteristic. When the acoustic signal strikes a surface such as the sea bottom, a portion of this energy is reflected back to the transducer as an echo. The shape and diameter of a transducer determines the cone angle. The acoustic signal strength is at maximum along the center axis of the cone and decreases away from it. The cone angle is based on the power at the center to a point where the power decreases to –3dB, with the total angle being measured from a –3dB point on each side. The echo that is received back at the transducer converts it into a low-voltage signal for amplification and processing. As sound travels at a fixed speed in water—4,800 feet per second—the depth can be calculated by measurement of the time difference between the transmitted pulse and the received echo. Piezoelectric transducers are used and are typically constructed of a crystal composed of various elements, including lead, zirconate, barium, titanate, and conductive coatings.

29.34 **Cone Angles.** The acoustic signal strength is at maximum along the center axis of the cone and decreases away from it. The cone angle is based on the power at the center to a point where the power decreases to –3dB, with the total angle being measured from a –3dB point on each side. Most manufacturers offer models with a variety of cone angles. Wide cone angles have less depth capability with wider coverage; small cone angles give greater depth penetration with reduced area coverage. High-frequency transducers (190kHz) are available in either wide or narrow cone angles. Low-frequency transducers used in fish-finders have cone angles in the range of 30 to 45 degrees. The further away from the centerline of the cone, the weaker the return echoes. This can be improved by increasing the sensitivity control.

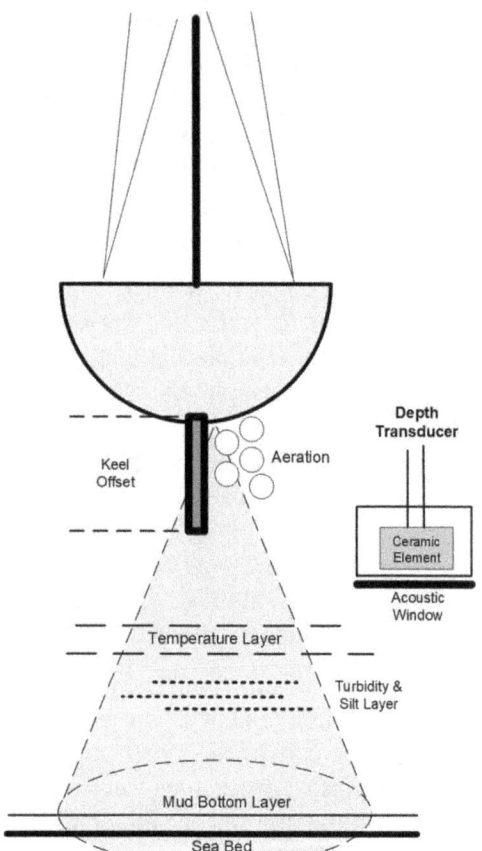

Figure 29-6. Typical Depth-Sounder

29.35 **Transducer Cavitation.** Cavitation affects transducer performance and is caused by water turbulence passing over a transducer head. At slow speeds, the laminar flow of the water is smooth, without any interference; however, at higher speeds, air bubbles are created over the transducer face, affecting acoustic signal transmission and reception. The effect is to interfere with transmitted acoustic signals that reflect back off the bubbles, effectively creating noise and masking the signal. Turbulence is also caused by the boat hull form or any obstructions, water flow over the transducer, turbulence from the propulsion, and other influences. Transducers must be mounted in areas of little turbulence or clear of hull flow areas, which is often a challenge.

29.36   Frequencies and Power Output. Transmission frequency affects both the depth range and cone angle. The speed of sound in water is a constant of 4,800 feet per second, and the time between the transmission and reception of the returned signal is measured to give a range or depth figure. Shallow waters less than 300ft (91m) give the best results with high-frequency transducers of 200kHz and wide cone angles up to 20 degrees. In depths greater than 300ft (91m), low-frequency transducers of 50kHz with small cone angles of 8 degrees are the best option. Power outputs are quoted in watts; although some quote peak-to-peak, watts RMS (root-mean-square) are the more accurate measure and are typically within the 100W to 600W range.

29.37   Depth-Sounder Maintenance. The transducer is the only item that can be maintained; if not, it will dramatically reduce the sounder performance. Inspect the transducer for damage, marine growth, and excess antifouling paints. Clean the transducer surfaces using soapy water; do not use heavy abrasives or chisels to clean the transducer aperture face. Do not bump the transducer or apply any impact to the surface. Avoid applying antifoulant to the transducer surface; it includes small voids and air bubbles, which will reduce sensitivity. If necessary in high-growth areas, smear on a very thin layer with your gloved finger. Troubleshooting entails reading the manual and determining whether settings and operating procedures are correct. Go into the settings or options menu and verify that settings are set on "auto" or defaulting to factory settings. If possible, when at a mooring or alongside, remove the log and depth transducers and replace them with the dummy plug, blanking cap, or blanking dummy.

a.   **Electrical Cables and Connections.** Check all connectors and connector pins for damage, and verify that they are straight and not bent. If straightening the pins, there is a risk of breakage, as they are brittle. Connectors not properly inserted or tightened are prone to saltwater ingress and corrosion. Inspect all cables for damage, cuts, chafing, or degradation.

b.   **Electrical Power Supply.** Bad connections are the major cause of problems. Check the power at the plug using a multimeter; a check should be made with the engine on or off. If the engine voltmeter shows normal charge voltages and the battery checks out, the problem is probably within any intermediate connections or junction boxes.

c.   **Electrical Interference.** If the depth-sounder shows interference, turn off all other equipment and then turn the engine off. Progressively start up the engine and then sequentially all other equipment to determine the source. Once the interference is identified, the power device supply may require suppression.

29.38   Depth-Sounder Installation. Be very careful not to bump the transducer and possibly damage the crystal element. Most installations are thru-hull mounted on a fairing block to ensure that the beam is facing directly down at even keel and within the recommended dead rise angle requirements to reduce any water flow turbulence. Locate the transducer in an area of minimal turbulence. Water bubbles from turbulence are a common cause of problems. In some cases transducers are mounted inside the hull within an oil bath or epoxy fastened to the hull on FRP/GRP boats. There is a sacrifice in maximum depths, which can reach a 60% to 70% reduction in range, and therefore should be avoided where possible. Always ensure that cables are installed clear of heavy current–carrying cables or radio aerial cables. Never install next to log cables as is generally done; the interference problem can be significant in older generation devices, but this is not an issue with new multisensors such as the Triducer from Airmar, which integrates speed, depth, and temperature in one unit. Follow the instructions that come with your transducer. The following installation recommendations should be observed when installing an instrument system.

a. **Central Processing Unit (CPU) Location.** If your system has a CPU, install the CPU or data box in a clean, dry area that permits easy access to transducer cables. Mount the CPU unit well away from fluxgate compasses and GPS receivers, as well as VHF, SSB, and AM/FM radios. The CPU must be a minimum of 3ft (1m) from a magnetic compass.

b. **Transducer Cables.** Transducer cables should not be lengthened or shortened. Coil up the extra length at the transducer end.

c. **Instrument Covers.** Do not cover or mount your instruments behind Perspex or plastic sheeting. This magnifies the heat from the sun and burns them out. When the instruments are not in use, always install the covers provided to prevent sun damage and weathering.

d. **Instrument Cables.** Do not stress or bend the cables sharply. All cables must be run through proper deck transits to connection junction boxes. Always run cables well away from radio antennae and heavy current–carrying cables.

Table 29-3. Depth-Sounder Troubleshooting

Symptom	Probable Fault
No depth indication	Transducer connection failure Transducer damaged Transducer fouled
Intermittent shallow indication	Fouled transducer Water aeration over transducer Weed beds or bottom issues
Shallow readings in deep water	Outside depth range Check your charts!
Inconsistent depth readings	Muddy or silted bottom Fouled transducer Poor transducer interface (in hull only) Processor fault

Electronics Interference

30.1   Interference Principles. Interference is the major enemy of boat electronics systems, corrupting position fixes and causing general performance problems and, in worst cases, causing electronics equipment damage. Interference and noise are the superimposing of a disturbance or voltage transient onto power, signal, and data lines. This is processed along with data to corrupt or degrade the processed information. The Federal Communications Commission (FCC) label or the FCC mark is a certification mark used on electronic products manufactured or sold in the United States. It certifies that the electromagnetic compatibility and interference from the device is under the limits approved by the FCC. The operation of any electronic device is subject to a number of criteria.

The device must not cause harmful interference. The device must accept any interference received, including interference that may cause undesired operation. This means that all electronics devices capable of emitting and transmitting radio frequency (RF) energy for the purpose of communication must adhere to US Code of Federal Regulations (CFR) FCC Part 15. There are classifications, the first being intentional radiators. These are devices that both intentionally transmit and receive radio energy and include wireless technologies that include Zigbee and Bluetooth, radio frequency identification (RFID), and alarm systems. Unintentional radiators have the ability to generate radio frequencies within the device and subsequently and unintentionally radiate through conduction, causing interference or upset with other electronic devices. Incidental radiators generate RF energy and include DC motors, mechanical light switches, and other sources. Finally, we have conducted emissions. All electronic devices will generate electromagnetic energy, and some of this energy will be conducted into power wiring and device connections. Sources include AC or DC power supplies, wired network communications, universal serial bus (USB) peripherals, multimedia devices, and personal computers, along with many others.

Industry Canada (IC) is similar to the FCC for equipment within the Canadian market. In Europe the Radio Equipment Directive (RED)—Conformité Européenne (CE RED)—is the prime approval mechanism. Many instrument systems designed for the European market are CE marked to IEC60945, which is a boat-specific electromagnetic compatibility (EMC) standard. Noises can be basically classified as radio frequency interference (RFI) or electromagnetic interference (EMI). Noise occurs in different frequency ranges; similarly, equipment may be prone to problems only within a particular frequency range. Multiple noise sources can cause a gradual degradation of electronics components, and when the cumulative effects reach a certain point, the devices fail. The following chapters describe the various sources of interference.

30.2   EMI Basics. Electromagnetic interference (EMI) has numerous sources, and all may impact other equipment. These include but are not limited to motor brush sparking, inductive and resistive load activation, switch activation, circuit breaker activation, fluorescent light operation, heater units, microwave cookers, mobile communications devices, atmospheric discharges, electrostatic discharges between people and equipment, and any electrical item that is switched on and off. All of these events and sources have the capacity to cause overloads, voltage peaks, transients, subvoltage dips, and other problems, and all can affect electronic equipment, including communication radios, AIS, and GPS. Prime areas where noise is observed include noise, hum, or hiss on audio systems and white lines or "snow" on television and radar screens. A big effect is degradation in data network performance, from error rate increases to data corruption and total loss of data. Many of these impacts can be mitigated by the use of various grounding techniques, shielding methods, twisted wires, and filters. While we discuss space weather, it should be remembered that electrical problems can arise from solar magnetic storms—coronal mass ejections (CMEs) that release plasma and magnetic field disruption—along with other various cosmic noises. You can experience atmospheric noises and those noises emanating from Earth's magnetic field flux. The sun irradiates the Earth continuously, and variations in intensity occur with sunspot and corona discharge activity cycles; with that comes noise. Atmospheric noises, also known as static or white noise, are caused by lightning discharges and other natural sources.

30.3   RFI Basics. Radio frequency interference (RFI) is a subset of EMI and encompasses a large spectrum of frequencies, although RFI covers a relatively narrow slice of the radio wave spectrum. It is a disturbance that has an external source. RFI is the conduction or radiation of radio frequency that causes an electronic or electrical device to produce noise that typically interferes with the function of an adjacent device and couples with these devices to create noise. Equipment that creates noise includes switch mode power supplies, radar, VHF, AIS, televisions, stereos, mobile phones, DVD players, satellite receivers, computers, 12V lighting (LED and fluorescent), variable speed drives, water pumps, air conditioners, and home appliances. To this list you can add motor-driven equipment, drills, food processors, vacuum cleaners, and microprocessor-controlled equipment. To mitigate and eliminate RFI, prevent coupling by use of filters and arc snubbers at the source. You can relocate equipment and reroute and segregate cable and wiring. You can add shielding and also ferrite chokes to cables. General rules covered elsewhere include keeping cables as short as practicable and adopting a routing segregation strategy. Long cables increase power line common impedance coupling but make the cable a larger antenna. Be careful about keeping large cable coils in a circuit, as they are prone to RFI reception. If using shielded cables, use those with heavy gauge shielding. Cables with foil and drain wire shields have a much higher common impedance coupling than cables with braided copper shields. Keep an eye on cable and wire connections; loose terminations or those with high contact resistance create noise. You can install RFI filters and also ferrites. Do not add additional grounds unless the device's manufacturer specifically recommends it. Too many grounds can increase circulating ground noise rather than reduce it. Never disconnect an AC safety ground as part of noise-reduction activity. In 2022 the USCG issued an alert about LED light RFI. Also in 2022, the Radio Technical Commission for Maritime Services (RTCM) adopted RTCM Standard 13700.0: "Electromagnetic Compatibility Requirements for Light Emitting Diode (LED) Devices and other Electrical and Electronic Equipment in the Vicinity of Shipboard Antennas for the Protection of On-Board Receivers." RTCM Standard 13700.0 is designed to protect GMDSS MF/HF and mobile satellite receivers, maritime VHF, AIS, and GNSS receivers from deck-mounted electrical equipment, primarily equipment installed near antennae and aerials. This was in response to LED navigation lighting and other LED lights causing interference.

30.4   Transients and Voltage Spikes. An electrical transient is a very fast, short-duration spike in voltage that could be as much as several thousand volts in magnitude. This voltage spike then produces a corresponding increase in current within the load, which is observed as a current spike. Consequentially, there is a momentary increase in transferred energy. Depending on the magnitude and duration of the transient, the resultant energy transfer to the load can be of minimal to no consequence—or the opposite, resulting in significant damage. Transients can occur as bursts rather than singular events. The most common source of transients within shore- and boat-based electrical systems is switching operations. Switching a circuit breaker or opening or closing a contactor creates a transient. Even a simple light switch operation creates a transient. The transient affects are amplified if the isolation devices have poor condition or faulty contacts where the momentary arcing is longer and more intense. If you encounter an electrical issue on board and the circuit breaker trips under high load, a larger arc and transient are generated. Spikes can have many negative effects on electrical systems and the devices and equipment connected to those systems. Frequent transient activity can gradually degrade equipment over time; however, lightning strike voltage spikes and the switching of large inductive loads can attain levels that can break down insulation and cause catastrophic damage. When a voltage transient exceeds the insulation value or dielectric strength, a flashover occurs. When this flashover happens, a low-impedance path is effectively created through the arc, through which the normal supply voltage will flow. The arc creates heat, burning occurs, and immediate insulation failure results. Either at home or aboard, if you have appliances, they will create disturbances, as will office equipment, including photocopiers and laser printers, and HVAC systems. Every time an inductive or capacitive load is switched on or disconnected from the power source, a surge impulse will be generated, and this will propagate back through the entire electrical system. The outcomes include automation disruption, variable speed drive (VSD) failure, computer failures, general IT crashes, and data loss. On the AC side in a marina, electrical spikes regularly cause nuisance tripping of RCD and GFCI protective devices. All new-generation electronic equipment is vulnerable to transient voltages. Everything has a microprocessor of some description installed; everything has microcontrollers and other internal processor components that comprise literally millions of active circuits in a small and compact package. It is fundamental electrical theory that the smaller the distance between conductors and terminals, the lower the transient voltage required to flashover. The outcome is that the transient voltage will stress these components, and repeated exposure will cause premature failure. Understanding and mitigating the various sources will increase longevity, reliability, and performance of devices.

30.5   Lightning Transients. Transients associated with boat lightning strikes have a very fast rise time to the maximum level, within 1 to 10 microseconds. The transient then decays at a rate of around 50 to 200 microseconds. As the current generated in a lightning strike is a transient one, several phenomena are involved, as discussed in the lightning chapter. Current spikes of short duration have a tendency to travel on the surface of a conductor; this is due to skin effect. Rapidly altering currents create an electromagnetic pulse (EMP), which radiates out from the point of the strike. When radiated pulses pass over conductive items, including power, communications, and instrument cables, they can induce a transient current into those items that runs along the surface to the point of termination.

30.6   Surge or Electromagnetic Pulse. Surge is defined as a very fast, short-term, high-voltage variation that exceeds 110% of nominal value. Surge sources often result from lightning, line or capacitor switching, or the disconnection of heavy loads. Often called transient voltages, these are random, high-energy electrical disturbances with a short time duration of 1 to 10 microseconds. Surges are different from long-duration events, which include swells or temporary overvoltages.

30.7   Arcing Noise. Arcing causes EMI. The average arc generates a lot of current over a very short time. This is a spike-shaped EMI disturbance of a very short time duration, and it possesses a lot of power due to the high current. The frequency spectrum of the average spike is close to uniform, from zero to infinity, and so is registered by everything. There are repetitive spikes caused by commutators in electric motors and sparking brushes. Charging systems or where loose connections exist commonly cause this. One common cause is loose or poor engine return paths for alternators, when the negative path arcs across points of poor electrical contact. Shielding and filtering are the most common mitigation methods. Static charge buildup on alternator rubber drive belts can cause noise as accumulated charges dissipate and arc to ground.

30.8   Induced or Inductive Coupling Interference. Inductive coupling may be intentional or unintentional. Unintentional inductive coupling arises when signals in one circuit are induced into an adjacent circuit. This is often referred to as "cross-talk" and is a type of electromagnetic interference. The "disturbing wire" and "victim wire" are accompanied by a magnetic field. The level of disturbance is totally dependent on the variation of the current (di/dt) over time and the mutual inductance coupling. Inductive coupling increases with frequency, and inductive reactance is directly proportional to the frequency. Factors include the distance between the two cables and the parallel cable distance; the greater the distance, the greater the exposure. The load impedance of the disturbing circuit cable, low ground impedances, and unbalanced circuits are the most problematic, with serial data, multicable control, and coaxial cables being the most susceptible. To mitigate inductive coupling, you should limit the length of parallel run cables. Use segregation, and increase the distance between the two cables. You can ground one shield end of both the cables. You can reduce the dv/dt of the disturbing cable; this increases the signal rise time, and ferrites are often used. What is dv/dt? Back to calculus, it means the rate of change (V) over time (T). These are the steep fronted voltage pulses that travel down the long cables to the motor and then come back as a reflective wave. Twisted pair wiring techniques can reduce crosstalk between a pair of wires and will consequentially decrease the level of EMI and RFI. Twisted pair cables are reasonably efficient as long as the induction within each twist area is approximately equal to the adjacent induction. The quantity of twists may have to vary to reduce the electrical coupling. The construction provides a capacitive coupling between the pair conductors.

30.9   Capacitive Coupling Interference. Capacitive coupling is also known as electrostatic coupling or electric field coupling. It is the process where energy is able to move between conductive elements that are separated by insulators. This happens when energy is coupled from one circuit to another through an electric field. Capacitive coupling causes crosstalk, where the signal from one conductor is coupled into an adjacent one.

30.10   Turn-On and Turn-Off Spikes. Turn-on spikes result from the initial charging of power supply input filters on power supplies. Turn-off spikes arise when reactive loads are switched and the magnetic fields collapse on inductive loads, such as transformers, relay or contactor coils, solenoid coils, pump motors, and other loads. Spikes can be as much as 500V peak-to-peak.

30.11   Ripple Noise. Ripple is a low-frequency component, created in rectifier bridges from diodes and silicon controlled rectifiers (SCRs), such as those found in alternator diode bridges, battery chargers, fluorescent lights, and inverters. It usually presents as a high-pitched whine. Ripple badly degrades communications audio quality. Good equipment has well-designed component layouts and suppressed electronics.

30.12   Static Charges. These charges have a number of sources. External charges and interference can arise due to static buildups in rigging. On reaching a certain voltage level, the static charge discharges to ground, causing interference. Another common cause is dry, offshore winds; a static charge builds up on FRP/GRP decks, and I have seen this on catamaran decks. The problem is prevalent on larger FRP/GRP vessels with large deck areas. A lightning protection system can help ground out these charges (see my note in the lightning protection chapter). Another is engine and propellor shaft charges. This type of interference can arise due to static buildup, both induced and due to moving parts in the engine. The static charge builds up and discharges to ground, causing the interference. Propeller shaft interference can arise due to static buildup on the shaft during motoring periods. The static will discharge to ground when it reaches a high voltage level and create interference. Typical cures are grounding of the shaft with a brush system. (See the corrosion chapter.)

30.13   AC Transients. All of the above can be applicable to both AC and DC systems. Surges and transients on an AC shore power system can be carried through to your own AC shore power system. Every time something happens on another boat in a marina you can end up seeing a surge or spike or disturbance on your boat systems. That is a reality that is rarely understood or considered. If you keep getting unexplained issues best check out the neighbors and do it nicely.

30.14   Harmonics. Often associated with shore-based industrial sites, harmonics are actually a big deal on commercial ships and oil rigs. Many do not realize that harmonics are ever present in your house and impact your power supply quality. If harmonic levels become severe, it often results in computer reliability issues, or circuit protection devices that operate in a seemingly nuisance or no-protective state. Motors can be noisy when running; cables can run at increased temperatures. Harmonics have become pervasive in our modern electronic world. This is due to the connection of a great number of nonlinear loads. Virtually all electronic devices are nonlinear loads. This starts with your computers, modems, printers, monitors and televisions, smartphone chargers, microwave ovens, washing machines, and refrigerators, LED lighting, inverter air-conditioning units, and so many more devices. This means there are harmonics everywhere. When a linear electrical load is connected to the electrical system, it draws a sinusoidal current with the same frequency as the voltage. Nonlinear loads draw currents that are not precisely sinusoidal. Current waveforms tend to get very complex, and these nonlinear loads increase current and, in the worst situations, cause voltage distortion of the supply. Waveform distortion is mathematically analyzed, and one of the measures used is to indicate total harmonic current distortion (THDi). This is a ratio of the sum of all the harmonic currents to the current at the fundamental frequency.

30.15   Voltage Dips or Sags. A voltage dip, or sag, is a sudden reduction in the supply voltage of between 10% and 90%, with a short time recovery period. The duration of a voltage dip is in the range of 10 milliseconds and 1 minute. The actual depth of a voltage dip is the difference between the minimum RMS voltage during the dip and the nominal voltage. Voltage changes that do not reduce the nominal supply voltage by less than 10% are not considered to be dips. Voltage dips may be caused by external or internal factors and can exist as random singular events or a series of repeated occurrences. External factors, which are normally singular events, include short-term reductions in supply voltage caused by load switching. Similar effects arise when switching between a shore power supply and an onboard inverter or generator supply. For example, using a high-current deck winch motor, windlass, or thruster as it switches off creates a large voltage dip. A supply voltage that dips can create issues with AC induction motors, and the results vary in severity. As the supply voltage decreases, the motor speed also decreases and, dependent on the severity of the dip, the motor speed may recover to nominal value as voltage amplitude recovers. If the magnitude and time duration of the voltage dip exceed nominal limits, the motor may stall and trip undervoltage protection, a motor contactor or relay may drop out, or a variable speed drive (VSD) or variable frequency drive (VFD) may shut down.

30.16   Voltage Swells. A voltage swell is the opposite to a voltage dip. It is defined as a sudden increase in the nominal supply voltage of 10% or more, then followed by a recovery in voltage after a short time duration. This has a typical time duration of 10 milliseconds up to 1 minute. A voltage swell is almost always caused when a heavy load is de-energized or switched off. Voltage swells have more destructive effects than voltage dips. Regular swell events can lead to premature insulation failure within induction motors that result from higher current and heating effects. Components within power supplies get stressed over time, and they have a cumulative impact. The effects can cause failure of electronic components and sensitive devices. Mitigation measures include supply segregation and for electronics equipment to be supplied through a regulated power supply.

30.17   Suppression Methods. A number of methods can be used to reduce or hopefully eliminate interference. The use of shielded cables along with proper grounding is important; however, the use of proper equipment enclosures is critical, as this minimizes electromagnetic radiation. A filter or capacitor is installed close to the "noisy" equipment and effectively short-circuits noise in the protected frequency range. Filters may take a number of forms.

    a.   **Filters.** The filter consists of either a capacitor or a combination of capacitor and inductor (LC filter) connected across the power supply lines. Some use filters with a very low ground impedance, typically lower than 20 milliohms at 1kHz, which cleans up ripple. An option is to supply sensitive equipment through a Navpac NP-12 from NewMar. This is a supply-conditioning module that filters out spikes and noise, regulates supply voltage, and has an internal power pack to ensure supply continuity. NewMar's StartGuard also protects against surges that occur when the voltage drops when starting engines. This device is connected in parallel with the equipment, and the sense circuit wire is connected with the starter switch or solenoid. The internal battery supplies the load when starting and recharges when in standby mode. The units are rated at 20A. Check out www.newmarpower.com.

b.   **Suppressors.** Suppression modules often use metal oxide varistor (MOV) technology and are available in AC and DC types. Many alternators do not have these fitted, so install them. Normally you will have noticed radio noise or interference on electronics equipment. A 1.0 microfarad suppressor is a starting point, but even experimentation with a couple of automotive types is simple and inexpensive.

c.   **Ferrite Beads and Chokes.** A ferrite bead has many names, including ferrite block, ferrite core, ferrite clamp, ferrite ring, EMI filter, ferrite collar, ferrite ring filter, and ferrite choke. They are basically electrical conductors surrounded by a magnetic material called ferrite, which is a ceramic material made of iron. The ferrite bead is a form of choke used to suppress or attenuate high-frequency (HF) electronic noise within electronic circuits and dissipate them as heat. There are many types of ferrite beads; the main type is the axial ferrite bead or clamp, which is cable mounted. The beads allow differential mode signals to pass but block common mode currents, as they interrupt RF ground loops and prevent any RF from coupling into the cables. They are ideal for eliminating problems in email connections to notebook, HF modem, and SSB connections and are recommended by SailMail in their installations. They can be used on any cables, such as autopilot cables or others exposed to interference. It is important when clipping on ferrites that no air gaps are left between the ferrite halves. Coaxial ferrite line isolators are available from various sources and are used on the coaxial cable and placed on the cable near the tuner unit. These block the stray RF ground path from the coaxial shield and transceiver grounds.

d.   **Power System Stabilization.** In cases of high-voltage induction, it is necessary to clamp voltages to a safe level, typically around 40V. One of the major causes of lightning strike damage is the failure of equipment power supplies to cope with high-voltage transients. The most common method of achieving this is to connect an MOV across the power supply. As the voltage rises, the resistance alters to shunt the excess voltage. A second method is to use an avalanche diode across the supply. MOVs are designed for AC systems, and DC surges tend to have longer time durations. An MOV can be blown without any indication on a single event.

30.18   Practical Suppression Methods. A number of methods can be used to reduce or eliminate interference; these are shown in the circuit diagrams below.

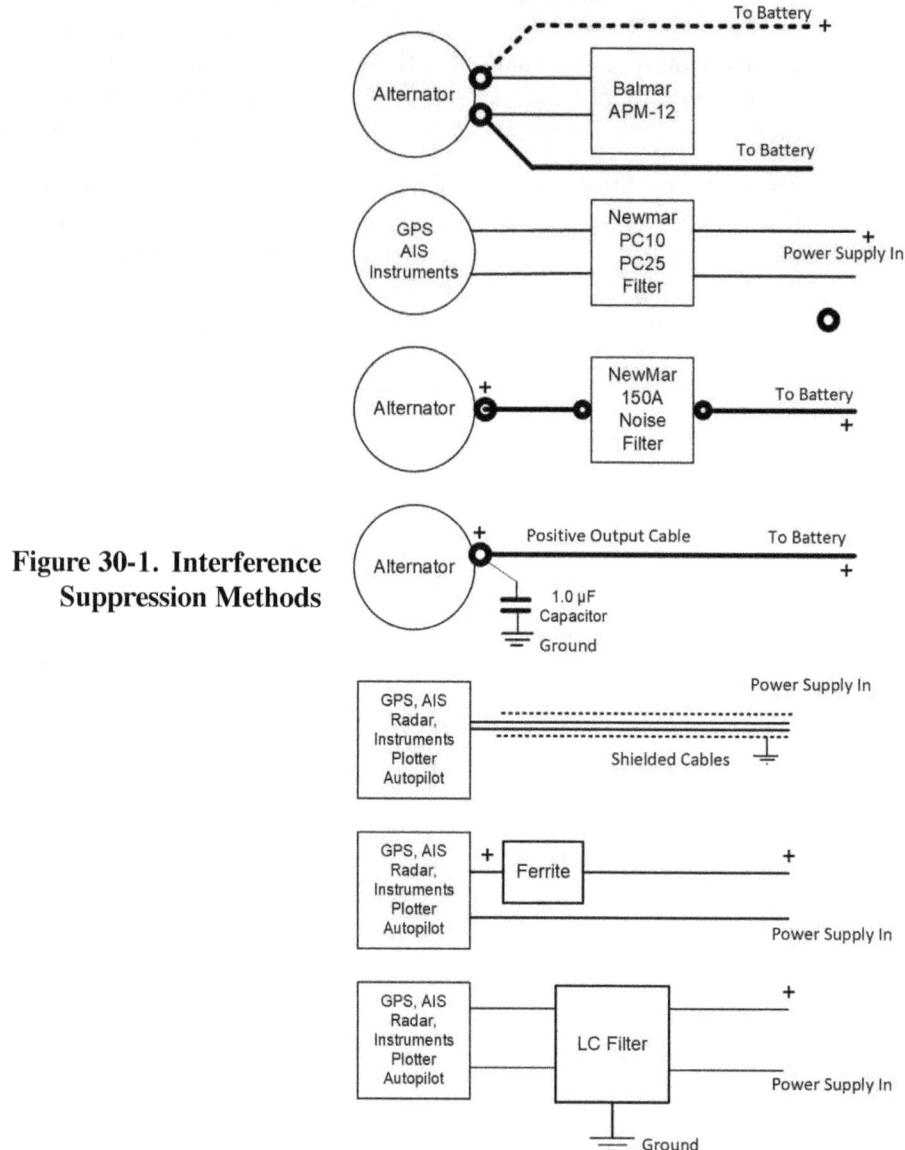

Figure 30-1. Interference Suppression Methods

30.19   Screening. Screening is used to mask sensitive equipment from radiated interference. Common sources included radio equipment and high current–carrying cables. The equipment or cables are covered and grounded by metal covers or screens (commonly called the Faraday cage principle). This may be a simple aluminum cover grounded to the RF ground point.

> a.   **Cable Covering.** Noisy power cables can be wrapped in noise tape, a flexible copper foil with an adhesive backing such as that from NewMar. I have successfully used this product and method on previous boats.

b. **Cable Shields.** Shields are designed to protect against interference from unknown or unspecified sources. The effectiveness of shields is measured in terms of transfer impedance—a measure of effectiveness in capturing the interference field and preventing it from reaching the conductor pairs inside. Data cables have shields made from a foil/polymer laminate tape or layers of braiding. They may have a drain wire installed to enable termination of the screen. Most manufacturers also specify the termination of shields. Never ground at both ends; always ground one end only, typically at the equipment end. In many cases shields are not connected at all, so check and connect them if necessary.

c. **Grounding.** The ground must be clean, which means it should have a ground potential between equipment no greater than 1V peak-to-peak. A ground is capable of conducting transients and emissions, so it must be sound. Another grounding source on boats is the grounding of static-causing equipment, such as shafts and engine blocks. As discussed in the alternators section, the negative connections to the engine block are a common source of problems. Ensure that the starter motor negative is attached close to the starter itself. Add an additional negative to the alternator. In many cases, interference is caused by arcing and sparking within the engine, as it is effectively part of the negative return conductor. Modifying the negative system eliminates this problem. Ensure that all ground connections are clean and tight.

d. **Twisted Pair Noise Cancellation.** The wires to a piece of equipment can be twisted together. This effectively causes cancellation as the electrical fields are reversed. When a current passes through a wire, an electromagnetic field is generated around the wire. The twisted pair concept employs the field that forms between the twists to cancel out the noise induced into one of these wires. If the wire is twisted, each twist of the wire can cancel out the noise generated in the twist just prior to it.

30.20 Noise Troubleshooting. Tracing the sources of noise is a matter of logic and systematic switching off of equipment and circuits to identify the source. In some cases, it may consist of two or more sources causing a cumulative effect. Some noise will be simply intermittent, such as static discharges, which may be synchronized with hot, dry wind conditions, or lightning pulses, which may not even be visible locally. It used to be that a cheap battery-powered AM radio was a good tool for tracking down radiated sources on board, with static being easily picked up by passing the radio close to the equipment (although these are rather rare devices these days). Some noise is simply related to the time of day. Interference from ionosphere factors on HF/SSB radios is well described in that chapter. This may affect GPS, HF/SSB, and satellite communications, all simultaneously, giving the appearance of some greater problem.

30.21 Cable Installation Planning. Proper planning and installation of cables can significantly reduce problems since cables are a major source of interference. If you experience problems, cable routing will have to be assessed and possibly require rerouting of the sensitive cables. Care must be taken with segregation of power, data, and signal cables. Where transducer or network cables must cross a power cable, they should be done at a 90-degree angle. Ensure that recommended grounding recommendations and practices are adhered to.

Autopilots

The autopilot is one of the few indispensable marine electronic devices. It is often referred to as the non-complaining, non-eating, extra crewmember. The advances in autopilot technology are powerful microprocessors and equally complex software algorithms that provide "intelligent" control. Some years ago, Raymarine introduced AST (advanced steering technology), which used a rate gyro that monitored vessel pitch and roll and adapted to the boat's sailing characteristics. Now Raymarine has the innovative Evolution system that uses AI control algorithms. This system uses the EV-1 sensor core—a combination of solid-state sensors that input data to the processor. The EV-1 takes data for pitch, roll, yaw, and heading from a nine-axis monitoring system that monitors vessel motion in three dimensions. This comprises a three-axis digital accelerometer, a three-axis digital compass, and a three-axis gyro digital angular rate sensor. The software uses complex AI control algorithms, so the autopilot reacts almost instantaneously to both sea conditions and vessel dynamics accurately to within 2 degrees in all conditions. The Garmin Reactor 40 autopilot follows a similar route. It also incorporates an attitude heading reference sensor (AHRS) system, which is a nine-axis solid-state device. With the programed algorithm, it processes data from nine sensors for heading, pitch, roll, and yaw.

31.1   Autopilot Basics. Most autopilot problems occur as a result of incorrect installation, improper matching to the vessel, or improper operation rather than personality conflicts. The basic function of an autopilot is to steer the vessel on a predetermined and set course, to a position or waypoint, or to wind angle. The pilot makes course corrections at a value corresponding to the course error, usually correcting to eliminate any overshoot as the course is met. All autopilots are microprocessor based and use the proportional rate system of operation. Correction is based on the amount of course deviation and the rate of change. Autopilots vary depending on the type of steering system used. The factors affecting autopilot selection are as follows:

a.   **Autopilot Selection.** An autopilot is selected on the basis of a number of important criteria:

(1)   **Steering Type.** The installed steering system, either tiller or wheel hydraulic, wire or direct drive.

(2)   **Displacement.** The loaded vessel displacement, as it is more valid than boat length, which has wide variations in beam, draft, and displacement. A common recommendation is that 20% be added to design displacement to get realistic cruising displacement.

(3)   **Sailing Style.** The type or style of sailing is important. For cruising you must base all factors on the worst weather possible, which means that power ratings must be capable of coping with prevailing conditions.

b.   **Power Consumption.** Always compare the current consumption at full-rated load, not average consumption. Many find that the autopilot uses far more power than expected, although much of the heavy consumption relates to excess weather helm activity and overworking of the pilot. There is no significant difference between the average consumption of the various drive types for a specific vessel size.

c.   **Factors Affecting Performance.** The following factors must be considered when selecting an autopilot, including the speed of rudder travel, rudder size, required number of turns lock-to-lock, and expected wind and sea conditions.

d.   **Autopilot Torque.** Torque is the force required to hold the rudder in position due to the pressure of water on the rudder, and to overcome the steering gear resistance of bearings and steering system drives. The vast majority of people underestimate this, and while the pilot works well in average conditions, it fails to keep course in bad weather.

e.   **Sail Trim.** Overloading and burnouts are commonly due to excessive weather helm. Weather helm is the tendency of a sailing vessel to turn toward the wind and create an unbalanced helm that requires pulling the tiller or steering toward the wind to counteract. If the helm is constantly held over by the pilot, trim the sails. To improve performance and reduce electrical and mechanical loads, reduce or minimize vessel heeling, ease the mainsheet, or travel to leeward. Weather helm will be reduced whenever the angle of heel when reaching or going to windward is reduced. Weather helm results from the leeward and aft shift of the vector center of effort, and this is caused by excess mainsail pressure. It is a good idea to reef the first time you think about it. The power saving is worth the effort.

31.2   Drive Types. Several types of drive are in use, depending on the steering configuration of each vessel. These are very briefly summarized below.

a.   **Mechanical Rotary Drives.** These drives are usually fitted on vessels where linear drives cannot be installed, where there are space restrictions, or where an inaccessible or small steering quadrant cannot accommodate any other drive. They can be powered from the helm position through a chain and sprocket.

b.   **Electrical Rotary Drives.** The motors on these systems consist of an electric motor coupled to a precision manufactured epicyclic gearbox.

c.   **Wheel Drive.** Wheel pilots are usually located on the steering pedestal. The drive unit is mounted in line on a cockpit side and consists of an integrated gearbox and motor, rotating the wheel via a belt. Vessel steering characteristics can be programmed into the control system, and a simple clutch lever enables instant changeover to manual steering. Belts must be correctly tensioned to avoid premature breakage or wear. These are generally suited for vessels for smaller day sail, weekending, and limited coastal sailing boats.

d. **Tiller Drive.** These are used on smaller boats and are not suited for offshore use—something I can attest to from experience. Larger tiller-steered boats just cannot handle it.

e. **Mechanical Linear Drive.** The linear drive unit is either an integrated hydraulic ram and pump system or a motor and gearbox drive directly connected to the rudder quadrant. The mechanical linear drive will move the rudder by pushing the tiller arm or the rudder quadrant. The linear drive has a minimal effect on helm "feel." It is relatively low cost, and the hydraulic units are very reliable. There is the advantage of a backup steering if some part of the steering drive or pedestal fails. Typical power consumption is relatively low, in the range 1.5A to 3A, 2.75A to 6A for larger vessels.

f. **Mechanical Hydraulic Linear Drives.** These drives are used on larger and heavier displacement boats. They generally comprise a reversing pump, a hydraulic oil reservoir, and a hydraulic ram. These drives are attached to the rudder quadrant or stock using a tiller arm. In these installations, the boat steering system must be capable of being back-driven from the rudder.

31.3   Hydraulic Steering Systems. Many steering systems are hydraulic; there are direct-connected units as well. The typical yacht hydraulic steering system is composed of several elements that make up the hydraulic actuator, which in most cases is a cylinder. The steering wheel operates the helm pump. When the wheel is turned, the helm pump also turns and pumps oil through the steering hydraulic circuit. For example, if the helm is turned to starboard, the hydraulic oil is pumped into the starboard hydraulic steering line. The pressurized hydraulic oil is input into the starboard side of the steering cylinder. As the cylinder moves to starboard, the oil from the port side of the actuator or cylinder then returns to the helm pump through the port hydraulic lines. The end of the cylinder is fixed and the opposite end connected to the tiller arm; this is connected to the rudder shaft. The tiller arm connection linear cylinder movement converts this force to a rotary movement, pushing the rudder over to starboard. If you turn the wheel to port, the same process occurs; the oil is pumped to the port line and returns through the starboard line. There are three basic types of hydraulic steering systems: the two-line system where pressurized fluid is pumped into the ram from either end, depending on the direction required and which has an external pressurized reservoir. In the three-line system system pressurized fluid flows in one direction only. A uniflow valve is installed within the system to direct all fluid back to the reservoir.

31.4   Hydraulic Reversing Pumps. Principal manufacturers include Octopus, Accu-Steer, Raymarine, and Simrad. Some are gear pumps, others piston pumps, and others gerotor. When an autopilot is installed, a hydraulic reversing pump unit is connected and has the same port, starboard, and return or fill line configuration. The Accu-Steer reversing pump is a good example. It employs a gerotor pump—a low-pressure, positive and fixed displacement pump that is both compact and has great power. The pumps have a very low operational noise footprint, and they suit a broad speed and oil viscosity range, with very good suction characteristics. "Gerotor" is derived from "generated rotor." A gerotor pump is composed of an outer and an inner rotor and works on the orbit principle. The gerotor motor consists of seven lobes and six teeth, with voids or spaces between them acting as pressure chambers. The pressurized hydraulic oil flows into these chambers and creates high pressure within one chamber and low pressure in the other. This imbalance of forces creates movement of the rotor, which starts rotating or orbiting. The output flow rate of the pump determines rudder speed. The Accu-Steer HRP35 model is a good example. The pump assembly comprises the reversing gerotor gear pump, flow control, pressure relief, hydraulic lock valves, suction makeup check valves, the valve housing manifold, and the permanent magnet drive motor, which is brushless. The suction makeup valves are installed to facilitate pump breathing or venting if air enters the steering lines. The output check valves are part of the lock valve assembly. The lock valve assembly comprises two output check valves and a lock valve spool. The rudder speed, called the hard-over to hard-over (HOH) speed, is calculated by dividing the volume (cubic inches) of the cylinder by the output of the pump unit (cubic inches per second). Under normal conditions the rudder speed is in the range of 8 to 16 seconds, so the pump output and cylinder volume must be matched correctly. If self-calculation of a cylinder volume is required, the formula is given as Volume = L (D2 –d2) $\pi \div 4$, where L = length of stroke of cylinder; D = internal diameter of cylinder; d = diameter of piston rod; and $\pi = 3.14$ (pi).

31.5   **Hydraulic Constant-Running Pumps.** These are solenoid valve–controlled steering systems. The pump-pressurized oil pressure for directional activation of the hydraulic ram is controlled by opening electric-controlled solenoid valves. Constant-running pumps, as the name implies, run continuously while the autopilot is on. They use an electric solenoid to control the flow of hydraulic fluid to the steering ram, which drives rudder movement to port or starboard. Some pumps are dual speed to save power in lighter conditions. Raymarine and others use the pumps, which are primarily on larger boats. Octopus pumps are geared type and incorporate a fluid reservoir and solenoid control valve. They have an integral motor starter relay and running light. Some pump manufacturers now have brushless motors.

31.6   **Hydraulic Oil.** Cleanliness is absolutely mandatory. Any dirt and debris or particles lodged in a solenoid valve will create excess electrical power consumption and solenoid valve failure. The condition is often characterized by a slight rise in the operating pressure when the motor is running in idle. The recommended hydraulic oils vary between manufacturers, so you should verify what your system uses; ISO15 is the nominal standard. Always carry spare oil when cruising.

31.7   **Hydraulic Installation.** There are a number of important considerations when installing pumps.

      a.   **Hydraulic Pump Mounting.** The pump must be mounted in a horizontal position. I have seen units mounted in the vertical when it was more convenient. The pump should be mounted adjacent to the steering cylinder and must be securely mounted to prevent vibration.

      b.   **Hydraulic Check Valves.** Check or nonreturn valves must be fitted to the helm pump to prevent the autopilot pump from driving it instead of the ram.

      c.   **Hydraulic Joints.** Check and tighten all hydraulic joint and hose connections. Do not use Teflon thread tape or thread compounds on hydraulic joints. Wrap all connections with grease impregnated tape, such as Denso Tape, to prevent corrosion. Try and stick with the same brand or type of fittings, as they are manufactured to the same tolerances and therefore tend to leak less than mix-and-match fittings.

d. **Hydraulic Hoses.** Make sure that hoses are not installed in contact with any part that can cause chafe and wear. Hoses move a lot when they are under pressure and also with boat motions.

31.8   Hydraulic Pump Maintenance and Testing. Perform maintenance and testing as follows:

a. **Test Rudder Operation.** Drive the rudder lock-to-lock (hard-over to hard-over) using the pilot control unit. Ensure that the rudder moves to the same side as the required command signal. If reversed, the motor terminal connections require reversal at the autopilot control box, and this is a new installation issue. The oil expansion reservoir, if fitted, may require topping up. Make sure the rudder stops before reaching the mechanical stops; driving into mechanical stops at full speed may result in damage.

b. **Maintenance.** Before maintaining the autopilot, switch off the power supply and relieve any residual system hydraulic pressure. Dismantle the pump after 1,000 hours of operation. Examine oil seals and replace them (I always do this regardless of the condition). Check the motor brushes and replace if excessively worn. Clean the motor brush gear with an electrical cleaner, and make sure the brushes move freely in the brush holders. Change the oil and flush through old oil.

c. **External Cleaning.** Clean with a damp rag, and never use any abrasive materials, acids, or solvents. Absolutely DO NOT use a high-pressure jet wash-down unit to clean the autopilot components as the seals will be damaged.

d. **Software Updates.** Software updates are very important, just as they are on your mobile phone. All manufacturers regularly release updates. These can include improvements to performance or adding new features and functions. Before updating, make sure you back up any important files. Do not interrupt the power supply when downloading.

31.9   Hydraulic System Troubleshooting. The following faults and symptoms are applicable to most pump systems.

a. **Spongy Steering.** Air in the system is the primary cause of poor performance; it degrades autopilot performance and affects steering accuracy. Understand how to properly bleed your system. The system must be bled according to the manufacturer's instructions. When bleeding, ensure that the steering is operated stop-to-stop, or hard-over to hard-over, to expel air in the pump and pipework. This problem will greatly affect the performance of the autopilot.

b. **System Cleanliness.** The system must be absolutely clean, and no particles of dirt should be introduced into it. This means clean hands, clean tools, and clean oil. Particles commonly lodge in check valves, causing loss of pressure and back-driving of the steering wheel. Oil quality should be monitored; it can degrade with contamination or moisture absorption.

c. **Hydraulic System Bleeding.** System bleeding is essential after installation or at any time after breaking out hydraulic lines or pumps. Air within the system will affect the responsiveness of the steering and increase pump noise. Turn your wheel hard over to port and, while pressure is applied, check all joints for leaks. Repeat for the starboard side. Monitor the oil level; if the level drops, you have leaks or air in the system and need to repeat the process. Most manufacturers recommend an ISO #32 or ISO #10 oil type.

31.10   Direct Drive Systems. Simrad offers direct drives as part of the autopilot lineup. Their DD15 direct drive unit is both compact and rugged, and the drive efficiency is greater than hydraulic and electromechanical units. These units are designed for continuous operation and have a maximum output torque of 150kgm. The units comprise a pancake-flat wound motor that drives a planetary and spur gearbox, resulting in relatively low power consumption. The Simrad direct drive autopilot can be divided into five main parts: the electric motor, two-step spur gearbox, planetary gearbox, electromagnetic clutch, and final spur reduction gearbox. The Simrad direct drive has multiple advantages over existing integrated-drive units, as explained below:

a. **Electric Drive Motor.** The Simrad direct drive autopilot uses a flat-wound, or pancake, electric motor. They are also called disc armature and printed armature motors. These motors are characterized by flat rotors that are driven by an axially aligned magnetic field and have many advantages over standard DC motors. These motors have a high starting torque, low inertia, quick start-up, and rapid acceleration, which is required when a speed command is transmitted. The efficiency of these motors is around 72% compared to 50% for traditional DC motors.

b. **Planetary Gearbox.** The electric motor output is speed reduced by a 750:1 reduction gearbox. This is to obtain the correct rudder travel speed or hard-over time. Unlike worm-drive systems on other pilots, this uses a combination of a planetary gearbox and spur gear sets—one small gear and one big gear. The planetary gearbox has many advantages and increased efficiency over other gearbox types.

c. **Electromagnetic Clutch.** When the steering is moved manually, the autopilot should disconnect automatically. This system has a patented electromagnetic engagement clutch, which is controlled by the autopilot computer. This is much more efficient than friction-plate clutches.

Table 31-1. Autopilot Hydraulic Troubleshooting

Symptom	Probable Fault
Excessive pump noise	Air in hydraulic system
No piston movement on command	Oil valve closed Pump sucking in air Nonreturn valve leaking
Rudder moves back to amidships	Nonreturn valve leaking
Wheel moves with pump operation	Lock valve leaking
Piston moves erratically	Air in hydraulic system
Rudder movement stops with load increase	Pump underrated
Motor runs, solenoid valve does not operate	System not filled with oil Shut-off valve closed Vent or drain line not connected No power to solenoids Dirt in solenoid Solenoid coil failure
Pump motor does not run	No voltage to motor Check start relay Autopilot not powered up Motor terminal loose or off Motor brush stuck if fitted

31.11   Autopilot Installation. There are a few fundamental points to observe when installing autopilots. Appraisals and postmortems of many offshore incidents revealed that many problems were directly attributable to improperly installed autopilots. The following factors should be considered, as they are the major causes of problems:

a.   **Mechanical Installation.** Always ensure that the drive units are mounted and anchored securely. It is sensible to mount a strong pad at anchoring points, as it is quite common on fiberglass vessels to see the hull flexing because the inadequate mounting points are unable to take the applied loads. Autopilot drive motors can cause vibration and noise, and this is amplified by both hull and deck. Most manufacturers supply special bolts, rubber washers, and bushings that can significantly reduce these effects. If you have vibration on an existing unit, consider the same measures.

b.   **Autopilot Wiring.** There are a number of important points to consider, and all must be carefully monitored when installing an autopilot:

(1)   **Power Cables.** Make sure that power cables to drive units are rated for maximum current demand and voltage drops, as cable runs are normally long. As standard, I install a minimum 10AWG (6mm²) twin tinned-copper cable to the motor and computer unit.

(2) **Control Cables.** Observe the safe distance advice when installing special autopilot cables. Do not cut, shorten, extend, or alter the cables supplied.

(3) **Circuit Protection.** The general circuit breaker protection should be rated for the supply cable. The EV-1/2 has additional levels of internal protection. The drive motor circuit and cables have current-sensing and stall-protection monitoring within the ACU hardware and software. Additional protection is installed by an internal power fuse.

(4) **Power Source.** Do not connect an autopilot to an engine start battery; data loss and control disruptions occur when the engine starts, causing a surge condition on the supply.

(5) **Radio Cable Clearances.** Make sure that all wiring is routed well away from radio aerial cables, since interference is a major cause of problems during radio transmission. Ensure that a ground cable is run from the computer unit to your RF ground. In rare cases, you may have to put a foil shield on SSB tuner unit interconnecting cables as well.

(6) **Suppression Ferrites.** These are often installed or supplied with autopilots. Installation is important for EMC mitigation. Install in accordance with installation instructions; do not leave them off. Read about ferrites in the chapter on interference. If you have to remove the ferrites for maintenance or installation purposes, make sure to reinstall them in the same location. Use only the specified ferrites and do not substitute another type, as each has its own characteristics. As extra ferrites can create weight on a cable, ensure that cable is properly supported. Raymarine specifies the use of ferrites on all cables that are not supplied by them.

(7) **Grounding.** Many autopilot units have cables with dedicated drain conductors for the cable screen. This drain wire should be connected to the boat's RF ground. (Read the chapter on grounds.) Raymarine has an implementation guideline that specifies a flat, tinned-copper braid with a rating of 30A, ¼-inch dimension or greater. An equivalent wire conductor is 8AWG (8mm^2) for a 3ft (1m) run or greater. It is always recommended to keep connecting braids or cables as short as practicable.

31.12   Input Sensors and Devices. Optimum use of autopilots depends on interfacing or installing wind, heading, and speed data, and other information. The Raymarine Evolution autopilot system incorporates autopilot response levels. The Evolution allows quick configuration for optimum performance to suit the circumstances. The "Leisure" setting is for passage making, where tight heading control is not required. The "Cruising" setting is for good course-keeping without stressing the autopilot. The "Performance" setting prioritizes accurate and concise heading control.

a. **Course Computers and Actuator Control Unit (ACU).** The arrow on the front of the EV sensor must be parallel to the centerline of the vessel and pointing toward the vessel's bow. It houses the main power and drive electronics for direct connection to a vessel's steering system. Control units must be mounted in an accessible and dry location and at least 3ft (1m) from any source of electromagnetic interference such as cables; make sure you mount on a vibration-free vertical surface. As many controllers have a diagnostic LED status indicator, make sure you can access it. Some control units have a specific orientation requirement and are marked showing the forward bow-facing point; they must be mounted in parallel alignment with the longitudinal axis or centerline of the boat. Check your manual for specific details.

b. **Heading Sensor.** This is an external electronic or fluxgate compass. New-generation autopilots such as the Raymarine Evolution EV-1 no longer require an external electronic or fluxgate compass, as the course computer incorporates a nine-axis sensor called the attitude heading reference sensor (AHRS), which is the primary heading sensor. If a non-Evolution heading data input is designated, the unit will combine this with the inbuilt gyro and accelerometer data for improved heading data.

c. **Rudder Angle Reference Sensor.** This inputs rudder angle data to the course computer in degrees. This feedback is part of the course control calculation in conjunction with heading data. In some situations where crosscurrents exist, this may result in a port or starboard steering bias when the rudder is midships and centered. Some pilot controllers allow input of offset angles for compensation in these conditions. Accurate rudder angle data is a prerequisite for good course-keeping.

d. **Speed Reference Input.** The autopilot uses speed data when making calculations relating to navigation. At a minimum, this information is supplied from a GNSS (GPS) receiver providing SOG (speed over ground) data.

e. **Wind Speed and Direction Data.** The autopilot uses wind vane data to steer relative to a specified wind angle. This data is supplied from an analog wind transducer.

f. **GPS Input.** GPS position data is used to follow routes and for optimum steering directions, such as steering to a waypoint.

31.13 Input Device Installation. Optimum performance depends on installing input data devices and control units correctly.

a. **Course Computer Location.** This should be located clear of magnetic influences and away from radio aerial cables. While older units were prone to induced interference, newer units are generally made to strict international noise-emission standards. They must follow the same rules as compass installations and be well away from AIS, radar, VHF, or HF/SSB antenna or aerial cables.

b. **Compass Location FRP/GRP and Timber Vessels.** The fluxgate or electronic compass should be installed in an area of least magnetic influence, and close to the center of the boat's roll to minimize heeling error. Turning errors can arise if the compass is not properly compensated. Southerly and northerly turning errors increase with distance from the equator. This causes slow wandering and slow correction; compensation normally reduces this problem. Make sure it is aligned correctly.

c. **Steel and Alloy Vessel Locations.** Steel and alloy vessels pose problems due to the inherent magnetic field in the hull. Mount the fluxgate sensor at a minimum of 5ft (1.5m) above the deck. As this is often on the mast, it may become disturbed when radar or radio cables passing through the mast are carrying current or signal. Cable location is critical in these installations.

d. **Cable Installation.** Ensure that the compass is mounted clear of cable looms or any other metallic equipment. As fluxgates are invariably installed under saloon bunks, do not store any metallic items such as tool boxes or spares there. Make sure no AIS, radar, VHF, or HF/SSB antenna cables pass close to the compass and attached cables. The recommended minimum safe distance is 3ft (1m) from any transmission equipment or signal-carrying cables, including the AIS, VHF, GPS, and radar; for HF/SSB this should be 6ft (2m).

31.14   Wind Vane Mode. When the autopilot is in wind vane mode, it uses the apparent or true wind angle as the primary heading reference. As changes in the true or apparent wind angle occur, it adjusts the locked heading to maintain the original wind angle. Most pilots have several settings to choose from. In the Raymarine EV-1 and EV-2, when there are changes in either the true or apparent wind angle, the pilot adjusts the locked heading to maintain the original wind angle. Adjustments to selected wind angle are simple push-button commands. The Raymarine Evolution has wind shift monitoring; and if a change of greater than 30 degrees for 60 seconds is detected, a wind shift alarm is activated. Check your manual for information on using this function correctly. This unit also has wind trim response, which controls the response time to wind change directions. Higher settings give faster responses.

a. **Automatic Tacking Mode.** Called the AutoTack function, this is ideal for shorthanded sailing. AutoTack is a programmable turn through to the same relative wind angle on the opposite relative wind angle. Raymarine advises to trim sails to reduce standing helm. In wind vane mode, the autopilot will respond to long-term wind shifts but is unable to react to short-term wind gusts. They advise against using this function where sudden wind shifts are possible. In addition, they advise to sail a few degrees further off the wind in gusty conditions to cope with wind directions changes.

b. **Accidental Gybe Function.** This gybe inhibit feature prevents the vessel from turning away from the wind if AutoTack is accidently in the wrong direction.

31.15   Autopilot Controls and Alarms. Many adjustments can be made to achieve optimum autopilot operation. DO NOT use an autopilot to steer the vessel in any channels, fairways, confined areas, or heavy traffic zones, as VHF and HF/SSB operation has been known to cause sudden course changes. Wireless remote controls are now available for autopilot control, and this makes quick course changes easy from anywhere on deck. The latest autopilots, such as the Raymarine Evolution systems, have comprehensive alarm and monitoring capabilities. These include the "clutch short" alarm, which indicates that a short circuit has occurred in the drive unit clutch; the autopilot will power down. The "current limit" alarm indicates that the drive overload current was exceeded; the autopilot will drop to standby condition and the alarm will time out after 10 seconds. To resolve this condition, check the drive unit and connections for stall or short-circuit conditions. A "drive short" alarm indicates a short circuit within the drive unit; the autopilot will power down. A "drive stopped" alarm indicates a rudder stall condition has occurred or the power has been removed from the drive unit; this is triggered in auto, track, and wind modes. An "EEPROM corruption" alarm indicates that corruption of critical configuration data has occurred. A "rate gyro fault" alarm indicates that the gyro sensor has failed. A "rudder reference unit failure" alarm indicates that the rudder reference connection has been lost or has exceeded the limits; this includes the rudder reference transducer failing while in auto, the angle being more than 50 degrees, or the connection to rudder reference being lost. The "solenoid short" alarm indicates a short circuit within the solenoid; the autopilot will power down. Device connection or data source alarms indicate that a device is no longer connected or a data source is no longer inputting. Most pilots have very effective self-test functions; become familiar with them, along with the menu structure.

31.16   Rudder Settings. You need to understand a number of rudder-related settings and controls when operating your autopilot.

    a.   **Rudder Gain.** This relates to the amount of rudder to be applied for the detected heading error and must be calibrated under sail. It is inextricably linked to proper compass setup and damping.

    b.   **Rudder Feedback.** Rudder feedback instantaneously provides the precise rudder position information to the pilot. It is essential that the feedback potentiometer be properly aligned. Most new autopilots have a high-resolution potentiometer that offers more precise feedback than earlier, coarser units.

    c.   **Rudder Limits.** This controls the limit of rudder travel. The autopilot must stop before reaching the mechanical stops, or serious damage may result.

    d.   **Rudder Damping.** This calibration is used where a feedback transducer is installed and minimizes hunting when the pilot is trying to position the rudder. In the Raymarine Evolution system, rudder damping is used to prevent autopilot overactivity, which is observed as hunting. This is addressed by a range of damping levels. Rudder damping relates to dead band angles, which are configurable on the pilot control head. Finding the lowest acceptable value entails some experimentation.

    e.   **Deadband.** This is the area in which the heading may deviate before the pilot initiates a correction.

f. **Rate of Turn.** The rate of turn limitation is typically 2 degrees per second.

g. **Dodge Function.** This function allows a manual course change to avoid debris and so on and then automatically return to the original course.

h. **Auto Speed Gain.** This allows adjustment of the amount of helm applied at varying boats speeds.

i. **AutoTrim and AutoSeastate.** These functions were pioneered by Raymarine (Autohelm) and are now standard on many autopilots makes:

  (1) **AutoTrim.** This function automatically compensates for alterations in weather helm and applies the correct level of standing helm.

  (2) **AutoSeastate.** This function enables the pilot to automatically adapt to changing sea-state conditions and vessel responses. It alters automatically the deadband settings and controlled by the pilot software. The pilot does not respond to repetitive vessel movements, only to true course variations.

31.17   Compass Inputs. The basis for good autopilot performance is proper setting of compass damping. You should start with minimum damping and increase according to conditions. Failure to get this right will cause either lagging or overshooting as rudder is applied to maintain course. This of course has detrimental effects on power consumption rates, as well as making you sail a lot farther than you have to. Variation is the local difference between magnetic and true north, and this alters fractionally every year. The Evolution autopilot control has options for variation input and control.

a. **Heading Error Correction.** This correction compensates for northerly and southerly heading errors. Failure to do this will cause amplification of rudder responses on those headings.

b. **Compass Linearization.** The process on a new-installation Raymarine Evolution system is automatic, subject to some testing criteria. The process will commence automatically after the boat has turned approximately 100 degrees at a speed range of 3kts to 15kts. For linearization completion, a 270-degree turn is required. Completion times for this process vary according to the vessel characteristics, the EV installation, and the magnetic interference levels. If the deviation is above 45 degrees, the EV unit will require relocation to an area with less interference.

c. **Compass Swing.** This calibrates the autopilot compass by turning the boat in slow circles so that automatic adjustments are made for compass deviation. Each 360-degree circle should take 2 minutes or more to complete, and a speed of around 2kts is optimum. If the deviation is greater than 15 degrees, the process should be repeated; if similar to the first run, consideration to relating the compass is required. Once calibration is complete you can lock the compass to prevent automatic linearization.

d.   **Magnetic Variation.** The variation must be entered into the autopilot. Newer units have automatic compass linearization to correct for compass deviation errors, as described above:

    (1)   **Fluxgate Compass.** Input from the fluxgate compass gives accurate heading data to the course computer.

    (2)   **Rate Gyro.** This allows rapid real-time sensing of vessel yawing prevalent in lightweight vessels and multihulls in following and quartering seas. The data input supplements the fluxgate signal and allows fast correction to counter the rapid heading changes the fluxgate cannot compensate for.

31.18   GPS Data Input. Interfacing of GPS receivers is standard. Input from the GPS allows route and waypoint navigation. It is important to remember that position-fixing systems are subject to errors, sometimes extremely large. This will have obvious effects on the steering, so it is important to keep a regular plot, as the autopilot will not be able to recognize the errors. Always be aware that if the GPS passes corrupt or faulty data, the pilot may initiate a course and put your boat in danger. As mentioned before, do not steer with an unsupervised pilot in confined waterways or close in to the shore. There is little time to react if a large GPS error is transmitted.

31.19   Track Mode Control. Track mode or control enables a pilot to steer from waypoint to waypoint in conjunction with a GNSS (GPS) receiver. The autopilot automatically adjusts to take account of tide and leeway. To do so it takes cross-track error data and uses this to compute and initiate course changes to maintain the designated track. New-generation pilots such as those from a Raymarine allow this to be set up from an MFD or chart plotter display. Most pilots will activate an "off-course" alarm when the course error exceeds 15 degrees for 20 seconds. Alarm angles can be programmed in.

    a.   **Cross-Track Error (XTE).** This is the distance between the current position of the boat and the planned track line. Cross-track errors arise when there is course change to avoid something and then resume course for waypoint arrival with some conditions. If the XTE exceeds 0.3nm, an alarm is activated, Large XTE, and which side of the planned track line the error is on, port or starboard.

    b.   **Waypoint Arrival.** The waypoint arrival circle is a virtual circle around the target waypoint. When the boat enters this circle, an alarm sounds to indicate arrival. Be aware that the vessel may still have distance to run to the actual waypoint and should prepare for the autopilot to turn onto the next leg. You should prepare for the course change if sail handling or trimming is required.

31.20   Autopilot Maintenance. Use the following advice to increase the reliability and performance of your autopilot.

    a.   **Electronics Temperature.** This applies to tiller units. Install the pilot out of direct sunshine if possible, and keep it cool. Units are made of black plastic to facilitate heat transfer from internal components. While this aids in heat dissipation, the casing will absorb heat. To reduce this, cover the unit with a lightweight white sailcloth cover in hot sunny weather; use Velcro for easy removal. I did this on an older boat of mine, and it worked well.

b. **Corrosion Control.** Ensure that systems are not exposed to excessive salt water and that seals are intact. Exposed units will be protected by the additional cover.

c. **Plugs and Sockets.** Regularly check plugs and sockets for water and moisture. Make sure they seal properly.

d. **Cleaning.** Clean using a damp cloth. Do not use any solvents or abrasive materials. Do not use a high-pressure hose on the unit; the seals are not designed for this.

Table 31-2. Autopilot Troubleshooting

Symptom	Probable Fault
Display is blank	Loss of main power, fuse failure
No rudder response	Autopilot unit fuse failure Rudder jammed Plug or connection fault Control unit fault Radio interference
Rudder drives hard over	Loss of feedback signal Rudder limit failure Fluxgate compass failure GPS data corruption Control unit failure Wind data corruption Calibration settings incorrect
Wandering course	Overdamped compass Rudder gain setting incorrect Feedback transducer linkage loose Control unit fault Drive unit fault Gain setting incorrect

Global Maritime Distress and Safety System (GMDSS)

GMDSS was fully implemented in 1999 for all commercial vessels exceeding 300 GRT (gross register tonnage). In many areas VHF Channel 16, or MF/HF 2182kHz, is not monitored, and requirements are being dropped. The primary function of GMDSS is to coordinate and facilitate search and rescue (SAR) operations, by both shore authorities and vessels, with the shortest possible delay and maximum efficiency. It provides efficient urgency, safety communications, and broadcast of maritime safety information (MSI), such as navigational and meteorological warnings, forecasts, and other urgent safety information. MSI is transmitted via NAVTEX, International SafetyNet on INMARSAT C, and some narrow band direct printing (NBDP) radio telex services.

32.1   GMDSS Operational Details. Worldwide communications coverage is achieved using a combination of INMARSAT and terrestrial systems. All systems have range limitations that have resulted in the designation of four sea areas that define communications system requirements.

a.   **Area A1.** Within shore-based VHF radio range. Distance is in the range of 20nm to 100nm. Radio required is VHF operating on Channel 70 for DSC and Channel 16 radiotelephone. EPIRB required is 406MHz or L-band unit (1.6GHz). Survival craft require a 9GHz radar transponder and portable VHF radio (with Channel 16 and one other frequency).

b.   **Area A2.** Within shore-based MF radio range. Distance is in the range of 100nm to 300nm. Radios required are MF (2187.5kHz DSC) and 2812kHz radiotelephone, 2194.5 NBDP, and NAVTEX on 518kHz. Also needed are the same VHF requirements as Area A1. EPIRB required is 406MHz or L-band (1.6 GHz). Survival craft requirements are the same as in Area A1.

c.   **Area A3.** Within geostationary satellite range (INMARSAT). Distance is in the range of 70°N–70°S. Radios required are MF and VHF as above and satellite (with 1.5–1.6GHz alerting), or as per Areas A1 and A2 plus HF (all frequencies). Survival craft requirements are the same as in A1.

d.   **Area A4.** Other areas (beyond INMARSAT range). Distance north of 70°N and south of 70°S. Radios required are HF, MF, and VHF. EPIRB required is 406MHz. Survival craft requirements are the same as in Area A1.

32.2   GMDSS Radio Distress Communications Frequencies. The frequencies designated for use under GMDSS are as follows:

a.   **VHF**   DSC Channel 70, Channel 16, Channel 06 Intership, Channel 13 Intership MSI.

b.   **MF**   DSC 2187.5kHz and 2182kHz.

c.   **HF4**   DSC 4207.5kHz and 4125kHz.

d.   **HF6**   DSC 6312kHz and 6215kHz (CH421).

e.   **HF8**   DSC 8414.5kHz and 8291kHz (CH833).

f.   **HF12**   DSC 12577kHz and 12290kHz (CH1221).

g.   **HF16**   DSC 16804.5kHz and 16420kHz.

32.3   Digital Selective Calling (DSC). A primary component of GMDSS, DSC is used to transmit distress alerts and appropriate acknowledgments. DSC improves accuracy, transmission, and reception of distress calls; VHF Channel 70 is the nominated DSC channel. DSC has the advantage that digital signals in radio communications are at least 25% more efficient than voice transmissions, as well as significantly faster. A DSC VHF transmission typically takes about 1 second, and MF/HF takes approximately 7 seconds, both depending on the DSC call type. A dedicated DSC watch receiver is required to continuously monitor the specified DSC distress frequency. VHF DSC radio equipment is a priority for small vessels, and some great units are available from Horizon, Raymarine, Icom, Standard, Simrad, and others. DSC equipment enables the transmission of digital information based on four priority groupings: Distress, Urgency, Safety, and Routine. The information can be selectively addressed to all stations, to a specific station, or to a group of stations. To perform this selective transmission and reception of messages, every station must possess a Maritime Mobile Service Identity number (MMSI) code. Note that distress "Mayday" messages are automatically dispatched to all stations. A DSC distress alert message is configured to contain the transmitting vessel identity (the nine-digit MMSI code number), the time, the nature of the distress, and the vessel position when interfaced with a GPS. After transmission of a distress alert, it is repeated a few seconds later to ensure that the transmission was successfully transmitted.

32.4   GMDSS Distress Call (Alert) Sequence. It is important to explain the various elements of GMDSS in an emergency situation.

a.   **Distress Alert.** This is usually activated from a vessel to shore. For sailing yachts, this is usually via terrestrial radio; larger vessels use satellites. Ships in the area may hear an alert, although a shore-based rescue coordination center (RCC) will be responsible for responding to and acknowledging receipt of the alert. Alerts may be activated via an INMARSAT A, B, or C terminal; COSPAS/SARSAT EPIRB (243/406 MHz); or an INMARSAT E EPIRB. DSC VHF or MF/HF can also activate alerts.

b.   **Distress Relay.** On receipt and acknowledgment of an alert, the RCC will relay the alert to vessels in the geographical area concerned, which targets the resources available and does not involve vessels outside the distress vessel area. Vessels in the area of distress can receive appropriate alerts via INMARSAT A, B, or C terminals; DSC VHF or MF/HF radio equipment; or NAVTEX MSI. On reception of a distress relay, the vessels concerned must contact the RCC to offer assistance.

  c. **Search and Rescue.** In the SAR phase of the rescue, the previous one-way communications switch over to two-way for effective coordination of both aircraft and vessels. The frequencies used are as outlined in the previous chapter.

  d. **Rescue Scene Communications.** Local communications are maintained using short-range terrestrial MF or VHF on the specified frequencies. Local communications take place using either satellite or terrestrial radio links.

  e. **Distress Vessel Location.** A search and rescue transponder (SART) and/ or the 121.5MHz homing frequency of an EPIRB assist in determining the precise location of the vessel in distress.

32.5 GMDSS False Alerts and System Coverage. GMDSS has commonly suffered a false-alert rate as high as 95%. False alerts are not desirable simply because of the load placed on SAR services. They are generally caused by operator errors and incorrect equipment operation. Another cause of false alerts is the improper acknowledgment of distress alerts, leading to excessive DSC calls. Training and experience in equipment operation is essential to resolve these problems.

32.6 GMDSS and Yachts. The installation of GMDSS is not compulsory for pleasure boats, but due to its universal implementation on commercial vessels, most yachts will need to be installed with GMDSS-compliant equipment simply to remain "plugged in" to the safety system. GMDSS will certainly maximize SAR situations for yachts, so in most cases it will dramatically enhance offshore safety. GMDSS equipment will accurately identify your yacht, current position, and type of emergency, and this information will be broadcast automatically. What you get is automatic activation of alarms at coast stations and on other vessels simply by pushing one button.

32.7 Satellite Communications Systems. Satellite systems play a major role in GMDSS, and prices have become more affordable for sailing yachts. INMARSAT was established by the IMO to improve distress and safety of life at sea communications and general maritime communications. INMARSAT is based on satellites placed in a geostationary orbit. Under GMDSS all commercial vessels operating in areas outside designated areas of international NAVTEX coverage require a receiver for reception of INMARSAT SafetyNET maritime safety information (MSI). The Standard-C SES is a GMDSS-compliant system that offers compact and lightweight terminals. These systems are designed to support data-only services, not voice.

EPIRBs

32.8 COSPAS/SARSAT System. GMDSS incorporates the COSPAS/SARSAT system as an integral part of the distress communications system. The acronym is based on the former Soviet "Space System for Search of Distress Vessels" and the American "Search and Rescue Satellite Aided Tracking." The COSPASS-SARSAT system consists of satellites that are configured in three constellations. The first is geostationary (GEOSAR) satellites. This is supported by low-earth-orbit (LEOSAR) satellites and medium-earth-orbit (MEOSAR) satellites. The MEOSAR satellites allow much faster response times, as they have overlapping footprints and enable almost instantaneous trilateration on an activated EPIRB. This currently stands at nine active GEOSAR satellites and five LEOSAR satellites. Ultimately there will be a network of 75 MEOSAR satellites. The satellites are maintained and monitored by strategically located ground stations that comprise local users terminals (LUTs), mission control centers (MCCs), and the various local rescue coordination centers (RCCs). Under GMDSS, if a vessel does not carry a satellite L-band EPIRB in sea areas A1, A2, and A3 (described earlier), a 406MHz EPIRB operating in the COSPAS-SARSAT system is required. This unit must have hydrostatic release and float-free capability. The system is a worldwide satellite-assisted SAR system for location of distress transmissions emitted by EPIRBs on the 121.5/243MHz and 406MHz frequencies, where 121.5kHz is an aircraft homing frequency and 243MHz is a military distress frequency that enables military aircraft to assist in SAR operations. The emergency position indicating radio beacon (EPIRB) is an essential piece of safety equipment for any offshore vessel. Earlier EPIRB units relied solely on overflying aircraft for detection of signals and relay of the position to appropriate SAR authorities; the new systems utilize satellites. Note that the 406MHz units are far more effective at lower latitudes than the 121.5/243MHz units, and the latter have been effectively phased out. In late 2022 a massive rescue effort was undertaken for a missing yacht by the US Coast Guard, commercial shipping, and the US Navy. The 30-foot yacht was somewhere on the East Coast between New Jersey and Florida. Eventually an oil tanker sighted the missing boat and rescued the crew. The yacht had been dismasted, had no fuel, no electrical power, and, even worse, no EPIRB.

32.9   Satellite (L-Band) EPIRBs. This system, developed by the European Space Agency (ESA), will alert rescue services in distress within 2 minutes rather than hours. The system combines position determination along with a distress signal using the INMARSAT geostationary satellites. The system uses special EPIRBs that incorporate GPS receivers and ensure a position fix within 650ft (200m). The distress signal transmits via one of the LUTs and landline links with appropriate rescue coordination centers.

32.10   406MHz EPIRBs. The 406MHz units have a unique identification code, and information is usually programmed at time of sale with MMSI or registered serial numbers. Some units have integral strobes and all incorporate 121.5MHz for aircraft homing signal purposes. Float-free units are called Category 1; manual bracket units are Category 2. Orbit time is 100 minutes using COSPAS satellites, so a delay in transmission up to 4 hours near the equator can occur; accuracy is 1nm (2km) and uses the Doppler effect, so two satellite passes are usually required. The system uses a store-and-forward system, so the satellite stores and downloads distress data when in view of a local user terminal (LUT). New technology units are the ACR Electronics GlobalFix V5 AIS EPIRB, which is an EPIRB incorporating near field communication, GNSS positioning, 406MHz COSPAS-SARSAT distress signal with return link service and 121.5MHz local homing. EPIRBs are now GPS enabled and allow location information to be transmitted with the distress transmission.

32.11   406MHz EPIRB Registration. If you acquire a yacht with a 406MHz EPIRB, you must register the EPIRB unit properly and provide all the appropriate data, including its unique identification number (or MMSI). Registration should be done immediately upon purchase. Failure to do this can cause absolute havoc if you use the EPIRB, because a vessel may be incorrectly identified or, worse still, not identified at all, which could seriously jeopardize your rescue. Bad information means bad rescue problems for everyone. If you have not registered, contact the organizations listed and do so.

a. **United States of America.** You can register online at https://beaconregistration.noaa.gov/RGDB/forms; there are also email and paper-based options.

b. **Canada.** Register online at www.cbr-rcb.ca/cbr/.

c. **United Kingdom.** Visit the UK Beacon Registry at www.gov.uk/register-406-beacons. This allows free registration of a UK 406MHz beacon, updating of your beacon registration details, and updating of your vessel details. You can register a personal locator beacon (PLB), emergency position indicating radio beacon (EPIRB), or simplified voyage-data recorder (VDR) for maritime use or an emergency locator transmitter (ELT). You can also email for advice at ukbeacons@mcga.gov.uk or telephone +44 (0)20-3817-2006.

d. **Australia.** Go to beacons.amsa.gov.au.

e. **New Zealand.** Go to beacons.org.nz.

32.12   EPIRB Activation Sequence. On activation of an EPIRB, the following sequence occurs.

a. A satellite detects the distress transmission. With 243/121.5MHz units, a satellite and the EPIRB must be simultaneously in view of the LUT.

b. The detected signal is then downloaded to an LUT. In 406MHz units the satellite stores the message and downloads to the next LUT in view.

c. The LUT automatically computes the position of the distress transmission. The distress information is then passed to a mission control center (MCC) before going to a rescue control center (RCC) and then to SAR aircraft and vessels.

32.13   EPIRB Operation. Do not operate an EPIRB except in a real emergency, because you could initiate a rescue operation. Do not even operate it for just a short period of time and then switch it off; authorities may assume your vessel went down quickly before circumstances stopped transmission. With current attitudes changing toward false alarms, it may reflect very badly on boaters as a whole in terms of wasting taxpayers' money. If you activate your EPIRB during an emergency, once rescued, do not leave the EPIRB in the raft or floating; the beacon may continue to transmit for some time.

32.14   Rescue Reaction Times. There is a mistaken belief that rescues are instantaneous after activation of an EPIRB. The reality, however, is a time lag that can average up to 6 hours or more from detection of a signal and physical location, although position is usually confirmed in less than 2 hours. This is dependent on suitable aircraft, weather conditions, and SAR coordinator response times. Every LUT has a "footprint" coverage area, and the closer you are to the edge of that footprint, the longer the delay. Time lags depend on intervals between satellite passes over a given location. There are six polar orbiting satellites and, although random in orbit, their tracks are predictable. If you have to activate, be patient and wait. Remember, you are not a survivor until you're on the deck of a rescue vessel or in the helicopter. Priority one is a survival training course. Have you taken one? Have you evaluated and planned a helicopter evacuation procedure?

32.15   Battery Life and Transmit Times. Much concern has been raised over battery transmit life after activation. Always ensure that the battery pack is replaced within the listed expiration date. Nominally a lithium battery has a life of 4 to 5 years, depending on the manufacturer. Typical transmit times are 80 to 100 hours at 5W output. Standards require a minimum of 48 hours.

32.16   EPIRB Maintenance. The only maintenance required is to test the EPIRB using the self-test function every 6 months in accordance with the manufacturer's instructions. Do not self-test by activating the EPIRB distress function. Do not drop the unit unless it is in the water.

32.17   Personal Locator Beacons (PLBs). These are not GMDSS equipment. PLBs are essentially miniature EPIRBs. They integrate 121.5MHz, the frequency used for homing in by SAR vessels and aircraft. Due to their small size, they can be attached to your wet-weather gear or life vest or carried in a pocket or panic pack. Some units are configured to activate in water; most operate for at least a 24-hour period, and some work for up to 48 hours. The PLB is not a substitute for a 406MHz EPIRB. The Ocean Signal RescueME PLB3 has a communication link to emergency services in a man overboard (MOB) emergency, even if the survivor is incapacitated. This mobile connected compact unit integrates GNSS positioning, 406MHz, 121.5MHz signals, and AIS transmissions, as well as the new Galileo return link service (RLS). It's an essential kit for blue water cruising, I think, given the difficulty in finding and recovering people. ACR has the ResQLink AIS PLB. This PLB has both near field communications (NFC) and RLS.

32.18   Radar Search and Rescue Transponders (SARTs). Under GMDSS these units are required on all vessels over 300 GRT. These devices are designed for use in search and rescue. An EPIRB will put potential rescue vessels in the area, but the transponder will accurately localize your position to search radars. Units typically have the following characteristics:

> a.   **Signal Transmission.** The transponder is omnidirectional and responds automatically when triggered or scanned by a radar emission or pulse. It emits a 9200GHz to 9500GHz high-speed frequency sweeping signal that is synchronous with the scanning radar pulse.

> b.   **Signal Reception.** On reception of the signal, the position is indicated on radar screens as a line of twelve blips giving range and bearing.

> c.   **Transponder Receiver.** The transponder gives an audible alarm when the radar emission of a search and rescue vessel is detected.

32.19   AIS-SARTs. These are part of GMDSS. A self-contained radio device like an EPIRB, they transmit position information using the Automatic Identification System (AIS). Both the boat position and time synchronization are taken from an integral GNSS receiver, such as a GPS. They have all your boat data, such as an MMSI number, programmed in. The transmitted distress signal is detected by both Class A and Class B AIS receivers. The ACR GlobalFix V5 is the world's first mobile connected AIS EPIRB with RLS. This innovative EPIRB combines AIS, NFC technology, GNSS positioning, and 406MHz COSPAS-SARSAT distress signals with RLS and 121.5MHz local homing capability. When your mobile phone is held over the designated area of the device, the latest data is automatically transferred into the ACR phone app for review. If you want to maximize survival, this would be the device to have. Visit ACR at www.acrartex.com.

NAVTEX and Weatherfax

33.1   **About NAVTEX.** NAVTEX is an integral part of GMDSS as well as the Worldwide Navigational Warning Service (WWNWS). It is an automated information system that provides meteorological, navigation, and maritime safety information (MSI). When NAVTEX became available, I was an early adopter and installed an ICS Electronics NAVTEX receiver in the cockpit. While crossing the Bay of Biscay, it was a great source of weather information during what became a very difficult and challenging solo crossing.

33.2   **NAVTEX Basics.** The world is broken down into twenty-one NAVAREAs, and each of these is allocated responsibility for transmission of marine and safety information, such as navigational warnings, to vessels at sea. Within each of these areas are transmission stations. INMARSAT enhanced group calling (EGC) provides long-range information. Each of the NAVAREAs is divided into four groups, each with up to six transmitters with an allocation of 10-minute transmissions each 4-hour period. The exceptions are Australia and New Zealand, which regrettably decided against implementing the service. The transmissions are time shared to prevent interference on adjacent areas, and they have limited power outputs. Message reception requires a dedicated receiver on your boat. Broadcast times are included within the frequency listings.

33.3   **NAVTEX Broadcasts.** Messaging is broadcast every 4 hours in English on a pre-tuned and dedicated frequency of 518kHz and 590kHz from stations around the world. Another frequency used for "tropical and long-range use" is 4209.5MHz. These stations are part of each NAVAREA. Broadcasts are made in local languages. Effective operating range is typically around 250nm; however, some stations can extend out to 550nm.

Table 33-1. NAVTEX Broadcast Codes

Code	Message Class
A	Navigational warnings (cannot be rejected)
C	Ice reports
D	Search and rescue information and piracy warnings
E	Meteorological forecasts
F	Pilot service messages
G	DECCA messages (discontinued)
H	LORAN messages (discontinued)
I	OMEGA messages (discontinued)
J	Satnav messages (GPS, Galileo, GLONASS)
L	Navigational warnings (additional to letter A)
V	Notice to fishermen (US only; not in use)
W	Environmental (US only; not in use)
X	Special services (allocation by IMO NAVTEX panel)

33.4 **NAVTEX Message Priorities.** Prioritization is used to define message broadcasts. Vital messages are broadcast immediately, usually at the end of any transmission in progress. Those classed as "Important" will be broadcast at the first available period when the frequency is not in use. Routine messages are broadcast at the next scheduled transmission time. Those messages classified as "Vital" and "Important" will be repeated if still valid at the following scheduled transmission times. Messages incorporate a subject indicator code (B2 character), which allows acceptance and rejection of specific information. Navigational and meteorological warnings and SAR information are nonselective so that all stations receive important safety information. B2 codes include navigational warnings (buoy positions altering, wrecks, floating hazards, oil rig moves, naval exercises, and meteorological warnings such as gales, etc.); ice reports, SAR, and anti-piracy info (cannot be rejected); met forecasts shipping and synopsis; pilot service messages; other NAVAID messages; navigation warnings additional to A (cannot be rejected): A = Nav warnings, B = Gales, D = Distress information, E = Forecasts.

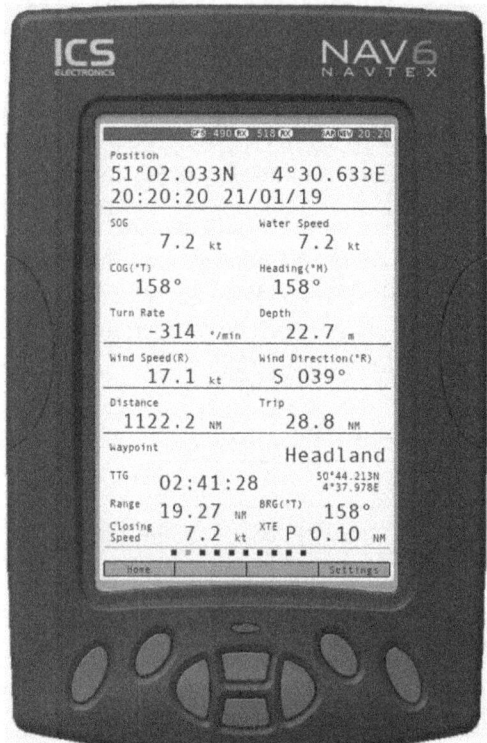 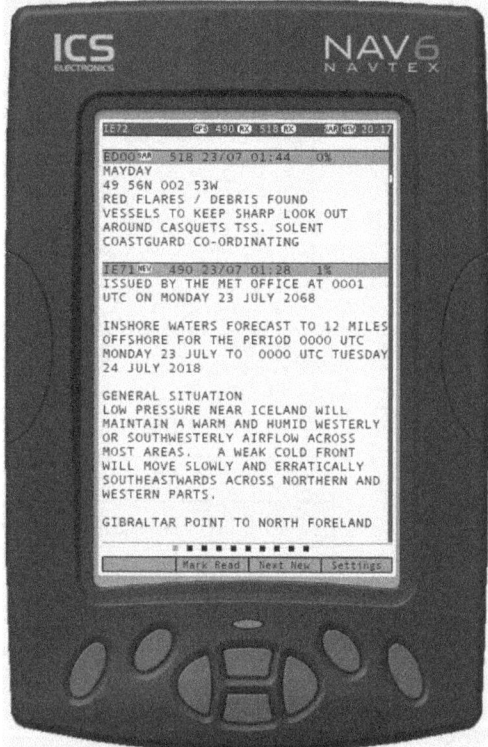

Figures 33-1.1 and 33-1.2. ICS NAV 6 NAVTEX Screens (*Courtesy of ICS Electronics*)

33.5 **NAVTEX Station Identification.** Navigation information is broadcast from a number of stations located within each NAVAREA, and broadcast times as well as transmitter power outputs are carefully designed to avoid interference between stations. Each station is assigned an identification code (B1 character). This is essential so that specific geographical region stations can tune in. Stations are selected by letter designation, such as M–Casablanca, I–Las Palmas; there two 24-hour forecasts each day for each sea area, and the message format contains the following. "Z" indicates there are no messages to transmit, checks system, and is an operational check message:

- Format nine characters, header code followed by technical code

- ZCZC B1 (transmitter ID), B2 (subject ID), B3, B4 (consecutive number)

- Time of origin

- Series ID and consecutive number

- The message text

- NNNN (end of message group)

33.6   US NAVTEX. There is a relatively reliable and continuous coverage on the West, Gulf, and East Coasts. There is some coverage near Kodiak in Alaska, Guam in the western Pacific, and Puerto Rico. The Great Lakes have no coverage, although a good proportion of them are covered by the Canadian Coast Guard. The following are the US NAVTEX stations and weather broadcast times. In 2019 the USCG proposed that US NAVTEX should be discontinued, citing aging MF NAVTEX equipment and replacement costs. They proposed replacement with IMO-recognized satellite services. No decision has been made at the time of this edition. (Data sourced from the ITU and NOAA.)

NAVTEX Eastern Pacific (NAVAREA XII). These are useful stations and broadcasts when cruising. Always check official schedules for broadcast time changes.

Table 33-2. Eastern Pacific NAVTEX–518kHz. NAVAREA XII

Station	B1 Character	Call Sign	Transmission Times (UTC)
Point Reyes, CA	C	NMC	00:20, 04:20, 08:20, 12:20, 16:20, 20:20
Cambria, CA	Q	NMQ	02:40, 06:40, 10:40, 14:40, 18:40, 22:40
Astoria, OR	W	NMW	03:40, 07:40, 11:40, 15:40, 19:40: 23:40
Kodiak, AK	J X	NOJ	J – 01:30, 05:30, 09:30, 13:30, 17:30, 21:30 X – 03:50, 07:50, 11:50, 15:50, 19:50, 23:50
Honolulu, HI	O	NMO	
Guam	V	NRV	01:00, 05:00, 09:00, 13:00, 17:00, 21:00 03:30, 07:30, 11:30, 15:30, 19:30, 23:30

33.7   NAVTEX Western Atlantic (NAVAREA IV). These are useful stations and broadcasts when cruising. Always check official schedules for broadcast time changes.

Table 33-3. Western Atlantic NAVTEX — 518kHz

Station and Character	Transmission Times (UTC)
Miami [A], USA	00:00, 04:00, 08:00, 12:00, 16:00, 20:00
Bermuda Harbour [B], BER	00:10, 04:10, 08:10, 12:10, 16:10, 20:10
Riviere au Renard [C], CAN	00:20, 04:20, 08:20, 12:20, 16:20, 20:20
Boston [F], USA	00:50, 04:50, 08:50, 12:50, 16:50, 20:50
New Orleans [G], USA	01:00, 05:00, 09:00, 13:00, 17:00, 21:00
Wiarton [H], CAN	01:10, 05:10, 09:10, 13:10, 17:10, 21:10
Curacao [H], NLD	01:10, 05:10, 09:10, 13:10, 17:10, 21:10
Portsmouth [N], USA	02:10, 06:10, 10:10, 14:10, 18:10, 22:10
St. Johns [O], CAN	02:20, 06:20, 10:20, 14:20, 18:20, 22:20
Thunder Bay [P], CAN	02:30, 06:30, 10:30, 14:30, 18:30, 22:30
Sydney (Nova Scotia) [Q], CAN	02:40, 06:40, 10:40, 14:40, 18:40, 22:40
Isabela [R], Puerto Rico	02:50, 06:50, 10:50, 14:50, 18:50, 22:50
Iqaluit [T], CAN	03:10, 07:10, 11:10, 15:10, 19:10, 23:10
Saint John (Yarmouth) [U], CAN	03:20, 07:20, 11:20, 15:20, 19:20, 23:20
Kook Island (Nuuk) [W], CAN	03:40, 07:40, 11:40, 15:40, 19:40, 23:40
Labrador [X], CAN	03:50, 07:50, 11:50, 15:50, 19:50, 23:50

Table 33-4. Western Atlantic NAVTEX — 490kHz

Station and Character	Transmission Times (UTC)
Saint John (Yarmouth) [V], CAN	03:35, 07:35, 11:35, 15:35, 19:35, 23:35
Sydney (Nova Scotia) [J], CAN	02:55, 06:55, 10:55, 14:55, 18:55, 22:55
Riviere au Renard [D], CAN	00:35, 04:35, 08:35, 12:35, 16:35, 20:35
Iqaluit [S], CAN	03:00, 07:00, 11:00, 15:00, 19:00, 23:00

33.8 **NAVTEX UK and Europe.** Having used NAVTEX sailing around the UK and Europe down to the Mediterranean Sea, I would highly recommend installing a NAVTEX unit. UK stations are Niton, Cullercoats, Portpatrick; Republic of Ireland are Malin Head and Valentia.

Table 33-5. NAVTEX UK/Ireland/Belgium/Norway/Netherlands—518kHz

Station & B1 Character	Transmission Times	Weather Information
Niton [E]	00:40 04:40 08:40 12:40 16:40, 20:40	00:40, 08:40, 20:40
Cullercoats [G]	01:00, 05:00, 09:00, 13:00, 17:00, 21:00	01:00, 09:00, 21:00
Niton [K] *Note 1*	01:40, 05:40, 09:40, 13:40, 17:40, 21:40	None
Portpatrick [O]	02:20, 06:20, 10:20, 14:20, 18:20, 22:20	02:20, 06:20, 18:20
Malin Head [Q]	02:40, 06:40, 10:40, 14:40, 18:40, 22:40	10:40, 22:40
Valentia [W]	03:40, 07:40, 11:40, 15:40, 19:40, 23:40	07:40, 11:40, 19:40, 23:40
Cross Corsen [A], France	00:00, 04:00, 08:00, 12:00, 16:00, 20:00	00:00, 12:00
Rogaland [L], Norway	01:50, 05:50, 09:50, 13:50, 17:50, 215:0	01:50, 13:50
Den Helder [P] *Note 2*	02:30, 06:30, 10:30, 14:30, 18:30 22:30	02:30, 14:30
Ostend [T], Belgium	03:10, 07:10, 11:10, 15:10, 19:10, 23:10	07:10, 19:10

Note 1: The Niton [K] broadcast is MSI information by France for the English Channel south of the median line along the Channel and east of 3°W. To the west of 3°W, MSI information for the waters off Brittany is included in the Cross Corsen [A] broadcast.

Note 2: The Den Helder [P] broadcast includes gale warnings; forecasts (12-hour); and a 12-hour outlook for *Thames*, *Humber*, *German Bight*, and *Dogger*; and gale warnings for sea area *Dover*.

Table 33-6. NAVTEX UK—490kHz

Station & B1 Character	Transmission Times	Inshore Weather Forecast
Niton [I]	01:20, 05:20, 09:20, 13:20, 17:20, 21:20	05:20, 17:20
Cullercoats [U]	03:20, 07:20, 11:20, 15:20, 19:20, 23:20	07:20, 19:20
Portpatrick [C]	00:20, 04:20, 08:20, 12:20, 16:20, 20:20	08:20, 20:20

33.9   NAVTEX UK Inshore Waters Forecast. The sixteen-area Inshore Waters Forecast is available through the national NAVTEX service on 490kHz. The text comprises a 24-hour forecast for a subset of the sixteen areas relevant to the area of reception of each of the NAVTEX transmitters, as shown in the table. A general 3-day outlook for all areas is added at the end of the text.

Table 33-7. NAVTEX UK—Inshore Waters Forecasts

Station	Areas Included in Broadcast
Niton	North Foreland to Selsey Bill Selsey Bill to Lyme Regis Lyme Regis to Lands End, including the Isles of Scilly Lands End to St. Davids Head, including the Bristol Channel
Portpatrick	Lands End to St. Davids Head, including the Bristol Channel St. Davids Head to Colwyn Bay, including St. Georges Channel Colwyn Bay to the Mull of Galloway, including the Isle of Man Lough Foyle to Carlingford Lough The Mull of Galloway to Mull of Kintyre, including the Firth of Clyde and the North Channel Mull of Kintyre to Ardnamurchan Point Ardnamurchan Point to Cape Wrath, including the Outer Hebrides Shetland Isles
Cullercoats	Shetland Isles From Cape Wrath to Rattray Head, including Orkney Rattray Head to Berwick on Tweed Berwick on Tweed to Whitby Whitby to the Wash The Wash to North Foreland North Foreland to Selsey Bill

Table 33-8. Mediterranean NAVTEX—490kHz

Station and Character	Transmission Times (UTC)
Istanbul [B], TUR	00:10, 04:10, 08:10, 12:10, 16:10, 20:10
Samsun [A], TUR	00:00, 04:00, 08:00, 12:00, 16:00, 20:00
Antalya [D], TUR	00:30, 04:30, 08:30, 12:30, 16:30, 20:30
Iraklio [Q], GRC	02:40, 06:40, 10:40, 14:40, 18:40, 22:40
Izmir [C], TUR	00:20, 04:20, 08:20, 12:20, 16:20, 20:20
Kerkyra [P], GRC	02:30, 06:30, 10:30, 14:30, 18:30, 22:30
Limnos [R], GRC	02:50, 06:50, 10:50, 14:50, 18:50, 22:50
Cross La Garde [S], FRA	03:00, 07:00, 11:00, 15:00, 19:00, 23:00
Cabo de la Nao [M], ESP	02:00, 06:00, 10:00, 14:00, 18:00, 22:00

33.10 **NAVTEX Mediterranean.** I have used NAVTEX in the Mediterranean, and it is a very useful weather source. I have not included all stations, just some useful ones to use when cruising. Always consult the updated official broadcast listings, which are subject to revision and change.

Table 33-9. Mediterranean NAVTEX—518kHz

Station and Character	Transmission Times
Istanbul [D], TUR	00:30, 04:30, 08:30, 12:30, 16:30, 20:30
Samsun [E], TUR	00:40, 04:40, 08:40, 12:40, 16:40, 20:40
Antalya [F], TUR	00:50, 04:50, 08:50, 12:50, 16:50, 20:50
Iraklio [H], GRC	01:10, 05:10, 09:10, 13:10, 17:10, 21:10
Izmir [I], TUR	01:20, 05:20, 09:20, 13:20, 17:20, 21:20
Kerkyra [K], GRC	01:40, 05:40, 09:40, 13:40, 17:40, 21:40
Limnos [K], GRC	01:50, 05:50, 09:50, 13:50, 17:50, 21:50
Cyprus [M], CYP	02:00, 06:00, 10:00, 14:00, 18:00, 22:00
Malta [O], MLT	02:20, 06:20, 10:20, 14:20, 18:20, 22:20
Split [Q], HRV	02:40, 06:40, 10:40, 14:40, 18:40, 22:40
La Maddalena [R], ITA	02:50, 06:50, 10:50, 14:50, 18:50, 22:50
Mondolfo [U], ITA	03:20, 07:20, 11:20, 15:20, 19:20, 23:20
Sellia Marina [V], ITA	03:30, 07:30, 11:30, 15:30, 19:30, 23:30
Cross La Garde [W], FRA	03:40, 07:40, 11:40, 15:40, 19:40, 23:40
Cabo de la Nao [X], ESP	03:50, 07:50, 11:50, 15:50, 19:50, 23:50

33.11 **NAVTEX Installation.** It is worthwhile remembering that NAVTEX is a medium-wave transmission, so a proper and quality aerial is required. Signal strength in harbors and some near-coastal areas, such as bays and inlets, may be very poor. The better quality the aerial, the better the reception.

WEATHERFAX

33.12 **Weatherfax Receivers.** Weather facsimile gives access to many stations that transmit weather charts, and the charts are a lot easier to interpret than foreign language voice forecasts. Transmitted data is varied and includes ocean current positions, sea temperatures charts, and current weather charts every 6 hours; forecasts up to 5 days in advance; sea state and swell forecasts; and ionosphere propagation forecasts. The technology used for sending weather faxes is the comparatively slow analog method. Images can subsequently take several minutes to download, and image quality is dependent on radio propagation. Weatherfax transmissions, such as those from the NOAA, are a result of analysis of GRIB (GRIdded

Binary) data, buoy data, ship observations, and other data sources, and the result of much knowledge, forecast modeling, and other information to arrive at something to send out.

a. **Weather Facsimile Components.** A facsimile transmission consists of a number of distinct components:

   (1) **Continuous Carrier.** This single tone is emitted before the start of any broadcast to allow the receiver to be tuned to maximum signal strength prior to data reception.

   (2) **Start Tone.** Also called the Index of Cooperation (IOC) select tone, this enables receivers to recognize the start of a transmission and to select the appropriate IOC drum speed. This governs the image resolution. The usual speed is 120 lines per minute (LPM); tone duration is 5 seconds.

   (3) **Phasing Tone.** This tone is used to synchronize the edge of the transmitted image. The tone has a duration of up to 30 seconds.

   (4) **Scale Tone.** Some systems enable the tone variations within the broadcast to be selected or varied.

   (5) **Body of Transmission.** This characteristic rhythmic "crunching" tone is the facsimile data being transmitted for decoding into an image.

   (6) **Stop Tone.** The stop tone is similar to a start tone and indicates the end of the transmission. The tone has a duration of around 5 seconds.

   (7) **Close Carrier.** This tone follows conclusion of the transmission; it has a duration of around 10 seconds.

b. **Decoders.** To obtain weatherfax data, it is necessary to obtain signals via a SSB or short-wave radio and decode it so that it can be displayed on your notebook computer. The basic function of a decoder is to convert transmitted audio signals into data. The audio signal is taken from the audio jack, if fitted, or a terminal on the rear of the SSB set.

c. **Discrete Systems.** On larger vessels, an integrated decoder and printer system is often used. A paper roll lasts a considerable period. The unit has a number of useful features. An additional aerial can be added for full NAVTEX reception, and the reception of RTTY and FEC signals is possible. Like most weatherfax units, programming of specific reception times is possible, which takes the worry out of looking up and catching broadcast times.

d. **Power Consumption.** The power consumption rate is relatively low, although it should take SSB consumption into account as well if a unit is left on permanently to capture programmed transmissions. If power consumption is an issue, you will have to power up before the required broadcast and after receiving shutdown again. Power consumption for a combination decoder and SSB over 24 hours can be at least 25Ah to 30Ah, which is considerable.

33.13   GRIB Broadcasts. This term is often seen when discussing access of weather data. The acronym "GRIB" is derived from "GRIdded Binary" or "General Regularly-distributed Information in Binary form." This is a concise data file format used to store and transmit gridded meteorological data, such as numerical weather prediction (NWP) model output. This was designed and is maintained by the World Meteorological Organization's Commission for Basic Systems. It is also denoted by the code GRIB FM 92-IX. While there are three versions of GRIB, GRIB Edition 0 is now obsolete. GRIB Edition 1 is still in use as part of the World Area Forecast System of the International Civil Aviation Organization (ICAO) and is still recognized by the WMO. The current GRIB Edition 2 format is a major upgrade of the GRIB standard. Edition 2 requires Acrobat Reader to view and is being phased in by European Centre for Medium-Range Weather Forecasts (ECMWF) along with other US and European NWP agencies. GRIB data is used to access raw weather model numerical data for decoding and data processing into visual images or any application using gridded data. Each GRIB file will contain one or more data records, arranged as a sequential bit stream, each with a header and packed binary data. Windy (www.windy.com) and PredictWind (www.predictwind.com/grib-files/) use GRIB files.

33.14   Meteorological Forecast Models. Phone apps for weather and software use a large variety of meteorological models, mostly identified by unrecognizable acronyms. Here's an explanation.

    a.   United States:

        (1)   Climate Forecast Model (CFS). This models the various interactions between ocean, atmosphere, and land at a global level. This comes from NOAA, with look-ahead up to 9 months.

        (2)   North American Mesoscale (NAM) Forecast System. This is one of the National Centers for Environmental Prediction (NCEP) principal models for weather forecast production. Resolution is approximately 12km in the United States.

        (3)   Global Forecast System (GFS). This is the NCEP weather forecast model. Data is generated from a large number of variables and has a resolution of approximately 27km. There are four updates every day.

        (4)   High-Resolution Rapid Refresh (HRRR). This is NOAA real-time hourly updates with a 3km resolution. It uses radar data and other inputs.

        (5)   Spire. This utilizes an extensive constellation of nano satellites that use a technique called radio occultation to collect atmospheric data, including air pressure, humidity, and temperature, globally. Resolution is 12km.

        (6)   Weather Research and Forecasting (WRF) Model. This mesoscale numerical weather prediction system is designed for atmospheric research and operational forecasting applications.

b. **UK and Europe:**

(1) **European Centre for Medium-Range Weather Forecasts (ECMWF).** This is the major European model; it issues twice-daily updates and has a resolution of around 9km. It is considered the most accurate global model available (www.ecmwf.int).

(2) **Icosahedral Nonhydrostatic Model (ICON).** This is a joint project of the German Weather Service and the Max Planck Institute for Meteorology. They are in the process of developing a next-generation global numerical weather prediction and climate modeling system. It has a 7km resolution in Europe and four daily updates.

(3) **AROME-Arctic.** This is a Meteo France–based model with a geographic resolution of 1.25km. It was primarily developed for the improvement of short-range forecasts, including high-intensity Mediterranean Cévenole events, severe storms, and other weather phenomena.

(4) **UK Meteorological Office (UKMO).** Also known as a unified model, it has an approximate resolution of 10km globally and just 1.5km in the UK. The model runs every 12 hours.

(5) **High Resolution Limited Area Model (HIRLAN).** This synoptic and mesoscale weather prediction model is under Finnish Meteorological Institute management. Developed by Sweden, Finland, Norway, Denmark, Iceland, Netherlands, Ireland, Spain, Lithuania, and Estonia, the model has good short-range accuracy.

(6) **Open Skiron (OS) Model.** This comes from Greece and is very good in the Mediterranean.

33.15   Weather Facsimile Frequencies. The following are selected weatherfax broadcast frequencies. You can go online and get full listings of broadcast times and content. *Note:* In upper side band (USB) mode, tune frequency 1900Hz (1.9kHz) lower; if in lower side band (LSB) mode, tune 1900Hz (1.9kHz) higher.

Table 33-10. Weather Facsimile Frequencies

Station & Call Sign	Frequencies
Kodiak (USA), NOJ	2054, 4298, 8459, 12412.5
Point Reyes (USA), NMC	4346, 8682, 12786, 17151.2, 22527
New Orleans (USA), NMG	4317.9, 8503.9, 12789.9, 17146.4
Boston (USA), NMF	4235, 6340.5, 9110, 12750
Halifax (Canada), CFH	122.5, 4271, 6496.4, 10536, 13510

Station & Call Sign	Frequencies
Sydney, NS (Canada), VCO	4416, 6915.1
Northwood (UK), GYA	2618.5, 4510, 8040, 11086.5
Charleville (Australia), VMC	2628, 5100, 11030, 13920, 20469
Wiluna (Australia), VMW	5755, 7535, 10555, 15615, 18060
Wellington (New Zealand), ZKLF	3247.4, 5807, 9459, 13550.5, 16340.1
Honolulu (Hawaii, USA), KVM70	9982.5, 11090, 16135
Tokyo (Japan), JMH/JMH2/JMH4	3622.5, 7795, 13988.5
Seoul (South Korea), HLL2	3585, 5857.5, 7433.5, 9165, 13570
Bangkok (Thailand), HSW64	7395
Rio de Janeiro (Brazil), PWZ-33	12665, 16978
Athens (Greece), SVJ4	4481, 8105
Hamburg (Germany), DDH3/DDK6	3855, 7880, 13882.5

33.16 **Computer-Based Weather Systems.** The onboard laptop computer offers a range of weather information options; some of these are as follows:

a. **Xaxero Weather Fax 2000.** A simple package, this is an easy-to-use hardware and software system for PCs. It offers a multitasking system with weatherfax frequency database, NAVTEX, Morse code, and TELEX capability. Check them out at www.xaxero.com.

b. **HF Fax.** This leading package uses a demodulator and has auto-tuning; signal tracking is possible. Once an image is received, you can zoom in, scroll, save to disc, or print out if required. The programs are capable of receiving and printing RTTY, FEC, and CW (Morse code) modes, as well as NAVTEX transmissions. The system has mouse control capability and can automatically control frequency of Lowe and Icom receivers. For more information check out www.psicompany.com/weather-fax/.

33.17. **Computer Fax Problems.** HF/SSB radio reception criteria apply to receiving weatherfax quality signals. The greatest problem with weatherfax reception is noise generated from the computer and picked up through the audio demodulator line. When troubleshooting, ensure that the frequency is accurately tuned; if instability and drifting occur, the signal will be inconsistent. Often weatherfax reception problems can indicate that aerial and earth connections in the HF/SSB system are defective. Check these out first. The frequencies being used may be affected by adverse propagation conditions, which affect all HF transmissions. It is advisable to retune to an alternative frequency and try again or wait until conditions improve. If this is a regular problem, obtain propagation forecasts. Check all sources of onboard electrical noise and interference.

VHF Radio

Very high frequency (VHF) radio is probably the most useful and essential radio communication equipment device available. It allows simple and easy ship-to-ship or ship-to-shore communications. The one disadvantage is that the range of a boat VHF radio is line of sight, and this is typically around 35nm. (See more about VHF aerials and radio range below.) All countries have licensing regulations that must be adhered to. Failure to comply may result in prosecution and fines. Major manufacturers include Icom, Uniden, Standard, Raymarine, GME, JRC, B&G, Garmin, Simrad, Lowrance, and Sailor. VHF has advanced considerably over the past few years, not only with GMDSS DSC VHF radios but also convergence with the inclusion of AIS and smartphone app capability. Some are compatible with and can be upgraded for NMEA 2000. The ICOM M510E is a Class D DSC transceiver with AIS. The large full-color TFT LCD display has near 180 degrees wide angle screen viewability. Command and control is available with iOS and Android smart devices. The hand mike has AquaQuake, which emits a buzzing vibration when water enters the speaker grill and sheds water from the speaker. Go to www.icomamerica.com for details. The Garmin Cortex V1 is similar with an AIS transponder, anchor watch, MOB functions and multiconstellation GNSS and smart phone remote monitoring and connectivity. Handheld VHF technology has also advanced. Many units have all the features of fixed mount types and are GMDSS compliant. A number of companies including ICOM, Standard Horizon, Uniden, Entel, Cobra, GME, Lowrance, McMurdo, and Midland manufacture VHF radios.

Many units have DSC including the Distress Alert red button feature and also MMSI calling. Some models have integrated AIS and GPS receivers. A MOB function is also a feature. Due to the lower power outputs and the low antenna height, the range is much less than the fixed units. Some have both AC and 12VDC charging capability and most now have lithium-ion (Li-ion) batteries. DSC radios also require MMSI numbers and there are licensing requirements. The Standard Horizon HX320 has features that include USB charging capability of the lithium-polymer (Li-Po) battery, Bluetooth for hands free operation, a submersible IPX7 rating; and it floats. It has a water activated strobe light, and three power settings: 1W, 2.5W and 6W. The HX890 is a Class H DSC VHF/GPS with a 66 Channel WAAS GPS receiver, with two inbuilt scrambler systems for secure communications and integrating an FM broadcast band receiver.

34.1   VHF Radio Licensing. All countries have boat VHF radio licensing regulations that must be adhered to. Failure to comply may result in prosecution and fines. All boat VHF installations must possess a station license issued by the appropriate national communications authority, e.g., FCC. At least one operator, normally the person registering the installation, should possess an operator's license or certificate, such as a Restricted Radiotelephone Operator Permit (RROP). See details below for the United States, Canada, UK, Australia, New Zealand, and Ireland; check the specific requirements of your local jurisdiction. There are also requirements for handheld VHF units.

34.2   VHF Radio Licensing, United States. In the United States the Federal Communications Commission (FCC) is the regulator. Criteria required for a radio operator's license or permit include the vessel type and size, the area the boat is operating in, who the communications will be with, and the radio equipment to be used. Most recreational boaters do not require a license. If your boat is under 300 tons, does not carry more than six passengers for hire, does not have an MF or HF radio, and is not in communication with foreign radio stations, you normally do not need a license—but check! Note that DSC VHF, along with SatCom and HF/SSB, radios have different requirements. You can obtain licenses online at www.fcc.gov. If your plan includes cruising overseas, you will need to get several licenses. You also will have to obtain an FCC registration number (FRN). The ship's station license (SSL) will assign you a unique call sign, which covers the VHF radio, radar, EPIRB, and all other radio transmission equipment. At this time, make sure you obtain your MMSI number, which is required for your satellite EPIRB and the GMDSS DSC VHF radio. In 2022 the FCC started issuing handheld VHF MMSI numbers on request from operators of licensed ship stations.

34.3   VHF Radio Licensing, Canada. The ROC license is a Canadian federal government–issued card that allows you to operate a marine VHF radio. Check out requirements at tc.canada.ca/en.

34.4   VHF Radio Licensing, UK. In the UK you will need a ship radio license. The handheld VHF/DSC radio requires a special license. A Short Range Certificate (SRC) is required for the operation of a VHF and a VHF DSC radio. The next level license is the Long Range Certificate (LRC), which allows you to operate MF/HF and VHF equipment that is required for GMDSS (Global Maritime Distress and Safety System) when venturing beyond Sea Area A1. These services are administered by AMERC in the UK. Check www.amerc.ac.uk.

34.5   VHF Radio Licensing, Republic of Ireland. Certificates of Competency are handled by the Maritime Radio Affairs Unit (MRAU) of the Department of Transport. The Irish SRC Operating License is required for operating a VHF radio. Visit www.gov.ie/en/service/1d6ea4-maritime-radio/ for details.

34.6   VHF Radio Licensing, Australia. Australia has the Australian Waters Qualification (AWQ). Licenses include the Long-Range Operator Certificate of Proficiency (LROCP), the Short-Range Operator Certificate of Proficiency (SROCP), and the Global Maritime Distress and Safety System (GMDSS) General Operators Certificate of Proficiency (GOCP). Check out details at ACMA (www.acma.gov.au).

34.7   VHF Radio Licensing, New Zealand. New Zealand has the Maritime VHF Operator's Certificate (MVOC). Visit these websites for further information, as well as a downloadable Marine Radio Handbook: maritimenz.govt.nz; www.boatingeducation.org.nz.

34.8   VHF Theory. The spectrum consists of fifty-five channels in the 156MHz to 163MHz band.

> **a.   Range.** As VHF is line of sight, the higher the antenna is mounted, the greater the distance. There are theoretical ways to work out the range, but for

simplicity I will leave them out. Factors such as atmospheric conditions and the installation itself also affect the actual range. Typical range with a coast station can be 35nm to 40nm. Handheld VHF radios have a much reduced range due to lower output power and aerial heights.

b. **Power Consumption.** Typically, it is 5A to 6A consumption on transmit. Reception-only consumption can add up, as the set is on virtually 24 hours. This is in the range of 1A to 7A. In a day, that can add up to 12Ah to 19Ah, depending on the set. However, VHF is one piece of equipment that should be left on regardless of power consumption, especially if it has integrated AIS function. The merchant ship that sights and tries to communicate with you will do so well before you may be aware of it.

34.9   VHF Signal Propagation. VHF signals penetrate the ionosphere rather than reflect. There are circumstances in which VHF signals can reflect back from the ionosphere to give "freak" long-distance communications, such as during very strong solar cycles. This occurred during Cycle 19 in 1957–58, Cycle 21 in 1980, and Cycle 22 in 1990. During these peaks, the monthly sunspot average rose to extremely high values and the ionosphere reflected higher frequencies than normal. VHF can also be reflected from clouds of increased ionization in the E layer of the ionosphere and during auroras, those spectacular light curtains caused by charged particles from the sun.

34.10   VHF Operation. As VHF is widely used by official and commercial operators, it is essential that you use your set properly for optimum performance.

a. **Channel Selector.** Select calling or working channel.

b. **Power Setting.** Select emission power. Always use the 1W Low (minimum) power setting for local communications and for stations or vessels within sight. Select the 25W high (maximum) power for distant contacts. For some designated channels, maximum power setting is mandatory.

c. **Squelch Setting.** Squelch, or noise limiter, reduces the inherent noise in the radio. Do not reduce the squelch too far, or you will lose the station.

d. **Volume Control.** Selects the sound level required.

e. **Simplex and Duplex.** Simplex means that talk is carried out on one frequency. Duplex is where transmit and receive are on two separate frequencies.

f. **Dual Watch/TriWatch.** This facility enables continuous monitoring on Channel 16 and the selected channel. Some now allow three-station monitoring.

g. **Talk Technique.** Hold the microphone clear of the mouth, approximately 2in (50mm), and speak at a volume only slightly louder than normal. Be clear and concise, and don't waste words. Many newer sets incorporate noise-canceling microphones.

34.11   VHF Radio Procedure. After selecting the required channel, use the following procedures (procedures are valid for coast stations or other vessels).

a. **Operating Procedure.** Wait until any current call-in-progress is terminated. Even if you do not hear speech, listen for dialing tones or other signals. Do not attempt to cut in or talk over conversations. Sometimes traffic may be busy and patience is required:

    (1) Always identify your vessel and call sign both at the beginning and end of transmission.

    (2) Keep conversations to a minimum, ideally less than 3 minutes.

    (3) After contact with other vessels, allow at least 10 minutes before contacting them again.

    (4) Always observe a 3-minute silence period on the hour and half hour. While not essential, it is good practice.

b. **Coast Station Calls.** Operate your transmitter for at least 7 to 8 seconds when calling, and use the following format:

    (1) Call the coast station three times.

    (2) State "This is <vessel name and call sign>" and repeat three times.

    (3) Response will be "Vessel calling <station name> this is <station name> on Channel <No.>." This is usually on VHF 16 or the nominated call channel.

    (4) Response will be "This is <call sign>; my vessel name is <name>."

    (5) State purpose of business, link call, and request information or advice.

    (6) On completion of business, "Thankyou <station>; this is <vessel name> over and out, and listening on Channel 16 or <Channel No.>."

34.12 VHF Distress, Safety, and Urgency Calls. Channel 16 should be used only for the following:

a. **Mayday.** Use of this distress call should be used only under the direst of circumstances, "grave and imminent danger." Use of the call imposes a general radio silence on Channel 16 until the emergency is over:

    (1) MAYDAY, MAYDAY, MAYDAY.

    (2) This is the vessel <NAME>.

    (3) MAYDAY, vessel <NAME>.

    (4) My position is "<Lat and Long>." True bearing and distance from known point.

    (5) State <Nature of Distress> calmly, clearly, and concisely.

(6)   State type of assistance required.

(7)   Additional relevant information, including number of people on board.

b.   **Pan-Pan** (pronounced PAHN-PAHN). Use of this call is to advise of an urgent message regarding the immediate safety of the vessel or person. It takes priority over all traffic except Mayday calls. The call is used primarily in cases of injury, serious illness, or man overboard:

(1)   <All Ships>.

(2)   PAN-PAN, PAN-PAN, PAN-PAN.

(3)   This is the vessel <NAME>.

(4)   Await response, and transfer to working channel.

c.   **Security** (pronounced SAY-CURE-E-TAY). For navigational hazards, gale warnings, and other warnings, as follows:

(1)   SAY-CURE-E-TAY, SAY-CURE-E-TAY, SAY-CURE-E-TAY.

(2)   This is the vessel/station <NAME>.

(3)   Pass the safety message.

d.   **Medical Services.** This call is used to advise of an urgent medical emergency. It takes priority over all traffic except Mayday calls:

(1)   PAN-PAN, PAN-PAN, PAN-PAN.

(2)   RADIOMEDICAL or MEDICO.

(3)   This is the vessel <NAME, CALL SIGN, NATIONALITY>.

(4)   My position is "<Lat and Long.> Diverting to <Location>."

(5)   Give patient details, name, age, sex, and medical history. Give present symptoms, advice required, and medication on board.

e.   **Phonetic Numbers.** When you need to spell out numbers, use the following:

No.	PHONETIC	SPOKEN	No.	PHONETIC	SPOKEN
0	Nadazero	NAH-DAH-ZAY-ROH	6	Soxisix	SOK-SEE-SIX
1	Unaone	OO-NAH-WUN	7	Setteseven	SAY-TAH-SEVEN
2	Bissotwo	BEES-SOH-TOO	8	Oktoeight	OK-TOH-AIT
3	Terrathree	TAY-RAH-TREE	9	Noveniner	NO-VAY-NINER
4	Kartefour	KAR-TOW-FOWER	**Decimal Point**	Decimal	DAY-SEE-MAL
5	Pantafive	PAN-TAH-FIVE	**Full Stop**	Stop	STOP

f. **Phonetic Alphabet. Boldface** indicates the pronunciation emphasis.

LETTER	PHONETIC	LETTER	PHONETIC
A	Alfa (**AL**-FAH)	N	November (NO-**VEM**-BER)
B	Bravo (**BRAH**-VOH)	O	Oscar (**OSS**-CAR)
C	Charlie (**CHAR**-LEE)	P	Papa (PAH-**PAH**)
D	Delta (**DELL**-TAH)	Q	Quebec (KEH-**BECK**)
E	Echo (**ECK**-OH)	R	Romeo (**ROW**-ME-OH)
F	Foxtrot (**FOKS**-TROT)	S	Sierra (SEE-**AIR**-RAH)
G	Golf (GOLF)	T	Tango (**TANG**-GO)
H	Hotel (HOH-**TELL**)	U	Uniform (**YOU**-NEE-FORM)
I	India (**IN**-DEE-AH)	V	Victor (**VIK**-TAH)
J	Juliett (**JEW**-LEE-ETT)	W	Whiskey (**WISS**-KEY)
K	Kilo (**KEY**-LOH)	X	X-ray (**ECKS**-RAY)
L	Lima (**LEE**-MAH)	Y	Yankee (**YANG**-KEY)
M	Mike (MIKE)	Z	Zulu (**ZOO**-LOO)

g. **DSC VHF Distress and Emergency Calls**. The red distress key or button is located under a flip-up key cover, and most DSC VHF radios are similar. In any distress situation, you lift the cover and press the button. If time allows you can scroll through distress description options on the menu and select the correct situation before pressing the button. This might be fire, collision, sinking, or MOB. You must PRESS and HOLD the BUTTON IN for at least 5 SECONDS or longer. The required depress time varies between makes; Icom radios are 3 seconds. A single button press will not transmit the distress message. Some radios will emit an audible countdown time with a series of "beeps." If time permits, stand by for an acknowledgment call. It is always advisable to then broadcast your distress or MAYDAY message on Channel 16 as well after that. Standby on Channel 16 for emergency contact. Also note that your DSC radio must be registered with your MMSI and also have a GPS data input. If you make a distress call in error, you should immediately notify the Coast Guard to avoid unnecessary rescue mobilization. Radios have a Distress Cancel call function. Do not forget to activate your EPIRB as well. For further details refer to Chapter 32, GMDSS, and read your DSC VHF radio product manual and be familiar with all the functions.

34.13 **VHF Antenna.** The majority of sailing yachts use VHF whip antenna. The antenna length is directly related to the antenna gain; the higher the gain, the narrower the transmission beam. A high-gain antenna has a greater range, but during vessel rolling and pitching, the lower gain antenna is more reliable with a greater coverage pattern. Half-wave omnidirectional whip antenna are typified by the stainless-steel rod construction. The radiation pattern has a large vertical component, which suits boats under heel conditions; most now are suitable for AIS applications as well. These VHF antennae can come in the form of a whip, with lengths varying from 3ft to 10ft (1m to 3m). The fiberglass whip effectively increases the height—about 8ft (2.5m)—and therefore the range of the radiating element; the gain is typically 3dB. Helical antenna have a gain of around 2.5dB but have a characteristically wider signal beamwidth. The higher the gain, the more directional the emitted signal becomes. For day coastal and weekend sailors the Shakespeare 5912-DORSAL VHF, AIS, and AM/FM antenna is an interesting alternative. The unique shark fin shape eliminates rope snagging and provides a range of 10nm at 25W or 4nm at 1W. I always carry an emergency antenna, and new technologies require new contingency plans. With AIS you need a high-performance emergency VHF antenna with AIS compatibility. The Revolve-tec emergency antenna is based on the Rolatube design and can be deployed quickly should you dismast or lose your antenna. The antenna is integrated into a rollable composite material. The antenna is ultra-light, 8ft (2.5m) long, and compact, with a high-visibility cover and an integrated SOLAS-approved LED strobe for location assistance. It has a 20ft (6m) cable with a PL259 connector. Check out revolve-tec.com.

34.14   VHF Antenna Coaxial Cables and Connections. Cables and bad connections are the principal causes of degraded VHF antenna performance. Avoid using thin RG58U coaxial cable where possible, as the attenuation is increased and large signal loss can occur. The amount of signal that gets out is dependent on low losses within the cable and the connections. For cabling antenna, always use RG213/U or RG8/U 50Ω for minimum attenuation, or switch to LMR-400-DB for optimum performance. Ensure that the cable has no sharp bends that may affect the attenuation of the cable. The typical cable attenuation of both types for a 100ft (30m) run is as follows:

a.   **About Gain.** The antenna gain is the principal parameter with a VHF antenna. This is defined as the quantity of energy the antenna is able to transmit and receive in the main direction when compared to a reference antenna. A reference antenna is considered as isotropic in that it will distribute energy in all directions. Gain is denoted as "dBi"; a reference antenna is a half-wave dipole and is denoted as "dBd." The formula for conversion between the two is dBd = dBi +2.15 dB. Antennae generally quote the gain as "marine dB"; however, this isn't a true value. If an antenna has a dBd of 0, the dBi will be 2.6 and the marine dB will be 3. If it has a dBd of 3, the dBi will be 5.1 and the marine dB will be 6. For a dBd of 4.5, the dBi will be 6.6 and the marine dB will be 9. For a dBd of 5, the dBi will be 7.1 and the marine dB will be 10.

b.   **RG58/U.** This is a nominal 7.1dB, which is a signal loss of 75% to 80%.

c.   **RG8/U, RG8/X (RG213/U), RF400.** This is a nominal 2.6dB, which is a signal loss of approximately 45%. When a signal is transmitted via the cable and antenna, a portion of that signal energy will be reflected back to the transmitter. The effect is that coverage is reduced due to the reduced power output. Measure the VSWR with a meter. The RF400 cable is 50Ω low-loss coaxial cable and is the general replacement for RG213U coaxial cable.

RG214U cable has a double screen and is recommended for long coaxial connection distances.

d. **LMR-400-DB Coaxial Cable.** Manufactured by Times Microwave Systems, this is now the industry-standard ultralow insertion loss cable. It is flexible, very light, and water blocked. If the coaxial cable outer jacket is damaged, a blocking agent flows to the area and reseals the cable. Any damage normally results in attenuation issues. The low-loss figures are possible due to the cable construction specifications. The center conductor is high-conductivity copper clad with a sophisticated insulation process. As a result, the attenuation is significantly reduced in comparison to single-shielded cables. The RF shielding is > 90dB in comparison to 40dB for RG8. The cable has an inner conductor, a nitrogen-injected foam closed-cell dielectric, an outer aluminum foil tape conductor, a tinned-copper overall braid, and a black PE jacket. It is applicable for VHF, GPS, VSAT, or any antenna or aerial requiring coaxial cables. It is a direct replacement for RG8, RG213, and RG214 air dielectric cables with 40% lower attenuation. I installed this on my own boat at time of publication.

e. **LMR-195-DB Coaxial Cable.** This is a replacement for RG58 and is for applications such as AIS with a nominal 50Ω impedance. These cables have much lower losses than RG58.

34.15   VHF Installation Testing. Many VHF installations operate poorly, with often-undiagnosed problems. Many boaters install their own cables, connectors, and antenna, but in the majority of cases, the installation is never tested. If the maximum range is to be realized, the installation requires proper testing. With the increasing reliance on new technology, in particular with DSC VHF units, reliability is of crucial importance. In an earlier chapter I highlighted the importance of installing the correct coaxial cable to reduce losses. The attenuation inherent within the cable is only part of the loss equation, and the following should be observed:

a. **Voltage Standing Wave Ratio (VSWR).** When a signal is transmitted via the cable and aerial, a portion of that signal energy will be reflected back to the transmitter. The effect is that coverage is reduced due to the reduced power output. Measure the VSWR with a meter. Until recently, you had to hire a technician to bring along an expensive meter. I am fortunate to possess a Bird wattmeter (see the image in the color insert) but also use a Shakespeare meter, the ART-1. This allows easy fault diagnosis and timely repairs and is highly recommended. A number of problems can reduce the VSWR. Regular testing of reflected power and detection of excessive values will alert you to potential installation problems. It may even save your life:

(1) **Damaged or Cut Ground Shields.** This is common where the cable has been jointed, or improperly terminated at the connector. Make sure the shield is properly prepared and installed.

(2) **Dielectric Faults.** This problem occurs when cables are run tightly around corners, through bulkheads, and through cable glands. Make sure that cables are bent with a relatively large radius; the tighter the bend, the more dielectric narrowing will occur, with increased reflected power.

(3) **Pinched Cable.** This common problem occurs where a cable has not been properly passed through a bulkhead, with the gland or connector impinging on the cable and reducing its dielectric diameter. Radio waves pass along the outside of the central core and along the inner side of the braiding, so any deformation will alter the inductance and reduce power output.

(4) **Connector Faults.** The most common problem is that of connectors not being installed or assembled correctly. Ensure that connectors are properly tightened, that pins are properly inserted, and that the pin-to-cable solder joint is sound and not a dry joint. Ensure that shield seals are properly made. Many connectors appear good at time of assembly but deteriorate very quickly when exposed to rain, salt spray, and corrosion. Check the status with a multimeter between the core and screen for short circuits.

(5) **Antenna Faults.** If an antenna is out of spec or suffered storm damage, or if a new antenna has been damaged in transit, functional efficiency will decrease and losses will increase. Inspect the antenna and connectors regularly. I always wrap the aerial connection with self-amalgamating tape to reduce ingress of moisture and salt air.

(6) **Surge and Lightning Protection.** Consider installing surge arrestors on your VHF coaxial cables. Manufactured by Times Microwave Systems, they are worth looking at in lightning hot spots.

VHF RADIO CHANNELS

34.16   US VHF Radio Channels. These are the nominated US VHF channel designations in 2023. Know and use them correctly. A four-digit channel number starting with "10" indicates simplex use of the ship station transmit side of what was previously an international duplex channel. Improper use of VHF channels can lead to fines by the FCC. You have been warned! These new channel numbers should eventually begin to be displayed on new models of VHF marine radios. Boaters should normally use channels listed as noncommercial. Channel 16 is used for calling other stations or for distress alerting; Channel 13 should be used to contact a ship where there is danger of collision. The Great Lakes and approaches operate on various frequencies: Ch16, Ch24, Ch26, Ch27, Ch28, and Ch85 are the most often used. Traffic, harbor, port, and bridge control are usually on Ch11, Ch12, and Ch13; and Ch14 is used for locks. Weather is broadcast continuously on WX1, WX2, and WX3.

34.17   About VDSMS. VDSMS (VHF Digital Small Message Services) is a relatively new VHF innovation and is intended to provide short messaging service between vessels and from vessel to shore. VDSMS operates on international VHF marine band frequencies and is defined in appendix 18 of the "International Radio Regulations" (RR Ap 18). VDSMS shares VHF channels with voice services and does not interfere with other communications.

Table 34-1. US VHF Channel Designations—Part 1

Channel	Designation
1001 (Old 01A)	Port Operations, VTS, New Orleans/Lower Mississippi Area
02	Harbor—Ship-to-Ship
03	Harbor—Ship-to-Ship
04	Harbor—Ship-to-Ship
1005 (Old 05A)	Port Operations or VTS; Houston, New Orleans, and Seattle Areas
06	**Safety—SAR Communications—Intership Safety**
1007 (Old 07A)	Commercial; VDSMS
08	Commercial (Intership Only); VDSMS
09	**US Calling Channel (Ship-to-Ship); VDSMS**
10	Commercial; VDSMS
11	Commercial (VTS in some areas); VDSMS
12	Port Operations (VTS in some areas)
13	**Bridge-to-Bridge (1 watt only); Intership Navigation Safety; Intracoastal Waterway (ICW);** Commercial Vessels. Do not use call signs. Use abbreviated operating procedures only. Maintain dual watch on Channels 13 and 16.
14	Port Operations (VTS in some areas; Ship Movements in San Francisco Bay)
15	Environmental Receive Only. Used by Class C EPIRBs
16	**International Distress, Safety, and Calling**
17	State and Local Government Maritime Control

Table 34-2. US VHF Radio Channel Designations—Part 2

Channel	Designation
1018 (Old 18A)	Commercial; VDSMS
1019 (Old 19A)	Commercial; VDSMS
20	Port Operations (Duplex)
1020 (Old 20A)	Port Operations
1021 (Old 21A)	US Coast Guard Only

(continued)

Table 34-2. *Continued*

Channel	Designation
1022 (Old 22A)	**USCG and Marine Safety Information Broadcasts**
1023 (Old 23A)	US Coast Guard Only
24	Public Correspondence (Marine Operator); VDSMS
25	Public Correspondence (Marine Operator); VDSMS
26	Public Correspondence (Marine Operator); VDSMS
27	Public Correspondence (Marine Operator); VDSMS
28	Public Correspondence (Marine Operator); VDSMS
1063 (Old 63A)	Port Operations & Commercial VTS; only New Orleans/Lower Mississippi
1065 (Old 65A)	Port Operations
1066 (Old 66A)	Port Operations
67	Commercial, Bridge-to-Bridge Communications; Lower Mississippi River Only
68	Noncommercial; VDSMS
69	Noncommercial; VDSMS
70	**Digital Selective Calling (DSC) Only (NO VOICE ALLOWED)**
71	Noncommercial; VDSMS
72	Noncommercial Intership Only; VDSMS
73	Port Operations
74	Port Operations
77	Port Operations Intership Only
1078 (Old 78A)	Noncommercial; VDSMS
1078 (Old 78A)	Noncommercial; VDSMS
1079 (Old 79A)	Commercial. Noncommercial in Great Lakes Only; VDSMS
1080 (Old 80A)	Commercial. Noncommercial in Great Lakes Only; VDSMS
1081 (Old 81A	US Government Only. Environmental Protection Operations
1082 (Old 82A	US Government Only
1083 (Old 83A)	US Government Only
84	Public Correspondence (Marine Operator); VDSMS
85	Public Correspondence (Marine Operator); VDSMS
86	Public Correspondence (Marine Operator); VDSMS
87	Public Correspondence (Marine Operator); VDSMS
88	Commercial Intership Only; VDSMS
WX 1	NOAA Weather Broadcasts Receive Only

Channel	Designation
WX 2	NOAA Weather Broadcasts Receive Only
WX 3	NOAA Weather Broadcasts Receive Only
WX 4	NOAA Weather Broadcasts Receive Only
WX 5	NOAA Weather Broadcasts Receive Only
WX 6	NOAA Weather Broadcasts Receive Only
WX 7	NOAA Weather Broadcasts Receive Only
WX 8	NOAA Weather Broadcasts Receive Only
WX 9	NOAA Weather Broadcasts Receive Only
WX 10	NOAA Weather Broadcasts Receive Only

34.18   US VHF Weather Broadcasts. All national weather broadcasts for coastal waters are announced on Channel 16 followed by the broadcast on Channel 22A. Visit the National Weather Service website at www.weather.gov.

Table 34-3. US VHF Radio Weather Broadcasts

USCG District/Location	Broadcast 1	Broadcast 2
First CG District		
Northern New England	11:05z	23:05z
Boston	10:35z	22:35z
Southeastern New England	10:05z	22:05z
Long Island Sound	11:20z	23:20z
Northern New England	11:05z	23:05z
Fifth CG District		
Delaware Bay	11:03z	
Baltimore	01:30z	23:03z
Hampton Roads	02:30z	12:05z
North Carolina	Warnings only	11:20z
Seventh CG District		
Charleston	12:00z	
Jacksonville	Warnings only	
Miami	Warnings only	22:00z
Key West	12:00z	22:00z
San Juan	12:00z	22:10z
St. Petersburg	13:00z	23:00z
Eighth CG District		
Mobile	10:20z, 12:20z	16:20z, 22:20z
New Orleans	10:35z, 12:35z	16:35z, 22:35z
Houston-Galveston	10:50z, 12:50z	16:50z, 22:50z
Corpus Christi	10:40z, 12:40z	16:40z, 22:40z

(continued)

Table 14-2. *Continued*

USCG District/Location	Broadcast 1	Broadcast 2
Ninth CG District		
Buffalo	02:55z	14:55z
Detroit	01:35z	13:35z
Lake Michigan	02:55z	14:55z
Sault Ste. Marie	00:05z	12:05z
Eleventh CG District		
Humboldt Bay	16:15z	23:15z
San Francisco	16:30z	19:00z, 21:30z (winter)
Los Angeles/Long Beach	02:00z	18:00z
San Diego	01:00z	17:00z
Thirteenth CG District		
Seattle		18:30z
Columbia River	06:30z	17:45z
North Bend	0603z	18:03z
Fourteenth CG District		
Honolulu	05:00z	17:00z
Guam	09:00z	21:00z

34.19   International (and UK) VHF Channels. The following table is adapted from appendix 18 of the "International Telecommunications Union Radio Regulations." Some UK administration allocation changes are added. Changes are always occurring, so check locally.

Table 34-4. International VHF Radio Channel Designations—Part 1

Channel	Designation
00	UK Search and Rescue (156 MHz)
01	Public Correspondence, Port Operations, Ship Movement
02	Public Correspondence, Port Operations, Ship Movement
03	Public Correspondence, Port Operations, Ship Movement
04	Public Correspondence, Port Operations, Ship Movement
05	Public Correspondence, Port Operations, Ship Movement
06	Intership (Simplex). SAR (Solent Leisure Craft Working Channel)
07	Public Correspondence, Port Operations, Ship Movement
08	Intership (Simplex)
09	Intership, Port Operations, Ship Movement (Simplex)
10	Intership, Port Operations, Ship Movement (Simplex)

Channel	Designation
11	Port Operations, Ship Movement
12	Port Operations, Ship Movement
13	Intership (Simplex). Bridge-to-Bridge Navigational Safety
14	Port Operations, Ship Movement
15	Port Operations, Intership, Ship Movement (**1 watt only**)
16	**International Distress, Safety, and Calling**
17	Port Operations, Intership, Ship Movement (**1 watt only**)
18	Public Correspondence, Port Operations, Ship Movement
1078	Port Operations, Ship Movement (Simplex)
19	Public Correspondence, Port Operations, Ship Movement
1019	Port Operations, Ship Movement (Simplex)
20	Public Correspondence, Port Operations, Ship Movement
1020	Port Operations, Ship Movement (Simplex)
21	Public Correspondence, Port Operations, Ship Movement
22	Public Correspondence, Port Operations, Ship Movement
23	Public Correspondence, Port Operations, Ship Movement.
24	Public Correspondence, Port Operations, Ship Movement
1024	For use under the VHF Data Exchange System VDES
1084	For use under the VHF Data Exchange System VDES
25	Public Correspondence, Port Operations, Ship Movement (Jersey CG)
1025	For use under the VHF Data Exchange System VDES
1085	For use under the VHF Data Exchange System VDES
26	Public Correspondence, Port Operations, Ship Movement
1026	For use under the VHF Data Exchange System VDES
1086	For use under the VHF Data Exchange System VDES
27	Public Correspondence, Port Operations, Ship Movement
1027	Port Operations, Ship Movement
28	Public Correspondence, Port Operations, Ship Movement
1028	Port Operations, Ship Movement
60	Public Correspondence, Port Operations, Ship Movement

Table 34-5. International VHF Radio Channel Designations—Part 2

61	Public Correspondence, Port Operations, Ship Movement
62	Public Correspondence, Port Operations, Ship Movement
63	Maritime Safety Information (MSI), UK
64	Maritime Safety Information (MSI), UK
65	**National Coastwatch Institution, England and Wales**
2006	Reserved
66	Public Correspondence, Port Operations, Ship Movement
67	CG UK Small Craft Safety; Irish SAR Control
68	Port Operations, Ship Movement
69	Port Operations, Intership, Ship Movement
70	**Digital Selective Calling (DSC) Only (NO VOICE ALLOWED)**
71	Public Correspondence, Ship Movement
72	Intership (Simplex)
73	UK Coast Guard Working
74	Public Correspondence, Ship Movement
75	Guard Band Protecting Channel 16, 156.7625–156.7875 MHz
76	Guard Band Protecting Channel 16, 156.8125–156.8375 MHz
77	Intership (Simplex)
78	Public Correspondence, Port Operations, Ship Movement
2078	Port Operations, Ship Movement (Simplex)
2019	Port operations, Ship Movement
79	Public Correspondence, Port operations, Ship Movement
2079	Port Operations, Ship Movement
2020	Port Operations, Ship Movement
80	International Marina and Port Operations
81	Public Correspondence, Port Operations, Ship Movement
82	Public Correspondence, Port Operations, Ship Movement. (Jersey CG)
83	Public Correspondence, Port Operations, Ship Movement
2024	Intership (Digital Only)
84	Public Correspondence, Port Operations, Ship Movement
2084	Intership (Digital Only)
2025	Intership (Digital Only)
85	Public Correspondence, Port Operations, Ship Movement

2085	Intership (Digital Only)
2026	Reserved
86	Public Correspondence and HMCG Medilink; Solent Coastguard
2086	For use under the VHF Data Exchange System (VDES)
2027	ASM (Application-Specific Messages)
87	AIS Channel 1
2028	ASM (Application-Specific Messages)
88	AIS Channel 2
M1/37A	Yacht Clubs, Race Committees, and Marinas
M2	Yacht Clubs, Race Committees, and Marinas
AIS 1	161.975/161.975 (161.975 MHz may be used for SAR)
AIS 2	162.025/162.025 (162.025 MHz may be used for SAR)

34.20   Canada VHF Radio Channels. Canadian marine weather forecasts are available on VHF Ch21B and Ch83B (Atlantic coast and Great Lakes) and VHF Ch21B and WX1, WX2, and WX3 (Pacific coast). Consult the Canadian Coast Guard publication "Radio Aids to Marine Navigation" (RAMN) for an up-to-date list of channels for your location. There are many allocations for Canadian Coast Guard, commercial traffic, SAR, pilotage, VTS, commercial shipping, and other operations. Channel 68 is reserved for yachts and marina communications. Ch21B and Ch83B carry continuous marine broadcast (CMB) service in English and transmit severe weather warnings, ice information, hazards to navigation, and other safety warnings. Visit the website at www.ic.gc.ca.

34.21   Mexico VHF Radio Channels. Chetumal Ch26, Cozumel Ch26, Ch27, Cancun Ch26, Ch27, Veracruz Ch26, Ch27. Ch68 is used for local cruiser nets. If in Baja, then La Paz. DO NOT use the following channels: Ch09 (Pemex), Ch10, Ch11 (Mexican Navy), Ch14 (La Paz Port captain), Ch16 (hailing only), Ch22a (La Paz cruiser hailing channel), Ch72 and Ch74 (local fishermen), Ch74 (La Paz ferry terminal), Ch83 (Immigration), or Ch88 (Customs).

34.22   UK Weather—BBC Radio 4. The UK government downgraded long-wave (LW) services mid-2022. Twenty-four hour forecasts for coastal areas are at 00:48, 05:20, 12:01, and 17:54 on LW 198kHz, providing a summary of gale warnings in force, a general synopsis, and area forecasts for specified sea areas around the UK. Summary of broadcasts and times are local. LW Shipping Forecasts will cease in 2024.

    a.   BBC4 Weekday Broadcasts:

        (1)   00:48: gale warnings, shipping forecast, weather reports from coastal stations, and the inshore waters forecast

        (2)   05:20: gale warnings, shipping forecast, weather reports from coastal stations, and the inshore waters forecast

(3)   12:01: gale warnings, shipping forecast

(4)   17:54: gale warnings, shipping forecast

b.   **BBC4 Weekend Broadcasts.** Not VHF but a valuable weather source. The famous forecast zones are Viking, North Utsire, South Utsire, Forties, Cromarty, Forth, Tyne, Dogger, Fisher, German Bight, Humber, Thames, Dover, Wight, Portland, Plymouth, Biscay, Trafalgar, FitzRoy, Sole, Lundy, Fastnet, Irish Sea, Shannon, Rockall, Malin, Hebrides, Bailey, Fair Isle, Faeroes, and Southeast Iceland.

(1)   00:48: gale warnings, shipping forecast, weather reports from coastal stations, and the inshore waters forecast

(2)   05:20: gale warnings, shipping forecast, weather reports from coastal stations, and the inshore waters forecast

(3)   12:01: gale warnings, shipping forecast

(4)   17:54: gale warnings, shipping forecast

34.23   UK VHF Radio Channels. The UK uses the international frequency allocation. There are localized variations to channel usage. Weather and navigation safety information is broadcast on Ch23, Ch84, and Ch86. These broadcasts are every 3 hours starting at 07:10 local time. Initial announcement is broadcast on Ch16. Strong wind warnings are always issued on receipt from the Meteorological Office if the wind within an inshore forecast area is predicted to exceed Force 6. MSI information is broadcast on Ch10 and Ch73. Ch6 is used for ship-to-aircraft communications in SAR activities. Ch67 is used for SAR and HM Coast Guard safety communications. Forecast areas are Swansea, Thames, Clyde, Yarmouth, Solent, Brixham, Dover, Shetland, Stornaway, Falmouth, Forth, Liverpool, Portland, Holyhead, Belfast, Aberdeen, Milford Haven, and Humber.

34.24   Channel Islands (UK). Jersey Coast Guard MSI broadcasts on Ch25 and Ch82. Gale or navigational warnings are broadcast at 06:45, 07:45, 08:45, 12:45, 18:45, and 22:45. Gale warnings are broadcast on receipt and at 03:07, 09:07, 15:07, and 21:07. Wind information, speed, and gusts at St. Helier VTS are transmitted on Ch18 every 2 minutes.

34.25   Ireland VHF Channels. The Irish Coast Guard provides weather broadcasts on Ch16 and then on a designated VHF working channel. Met Éireann Sea Area forecasts are broadcast every 3 hours, commencing at 01:03, 04:03, 07:03, 10:03, 13:03, 16:03, 19:03, and 22:03. Forecast on Valentia at 01:03 and every 3 hours to 22:03 on Ch24. Malin Head on Ch23 for Fastnet, Shannon, and Irish coastal waters. Gale warning broadcasts are preceded by an announcement on Ch16, broadcast on receipt, and repeated at 00:33, 06:33, 12:33, and 18:33.

34.26   France VHF Channels. Meteo France broadcasts weather bulletins in French. Storm, gale, and navigation warnings are in English and French.

a.   **Gris-Nez, Ch79:** Belgian border to Baie de la Somme and Baie de la Somme to Cap de la Hague. Broadcast times: 07:10, 13:10, 19:10.

b. **Joburg, Ch80:** Baie de la Somme to Cap de la Hague and Cap de la Hague to Pointe de Penmarc'h. Broadcast times: 07:33, 13:33, 19:33.

c. **Corsen, Ch79:** Cap de la Hague to Penmarc'h (Stiff/Ouessant). Broadcast times: 05:03, 07:15; 13:15, 19:15.

d. **Etel, Ch80:** Pointe de Penmarc'h to l'Anse de l'Aiguillon (Les Sables-d'Olonne). Broadcast times: 08:15, 14:15, 20:15.

e. **L'Anse, Ch79:** L'Anse de l'Aiguillon to the Spanish border. Broadcast times: 08:15, 14:15, 20:15.

34.27 **Spain VHF Channels.** Navigation warnings are broadcast at different times.

a. **Bilbao, Ch10:** 00:33, 04:33, 08:33, 12:33, 16:33, 20:33

b. **Santander, Ch11:** 02:45, 06:45, 10:45, 14:55, 18:45, 22:45

c. **Coruña, Ch13:** 00:05, 04:05, 08:05, 12;05, 16:05, 20:05

d. **Finisterre, Ch11:** 02:33, 06:33, 10:33, 14:33, 18:33, 22:33

e. **Vigo, Ch10:** 00:15, 04:15; 08:15:12:15, 16:15, 20:15

f. **Huelva, Ch11:** 04:15, 08:15, 12:15, 16:15, 20:15

g. **Tarifa, Ch10 & Ch67:** Even hours + 15

h. **Algeciras, Ch74:** 03:15, 05:15, 07:15, 11:15, 15:15, 19:15, 23:15

i. **Almeria, Ch74:** Odd hours + 15

j. **Valencia, Ch10:** Even hours + 15

k. **Castellón, Ch74:** Summer: 05:03, 09:03, 15:03, 19:03; winter: 06:03, 10:03, 16:03, 20:03

l. **Tarragona, Ch13:** 06:30, 10:30, 16:30, 21:30

m. **Barcelona, Ch10:** 07:00, 11:00, 16:00, 21:00

n. **Palma, Ch10:** Summer: 06:35, 09:35, 14:35, 19:35; winter: 07:35, 10:35, 15:35, 20:35

o. **Palma, Ch20:** 09:10, 14:10, 21:10

p. **Menorca, Ch85:** 09:10, 14:10, 21:10

q. **Ibiza, Ch03:** 09:10, 14:10, 21:10

r. **Tenerife, Ch65:** 03:40, 13:40, 19:03

s. **Las Palmas, Ch61:** 03:40, 13:40, 19:03

t. **Fuerteventura, Ch22:** 03:40, 13:40, 19:03

34.28   Portugal VHF Channels. NAVTEX (and internet) is the primary way of broadcasting maritime safety information (MSI). Interesting warnings on the website and valid for everyone getting weather from the internet. Note that the internet is not part of the Maritime Safety Information System and should never be relied upon as the only means to obtain MSI navigational warnings, Notices to Mariners, and weather forecasts and warnings. These were previous listings and the best I could get at time of publication: Arga, Ch25, Ch28, Ch83; Arestal, Ch24, Ch26, Ch85; Monsanto, Ch11 at 02:50, 06:50, 10:50, 14:50, 18:50, 22:50; Montejunto, Ch23, Ch27, Ch87; Lisboa, Ch23, Ch25, Ch26, Ch27, Ch28; Atalaia, Ch24, Ch26, Ch85; Picos, Ch23, Ch27, Ch85; Estoi, Ch24, Ch28, Ch86; Sagres, Ch11 at 08:35, 20:35; Acores, forecast on Ch16, Ch23, Ch26, Ch27, Ch28 at 09:35, 21:35; Madeira, Ch25, Ch26, Ch27, Ch28.

34.29   Italy VHF Channels. Weather bulletins are broadcast on Ch68 at 01:35, 07:35, 13:35, 19:35. Navigation and gale warnings are broadcast on receipt H+03 and H+33, and continuous in the northern Adriatic. The issue for all sea areas: (1) Corsican Sea, (2) Sardinian Sea, (3) Strait of Sardinia, (4) Ligurian Sea, (5) North Tyrrhenian Sea, (6) Central Tyrrhenian Sea West, (7) Central Tyrrhenian Sea East, (8) South Tyrrhenian Sea West, (9) South Tyrrhenian Sea East, (10) Strait of Sicily, (11) South Ionian Sea, (12) North Ionian Sea, (13) South Adriatic Sea, (14) Central Adriatic Sea, and (15) North Adriatic Sea.

34.30   Greece VHF Channels. Olympia Radio broadcasts prerecorded weather bulletins in Greek and English. Bulletins are broadcast at 06:00, 10:00, 16:00, and 20:00 UTC. Chios, Ch85; Kefallinia, Ch27; Kerkyra (Corfu), Ch02; Knossos, Ch83; Kythira, Ch85; Limnos, Ch82; Moustakos, Ch04; Mytilini, Ch01; Parnis, Ch25; Patra, Ch85; Petalidi, Ch8; Pilio, Ch60; Rodos (Rhodes), Ch63; Sitia, Ch85; Syros, Ch04; Sfendami, Ch23.

34.31   Croatia VHF Channels. Weather is broadcast on the following stations and channels: harbormaster on Ch10; marinas on Ch17. Continuous weather broadcasts every 10 minutes and updated at 07:00, 13:00, and 19:00 local time on Ch73 and Ch69 for Northern Adriatic Eastern Part; Ch67 for Central Adriatic Eastern Part; Ch73 for Southern Adriatic Eastern part. Weather reports are broadcast at the following times in UTC; UTC + 1.

    a.   **Rijeka Radio, Ch24:** 05:35, 14:35, 19:35

    b.   **Split Radio, Ch21, Ch23, Ch07, Ch28:** 05:45, 12:45, 19:45

    c.   **Dubrovnik Radio, Ch04, Ch07:** 06:25, 13:20, 21:20

34.32   Türkiye VHF Channels. Weather observations and then weather bulletins.

    a.   **Istanbul, observations, Ch67:** 07:30, 09:30, 11:30, 13:30, 15:30, 17:30, 19:30. West Black Sea, Sea of Marmara, Aegean Sea. **Forecasts on Ch67** at 07:00, 19:00. General marine weather forecast bulletins for the Black Sea, Sea of Marmara, Aegean Sea and Mediterranean Sea (in Turkish and English).

    b.   **Antalya, observations, Ch67:** 07:30, 09:30, 11:30, 13:30, 15:30, 17:30, 19:30. Mediterranean and Aegean Seas. **Forecasts on Ch67** at 07:00, 19:30. Seventy-two-hour marine weather forecasts for the Black Sea, Sea of Marmara, Aegean Sea, and Mediterranean Sea (in Turkish only).

c.   **Samsun, observations, Ch67:** 07:30, 09:30, 11:30, 13:30, 15:30, 17:30, 19:30. Black Sea.

34.33   Cyprus VHF Channels. Weather bulletins from the Cyprus Meteorological Service after initial announcement on Ch16 and then nominated VHF channel at 06:00, 10:00, 16:00, and 22:00 UTC.

34.34   Malta VHF Channels. Weather and MSI are broadcast by Valletta VTS on Ch11 at 07:03, 11:03, 17:03, and 23:03.

34.35   Caribbean VHF Channels. There used to be many VHF channels in Caribbean countries providing weather information, but that has changed over recent years. VHF weather services have steadily declined over the past 15 years, and I have added other sources, such as radio and internet. Ignoring that side trip and long offshore passage, our start point of this review is the Bahamas, Cuba, Cayman Islands, Jamaica, Turks and Caicos, Haiti, Dominican Republic, Puerto Rico, US Virgin Islands, and the British Virgin Islands. Next up are Anguilla, Saint Kitts and Nevis, Montserrat, Sint Croix, Sint Maarten/Saint Martin, Saint Barthélemy (Saint Bart, Saint Barth), Antigua and Barbuda, Guadeloupe, Marie Galante— and that's not the end of it, when too many islands are never enough. Dominica, Martinique, Saint Lucia, Saint Vincent and the Grenadines, Bequia, Barbados, Carriacou, Grenada, Trinidad and Tobago, Margarita, and Bonaire, Curacao, and Aruba (Netherlands Antilles).

a.   **Bermuda.** Ch27, Ch28. Ch10, Ch12, Ch16, Ch27, Ch38 at 12:35 and 20:35. Bermuda Weather Radio has one station on WX2 providing marine weather and forecasts. Operated by the Bermuda Maritime Operations Centre. Coastal forecasts available at www.weather.gov.ky.

b.   **Bahamas.** Uses US frequencies: Nassau, Ch27; Exuma, Ch22. Weather forecasts every odd hour on Ch27. Cruisers Net operates on Ch68 at 08:15 with weather forecasts. Internet sources include Barometer Bob (barometerbob .org) and SSB Marine Weather Center (www.mwxc.com). Bay Street Marina is on Ch08. Ch09 and Ch68 in the Abacos and George Town areas. Net at Nassau 07:15 on Ch14.

c.   **Cuba.** Requires communication before arrival on Ch16 and Ch68.

d.   **Cayman Islands.** Weather forecasts by Radio Cayman 89.9 and 103.5 at 07:14, 08:08, 12:09, 18:22, and 22:18.

e.   **Jamaica.** Kingston Ch26 and Ch27. Forecasts for southwest, northwest, and eastern Caribbean and Jamaica coastal waters at 01:30, 14:30, and 19:00 on Ch13.

f.   **Turks and Caicos.** Ch16 used when arriving then switch to nominated channels. No weather forecasts. Radio Turks and Caicos, FM 89.1.

g.   **Haiti.** No identifiable broadcasts; if visiting use Ch16 and local radio stations for weather information. Very affected by massive earthquakes and hurricanes.

h. **Dominican Republic.** No weather forecast identified. Several ports, so usual VHF channel communications. Monitor local radio stations for forecast. The Luperon Cruisers Net reportedly has weather forecasts on Ch68 then Ch72 every Wednesday at 08:00 and on Sunday at 09:00.

i. **Puerto Rico (US Coast Guard).** NOAA forecasts broadcast continuously on WX1, WX2, and Ch22. NAVTEX forecasts available online at www. weather.gov/sju/.

j. **US Virgin Islands (USVI).** Forecasts for western North Atlantic, Caribbean, and Gulf of Mexico on Ch28. Weather on Channel WX6 and WX1. WAH (Virgin Islands Radio) broadcasts weather on Ch16 then on Ch28 and Ch85 at 06:00, 14:00, and 22:00. WOJO 1030kHz AM, every day at 6 minutes past the hour. Saint Thomas, Radio WIVI FM 99.5, with forecasts at 07:30, 08:30, 15:30, 16:30 daily. WVWI 1000kHz AM Radio One (Charlotte Amelie). Christiansted on WSTX 970 AM News Talk.

k. **British Virgin Islands (BVI).** Use VHF weather channel WX3 or WX4. NOAA forecasts are broadcast continuously on Saint Thomas Radio ZBVI 780 AM. Detailed marine weather forecast daily at 08:05; updates at 07:30, 08:30, 09:30, 10:30, 11:30, 13:00, 14:15, 14:55, 16:00, 18:27, 19:00, 20:00, and 20:30. Visit Bitter End Yacht Club and Quarterdeck Marina in Virgin Gorda; call on VHF Ch16 on the approach.

l. **Anguilla.** No VHF weather; Radio Anguilla is best weather source at 95.5 FM. Weather follows the news at 07:05 Monday to Saturday and 19:05 Monday to Friday. Forecast available online at www.antiguamet.com. FM radio 1000 FM, with news and weather on the hour and/or the half hour.

m. **Saint Kitts and Nevis.** Arrival into Charlestown requires Ch16 contact with Nevis Air and Sea Ports Authority (NASPA). FM Radio 92.7 and 93.2 FM, with news and weather on the hour and half hour. Internet radio stations with weather: ZIZ radio (zizonline.com) and 2020 Vision Radio (www.2020visionradio.com). Online at www.antiguamet.com.

n. **Montserrat.** No VHF weather but ZJB Radio Monserrat (www.zjbradio.com).

o. **Saint Martin/Sint Maarten.** Port Sint Maarten commercial, Ch12. Sint Maarten Cruisers net has weather Monday through Saturday on Ch14 at 07:30. Weather on Shrimpy's net, Monday through Saturday on Ch10 at 07:30. Radio Island 92 (91.9 FM) marine weather forecast at 09:00 and 12:00. Simpson Bay Marine contact on Ch12 and Ch79A. VHF radio, Ch10 at 07:30, the Cruisers net. FM Radio 88.9 FM, news and weather on the hour and half hour. Voice of Saint Martin, 102.7 FM, Sundays at 09:00. Weather channel WX-1 all day.

p. **Saint Barthélemy** (Saint Bart, Saint Barth). Ch12 to call Port de Gustavia and Port Control. Weather via online radio stations.

q. **Antigua and Barbuda.** Antigua, Ch06 at 09:00 (English and Falmouth Harbours Radio). Jolly Harbour, Ch74, Monday through Friday at 09:00. Carlisle Bay, 09:00 on Ch72. FM radio 620 AM, with broadcast at 06:50. FM radio 93.9 FM, with news and weather on the hour and half hour. Malones Food Store and Creole Restaurant (English Harbour). Contact them on Ch68 for immediate free delivery. Visit www.antiguamet.com for Meteo information.

r. **Guadaloupe.** Ch06 and then Ch11. Warnings on receipt, odd hours +33 and Ch26 and Ch27 every odd hour +30. Weather messages Ch26 and Ch27 at 03:30 and 14:30.

s. **Dominica.** Portsmouth Ch72 at 07:30. Then Net in Spanish and French. Government Met information at www.weather.gov.dm.

t. **Martinique** (Fort-de-France). VHF French language forecast on Martinique Cruisers Radio Net on Ch08 and Ch11 at 08:30 and 18:30, Monday, Wednesday, Friday.

u. **St. Lucia.** VHF Ch69 at 08:00, Monday to Saturday. Weather on FM Gem Radio, with marine weather broadcasts on FM 93.7 at 07:30 and 09:30 each day. If you tune to FM 94.5, news and weather is every hour and half hour. Internet forecasts at Rodney Bay Marina. Check met.gov.lc.

v. **Saint Vincent and the Grenadines.** No VHF, but weather available on AM 790kHz, with marine weather broadcast at 08:50 and 18:30 each day. Net broadcasts on Ch68 in Saint Vincent Blue Lagoon Hotel & Marina Monday, Wednesday, and Friday at 09:00.

w. **Bequia.** Ch68 at 08:00 daily. In Port Elizabeth, Caribbean Diesel on Ch68. Trash collection on Ch67 by Daffodil Marine. Frangipani Yacht Services on Ch68.

x. **Barbados.** Ch16, Ch26. Forecasts at 00:50, 12:50, 16:50, 20:50. Warnings on receipt and every 4 hours for Caribbean, Antilles, and Atlantic waters on Ch26. On radio AM 900, hourly weather updates and after the news broadcast at 07:00.

y. **Carriacou.** Ch69 at 07:30 on Monday, Wednesday, and Friday. German net daily on Ch71 at 08:30. Carriacou Marine, Ch16 and Ch68 for customs entry Grenada and services/provisioning.

z. **Grenada.** Ch 66 at 07:30 Monday to Saturday. Listen on Ch69. Saint George's Ch06, Ch11, Ch12, Ch13, Ch22A. Forecast on request.

aa. **Trinidad and Tobago.** Chaguaramas, 08:00 daily on Ch68. North Post Radio, Ch24, Ch25, Ch26, Ch07 daily. Forecast at 13:40 and 20:40; metoffice.gov.tt/forecast.

bb.   **Bonaire** (Netherlands Antilles). Ch77 as hailing channel, then to Ch71 or Ch88A for traffic. All remaining channels are assigned. Bonaire has two all-weather marinas: Harbour Village Marina, Ch17; Plaza Marina, Ch18 (www .meteo.cw) and also for Sint Maarten and Saint Eustatius.

cc.   **Curacao** (Netherlands Antilles). Busy commercial port in Willemstad and VTS Control on Ch12 (www.meteo.cw) and also for Saint Eustatius. Curacao Yacht Club contact on Ch68 (www.curacaoyachtclub.com).

dd.   **Aruba** (Netherlands Antilles). Use Ch16 for Coast Guard emergencies, Ch16 for Aruba Port Control, and then switch to Ch11 or Ch14. Ch06 used for local charter fishing boats. Ch16 and then Ch69 for the Renaissance Marina. Ch16 and then Ch68 for Varadero Marina. Ch68 is used for local pleasure and fishing boats, Bucuti Yacht Club, and Aruba Nautical Club. Local watersports use Ch19, Ch21a, Ch23a, Ch70a, Ch71a, Ch72a, Ch80a, and Ch81a. Cruisers occasionally have a net on Ch18a.

ee.   **Venezuela.** Puerto La Cruz, 07:45 on Ch72; intermittent net.

ff.   **Colombia.** Cartagena, 09:00 daily on Ch71; Santa Marta Marina, Ch72.

gg.   **Panama.** Bocas Del Toro emergency net, 07:45 daily on Ch68. Shelter Bay in Colon daily, 07:30 on Ch77. Vista Mar Marina in Colon, 08:30 on Ch74 daily.

AUSTRALIA AND NEW ZEALAND

34.36   **Australia VHF Channels.** Australia uses the international frequency allocation. There are localized variations with usage. VHF weather and other services vary between each state jurisdiction.

a.   **Queensland.** Marine weather is broadcast to limited areas in the state of Queensland by the Bureau of Meteorology (BOM). Alert announcements precede these on Ch16 and then the forecast on Ch80 at 07:15Z and 22:15Z every day. Monitor weather forecasts by the many volunteer rescue organizations in Queensland. Again, announcements are made on Ch16 then broadcast on Ch67.

b.   **New South Wales (NSW).** In NSW, Ch67 is used in the Newcastle, Sydney, and Port Kembla areas to broadcast weather at 07:33 and 17:33. Severe weather warnings are broadcast every hour. The state has repeaters along the coastline, including Ch21, Ch22, Ch80, Ch81, and Ch82.

c.   **Victoria.** Broadcasts forecasts for all Victorian coastal waters. This includes Bass Strait (four zones) and Port Phillip, Western Port, and Gippsland Lakes. Broadcasts are on Ch67 at 06:48 and 18:48 EST. Weather warnings on Ch67 on receipt. Broadcasts of current weather warnings are made on Ch67 at 00:48, 02:48, 04:48, 06:48, 08:48, 10:48, 12:48, 14:48, 16:48, 18:48, 20:48, and 22:48 EST. Visit transportsafety.vic.gov.au/maritime-safety.

d. **Tasmania.** Weather is broadcast at 08:03 and 17:33 EST. North Tasmania, Ch28; Hobart, Ch07; Bruny Island, Ch24/Ch27; St. Marys, Ch26; Ch68, Central North Coast, North East Coast, Central West Coast, and Maatsuyker Island; Ch69, Eastern Bass Strait and Lower East Coast and Far North East Coast.

e. **Western Australia (WA).** VHF marine forecasts are broadcast by volunteer sea search and rescue groups on Ch16 and Ch67. WA Water Police weather and navigation warnings (within 20 nautical miles of Perth metro area) at 07:18 and 19:18 hours on Ch16 and Ch67. Carnarvon, Ch73, at 21:45, 23:45, 04:15, 08:15, 05:45, 07:45, 12:15, 16:15 WST); Esperance, Ch72, at 22:35, 04:35, 08:35 (06:35, 12:35, 16:35 WST); Geraldton, Ch73, at 22:15, 00:15, 04:15, 09:15, 06:15 (08:15, 12:15, 17:15 WST); Broome, Ch72, at 22:40, 09:10 (06:40, 17:10 WST); Weather (06:33, 17:03 WST); Rottnest Island, Ch60; Jurien Bay, Ch62; Geraldton, Ch28; Carnarvon, Ch24; Dampier, Ch26; Port Headland, Ch27; Broome, Ch28.

f. **Northern Territory.** Weather broadcasts for Darwin on Ch67 (08:03, 18:03 CST); Gove, Ch28 (08:03, 18:33 CST); Frances Bay Lock, Ch06; Cullen Bay Lock, Ch11; Darwin Port Working, Ch10; Gove Harbor, Ch12; Groote Eylandt, Ch06. Visit nt.gov.au/marine/.

g. **South Australia.** Weather broadcasts (07:48, 17:18 CST); Adelaide, Ch23/Ch26; Kangaroo Island, Ch61; Port Lincoln, Ch24 and Ch27.

34.37   New Zealand VHF Channels. Weather information is broadcast by Maritime Radio simultaneously through all coastal VHF channels. Pre-broadcast announcements are made on Ch16 and then on the channel announced. Local area weather forecasts broadcast continuously by the NZ Coast Guard using Ch19, Ch2019, Ch20, Ch2020, Ch79, and Ch2079. Chatham Islands coastal VHF stations transmit weather forecasts simultaneously on Ch60 and Ch62 at 06:03, 14:03, 18:03, 22:03 (Chatham Islands local time). Ch16 and working channels Ch25, Ch67, Ch68, Ch69, or Ch71. Safety information and weather at 01:33, 05:33, 07:33, 10:33, 13:33, 17:33, 21:33. Cape Reinga, Ch68; Kaitaia, Ch71; Whangarei, Ch67; Great Barrier Island, Ch25; Plenty, Ch68; Runaway, Ch71; Tologa, Ch67; Napier, Ch68; Wairarapa, Ch67; Wellington, Ch71; Picton, Ch68; Kaikoura, Ch67; Akaroa, Ch68; Waitaki, Ch67; Chalmers, Ch71; Bluff, Ch68; Stewart Island, Ch71; Puysegur, Ch67; Fiordland, Ch71; Fox, Ch67; Greymouth, Ch68; Westport, Ch71; Farewell, Ch68; D'Urville, Ch67; Wanganui, Ch69; Cape Egmont, Ch71; Taranaki, Ch67; Auckland, Ch71; Russell Radio (BOI), Ch63.

34.38   Pacific Islands. Society Islands Ch13 Weather in French: 06:30, 12:00, 16:00, 20:00 VHF Tahiti Cruiser Net Ch68 M-F 07:30. **Tonga**, Vava'u. Cruisers Net Ch26 at 08:30. **Fiji** Musket Cove Marina Ch64/Ch68. Port Denarau Marina Ch14. **New Caledonia**. Noumea Radio weather daily Ch16, 06:30, 09:30, 15:15, 18:30 and Ch26, Ch28 Main Island South, Ch25 Île des Pins. **Vanuatu.** Net daily at 07:30 and 20:30 UTC in cruising season 8.230kHa/8.188 kHz. **Marshall Islands**. Majuro. VHF Net Ch71 Mon-Sat at 07:30 and HF daily at 07:45 on 6.224kHz. **Wallis & Fortuna.** VHF Ch09 and Ch22.

MF/HF-SSB Radio

35.1   Marine MF/HF-SSB Radio. Medium and high frequency (MF/HF) radio is part of GMDSS; Marine MF/HF-SSB radios operate within the frequency range of 1.6 MHz to 30 MHz. Medium Frequency (MF), High Frequency (HF) and Single Sideband (SSB) or MF/HF-SSB. Channel pairs are designated by the International Telecommunications Union (ITU), which defines standardized pairings for transmit and receive frequencies. Simplex is where transmit and receive are on the same frequency such as 2182.0 kHz and are generally used on ship-to-ship communications. Duplex is where transmission and reception are on two different frequencies. Many of these frequency pairs have ITU designated channel numbers and are used in ship-to-shore communications. There are a number of emission settings on a HV radio that include Upper Sideband (USB), Lower Sideband (LSB), and Amplitude Modulation (AM). Single sideband or single sideband suppressed carrier modulation refines amplitude modulation to use bandwidth and transmitter power more efficiently. The terms HF and SSB are generally used interchangeably. MF-HF/SSB allows long range communications and access to the various nets but also for email and weatherfax access.

a.   **Signal Propagation.** Sky waves travel up until they reach the ionosphere and are bent and reflected back over a wide area. The ionosphere extends from about 50km to 500km into the atmosphere and is formed by the ionization of air atoms by incoming UV ionizing radiation from the sun. The ionosphere is weakly ionized plasma that is constantly changing, and the changes alter the propagation characteristics of the radio waves. This is typified by the differences in night- and daytime transmission characteristics. Good HF communications depend on utilization of the changing conditions with use of optimum frequencies. The ionosphere structure is divided in to layers D, E, F1, and F2, in order of increasing height:

(1)   **F Layer.** The main reflecting layer is called the F layer and is approximately 320km high. The layer is permanently ionized, but during daylight hours energy from the sun causes the intervening layers E and D to form; at night it reflects the highest radio frequencies in HF bands.

(2)   **E & D Layers.** The signals reflected from these layers have lower ranges. Frequencies of 3MHz or less are absorbed by the D layer and eliminate sky wave propagation; therefore, 2MHz is not favored.

(3)   **Ground Wave.** Ground wave signals travel along the Earth's surface but are absorbed or masked by other radio emissions.

(4)   **Skip Zone.** The skip zone is the area between the transmission zone and the zone where the signal returns to Earth, with generally negligible signal within the skip zone.

b.   **Propagation Changes.** The ionosphere will affect each frequency differently. Extreme ultraviolet (EUV) radiation is responsible for forming and maintaining the ionosphere and is dependent on solar sunspot activity. When sunspot activity is lowest, the solar cycle EUV radiation is weak and the density of charged particles in the F region is lowest. In this state, only lower frequency HF signals can be reflected. At sunspot cycle peaks, the EUV and ionosphere density are high and higher frequencies can be reflected. The season, time of day, and latitude also affect HF radio communications. Solar flares produce high levels of electromagnetic radiation, and the X-ray component increases the D layer ionization. As HF communications use the F layer above, they must transit the D layer twice during any signal skips. During a major solar flare, the increased ionization results in a higher density of neutral particles and absorption of signal in the D layer; this is called sudden ionosphere disturbance (SID). Characterized by increased attenuation of HF signals at lower frequencies, it is also called short wave fadeout—SSWF for sudden and GSWF for gradual. These events are synchronized with solar flare patterns and are characterized by rapid onsets of just several minutes and declines of up to 1 hour or greater.

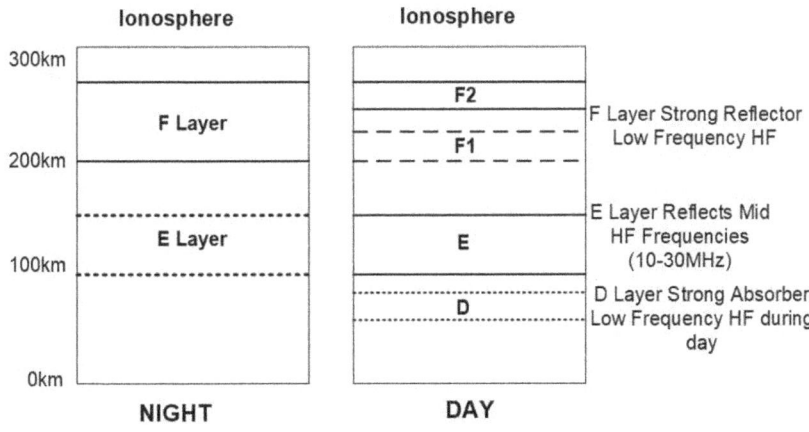

Figure 35-1. Atmospheric Layers

35.2   Space Weather Effects and HF/SSB Radio. Space weather has become very important in the satellite communications age. The underlying factor ruling space weather, at least in this end of the galaxy, is our sun. The sun is by nature prone to dramatic and violent changes, with events such as solar flares and the resultant blast streams of radiation and energized particles that stream toward Earth. Space weather is caused by changes in the speed or density of the solar wind, and this is the continuous flow of charged particles that flow from the sun past Earth. The flow tends to distort the Earth's magnetic field, compressing it in the direction of the sun and stretching it out in the opposite direction. The solar wind fluctuations cause variations in the strength and direction of the magnetic field near the Earth's surface; sudden variations are called geomagnetic disturbances, and the electrical layers of the ionosphere are disrupted.

35.3   Solar Cycles. Space weather depends on an 11-year solar cycle. Cycles vary in both intensity and length, and the solar activity is characterized by the appearance of sunspots on the sun. Sunspots are regions of stronger magnetic field, and the solar maximum is the time when maximum spot numbers are visible. Sunspot numbers are quoted for average numbers over a 12-month period and are the traditional measure of solar cycle status. Peak sunspot activity, like highest rainfall, is that for recorded worst cases, and five of the last six cycles have been high magnitudes. Cycle 19 in 1957 had a peak sunspot number of 201, the largest on record. Cycle 21 in 1979, with a peak sunspot number of 165, was the second largest. Cycle 22 in 1989 was the third largest. The current cycle, 25, is expected to peak in July 2025 with 115 sunspots, although some academics forecast that it will be one of the strongest ever, with 210 to 260 sunspots.

a.   **Frequency Preferences.** The best ocean frequencies are on 4MHz, with ranges of up to 300 miles in day conditions and thousands of miles at night without static at 2Mhz. Characteristics are as follows:

(1)   **Sunset.** At sunset, the lower layer ionization decreases and the D layer disappears.

(2)   **Dusk.** At dusk, range increases on 2MHz over thousands of miles almost instantaneously. Interference levels are dramatically reduced.

(3)   **Night.** The reflecting layer of the ionosphere rises at night, increasing the ranges for 4MHz to 6MHz, so lower frequencies are best at night. High frequencies are no good at night.

(4)   **Day.** Low frequencies are weak during the day. High frequencies are used in the daytime.

35.4   Typical Frequency Propagation. The following are derived from a number of sources. As most already know, propagation alters with seasons, time after sunset, and also after sunrise. I have used these for many years as a quick reference. The typical range is given for each band.

Table 35-1. HF Propagation Table—Spring and Summer

Sunset + hr	4MHz Miles	8MHz Miles	12MHz Miles	16MHz Miles	22MHz Miles
1	50–300	250–1,100	550–3,500	750–6,500	1,500–7,500
2	100–650	250–1,500	500–3,500	750–6,000	
3	100–650	250–2,000	500–3,500		
4	100–850	250–2,600			
5	100–1,100	250–2,600			
6	100–1,600	400–3,200			
7	100–1,700	500–3,600			

Sunset + hr	4MHz Miles	8MHz Miles	12MHz Miles	16MHz Miles	22MHz Miles
8	250–2,100	750–4,200			
9	250–2,600	750–4,200			
10	250–2,600	750–4,200			
11	250–2,600	750–4,200			
12	100–1,100	500–2,700			
Sunrise + hr					
1	100–550	410–2,100			
2	0–100	410–2,100			
3	0–100	260–1,600			
4	0–100	260–1,600	500–1,100		
5	0–100	260–1,600	500–1,600		
6	0–100	260–1,600	500–2,600	700–4,200	
7	0–100	260–1,600	500–3,600	700–4,200	1,400–7,200
8	0–100	260–1,600	500–3,600	700–4,200	1,400–7,200
9	0–100	260–1,600	500–3,600	700–4,200	1,400–7,200
10	0–100	260–1,600	500–3,600	700–4,200	1,400–7,200
11	0–100	160–600	500–3,600	800–6,200	1,400–7,200
12	0–200	160–600	500–3,600	800–6,200	1,400–7,200
13	50–270	160–850	500–3,600	800–6,200	1,400–7,200

Table 35-2. HF Propagation Table—Fall and Winter

Sunset + hr	4MHz Miles	8MHz Miles	12MHz Miles	16MHz Miles	22MHz Miles
1	100–650	400–2,100	500–3,600	700–6,200	1,400–7,200
2	100–650	400–2,100	500–4,200	700–6,200	
3	100–1,100	400–2,100	500–4,200		
4	100–1,100	400–2,600	500–4,000		
5	100–1,100	400–3,200	500–4,200		
6	100–1,600	400–3,600			
7	250–2,100	400–4,200			
8	250–2,600	500–4,200			
9	500–3,100	500–4,200			

(continued)

Table 35-2. *Continued*

Sunset + hr	4MHz Miles	8MHz Miles	12MHz Miles	16MHz Miles	22MHz Miles
10	500–4,200	500–4,200			
11	500–3,200	750–5,200			
12	250–2,600	750–5,200			
13	250–1,600	500–2,600			
Sunrise + hr					
1	100–1,100	400–2,200			
2	100–600	400–2,200			
3	0–100	400–2,200			
4	0–100	400–2,200	500–3,600	700–4,100	1,500–3,200
5	0–100	250–1,600	500–3,600	700–4,100	1,400–4,200
6	0–100	250–1,600	500–3,600	700–4,100	1,400–4,200
7	0–100	250–1,600	500–4,200	700–5,100	1,400–6,100
8	0–100	250–1,600	500–4,200	700–5,100	1,400–7,200
9	0–100	250–1,600	500–4,200	700–6,100	1,400–7,200
10	0–100	250–1,100	500–3,600	700–6,100	1,400–7,200
11	0–250	250–1,600	500–3,600	700–6,100	1,500–7,200

Table 35-3. Quick Reference Optimum Transmission Times

Frequency MHz	Sunrise (0600)	Noon (1200)	Sunset (1800)	Midnight (2400)
22000	Average 100–2,000nm	Good 2,000nm plus	Good 2,000nm plus	Average 100–2,000nm
12000	Good 2,000nm plus	Good 2,000nm plus	Good 2,000nm plus	Good 2,000nm plus
8000	Good 2,000nm plus	Average 100–2,000nm	Average 100–2,000nm	Good 2,000nm plus
6000	Good 2,000nm plus	Average 100–2,000nm	Average 100–2,000nm	Good 2,000nm plus
4000	Average 100–2,000nm	Bad 50nm	Bad 50nm	Average 2,000nm plus
2000	Good 2,000nm plus	Bad 50nm	Bad 50nm	Good 2,000nm plus

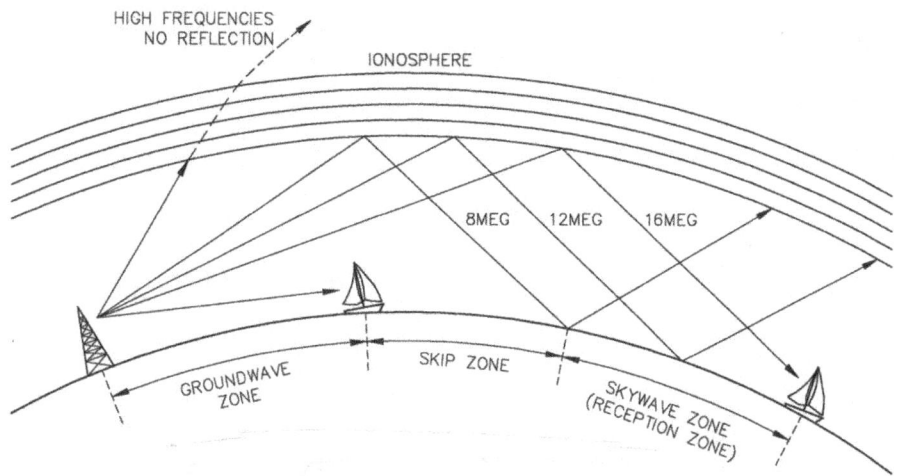

Figure 35-2. HF Radio Wave Behavior

35.5   Operation Requirements. Certain legal requirements and operational procedures must be observed.

a.   **Ship Station Licensing.** Every vessel must have a license issued by the relevant communications authority. Transmitters must be type-approved by the appropriate communications authority. The issued call sign and vessel name must be used with all transmissions. Go to the official authorities listed in the VHF chapter.

b.   **Operator Licensing.** An operator's certificate is required, and a test is given that covers knowledge of distress and safety procedures and related marine communications matters.

35.6   HF Radio Frequencies and Bands. Always consult a current list of radio signals. The UK Admiralty List of Radio Signals (ALRS) is by far the most accurate; invest in the relevant volume for any world location, or do some online research before you go.

a.   **Listen to Station.** If you can hear traffic clearly on the band, you will probably have relatively good communications on that band.

b.   **Monitor Bands.** Monitor the various bands and channels and determine the best peak period for communications. If the signal strength is good but the channel is busy, use a second channel if available, or wait. Do not tune while a call is in progress.

c.   **Station Identification.** Have name, call sign, position, and accounting code ready for the operator if required.

35.7   US Coast Guard Broadcasts. US Coast Guard HF voice broadcasts are on upper sideband mode and are broadcast by "Iron Mike," a synthesized voice. Visit www.weather.gov.

a. **Chesapeake (NMN):**

Frequencies: 4426, 6501, 8764kHz; WX forecasts at 03:30, 05:15, 09:30

Frequencies: 6501, 8764, 13089kHz; WX forecasts at 11:15, 15:30, 21:30, 23:15

Frequencies: 8764, 13089, 17314kHz; WX forecasts at 17:15

b. **New Orleans (NMG):**

Frequencies: 4316, 8502, 12788kHz; WX forecasts at 03:30, 05:15, 09:30, 11:15, 15:30, 17:15, 21:30, 23:15

c. **Point Reyes (NMC):**

Frequencies: 4426, 8764, 13089kHz; WX forecasts at 04:30, 10:30

Frequencies: 8764, 13089, 17314kHz; WX forecasts at 16:30, 22:30

d. **Kodiak (NOJ):**

Frequencies: 6501kHz; WX forecasts at 02:03, 16:45

e. **Honolulu (NMO):**

Frequencies: 6501, 8764kHz; WX forecasts at 06:00, 12:00

Frequencies: 8764, 13089kHz; WX forecasts at 00:05, 18:00

f. **Guam (NRV):**

Frequency: 6501kHz; WX forecasts at 09:30, 15:30

Frequency: 13089kHz; WX forecasts at 03:30, 21:30

35.8   Caribbean Marine Weather Center. This service was established and is operated by Chris Parker. If you sail within the Caribbean area, I strongly suggest that you invest in one of many subscriber-based weather services he offers. The decline of Caribbean SSB and VHF weather information makes the service indispensable. The range of Chris's services includes all of the Caribbean, the Gulf of Mexico, the Pacific coast of Central America, the Bahamas, the entire US East Coast, and out to Bermuda; also, the Atlantic waters north of the equator in the area west of about 50W longitude. He uses omnidirectional and directional beams to target a yacht location. Summer net for the Eastern Caribbean is at 08:30 AST/EST (12:30 UTC) on 8.137 USB and 12.35 USB. At the start of each net, emergency traffic to and from boats is passed, followed by the weather, comprising a 5-minute broadcast of current conditions and weather forecast for the target geographical area. Check out and communicate through www.mwxc.com.

35.9   UK Coast Guard Broadcasts. The following are selected frequencies for navigational warnings, weather forecasts, and working frequencies for the UK Coast Guard.

Table 35-4. UK HF Broadcast

Frequency kHz	MRCC	Time
2226	Aberdeen	07:30, 19:30
1883	Belfast	08:10, 20:10
1880	Falmouth	07:10, 19:10
1925	Humber	07:50, 19:50
1770	Shetland	07:10, 19:10
1743	Stornaway	07:10, 19:10

35.10 **Mediterranean Radio Frequencies and Weather Forecasts.** The following frequencies are current:

    a. **France,** Cross Med La Garde: 1696, 2677 at 10:00 and 16:00; 22900kHz USB

    b. **Monaco Radio:** 4363, 8728, 13146, and 1726kHz USB

35.11 **Australian Frequencies and Weather Forecasts.** The Bureau of Meteorology (BOM) broadcasts marine weather warnings, forecasts, and observations. Marine weather warnings broadcast every hour. Weather forecasts and coastal observations and high seas are repeated every 4 hours. MSI for each state or territory. The service provides 24-hour distress, safety, and urgency watches on 4125kHz, 6215kHz, and 8291kHz and navigation warnings on 8176kHz twice daily.

    a. **Charleville VMC (Australian Weather East):** This service covers coastal waters to around 200nm offshore. Daytime (07:00–18:00 EST) on 4426, 8176, 12365, and 16546; nights (18:00–07:00 EST) on 2201, 6507, 8176, and 12365. Coastal waters, high seas areas, and marine weather warnings broadcast on the hour for Northern Territories, Queensland, New South Wales, Victoria, and Tasmania. MSI at 25 minutes past each hour. Navigation Warnings are broadcast on 8176kHz at 3 minutes before each hour. They complete prior to the weather broadcast start on VMC. Times are EST (UTC+10).

    b. **Wiluna VMW (Australian Weather West):** This service covers coastal waters to around 200nm offshore. Daytime (07:00–18:00 EST) on 4149, 8113, 12362, and 16528; nights (18:00–07:00 EST) on 2056, 6230, 8113, and 12362. The best reception is generally on 8113kHz from Kimberley down to South Coast. Coastal waters, high seas areas, and marine weather warnings broadcast on the hour for Queensland Gulf, Northern Territories, Western Australia, and South Australia. MSI at 25 minutes past each hour. Times are WST (UTC+8). Forecasts for South Australia at 03:00, 07:00, 15:00, 19:00, 23:00 CST. Forecast for WA North of NW Cape 03:30, 07:30, 11:30, 15:30, 19:30, 23:30 WST. Forecasts for WA South of NW Cape 00:30, 04:30, 08:30, 12:30, 16:30, 20:30 WST. Forecasts for Northern Territory 01:00, 05:00, 09:0, 13:00, 17:00, 21:00 CST.

35.12   New Zealand (Taupo Maritime Radio—ZLM). This service operates on 2207, 4146, 6224, 8297, 12356, 16531. Weather (NZST) coastal navigation warnings, synopsis, and warnings for the New Zealand coast at 01:33, 05:33, 08:03, 12:03, 13:33, 17:33, and 20:03 on 2207, 4146, and 6224. Oceanic, navigational, and meteorological warnings in force for NAVAREA XIV; synopsis and forecast for high seas area Southern. This covers Norfolk Island, Cook Islands, Samoa, Tonga, Tokelau, Niue, and American Samoa. Oceanic warnings at 03:03, 03:33, 09:03, 10:03, 15:03, 15:33, 21:03, and 22:03 on 6224 and 12356.

35.13   HF Radio Tuner Units. The tuner unit function is to match the antenna length to the frequency being used.

    a.   **Manual.** There are still manual tuner units around, although they have been largely phased out by fully synthesized systems with automatic tuner units (ATUs). These entailed the matching of the antenna by adjusting tune and load controls using an inbuilt tune meter.

    b.   **Fully Synthesized Units.** The new synthesized radio sets with automatic tuner units enable nontechnically oriented people to easily communicate. Units consist of a full range of International Telecommunication Union (ITU) frequencies. The tuner unit essentially consists of inductors and capacitors that are automatically switched in series or parallel with the antenna to achieve the correct tuned length.

35.14   HF Radio Aerials. The aerials are crucial to proper performance of the HF radio. The whip generally operates over a wider frequency range than the wire line aerials seen on sailing boats.

    a.   **Loaded Whip.** These aerials have loading coils and are generally very long.

    b.   **Unloaded Whip.** These whips have a similar performance to long wire backstay aerials, and the ATU provides the required aerial length. As the voltage and currents can be significant at the base, it is essential that high-quality insulators and well-insulated feed line cables be used to minimize losses. A very low-resistance ground system is required. Unloaded whips are a viable alternative to backstay aerials. The smallest is about 15ft (4.5m) long for 500W input and is rated to 9kV and 1.4MHz to 30MHz; length increases with power input.

    c.   **Backstay.** Insulated backstays are the most common type on sailing boats. They do find some use on trawler motor yachts as well, often in a triatic stay arrangement, and are most efficient in the 2MHz to 8MHz range. Losses can occur here as well, as signal radiates into the mast and rigging. The backstay should be at least 35ft (11m) long for an effective aerial. The insulators should be free of chips and have long leakage paths.

    d.   **Split Lead Antenna.** The GAM/McKim split lead antenna is a very good option from GAM Electronics. The antenna simply press-fits onto and over the backstay—easy to install; no need for backstay insulators. The active elements are specifically designed for RF, so they're more efficient than a stainless-steel backstay. The backstay needs to be 34ft (10m) or longer. Check them out at gamelectronicsinc.com.

e. **Aerial Feed Line.** The feed line to the aerial is very important, as resistance causes attenuation of the transmission signals.

(1) **Feed Line Cables.** Problems with feed line cables are mainly caused by thin conductors and bad joints, which result in conductor heating and losses. Ideally the cable should not run close to metal decks or hull.

(2) **Insulation Quality.** Insulation losses occur through conductors and deck feed insulators. Use cables with good insulation values; ideally, a silicon insulated high-voltage cable should be used.

(3) **Deck Transits.** Poorly insulated leads close to metal decks and hull can cause arcing or induction losses. External cables also cause leakages as cable breaks down the insulation, causing cracks due to the UV effects of the sun. The best system in steel vessels is the use of through-deck insulators. These offer long leakage paths and therefore less signal loss; they must, however, be kept clean.

(4) **Backstay Connections.** It is imperative that the feed line to aerial connection be made properly.

35.15 HF/SSB Radio Grounds. HF/SSB radio problems of transmission and reception are more often than not attributable to poor antenna systems. Remember that the so-called "ground" is an integral part of the aerial system; it is more accurately referred to as the counterpoise and not a ground. If it is poor, you may not be able to tune properly to required frequencies.

a. **Ground Plates/Shoes.** Ground plates or shoes are the most effective method of providing an RF ground plane or counterpoise on FRP/GRP and wooden boats. They provide half the required aerial length and are an integral part of the radiating system. Copper plates are externally attached to the hull.

b. **Internal Copper Mesh.** Glass and timber vessels try to avoid the installation of ground plates by glassing in a large sheet of copper mesh.

c. **Copper Straps.** The interconnecting copper strap from tuner unit to ground plane is essential. It must be a strap, not a cable, and the surface area is the critical factor. To be effective, a low resistance is required; higher resistance is the cause of many performance drops and interference. The ground strap should be at least 2in (50m) wide. The copper strap should be run clear of bilge areas.

d. **KISS-SSB.** This is a US patented marine SSB counterpoise system. It is very simple and easily self-installed; no external thru-hull connections are required. The system comprises many wire radials that resonate in the entire frequency range, 2MHz to 28MHz. A 4-foot lead is connected to the ATU and then stretches out the remaining 10 feet of 1-inch-diameter tubing, which encloses the array of specific-length radiating copper radials. Switch to this system, and forget about the ground plate. Visit www.kiss-ssb.com.

35.16   HF/SSB Radio Maintenance. Regular maintenance tasks will ensure good radio performance.

    a.   **Aerial Connections.** The lead wire aerial connections should be regularly checked for deterioration. If exposed, the wire may degrade and introduce resistance into the circuit. Always tape up the connection with self-amalgamating tape.

    b.   **Insulators.** Always clean the insulators to remove salt deposits that encrust and cause surface leakages. This should include the upper insulator on wire antennae. A damp rag is the best cleaning tool.

    c.   **Ground Connection.** Check the RF ground connections. Clean and tighten the bolts and connection surfaces. After this, apply a light smear of petroleum jelly to prevent deterioration if in a bilge area. Always check and keep this area clean and dry if in a bilge area, as reaction between the copper strap and metalwork can cause corrosion problems.

35.17   Standard Time Frequencies. The National Institute of Standards and Technology (NIST) broadcasts time and frequency from WWV and WWVH in the United States. This is also known as the "time tick." Since November 2021 a test signal has been broadcast on minute 8 of every hour on WWV and minute 48 on WWVH. This is being initiated as part of ionospheric research and is a venture of the Ham Radio Citizen Science Investigation (HamSCI) and NIST. This test signal comprises various tones, chirps, and Gaussian noise bursts and may be modified occasionally.

    a.   **WWV (Fort Collins, Colorado).** Times 8th, 9th, and 10th minute past the hour on 2.5, 5, 10, 15, and 20MHz. NWS Atlantic high seas broadcasts were discontinued in 2019.

    b.   **WWVH (Kauai, Hawaii).** Times 48–51 minutes past the hour on 2.5, 5, 10, and 15 MHz. NWS Pacific high seas broadcasts were discontinued in 2019.

    c.   **DCF-77 (Mainflingen, Germany).** The data signal is broadcast every 60 seconds on 77.5kHz.

    d.   **MSF (Cumbria, UK).** The times signal is broadcast every 60 seconds on 60kHz.

35.18   HF/SSB Radio Troubleshooting. Basic HF/SSB radio system troubleshooting is described in the following table.

Table 35-5. HF/SSB Radio Troubleshooting

Symptom	Probable Fault
No reception	Wrong channel selected Propagation problems Aerial lead wire broken Aerial connection corroded Tuner unit fault
Poor reception	Propagation problems Aerial connection corroded Insulators encrusted with signal leakage Aerial grounding out
No transmission	Tuner unit fault Aerial connection corroded Insulators encrusted, with signal leakage Aerial grounding out Aerial lead wire broken Ground connection corroded Low battery voltage Transceiver fault
Poor transmission	Propagation problems Aerial connection corroded Insulators encrusted, with signal leakage Aerial grounding out Tuner unit fault Ground connection corroded

Amateur (Ham) Radio

Amateur, or ham, radio is the realm of a worldwide group of radio enthusiasts. Although ham operators have been involved in many lifesaving efforts with sailors, regrettably, ham operators and the system have been somewhat abused. Ham operators were crucial in the more recent tsunami and earthquake events and assisting after many other disasters. Ham radios are a major communication source in the cruising world, and we all appreciate the efforts. Ham radio use is prevalent in the United States, with many long-distance cruisers equipped; however, they're not as prevalent elsewhere in the world.

36.1   Licensing Requirements. There are a number of important factors to consider when planning to install or use ham radio on a yacht. The first step is reviewing requirements. Check out the American Radio Relay League (ARRL) and invest in a copy of the *ARRL Handbook*, now in its 100th edition. Visit www.arrl.org.

> **a.   Operator Licensing.** It is the operator that is licensed, not the station. There are a number of levels that give either partial or full access to frequencies. Levels used to require an examination in Morse code, although that has changed. An understanding of radio theory and rules and regulations with respect to operations is still required. Fear of technical matters and theory, as well as the Morse test, frightened off many would-be amateurs; again, that's changed. There are three levels of ham radio licensing. First is the entry-level Technician License. Second is the General Class License, which is the best for cruising and will be required for access to Maritime Mobile Nets in the 15-, 20-, and 40-meter bands. The top level is the Extra Class License.

> **b.   Penalties.** You must be licensed for the country of operation. Beware in some third-world countries, where communications are controlled; prison time and vessel loss can occur if ham radio is used in port and without authorization. In many cases you will not be acknowledged on ham bands unless you are licensed and have a call sign.

> **c.   HF/SSB radio versus Ham radio.** This is a perpetual argument, with both systems having a place. Carry both, or install a combined ham-HF/SSB unit:

> > **(1)   HF/SSB.** Radio sets are generally easier to operate for nontechnical people, and with automatic tuning, it is simple to punch in an ITU channel number and talk. Additionally, radios have automatic emergency channel selection and are type approved for marine communications. Only a restricted license or permit is required. It is allowable to operate a HF/SSB radio on amateur frequencies if you have a ham license. One of the disadvantages of HF/SSB on ham frequencies is that synthesizers are programmed in 0.10kHz steps. Ham communications may be at frequencies outside of that, so HF/SSB sets can be marginally off frequency. Most HF/SSB

sets operate on upper side band (USB), while most frequencies below 40 meters are lower side band (LSB).

(2) **Ham.** The ham operator must have a license appropriate to the frequency band being worked. Access to GMDSS emergency frequencies is illegal except in emergencies. It is illegal to operate non-type-approved radios such as ham radios on marine frequencies. Ham allows the use of casual conversation, which marine SSB does not. Ham allows full access to information packed nets and a worldwide communications network.

36.2   Ham Nets. The following Maritime Mobile Net times could vary an hour either way, depending on summer time changes in respective countries; 14.314MHz is monitored virtually 24 hours and is the de facto maritime mobile international calling frequency. This list has evolved over 30 years. Nets come and nets go, so if it's not in operation, apologies! Thanks to all who operate these and provide service; it's much appreciated! These are USB frequencies unless noted.

Table 36-1. Caribbean Nets

Time (UTC)	Frequency (MHz)	Net Name and Area
00:00–22:00	14.300/14.313	Maritime Mobile Net
02:00	14.334	East Coast/Brazil Net
10:00–12:00	6.215	Daily Caribbean Weather
10:00	4.045/8.137	Eastern Caribbean Weather (Chris Parker)
10:30	3.855	Trinidad Emergency Net (LSB/Ham)
10:30	3.815	Caribbean Emergency/Weather Net (LSB/Ham)
10:30	4.045/8.137	Bahamas Weather (Chris Parker)
11:00	7.250	Caribbean Net (Mon–Sat) (LSB/Ham)
11:00	14.300	Intercon Net (Caribbean/Pacific)
11:10	3.930	Puerto Rico Weather Net
11:20	7.096/3.696	Bahamas Weather (LSB/Ham)
11:30	8.137/12.350	US East Coast, NW Bermuda Atlantic (Chris Parker)
11:45	7.268	Waterway Cruising Club (LSB/Ham)
12:00	4.060	Coconut Telegraph
12:15	8.104	SSCA KPK Cruising Safety (Migration Season)
12:30	8.137/12.350	Bahamas and East Caribbean WX (Chris Parker)
12:30	8.152/8.146/8.164	Cruizheimers Net
12:30	7.185	Barbados Information Net

(continued)

Table 36-1. *Continued*

Time (UTC)	Frequency (MHz)	Net Name and Area
13:00	8.137/12.350	Western Caribbean Weather (Chris Parker)
13:00	7.083	Central American Breakfast Club
13:15	6.209/6.212/6.516	SW Caribbean Net
13:30	8.107/8.167	Panama Connection Net
13:30	6.212	Picante Net (Oct to July)
13:45	8.152	SSCA Net Caribbean Season
14:00	7.292	Florida Coast Net
14:00	6.209	Northwest Caribbean Net
14:30	6.209/6.212/6.516	Caribbean NW Net
17:00	14.300	USCG Amateur Net
17:00	14.300	Maritime Mobile Net
20:30	7.086	Caribbean Cocktail and Weather (LSB/Ham)
21:00	8.152	Doo-Dah Net
21:10	12.350	Trans-Atlantic Cruisers Net (Migration Season)
22:00	21.404	Central American Net
22:00	8.137/12.350	Caribbean and Atlantic Weather (Chris Parker)
22:30	3.815	Caribbean WX and Emergency (LSB/Ham)
23:30	3.815	Barbados/Trinidad Net

Table 36-2. Pacific and Mexican Nets—Part 1

Time (UTC)	Frequency (MHz)	Net Name and Area
00:00	3.968	Happy Hour Mexico (Baja, West Mexico)
00:00	14.135	Pacific Island Net
00:00–22:00	14.300/14.313	Maritime Mobile Net
00:00	14.320	Sea Maritime Mobile Net (South/West Pacific)
00:00	14.135	Pacific Island Net
00:25	14.323	Mar Mobile SE Asia (Hong Kong to Australia)
01:00	6.516/8.122	Southbound Net (West Coast Baja)
01:00	3.855	BC Boaters Net (Straits of St. George)
01:00–03:00	14.305	California Hawaii Cocktail Hour (Pacific)
02:00	6.516	Bluewater Pacific

Time (UTC)	Frequency (MHz)	Net Name and Area
02:00	21.492	Gerri's Happy Hour (Pacific to Baja)
02:30	4.051/4.060	North Sea of Cortez
03:00	14.300	Pacific Seafarers Net (LSB/Ham)
03:00	14.116	Travelers Net (West Pacific/Australia)
03:30	7.294	Baja and West Coast
04:00	3.856	Taco Net (Baja)
04:00	14.115	Canadian DDD Net (Pacific)
04:00	14.318	Arnolds Net (South Pacific)
04:30	4.030/4.024	Papagayo Net (Pacific)
05:00	14.313	Pacific Maritime Net
05:05	12.353	Pacific Maritime Radio
05:15	8.752	Gulf Harbour Radio (NZ/SW Pacific)
07:15	3.820	Bay of Islands Net (NZ, Australia, and Pacific)
08:00	14.315	Pacific Inter-Island Net (West and South Pacific)
10:00	14.320	South China Sea Net
11:00	14.300/14.313	Intercon Net (Caribbean/Pacific)
12:00	14.320	South East Asia Net (Pacific)
13:00	7.085	Central American Breakfast Club
13:30	6.212	Picante Net (Oct to July) (Western Mexico)
13:30	3.968	Sonrisa Net (Sea of Cortez/Pacific Mexico) (LSB)
14:00	8.143/8.137/8.155/6.230	Panama Pacific Net (Weather to Galapagos)
14:00	4.149/4.146/6.224/6.270	Amigo Net (Mexico and Pacific) (CLOSED)
14.45	7.192	Chubasco Net (US Pacific/Mexico) (LSB)
15:15	7.233.5/7.238	Baja California Net (LSB)
15:30	3.865	NW Boaters Net
16:00	7.200	Taco Net—Baja
16:00	3.870/7.285	Great Northern Boater's Net
16:00	8.104	Westbound Pacific Net
16:00	14.320	California Hawaii Net
17:00	14.300	USCG Amateur Net
17:00	7.240	Bejuka Net (Central America)
17:00	14.300	Maritime Mobile Net

(continued)

Table 36-2. *Continued*

Time (UTC)	Frequency (MHz)	Net Name and Area
17:30	8.173/8.137/8.294	Pacific Magellan Net (Tahiti, Marquesas)
17:30	8.188	Coconut Breakfast Net (French Polynesia)
18:00	8.173	PolyMagNet (Polynesia)
18:00	6.516/4.417	Far North Radio (NZ and Pacific)
18:00	7.076	South Pacific Cruising Net
18:30	12.353	Coconut Breakfast Net (Polynesia)

Table 36-3. Pacific, Mexican Nets—Part 2

Time (UTC)	Frequency (MHz)	Net Name and Area
19:00	14.340	Manana Net (Mexico, Eastern Pacific)
19:00	14.305	Confusion Net (Pacific)
19:00	7.285	Hawaii Net
19:00	14.329	Bay of Islands Net (NZ and South Pacific)
19:00–19:30	4.417	Far North Radio (NZ/Pacific)
19:00/22:00	14.340/14.305	Hawaii Net
19:45	6.224	Yokwe Net (Marshall Islands)
20:00	7.095	Harry's Net (Pacific)
20:00	7.080	New Zealand Weather Net
20:30	8.230/8.188	Vanuatu Net
20:40	7.087	Comedy Net (Pacific and Australia)
20:40	7.190	Admirals Net (US West Coast)
21:00	14.315	Tony's Net (South Pacific, Australia, and NZ)
21:00	21.402	Pacific Maritime Mobile Net
21:15	8.101	Namba Net (Solomons, Vanuatu, New Caledonia)
22:00	21.404	Central American Net
22:00	14.305	Hawaii Net
22:00	8.161	Sheila Net (Western Pacific, NE Australia)

Table 36-4. UK/Europe/ Mediterranean/African Nets

Time (UTC)	Frequency (MHz)	Net Name and Area
01:00	21.407	Pacific Indian Ocean Net
04:45/04:50/05:00/05:10	21.300/14.300/7.115	Durban Maritime Net (South Africa)
05:00	14.303	Tony's Net (Indian Ocean)
05:00	8.101/12.353	Peri-Peri Radio East Africa
05:00	21.200	UK, NZ, African Net
06:30	14.316/7.045	South African Maritime Mobile Net
07:00	7.085	Mediterranean Sea Cruisers Net
07:00	14.313	International Maritime Mobile Net
08:00/18:00	14.303	UK Maritime Mobile Net
09:00	14.313	Mediterranean Sea Net
10:00	14.313	German Maritime Mobile Net
11:15	14:316/14.341	Indian Ocean Maritime Net
11:30	14.316/7.045	South African Maritime Mobile Net
12:45	14.121.5	Mississauga Net (Atlantic, Mediterranean)
13:00	21.400	Trans-Atlantic Net
13:00	6.516/8.134/12.359	Med Net
13:00	21.400	Transatlantic Net
15:00	8.101/12.353	Peri-Peri Net
16:30	14.303	Swedish Maritime Net
16:30	14.313	German Maritime Net
18:00	14.303	UK Maritime Net
19:00	14.297	Italian Maritime Net
20:00	21.390	Inter-American Traffic Net
20:30	14.303	Swedish Maritime Net

36.3 **Email Services.** Snail mail is a thing of the past as we use email for most every-thing. For most boaters an INMARSAT terminal is not a viable economic alternative, although GMDSS-inspired changes make communications improvements essential. If you have a quality HF/SSB radio on board, that valuable piece of equipment is your means to get connected to the world.

    a.   **HF E-Mail System Components.** The basic components are as follows:

        (1)   **SSB Radio.** Not all SSB radios are configured for email and may require modification to operate and require an audio output jack.

This should provide a line level output signal of 100mV RMS. A good power supply is essential to maintain constant transmission; battery voltages must be up and power supply connections sound. Aerials and ATU grounds must be good to ensure optimum transmission and reception. Refer to the SSB sections for details.

(2) **HF/SSB Modem.** Modems are generally part of the service providers' systems, although those using non–service company systems, such as packet radio enthusiasts, use a terminal node controller (TNC). The most common modems are those from Kantronics (kantronics.com). A modem has a power input, data port, and radio port, along with operating software. The audio cable to the SSB consists of four wires; transmit audio (TxD); receive audio (RxD), push-to-talk (PTT), and the audio signal ground. The audio cable must be shielded, with the shield being connected at both ends. Prewired cables are available from Kantronics. Clip-on Ferrites must be fitted at both ends to reduce RF interference. See the chapter on interference for ferrite principles. Software is required; check sailmail.com for details.

(3) **PacTOR Modems.** Made by SCS in Germany, these modems use a two-tone signal and are far more effective and reliable, with data transfer in noisy or weak signal environments and transmission speed adapting to the link quality. Effectively a hybrid Packet/Amtor modem, they are becoming the favored modem type for most marine HF email systems. PacTOR is a radio modulation mode that is utilized by many amateur radio operators. Modems have an optional advanced encryption standard (AES), which provides data encryption. The term "PacTOR" is the combination of "amateur teleprinting over radio" (AMTOR) and "packet radio." There have been a number of generations, with PacTOR IV the current unit. PacTOR is used on frequencies from 1MHz to 30MHz. Check www.pactor4.com.

36.4   Email Service Providers. There are several ways to connect to email via SSB/HF, SailMail and WinLink being the best. They offer comprehensive services that cannot be afforded without installing satellite systems, such as weather and navigational warnings, along with the very important email.

a. **SailMail.** The SailMail Association is a nonprofit organization comprising yacht owners that operates an email system for members. Check out sailmail.com for complete details. SailMail email can be sent out via satellite services, but its great strength for yachts is the worldwide network of SSB-PacTOR radio stations. I highly recommend subscribing to this service.

b. **WinLink 2000 Network.** This global radio messaging system utilizes amateur-band radio frequencies that can broadcast email with attachments, position reporting, weather forecasts, and emergency communications. It is available to amateur radio operators by linking their radio to the internet (winlink.org).

c.   **SeaMail.** Xaxero provides unrestricted service to the Pacific, free of charge. Obtain free software at www.seamail.org. This service operates through Kumeu Radio ZMH302 Auckland, New Zealand, out into the Pacific. The USB assigned frequencies are 6382.5kHz, 8488kHz, 12708.5kHz, 16954.53Khz; USB 6380.3, 8485.8, 12706.3, 16952.3. The service uses PacTOR and PacTOR II.

36.5   Morse Code. The final commercial Morse code transmission was made on July12, 1999, from the Globe Wireless master station at Point Reyes near San Francisco. The transmission signed off with Samuel Morse's first message of 1844, "What hath God wrought?" and the prosign "SK," which means "end of contact." Although commercial Morse code transmissions have ceased worldwide, radio amateurs continue to use and keep the "language" alive. My earliest seagoing experiences on old tramp freighters was all Morse communication. Morse code still has its uses and is most practical when used with a signal lamp. In the interest of tradition, I have included the Morse alphabet and recommend at least learning the Morse "SOS" for distress and emergencies. SOS was devised as it was easy Morse code to remember; however, SOS also acquired the mnemonic of "Save Our Souls" or "Save Our Ship" to many seafarers. The definitive reference for international Morse code is "Recommendation ITU-R M.1677-1." There are Morse practice transmission schedules to maintain skills. These include some Australian ones such as VK2RCW, continuous on 3699kHz and 144.950MHz 5wpm, 8wpm, and 12wpm; VK3COD, nightly (Monday–Friday) at 10:30 UTC on 28.340 MHz and 147.425 MHz; and VK4WCH, Wednesday at 10:00 UTC on 3535 kHz. Visit morsecode.world/; rsgb.org.

SOS (Save Our Souls)
. . . Dit Dit Dit
- - - Dah Dah Dah
. . . Dit Dit Dit

Morse Alphabet A - M		Morse Alphabet N - Z	
A . -	Dit Dah	N - .	Dah Dit
B - . . .	Dah Dit Dit Dit	O - - -	Dah Dah Dah
C - . - .	Dah Dit Dah Dit	P . - - .	Dit Dah Dah Dit
D - . .	Dah Dit Dit	Q - - . -	Dah Dah Dit Dah
E .	Dit	R . - .	Dit Dah Dit
F . . - .	Dit Dit Dah Dit	S . . .	Dit Dit Dit
G - - .	Dah Dah Dit	T -	Dah
H	Dit Dit Dit Dit	U . . -	Dit Dit Dah
I . .	Dit Dit	V . . . -	Dit Dit Dit Dah
J . - - -	Dit Dah Dah Dah	W . - -	Dit Dah Dah
K - . -	Dah Dit Dah	X - . . -	Dah Dit Dit Dah
L . - . .	Dit Dah Dit Dit	Y - . - -	Dah Dit Dah Dah
M - -	Dah Dah	Z - - . .	Dah Dah Dit Dit

Figure 36-1. Morse Code Alphabet

Satellite Communications

37.1 **Satellite Services.** There is an ever-increasing number of satellite services, and these are described along with their salient features. The system you select will depend on many issues, including the services required, coverage area, initial installation costs, antennae sizes, and the all-important call costs.

a. **Viasat/INMARSAT.** In 2023 Viasat acquired INMARSAT, the cornerstone communications network that supports the GMDSS. This is based on the International Maritime Satellite Organization. The system comprises fourteen satellites in geostationary orbit 22,236 miles (35,786km) above the Earth's surface, and the satellites remain in the same position relative to the Earth. There are four Ocean Regions: Atlantic East, Atlantic West, Pacific, and Indian. These satellites operate in the L-band, Ka-band, and S-band. Currently, new generation satellites, including the Inmarsat-6 (I-6), are entering service; these hybrid satellites incorporate ELERA (L-Band) and Global Xpress (K-Band). Global Xpress is a high-speed mobile broadband network with global coverage. The next generation of GX7, 8, and 9 satellites are due for launch in 2024 and will significantly increase capacity and capability. Check them out at www.inmarsat.com.

b. **Inmarsat C & Mini-C.** This is two-way, store-and-forward packet data communication via a lightweight, low-cost terminal. This is the only system that allows ships to meet most GMDSS communications requirements. Antennae are small at 10in (25cm) and have gyrostabilized omnidirectional antennae. This allows the transmission and reception of data, including email, SMS, weather updates, and more.

c. **Fleet One Global and Fleet One Coastal.** These services offer email, electronic charting, weather access, and voice calls worldwide. You can check out coverage on the INMARSAT website. The antenna size, always an issue on yachts, is 7lb to 9lb (3kg to 4kg) and the radome size is 12½in × 11in (319mm × 277mm) for the smallest option. An important feature is that Fleet One also supports the free "505" safety service, which allows a direct call to the nearest Maritime Rescue Coordination Center (MRCC).

d. **Viasat.** This system uses six low earth orbit (LEO) small-sat satellites that will increase to twelve with future launches. They offer sea-based systems and cover the United States and Europe, including the Mediterranean, and when fully implemented will cover everywhere. Visit www.viasat.com.

e. **Globalstar.** This system replaced their twenty-four LEO 930m (1,500km) satellites. A spare satellite was placed into orbit in mid-2022. This uses a handset similar to a cell phone with larger antenna or a remote one. Services include high-quality voice, short messaging services (SMS), and roaming. Primary coverage areas are very good, with virtually the entire United States,

516

most of Europe and the Mediterranean, the majority of South America, and also Australia and New Zealand. This is not a GMDSS compliant system, but they do have an SOS facility. Iridium SafetyCast is an automated satellite system for promulgation of weather warnings and marine navigation warning. Check out their services at www.globalstar.com.

f.   **Iridium.** This constellation consists of 75 satellites in LEO at a height of 480 miles (780 kms). Sixty-six are operational and the rest are spares. The system runs on L-band frequencies and offers global voice, data, fax, and paging services. In 2020 the system was certified for use in the GMDSS. With a computer, this can provide internet access and email, as well as fax and enhanced messaging services. Iridium CertusSM is a range of global satellite services that is powered by Iridium NEXT. This advanced L-band global satellite constellation will create cross-linked LEO architecture that will cover 100% of the planet. They also offer SafetyCast, which broadcasts MSI via an EGC service. The latest service is Iridium GO! Exec, which offers more affordable service with improved data speeds and functionality. It is available through PredictWind. Log on to www.iridium.com.

g.   **Starlink.** SpaceX will operate a LEO constellation of 2,400 small satellites, with up to 12,000 more authorized, and an FCC application lodged for 30,000 more. The services offer high-speed, low-latency internet and was approved in 2022 for mobile applications. Starlink uses the Dishy McFlatface flat-panel antenna. Unlike installations on relatively stable cruise ships, they are not as effective on rolling and pitching boats. Mounting the dish is

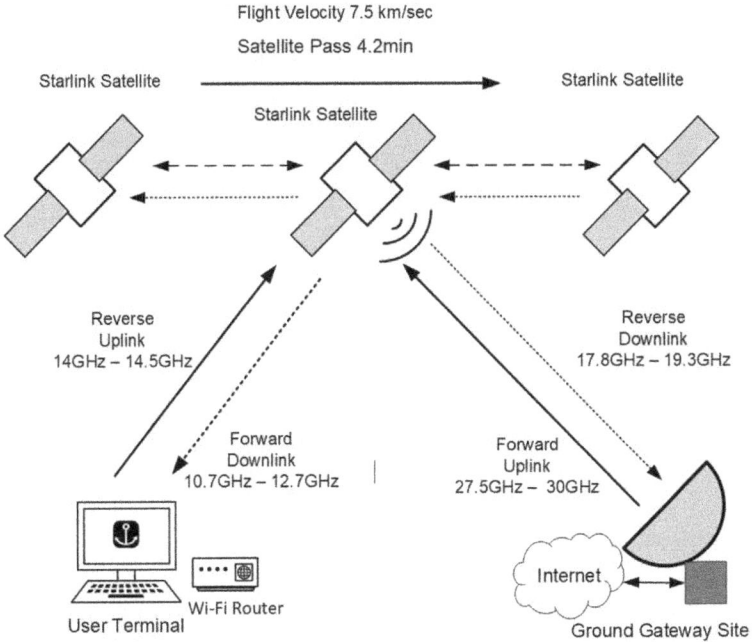

Figure 37-1. Starlink Architecture

problematic, and dropouts are common when swinging on the anchor, unlike satellite systems with a parabolic antenna. In rough weather the issue is that it operates at the lower end of the Ku-band, which is a common frequency range for high-speed data. Although it has higher bandwidth, the frequency is affected by heavy rain. Check out www.starlink.com.

h. **OneWeb.** This UK-based company launched 36 satellites in 2022, with a total of 146 in orbit out of a planned constellation of 648 satellites. They will operate at an altitude of 750m (1,200km). OneWeb will launch 40 satellites in early 2023. OneWeb uses a parabolic antenna for satellite communications.

i. **Thuraya.** This system has two geosynchronous orbit satellites. Satellite 4-NGS is due to launch in 2023. The handsets offer voice, fax, data, and short messaging services. Go to www.thuraya.com.

j. **Eutelsat–VSAT (Very Small Aperture Terminals).** They use a fleet of thirty-six Ku-band geostationary satellites and have plans to launch a new-generation very high throughput satellite (VHTS) , Eutelsat Konnect VHTS, which will have 230 spot beams and 500gps capacity to provide two-way broadband connectivity. Go to www.eutelsat.com.

k. **HughesNet.** This popular service in the United States uses geosynchronous orbit satellites. The system uses their EchoStar XIX satellite, which they claim is the highest broadband capacity satellite in the world. This service is home based, and I am unsure if a mobile service for yachts is available; you will have to check at www.hughesnet.com.

l. **Astranis.** This system utilizes geostationary orbit satellites. Ultimately, the company plans hundreds of them. They are called microGEO, as they are very small satellites when compared to many others. The first satellite is positioned above Alaska. This system uses a spot beam, and satellites will be inserted into geostationary orbits. Check them out at www.astranis.com.

m. **AST SpaceMobile.** This is a cellular broadband direct-to-mobile phone service with equivalent 4G and 5G cell phone speeds. The BlueWalker 3 satellite, a low-latency phased array unit, has been undergoing testing, and plans are to launch the first five commercial Block 1 BlueBird satellites in late 2023. Visit ast-science.com.

n. **Omnispace.** This hybrid mobile network uses 5G mobile as part of the system. Omnispace use non-geosynchronous orbit satellites. They have two LEO satellites for testing and will ultimately deploy 200 satellites. Visit omnispace.com.

o. **Telesat Lightspeed.** This system will have a constellation of 188 LEO satellites offering low-latency broadband. Satellites will operate in hybrid polar and inclined orbits. Innovative optical inter-satellite communication is part of this system. Telesat also has a GEO satellite system comprising thirteen satellites operating on C, Ka, and Ku bands using the Nimiq, Anik, and Telstar series of satellites. Visit www.telesat.com.

p. **Kuiper.** This system is backed by Amazon and is aiming to offer the same satellite broadband services as Starlink. They are due to launch their initial prototype satellites—KuiperSat-1 and KuiperSat-2—for testing in 2023. The constellation will consist of 3,236 satellites. Services will be available when they have successfully launched 578 satellites into orbit. The Project Kuiper FCC4 license requires half the satellite constellation to be launched by 2026 and the complete constellation by 2029. They have partnered with Verizon.

37.2   Satellite System Installation. Systems consist of the radome, which encloses a stabilized antenna dish, a pedestal control unit (PCU), and the RF Unit. Follow installation instructions in the manufacturer's user manual precisely, as warranty may be voided if installation is incorrect.

a. **Radome Installation.** The radome must be located as far as practicable from any HF and VHF antennae, and preferably a minimum of 15ft (5m) from all other communications and navigation receiver antennae. Do not mount the unit in any location subject to vibration. As some are mounted on stern posts, make sure the unit is well supported. Obstructions will create blind spots and disrupt communications. Obstructions less than 6in (15cm) are acceptable within 10ft (3m) of the antenna; however, note that marginal signal strengths are vulnerable to them. Safe compass distance is a minimum of 3ft (1m).

b. **Emissions.** Systems require radiation precautions and should be installed 6ft (2m) or above to avoid excessive microwave radiation. The radome should be outside the beamwidth of radar antennae, typically 10 degrees each side of the central plane.

c. **Visibility.** The radome should be properly aligned parallel with the boat's axis. As beamwidth is around 10 degrees, a clear line of sight is required from 5 degrees elevation and above. The azimuth and elevation angles must be considered at all times.

d. **Cable and Wiring.** Normal cable installation rules apply, and they must be installed to prevent mechanical damage. The antenna unit uses double-screened 50Ω coaxial cable. This is usually RG223/U and RG214, with maximum lengths of 43ft (13m) and 82ft (25m), respectively, to achieve 10dB/0.6Ω maximum losses and attenuation. All cables must be shielded and the shield grounded. Peripheral equipment must be grounded. Coaxial connectors must be installed correctly; poor installation is a frequent cause of problems.

37.3   System Operation. The nominated satellite is based on the boat's position and then selection of a relevant satellite within area coverage. The vessel heading is required to give correct azimuth heading, and the gyro or fluxgate input provides this. The azimuth angle is the angle from north and the horizontal satellite direction. The elevation angle is the satellite height above the horizon with respect to the vessel. At power-up, the system must locate a satellite and synchronize with it. This is by either automatic or manual-initiated hemispheric scan for the selected satellite. The dish performs a search pattern until the satellite signal is located in the relevant Ocean Region. Systems carry out a self-test at initialization. Systems default to the last settings on gyro, azimuth, and elevation; if the vessel position is lost, this data is required. During operation, displays on handsets show signal quality and signal strength signal/noise (S/N) ratio. Bit error rates (BER) decrease with increased signal quality. Log on to www.heavens-above.com for information on satellite orbits and tracking.

37.4   OCENS OneMail. This is a viable solution for email access through satellite or low-bandwidth connections. These include Iridium GO, Globalstar, Sat-Fi, Thuraya IP, and others. The system is able to send and receive attachments, including images, video, and GRIBs. The service is also available for iOS and Android smartphones and tablets. Check www.ocens.com.

37.5   Small Boat Systems. KVH has been a leader in this space for a long time. One offering is the TracPhone V30, a small, lightweight VSAT antenna with dimensions of 15.5in (40cm) by 17.6in (45cm). Inside is a 14.5in (37cm) antenna dish and modem. At just 23.6lb (11kg) with a variable 10V to 36V power supply range, this unit operates on the KVH mini-VSAT BroadbandSM network. As with all service providers, there is a range of plans. Another leading supplier, Intellian, has the v45C unit with a very compact 18in (45cm) antenna. These units use a three-axis stabilization system to stay locked onto a satellite even in rough conditions. KVH has the innovative KVH ONE hybrid system, which integrates satellite, cellular, and wireless technology with automatic switching between sources. KVH uses a layered HTS Ku-band satellite network that is powered by Intelsat and gets cellular support in 150+ countries as far as 20nm (32km) offshore. It is enabled for 5G and LTE-A cellular speeds and is compatible with the KVH or user-supplied subscriber identity module (SIM) cards along with support for shore-based wireless and high-quality enhanced voice over internet protocol (VoIP) service. Go to www.kvh.com for more technical information and services. The KVH Trac-Phone LTE-1 is worth a mention. TracPhone LTE-1 is an ultracompact, marine-grade system for use in US waters. It is much faster than 4G LTE and has a high-gain, dual-antenna array, modem, GPS, and wireless router in the dome, perfect for streaming video and music, wireless calling, web browsing, email and social media access, and a lot more applications. Satellite performance and reception on sailing yachts is often affected by masts or other onboard structures. KVH has created Tracking Avoidance Zone (TAZ) functions, which give optimal reception data on their TracPhone VSAT system.

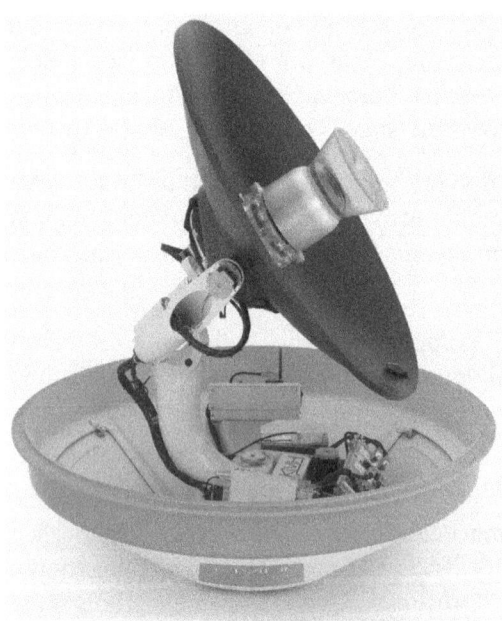

Figure 37-2. KVH TracPhone V11-HTS Interior (*Courtesy of KVH Industries, Inc.*)

Mobile Phones

Cell phone, handheld phone, mobile phone, smartphone, or whatever you call it, wherever you are in the world, the smartphone has permeated our existence at almost every level. It really is the best thing since sliced bread! The impact has been just as great within boating circles as it has been ashore. Mobile phone technology is the convergence of so many technologies, from high definition (HD) photo and video to messaging and telephony, data and internet access, games and music; from banking to payments systems, supermarket orders and fast food delivery, COVID tracking apps, online access for almost all suppliers of goods and services, and ride share. There is an application (app) for almost everything you can think of. Some of us can still recall the now-defunct analog devices and the early GSM technology and the challenges they created. At the time, however, they were great when cruising. We are now in a transition phase from 4G to 5G that will revolutionize the space. Finding a wireless hot spot is always top of the agenda. Do not use your phone when piloting in channels or port environs, or when sailing in congested or busy traffic areas. Vessel pilots in one US state were recently forbidden to use phones when piloting commercial vessels after a container ship ran aground while the ship pilot was preoccupied for an extended period on his phone. Phone use degrades situational awareness and poses a safety threat, so use yours responsibly.

38.1   Distress Calls and Mobile Phones. There are many reasons mobile phones are not good for distress calls except as a last resort. A vessel in distress cannot communicate with other potential rescue vessels in the area. This has the potential to considerably delay rescue, uses greater resources, and increases the risks to all involved. If you are in distress, you may not get through to an appropriate authority. As documented, you can be placed on hold in a queue, or you may be at the outside of the cell range and drop out repeatedly. Vessels in distress that cannot provide exact position information cannot be located using VHF radio direction-finding (RDF) equipment, and a VHF radio facilitates this important function. Vessels in distress cannot activate priority distress alerts using mobile phones. Rescue scene communications can be severely disrupted because normal mobile phone communications can occur only between two parties. Most rescue vessels and SAR aircraft do not have mobile phones. These communications problems and the resultant message chains have the potential to cause disruptions or delays to the extent that a safe rescue opportunity is lost or abandoned, with catastrophic results.

38.2   Phone Coverage. If you use your mobile phone on coastal trips, you will have occasional signal dropout problems. The problems will occur at the outer range of the transmission cell, and this is more pronounced at sea. Range is typically 8nm, which is a lot less than a VHF radio. If constant connectivity is important, install an external aerial or booster to increase range.

38.3   Foreign Voyaging and SIMs. Global roaming rates can be very expensive. If you are moored in one place for any length of time and make lots of calls and short message service (SMS) texts, it is a no-brainer to purchase a pay-as-you-go SIM card and number. This offers real long-term savings in most countries, and I have done this on a number of occasions. Like most folk, I get onto any free wireless network available. I use WhatsApp for most voice and messaging communications; its encrypted and free. You can really cut costs if you work at it. Data roaming is the killer, and I switch my phone off in order to maximize my mobile top-up or prepaid credit load (data sure chews up the credit!). SMS text messaging is a useful low-cost way of staying in touch. Again, that is why I use WhatsApp.

38.4   Mobile Phone Installation. There are solutions to improve signal reliability and range. Many install a cell phone signal booster. It is highly dependent on there being a cell with enough range to start with. Boosters create some issues, including signal oscillation and overload. Oscillation can cause feedback issues between the external aerial and the cell phone. This is best resolved by maximizing the distance between each aerial. Signal amplifiers can get overloaded, and this is within a strong cell. A good shore signal can reach as far as 40nm offshore, and the booster will amplify that. Cell boosters need a quality high-gain omnidirectional antenna, such as those from www.poynting.tech. There are several suppliers on the market; contact them and work out what you need. The weBoost Drive Reach system (www.weboost.com) seems to be popular on many boats; another is from leading aerial supplier Shakespeare. The good thing is that they will function everywhere you sail, from the US coast to the mid-Pacific and the Mediterranean. Read the user manual and installation instructions before installing a system.

38.5   Mobile Phone Boating Apps. There so many apps out there, and these are just a small selection of those available. The number 1 app is WhatsApp. More than 2 billion users can't be wrong! I use this extensively, as do many business operations. You can message, make voice calls from and to any country, and make video calls for free (although data charges may apply with your service provider). This app uses the phone internet connection, so no SMS charges. One good function is the group chat feature—you can update everyone about where you are and how life afloat is going. Another key feature is that all communications are end-to-end encrypted. See the various meteorological models in use for weather apps in the weatherfax chapter. I have curated the following selection of other useful apps.

a.   **Safety.** The following are essential apps when in the United States:

(1)   **United States Coast Guard (USCG) app.** With this free app, you not only get instant access to state-by-state boating information but also can request a safety check of your vessel and get a checklist of the federally required (and recommended) equipment you should have on board depending on the size of your vessel. With the touch of a button, you can file a float plan, get "rules of the road," pinpoint nearby NOAA buoys, and report a hazard, pollution, or suspicious activity. A big bonus is that this app also features an "Emergency Assistance" button which, when location services are enabled, will connect you with the closest US Coast Guard command center if you get into trouble.

(2)   **First Aid by the American Red Cross app.** The official American Red Cross app provides expert advice for common emergencies. The app features videos and simple step-by-step advice, from common

allergies, burns, sprains, and bleeding, to broken bones to stroke. Using this app, you can quickly assess a person's condition and then use the USCG app if you need assistance.

b. **Weather.** The most popular of all apps are anything to do with weather:

(1) **NOAA Weather Radar Live app.** If you are super careful about weather checks, the National Weather Service forecast app is a very useful information source. All that is required is your zip code, city, or town for local forecasts, radar charts, tide information, astronomical data, high seas warnings, and more. Important National Weather Service watches, warnings, and other alerts are shown on the map as interactive polygons covering the area you are in; simply tap to get the full alert text.

(2) **MyRadar app.** This popular app has had over 50 million downloads. It is fast, easy to use, and displays weather radar around your current location. Just start the app, and your location comes up with animated weather. MyRadar can show when a squall or storm is coming, how long it will last, and how heavy it will be, assisting in the decision-making process of either run for home or sit out the squall.

(3) **The Marine Weather app.** From Accuweather, this app is for advising local weather situations. Along with detailed weather forecasts, it includes GPS navigation.

(4) **SailGribWR app.** This app provides routing along with weather forecast models. Optional tidal stream data can be factored into the routing calculations.

(5) **PredictWind app.** This weather forecasting app includes route optimization options. The app runs its own models for popular sailing locations around the world. Popular with racing sailors, it's also great for cruising and provides powerful weather routing, along with isobar and rain maps. There is much more to this package, which delivers extremely accurate weather forecasts supported by high-resolution modeling. They have their own ultrahigh 1km resolution coastal forecasts that utilize PWG & PWE along with NAM, HRRR, and AROME regional models, which accurately predict sea breeze and geographic wind effects. One feature is termed a GMDSS forecast; while not being strictly true, it innovatively uses artificial intelligence (AI) to decode or derive a weather forecast or warning from a NAVTEX transmission. GPS tracking is available, as is a satellite version for offshore forecasting that requires the Predictwind Offshore app. Also available is the Iridium Go Predictwind app. Check out www.predictwind.com.

(6) **Windy app.** This weather app uses a range of models to provide data. This includes the European Centre for Medium-Range Weather Forecasts (ECMWF), a European global forecast model that's often regarded as the most reliable model there is. For those located in the southern portion of the UK and France, the app uses the fine-grained AROME model. This model was developed to improve short-term forecasts of dangerous weather events, including intense Mediterranean precipitation, heavy storms, and fog. The app provides wind and wave data for thousands of locations, as well as cloud cover, precipitation, and other weather data. The free version gives tidal times and heights. Check out this highly rated app at www.windy.app.

(7) **Buoyweather app.** This very easy-to-use weather app uses a features map and provides location-specific forecasts. Detailed wind and wave forecasting along with simple navigation are available.

(8) **GRIBView aap.** This app utilizes seventeen different forecast models, some of which are highly accurate wind models of the English Channel and 5 miles globally. The free subscription service provides 6nm resolution; for an extra cost you get 6nm and wave data.

(9) **Nova Buoys app.** Provides real-time weather data from more than 2,000 NOAA buoys around the US coast. Some are located in the Atlantic and the UK. Data is varied depending on buoy types but can include wind speed and direction, air pressure, wave heights, periods, and direction and other information. Visit www.flytomap.com.

(10) **Squid Mobile app.** This app provides a wide data set, including wind speed, direction, precipitation, cloud base heights, wave data, air pressure, and temperature. It uses cloud-based weather routing calculations and has GRIB file weather download options (www. squid-sailing.com).

(11) **Windfinder app.** Offers global wind conditions, and the "Plus" subscription level provides greater detail. Resolution is 7km in Europe and 5km in the United States. Coverage is available in Europe and the mid-Atlantic in the Canary Islands and South Africa. The app sources data from over 20,000 weather stations and has access to 63,000 web cams. Check www.windfinder.com.

c.   **Tides.** The ability to make tidal predictions is always an advantage:

(1) **Absolute Tides app.** Requires a license for a year. Uses UKHO data for the UK and Ireland and has tidal curves and a tidal stream atlas. It also has inshore water forecasts for UK coastal areas.

(2) **Imray Tides Planner app.** If you still do tidal calculations the manual way, this free app can provide both tidal stream and tidal height data. This data is available in a graphical form, and you can simply scroll forward in time for an immediate height of tide at any time and location. The result is that the secondary port tidal height calculations are very quick and extremely simple. Information encompasses some 8,000 global locations. A small fee gets you forward tidal data—up to 6 days for the UK, the current day in France, and any day for Italy; also any day for the United States, Australia, New Zealand, and elsewhere (www.imray.com).

(3) **Tides Pro app.** Small subscription for this, with global tidal descriptions heights, times, and charts. Doesn't cover all areas, primarily United States and US territories; details at www.tidespro.com.

(4) **Tidal Charts Near Me app.** Uses over 7,000 tidal stations around the world. The app has a weather forecast screen.

(5) **Aye Tides app.** This can provide tide times and heights for 12,500 locations around the world. Display provides that day's high and low water, along with heights. Sun and moon heights are available, along with tidal curves (www.ayetides.com).

(6) **UK Tides app.** Small subscription fee. Uses UKHO data and can access a large amount of data. The app has very intuitive screen and includes tidal streams, tidal set, and drift.

d. **AIS.** There are some useful AIS apps available; here are two:

(1) **Boat Beacon app.** This is an AIS-based collision detection system for smartphones and tablets. The app is able to share the AIS feed with other apps on your device that include Navionics and then network to the internet. The app can provide the bearing, range, and closest point of approach (CPA) calculations as well as normal AIS information. This is the only app that continuously monitors CPA and issues a notification when potential collisions are detected.

(2) **Marine Traffic app.** I use the web version of this AIS app regularly. I recommend that every boater either checks out the website or the app to get a reality check on just how many ships are out there. The app incorporates wind data and weather forecasts. Highly recommended, but it's no substitute for AIS.

e. **Navigation and Management.** Here is an assortment of apps from among the many:

(1) **Seapilot app.** This Raymarine app offers navigation along with automatic tracking and includes weather forecasts and charts. It allows downloading of charts for later review.

525

(2) **iNavX Navigation app.** This app has had many positive reviews and assessments. It offers professional-level chart plotter navigation capability. You can call up required charts and display bearing and speed instrument data.

(3) **SeaNav app.** This app offers augmented reality; you connect your camera and view the route. The app includes weather and chart information and updates.

(4) **Navionics Boating app.** This app offers full chart plotting capability along with tidal height and tidal stream data. The Navionics charts can be displayed with contours showing actual depths, compensating for height of tide. Vector charts are used and require a tight zoom to see all the charts' fine details. The app can upload and download sonar data with the SonarChart feature and is able to synchronize with boat navigation systems. Navionics is considered the world's number-1 boating app.

(5) **Raymarine Rayview app.** This app turns your smartphone into both a radar and a chart plotter. There are two multifunction display (MFD) types, the C-series and the E-series. You can then stream the video images directly to your phone or tablet. Simply enable wireless streaming on the E and C series displays and then the wireless network.

(6) **Simrad app.** This app allows easy activation or registration of Simrad devices. It has a suite of other information, including free updated charts, waypoints and routing, real-time AIS, weather fore-casts, and other useful information such as marinas and harbors.

(7) **Snag-A-Slip app.** This app allows quick identification of all marinas along the US coast. It has all the marine services and facilities and allows you to make a direct booking. The app is for US, Canadian, and Caribbean users, with a comprehensive listing of all the marinas within these areas. This app will store your boat information to enable simple future bookings.

(8) **AnchorChainCalculator app.** This very useful app allows quick calculation of anchor loading along with required chain lengths. Just punch in your boat particulars and out comes the answers. It does remove some of the guesswork when anchoring. Highly recom-mended, and I found it easy to use. For practice, just go to the free online version and try it out. I keep this on my partner's tablet.

(9) **SailTimer app.** This app is for those who want to sail efficiently. It allows calculation of the optimum tacking angles, along with distance and times to each tack. The app displays your optimized tacks as a chart overlay, and you can observe all the tacks along an entire route.

(10) **Beneteau Seanapps app.** For boat owners, Beneteau has an app that allows simple checking of fuel levels, battery charge status, bilge level, water tank level, GPS position, engine information, water depth, wind and log speeds, internal and external temperatures, and other relevant information. It incorporates a maintenance logbook.

(11) **CZone Mobile app.** This mobile and cloud-based app from Navico for boatbuilder client services monitors boat location, engine data, battery condition, fuel levels, and bilge level. Weather forecasts are also available.

(12) **C-Map app.** This digital cartography app employs Genesis, an outsourced mapping data application that incorporates more than 30,000 user-generated elements to the charts. Visit www.c-map.com.

38.6 **Smart Phone Cybersecurity.** The bad guys are after your phone and all its treasures, such as banking data and payments software, and most every other bit of data to steal your identity.

Android and iOS operating systems are susceptible to many types of malware; however, many good phone malware protection software packages are available. Do not download any apps other than from Apple's App Store or Google Play. Avoid using antivirus apps, and Samsung does not recommend third-party antivirus products. Make sure to update your phone applications and run the latest software version and operating system, which often contain vulnerability fixes with updates and security patches. Use screen-lock protection in the event you lose your device.

a. **Spyware and Madware.** Spyware and madware infect by installing a script or program, usually without owner consent. Madware is used to harvest data so that you get spammed with adverts. Spyware is used to collect information on internet usage and sell or on-pass to a third party. Data includes your location, passwords, contact list, and other personal information.

b. **Drive-by Downloads.** Opening the wrong email or visiting a malicious website exposes you to malware known as drive-by downloads. These variants are automatically installed on your device. Threats include spyware, malware, adware, or, even worse, a bot. A bot utilizes your device to perform malicious tasks, such as virus transmission.

c. **Viruses and Trojans.** Some apps can contain a virus or trojan, which in worst cases can mine your phone for passwords and banking information.

d. **Mobile Phishing.** Mobile phishing uses applications to deliver mobile malware. Users cannot differentiate between legitimate and fake applications; the result is your bank account numbers, passwords, and personal data being stolen.

e. **Phone Scams.** Be careful when you answer an unrecognized number. Missed the call? Don't call back, and certainly don't call back if it's a foreign number. Repeat: Don't call back! No message, no callback! There are imposter, credit repair and debt relief, loan and investment, "you won't get arrested," IRS and tax office demand threats, fake prize or lottery winner, fake charity, false vehicle accident, fake debt collector scams, and new scams appear daily. Hang up immediately, including robocalls. If they use the word "urgent," it's a scam. Block the number! Fake technical support scams, hang up. Do the same with fake family in peril emergency scams. Scammers are able to use spoofing to make any name or number show on your caller ID. Beware of "Can you hear me" calls. If you answer and they ask, "Can you hear me?" they want to record your voice saying "Yes" for use in fraudulent purchases. Say nothing; just hang up. One-ring and hang-up calls are robocalls to verify the number. Jury eligibility scams are used to collect your social security number. Area code missed call scams; if you call back, premium call rates apply on your bill. Then there are the "Hi Mum" text scams on phone and WhatsApp. Don't be tempted to call back. The takeaways are not to open suspicious SMS text messages, pop-ups, or any email attachment or link in an SMS. Review your privacy settings and choose strong passwords for all mobile devices and computers to make sure your personal data is secure. Review your privacy settings on social media. Absolutely do not respond to any payment requests of any type, from any source. If you shop online, make sure it's a trusted source. If it sounds too good to be true, it absolutely is. It's hard to believe that scammers get hundreds of millions and now billions of dollars every year. Okay, you got the message: Be careful out there!

38.7 **Phone Charging.** Charging a phone is a priority. You can keep your phone plugged into an inverter output with an integral charger or look at some of the alternatives. Scanstrut has come out with the waterproof Rokk Nest SC-CW-06E wireless inductive charger. It is easily installed and has a 10W fast-charge rating.

38.8 **Wireless Technologies.** It is common to hear people use the term Wi-Fi when looking at connectivity options. The registered name of Wi-Fi has become an almost generic description. There are 4 basic wireless technologies. The first is wireless LAN (WLAN), which is used within buildings and small outdoor locations and is the system you use at cafes and restaurants. The second is wireless metropolitan networks, or wireless MAN, which is used in larger outdoors scenarios. The third is wireless personal area networks, or PAN. These are relatively small area networks and use protocols such as Zigbee and Bluetooth. The fourth is wireless wide area network, or WAN, which use the cellular phone network. The rollout of 5G technology is opening up new wireless solutions that are still in development.

Radar

39.1   Radar Basics. The word radar is derived from "**RA**dio **D**etection **A**nd **R**anging, which originated with the US Navy in 1941. Radar is a method of locating the presence of a target and then calculating its range and angular position with respect to the radar transmitter. Radar systems make close-in navigation a lot easier for making landfalls, navigating channels, or navigation in reduced visibility. With the advent of GPS, some mistakenly see radar as redundant, but there is no substitute for radar as a navigational aid. Radar indicates where things are. GPS indicates where *you* are. Many times I have wished for radar when closing on a shore or port in bad visibility at night and, worst of all, in fog. Radar offers many very useful functions, including position fixing from geographical points, and the positions, speeds, directions, and close approaches of other vessels. Radar indicates buoy and navigation mark positions, rain and squall locations, land formations in poor visibility, the all-important collision avoidance at night and in poor visibility, AIS functionality, and more. There have been rapid advances led by Raymarine, Furuno, and Navico/Simrad. Among the various products are the innovative Raymarine Quantum series, Furuno 1st Watch Wireless Radome models, Furuno DRS4DNXT, B&G Broadband 4G, Simrad/Lowrance Halo20 Pulse Compression Radars, and similar offerings, such as the Garmin Fantom and Humminbird CHIRP radars. Many new radars using multifunction displays are capable of overlaying an electronic chart with the radar display—a great innovation coming out of high-speed networks. Anything that improves a sailor's situational awareness is welcome.

39.2   Pulse Radar. This is the traditional magnetron-based pulse radar technology. Radar transmits a pulse of microwave radio frequency (RF) energy, which is radiated from a highly directional rotating transmitter called the scanner. Any reflected energy is then received and processed to form an image on a screen. The time interval between transmission of the signal and reception of reflected energy can be calculated to give target distance and direction.

39.3   Broadband Radar. Frequency modulated continuous wave (FMCW) radar is also referred to as broadband radar. In this continuous-wave transmission technology, the transmitted signal is a range of frequencies, so it gets the broadband tag. The signal is transmitted from the rotating scanner; however, the return signal is processed differently. The receiver processes the change in frequency, or frequency difference, and is able to calculate the target range and bearing. One key change is that broadband radar has two antennae, one transmitting and one receiving. The big difference over conventional pulse radar technology is much clearer definition of targets. Improvements in software processing now enable dual-range viewing capability and increased target definition. Simrad models have high-speed antenna rotation speeds of 60rpm for short-range tracking of high-speed targets, providing collision avoidance advantages. The Halo pulse compression radar is a combination of pulse and FMCW broadband systems.

39.4   CHIRP Radar. This is the foundation of the latest radar technology as well as sonar and fish-finder applications. Compressed high-intensity radiated pulse (CHIRP) is sometimes referred to as linear frequency modulation (LFM) and frequency modulation on pulse (FMOP). While relatively new on small vessels, pulse compression technology has been used in military applications for a long time. This technology is a combination of frequency modulation and pulse radar. The basis for CHIRP is the addition of a time delay to the receive signal and transmitted swept frequency so that the transmitted signal is effectively compressed in time to create a pulse and reception signal that is shorter. A chirp signal where the frequency increases is called an "up-chirp," and one that decreases is called a "down-chirp" with time. The advantages are enhanced range, enhanced range resolution, lower peak power requirements, greater noise immunity, and higher performance on short- and long-range targets. Raymarine was the first in this space with its innovative Quantum radar. This has dual-range viewing capability and, when using the MFD, can allow split screens showing two different ranges. This series has integrated Wi-Fi. Quantum chirp lengths are 400 nanoseconds (Ns) to 20 microseconds (µs) with a CHIRP bandwidth up to 32MHz.

39.5   Doppler Radar. Furuno leads the market with inclusion of Doppler technology. When a radar signal echoes off a target, the frequency of that signal return varies depending on target travel direction. These frequency changes allow differentiation between targets either moving away from or headed toward your own boat. This is displayed depending on threat level; targets moving away are green, and targets moving toward are denoted in red. I am installing the Raymarine Quantum 2 with Doppler on my boat at time of publication.

39.6   Radar Scanners. In practice, it used to be said that the larger the scanner, the narrower the beamwidth, and the better the target discrimination. Of the two main scanner types, the beamwidths of enclosed scanners are always generally larger than the open-array types. An open-array scanner has a beamwidth nearly half that of enclosed units, which gives far better target discrimination. This factor is one of the trade-offs that need to be considered when selecting a radar scanner unit. If it can be accommodated, an open-array scanner performs far better. Older generation pulse radars used to use center-fed slotted waveguide antenna technology, and it is still used in commercial and naval applications. Some radar scanners have an integral safety switch that disables the antenna from rotation when servicing or maintaining it. Do not forget to switch "OFF" before installing and back "ON" after completing the work. Furuno has the 1st Watch Wireless radar units, resolving cable issues such as the following:

   a.   **Patch Arrays.** Radar antennae are now patched or printed antenna arrays. They are also termed microstrip narrowband or wide-beam antennae, where the beam can be electronically steered. These are horizontally polarized arrays specifically designed for FMCW marine radars. The array comprises ninety-six series-parallel-fed elements printed onto a microwave substrate, with the configuration being optimized for minimum line loss and maximum aperture efficiency.

   b.   **Sidelobe Attenuation and Suppression.** Beamwidths are not precisely cut off. There are zones outside the main beam where power is wasted and dissipated, as the transmitted radar energy cannot be precisely focused into a single beam. A small portion of energy escapes and is transmitted in other directions; this is known as sidelobe energy and appears as arcs on the radar screen. False target returns do occur adjacent to strong targets such as coastal formations and merchant vessels. Radar systems such as Broadband 4G radar have improved sidelobe rejection when dealing with large targets to make sure the option is switched on. Adjusting sensitivity and STC can reduce this issue.

c. **Antenna Speed.** A pulse radar scanner revolves at 24rpm. On new-generation radars, 24rpm is the standard nominal rotation speed, with 48rpm and up to 60rpm on a Navico Halo unit, with faster speeds depending on the range setting.

d. **X Band Radar Frequencies.** All small boat radars operate on microwave frequencies in what is termed the X band. Frequency ranges are 9200MHz to 9500MHz, a wavelength of around 3 centimeters. Transmit frequencies for the Quantum are in the 9354MHz to 9446MHz range.

e. **Radar Output Power.** Peak power ratings for standard pulse radars are given for the actual microwave output power. A kitchen microwave operates on the same principle. Given the effect microwaves have on food, always follow the warnings on eye protection with pulse radars. It is quite common for naval vessels with high power radars to incinerate any birdlife in the rigging at start-up. This is changing as small boat radars now go broadband and CHIRP, with significant power reductions.

f. **Range Discrimination.** Range discrimination or resolution is a function of transmission pulse length. When the distance between two targets on the same bearing is longer than the pulse length, the targets are shown as separate. When the distance between targets is less than the pulse length, they appear as one target. Most radar sets automatically alter pulse length with a change in range settings.

39.7 Beam Angles. Radar transmissions are similar to the light beam from a lighthouse in that a radar beam has a defined angle in both vertical and horizontal planes. The beamwidth is normally defined as the angle over which the power is at least half of maximum output; this is usually rated at −3dB. Target discrimination or resolution is a function of beamwidth. A scanner with a narrow beamwidth is effectively slicing and sampling sectors of approximately 2.5 degrees around the azimuth. Large targets will be sampled a number of times and their size quantified. A wider beamwidth will sample an area twice that size but will not always discriminate between two or more targets. If a harbor entrance is narrow, the radar beam may in fact see it as part of the breakwater until the range has closed up. At longer ranges, two targets at the same distance and close together may appear as one target.

a. **Horizontal Beamwidth.** This is width of the horizontal beamwidth transmitted by the antenna. It defines the capability to discriminate between two or more detected targets on similar but different bearings located at a similar range. Narrow beamwidths have better performance than wide beams. Generally, the wider the length of an antenna, the narrower the beamwidth, and radome antennas have wider beamwidths than open scanners. Quantum nominal beamwidths are horizontal at 4.9 degrees and vertical at 20 degrees. Raymarine and Navico also have what they call "beam sharpening," a software algorithm for improving target resolution. The Raymarine Cyclone radars for 72nm and 96nm and open scanners have beamwidths as low as a range of 1.8 to 1.15 degrees.

b. **Vertical Beamwidths.** Open-array scanners have a beam angle of around 25 degrees for all open-array scanners. These can vary between 20 and 25 degrees on both open and radome scanners.

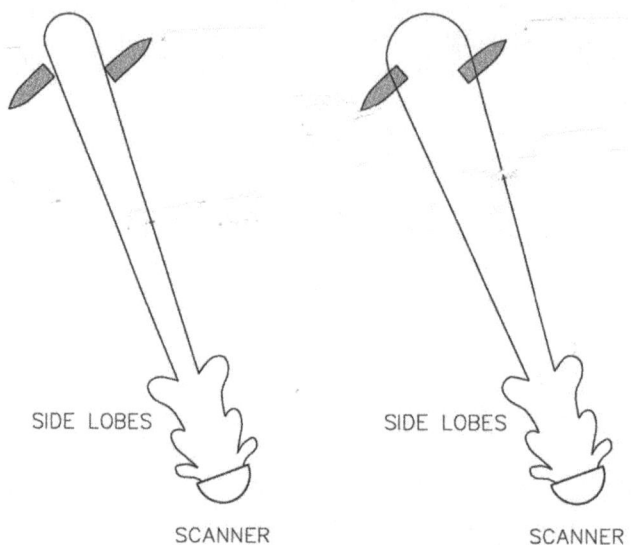

Figure 39-1. Radar Target Discrimination

39.8 **Radar Heel Angles.** The heel of a vessel, and therefore of the scanner, has an adverse effect on performance. Most radars have vertical beam angles of between 20 and 25 degrees, so at a heel angle of approximately 15 degrees, anything to windward is virtually invisible and there is a significant blind spot to leeward. That doesn't take into account the additional masking of the signal by waves. This problem is more pronounced on stern post–mounted units. One solution is to alter the attitude of the scanner using one of the several self-leveling systems available. These include gimbal systems, or manually operated hydraulic systems. The Questus system can be mounted around a backstay, mast, or stern post. The system uses a hydraulic pivot damping system on the leveling mechanism to prevent oscillation during all sea states. The hydraulic system employs heavy gear oil that is devoid of air and is able to withstand wide temperature gradients and thermal expansion. The cheaper alternative is to level up the boat periodically and have a look, which shouldn't be a problem on a cruising yacht. Go to www.questus.com for details.

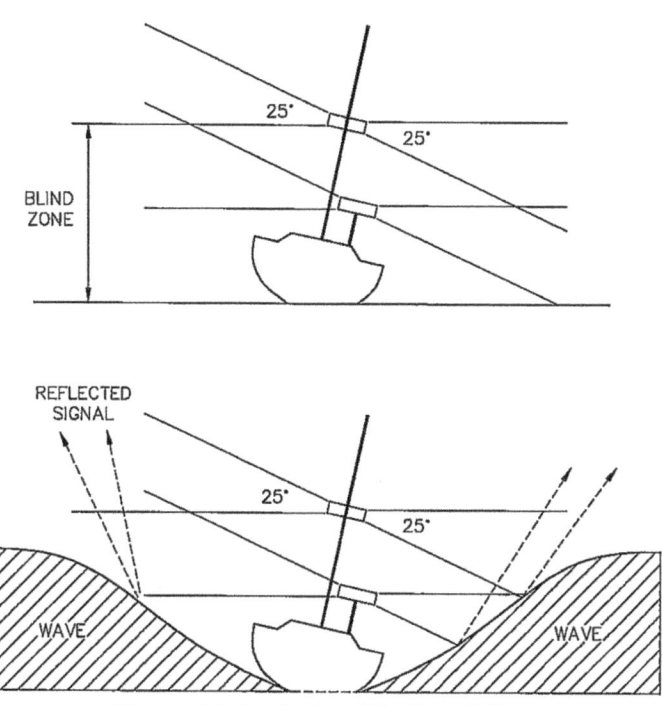

Figure 39-2. Radar Heeling Effects

39.9   Radar Ranges. Maximum radar ranges are a function of scanner height. Table 39-1 gives the approximate horizon ranges for radar under standard conditions for targets of a known height. Conditions at sea may give better results, but it should be noted that atmospheric conditions affect the ranges as well. Radar signals travel in a straight line but are subject to bending under normal atmospheric conditions. This bending increases the radar's horizon approximately 6% over optical horizons. The maximum detecting range of the radar, Rmax, is variable and is dependent on factors that include antenna height above sea level and the target height, along with the shape, size, and material of the target. Under normal atmospheric conditions, the maximum range for radar is equal to the radar horizon, or a fraction less. The radar horizon is longer than the optical one by about 6% due to the diffraction properties of a radar signal. Rmax is denoted in the equation Rmax= 2.2 × (h1 + h2), where Rmax = radar horizon in nautical miles; h1 = the antenna height (m), and h2 = the target height (m). Look at the following example: The antenna height above the waterline is 9m, and the height of the target is 16m. Rmax is denoted in the equation Rmax = 2.2 × ($\sqrt{h1}$ + $\sqrt{h2}$), where Rmax = radar horizon in nautical miles; h1 = the antenna height (m), and h2 = the target height (m). Look at the following example: The antenna height above the waterline is 9m, and the height of the target is 16m. The maximum radar range is Rmax= 2.2 × ($\sqrt{9}$ + $\sqrt{16}$ = 2.2 × (3 + 4) = 15.4nm. Note that the radar detection range is reduced by precipitation, which absorbs the radar signal. Targets such as large vessels or landmasses, for instance, may appear at much greater ranges, and the known height of these should be added to the scanner height. Ranges can increase or decrease depending on the prevailing atmospheric conditions. Another factor that comes into play is super-refraction—when an upper layer of warm dry air is laying over a surface layer of cold moist air. The radar waves tend to bend downward and increase target detection waves. Subrefraction is the opposite, with cold air over warm air layers. The result then is that waves bend upward and decrease radar ranges.

Table 39-1. Radar Horizon Table

Target Height ft (m)	Scanner Height 16.5ft (5m)	Scanner Height 33ft (10m)	Scanner Height 49ft (15m)	Scanner Height 65.6ft (20m)	Scanner Height 82ft (25m)
Zero	5.0nm	7.0nm	8.5nm	10.0nm	11.0nm
16.5ft (5m)	10.0	12.0	13.5	15.0	16.0
33ft (10m)	12.0	14.0	15.5	17.0	18.0
49ft (15m)	13.5	15.5	17.3	18.5	19.8
65.6ft (20m)	14.8	17.0	18.5	19.8	21.0
82ft (25m)	16.0	18.2	19.8	21.0	22.3
98.4ft (30m)	17.3	19.0	20.8	22.0	23.3
115ft (35m)	18.0	20.0	21.8	23.0	24.3

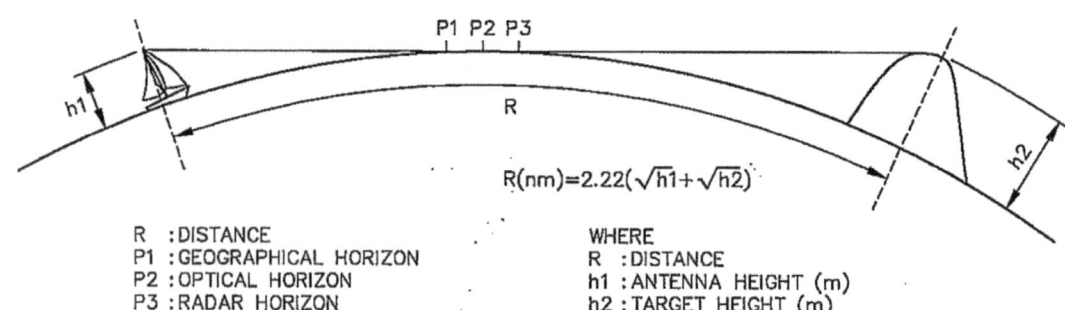

P1 P2 P3

R

$$R(nm)=2.22(\sqrt{h1}+\sqrt{h2})$$

R : DISTANCE
P1 : GEOGRAPHICAL HORIZON
P2 : OPTICAL HORIZON
P3 : RADAR HORIZON

WHERE
R : DISTANCE
h1 : ANTENNA HEIGHT (m)
h2 : TARGET HEIGHT (m)

h1

h2

Figure 39-3. Radar Horizons

39.10   Radar Displays. Display used to be called plan position indicators (PPIs) and are still referred to as such by some manufacturers. The trend for radar displays as part of an integrated system rather than a standalone system has advanced significantly. There is now a single MFD for all radar, chart plotter, video, instrument, and fish-finder data. The various display types available on radar systems are as follows:

a.   **Rasterscan Displays.** These older-generation displays use the same technology as computer monitors. Essentially, the radar screen consists of many dots called pixels. The status of the pixels installed in the memory is altered in response to signal processing changes. Resolution is quoted as the number of pixels on the screen, such as 480×640 pixels. Unlike the CRT display, the rasterscan display is a result of complex digital-signal processing and allows the use of numerical information on the screen. Digitally processed signals usually have to be above a minimum level to be displayed, so weak echoes are often rejected. For this reason, proper tuning and operation are essential if all targets above that threshold are to be displayed.

b. **CRT Displays.** The cathode ray tube (CRT) was the primary display type until technological advances made them redundant, although some boats still have them on board. The radial display was synchronized with the scanner and effectively displayed every return, with target brightness being relative to target strength. These displays required hoods for daylight operation and consumed large quantities of power.

c. **Liquid Crystal Displays (LCDs).** Earlier LCD radars were the ideal solution for most yachts, and some are still in use. They were low on overall power consumption as well as compact, with low profile and lightweight radomes, which means less weight aloft.

d. **Multifunction TFT LCD Displays.** A thin film transistor (TFT) LCD is an active-matrix LCD that improves contrast by utilizing thin film transistor technology. The displays are sunlight viewable, with 256 color capability, and some have LED backlighting.

39.11   Radar Display Presentation Modes. A radar display can be aligned to orient to one of three basic configurations along with several more on the latest broadband and CHIRP radars. There has been an increase of modes with technology and mainstay modes such as North-up, Head-up, Course-up, and True Motion. Today we are on a Simrad Halo unit, which includes Harbor, Offshore, Weather, Bird, and Dual Range modes. These various modes are available on various radar models as below.

a. **North-Up (N-UP).** North is always at the top of the display. Interfacing a gyro or fluxgate compass puts true north at the head of the screen. One of the advantages of this display is that both chart and display correspond and bearings are easily transferred for plotting purposes. Many plotting and navigating errors are created by incorrect transfers of screen information. As the heading changes, the heading marker also changes.

b. **Head-Up.** The top of the screen is the same as the vessel heading, and all bearings are relative to the boat. You can select a ship heading marker on the screen; as the heading changes, the display follows and rotates.

c. **Course-Up.** The top of the screen is aligned to the selected course using an interfaced electronic fluxgate compass. The radar image is stabilized and displayed with the course at the screen top. As the boat heading changes, so does the heading marker. When you alter course, the image adapts to the new course.

d. **True and Relative Motion.** A radar display can be configured for true motion or relative motion. In relative motion mode the yacht's position remains fixed on the radar screen and all targets are relative to the yacht. In true motion mode all the targets are in a constant position on the display. The yacht moves across the display on the course heading, and the display is similar to a chart. All other boats appear in true relationships to other targets and to displayed landmasses. As the vessel comes to the edge of the screen, the radar automatically resets to display the area in front of the boat.

e. **Weather Mode.** Settings are optimized for good detection and presentation of rain clutter. The update rate is reduced and color imaging enhanced. This automatically increases gain and reduces clutter and tunes the inbuilt filters for detection of weather, such as precipitation and squalls.

f. **Bird Mode.** This mode automatically increases gain and reduces clutter and tunes the inbuilt filters for detection of bird flocks. Sensitivity is maximized. This allows easy bird flock detection, so it is popular in the fishing fraternity.

g. **Dual Range Mode.** This allows you to monitor two ranges simultaneously; you can select a long range and a short range, depending on circumstances.

h. **Harbor Mode.** On some radar sets the scanner operates at high speed to give faster target updates when in close or around areas with high-speed traffic. The various radar settings are optimized when quality target discrimination and rapid update are required.

i. **Offshore Mode.** The settings are optimized for sea conditions offshore. The settings expand targets to ensure visibility.

j. **Waypoint Up.** This function on some radars is an azimuth stabilized display where the centerline to display top indicates waypoint bearing. The waypoint marker indicates the waypoint position, which is output from a GPS input.

39.12 MARPA. Mini automatic radar plotting (MARPA) is essentially multiple-target tracking and allows tracking of 10 targets and up to 25 or more simultaneously and the automatic computing of target course, target speed, and the closest point of approach (CPA) and time to closest point of approach (TCPA) of each. A heading sensor is required for these functions. Simrad radars have oneTrack, which is used to automatically detect and track targets within a designated zone, and VelocityTrack, which allows for instantaneous identification and automatic checking of every target using Doppler technology. ZoneTrack is an automatic target tracking feature for detection and tracking within one or two user-defined designated zones. This uses a color-coded system with vessels approaching, stationary, or moving away. These functions require good quality heading and GPS data. MARPA and AIS targets that are moving also have graduated trails on the PPI that show past position history. This is a function for target tracking and risk analysis. The result is accurate and continuous situation evaluation. MARPA effectiveness may be reduced or ineffective when target echoes are weak or when targets are close to landmasses or other large targets. If the target is engaged in rapid maneuvers or in rough or very choppy sea state conditions, the target is obscured by sea clutter or large swell conditions. This applies when your own yacht stability is affected in big sea conditions or the heading data is degraded.

39.13   Radar Installation. The two most common scanner mountings are mast mounted and stern post mounted. Each has advantages and disadvantages.

    a.   **Mast Mounted.** A number of factors affect a scanner mounted on the mast:

        (1)   **Radar Range.** Mast mounting increases radar range. This is clearly illustrated in the radar horizon table. You will have to work out what suits your particular requirements.

        (2)   **Weight and Windage.** Scanners used to be quite heavy, but new-generation scanners are lightweight and radomes have little windage.

        (3)   **Scanner Height.** Consider sail interference and fouling of the scanner. Scanstrut makes radar guards that do help with wear and tear on sails, but it is worth checking out the risks of fouling before installation. A sail can generate enough force to rip a scanner and mount off the mast. Secondary retention should be considered. This comprises a stainless wire attached to mount and mast so that should it come loose or be torn off, you don't lose the scanner or have it end up on your head. Check out www.scanstrut.com.

        (4)   **Blind and Shadow Sectors.** Masts and even rigging can obstruct and reduce the intensity of a radar signal. The position of the scanner is important. Install the bracket above or below the spreaders to minimize obstruction. There will be a small blind sector astern. Where scanners are mounted on a ketch mizzen mast, you have both forward and stern shadows. There has always been a blind spot of around 150ft to 200ft (30m to 50m) around a boat. This is related to the time difference between signal transmission and reception. With broadband radars, that has come down to less than 20ft (6m). The result is that close-range targets may go undetected. Always check and determine the angular width and relative bearing of any shadow sectors and how they affect your radar as soon as possible after installation.

    b.   **Stern Post Mounted.** The stern post mounting arrangement is more accessible than mast mounting and has become a very popular alternative in recent years. Some stern posts are hinged to allow easy lowering. I can never understand why 24nm radars or above are used in these installations, as their range is limited by the post's height. It is preferable to use an open-array antenna to improve the resolution if halyard fouling isn't an issue:

        (1)   **Radar Range.** The radar range is reduced by a couple of miles depending on the target height. Typical height is around 10ft (3m), compared with around 20ft to 28ft (6m to 8m) for a mast mount.

(2) **Scanner Leveling.** When the boat heels, the scanner also tilts, leading to significant loss of performance and range. Some innovative self-leveling mountings are now available.

39.14 Radar Emission Safety. Emission safety hazards have dramatically reduced with radars having a lower power radar transmitter. Emissions are down to around that of a cell phone output these days, and health risks and the risk of eye damage are virtually gone. This was often an important consideration in stern-mounted units. The Broadband 4G and Broadband 3G radars have emissions 80% lower than a smartphone, and transmit just one-fifth of a watt. On older-generation pulse radars, this was over 2kW, which meant a minimum safe distance of 4.6ft (1.4m); a 4kW unit required 9.3ft (2.8m) clearance. The following important safety factors must be considered when working with radar.

a. **Eye Damage.** When working with older-generation pulse radar units, direct exposure to an operating radar transmission can permanently damage the retina or cause blindness. Safe distances are normally given as around 3ft (1m), but recent medical research has recommended an absolute minimum of 6ft (2m). In this respect, stern post radars used to represent a real health hazard, especially when powerful output units were installed.

b. **Electric Shock.** Magnetron pulse radar scanner units have high internal voltages. Owners should never open an operating scanner and attempt repairs or measurements; call an authorized service person.

39.15 Radar Cables. Cutting and rejoining cables should be avoided; however, this is sometimes required when a radar scanner is mast mounted on a sailing boat. On older-generation pulse radars, it was good practice to install a ferrite on each side of the break, and you should consult your supplier for specific details. Wires should be reconnected correctly, and that includes the screen. In general you will need to use a junction box, and radar-specific units are available. Always keep unscreened coaxial cores to less than 1.2in (30mm) to preserve electromagnetic compatibility (EMC) integrity.

a. **Cable Installation.** Always ensure that radar cables are well protected from chafing where they enter the mast. Ensure that cables are not bent too tight or kinked; minimum bend radii for small cables are 2.5in (60mm) and 3.75in (82mm) for larger cables. On new-generation broadband and CHIRP radars, standard scanner cable length is typically 50ft (15m); optional lengths are 15ft (5m), 35ft (10m), and 80ft (25m). For through-deck installations there are cable glands with a split-seal arrangement to allow larger plug assemblies to be put through. Have drip and service loops so that water cannot migrate down the cable into connectors. Do not install radar cables next to high current–carrying cables or radio aerial cables. Make sure all connectors are properly installed and pushed in. If they have a lock on them, ensure that it is engaged.

b. **Mast Cable Pulling.** When pulling a cable down through the mast with a mouse or messenger line, avoid excessive force. Do not tie off around the connector; instead tie off around the cable insulation jacket. Make sure that where cables exit a hole on the mast, all edges are covered to prevent chafe.

c.   **Radar Interface Box.** Where your new radar has a radar interface module or box, install it vertically in an easily accessible location where it is protected from water, condensation, spray, and water drips. Make sure you can access it easily to connect or disconnect power and network cables as well as display cables. Always make sure you have a drip loop when installing cables.

39.16   Radar Power Consumption. Radars are now all solid-state microprocessor based and consume a lot less power than previous magnetron pulse radar technologies, which used to consume 4A or more. There has been around an 80% to 90% reduction in energy requirements, which is good news on a sailing yacht. The Quantum radar consumes 17W maximum in transmit mode and 7W in standby. Open scanners typically have a power consumption 50% greater than enclosed types because they have a heavier scanner and require a more powerful motor to rotate it. Many new radars use what is called a pancake motor, which is a compact brushless motor.

39.17   Grounding. As a transmitter, radar scanners require proper grounding, usually at the scanner and at the rear of the display unit. When mounted on a stern post, a grounding cable is attached from the base to the RF ground on some models. The latest CHIRP and broadband radars do not require grounding; however, read the installation manual. The requirements vary between different radar makes and models. For example, the Simrad Halo installation manual states that the chassis ground is DC isolated from power negative to reduce the risk of galvanic corrosion. They recommend connecting the ground to either the vessel's bonded ground or a nonbonded RF ground at the closest possible location.

39.18   Radar Operation. Correct operation of radar is essential if you are to derive the maximum benefit. Don't be one of the all-too-common radar-assisted casualties. A radar has a number of controls, but they all have clearly defined functions that are easily learned. *Read the manual!*

a.   **Powering Up.** It used to take several minutes for an earlier generation magnetron-based pulse radar set to warm up and be ready for use. The new-generation CHIRP and Broadband units are ready to use in a matter of seconds.

b.   **Range Selection.** Older pulsed radars used to have just four or five ranges; however, new radar technology such as the Raymarine Broadband 4G radar has up to 18nm range scales that include short range. Always set the range you wish to work on. Typically, the 12nm range is ideal for the average yacht, given the radar's horizon. On a mast-mounted scanner, a greater range will enable you to detect a large vessel on or just over the horizon. Selecting a range automatically sets the appropriate range ring intervals, the pulse length, and the pulse repetition rate.

c.   **Screen Brilliance.** Adjust the brilliance control to suit your requirements. Don't make it too bright at night or so dim that targets are not clearly displayed. New-generation radars have night and daylight settings.

d. **Gain Control.** For those with legacy model pulse radars, the gain control removes background noise, which shows as large areas of irregular speckles on the display. Adjust the gain control so that screen speckling just starts to appear. Gain controls the signal amplification, so be very careful not to over adjust, as smaller echoes can be masked or, if under the required threshold, not appear at all. The gain sensitivity is normally set high for long ranges and reduced for low ones. Menu selections on some older radar units had a "normal" setting, with "SEA" reducing sea returns so that background noise is just visible on the screen. "RAIN" reduces close rain or snow returns; "FTC" reduces distant rain or snow returns. The latest generation broadband and CHIRP radars do this automatically. The Garmin GMR incorporates dynamic auto gain to optimize quality in poor conditions.

e. **Fast Time Constant (FTC).** For those with legacy model pulse radars, this control reduces the rain clutter typically associated with storms. Rain clutter is proportional to the density of the rain, fog, or snow. Although the control is useful in tracking squalls and rain, caution should be used so that targets are not obscured. Heavy rain may cause total loss of target definition and cannot be adjusted for. Echoes from waves can be troublesome, covering the central part of the display with random signals known as sea clutter. The higher the waves, and the higher the scanner above the water, the farther the clutter will extend. Sea clutter appears on the display as many small echoes that might affect radar performance. The FTC function separates these echoes into a speckled pattern, resulting in better solid target definition.

f. **Noise Rejection.** The Navico Broadband 4G radar has enhanced digital signal processing capabilities and allows user-definable settings from "High" to "Low" to control or reduce the noise levels that are detected. This can increase the range by up to 50% along with improving the sensitivity of target detection.

g. **Watchman Mode.** In this function setting, the radar automatically transmits for 1 minute to check for targets within a preset guard zone. If any change is detected from the previous transmission, an alarm is activated, canceling the mode, and transmits continuously. This is a very good option for sailing yachts. When selected, the scanner will automatically power up at the selected period and then power down to low-power standby mode. You get to do a long-range target watch and save valuable power at the same time.

h. **Clutter.** Sea clutter has always been a problem to tune out. New-generation broadband and CHIRP radar systems with greater processing power are able to discriminate between small targets in heavy seas by employing directional sea clutter processing, which improves small target resolution. On legacy older generation pulse radars, this control is often referred to as the sensitivity time constant control. Sea clutter, most apparent at the screen center in that region closest to the vessel, is interference caused by rough seas or wave action where some of the transmitted signal is reflected off wave faces. Three-centimeter (3cm) radars transmit at a very low signal angle, which grazes the water surface. On short ranges, clutter can mask targets, especially weak ones. The effect decreases at long ranges. Sea clutter always appears stronger on the lee side of the vessel because vessel heel in that direction exposes the beam to larger water areas. Sea clutter filters the random echo returns from waves. Furuno has a dynamic sea filter that automatically adapts to changing sea conditions to maintain good signal quality. Many radar units have preset sea state modes such as "calm," "moderate," and "rough." On new radars, it is easier to simply leave it on "auto." Auto sea clutter mode dynamically adjusts to increase suppression when looking into waves and reduce suppression when looking down the waves. Downwind signals will deflect with less clutter, while upwind signals will reflect with increased clutter. When a radar has a rain control setting to reduce clutter to differentiate targets, use caution.

i.   **Tuning.** The majority of radars are self-tuning, and adjustment will be indicated on a small bar readout on the screen. Most radars can be manually tuned, but this should be done carefully and according to your user manual.

j.   **Pulse Lengths.** Pulse length on new-generation radars is automatic, with range changes on modern small boat radars. At short ranges, pulses are at 0.05 microsecond to give better target resolution. At long ranges, pulses increase to 1.0 microsecond. Beam pulse lengths are long for detection of long-range targets and shorter for good target discrimination.

k.   **Pulse Repetition.** Repetition rates vary across ranges from 200 to 2,500 per second. Rates determine the size of the area around the vessel where there is a dead zone. At 0.05 microsecond, this is around 300ft (150m). At 1.0 microsecond, this reduces to 100ft (30m).

l.   **False and Multiple Echoes.** False echoes are sometimes observed on the screen where there is no actual target. They can be reduced or eliminated. You should be aware of how they appear on the screen and know the difference between legitimate contacts to avoid confusion. Multiple echoes occur when the radar is set for short range and a strong echo is received off a large commercial vessel, bridge, or a breakwater. Sometimes a second or third echo is observed; these can be reduced by varying the sensitivity or adjustment of the STC.

m.   **Interference Rejection (IR).** Interference can come from a number of sources:

(1)   **Other Radars.** Mutual interference from other radars operating in the area can cause interference on the display. This is particularly apparent near major shipping routes where powerful commercial

vessel radars operate. If available, use the IR function to remove these unwanted signals. Many new radars automatically correct this.

(2)   **Mast Clutter.** When a radar is installed, there will be a blind spot abaft the scanner due to the mast. No targets will be detected in this area at close ranges. On a stern post– or mizzen mast–mounted radar, the area in front will be masked for the same reason. Caution must be exercised, as this is the normal collision risk sector.

39.19   **Radar Plotting.** The main purpose of a radar is to detect stationary and fixed targets. A number of basic features facilitate this.

a.   **Range Rings.** The concentric range rings automatically change with the selected radar range; user-definable range ring settings are possible at setup.

b.   **Variable Range Maker (VRM).** This function uses the range rings and the marker. The readout appears on the screen, but as with all navigation exercises, make sure you are measuring the correct target. Many errors are made this way, which is why radar should be used in conjunction with other position-keeping systems, principally the charts and eyeball.

c.   **Electronic Bearing Line (EBL).** The most commonly used function in conjunction with the VRM enables easy plotting of a target. Be careful; many unfortunate incidents have occurred because a bearing was taken without checking which head-up display was being used. Most radars enable selection of either actual (magnetic/true) or relative (relative to boat heading) bearing readout in the EBL data box.

d.   **Target Expansion or Boost.** This function on many radars allows short- or long-range contacts to be expanded. This increases the pulse length or reduces the bandwidth to increase target images and increase radar sensitivity. It is useful when closing on low-altitude landfalls such as atolls and islands. This has become automatic on new-generation radars.

e.   **Off-Centering.** A number of radar sets have an offset function that alters the screen center 50% to 75% down the screen. This makes forward long-range observations possible in the same radar range. New-generation radars have many display options.

f.   **Safe (Guard) Zones.** Guard zones offer real safety advantages. They can be set for complete circular coverage or for specific sectors. It is a big error to rely on this function when sailing shorthanded; proper observations should be regularly made.

g.   **Target (Wake) Plotting.** This feature, now part of most radar models, allows a trail or wake of the target of multiple targets to be plotted. Target plotting is time related and can be continuous or set at a number of seconds. A clear plot of the target is invaluable for ensuring that collision risks do not arise.

h.   **Sector Blanking.** Some radars have blanking sectors to stop transmission in areas with poor reflection, such as masts. This can lead to false echoes, which can be confusing.

39.20   Radar Software Upgrades. All new radar sets require regular software updates, so make sure to check the website of your chosen radar unit; most have a download facility. This will improve your radar functionality and address issues that have been discovered by other users with patches and fixes implemented. It is an essential maintenance task. Prior to downloading and then installing to the radar, back up any valuable user data. Your manual will have clear instructions on system data backup procedures. You should do regular back-ups of your system-setting files.

39.21   Radar Maintenance. Not much maintenance is required on a radar unit, but taking the following steps will ensure long-term reliability. A mast suffers a lot of vibration, so loosened bolts are a possibility.

a.   **Scanner Electrical Connections.** Once a year, open the scanner and tighten all the terminal screws.

b.   **Scanner Check and Cleaning.** Clean the scanner with warm soapy fresh water to remove salt and dirt. Do not scour or use harsh detergents. Do not pressure wash the scanner. Check the radome for signs of cracking; cracks can be repaired with a sealing compound. Never paint a scanner radome or apertures, as it will affect the radar wave emission quality.

c.   **Scanner Fixing Bolts.** Check and tighten the scanner fixing bolts. Coat each bolt with silicone sealant. Remove the bolts and check for corrosion; coat with an anticorrosive compound or grease. Torque the bolts correctly—don't overtighten.

d.   **Scanner Gaskets.** Check that the scanner's watertight gaskets are in good condition and seal properly.

e.   **Scanner Drive Motors.** Some older scanner motors have brushes. Check the brushes every 12 months. Manufacturers sometimes provide a spare set taped to the motor. Many new radars have brushless pancake motors that have solid-state commutation along with sealed gearboxes that are maintenance free. Be careful if the screen has residues of crystallized salt or sand. Moisten the screen to dissolve residues, as screen scratching is possible when wiping. Some scanners have a drive belt, and it is important to check it periodically. If your scanner gets a lot of use, the belt may deteriorate or elongate and require replacement.

f.   **Display Unit.** Clean the screen with a clean soft cotton or microfiber cloth. On new screens, a screen cleaner such as Raymarine's works well. Check and clean the media port door, as salt crystal can partially leave it unsealed and make ingress of water possible. Do not use abrasive cleaners. Do not use any solvent- or other chemical-based cleaner, acetone, or alcohol.

39.22   Radar Troubleshooting. The following table gives typical faults that can be investigated and rectified before calling a technician. Many new radars have error messages or LED indication and provide a list of what they all mean. Some units have alarm messages and equipment status menus; learn how to use them.

Table 39-2. Radar Troubleshooting

Symptom	Probable Fault
Scanner stopped	Motor brush stuck (if fitted) Bearing seized Scanner motor failure Scanner motor control failure Communication issue
No display	Power switched off Brightness turned down Fuse failure Loose power plug
Display on, no targets	Scanner stopped Scanner plug not inserted correctly Software issue Ethernet plug issue if installed

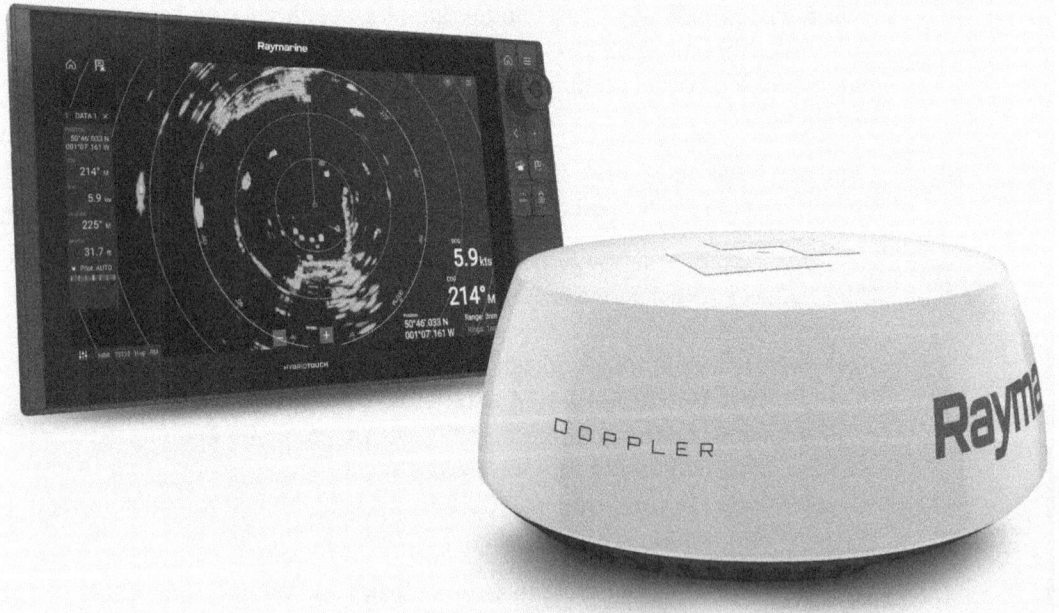

Figure 39-4. Raymarine Quantum 2 Doppler Radar (*Courtesy Raymarine*)

Radar Reflectors

40.1　Merchant Vessel Visibility and Radar Reflectors. The subject of radar visibility and radar reflection has sparked ongoing controversy over the years. A constant stream of so-called radar reflective safety devices have been marketed to yachtsmen. Not to have an effective reflector mounted at all times is, in my judgment, negligent in the extreme. While AIS is rapidly spreading onto small boats, being radar visible is another issue. Having been on the bridge of a fast merchant vessel steaming at 22 knots while trying to observe radar-invisible vessels, it is a challenge. Whether you sail around port approaches, vessel traffic separation zones, or deep ocean, you still need to be seen. The image in the color insert is of a Maersk container vessel transiting the Singapore Strait, courtesy of Piet Sinke at Maasmond Maritime. This vessel can carry 13,100 TEU (twenty-foot equivalent unit) and has a service speed of about 18kts to 20kts. In mid-2022 a vessel with 24,346 TEU capability was launched. They all move at high speeds, and they are double the size illustrated.

If you want to keep track of what is happening in the maritime world, subscribe to the free "Daily Newsclippings" and join more than 44,750 other subscribers at maasmondmaritime.com/en/subscribe/.

40.2　Maritime Traffic. While the inaptly termed "shipping lanes" may constitute areas of heavy commercial traffic, commercial vessels ply waters everywhere. I strongly suggest visiting this website for the ultimate reality check so you understand how many ships are out there and how effective AIS is: ship-tracking.net. The commonly adopted attitude—that no one is keeping a lookout anyway, so why bother—is fatally flawed. Today all vessels have AIS and radar with collision avoidance tracking systems. If the vessel's radar cannot lock onto a good, consistent signal, it cannot identify and track a target. I have sailed under many flags commercially, from general cargo to large tank ships and reefer vessels under much-maligned "flags of convenience" such as Liberia, Marshall Islands, Panama, and others. Contrary to popular misinformed opinion, I can verify that officers were all qualified and competent. With large and fast vessels, the earlier you are detected on radar and your course and collision risk are assessed, the earlier action can be taken to change course and avoid a close-quarter situation.

    a.　Search and Rescue. Besides the collision risk problem, the passive radar reflector has an important role during search and rescue (SAR) operations that cannot be overstated. Many SAR operations are called off at night. Much valuable helicopter airtime and aviation fuel are wasted in aerial search patterns under poor conditions and low cloud cover simply because there is no effective radar reflector on board. Reaction times, rescues, and survival prospects, even with EPIRBs, are decreased in the localization and visual identification phase of a SAR operation.

    b.　Mast Weight and Windage. One of the main reasons given by many for not having a reflector is that reflectors are too bulky or create windage. Also cited is that they are too heavy when mounted on the mast, or they foul halyards. Given that the mast will often carry the radar scanner and lights, it's a rather weak excuse.

c.   **Mast Shadowing.** Wherever you mount your radar reflector, there will be some shadowing from the mast. When reflectors are mounted directly in front of the mast, there is a typical 10-degree blind spot directly aft, the lowest collision risk sector of all. A yacht's track is far from straight, whether under autopilot or hand steering. The variation is in the range of 10 to 25 degrees, although some reflective surface will be "seen" overhanging the mast. This movement will expose a substantial number of reflective corners, enough to offer a reasonably consistent return at a range of at least 5nm.

40.3   Radar Reflector Theory. To understand reflectors, a basic understanding of radar signal behavior is required. The following information will prove useful when using a radar unit.

a.   **Radar Beam Behavior.** When a radar beam reaches a target, in theory it reflects back on a reciprocal course to be processed into a range and bearing for display on the radar screen. In practice, a beam does not simply bounce back off an object. Some materials are more reflective than others, while others absorb the signal.

b.   **Radar Reflective Materials.** The best reflective structures are made of steel and aluminum. Wood, fiberglass, and sailcloth do not reflect at all. In fact, fiberglass absorbs some 50% of a radar signal. There will almost always be some reflection, but the direction of the reflected beam will be erratic and so minimal that no consistent return can be monitored on the screen.

c.   **Radar Reflection Consistency.** Consistency is one of the major requirements of a good reflector. A good reflector consists of a metallic structure, normally aluminum, with surfaces placed at 90 degrees to each other. If a beam is directed to the center of a reentrant trihedral parallel to the centerline, it will reflect on a reciprocal course back to the scanner. A reentrant trihedral is simply a corner with three sides, such as the corner made up of two walls and a ceiling. The centerline of the corner points in a direction approximately 36 degrees to each of the sides making up the trihedral. The more the angle increases away from the centerline from a radar beam, the less radar signal returns. This simple principle forms the basis of radar reflectors.

d.   **Radar Reflection Standards.** The basic standards include a number of specifications. Never buy a reflector that does not comply. A peak echoing area of $10m^2$ is defined as the equivalent of a metal sphere of approximately 12ft (3.6m) diameter. International requirements and standards are as follows:

(1)   **ISO (8729-2 2009).** This IMO-sponsored standard specifies a radar cross section (RCS) of $2.5m^2$ as the minimum threshold of radar visibility. The standard specifies the minimum requirements for a radar reflector intended to enhance returns from small vessels as

required by IMO Resolution MSC.164 (78). There are two types of radar reflector: passive and active. Passive reflectors are mechanical, whereas active reflectors have an electronic element.

(2)   **USCG.** A standard is set for survival craft reflectors. Manufacturers are required to demonstrate a range of 4nm in a calm sea.

(3)   **MCA (UK).** General compliance required with ISO 8729-1:2010 for passive devices. This requires a peak Radar Cross Section (RCS) of at least 10m².

(4)   **Royal Ocean Racing Club (RORC) (UK).** An equivalent echoing area of not less than 10m² is required.

(5)   **Australian Yachting Federation. (AYF) (Australia).** A minimum equivalent area of at least 10m² is required.

40.4   Radar Reflector Types. There are a variety of reflectors on the market. The various manufacturers have challenged other reflector makers' numbers and performance standards.

a.   **Octahedrals.** The standard octahedral is a structure consisting of eight reentrant trihedrals. It was developed in the early 1940s, when radar was under development. For maximum effect, the octahedrals must be mounted in the proper orientation, called the "catchrain" position. Many are hoisted up by a corner. One magazine survey had a figure approaching 70%; my survey was closer to 80%. The structure has only six effective corners, pointing alternately up and down, the remaining corners being of little use. The effectiveness of the radar reflector is shown in a polar diagram, which shows peak reflection areas and large areas between the lobes where no reflection occurs or is so minimal it is under the minimum standards (2.5m²) set down by IMO. The total number of blind spots on a correctly hoisted octahedral total nearly 120 degrees, which is not ideal. The bad news is that when heeled to 15 degrees, the blind spots increase to nearly 180 degrees. Under sail, you have a 50% chance of being seen on radar, in most cases intermittently. This can be further reduced when part of the signal, after reflecting off the sea surface, cancels out another beam traveling directly to the reflector.

b.   **Optimized Arrays.** The Firdell Blipper and the Echomax are representative of these reflector types and are the most common on masts. The reflector consists of an array of precisely positioned reentrant trihedrals designed to give consistent 360-degree coverage, and through heel it angles up to 30 degrees. As a vessel moves around in a typical three-dimensional motion, each of the corners moves in and out of "phase" to the radar signal, with one corner sending back signal directly and others giving partial returns, resulting in a consistent return at all times. The units are rotationally molded inside a radar-invisible plastic case; the windage is only 15% of an 18in (45cm) octahedral, and the unit weighs less than 4.4lb (2kg). These reflectors meet and exceed all published standards. On a personal note, I have installed a Blipper on my two previous boats and done my own impromptu ad hoc testing of my radar visibility; all those tests were positive.

c. **Stacked Arrays.** These are typified by tubular reflectors that resemble a fluorescent tube or rolling pin, such as the Mobri and Slim Jim units. I have seen many of these taped to a backstay or stay, sometimes three or four on a yacht. They consist of an array of tiny reflectors housed in a see-through plastic case. These reflectors are purchased because they are cheap and small, not for the radar-visibility factor, which is the primary safety requirement. The analysis I have looked at shows that such reflectors can only effectively return the amount of signal required in a perfectly vertical position. At any angle of heel of 1 degree or more, the unit return falls away to virtually nil. At best tabulated positions, at 0-degree azimuth, the RCS is 6.05; heeled to 1 degree it falls to 1.46, and at 2 degrees to 0.18. You can draw your own conclusions, but I am yet to be convinced. If you have one taped to a backstay, it probably isn't radar visible.

18" OCTAHEDRAL IN CORRECT "CATCHRAIN" POSITION

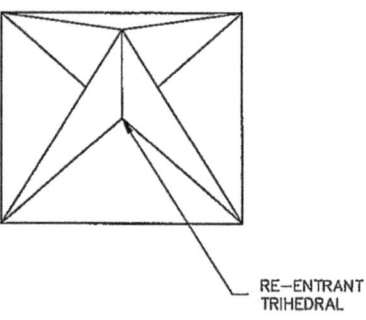

RE–ENTRANT
TRIHEDRAL

MOBRI S2 REFLECTOR

BLIPPER 210–7 REFLECTOR
(CUT AWAY)

OUTER
PVC
CASING

RE–ENTRANT
TRIHEDRALS

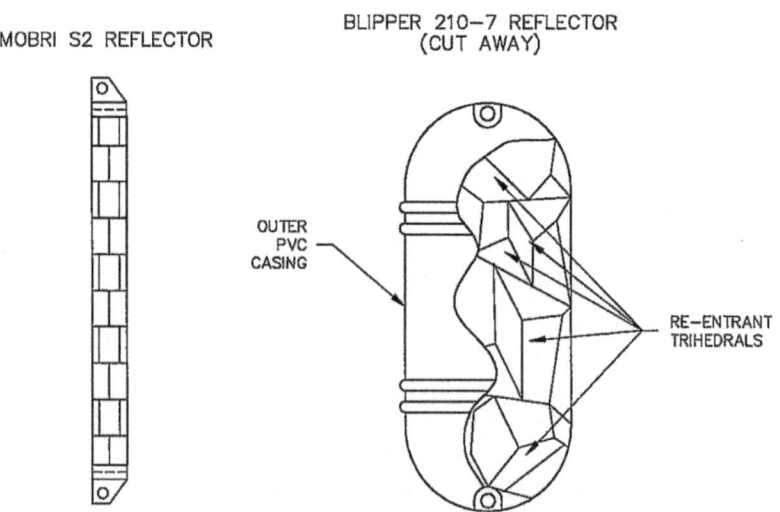

Figure 40-1. Radar Reflector Types

   d.   **Luneberg Devices.** These devices resemble two half spheres mounted back-to-back. They are normally fitted to the masthead in a fore-and-aft configuration. They are very heavy. The main criticism of the reflector is that the returned echo is only fore-and-aft and not athwartships, permitting a large and dangerous blind sector.

   e.   **Foil Devices.** I have heard many people advocating filling the mast with foil or simply hanging a pair of pantihose full of aluminum foil in the rigging. A case was heard in the UK courts some years ago regarding the loss of a catamaran in a collision with a coastal vessel during a yacht race. The skipper did not hoist a reflector because he feared windage would reduce sailing performance; instead he inserted a foil-filled stocking into the mast. The Admiralty judge included the following in his decision against the catamaran skipper: "To leave an anchorage and proceed without radar into a shipping lane when the visibility is less than 75 yards, so that the navigator is blind, and without a radar reflector so that the yacht is invisible, is in my judgement seriously negligent navigation." That statement sums up the issue of passive radar reflectors and the necessity for having them.

40.5   Radar Detectors and Transponders. AIS is changing the radar visibility space, but many have still not invested in AIS, reinforcing the need for passive radar visibility.

   a.   **Radar Detectors.** These devices are omnidirectional units that activate an alarm when a radar signal is detected in the vicinity. Typical range is approximately 5nm. When an alarm is activated, the units can be used as a radar direction finder and a plot of the vessel track can be made. The disadvantage is that with more than one fast oncoming vessel, it is difficult to plot all targets and make judgments based on the plot. Vessels may have already made collision-avoidance alterations, and this only in the case that a good radar reflector has been fitted so that you are radar visible.

   b.   **Radar Target Enhancers (RTEs).** These are not GMDSS equipment. The operation of these devices works by the reception of an incoming radar signal, the amplification of that pulse, and the retransmission of the pulse back to the radar signal source. This has to occur simultaneously and at the same frequency. The returned signal is displayed in enhanced form, with the relatively small return of the boat appearing significantly larger than it actually is. This obviously has the advantage of displaying strong and consistent echoes on radar screens. Effectiveness depends on the incoming radar signal strength, height at which the RTE is installed, and the height of the other vessel's radar above sea level. These units operate in response to 3-centimeter X-band and S-Band radars. The effective range is typically around 12nm, but not less than 3nm.

40.6   Radar Reflection Polar Diagrams. Polar diagrams are the usual way manufacturers represent the performance of their radar reflectors. There are two types of polar diagrams:

a. **Horizontal Polar Diagrams.** Polar diagrams are essentially signal returns plotted for all points around the azimuth for a reflector in the vertical position. This is crucial to understanding test claims and actual onboard performance of the reflector.

b. **3D Polar Diagrams.** The more accurate test of a radar reflector is a three-dimensional polar diagram, which indicates performance under actual heel conditions. Given that performance under heel is the critical requirement, I would caution purchasers against buying a product that cannot produce such data, or without verifiable proof that it works under the normal heeled sailing conditions of a yacht.

40.7 Radar Fresnel Zones. In some cases, radar signals self-cancel in either the transmission or the return path. This is due to factors including radar height, target height, sea and earth surface conditions, and radar range. The regions where cancellation occurs are called Fresnel or extinction zones. In such conditions, the radar signal reaching the radar reflector may be relatively weak, with a weak return. The result is either no return to the radar or a return so weak that it is not processed.

a. **Reflector Mounting.** It is apparent from the Fresnel tables that the masthead is not the ideal place to put your reflector, as a relatively large cancellation zone exists. Reflectors are best mounted around the spreaders, or about 13ft to 16ft (4m to 5m) high.

b. **Fresnel Tables.** The following Fresnel tables are published courtesy of Firdell and cover the first Fresnel zone for 12ft (3.66m) and 16ft (4.9m) radar heights and 4ft (1.2m) to 22ft (6.7m) target heights. The tables are based on a radar frequency of 9.4GHz.

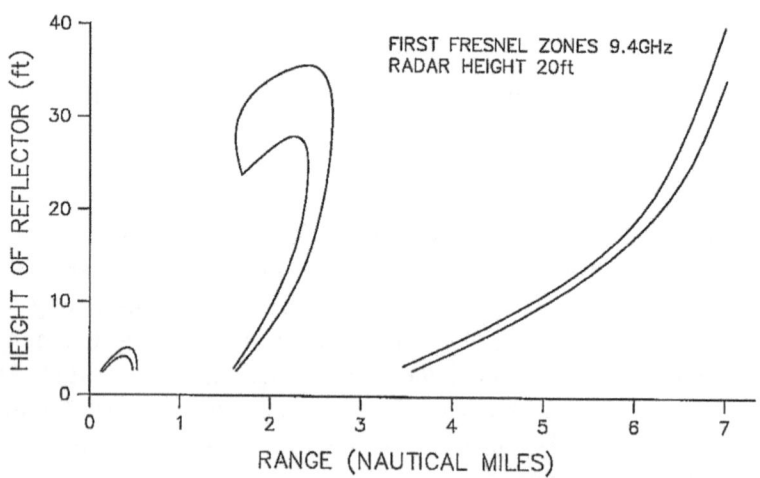

Figure 40-2. Radar Fresnel Zones

Table 40-1. First Fresnel Zone—Radar Height 12ft (3.66m)

Target Height	Zone (nm)	Zone (nm)	Zone (nm)
13.1ft (4m)	0.140–0.457	1.536–1.578	3.626–3.692
19.7ft (6m)	0.212–0.481	1.692–1.741	4.086–4.166
26.2ft (8m)	0.288–0.487	1.814–1.868	4.440–4.529
32.8ft (10m)	0.393–0.453	1.906–1.965	4.708–4.805
39.4ft (12m)	1.978–2.041	4.915–5.016	
45.9ft (14m)	2.033–2.102	5.078–5.182	
52.5ft (16m)	2.077–2.150	5.209–5.316	
59ft (18m)	2.122–2.190	5.318–5.427	
65.6ft (20m)	2.141–2.224	5.410–5.521	
72.2ft (22m)	2.164–2.253	5.492–5.604	
78.7ft (24m)	2.183–2.279	5.565–5.678	
85.3ft (26m)	2.199–2.301	5.632–5.747	
91.9ft (28m)	2.211–2.321	5.695–5.811	
98.4ft (30m)	2.221–2.339	5.755–5.872	

Table 40-2. First Fresnel Zone—Radar Height 16ft (4.88m)

Target Height	Zone (nm)	Zone (nm)	Zone (nm)
13.1ft (4m)	0.188–0.469	1.626–1.673	3.824–3.891
19.7ft (6m)	0.290–0.473	1.775–1.828	4.271–4.351
26.2ft (8m)	1.898–1.958	4.648–4.740	
32.8ft (10m)	1.997–1.064	4.957–4.058	
39.4ft (12m)	2.077–2.150	5.209–5.316	
45.9ft (14m)	2.140–2.220	5.414–5.526	
52.5ft (16m)	2.189–2.276	5.583–5.700	
59ft (18m)	2.227–2.322	5.723–5.844	
65.6ft (20m)	2.256–2.359	5.842–5.966	
72.2ft (22m)	2.276–2.389	5.944–6.071	
78.7ft (24m)	2.288–2.413	6.033–6.162	
85.3ft (26m)	2.292–2.532	6.112–6.243	
91.9ft (28m)	2.288–2.447	6.183–6.316	
98.4ft (30m)	1.574–2.457	6.248–6.382	

Onboard Computing

41.1   Notebook Computer Systems and Tablets. The notebook computer is a powerful mobile office work tool to incorporate into major yacht management systems. Several companies specialize in mobile computing as well as provide a full range of options and marine software, and I recommend that you contact them and discuss the options. With the COVID-19 pandemic, most of us have reassessed working from home or boat options, and the notebook is central to that. In previous editions, looking at computers and related options was required, but not anymore.

41.2   Notebook Selection. The criteria in selecting a suitable notebook computer for a vessel are the same as for any other environment, and manufacturers such as Lenovo, Dell, Toshiba, and HP all have acceptable models. Machines continue to get smaller, slimmer, and lighter due to processing circuit design along with the latest battery technology, as well as deletion of the obsolete CD and disc drive, and display sizes with massively improved resolution and quality. I have successfully carried notebooks around the world on and off helicopters and other aircraft and used them for most of my professional work on oil rigs and ships. Some of this edition was written offshore in the South China Sea on an oil drilling rig sea trial and some aboard my own ketch using a range of machines from a Dell (sadly the power supply and the fan failed) to my current Lenovo ThinkBook machine. In the end, if you treat them rough, they will fail; if you soak them, they will fail. It's often all a question of budget. If you are doing serious boating, a rugged version may be ideal, but the economics don't always stack up. Each notebook manufacturer has different processors, which affect price. Whether an Intel Core i7 or AMD processor, they are designed along with associated circuitry to maximize power efficiency and speed. When selecting random access memory (RAM), go for as much as your budget can afford and as much memory as you can. The more you have, the faster applications will operate, particularly some of the powerful graphics packages and chart-plotting software. If you are a gamer, you will need something designed for that. I would always opt for the largest hard drive possible within your budget, as so many software packages soon absorb capacity. The one thing driving up notebook prices is screen quality and resolution; the higher the nit (a unit of brightness), the more expensive the notebook.

41.3   Software Options. The following is a general guide to software options; choose your favorites.

    a.   Standard Software Packages. I use standard Microsoft Office with Word and Excel, which handles most things I need to do. Excel is used for almost all functions, from log to maintenance records, stores and parts inventory, and more.

    b.   Electronic Charting. Refer to the chapter on electronic charting.

    c.   Entertainment. I have a big file of movies and music loaded up. There are so many games apps to choose from.

d.   **Weather Information.** There are many weather software options, and these are listed in the weatherfax chapter.

e.   **Digital Photographs.** For many, the digital camera has pretty much been replaced by smartphones, many with very high-resolution still and video capability. If you use a drone, you need somewhere to hold your images.

f.   **Tidal Prediction Software.** These various software packages are really useful for tidal predictions. Most use databases of tidal harmonics. One package TidePredictor allows analysis of tide and current measurements and incorporates wind speed and barometric pressure data. Neptune Navigation offers another great package; check www.neptunenavigation.co.uk/tides.htm for details. User-friendly tidal height prediction software for UK and European sailors, features include daily curves, instantaneous rate of rise or fall of the tide, sun rise and set times, lunar phases, springs to neaps indicator, and monthly tide times and heights. Another package good for worldwide use is JTides; check them out at arachnoid.com/JTides/. A word of caution when using these packages: Actual tidal conditions may vary due to local conditions and weather.

g.   **Astro and Celestial Navigation.** Why include this? Because many choose to maintain these basic skills, and I am somewhat remiss in having forgotten how to do sights. Fortunately, companies still develop software packages that include the following and are way better than the DOS packages I used to use. A great package for those who want to keep their skills alive is Starpilot, a user-friendly celestial navigation program allowing easy sextant reading computations. Check them out at starpilotllc.com. Pangolin has another good astro/celestial package that includes very useful tutorials. Visit pangolin.co.nz/astronavigation/.

h.   **Performance Navigation Software.** There are several very popular packages available. Adrena has the very impressive Octopus cruising software. The package has several modules that include routing calculations, which allows analysis of optimum route plans. The mooring assistance module has graphical condition displays for the selected mooring location that include weather forecasts and tidal and sea level information along with sea condition data. The package can be networked for everything from MOB and AIS, anchor watch function, and a range of other capabilities. Check the system out at www.adrena-software.com. Another powerful package is the Expedition Deckman. The Expedition system has been utilized in many racing events, including the Volvo Ocean Race. Among the system's many capabilities are the display of marks, courses, laylines, track, and AIS targets, as well as polar functions. All the data allows the system to use multiple routing algorithms, multiple optimum routes, and a great deal more. Check this system out at www.expeditionmarine.com.

i. **Email.** There are a variety of methods for sending and receiving email. This is discussed in the ham radio section, in particular for HF radio and your computer. Smartphones are making email access very easy when coastal cruising, as are the plethora of satellite systems being launched.

j. **Internet.** Getting onto the internet is getting easier by the day, whether it's from a wireless hotspot or via satellite. Many use their smartphones for internet access, and like many, I do so almost daily. The Glomex Lite EVO is a compact plug-and-play 4G/Wi-Fi coastal internet single SIM system that allows web browsing up to around 15nm offshore.

41.4   Computer Maintenance. The following basics will improve reliability. Remove particles regularly, and wipe the board with a slightly damp cloth. Keep wet fingers, coffee mugs, and such well away. Small purpose-made, battery-powered vacuum cleaners are available to extract dust and particles. (My USB rechargeable compact fan/blower also extracts dust.) Screen maintenance is simple: Wipe using a lint-free cloth; I use a spectacle cloth. You can buy a pack of purpose-made screen cleaning wipes and use them to gently wipe off marks. Be careful when using touchscreens with salty fingers, as they can scratch. Making your battery power last longer between charging cycles is a constant challenge. Suggestions include using a dark background, resetting your options with respect to sleep, and shut down if not in use; all notebooks have power-saving options. Another hint is to use one application at a time, single-tasking; opening several applications at the same time uses more power. Go into your Windows performance management tool and set energy-saving levels. In the end, it is a trade-off between speed and battery run time. Remember that all those applications running in the background consume power. Another power-saving hint is to turn off the wireless and enable airplane mode.

41.5   Computer Troubleshooting. There are a few basic checks to make. The majority of problems with computers on boats can be attributed to poor-quality power supplies, both AC and DC, carrying damaging voltage transients. If you are using an alternator fast-charge device, disconnect the computer from the vessel DC system during charging as a precaution. If things are not working out, there is always the option of rebooting.

41.6   Computer Cybersecurity. The bad guys never give up, and you need to be on your guard if traveling around. Your email address has been harvested from some subscription service you signed up for or some retailer who then on-sold your data. Don't open the email; delete it. Don't click on any link and, worst of all, do not open any attached files. Invest in good virus software such as Norton, McAfee, Panda, or TotalAV. Beware of the following:

a. **USB File Sharing.** I know the files from the boat next to you have some interesting stuff, but transferring files is inherently risky. So bad is the problem of importing virus and malware that there are no USB sockets in the offshore oil industry. You need to scan and be very careful before inserting that borrowed USB stick, flash drive, thumb stick, jump drive, pen drive, thumb drive, or whatever it is called where you are. Be extra careful or risk suffering some serious problems.

b. **Malware.** This is malicious software, and is the overarching umbrella term used to describe any software that seeks to be deliberately harmful to digital devices such as computers and smartphones. Malware can be relatively harmless or be scary, such as pop-up adds that you can't close. Others can use your computer to run illicit activities without your knowledge. Use a pop-up blocker, and don't click on any links or pop-ups. Download software only from websites you know and trust. Signs you may have malware include slow operation, unexpected crashes, strange error messages, shutdown and restart problems, getting taken to web pages you didn't plan to visit, or new icons appearing.

c. **Computer Viruses and Worms.** A computer virus is a type of program that replicates itself to spread infection. Viruses are harmful and can destroy data, slow down system resources, and log keystrokes. A computer virus attaches bits of its own malicious code to other files or replaces files outright with copies of itself. A computer virus requires a host program and requires user action, such as opening an infected Word file, to transmit from one system to another. Worms, however, are able to spread across systems and networks independently and thus are very dangerous.

d. **Open Wireless Access.** Every time you go to a Starbucks or some other place for coffee and connect to a wireless hot spot, you are open to being hacked.

e. **Email Phishing.** Phishing expeditions through your inbox include generic email greetings, URLs that seem a little odd, and links and attachments that try to entice you and get the click. Resist the temptation!

f. **Scams.** There are so many junk emails and scams are out there, and people still get sucked in. Notifications of a lottery win is a favorite; simply delete. Then there are unsolicited job offers. If it seems too good to be true, it probably is.

g. **Trojans.** Trojans aim to trick you into downloading a seemingly legitimate software or file, but once downloaded, they run the malicious code.

h. **Ransomware.** When your system gets infected with ransomware, it allows hackers to lock you out of your computer or wipe out the entire system unless a ransom is paid. This is currently a major problem worldwide for many businesses.

Entertainment Systems

42.1 **Music Systems.** Without music, is your boat really ready for sea? On a new yacht, there is a definite psychological lift when the music goes on for the first time. When selecting or installing an entertainment system, there are a number of important factors to consider. Like radios, choose units that are designed for RVs and more-rugged shore duty. There are some really good-quality systems designed for boats, mostly the powerboat market but suitable for yachts as well. In many places in the Caribbean, radio stations are streamed, and getting an AM or FM station is somewhat harder these days. I used to love loading up my ten-disc Pioneer CD player, and then I got really used to my MP3 player. Now most folk have their favorite playlists on their phones. I like my simple radio with the USB port to plug in my favored USB memory stick for a day of Jimmy Buffet and "Margaritaville," Cool Change, or perhaps AC/DC and "High Voltage" rock and roll; it is simple to change music to suit your mood. Marine audiovisual (AV) systems continue to become more sophisticated. People want the same quality systems on board as they do at home. Systems have three basic elements: reception, which is getting data to the boat; distribution, which is moving files around the boat; and display, which is broadcasting entertainment through speakers and display screens. Networking of audiovisual systems is becoming the norm. It is now an online world, and many have moved to streaming with Spotify, YouTube, Amazon Music, and Netflix. Many choose to store music on their smartphones and connect with Bluetooth, given that terrestrial connectivity is challenged by range and internet access for getting entertainment on board is not easy. The transition to 5G is providing greater ability for downloading or watching AV content, depending on where you are. Here are some basics.

a. **Power Output.** Power output is rated in watts and is specified in either "watts per channel" (RMS) or "total power output" (PMPO). Watts per channel is the power through each speaker; total power output is combined power. There is no need for high-rated units with 60W per channel simply because the area involved is relatively small and the ear cannot distinguish between a 30W and a 60W system. Quality, not volume, is what counts. Check out clarionmarine.com.

b. **Speakers.** A speaker produces sound when the cone moves back and forth, compressing air in waves that produce the sound. Best sound is produced along a line projecting outward from the cone center. Sound radiates at around 45 degrees off the main axis in a cone configuration. If not listening in this zone, sound reflects and distorts off other surfaces. The average speaker requires air space of at least 1 cubic foot. When the cone moves forward, a vacuum is created; as it moves back, the air is compressed. Small cavities behind speakers reduce the cone movement, and you will get a lower bass response, as low frequencies require more air movement. Be aware that magnetics on the speaker cones can affect navigation instruments, so install them well away from compasses and GPS. It is very important to install the best quality speaker cable you can. Long runs are normal in boats, and quality suffers accordingly. For speakers mounted in the cockpit or on the stern arch, use tinned-copper cables. If you want good quality sound, this is only as good as the speaker placement. Don't mount speakers facing toward each other; angle them toward the listening position. Avoid having multiple speakers and subwoofers inside the same enclosed area. Consider an amplifier before upgrading speakers; tuning them for optimum performance is required. Manufacturers include Poly Planar, which makes quality waterproof speakers. Others include Fusion, James Loudspeaker, JL Audio, Rockford Fosgate, and Roswell Marine Audio.

42.2   Entertainment Systems. These are becoming fairly sophisticated and complex. There are so many options from manufacturers such as Clarion, Fusion, JL, and others. Systems now comprise amplifiers, multiple zone speaker arrangements, and much more than just the simple saloon and cockpit speaker setup like on my current boat. Add in subwoofers, smartphone Bluetooth connectivity, SiriusXM satellite radio, and remote control and the system becomes very versatile. A simple system will have a source unit, four speakers, an amplifier, and possibly a control head. NMEA 2000 connectivity and control is available on some MFD systems such as Raymarine's. Connected up to the source unit will be the AM/FM radio aerial. Units such as the Clarion CMS4 are more correctly termed "media hubs." They incorporate a seven-band graphic equalizer that offers control in the 50Hz to 16kHz range. The Bluetooth aptX feature allows streamed compressed audio, depending on compatible mobile phones. The unit has USB ports, which suits my personal practice of using memory sticks for my curated music mix. Multizone control for up to four zones is possible, along with volume control on the sources. Interactive Pandora internet radio works using the Pandora app on your smart device. SiriusXM satellite radio requires a tuner unit. To add to all those features is a global radio tuner with weather band.

42.3   Video and DVD Players. The video cassette, like the dodo, is long extinct. It was replaced by the DVD as the main audiovisual source on board. I still use a DVD player, as do many others. There's nothing like binge-watching DVDs while swinging on the hook with a nasty day outside. But the end of the road for DVDs is approaching, with entertainment so easy to download from Netflix and elsewhere. Common faults with DVDs are damaged or scratched discs that cause skipping, failure to play, or disc rejection. Vibration can also cause skipping. Failure to play can be caused by moisture on the laser pickup, which can be rectified by leaving the player on to warm up for an hour. Picture problems are usually caused by incorrect settings, poor cable connections, or the disc having incompatible copy-prevention guards. Sometimes powering off, removing the plug for 30 seconds, and restarting will solve operation problems, which are often caused by static, lightning, electrical noise, or simply improper operation. Remember that any switch off causes a loss in settings. Current demands are relatively small. It is a good idea to place a cover or bag over unused players with a bag of silica gel and perhaps a corrosion inhibitor, as players are not marine grade.

42.4   Satellite TV. There are some very sophisticated satellite TV systems available, including those from KVH and Raymarine. Systems operate on the C-Band (3.7GHz to

4.2GHz) and the Ku-Band (10.7GHz to 12.95GHz), and some are dual frequency. The units use a control unit and an auto-tracking gyrostabilized antenna dishes housed in a radome. Yacht-friendly radomes have sizes in the range of 24in to 59in (0.6m to 1.5m). As satellite tracking must be fast and accurate to remain locked on, units have pitch-and-roll sensors, with three-axis servos and track satellites to compensate for vessel pitch, roll, and yaw. These sensors use rate gyro sensors and inclinometers. Companies have invested heavily in research and development, producing major advances in parabolic reflector efficiency, reduced weights, and more precise tracking. They have a four-axis design compensating for elevation, pitch and roll, azimuth, and active skew to maximize signal quality. Smaller vessels usually install the radome on a dedicated stern post or mast. TracVision systems offer the ability to access digital HD TV, movies, and music; the range, TV1 through to TV10, depends on whether you cruise coastal or are ocean voyaging. The TV1 is suited for smaller vessels cruising around coastal United States or Europe. The TV6 has a 24in (60cm) antenna, allowing access to worldwide satellite TV. It allows the reception of standard definition and HD Ku-band satellite TV programming and incorporates their RingFire antenna technology for greater geographic coverage and enhanced reception. Tracking stability is improved as a result of advanced algorithms to enable good performance in rough sea states. Check out the KVH website for the range of systems and options available: www.kvh.com.

42.5   Terrestrial Television. Television aerials and their performance on yachts is a controversial subject along with often overoptimistic performance claims by some manufacturers, as well as being relatively expensive. Comparable performance with home aerials should not be expected, and attempting to get a reasonable picture while under way is generally out of the question. The off-watch should probably stick to Netflix downloads. The principal problem is obtaining a good picture at anchorages, without the continual ghosting that occurs in varying degrees of severity as you swing around. The ghosting problem largely depends on the path of the transmitted signal and the frequency characteristics of the transmission. Note the following issues.

a.   **Signal Distortion.** Transmission signals are essentially straight line and do not bend significantly when meeting obstructions. The effect is one of creating shadows and areas of low signal behind the obstruction. Reflection of the signal alters the direction of propagation, causing signals to arrive at the aerial from a direction other than the straight-line path from the transmitter. The receiver ultimately receives two signals that arrive at different times. The effect is the reception of a distorted signal pattern. The distortion of signal occurs from a number of sources, including hills, other boats, rigging, and reflection off the water's surface.

b.   **Signal Polarization.** Signal transmissions are generally horizontally polarized. When signal is reflected, the polarization is altered, causing distortion.

c.   **Transmission Systems.** The variety of TV transmission systems in use is confusing. The main systems are M/NTSC (United States, Canada, Guam, most of South America); BG/PAL (Australia, Netherlands, Spain, Portugal, Italy, Canary Islands); I/PAL (UK, South Africa); L/SECAM (France).

42.6   TV Antenna. There are various antenna types in use.

    a.   **Directional Antenna.** These types of antenna are used so that the antenna can be aligned with the transmitted signal. Antenna are of the domestic type, and may be of use if you live on board and rarely venture out from the marina. But on an anchorage they are nearly useless, as you must continually adjust the aerial.

    b.   **Omnidirectional Antenna.** These antenna are able to receive transmission signals consistently and are not affected by the vessel swinging at anchor or mooring. This type of antenna does not discriminate between directly transmitted or reflected signals. The most common type of antenna is the ring or loop, which is hoisted when required. These antenna do have a problem with reception of reflected signal from any masts and rigging, and they are relatively poor within marinas.

    c.   **Active Antennas.** These units are typically a fiberglass or plastic dome, with an integral omnidirectional loop aerial element inside. The signal is amplified to compensate for the smaller antenna, and performance is dependent on a good gain value within the amplifier unit. These antenna are designed for the reception of UHF signals as well as AM/FM radio transmissions. Glomex antenna are very popular in this space.

    d.   **Installation Factors.** The following factors should be considered when mounting and installing antenna:

        (1)   **Antenna Height.** Install the antenna as high as practicable.

        (2)   **Antenna Cables.** Cables should always be low-loss coaxial (RG59), which is normally 75Ω impedance.

42.7   Satellite Radio. The service offers over 150 channels of entertainment. SiriusXM satellite radio is available in the forty-eight contiguous states, the District of Columbia, and Puerto Rico (although limited), and around 200nm seaward into the Atlantic and Pacific Oceans, the Gulf of Mexico, the Caribbean, and the Great Lakes. Marine subscription plans include SiriusXM Coastal and SiriusXM Offshore, with extensive weather information that includes NOAA weather forecasts along with weather radar, storm cell attributes, tropical storm tracks and lightning strike data, offshore surface wind forecasts, and weather warnings. Of course, they also have great music, including Yacht Rock Radio, sports, and much more.

    a.   **SiriusXM Receivers.** You can purchase a Raymarine SR200 SiriusXM satellite weather receiver for Raymarine Axiom, Axiom Pro, and Axiom+ displays with LightHouse 3.9 or updated software. Raymarine and SiriusXM have partnered to enable transmission of real-time up-to-date weather information and forecasting. The weather information is delivered via the satellite receiver through SiriusXM's Marine Weather services. For Garmin multifunction displays, you will need the Garmin GXM 54 Satellite Weather Receiver. Simrad, B&G, and Lowrance systems require a WM-4 Satellite Weather receiver.

The graphical information is overlaid onto your charts so that observations of weather and fishing information relative to boat location are available. A 20.0 software update or later is required for WM-4 operation on Lowrance HDS Live, HDS Carbon; Simrad NSS evo3, NSS evo3S, NSO evo3, NSO evo3S; B&G Zeus3, Zeus3S, Zeus3 Glass Helm, and Zeus 3S Glass Helm. Again, check their website for updated information. For Furuno systems you will need the BBWX4 fourth-generation SiriusXM satellite weather receiver for Furuno NavNet TZtouch, TZtouch2, and TZtouch3 systems. A relatively new service, SiriusXM Fish Mapping, is available for those who go offshore looking for fish that don't get away. Check out www.siriusxm.com.

b.   **Hardware Installation.** The receiver comprises an antenna module and the receiver module. The antenna module receives signals from the satellite, then amplifies that signal before processing to filter out any interference and noise. For maximum range and optimal signal reception, you can install upgraded helical antenna that do not require a ground plane connection and have a dual-shield low-loss cable. Install the antenna where sheets cannot get tangled around them; if signal is degraded, always check the antenna and connectors first. Some practical installation notes: For the Furuno unit, the antenna and receiver are to be installed 5ft (>1.5m) clear of an AIS transmitter or radio and 6.5ft (>2m) away from a radar and 6ft (>1.8m) clear of a compass.

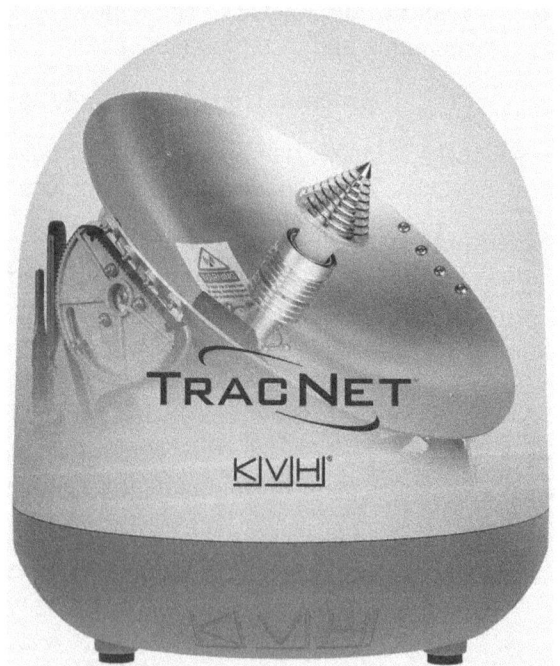

Figure 42-1. KVH TracVision H30 Hybrid Terminal (*Courtesy of KVH Industries, Inc.*)

Forward-Looking Sonar and Fish-Finders

43.1   Forward-Looking Sonar. Forward-looking sonar maps out the seabed ahead of the boat. Forward Scan is a great tool for nosing into a bay and even up closer to some idyllic shelving beach. It is invaluable when the water is muddy up that narrow tidal river or estuary, and a lot more informative and less stressful than looking at a simple depth-sounder digital readout. Entering challenging anchorages or new and unfamiliar harbors is easier when you can see what is ahead along with any potential underwater obstructions. I have always found piloting unfamiliar waterways rather stressful—even more so when it's a refuge from weather or at night. EchoPilot has been a leader in this space for many years and is now owned by Navico, which also owns Simrad, B&G, and Lowrance. They have moved ahead with a forward sonar transducer and updated software innovations for their SonarHub module and chart plotters.

43.2   Floating Containers. There was an unrealistic expectation that forward sonar would be suitable for detection of partially submerged shipping containers; however, they are unable to do this. Most containers sink reasonably quickly as water seeps into them, but it is the ones floating just under the surface that create the greatest risk of catastrophic collision for all shipping and yachts. The IMO's Sub-Committee on Carriage of Cargoes and Containers has agreed on a draft for new requirements relating to the mandatory reporting of freight shipping containers lost overboard at sea, usually in heavy weather. They highlighted the serious hazard to navigation and safety at sea in general, in particular to recreational sailing vessels, fishing vessels, and other small craft. The subcommittee also set out a specific set of amendments to the SOLAS and MARPOL treaties. The World Shipping Council (WSC) has reported that the international liner shipping industry transports approximately 226 million containers annually. The industry recognized the growing dangers after a series of high-profile incidents in 2020 and 2021. The WSC estimated that an average of 1,566 containers were lost at sea each year in a 15-year period (2008–2022). The WSC calculates that over the past 5 years, the average was more than 2,301 shipping containers lost each year. The draft amendments to SOLAS Chapter V add new paragraphs specifically addressing danger messages. This would require the master of every ship involved in the loss of freight shipping containers to communicate the particulars of an incident without delay, and to the fullest extent possible, to all vessels in the incident vicinity.

43.3   Forward Sonar Transducers. The transducer head scans an envelope vertical to horizontal and has a beamwidth and cone of 15 degrees either side of the bow, with a forward-looking range of 490ft to 656ft (150m to 200m). Range is dependent on the type of objects or bottom formations ahead, along with water depth and contours. A vertical seawall will reflect acoustic energy better than a muddy bottom, and shelving mud has a range of around 165ft (50m). The range is relatively short, and the transducer is typically installed around 6ft (2m) forward of the keel and extends some 1.2in (30mm) below the hull. Depth-sounders emit an acoustic signal and process the return echo. As described in the instrumentation chapter, many things affect accuracy, including water temperature. Shoaling waters are easier to image than ones that are deepening. Transducer heads must be free of turbulence across the head. Operating frequency is 200kHz. Like depth- and fish-finder transducers, they have piezoceramic elements. These ceramic elements are manufactured from polarized barium titanate or lead zirconate titanate. The transducer elements have an acoustic window on the outside; the remainder of the element is surrounded by acoustically matched material that facilitates emission and reception of acoustic pulses. EchoPilot's patented real-time forward-looking sonar uses an electronically controlled array to scan the seabed from straight ahead through 90 degrees to straight down and is not unlike a forward-looking underwater searchlight.

FISH-FINDERS

43.4   About Fish-Finders. The fish-finder started off as a submarine finder. Recent developments in computing power, signal processing, and microelectronics have resulted in major advances in sonar development. I spent a couple of years on a new submarine sonar systems project, and this is an interesting subject, with defense-level capability now part of the boating and fishing world. The term SONAR is derived from the words **SO**und **NA**vigation **R**anging and now simply called sonar. Why is the subject of fish-finders being included in a primarily yachting electrical and electronics book? A great question, and it is because an ever-increasing number of people are installing them on yachts to get clear information on what is below, especially those who like to fish as well. The depth determines the fishing technique, and what type of lure or bait to use. The fish-finder lets you see who may be at home down there, but it will not guarantee a catch. That part is up to your own skills. In most fishing, you will use your fish-finder to look for underwater structure such as ledges and significant bottom changes. Cover consists of underwater objects and structures such as tree stumps, weed beds, and logs. You may be looking for drop-offs like gullies and channels or break lines, which are sharp bottom drop-off points. You may just be looking for fish activity. If you locate a tight ball of baitfish, they are probably being chased by a predator. A fish-finder is not necessarily going to tell you precisely what the fish is, although some people are able to do so. Fish-finder interpretation involves understanding both the underwater features and fish behavior to distinguish between baitfish and the target fish. Distinguishing between the species is the angling part. Selection is challenging, and the choices range from networked and broadband finders to standalone and combination units.

43.5   Fish-Finder Principles. The fish-finder electrical signal is sent to the transducer. The electrical signal to energize the transducer crystal is generated by an amplifier. The energized piezoelectric element (crystal) then reverberates at a particular frequency, converting the electrical signal into mechanical acoustic or sound energy. This acoustic energy or sound wave creates oscillation of the water molecules through which the sound travels. The sound is pulsed out in a defined beam with a cone-shaped characteristic. These acoustic beams do not travel out in a straight line but in a wave pattern. The acoustic pulse travels through the water at a rate of approximately 4,800ft/s (1,463m/s) in salt water and 4,920ft/s (1,500m/s) in fresh water. When the energy strikes an object within that beam, such as a fish air bladder, the sea bottom, or a structure, some of that energy is reflected, or echoed, back to the transducer. As the processor is programmed with the rate of sound transmission in the water, it calculates the time difference between the transmission and reception of the returned signal to give a range or depth figure. This result is then displayed as a number or as an image on a screen. Fresh water and salt water tend to absorb and scatter sound signals, and the higher frequencies are more susceptible than lower frequencies. Water is frequently being mixed due to environmental factors such as wind and wave action. The water has air bubbles, suspended materials such as silt, minerals, and salts that vary in quantity. There are also microorganisms including plankton and algae, all which scatter, absorb, and reflect fish-finder sonar signals. StructureScan 3D imaging scans the underwater terrain and delivers a high-resolution, three-dimensional view that enables display of bottom formation and is able to determine composition such as mud, sand, gravel, or rock; it works by analyzing the return echo strength to determine hardness. Different bottoms support different fish species.

43.6   Transducer Basics. Transducers are typically constructed of a crystal composed of various elements, including lead, zirconate, barium, titanate, and conductive coatings. The shape and diameter of a transducer determines the cone angle. The acoustic signal strength is at maximum along the center axis of the cone and decreases away from it. The cone angle is based on the power at the center to a point where the power decreases to –3db, with the total angle being measured from a –3db point on each side. Most manufacturers offer models with a variety of cone angles. Wide cone angles have less depth capability with wider coverage; small cone angles give greater depth penetration with reduced area coverage. The farther away from the centerline of the cone, the less strong the return echoes are. This can be improved by increasing the sensitivity control. Resolution defines its measuring precision or detection capabilities. This is a direct function of several factors, including pulse duration, angle of incidence of the acoustic wave front on the target, the nature of the target, and the beamwidth of the transmission. The minimum threshold of distinguishability directly corresponds to one-half pulse length. If two targets are less than a half pulse length apart, they will reflect as a single target. If the two objects are separated by more than a half-pulse length, they will be seen as two separate echoes.

43.7   Transducer Operation. The standard transducer is a single ceramic element device that transmits and receives sonar acoustic pulses on a fixed frequency. Dual-band devices have two elements—one for high frequency (200kHz) and another for low frequencies (50kHz). The more elements, the greater the performance of the system. High-frequency transducers have the ability to differentiate smaller targets but do not have the depth range. Lower frequencies such as 50kHz devices have greater depth capability but lower resolution. In dual-frequency units with 50kHz/200kHz transducers, the fish-finder software processes both frequency returns of fish or bottom to create a composite water column image. The standard transducer has a high-power output signal in the range of 600kW to 3kW, and this output pulse is relatively short. The leading designer and manufacturer of transducers is Airmar, and they have labeled the three common frequency bands: Low band is the 40kHz to 75kHz spectrum, medium band is the 80kHz to 130kHz spectrum, and high band is the 130kHz to 210kHz spectrum. An equation used when calculating speed of sound in waters is: $C = T - T^2 + S + D$, where "C" is the speed of sound in meters/second; "T" is the water temperature in Celsius; "S" is the water salinity in parts per thousand; and "D" is the water depth in meters. The speed of sound in water is dependent on pressure (depth), temperature (a change of $1°C \sim 4m/s$), and salinity (a change of $1‰ \sim 1m/s$).

43.8 **CHIRP Transducers.** Compressed high-intensity radar pulse (CHIRP) transducers are the most transformational development in the sonar and fish-finder world. The CHIRP transducer emits a long time duration, linear pulse that sweeps across a frequency range. Unlike a standard transducer, the CHIRP transducer emits across a wide range of frequencies, typically 130kHz to 210kHz. The reflected signals are quite complex, so more complex signal processing and electronic circuits are required. CHIRP technology results in greater definition at greater water depths. The signal-to-noise ratio is improved, resulting in significantly improved target discrimination against the seabed so that individual fish can be detected with very detailed definition. CHIRP transducers are able to operate in either a single- or dual-frequency band. Low CHIRP is able to accurately track the seabed in deep water and is compatible with 50kHz single-frequency transducers. Medium CHIRP offers better resolution than low CHIRP, with relatively small loss of depth capability, and is compatible with 83kHz single-frequency transducers. High CHIRP offers high resolution with high image quality in shallow water and is compatible with 200kHz single-frequency transducers. Simrad has dual-channel CHIRP transducers that are compatible with their Simrad S5100 broadband sounder module; these are capable of dual frequency, sweeping on two ranges simultaneously. Lowrance developed its own CHIRP transducer, labeled the Med/High Skimmer. Garmin has a multibeam Med/High CHIRP sonar that produces photo quality high-resolution images. The Furuno CHIRP Side-Scan for NavNet TZtouch3 is able to scan port and starboard, and allows observation of bottom structure shape in high definition. It operates using lower frequencies. It has a range of 750ft (200m) on each side of the boat. The DFF-3D Multibeam sonar operates at 165kHz and provides visibility under the boat of 1,000ft (300m). The optional Deep Impact TruEcho CHIRP boosts output power. Head to www.furunousa.com for details of these impressive systems.

43.9 **Operating Frequencies.** The acoustic transmission frequency affects both the depth range and cone angle. The speed of sound in water is a constant of 4,800ft/s (1,463m/s), and the time between the transmission and reception of the returned signal is measured to give a range or depth figure. The lower frequencies are used in deeper waters and have lower power losses. They tend to have wider beam angles and cover wider viewing areas. High frequency (HF) transducers (190kHz to 200kHz) are available in either wide or narrow cone angles. Low frequency (LF) (50kHz) transducers have cone angles in the range 30 to 45 degrees. Many frequencies are used in fish-finder sonar systems; typical are 38kHz, 40kHz, 50kHz, 75kHz, 107kHz, 120kHz, 150kHz, 192kHz, 200kHz, 400kHz, and 455kHz. Some have a frequency of 192kHz transducers with either a wide (20-degree) or narrow (8-degree) cone angle and a deepwater one of 50kHz with a 35-degree cone angle. Some units have a user-selectable tri-frequency capability of 38kHz/50kHz, 38kHz/200kHz, or 50kHz/200kHz, a depth range up to 5,905ft (1,800m), and a maximum ping rate of fifteen pings per second. Shallow waters less than 300ft (90m) give the best results with high-frequency transducers of 200kHz and wide cone angles up to 20 degrees. In depths greater than 300ft (90m), low-frequency transducers of 50kHz with small cone angles of 8 degrees are the best option.

43.10 **Power Output.** Power is measured in watts root-mean-square (RMS) or overall peak power capability. Higher power ratings in watts means larger output power. Deepwater fish-finders are more effective with higher power. Many units now have transmission power ratings in the range of 1kW to 300W RMS. The average is around 250W to 500W RMS.

43.11   Multibeam Fish-Finders. While simple fish-finders have a single beam transducer, all the leading manufacturers are now introducing multibeam systems that have several sonar beams, ranging from two up to eighteen. Each of the narrow beams has a bottom resolution that equates to a narrow single beam. The accuracy measurement is not any better than single-beam units, and accuracy will decrease as the swath angle increases. Multibeam systems are either swath or sweep systems. This subsequently increases the coverage area and accuracy. The single-beam transducer has a typical cone angle of 20 to 24 degrees, which gives a depth range of about 600ft to 1,000ft (183m to 305m). The dual-beam system has a range up to 2,000ft (610m). The first beam is in the cone center, and a second beam surrounds it to increase the coverage area. The tri-beam systems have a 90-degree coverage area, with ranges up to 1,000ft (305m). The main beam is directed down, and two beams are configured to each side to provide a large coverage area. The wide side has three beams to view bank and bottom contours with the center beam directed down 120ft (37m) and port and starboard to 120ft (37m). The six-beam system gives a 3D contour display of the sea bottom, and some systems have a coverage of 53 degrees up to a depth of 240ft (73m).

43.12   Fish-Finder Technology. The technology has come a long way in a short time. Some of the impressive systems on the market include the Simrad NSS and NSO evo3s fish-finders. The NSS model incorporates a SolarMax IPS display, six-core processor, 1kW CHIRP sonar, and advanced networking options. The finder can be integrated with autopilot, radar, sonar, and smartphone devices. Simrad has C-MAPS and CMOR relief shading that can image the 3D structure of the bottom. The Simrad NSS evo3s has screens from 9in to 16in (230mm to 406mm); the NSO evo3s display is from 16in to 24in (406mm to 610mm). Some units, including a Garmin, have a 7in to 12in touchscreen. It's all about pixel power, and some Garmin units have a HD 1920 × 1080 display. The top-of-the-line Raymarine Axiom with RealVision 3D and RV-100 transducer package includes RealVision 3D image processing, SideVision, DownVision, a 3D fish-finder, and very powerful CHIRP sonar. All this is powered by a quad core processor that runs on the LightHouse 3 operating system. The multitouch interface with very high-resolution display makes this one powerful device. Lowrance has the new HDS Pro display with high resolution ActiveTarget 2 live sonar and the ultra–high definition active imaging FishReveal transducer. The Lowrance Ultimate Fishing System HDS Pro is able to network and control trolling motors, autopilots, engines, radar, communications, and PowerPole shallow water anchors. The system is able to display clear and crisp imaging from CHIRP, SideScan, and DownScan sonar transducers. The Furuno RezBoost uses advanced signal processing to boost targets by a factor of eight. This enables greater target definition and also the ability to identify and analyze bottom types.

43.13   Transducer Cavitation. Cavitation affects transducer performance and is caused by water turbulence passing over a transducer head. At slow speeds the laminar flow is smooth without any interference; however, at speed, air bubbles are created over the transducer face, affecting acoustic signal transmission and reception. The effect is to interfere with transmitted acoustic signals that reflect back off the bubbles, effectively causing noise and masking signals. Turbulence is caused by the hull form or obstructions, water flow over the transducer, and turbulence from propulsion. Transom-mounted units must be carefully mounted to avoid turbulence from water flow off the transom. The higher the speed, the greater the turbulence that is caused. Manufacturers are designing transducers that work better at higher speeds, including transducers with improved hydrodynamic shapes. Transducers must be mounted in areas of little turbulence or clear of hull flow areas, which is not always easy.

43.14   Fish-Finder Displays. The most common display type is the liquid crystal display (LCD). Displays must have both high resolution and good contrast. Displays are also sometimes quoted in pixels per square inch, and the more pixels the better the resolution. The LCD display is composed of a complex grid of pixels, which are small square display elements that make up a screen image. Pixels are turned on or off to form an image on the screen, and return echoes are processed and displayed as dark pixels. Grayscale-display images use several shades of gray to indicate signal strength variations, with strong signals being very dark and weak ones being light gray. Each successive return activates a new column of pixels so that a continuous image is displayed on the screen as each column is replaced. The display resolution quality is dependent on the number of pixels in each vertical column. The number of horizontal pixels determines the retention period that a displayed image is on the screen. This determines the ability of a system to support additional image windows in a split-screen mode. Some units have high-resolution displays in eight or sixteen colors on TFT LCD displays. Color displays use up to sixteen colors for different signal strengths; stronger ones are displayed in red, weaker signals as green or blue. For example, baitfish schools are generally in blue or green, with larger game fish showing as yellow, orange, or red. The seabed and wrecks are usually displayed as dark orange or red. LCD panels are constructed of grids of pixels—tiny dots that are individually color controlled when an electrical current passes through them. The more pixels an LCD display contains per square inch, the higher resolution the screen will be, meaning it will be easier to read, brighter, and have better picture quality. The number of vertical pixels will determine the depth resolution; the number of horizontal pixels will determine how much data is shown on the screen.

43.15   White Line Function. Fish-finders have a feature called white line or edge that assists in discriminating bottom hardness from the bottom contours. The thickness of the white line is indicative of hard or soft surface strength returns. A rocky seabed has a strong echo return and displays a thick white line; a soft or sandy bottom will show as a thin white line.

43.16   Zoom Function. The zoom function allows the magnification of a portion of the depth range to improve analysis and identification of targets in that area. The typical magnification scales are x2 and x4 the normal scale. This allows monitoring of a certain depth range such as 40ft (12m) to 50ft (15.25m), or zooming on the bottom and 10ft (3m) above it. The split-screen feature allows tracking of different features simultaneously, such as zoom segment and the bottom contour.

43.17   Sensitivity. Where installed, the sensitivity control is used to adjust the receiver to tune in or tune out signal returns. If the unit is set with a low sensitivity, it will not detect bottom details, fish, or obstructions. High sensitivity settings will return signals on everything and clutter the screen with spurious returns. Most fish-finders now have automatic sensitivity adjustment, which compensates for ambient water conditions and depth. Most also have sophisticated and powerful signal processing algorithms that uses complex software to process parameters. These include water conditions, noise and interference levels, and boat speed to automatically adjust control settings to optimize the images on the display.

This entails setting the sensitivity to the highest level possible without allowing noise to be displayed—a balance between noise rejection and sensitivity. Submarine sonar systems use extremely sophisticated signal processors and software, and this approach is very similar. The sea is a very dynamic environment, and sound is capable of traveling very great distances. The acoustic signals can be absorbed and reflected; the higher the frequency, the greater the scattering effects, and the lower the frequency, the greater the range. Wave action, microorganisms, varying salt densities, and suspended solids further enhance signal scattering.

43.18   Water Temperature and Thermoclines. Water temperature affects fish, as they are cold-blooded animals and have the same temperature as the surrounding water. Due to biological factors, fish feeding and spawning behavior is dependent on water temperature. Fish are generally found at locations where the water temperature suits them. Any body of water consists of layers, with the surface generally warmer than the middle or bottom layers. The interface between areas of different temperature is called a thermocline. Thermoclines are important to locating fish, as they tend to be found either just above or below them. Fish-finders can detect thermoclines, and the greater the difference in temperatures the more visible a thermocline becomes.

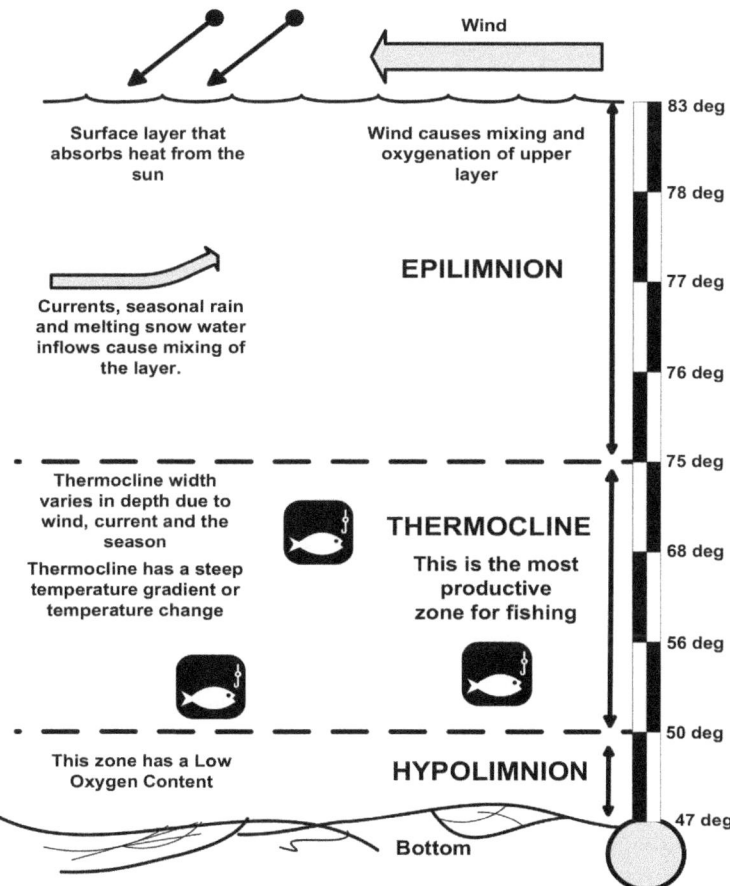

Figure 43-1. Thermoclines

43.19   Fish Arches. The display of fish arches on the screen is directly related to sonar acoustic characteristics. As a fish enters the acoustic cone, a display pixel is turned on, and as it moves toward the center of the cone, the distance between the transducer and fish decreases so that pixels are progressively turned on and display a shallower depth and therefore a stronger signal. When the fish reaches the cone center, this forms half the arch; the other half is completed as the fish moves toward the outer edge of the cone. Very small fish probably will not arch at all. Because of water conditions such as heavy surface clutter or thermoclines, the sensitivity sometimes cannot be turned up enough to get fish arches. For the best results, turn the sensitivity up as high as possible without getting too much noise on the screen. In medium to deep water, this method should work to display fish arches. If the fish does not pass through the cone center, the arch will either be partial or not be displayed. Arches are not formed in shallower waters, as the cone angle becomes too narrow. Arches are not formed when the boat is drifting or anchored. Fish schools vary in displayed shape, depending on how much of the school is within the cone. In deeper water, each fish, if large enough, may have an arch displayed. The size of fish arches depends on the sensitivity adjustments, boat speed, water depth and cone angle, and location of the fish within the cone.

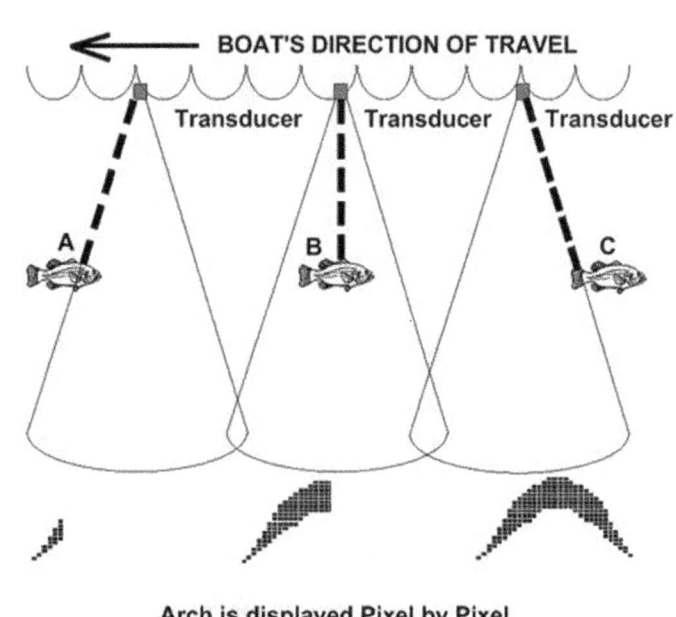

Figure 43-2. Fish Arches

Arch is displayed Pixel by Pixel
as fish passes through center of
sonar beam

43.20   Fish-Finder Installation. Be very careful not to bump the transducer and possibly damage the crystal element. Most installations are thru-hull mounted on a fairing block to ensure the beam is facing directly down on an even keel and to reduce any water flow turbulence. Locate the transducer in an area of minimal turbulence. Water bubbles from turbulence are a common cause of problems. In some situations they are mounted inside the hull within an oil bath or epoxy-fastened to the hull on FRP/GRP boats. There is a sacrifice in maximum depths, which can reach a 60% to 70% reduction in range and therefore should be avoided where possible.

a.   **Stern Mounting.** Many powerboats mount the transducer on a retractable bracket on the stern or transom. This arrangement, while effective and less work on smaller vessels, is not ideal on bigger ones, including sailing boats. Turbulence from the propeller and laminar water flow breaking away from the hull generally affect operation, and the transducer is only effective at very low speeds.

b.   **Cabling.** Always ensure that cables are installed clear of heavy current–carrying cables or radio aerial and antenna cables. Never install next to instrument cables as is often done; the interference problem can be significant.

43.21   Maintenance. Maintenance entails reading the manual and performing the tasks described or following recommendations. Perform the following maintenance checks.

a.   **Connections.** Check all connectors and connector pins for damage, and that they are straight and not bent. If straightening the pins, there is a risk of breakage, as they are brittle. Connectors not properly inserted or tightened up are prone to saltwater ingress and corrosion.

b.   **Cables.** Check all cables for damage, cuts, or fatigue. Transom-mounted transducer cables are prone to damage.

c.   **Power Supply.** Connection problems are the major cause of power supply issues, either at the supply panel or at the battery. Check the power at the plug using a multimeter, and a check should be made with engine on or off. If the engine voltmeter shows normal charge voltages and the battery checks out, the problem is usually in the intermediate connections.

d.   **Interference.** If the fish-finder has interference, turn off all other equipment and then turn the engine off. Progressively start up the engine and then other equipment to determine the source; the power supply may require suppression. Check that two fish-finders are not being run at the same time; also, two vessels in very close proximity may cause mutual interference if using similar acoustic frequencies. If the interference is present with all systems off, the fish-finder's automatic noise rejection facility maybe malfunctioning.

e.   **Transducer.** Inspect the transducer for damage, marine growth, and antifouling paints, and clean off the surfaces using soapy water. Do not use heavy abrasives or chisels to clean the faces.

43.22   Troubleshooting. Troubleshooting starts with reading the manual and determining whether settings and operating procedures are correct. Go into the settings or options menu and ensure settings are on auto or defaulting to factory settings. Many fish-finder problems are simply incorrect operation. Read the manual and practice using the fish-finder. Understand how to navigate through the various menus and options. If you have lost the user manual, you can get most off the manufacturer's website. Furuno, Lowrance, Raymarine, Garmin, Simrad, and Humminbird have most of their user manuals available for free download. The transducer is the only item that can be maintained, and if it isn't, it can dramatically reduce performance. Cleaning is essential, and regular removal of growth off the transducer should be undertaken. Do not bump it or apply any impact to the surface. Avoid applying antifouling paint to the transducer surface—it includes small voids and air bubbles that will reduce sensitivity. If absolutely necessary, smear on a very thin layer with your finger. Check out these troubleshooting hints when you have performance issues.

a. **Fish-finder Inoperative.** This is generally caused by loss of supply, so check supply switches and fuses. Check any supply circuit connections. Check for any corrosion at the in-line fuse, which causes high resistance, and at the power connection plug pins. The fuse may show as okay when testing, but when on load and the fish-finder is on, the high resistance creates a voltage drop. Check that the plug is properly inserted, and also check that pins are not damaged or broken. Check that supply cables are not damaged, and check for cable nicks or cuts. If all is found to be okay with power checked at the plug, the sounder switch is faulty. Check all connectors and connector pins for damage, and that they are straight and not bent. Use caution when straightening bent pins; they're brittle and break easily. Connectors not properly inserted or tightened are prone to saltwater ingress and corrosion.

b. **Poor Resolution or Range.** If bottom images are poorly defined or performance has degraded, or fish aren't being detected, check the transducer. If the transducer is angled and not directed straight down, this can reduce performance. Perform an interference check. Check that the water depth you are fishing in is within the sounder's nominal depth range. Check that the power supply is correct; lower voltages due to poor connections or low battery voltage can affect the unit. If the transducer is the inboard type, check that this has not deteriorated. Make sure some water surrounds the transducer to displace any air bubbles that may have formed.

43.23   Fish-Finding Satellite Services. Not exactly sailing related but important to know, and some of the information is useful to sailors like myself who like fishing. ROFFS and SiriusXM Marine Fish Mapping services are offered by subscription and provide very detailed oceanographic analyses for anglers and sailors. These companies provide detailed potential fish location services based on ocean circulation features. You can find out more about their impressive range of services at www.roffs.com and www.siriusxmcommunications.com. The services are summarized below.

a. **Sea Surface Temperatures.** Breaks in the seawater surface temperature are indicative of changes in normal conditions. This may be a 0.5- to 1-degree variation off the Florida Keys and a few degrees variation in the mid-Atlantic. When warm tropical water from the Gulf Stream, the Gulf of Mexico (GOM) Loop Current, or any other ocean eddy or current strikes a canyon wall, a hump, or any other subsurface geographical feature, the result is upwelling. These breaks denote the edges or boundaries of opposing or converging currents and rips. This cooler water then pushes up nutrients and creates a mini-ecosystem inhabited by everything from game fish to baitfish species.

b. **ROFFS Service.** The subscription ROFFS analysis incorporates an oceano-graphic map and an associated description. Data temperature breaks with the best prospect of fish inhabiting the area are identified. The data advises the time period when fish have been stationary over that bottom area. The eddies that rotate off the loop current in the Gulf of Mexico (GOM) and the mid-Atlantic Gulf Stream have a longer stationary period than similar conditions near Florida. The longer the temperature break remains stationary, the richer that ecosystem becomes. The ROFFS system updates every 90 minutes.

c. **SiriusXM Marine Fish Mapping.** This subscription service is able to input data and weather to the boat's multifunction display (MFD). Marine Fish Mapping categorizes temperature breaks as "SST Fronts" (sea surface temperature front strength) and "SST Contours." SST Fronts indicate higher temperature gradients; SST Contours track the temperature gradient. The fish-mapping feature updates every 3 hours.

d. **About Chlorophyll and Plankton.** Chlorophyll concentrations mirror the level of plankton in the water. These are major food sources for everything from whales, baitfish, squid, dolphins, porpoises, and pelagic species such as tuna, mahi-mahi, billfish, wahoo, and sharks. When high concentrations of chlorophyll and plankton arise, they give water an off-color appearance when compared to waters with low plankton levels. When a high chlorophyll-plankton concentration coincides with a temperature break, it has a multiplier effect with greater nutrient levels. The ROFFS chlorophyll zones are denoted with red for the highest levels to dark blue for low concentrations. The SiriusXM system lists the "Plankton Front Strength" and "Plankton Concentration Contours." These fronts denote the strongest plankton changes, and the contours data is related to plankton concentration tracking. This fish-mapping feature provides an update on a 24-hour cycle. ROFFS updates this data several times each day, with each satellite pass based on noon. If sailing, watch out for such events as a source for catching some lunch.

e. **About Weed Lines.** I have always kept an eye out when voyaging for telltale weed lines as a source for catching fish. ROFFS analyses of water temperature breaks often reveal weed lines. Like ROFFS, SiriusXM Marine Fish Mapping has a weed-line function. Satellite imaging identifies weed line locations and provides a 3-day series of images to track the movement.

f. **Thermoclines.** Thermoclines are subsurface convergence zones. Their depth is variable and they appear as horizontal lines on sonar. They are prime fishing areas for pelagic species. Fish Mapping has a feature called "30m (100ft) Subsurface Sea Temperatures" that provides readings at approximately 90ft (30m). Cold water welling up toward the surface has a cooling effect; some fish species don't like that, while others do.

g. **About Sea Height.** Sea surface heights can have areas that differ from surrounding waters. This can be attributable to strong upwelling, opposing countercurrents, rips, and eddies. These nutrient-rich areas are popular with various gamefish. ROFFS has an "Altimeter" feature; Fish Mapping has "Sea Surface Height Anomaly" feature.

43.24   Personal Bathymetric Generator (PBG). Advances in sonar technology have opened up other opportunities and innovations such as PBG. Multibeam sonar systems have evolved, and manufacturers such as Furuno, with their TZtouch3 and B165 series of sonar transducers and the DFF-3D sounder, typify these innovations. Every time a transducer emits an acoustic signal, it records the data points over a 120-degree-wide beam. This multibeam emission records fifty individual echoes or data points. This allows a permanent record of bottom mapping, also called swath mapping, to be taken in very high-resolution imaging in survey quality. This data topography includes structure, canyons, bottom data, ledges, wrecks, drop-offs, and depth contours, and this is recorded in 3D in color-rendered images that allow easy seabed interpretation. Furuno is recommending the use of a satellite compass with heave compensation when actively recording to increase accuracy.

Safety and Security Systems

44.1 **Inflammable Gas Detection.** Propane, or liquefied petroleum gas (LPG) is potentially lethal on a boat. If leaking propane gas (C_3H_8) accumulates in the bilges, once ignited it takes only a very small quantity of gas to explode and destroy a vessel. If LPG gas is installed, a quality gas detector is essential, preferably with a gas auto-stop solenoid linked to the detector. The Vetus GD1000 gas detector panel and sensor is able to monitor a diverse range of combustible gases, including include propane, butane, methane, and hydrogen. In addition, it will also detect the toxic and poisonous carbon monoxide. The operating principles are described below.

44.2 **Toxic Gas Detection.** The single sensor is able to detect flammable gases, including include bottled LPG or propane gas, and the poisonous carbon monoxide. If you do not have a carbon monoxide gas detector installed on your boat, go and buy one after you read this.

     a. **About Carbon Monoxide.** Carbon monoxide is an insidious killer that is colorless, odorless, and tasteless. Escaping gases from a faulty engine, generator exhaust, or head-installed water heater can kill you or all of your crew. In several countries, these heater types are illegal to install within enclosed spaces such as boats. Propane water heaters are potentially lethal, as are your LPG cooker and oven. Carbon monoxide poisoning is the leading cause of poisoning deaths in the United States, killing 300 to 500 people a year; 50,000 seek hospital emergency room assistance. The UK experiences around 120 deaths a year. In addition, there have been several yacht deaths caused by generators and heaters.

     b. **Toxic Effects.** Symptoms of carbon monoxide poisoning include nausea, vomiting, dull headache, dizziness, shortness of breath, confusion, agitation, and a raft of other signs. If you experience these symptoms while an engine or generator is running or you have a space heater, hot water heater, gas stove, or indoor charcoal grill running, stop them immediately. If you experience occasional symptoms, stop and do some serious troubleshooting to identify the cause.

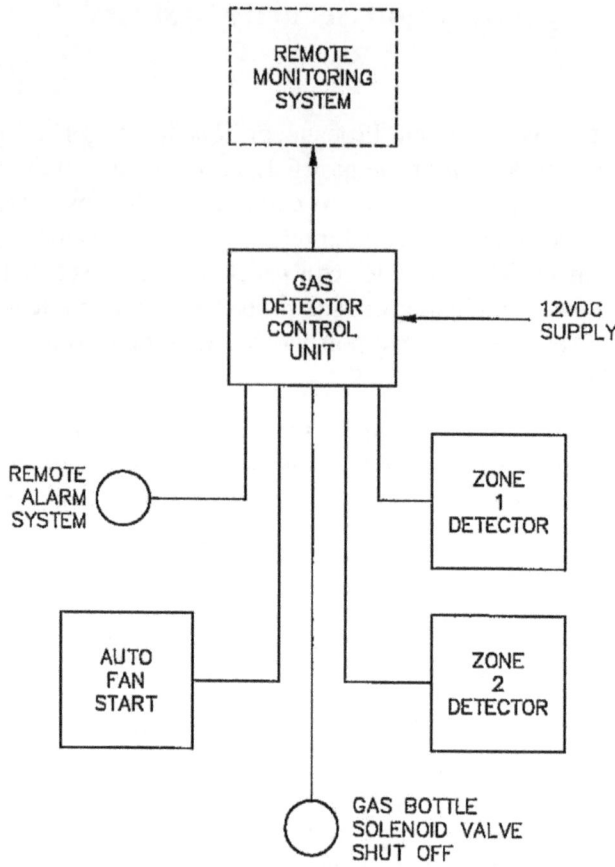

Figure 44-1. Gas Detection Systems

44.3 **Gas Detectors.** There are various flammable and toxic gas detectors on the market. These are described below, along with their operating principles.

a. **Theory.** All flammable gases have a lower explosion limit (LEL). As long as the gas-to-air ratio remains within this range, no explosion can occur. Once this level is exceeded, however, a significant explosion risk exists. A detector must indicate the presence of gas concentrations before the limit is exceeded, typically 50% of LEL. Better units have a sensitivity of 25% LEL. This applies to gasoline/petrol-powered engines.

b. **Detector Types.** Gas detectors are generally categorized by the type of gas they are designed to detect—either combustible or toxic. Further categorization is in the types of synonymous detector technology. Catalytic sensors are generally used to detect combustible gases. Electrochemical and metal-oxide semiconductor technologies are used to detect toxic gases:

(1) **Catalytic Bead Detector.** The detector uses a catalytic bead to oxidize the combustible gas. The resultant change in resistance is then converted into a sensor signal. A wire coil is coated with a

glass or ceramic catalyst material and then heated electrically to a temperature that allows catalyzation, or burning, of the gas. This releases heat and increases the wire temperature. The increase in the wire temperature alters the resistance; this is monitored by a Wheatstone bridge circuit, and the resultant measurement is converted to an electrical signal. A second sensor compensates for temperature, pressure, and humidity.

(2) **Metal Oxide Semiconductors (MOSs).** These are used to detect toxic gases such as carbon monoxide. They comprise a metal oxide that changes resistance in response to a gas, the change being measured and transformed into a measurement and concentration reading. The semiconducting metal (tin or tungsten) oxide is applied as a gas-sensitive film to a nonconducting substrate between two electrodes. The substrate is heated to a level at which the gas can create a reversible change in the semiconductor conductivity; the film produces free electrons that are able to flow uninhibited through the material and create current. If there is no gas present, oxygen is ionized onto the surface and the sensor becomes semi-conductive. When molecules of a gas, such as carbon monoxide are present, they displace the oxygen ions; this decreases the resistance between the electrodes and more current flows. This change is measured and the sensor resistance correlates and is proportional to the gas concentration.

(3) **Electromechanical Sensors.** These detectors measure a specific gas concentration, such as carbon monoxide, by oxidization or the reduction of the specific gas to an electrode; this generates a positive or negative current flow. An electrochemical detector has an ion conductor placed between a sensing electrode and a counter electrode. When a gas comes into contact with the sensing electrode, the gas oxidizes due to a chemical reaction with airborne water molecules. The reaction results in hydrogen proton flow through the ion conductor to the counter electrode, while electrons flow to it via a conductive path. The current is then measured to determine the toxic gas level and processed to create a measurement and alarm.

c. **Installation.** Sensor elements must be mounted in areas where gas may accumulate. The problem is that bilgewater or moist salt air can contaminate the element, causing degradation or failure.

d. **Testing.** Ideally, a precise gas-and-air mix of the appropriate LEL ratio would be used to calibrate the alarm level. In practice, however, this is never done. The simplest method to test whether the system is functioning is by activating a butane or disposable cigarette lighter at the sensor. Activation should be almost immediate.

e. **Alarm Outputs.** All flammable gas detectors should have a gas bottle sole-noid interlock that closes when gas is detected. This function should be fail-safe in operation. An external alarm or exhaust fan can be connected to the detector. Many integrated monitoring systems now have phone apps to receive alarms.

f. **Troubleshooting.** Note the following important factors:

(1) **Alarms.** If an alarm goes off, always assume it is real. If the alarm proves to be false or part of a sequence of false alerts, you can nor-mally readjust the alarm threshold. Do so only enough to compensate for the sensor drift causing the nuisance activation.

(2) **Sensor Element.** The principal cause of problems is a degraded sensor element. Carry a spare sensor for replacement if it is replaceable. If after replacing the sensor the alarm still causes problems, have the electronic unit tested. Sometimes it is easier and more economical to simply replace the unit.

44.4 Fire Detection Systems. Smaller yacht owners should invest in self-contained units that have an integral battery. Large yachts have a central control unit that processes sensor information and allows the setting of alarm thresholds and time delays that activate alarms. The various smoke types are different with respect to smoke particle sizes. Hot fires tend to produce very small, almost invisible particles; low-temperature, smoldering fires will produce larger, visible particles. Ion chamber detectors react quickly to small particles but are slower in detecting larger particles; the reverse is true for photoelectric detectors. Some detectors combine photoelectric and ionization sensing within the same unit.

a. **Photoelectric Sensors.** These detectors are ideal for low levels of smoke. They use the Rayleigh forward-scatter principle, which uses the scattering proper-ties of light from smoke particulates when they enter a light beam. The light sources use a narrow band gallium arsenide (GaAs) emitter and a silicon pho-todiode photodetector, with a lens installed in front of each. They are aligned so that the optical axes of each will cross in the center of the sampled volume. Baffles are installed within the narrow light beam so that no light reaches the detector. When smoke enters the chamber, some light will get scattered and reach the photodetector; the quantity of light at the detector is proportional to the smoke density. This is processed within an amplifier, and a 0 to 20mA analog signal is output to the control unit. Test response time is 6 to 22 seconds.

b. **Ionization Chamber Detectors.** These operate by the air within a chamber being ionized by a very small radioactive source of Americium-241. This allows a small current to flow between the source and a cover, which have a fixed voltage between them. The collector is a perforated electrode that has a nominal clean-air potential relative to the outer electrode. When combustion particulates enter the chamber, the collector potential increases, and the level of charge can then indicate smoke density. These units operate best with invisible smoke materials released by fast-burning fires. Test response time is 6 to 12 seconds. These detectors work well, as they are very sensitive to humidity and atmospheric pressure, along with airborne particles.

c.   **Heat Sensors.** There are two types of heat sensors. The first type activates when a set temperature is reached; the second activates based on the rate of temperature rise above a threshold level. Many units combine both functions. The heat sensor uses a bridge consisting of two matched thermistors, which are arranged to respond on absolute temperature and rate of temperature change and are fed to a differential amplifier. The thermistors are negative temperature coefficient types; one is exposed to air and the other is within the detector casing. The bridge voltage will track constant temperatures. When the temperature changes rapidly, the sense thermistor will be unable to follow and generates an analog output.

d.   **USCG Fire Extinguisher Update.** A new USCG regulation came into effect in 2022. The new regulation—Code of Federal Regulations (CFR) 33 Part 175, subpart E—is applicable to recreational boats less than 65 feet in length overall (LOA). The first important change is that no fire extinguisher can be older than 12 years. Check the date stamp on your extinguisher; if it is over 12 years or close, or there is no date stamp, you should replace the extinguisher. The second change is that all boats built from the year 2018 and after are required to carry fire extinguishers labeled "5-B," "10-B," or "20-B." Fire extinguishers labeled with "B-I" or "B-II" designations are not acceptable. Boats predating 2018 are allowed to carry extinguishers labeled "B-I" or "B-II," but that is conditional on being serviceable and not date-stamped greater than 12 years in age.

44.5   Security Systems. Trying to keep the villains off your boat is always a major undertaking. You can never keep out a determined thief, but my approach has always been to make the exercise as difficult as possible. The big trend today is wireless security systems, which means no wiring and even fewer trouble-causing connections. A variety of detectors can be coupled with control units and alarms. Some detectors combine IR and microwave detectors in one unit. Many providers supply remote sensing through the internet using an Internet Protocol (IP) address, mobile phone access to cameras, as well as alerts to your phone. One thing that needs to be checked is the electric power load on the system. There are two scenarios to consider: One is monitoring the boat when no one is on board; the other is when you are down below at night and want to monitor the deck for would-be thieves or others bent on causing harm. NMEA 2000 has changed the game, and you now can effectively monitor everything on your boat with up to thirty sensors and from literally anywhere on the planet. Systems now monitor and transmit alarms and status alerts with phone apps. They collect and transmit engine performance information and everything from tank levels to battery state of charge. Then we add the security part with constant video and camera surveillance. Actisense-I is one of these new-generation systems that offers systems diagnostics and total status monitoring. Check them out at https://actisense.com. Yacht Sentinel has the Sentinel Cam 2, which records HD video; the wireless camera is very economical on power, with a current draw of just 400mA. The CAM220 from Raymarine is a rugged high-definition video camera that employs infrared illumination for night vision. Power consumption is a very low 210mA to 370mA, which is dependent on IR operation. The OSCAR system has several variants; the Offshore 640 has a five-megapixel camera and provides eight times greater thermal resolution than the Flir M323. The system employs the latest generation of Nvidia graphics chips, and the company claims that it can see a small boat 1 mile (1.5km) in front and a man overboard (MOB) 150 yards (150m) away. This is the latest in imaging technology.

a.   **Ultrasonic Sensors.** Unsuited to vessel installation, ultrasonic sensors are easily set off by spurious signals and have a relatively high electrical power consumption.

b.   **Microwave Sensors.** These are often combined with passive infrared sensors (PIR) and use short K-band to reduce false-alarm rates.

c.   **Infrared (IR) Motion Detectors.** These dual-sensor units direct a pattern of infrared beams over a set area. When a heat source crosses a beam, the alarm is activated. Contrary to the theory that cats and other animals set them off, these detectors can be calibrated to react only to human-sized heat sources. One unit, properly located, can cover a typical belowdecks saloon or salon, but the installation site must be carefully selected so that it is not easily visible. These are now available in wireless models.

d.   **Magnetic Switches.** The most power-efficient security systems use magnetic switches on hatches and other access points. These are connected directly to the control unit. This system detects the thief before he or she enters the boat and alarms when the person is still on deck or in view. It is also fail-safe, so the alarm still activates if a sensor cable is cut. These are now available in wireless models. The installation must be concise, as rattling hatch boards can create false alarms.

e.   **Deck Pressure Pads.** Pressure-activated pads can be installed under carpets and mats. I have seen several installed under a mat in the cockpit.

f.   **Deck Sensors.** These are installed above and below decks and are effectively strain gauges that detect distortion when a person's weight is applied on the side decks or within the cockpit.

g.   **Photoelectric Beams.** These miniature beams activate an alarm when a beam is broken.

44.6   Security Alarm Indication Systems. Theft of boat contents, equipment, dinghies and outboard engines is rampant, and most boats employ traditional alarm systems. Once an intruder is detected, an alarm has to be activated to indicate their presence. The following alarm systems are recommended.

a.   **Strobe Light.** A high-intensity xenon strobe light mounted on the stern arch or mast is the most common indication method. Many install a blue light, but you simply cannot see it easily. That is why police vehicles worldwide now use a red/blue and white light combination.

b.   **Audible Alarm.** Install the highest output siren you can find. Put one outside in the cockpit and one below. A high-output unit in a cabin is very painful and will cut short any intruder's stay. A number of audible alarms may panic or disorient a thief. A system popular in the Caribbean uses several of the methods in this chapter, called "pirate lights." These have a long-range infrared sensor connected to a 130dB siren, which is the limit at which sound becomes painful and hazardous to the ear. The dual LED 5,000-lumen floodlights are enough to cause temporary blindness (even permanent, if the bad guy keeps looking at it). The system is controlled by a key fob unit. The aim is to make the intruder's presence untenable through light and sound. The remote function allows switching on spreader lights or any other lights from up to 3 miles away. The new V2 has bilge and smoke alarm interconnection. Check them out at www.piratelights.com.

c.   **Acoustic Barrier Alarms.** A viable method, these alarms emit a very loud noise that affects the inner ear and disorients the otolith organs and semicircular ear canal, causing dizziness, vertigo, and nausea. Check Global Ocean Security Technologies out at gostglobal.com.

d.   **Cloaking Systems.** The cloaking device will activate and flood the boat in a superthick, pea-soup type fog. This is created by vaporization of a specially formulated glycol solution. Visibility will decrease within seconds to just a couple inches, effectively blinding the thief. Pair it up with an acoustic barrier, and you limit the losses.

e.   **Interlocking Systems.** Connecting various systems to the alarm is another popular method. Spreader and foredeck spotlights, as well as any spotlight on stern post and arches, can be interlocked to come on with alarm activation.

f.   **Time Delays.** Entry and exit delays give you time to leave after you activate the alarm and time to disable the alarm when you return. I prefer to fit a remote isolator in a sail locker and have minimal delay. Generally, laws restrict alarm operation to 10 minutes; after that, the alarms must cease. Really ambitious thieves will set off the alarm and come back when the silence returns, so make sure yours resets automatically.

g.   **Remote Monitoring Systems.** This alarm method has transformed into something much greater and more comprehensive than simple back-to-base monitoring. Before internet and cell phone options and technology developed, these systems used to transmit a radio signal to a 24-hour monitoring station that could take corrective action. These systems can now monitor all vessel alarms, including bilge levels, smoke and fire, and gas, as well as security sensors and alarms. Systems include the Raymarine Yachtsense Link, a multipurpose marine mobile router that incorporates Raynet Ethernet ports, along with mobile broadband connectivity and onboard wireless. This allows connectivity of Axiom displays, mobile phones, tablets, and also PCs to a unified onboard network. The system is able to automatically switch between a marina wireless and mobile networks to maintain constant connectivity. Other systems include the Yacht Sentinel (yacht-sentinel.com); Garmin OnDeck (www.garmin.com); Nautic Alert X3 (nauticalert.com); and Blue Guard Innovations' Smart Skipper (www.bluebgi.com/smart-skipper). There are many other systems with similar remote monitoring capabilities.

h. **GPS Tracking.** You can install a GPS tracking device in your boat; if it detects any movement, you will receive an email alert. The C-POD Mini is a very effective tracking device. The tracker uses LTE communication and is user-friendly with Android and iOS apps; it is easy to hide and provides GPS positions four times per day and every 5 minutes when tracking. Powered by AAA batteries, it lasts about 6 months or a full season using patented technology. It incorporates a Guard Zone breach notification function. This activates within 30 minutes of breaching the guard zone, and tracking activation commences 10 minutes after the breach. It has integral GPS position monitoring using Google maps. Check out c-pod.com.

i. **CCTV and Video Surveillance.** These systems use mini-dome and ball cameras and have image backup for up to 3 months or more on a hard drive. You can log into an account and view your cameras from any device. Installing a dummy camera unit or two above decks just might deter would-be thieves.

j. **Phone apps.** There are a number of boat security phone apps. One popular app is the Siren 3 Pro app. It is able to monitor many critical boat parameters, including power status, shore power status, battery voltage, bilge pump activity, water levels, engine performance, temperature, location, and a lot more. The app is able to input both wired and wireless sensors and is NMEA 2000 compliant. Go to sirenmarine.com for more details. All the remote monitoring systems have associated phone apps as part of their access and monitoring functionality.

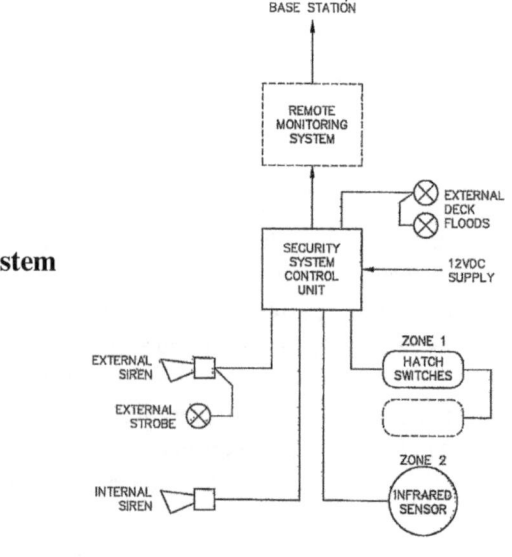

Figure 44-2. Security System

44.7   Collision Avoidance Camera. Camera technology has converged with artificial intelligence (AI). The SEA.AI system is able to detect floating objects using both high-resolution, low-light thermal and optical cameras to detect and identify other craft. It can detect floating debris, including shipping containers, navigation buoys, small boats, and a person overboard and also can detect wood, steel, organic, or plastic. The system uses powerful AI machine vision technology along with a database that contains more than 6 million annotated marine objects. It searches for anomalies and distinguishes water from non-water and compares these within a neural network. When the system detects a floating object, it processes and calculates the distance and its course; if a collision risk is calculated, the system sounds an alarm. The system's large database enables it to classify whether the detected object is a ship, yacht, shipping container, or 44-gallon drum. It has an integral inertial measurement unit (IMU), which allows calculation of direction and acceleration. The SEA.AI has a second reference point that is used for image stabilization and movement compensation through use of an AI-based horizon detection facility, which enables functioning in poor sea states. The SEA.AI unit should be mounted at the masthead so it can see over wave crests. This means no obstruction such as the tricolor and anchor light, wind instruments, or any antenna, and the minimum height should be 25ft (7m).

44.8   Marine Camera Systems. These are becoming very sophisticated and more common. The company FLIR has led the industry for many years, and FLIR now owns Raymarine. These camera systems are typified by the M300 series, which incorporates enhanced camera stabilization for optimum high-performance visible, thermal, and multispectral imaging. These systems are either single-sensor visible thermal models or dual-sensor multispectral. Each system has active gyrostabilization that copes with rough and choppy surface conditions and has a 360-degree panning capability. Thermal capability allows for imaging in darkness, glaring light, and light-mist conditions. The system has an integrated attitude heading reference sensor (AHRS) along with a two-axis mechanical stabilization system that eliminates pitch, heave, and yaw effects. It is also enhanced by active stabilization when integrated with electronic stabilization to reduce the image jittering that comes from engine rumble and boat vibration. Dual-sensor models employ the patented FLIR Color Thermal Vision (CTV) technology. This proprietary multispectral imaging technology for the FLIR M300 Series and FLIR Raymarine Axiom line of navigation displays integrates both thermal and high-definition visible color video. This enhances buoy, vessel, and other target identification at night. The M300 can be integrated with Raymarine's Axiom and ClearCruise Augmented Reality enhancements. Another impressive system is the Sionyx Nightwave D1 ultralow-light marine camera, which provides good imaging and can be combined with a chart plotter, AIS, and radar.

Troubleshooting and Maintenance

45.1 **Troubleshooting.** Troubleshooting has a clearly defined philosophy that should be understood and followed for successful outcomes. Troubleshooting is the logical process of evaluating a system, how it operates, and identifying the causes of low performance or failure. It involves the collection of evidence, such as visual burn or heat evidence, unusual sounds, acrid smells, temperature variations, abnormal vibrations, and other problems. This can be supported by the correct use of instruments and analyzing the data displayed on them, forming the basis for testing theories and assumptions so that the precise fault can be identified and subsequently rectified. I like this maxim that I came across recently; it should be adhered to as far as possible: Do not be neglectful, imprudent, irresponsibly inexpert, or incompetent on technical problems. The following factors must be considered in any troubleshooting exercise.

  a. **Systems Knowledge.** Understand the basic operations of the equipment. It is common to discover that "faults" are, in fact, improper operation of the equipment. If you have a basic understanding of the system, it is considerably easier to break down the system into functional blocks, which makes the process much simpler. Understanding how a system is designed to operate is rather different from how you think it should operate. A circuit diagram will show all the components in a system.

  b. **Systems Configuration.** Understand where all the system components are installed, where the connections and cables are, and where the supply voltages originate.

  c. **Systems Operation Parameters.** Understand what is "normal" during operation and the parameters or the operating range of the system. All too often, expectations are very different from the realities.

  d. **Test Equipment.** Understand how to use a multimeter. Be able to make the simple tests of voltage and continuity of conductors.

45.2 **Troubleshooting and Maintenance Safety.** When maintaining equipment or troubleshooting, the number-one priority is your safety and the use of safe work practices.

  a. **De-energize and Isolate.** De-energize all sources of stored energy—electrical, hydraulic, compressed air, equipment under tension, and any other source.

  b. **Risk Assessment.** Before you commence any maintenance tasks, please consider how to do so safely. Perform a simple risk assessment: What can go wrong? What are your risk factors? Consider boat movement, heat, accessibility, correct tools, skill sets required to do the task, and anything else that might affect your personal or a crew members safety. Perform a basic risk assessment prior to every job, and practice safety in every task. Be safe out there!

c. **Personal Protective Equipment (PPE).** Wear safety glasses or goggles when anywhere near pressurized systems, when blowing out dust, or when using hand drills.

d. **Personal Safety Factor Assessment.** Trips and falls are constant injury sources on deck and below, and foot trauma injuries are common, with many sources of lacerations. A surgeon friend once told me that shoulder rotator cuff injuries were very common and a major injury source on boats because when people fell, they put out one arm for support as they went down:

(1) **Hand Injuries.** Your hands are the most valuable tool on board, and fractures and severe trauma are common. I have experienced more than my fair share of hand scars. The fact is, most finger and hand injuries are avoidable. Sharp objects cause many injuries, so please be careful working with knives and cutting tools; one slip can sever a hand ligament. Tool injuries, such as whacking a finger with a hammer or slipping with a screwdriver and experiencing a puncture wound, are easy to do. Machinery injuries are common, so be careful where you insert your hand and fingers. The number-one exposure point is near engine drive belts, and the next is an operating windlass; many fingers are lost here. Be careful when tailing sheets onto a deck winch, as these have claimed many fingers.

(2) **Liquids and Chemical Exposure.** Be careful around hot and pressurized engine cooling systems and the galley stove as hot liquids can cause severe scalding and burns. Chemicals are a high-risk area. Be cautious when handling oils, acids, cleaners, and solvents; they can cause chemical burn damage to your hands.

(3) **Manual Lifting.** Gravity is a major injury source; be careful when handling heavy equipment or parts to avoid needlessly crushed fingers.

(4) **Electric Shock.** Electric shocks have the potential to burn your hands. The simple answer is to use gloves wherever you can when performing maintenance activities. You cannot sail your boat with a damaged hand, and the worst-case scenario is the need to call a USCG helicopter, a trip to the ER, or, even worse, expensive reconstructive surgery.

(5) **Biological factors.** Biological factors also injure hands, so be careful when dealing with toilet and holding tank work; exposed lacerations soon fester.

45.3 **Troubleshooting Procedure.** The following approach should be used when troubleshooting systems.

a. **System Inputs.** Verify that the system has the correct power input. Don't assume anything. For example, there may be a voltage input, but it may be too low. Check it with a multimeter.

b. **System Outputs.** Does the system have an output? Is the required voltage or signal being put out? If there is an input and no output, you have already isolated the main area of the problem.

c. **Fault Isolation.** In any troubleshooting exercise, split the system in two. This method is ideal when troubleshooting lighting and power circuits because it instantly isolates the problem into a specific, smaller area.

d. **Fault Complexity.** Most problems usually turn out to be rather simple, so start with the basics. Don't try to apply complex theoretical ideas you do not fully understand; the result is likely to be a lot of wasted time, as well as embarrassment. Stand back and think first.

e. **Failure Causes.** When a fault has been isolated and repaired, try to determine why the failure occurred, if possible. Context is everything, and problems often occur as a result of some other action or activity that often is not clear initially.

45.4   Troubleshooting Process Chart. The following process chart outlines a basic and logical fault-finding procedure. It is applicable to most equipment and systems on board.

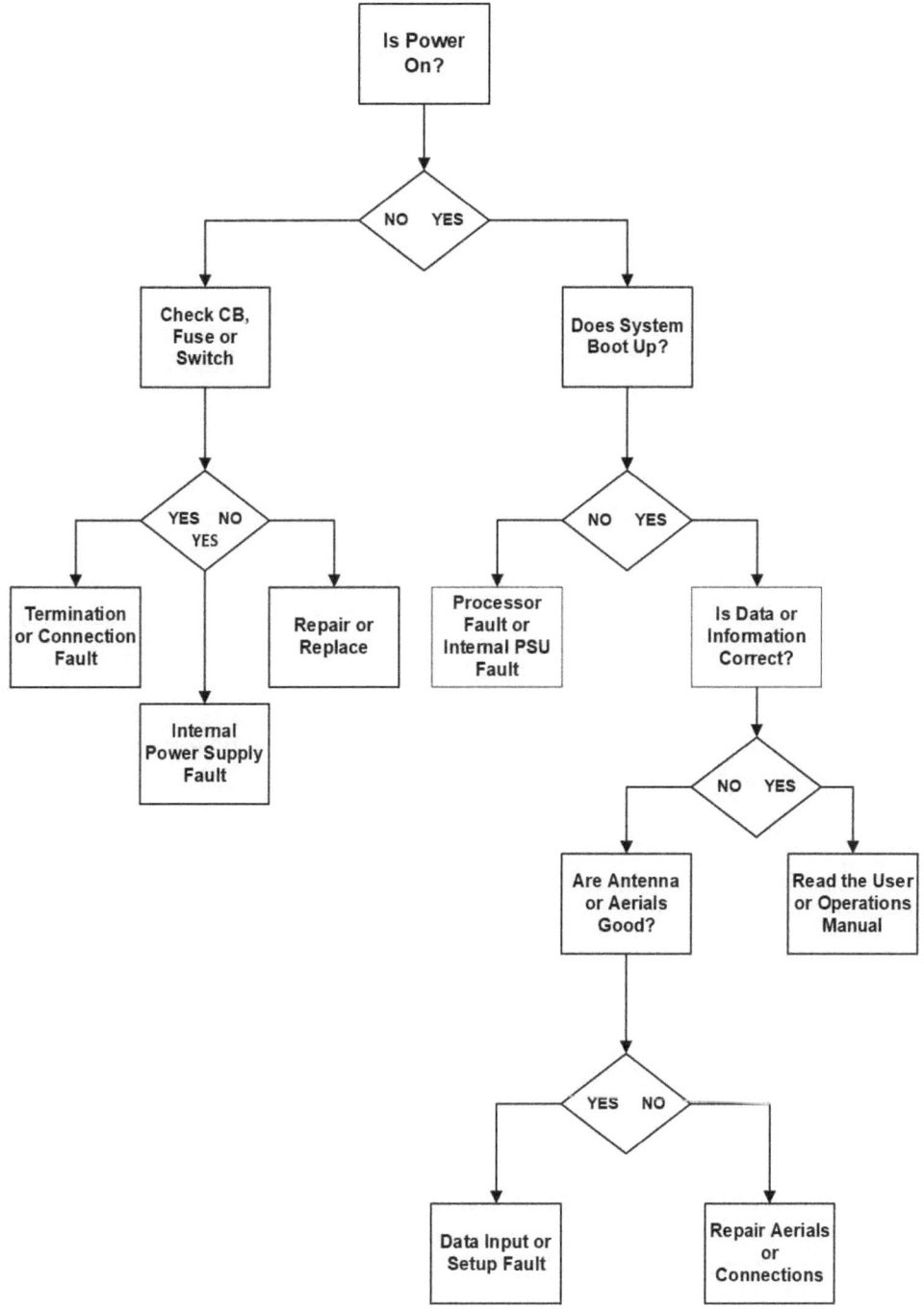

Figure 45-1. Troubleshooting Process Diagram

Failure Mode and Effects Analysis

46.1   Failure Mode and Effects Analysis (FMEA). I am actively involved in large ship and offshore oil rig systems, performing a Failure Mode and Effects Analysis (FMEA) and the subsequent testing to prove or disprove the assumptions. The FMEA and the Failure Mode and Effect Criticality Analysis (FMECA) are methodologies used to identify, assess, and address potential failure modes. The FMEA looks at single-point failures and hidden failure modes. This concept was developed by the US Armed Forces in the 1940s to classify failures according to mission success impacts and effects on personnel and equipment safety. The concept was behind the Apollo 11 moon landing in 1969, and NASA has adopted this methodology, as has much of industry, aviation, and the oil industry. I could never understand why this methodology was not being applied to smaller vessels such as sailing yachts and powerboats. I have adapted and scaled down some of these principles over the last 15 years.

46.2   Performing an FMEA. The FMEA exercise involves scoring the various attributes. Ranking severity (S) is the first metric and is for items and systems that require priority. That would include your engine, battery charging system, and mast and standing rigging. High severity scores effectively mean they are showstoppers if they fail. Low severity scores mean minimal effects on your activities. Detection (D) ranking is the probability of preventing a failure and intervention before the event. Low scores are for easily detectable conditions; high scores are for conditions that are hard to predict. Occurrence (O) ranks the probability of failure mode manifesting itself during the lifespan of a device or equipment. Low scores indicate failures are probably rare; high scores mean failures are highly probable. Each of the three factor areas assign a score between 1 and 10. One is for low probability and 10 is high probability. These three metrics are multiplied to derive a risk priority number (RPN) for each failure mode: RPN = Severity × Occurrence × Detection. Don't get bogged down with the number, as it's just an indicator. Failure modes are the various ways a system or equipment can stop functioning correctly or stop performing and functioning as designed. When looking at this, you need to brainstorm and try to determine all the ways and root causes that can bring your system down. A simple one is the seawater cooling inlet for your engine. If the pump impeller fails, the strainer might be blocked with jellyfish, the thru-hull inlet blocked with plastic, or the thru-hull seacock obstructed with debris. I have experienced all of these failure modes, which contributed to an overheating engine.

46.3   The Boat FMEA. The basic FMEA requires the analysis of boat engine starting and charging systems, as both are critical to propulsion or electric power. The exercise is to analyze the systems and identify single-point failures. In a single-engine boat, there is virtually no redundancy on any system. It is necessary to first identify all the points that, upon failure, will also fail the entire propulsion system, and then devise methods to improve redundancy. When assessing risk, it is important to consider the following factors and statistics: Approximately 80% of all electrical system circuit failures are due to faulty or failed connections. Approximately 70% of equipment and machinery failures are attributable to inadequate nonexistent, or improper maintenance. The following are examples of critical systems.

a.   **Battery Charging.** A failure in the alternator charging system results in no battery charging, no electrical power, and the eventual loss of all systems electrical power and main propulsion engine starting. See the battery charging section for an FMEA example.

b.   **Engine Starting.** A failure in the engine starting system means no propulsion, no charging of batteries, and eventual loss of all electrical power. See the diesel engine chapter for an FMEA example.

c.   **Engine Seawater Cooling System.** Any failure causes an overheating engine, possible loss of engine, no propulsion, and no battery charging.

d.   **Mast and Rigging.** This exercise can also be extended to the mast and standing rigging. If a corroded cap shroud fitting fails due to crevice corrosion, the rig fails. This results in loss of mast and sails in heavy weather, severely impacting boat and personal safety.

Multimeter Use

47.1   How to Use a Multimeter. As the name suggests, a multimeter is able to perform a range of electrical measurements, and the majority of tests can be carried out using a multimeter. There are two types of multimeters, analog and digital. An analog meter has a needle to display the readings. The digital multimeter (DMM) displays the test values numerically. Manual ranging meters require selection of measurement ranges, and auto-ranging types automatically select the best measurement range. Multimeters should be designed, constructed, and tested to international standard IEC 1010 and be verified as being tested by an independent laboratory. These standards ensure that a multimeter is tested to verify that its internal components are designed and constructed to protect the operator from hazards, including electric shock and burns, in specific high electrical risk environments. The measurement category rating of a multimeter will consider the working voltage and the maximum transient voltage that may be encountered. A momentary voltage spike or peak can have catastrophic results on an underrated multimeter. Observe and follow the AC electrical safety advice given in the AC Power Safety chapter, as there is an inherent risk of electric shock. When measuring AC circuits, even a small breakdown in a multimeter's internal insulation can initiate a substantial arc flash. Some countries have measurement categories. Cat 1 is for low-energy and low-voltage systems; Cat 2 is for domestic equipment with a potential fault current up to 5kA. Cat 3 includes wiring systems and wall sockets, outlets, and motors classified as medium-energy circuits with up to 25kA fault current capacity. Cat 4 covers measurements on sources of voltage and services classified as high-energy circuits with a potential fault current of greater than 25kA.

a.   **Voltage Test (AC and DC).** The volt is the unit of electrical pressure and is the force required to cause a current to flow against a resistance. The basic equation is $E = I \times R$ (see 47.1[d]). It is the most useful of all measurements, either to detect that voltage is present or to precisely measure voltage levels. I perform 95% of all my troubleshooting on complex oil rigs and commercial vessels with this function alone. The voltmeter is connected across the supply or equipment, which is negative probe to negative and positive to positive to measure the voltage potential between the two. Reversal of probes will simply show a negative reading. If the DMM is not auto-ranging, set the scale to the one that exceeds the expected or operating voltage of the circuit under test. To analyze results, consider the following:

(1)   **Voltage Is Absent.** Missing voltage indicates that the circuit supply is switched off, or that the circuit is possibly broken, such as a connection or wire (positive or negative) or a faulty switch or circuit breaker.

(2)   **Voltage Is Low.** Low voltage indicates that the supply voltage to a circuit from the battery is low, or that additional resistance is in the circuit, such as a faulty connection.

b. **Continuity Test.** The continuity test requires use of the "Resistance Ω" setting. It simply tests whether a circuit is closed or open. Many multimeters also incorporate a beeper to indicate a closed condition. Power must be switched off before testing. Touch the probes together to verify operation, and then place the probes on each wire of the circuit under test. What you are looking for is a simple overrange or OL reading if the circuit is open, and low or no resistance if it is closed.

c. **Resistance Test.** Resistance is resistance to the flow of electrons; the fundamental laws, formulated by Georg Ohm, are called Ohm's law. The ratio of a voltage through a conductor to a current flowing in it is constant, and is equal to the resistance of the conductor. The basic equation is $R = E/I$. If the DMM is not auto-ranging, set the range switch to the circuit under test; typically the 20Ω range is used. Turn off circuit power and discharge any capacitors. When testing, do not touch probes with fingers, as this may alter readings. Prior to testing, touch the probes together to see that the meter reads zero.

d. **Current Test.** Current is the rate of electron flow in a conductor in amperes. The basic equation is $I = E/R$, where "I" = current in amps, "E" = voltage, and "R" = resistance in ohms:

    (1) **Direct Current (DC).** This is the movement of electrons through a conductor in a single direction only. The ammeter function of a multimeter is rarely used or required, although some use it for measuring phantom leakage currents. The switchboard ammeter normally can be used for all measurements. The ammeter is always connected in series with a circuit, as it is a measurement of current passing through the circuit. The circuit should be switched off before inserting the ammeter into the circuit. Most DMMs have maximum DC measurement ratings of 10A only, and it is a little-used function.

    (2) **Alternating Current (AC).** This is the movement of electrons through a conductor in one direction, followed by a reversed movement in the other. At the start, the voltage is 0; at a quarter of the cycle, it reaches maximum; halfway through the cycle, it is 0; three-quarters though the cycle, it again reaches maximum; then, at completion of the cycle, it attains 0 again. The cycle is also referred to as a sinusoidal waveform. In normal practice, all references to voltage, current, and power values are root-mean-square (RMS) values. This is the maximum value multiplied by a constant, which in AC systems is 0.707. The frequency of AC is measured in cycles per second, and the measurement unit is the hertz (Hz). In most systems this is nominally 50Hz or 60Hz, although aircraft operate at 400Hz. AC current measurements are made using a clamp-on or clip-on ammeter.

e. **Power.** The watt is the unit of energy or power. The basic equation is $P = E \times I$. In AC systems, power factor is included so that $P = E \times I \cos \Phi$. Power can be tested using a wattmeter, as it is in domestic situations using watthour meters. Power meters may be installed on generator panels and display in kW. In most circumstances power is simply calculated using voltage and current readings.

f. **Capacitors.** When an insulator or electrolyte separates two metal plates, a potential difference exists. Excess electrons on the negative plate exert attraction on the positive plate when the potential difference is removed; this then reverses charge and then discharges. The unit of capacity is the farad (F) and microfarad (μF).

g. **Diode.** A diode is a semiconductor one-way valve consisting of an anode and a cathode. It allows electrons to flow one way only and has a high resistance the other way. To test a diode, where a diode test position is not included on the multimeter set, place a probe on each side of the diode. It will read low resistance in the forward direction and high resistance in the reverse, or blocking, direction. This is why they are often called blocking diodes. Diodes are used to form full wave bridge rectifiers that convert AC to DC. To test, each diode must be checked separately in the forward and reverse directions.

47.2   Multimeter Maintenance. Multimeter accidents are common, with many serious and some fatal. Look after your meter. Do not drop the meter or get it wet. Use it properly when testing live circuits. A few basics ensure reliability and safety.

a. **Meter Test Probes.** Ensure that probes and leads are in good condition. On many probes, the tips sometimes rotate out. A probe may come out and short across the terminals under test. Another problem is that the solder connections sometimes used in test leads break away due to twisting and movement.

b. **Meter Test Cables.** Cables should be kept clean and insulation undamaged. Cables can age and crack. If a cable is damaged, replace it. Do not attempt to test higher voltages, in particular AC voltages, if the cables are damaged. People have received severe shocks or been killed due to faulty leads.

c. **Meter Batteries.** Replace the internal battery every 12 months, or at least carry a spare. Many meters will have a low-battery warning function.

Boat Maintenance

48.1   Boat Maintenance. Boat maintenance is often considered an onerous chore, something one must begrudgingly undertake as the price for boating. Hourly rates can be quite steep for a marine engineer at the marina to do an oil change on your engine or other required maintenance tasks. Most of us can't really afford that, so some self-efficiency is useful. If you can afford to get people in, more power to you. Hopefully this section will enlighten you and stop the service people from smothering you with gobbledygook, as well as the common upselling attempts. You can mitigate the effects of maintenance or unpredictable failures—and understand them.

48.2   Maintenance Philosophies. In my professional career, I have been involved in performing and working up maintenance programs in commercial maritime and offshore oil rigs. Over the years, the philosophies behind this have changed or have partially merged. This has implications for how you maintain your boat, which is a valuable asset, so maintenance has personal safety implications.

a.   **Corrective or Reactive Maintenance.** Known as a reactive or "run to failure" strategy, it is based on the principle of allowing a device or equipment to fail and then be replaced or repaired. There is a dependence on the availability of spare parts or replacement equipment. If failure consequences have little or no impact on the environment or safety, this approach may be acceptable, but it is always about risk. The economics and time impact are low either before or after failure. One of the disadvantages is that failures tend to be unpredictable, as no condition monitoring is implemented. Surprise failures on a boat often tend to have consequential and cascading effects, and outages often have procurement impacts, with replacement parts requiring long lead times.

b.   **Preventive Maintenance.** Also known as preventative maintenance, its purpose is to perform maintenance tasks to reduce unexpected failures. This type of maintenance lies between corrective and predictive. Time-based maintenance is the most common methodology used. Tasks are performed at set intervals—weekly, every 6 months, or annually. For many years we had interval or time-based tasks on everything from diesel engines to deck winches and other auxiliaries. I am sure some of you are running spreadsheets to track your boat maintenance tasks. Usage-based maintenance is a subset of this type of maintenance and is based on hours run, distance traveled, or other parameters.

c.   **Predictive (Condition-Based) Maintenance.** Condition monitoring (CM) is based on continuous monitoring of motors, pumps, and other rotating machinery. Condition-based monitoring a boat owner can perform depends on the complexity of the boat and includes oil analysis, vibration analysis, and electrical testing, such as insulation resistance or megohmmeter testing.

The reality is that doing electric motor testing with an ohmmeter and insulation-resistance testing detects only around 10% of motor failures. Industrial sites tend to use Motor Circuit Analysis (MCA), which enables ground fault detection along with evaluation of terminations, cables, and internal stator winding faults, which enhances predictive maintenance, commissioning, and troubleshooting. These test units can be used on de-energized systems or on others that allow energized system testing with Electrical Signature Ansalysis (ESA), which looks at waveforms. Again, this is beyond the average boat owner, but it illustrates the extent technology has advanced in maintaining electrical equipment. Many other more-sophisticated condition-monitoring and testing techniques are well beyond those required for most boating. This strategy periodically measures or monitors one or more parameters and detects anomalies, significant changes, and various nonstandard operating conditions or out-of-parameter readings in machinery, including vibration, temperature, and other factors. This data can indicate impending failures that can become system-critical failures. The data is used to plan maintenance before failure. Given that faults often cascade and escalate, creating other system failures, predictive maintenance reduces reactive maintenance. Vibration monitoring and analysis was a cornerstone of this methodology; however, with technology advances, real-time monitoring using installed sensors is now becoming the norm.

d. **Reliability-Centered Maintenance (RCM).** This strategy is aimed at maximizing equipment life cycles and minimizing downtime as economically as possible or practicable. While a big-industry strategy, RCM can be reduced to elements that boat owners will or can use. The spread is less than 10% reactive maintenance, 25% to 35% preventive maintenance, and 45% to 55% predictive maintenance. RCM focuses on the use of technology for monitoring equipment, including vibration analysis, oil analysis, and more sophisticated methods such as infrared, acoustic (ultrasonics), and sound levels. The latter are normally beyond the average boater. The basic RCM paradigm is based on the proposition that the less maintenance you undertake, the better the outcome. Maintenance is performed only when absolutely necessary. RCM aims are the preservation of system functions; identification of failure modes that can affect system functionality; prioritization of failure modes in accordance with risk assessments; selection of the most effective tasks to control failure modes; assessment of the probable consequences of each failure; and, finally, the prediction and prevention of each specific failure. Complementary processes include the Failure Modes and Effects Analysis (FMEA), described earlier. There is also the Failure Mode and Effects and Criticality Analysis (FMECA), which analyzes causal links between failure modes, effects, and causes of failures. Other tools include hazard and operability studies (HAZOPs) and fault tree analyses (FTAs). The latter are used mainly at superyacht and commercial shipping levels.

48.3   Reliability-Centered Maintenance (RCM). Much research and analysis within the aviation industry has evolved maintenance into what is termed "reliability-centered maintenance" (RCM). A product of United Airlines and the US Department of Defense, it is now embraced by the US Navy and US Air Force. The philosophy is now standard with commercial aviation, mining, nuclear power, the offshore oil industry, health care, manufacturing, and transport. While this may appear as a big business system it can be scaled down and the core principles extracted for use on your own boat. Research revealed that for at least 70% of systems and equipment, time-based preventive maintenance was not making any real difference. The same equipment had an inherent probability of failure, so performing constant maintenance tasks, servicing, or replacement was actually pointless. All machinery will degrade over time, and some items degrade at a pace so slow that they are not of practical concern. The reality is that not all failures on your boat can be prevented by maintenance. Some equipment and system failures arise as a result of events and factors that are entirely outside our control. Sometimes they can be catastrophic, and I have used the analogy of a Black Swan event. This can be defined as a rare and unexpected or unpredictable phenomenon or high-impact event with a very low probability of occurring. The descriptors of freak and unprecedented are frequently and annoyingly used. While often linked to Wall Street and finance, the term is increasingly applicable to industry, systems, and equipment and certainly can apply to boats.

It doesn't matter how much maintenance you perform, the outcome cannot be changed, although the effects can be substantially mitigated. You can do all the maintenance in the world, but if the equipment is poor quality or the system has been poorly designed with inherent weaknesses, is poorly constructed, or has hidden failure points, you are probably on the losing end of the equation.

48.4   Yacht Maintenance. Properly planned and implemented maintenance uses the base assumption that not all failures can be prevented and, depending on the impacts or failure modes, are part of everyday life aboard. Research within the aviation industry and the US Navy came to the startling conclusion that between 70% and 90% of failure modes are not age related and that the likelihood of failure is random. It doesn't make much sense to use your time and money to maintain or replace equipment when the reliability hasn't actually been reduced, or where the nominated maintenance task will not improve the reliability. This introduces the concept of condition monitoring for up to 90% of equipment, with only a small percentage of time-based overhaul and replacement. Many boat owners actively consider the consequences of not performing maintenance and the possible impacts and consequences of allowing the failure mode to happen. Maintenance is about whether the consequences are severe enough or the outcomes tolerable. It must be noted that not all failures have the same probability, and not all failures have the same consequences. The things that have the highest propensity for failure, or where dominant failure modes are identified, get the priority. Low-frequency failures, even though they might have serious consequences, are not the priority. In other words, don't maintain things based on non-credible failure modes. You could go insane looking at every item on your boat and saying "What if?" this or that happened, even if the likelihood is extremely rare. Maintenance on your boat has to look at risk using this equation: Risk = Likelihood × Consequences. A typical boat example is looking at parts of the engine. How often does a drive belt fail? How often does an injector or fuel pump fail? How often does a water pump fail?

48.5   Systems Monitoring. Regular inspections can often indicate impending failure—for example, a noisy or warm to hot bearing; a fan belt with that burnt rubber odor from overheating. Heat, smell, noise, and vibration are great low-tech indicators. Most people react and troubleshoot when these signs appear and head off a failure. Many relatively simple bits of equipment wear out; once a certain age is reached, failure has a high probability. This brings in the philosophy of replacement before failure, or having a spare part on hand and allowing the failure as described above. This works for simple equipment, but when a system composed of several elements is considered, things change. Complex systems on a boat include an engine starting system or a charging system. Each element has its own failure mode, but when a system is considered, the various elements don't have a typical age-related failure pattern. One of the curses in boats is issues that are undetectable during normal operation; these are known as hidden failure points. Usually something fails after sitting unused and then switched on, also called failure on demand. Sometimes you stumble on a failure when testing a system or equipment. Typical examples are protective items like pressure switches. These protective function items don't get used until they have to work in a specific circumstance. Identifying hidden failure points is essential to reliability—some you can identify and some you cannot.

48.6   Reliability and Maintenance. You cannot maintain your way to reliability. No matter how much maintenance you perform, no matter how intrusive and rigorous you are, you are always going to be limited by the inherent design, material quality, and performance characteristics of the equipment. If it is poorly designed, the maintenance will not help much in reliability improvement. The fact is that boats are full of things like this. The only answer is when a piece of equipment fails and you know it's a design issue, try to eliminate it. That means choosing a replacement without the design deficiencies, or even reworking the overall system design at the same time. Most older boats that have been fortunate enough to have a series of hands-on owners will have gradually eliminated and worked out these design issues, usually as a result of frustration. You can learn to live with a deficiency and have it become the acceptable norm, or you can eliminate it. Don't waste your resources. Nearly everyone is on a budget, and you have to reserve valuable resources by not wasting money on needless tasks. Most people are time poor. If the maintenance on your boat does not add value and actually reduces overall reliability, why bother? If you don't really need to open up something and disturb everything, then don't; all you are doing is reducing reliability. Another important thing to remember is that much equipment is overrated for the job you are using it for. You do not have to over maintain to keep that level of performance. What you are seeking is reliable performance and functionality. To borrow that old maxim: "If it ain't broke, don't fix it!"

On the other hand, absolutely follow the manufacturer's maintenance recommendations. One example is a deck winch. The winch may sit with a cover on it for weather protection. During the sailing season, it gets used for a short summer cruise and several weekend trips and then is covered up again for the off-season. The question arises as to whether you dismantle the winch and perform maintenance. Intrusive and unnecessary maintenance generally leads to increases in failures by introducing human error into the equation. There might be mistakes in reassembly, use of defective parts, or lack of knowledge about lubrication, incorrect torquing or any other factor. Always review your maintenance regime on your boat, keep good records of your work, and be ready to adapt to changing circumstance.

48.7 **Flange Management.** I have lost count of the issues surrounding this, along with bolts and gaskets, both on commercial and smaller vessels, including cruising yachts, powerboats, superyachts, and trawler yachts. It is included here because I am frequently asked questions on the subject. If you have lots of hydraulic and fuel system or water system piping, then perhaps a refresher is required. Controlled bolting and flange management is another term for joint integrity and is important within any industrial or maritime installation. The performance of any pressurized, gasketed, and bolted flange joint relies on its ability to remain leak free through all operational ranges. If the joint does leak, you need to find the root cause and correct it. Don't let the leak turn into something normal and "standard"—something you just live with and tolerate. Flange management involves ensuring that the flange is acceptable without defects and is precisely aligned without flange stress. The flange must be parallel and axially aligned. Don't apply force to align flanges (I have seen people using chain blocks and more). The correct gasket type must be installed and aligned properly. The bolt must have the correct bolt preload and, along with nuts, be properly lubricated with the correct lubricant over the bolt thread length and nut face. So, job done and it still leaks? Depressurize the system and retorque again. If that doesn't cure the leak, disassemble and check the flange face quality for scores, dents, and scratches and for misalignment. Check the gaskets for damage and defects. Check the bolts for damaged threads. Bolts should have at least three threads showing outside the nut.

48.8 **Gaskets.** Gaskets should be of the required material and thickness and the correct size. Bolt holes should be just slightly larger than the bolt diameter. The inner diameter should not be smaller or larger than the flange inner diameter. Your spare ready-cut gaskets, whatever material they are made from, should have been stored properly—in a horizontal position and protected from heat and damage. They should never be bent, pinched, or crushed. If they reside in the bottom of a locker, perhaps rethink your storage options. Having damaged gaskets is the same as having no gaskets.

48.9 **About O-rings.** Like gaskets, O-rings are common on many systems and equipment, including linear pistons and rods, oscillating shafts, and rotating shaft seals. They can be found within pumps, cylinders, connectors, valves, and termination fittings in fluid systems such as hydraulics and pneumatics. O-rings are ring-shaped seals, essentially rubber sealing rings, and are placed within turned or milled grooves. When an O-ring is compressed, it is deformed and seals the gap between the two surfaces. The most common failure modes include abrasion, which occurs between the two surfaces and can result in lacerations and shearing. The fluids passing through the joints the O-rings protect can degrade them, causing them to harden and become fragile. Other signs of potential failure include cracking, brittleness, color changes, decreased flexibility, and blistering. O-rings are designed to tolerate high levels of thermal expansion, but if it's excessive, they can expand and fail. When thermal expansion occurs, the O-rings become out-of-round in shape. Other failure causes are incorrect selection and installation. The groove must be clean and lubricated correctly. Installation damage is often identifiable by visible cuts and notches. Degradation is also noticeable by ring swelling, softening, shrinking, and cracking. Eccentricity is caused by irregular clearance gaps. Either invest in an O-ring kit or carry spares.

48.10   Bolt Torquing. If flanges are aligned and gaskets inserted, it is time to torque up the fasteners—nuts and bolts. The correct bolt torque procedure must be used, as well as the use of a calibrated, or at least reliable, torque tool. Nuts or bolts should be tightened by hand initially using a cross-pattern sequence. The next step is to tighten to 30% using the same cross sequence; then to 60% using the same procedure. Finally torque at 100% using a cross-pattern sequence and then two passes in a clockwise sequence. You should be within plus or minus 5% of the required torque. Factors such as friction coefficients and nut or K-factors come into play, but they are beyond the scope of this book.

48.11   Nuts and Bolts. Are you installing the correct bolts for the job? Simple enough, but I have encountered so many issues with incorrect bolt grades, improper torquing, and other factors that a refresher is required and should be considered along with the previous chapter.

    a.   **Bolt Tightening.** Proper tensioning and clamping force are dependent on attaining and maintaining the correct tension level for each bolt. The flange pressure or clamping load must be sufficient to create and maintain a seal. The bolt must have the correct tensile strength values for the job. If you cannot determine the bolt's tensile value, usually denoted on the head, don't use it. When a bolt is tightened, it increases the tensile load or stretch until the required preload force is attained and overcomes opposing forces that are applied to the joint. Further tightening brings you to the proof load level, which is the maximum allowable within the safety margin. Further tightening brings you to the yield load point, where the bolt is stretched permanently and no elasticity remains. For most fasteners the preload is 75% to 85% of the proof load. The proof load is typically 90% to 95% of the yield strength and approximately 65% of the ultimate load. Bolt tensile strength properties are classified according to SAE grade designations and metric property classes. Where high-tensile bolts are used, do not reuse bolts, as they will have lost some of their properties. Nuts should have a specified proof load 20% greater than the bolts' ultimate strength rating. Always use the same material as the bolts for both nuts and washers. If reusing bolts, make sure they are brushed clean of corrosion or rust; and that nuts can be run down the threads by hand. A reminder of something we have all experienced: A stripped thread is where the thread has been damaged or has been removed as a result of cross threading or installing the wrong nut and incorrect threaded hole. This also can be caused by over-torquing; the threads cannot withstand the forces. Use caution when starting off; don't force a nut on if it is cocked and misaligned.

    b.   **Bolt Length.** How long should bolts be? The minimum theoretical length is based on the item to be bolted, along with the nut, washer, and bolt protrusion. The common rule is for three threads protruding from the nut, although two to five threads is the recommended range. If the bolt is flush with the nut, full engagement is not possible and a full thread cross section must be there. Excess protruding thread should be avoided; first, as it is an injury risk (been there and done that), and second, the excess exposed thread can be damaged making nut removal difficult, and it is just slower to fasten.

c.   **Locknuts.** Locknuts, also known as locking nuts and self-locking nuts, are an internally threaded fastener with the integral ability to lock and prevent loosening when subject to vibration or torque. Some use anti-seize or locking compounds. The most popular locknuts are those with nylon inserts, often referred to as a nyloc. They work by deformation of the nylon insert to prevent movement. The nylon insert creates tension or resistance between the nut and the threads. Repeated application and removal will decrease the effectiveness, and I only use nylocs twice. I have used them quite a bit, but they can come loose when the vibration occurs at the right resonant frequency. If you are having issues, you need to look at the vibration sources, including imbalances, misalignment, bent shafts, mechanical looseness, resonance, belts and pulleys, and bearings.

48.12   About Washers. Washers are a deceptively simple item; however, they have a distinct engineering purpose. The questions are frequent: What type do you use? What about spring washers? Do you put washers under bolt heads? The humble washer is used on almost everything, but often incorrectly. The washer has the prime purpose of distributing applied loads underneath bolt heads and nuts and does so by providing a larger surface area under stress. It is optional, but placing a washer under a bolt head helps prevent a bolt damaging or digging into the surface as it is torqued down. If you are inserting a washer plus a spring washer under the nut, ensure that you have a bolt long enough to include them and to allow a minimum of two or more exposed threads, as described above. The nut end installation is different and usually includes the common flat washer and the spring washer. On a boat a washer should always be made from 316 stainless steel. Never mix different bolt materials with different washers and nuts, which is more frequent than you might think. Washers reduce surface scratching, prevent the bolt from sinking, and evenly distribute force and pressure, as well as reduce heat and friction. Washers also prevent corrosion, maintain constant tension, and can serve as a spacer. Spring washers are a subset of flat washers and come in several forms. They have an integral role on the boat to maintain assembly tension, absorb shock loads, and, importantly, provide a controlled reaction for dynamic loads. Split-lock washers are the most common in use. As the bolt or nut is torqued up, the washer will flatten out, add tension, and prevent vibration-induced loosening. Each type of spring washer has specific advantages and benefits. Spring washer types include the single pattern, normal pattern, double pattern, and grip pattern, and all are subject to Hooke's law, or the law of elasticity.

48.13   Bolt Lubricants. Lubricants have friction coefficient values. The lower the value, the greater the energy transfer into the stretching of a bolt during torquing. If not lubricated, the torquing is transferred into overcoming bolt thread friction. There was a time when bolt lubricants were all about preventing seizure and easing disassembly; as a result, greases came to be simply called "anti-seize." Galling is a condition that arises on bolt threads and nuts; a form of surface damage that comes from two sliding metals, it is microscopic localized roughened surfaces and protrusions. This occurs when insufficient or no lubricant is used. There has been a big increase in the use of molybdenum disulfide lubricants, often called moly lubricants. Copper-based lubricants were the most common in prior years and still are in many areas, although there has been some transitioning to moly lubricants, which are environmentally friendlier and prevent galling. Some hints on applying thread lubricants: Always use clean hands or clean gloves, and use the same rules and precautions as bearing care. Don't contaminate the lubricant; it defeats the purpose if tiny particles of grit get mixed in with it. Take care not to contaminate the gasket or flange surfaces.

48.14   Vapor Phase Corrosion Protection. In winter, the moisture and salt-laden air, along with the constantly fluctuating humidity, create condensation within the boat electrical and electronics systems. It affects the rotating equipment such as engines, motors, and compressors, as well as hydraulic systems, bolts, valves, and a lot more. As a maintenance and anticorrosion measure, vapor-phase technology is hard to beat. It is used extensively within the offshore oil and gas and the maritime sectors. I have used the VpCI and MCI corrosion control technologies from Cortec for years. "VCI" is the generic term applied to volatile corrosion inhibitor, vapor corrosion inhibitor, or vapor phase corrosion inhibitor (VpCI). These products are specifically designed for corrosion protection, and while there are many products in the range, some are perfect for boat electrical and mechanical equipment. The technology works by forming a molecular barrier over metal components that separates them from oxygen and moisture, preventing commencement of the corrosion cycle. The most common protection for electrical and electronic cabinets is the insertion of a VpCI-101, VpCI-105, or VpCI-111 emitter. Close it up, stick on the label to show the date installed, and it will passively protect everything.

48.15   Lubrication and Zerk Fittings. Lubrication trivia time. A zerk is another name for a grease fitting or grease nipple. Invented by Oscar Zerk in 1929, it is also known as a grease zerk or an Alemite fitting. The zerk fitting has an internal spring-loaded ball that will compress when grease gun–induced pressure is applied and allows the one-way flow of grease through the channel down to the bearing. Clean the zerk before attaching your grease gun. When you finish greasing, clean off the nipple and then install a grease nipple cap or cover. These simple items are made from low density polyethylene (LDPE) in a range of colors. Nipple caps are designed to prevent the ingress of dirt, debris, contaminants, and salt water. If you finish greasing and leave grease residue on the nipple, the next time you apply grease, you force that contaminated grease into the bearing and risk causing damage.

48.16   Lubricants. Lubricants, both oil and grease, are designed to protect surfaces from degrading. Poor or no lubrication creates friction, and there is always some microscopic surface roughness, even in the best-machined surfaces. These various high points are known as asperities, and when they rub against each other they create abrasion and adhesion and assist in shedding metal particles and other residues. Lubricants reduce friction and so control wear, and they also have a function of corrosion protection, temperature, and contaminant control. Greases and oils separate various moving parts with a boundary film. Created by certain lubricant additives, this film formation is crucial to performance. These film-creating additives include molybdenum disulfide and graphite, which adhere to mating surfaces. Most common industrial lubricants are mineral, synthetic, or vegetable based, and the base oil has a thickening agent to form grease. These can be simple lithium soaps, lithium complex, or polyurea. The majority of performance-enhancing additives used in lubricating oils are part of grease formulations. A typical bearing grease will contain antioxidants, antifoam agents, anti-wear agents, and rust inhibitors. "Viscosity" is a term applied to a lubricant's internal flow resistance. The higher the viscosity, the less flow; the reverse for lower viscosity. Greases are rated for the temperature levels they can tolerate before becoming viscous. This is important in many applications, including bearings. *Caution:* Never mix two types of grease within a bearing.

48.17   Overgreasing. One statistic finds that 35% of all bearing failures are attributable to incorrect lubrication. The factors include overgreasing, undergreasing, the incorrect grease for the application, or a combination of all these. Overlubrication of rolling element bearings is one of the most common mistakes made with machinery and electric motors. It is stated that 40% of bearings never reach their engineered life cycle. This is a constant problem on commercial maritime, offshore oil industry, and pleasure vessels, as well as ashore in industry. Excessive grease leads to overheated bearings and a consequential reduction in service life. Excess grease insertion builds up pressure and pushes the rolling elements through the fluid film and up against the outer race. This means the bearing has to work harder to push the rolling elements through a mud bog of grease. The increased pressure and friction result in a temperature increase. Excess grease within the bearing cavity causes the balls or rollers to slide instead of turning and pushing and churning the grease out of the way. This results in decreased lubricant performance where the oil and grease thickener separate; this is known as oil bleed. When heat is coupled with oil bleeding along with time, this eventually cooks the grease thickener into a hard and crusty accumulation that will impair efficient lubrication performance and block any new grease. The general maxim is that for every 18°F (10°C) temperature rise above 150°F (65°C), the bearing life is reduced by 50%. The converse is true; under-greasing, either quantity or time interval, has adverse impacts.

48.18   Grease Application. The typical manual grease gun is able to develop 15,000psi of pressure when pumping in grease. Pumping a bearing with grease often damages the bearing seals and forces grease into parts of the bearing cavity where no grease is required. How much grease to use? Good question. First look at the equipment manual for guidance. If no information is provided, the commonly accepted formula is $G = 0.114 \times D \times B$. Take the outside diameter, "D," of the bearing in inches and multiply it by "B," which is the width in inches, then multiply this number by a factor of 0.114. This results in "G," which is the number of ounces of grease required. For example, a 3-inch outside diameter bearing with a ½-inch width requires 0.17 ounces, or 4.8 grams of grease. The math is $3 \times 0.5 \times 0.114 = 0.17$. When you grease and some of the old grease is forced out of a plug, any observed discoloration can be attributed to excessive heat, oxidation, or contamination. Slowly purge the old grease until the new grease is visible. Apply the grease only with the device running so that new grease is evenly distributed, and to avoid overpressurization of the seals. Observe grease application as outlined under the grease nipple section. How to avoid these problems? Understand the preventive maintenance requirements of each piece of equipment, and verify the correct amounts and frequency of lubrication.

48.19   Lip Seals. Lip seals are often treated badly, and many are unaware of the important and vital role they play on rotating machines. Lip seals can rupture when overgreasing or overpressuring and facilitate contaminate ingress such as dust and water into the bearing housing. Lip seals usually fail at around 500psi, so it is easy to destroy them with a grease pump. They are designed to keep grease in and contaminants out. Lip seals work by maintaining friction, and to be effective they must have proper contact with the rotating outer race of the bearing. Most lip seals are made from nitrile, which has a very broad temperature range. Most are damaged when removing them or installing them. Check that the elastomer is free of scratches, nicks, and cuts or any other visible damage. Inspect the case for damage, including dents and abrasions. If a garter spring is used, check that it is installed correctly. Clean the seal and shaft and ensure that it is truly clean; debris can get trapped and score the shaft. Make sure the assembly lube is compatible with the bearing grease being used. Make sure the elastomer and the case are of the correct specification. Check that the lip seal is installed and pointing in the correct direction. Inspect that the shaft is free of any defects, such as grooving and scratches. Check that the bore is defect free, with no debris, and check the roundness. When installing lip seals, make sure you apply pressure evenly when installing and that they are not cocked.

48.20   Bearing Basics. The bearing is in everything that rotates, and it is crucial in keeping the world's wheels turning. The common roller bearing is used to support and rotate with minimal friction rotating and oscillating machine elements such as shafts, axles, and wheels and the transfer loads between components. Bearings are high-precision components that enable high-speed rotation applications that also minimize heat, friction, energy loss, and wear. The two types in use are the ball bearing and the roller bearing.

   a.   **Bearing Parts.** The standard roller bearing comprises the following elements: the inner ring, the outer ring, the balls or rollers, and the cage. Bearings may have a seal or metallic shield on one or both sides, and bearings that are copped on both sides are factory pre-lubricated with grease; the code "ZZ" or "2Z" is used to denote this. Rings are manufactured from hardened steel to cope with fatigue caused by cyclic overrolling and the pressure that exists at the rolling contact area. The rolling elements, which are the balls or rollers, transfer the load between the inner and outer rings. While the same grade of steel is used for bearing rings and rolling elements, they are now being made from ceramic materials. These are referred to as hybrid bearings, and the rolling elements are made from bearing-grade silicon nitride (Si_3N_4), which makes the bearings electrically insulative.

   b.   **Bearing Cages.** The cage is used to separate the rolling elements to reduce the heat generated by friction within the bearing. It maintains precise rolling element separation and spacing for optimum load distribution. Cages are centered on the rolling elements, and this allows easy grease penetration. Cage types include the stamped metal and machined metal cages made from brass or light alloy that allow high speeds and temperatures, accelerations, and vibrations. There are also polymer cages, which are manufactured from polyamide 66 (PA66) or other polymer materials. Polymer cages have low friction and allow higher speed operation. There is a normally a relatively small gap between the inner ring and the shield. Bearings are used with shields when the operating conditions are clean. Where contamination risks are moderate, seals are used. Contact seals are used, and they contact the sliding surface of a bearing ring. Bearings are installed on a shaft or within a housing as an interference fit. The expansion of the inner or outer ring compression reduces the internal clearance. As bearings create heat when rotating, the differential expansion of the bearing along with mating components affect the clearances. The deep groove ball bearing is the most common bearing in use and consists of a row of ball bearings as the rolling element. These are enclosed and trapped between two annulus-shaped metal pieces known as races. The inner race rotates freely; the outer race is stationary.

48.21   About Vibration and Monitoring. Monitoring vibration is a core activity in condition monitoring and predictive maintenance. Vibration analysis devices are used to determine bearing inner and outer race defects, and to monitor ball, roller defect, and cage defects. Each of these conditions is characterized by harmonic peaks at the appropriate failing frequency. Excessive internal bearing clearances can be identified, as well as looseness between a bearing and the shaft or casing. Spectral signatures are characterized by vibration at various harmonics of the rotating frequency. Inadequate lubrication is characterized by high-frequency vibration between 1kHz and 20kHz. The level of oscillation in a vibrating object is known as the displacement. When an outer bearing ring vibrates, the outer surface will move upward to the upper limit, then down to the lower limit, and then return to the start point. The measurement between upper and lower limits is known as the peak-to-peak displacement. The entire oscillation movement from the start point through the upper and lower limits and back to the start point is called a cycle. This vibration cycle will continually repeat if the bearing maintains rotation. When these cycles are referenced to a time, a frequency is derived and expressed as hertz (Hz). When determining the origin of vibration, it is difficult to precisely pinpoint. Sources are diverse and include manufacturing tolerances, clearances, out of balance forces, and moving part friction. Vibration increases fatigue and shortens bearing longevity. Measurements of vibration velocity are made. and this is displacement times frequency, which indicates the vibration severity. The greater the velocity measurement, the more noise a bearing emits. While vibration velocity indicates the potential for bearing fatigue, vibration force results in deformation of the balls and the rings. To accurately assess condition, vibration acceleration is measured. Vibration acceleration is an indicator of force, where Force = Mass × Acceleration. This is measured in G (9.81 m/s^2), and these are often converted to decibels (dBs).

48.22   Bearing Failures. Removing bearings and pulleys is a task that requires both care and skill. More damage is done to motors and machinery because of improperly installed bearings than nearly any other cause. It is said that 30% of bearings are damaged prior to installation. This is due to storage in damp surroundings and contamination caused by unwrapping to check bearing numbers. A good bearing puller set is very useful, and I have a very good Proto set that caters to most pulling tasks. The most common causes of bearing failures are as follows:

a.   **Bearing Contamination.** The majority of bearing failures, estimated at over 90%, are caused by contamination introduced at installation. It is essential that bearings are handled using clean hands and tools. If you haven't got new gloves, use very clean hands. Don't unwrap the bearing until you are ready to install it. When regreasing using a grease gun, clean the grease nipple properly first so that dirt is not pumped in with the new grease. See the chapter on lubricants and overgreasing.

b.   **Bearing Mounting and Installation.** Improper mounting is the other major cause of premature bearing failures. Bearings are normally installed with a press fit on the shaft and the outer ring being either a press or interference fit. If a bearing is mounted using hammer blows or pressure to the outer race, it will cause dents or a true brinell. A loose fit on the shaft or housing will cause rotation and generate heat and metal particles that will also cause damage. Where the fit is too tight, the rings may be stressed, causing cracking and internal preloading, causing deformation leading to high temperatures and failures. The most common installation method on smaller bearings is the use of a tubular dolly that matches the inner ring. This is tapped using a metal (not timber) mallet. Where heating is used, this typically raises the bearing temperature to 158°F (70°C) using an oil bath or oven, then pressed on using a dolly. Motor end shields may be heated to around 104°F (40°C). Be aware that running a motor with roller bearings and no radial force applied to the shaft may damage bearings. Motors with angular contact bearings should not be run without axial force applied in the right direction to the shaft. Make sure that bearing rings are properly aligned onto the shaft. Cocking and forcing will result in distortion and damage. Never apply force to the rolling elements. Do not strike the outer ring as a way of forcing the inner ring. If an interference fit is required, use an oil bath to heat the bearing. Do not apply force with a mallet, hammer, screwdriver, drift, or anything else, and do not apply any sharp impact blows. Dirt and dust contamination contribute to increases in bearing vibration and noise levels. Improperly installed bearings or mishandling is often a cause, as described elsewhere. Shock loads on bearings also cause bearing damage to raceways.

c.   **Bearing Misalignment.** The main causes are due to bent shafts, out-of-square shaft shoulders, clamping nuts, and spacers, which lead to overheating and failure of separators.

d.   **Bearing Lubrication.** The purpose of bearing lubrication is to minimize friction at the various contact points within the bearing, protection of the finished surfaces from corrosion, dissipation of heat generated within the bearing, and protection from particles and dirt. The choice of lubricant depends on the bearing temperature, the size of the bearing, the operational speed, the load, the service conditions, the method of relubrication, and the method of sealing. Bearings with the notation "Z," "ZZ," or "2Z" are sealed and cannot be greased. If a greaseable bearing is removed and cleaned, the most common cleaning solvent is isopropyl alcohol for washing out the bearings Repack to only 50% with new grease. (See the previous chapters.)

e.   **Bearing Electrical Damage.** Electric currents can pass through motor shafts and the bearings, causing arcing at the contact points between balls and races. This leads to pitting and cratering, resulting in spalling and vibration with failure. The typical maximum voltage drop across a bearing is 0.4V.

f. **Bearing High Temperatures.** If motors are run within nominal ambient temperatures of 77°F (25°C), the operating life of bearings is typically around 16,000 to 26,000 hours for two- and four-pole machines. Vertical machines should have this reduced by 50%. Where temperatures increase, the heat will transfer from the shaft to the bearing and cause the grease to break down. The grease may liquefy and bleed off or cause oxidation and carbon formation, which may cause the balls to jam. Heating alters the metal temper characteristic and causes reductions in metal hardness, as well as changes to internal clearances and preloading.

g. **Bearing Fatigue.** The metal components in the bearing can fatigue due to the rotational forces and dynamic impacts between balls and races. The constant flexing and loading of the components eventually will cause flaking of the metal and wear, with bearing failure.

h. **Bearing Corrosion.** This is caused by moisture, acids, lubricant breakdowns, condensation, and water ingress. Corrosion of the finished surfaces creates abrasive particles that cause excess wear, pitting, and vibration. The particles are often absorbed into the grease, and the wear rate is increased.

i. **Bearing Vibration Brinell.** Motors that are not turned over regularly can suffer bearing damage called the vibration or false brinell effect. This is caused by the rapid movement of balls within the bearing race. As there is no rotation, the balls are not properly lubricated, causing wear and indentation at the metal contact points. If a motor is stationary for much of the time, the bearings should be maintained by turning the shaft by hand every month.

j. **Bearing Noises.** You can check the operation of bearings by creating a stethoscope from a large screwdriver. Place the tip on the bearing housing and then place the handle to your ear. The vibration will pass up the screwdriver shaft and denote the following:

(1) **Low noise or rumble sound.** The bearing is damaged due to poor installation.

(2) **Irregular rasping sound.** Balls and races are damaged.

(3) **High-pitched shrill sound.** Clearance problems or the bearing is running dry.

(4) **Intermittent sounds.** Bearing grease is contaminated with dirt.

(5) **Abnormal temperature rise.** There is an overload or lubrication failure.

(6) **Subdued humming sound.** The bearing is probably good.

Winterization

49.1 Winterization. Winter is coming, as John Snow says in *Game of Thrones*, and winter is as savage as the Wildlings. Winterization is preventive maintenance with an element of predictive maintenance. Having endured several harsh European winters, been iced and snowed in, and left my boat to escape to sunnier climes, I have learned that you need to be very thorough if you want a pain-free spring recommissioning and start-up. The extent of winterization depends on the expected environmental impacts, the length of the layup, whether you attend and run things over in the winter or not, and how big a boat and how many systems you have. If you are hauled out into a boatyard, the same rules mostly apply. This is my take on the minimum requirements to attend to, from many lived experiences; anything extra you consider is a bonus and will save you a lot of money. The following are more general tasks, and are from lessons learned over many years and learning by mistakes. Some are basic, but when you are busy shutting down, things get forgotten.

a. **Dehumidifiers.** My single best investment was a portable dehumidifier, which I leave running all winter. This uses a refrigerant system, and after 20 years it is still performing well. Caframo makes a unit called Stor-dry that warms damp air with an integral heating element above the dew point and circulates the air using a fan to reduce humidity. This reduces condensation and inhibits mold and mildew formation. The 230V and 120VAC units have a 70W power consumption.

b. **Stowage.** Move all life buoys, rescue gear, and other safety equipment down below; leave nothing on deck. Always check to see that your halyards won't wake the neighborhood when banging against the mast. Remove any sheets, both headsail and main, and stow below. While you are at it, put all your bedding and pillows into plastic bags. Books and paper charts soon absorb moisture; if they are valuable pilot books and guides, wrap them in plastic and store somewhere dry.

c. **Movement Prevention.** Tie off the boom to prevent movement; constant swinging doesn't help. If you don't use those neat canvas winch covers from Westmarine, wrap your winches in plastic. Lubricate the steering linkages; many prefer to move the wheel or tiller hard over and then lash to prevent water- and rudder-induced movement.

d. **Isolation.** Isolate the LPG gas bottle, and burn off the remaining gas on the stovetop. Verify that all isolation valves are closed. When you remove the sheets from the furler, make sure the furled sail cannot unfurl in the wind; wrap it up very tight. If you can take your gennaker or spinnaker home, do so; store in a dry place, as mildew and mold soon creep in during winter. If you are spending the winter in the water, make sure you are tied up correctly with properly tied-off lines and springs to prevent excess movement.

e. **Fenders.** Make sure your fenders are installed correctly; if they aren't, your hull may be damaged as it grinds against the pontoon.

f. **Moisture Prevention.** Install a closet desiccant module in your clothing and wet weather closets and storage area to absorb any moisture—it really works! It sure beats moldy and mildewed clothing and foul weather gear. I install these in almost every place on my boat.

g. **Refrigerator.** Empty the refrigerator icebox; clean and deodorize, and leave the lid open. If you have a water-cooled unit, drain down the water line and close the seacock if separate. This is a good opportunity to replace the anode if installed.

h. **Perishables.** On the food question, it is best to take it all out and either dispose of it or use it; some canned food can freeze solid and burst.

49.2   Water System Winterization. Drain down the water tank; clean and remove any accumulated sediment. Refill the tank and dissolve several water sanitizer tablets in it. Operate the water pump and pass the treated water through all outlets, shower, and galley; then hold for 12 hours to kill all bacteria. Empty the tank and drain all water from the water piping and pump so that none remains; open and drain water filters and strainers. Leave all valves and faucets open. Open the circuit breaker to isolate the water pump circuit. If you have an engine-heated calorifier or water heater tank, drain the tank. If you have an accumulator in the system or water filter, make sure you drain this as well. Some people like to use nontoxic propylene glycol antifreeze in the water system, but I prefer not to do this. Consider the following systems for winterization.

a. **Shower System.** Operate the drain pump and verify that all water has gone. Dry out the shower pump well, and clean any accumulated soap, hair, and other matter. If a separate pump float switch is installed, operate by hand to ensure it is clear of debris. Open the circuit breaker to isolate the shower pump circuit. Wrap the toilet paper in plastic wrap so you can actually use it when you come back.

b. **Toilet System.** Empty the holding tank; flush and deodorize same. Flush through the toilet and pump dry. Follow the manufacturer's instructions for your system. Close the overboard discharge valve and seawater inlet valve. Exercise them by operating several times, then lubricate. Toilet seals need lubrication, and you can buy additives to flush through. Open the circuit breaker to isolate the toilet circuit. I have been flushing some vinegar down the toilet every month. Adding some mineral oil or Head Lube from Westmarine before shutdown is also common, and don't forget to grease the piston rod. Consider using vapor phase protection such as Cortec ECO-SEPT or PORTA-TREAT 10X in the system.

c. **Bilge System.** Check each bilge pump and the associated bilge. Run the pump to remove any water. Check the float switch and verify that it is free and not frozen, stiff, or covered in debris. Close the overboard discharge valve and exercise several times; lubricate if necessary. If you are leaving the bilge pump energized and in auto mode, verify the float switch operation and do not close the overboard valve.

d. **Washdown Pumps.** Run the pump and ensure that the system is dry and the water within the line is drained out; close the seawater suction valve if required.

49.3   Engine Systems Winterization. There are several elements to winterizing an engine and its various subsystems. See the following options.

a. **Engine Freshwater System.** Replace the glycol in the cooling system or verify its quality. If it is old, it will have degraded, so consider changing it; it is cheap insurance. Consider adding vapor phase Cortec M-640L to the cooling system.

b. **Engine Seawater System.** Check the raw water strainer and clean out any debris.

Once you have completed the final engine shutdown run, drain the suction line of all water. Open and empty or drain the strainer. Close the seawater suction seacock. Exercise by operating several times, then lubricate and close same. Open the seawater pump cover and drain any water; leave open and coat the impeller with pump grease.

c. **Engine Lube Oil System.** Warm the engine and drain all the oil. Remove and replace the lube oil filter. Fill with new lube oil; depending on where you are, use the correct grade of oil for your winter conditions. Run the engine and allow oil to circulate through the engine properly. If a crankcase breather is installed, wrap this in plastic to prevent ingress of moist air. By the way, you should always change the oil again at spring commissioning. Consider adding Cortec M-531 to the system.

d. **Engine Fuel System.** Before winter always try to run down the fuel tank to a minimum level. There are two philosophies: full tank or empty tank. The full tank is with fresh fuel and added stabilizer. Popular stabilizers include PRI 32-D and STA-BIL; I use the latter. If you choose this method, open and check the fuel tank. If at a minimum level, check for accumulated sediment and water and remove through the tank drain plug. Add the stabilizer according to tank quantity ratio, and then fill with fresh diesel. I always use the highest quality and octane level diesel I can get, as it contains additives for cleaning and lubricating fuel systems. If you have water separators and filters, replace and drain them. Run the engine and allow the fresh treated fuel to pass through the fuel system. Consider adding Cortec VpCI-707 to the fuel tank.

e. **Engine Exhaust System.** Open and drain any water remaining in the exhaust system.

f. **Air System.** At the end of the engine running, wrap the air inlet with plastic wrap to prevent moist air getting in. Fogging is another option; it involves spray fogging through the air intake. This deposits a thin protective film on internal parts.

g. **Alternator.** After the end of the shutdown run, wrap the alternator in plastic wrap to prevent moist air getting into the windings.

h. **Start Battery.** Disconnect the start battery so that no electrical leakages can drain the start battery.

49.4 Electrical and Electronics Winterization. These hints and hacks should help you put the systems into hibernation until spring arrives.

a. **Desiccants.** For the main electrical panel, insert a vapor phase emitter behind the switchboard. Install a closet desiccant module behind the switch panel to absorb any moisture; it really works! These are available in supermarkets.

b. **Warm-up Devices.** Prior to the engine shutdown run, power up all electronic devices to warm them through and remove condensation. If you are removing the device, blank off the openings to seal the panel against moist air ingress.

c. **Instrument Removal.** If there is a security issue, remove all marine electronics and take them home to a warm, dry location. Most manufacturers guarantee their devices to very low temperatures of –20°F (–28°C), but check your device limits so they are able to withstand the winter. Don't forget to power down all electronic circuits, including the network power.

d. **Helm Instruments.** Remove any helm-mounted display units, and wrap any plugs with plastic wrap to stop moisture ingress. Make sure the sun covers are on. I wrap some tape around those to make sure they don't come off in the wind.

e. **Lighting.** Isolate the light switch circuit. Many people accidently leave one light on, and the battery is soon dead.

f. **Aerials.** Remove any antennae and aerials and place them down below; again, wrap up the cable plugs.

g. **Transducers.** All thru-hull transducers for depth sounder, speed/log, and fish-finder should be removed and blanking plugs installed.

h. **Wind Instruments.** Remove the masthead wind instruments and stow below. Ensure that the cable plugs are wrapped up; there is no point wearing the anemometer bearings out.

i. **Software.** If your devices have software, back up all the data. Check the software version; take home the micro SD card and prepare same with the latest software update to be ready for the spring commissioning. If you value all that data, back it up to an external drive. Most manufacturers enable website update downloads.

j. **Autopilot.** Operate the autopilot actuator several times and lubricate linkages.

k. **Batteries.** If a flooded cell type, top up the cells and charge for 24 hours before shutdown. If you are going dead ship, disconnect the batteries. I know some who are able to remove the batteries and take them home to the garage and charge.

l. **Shore Power.** If you are going to remain plugged into the marina power, remember that the GFCI on the pedestal can trip; if no one is checking, the battery charger will be off. Plan on worst case here. Make sure the cable has enough slack to accommodate boat movement.

49.5 **Windlass Winterization.** A piece of equipment that is so open and exposed to the elements requires special attention.

a. **Windlass Control.** Run the anchor up and down, and warm the electric motor.

b. **Corrosion Protection.** If you haven't done so, place a Cortec or similar vapor phase corrosion inhibitor inside the box. Coat the main terminals with petroleum jelly or similar product, or spray on Boeshield T-9, which is a really good protective product.

c. **Isolation.** Isolate and open the main circuit breaker for the windlass.

d. **Windlass.** Put a cover over the windlass if you have one; if not, wrap in plastic to keep water out—wind and water can get in anywhere.

Spring Commissioning

50.1   Spring Recommissioning. The winter layup period can and does take a heavy toll on many electrical and electronics systems. The insidious damp and moisture infiltrate literally everything, even the supposedly waterproof devices and equipment. In most cases it is not until you start switching equipment on that things start to go awry. This is even further aggravated by the corrosive effects that moisture and condensation have on equipment. A few months sitting idle with a coating of moisture, and everything starts to seize up as corrosion advances; 6 months is a long time.

50.2   Recommissioning Commencement. How do you prioritize starting up and commissioning things? It is always best to start the recommissioning on a system-by-system basis, and simply be methodical. This logically is based on the most important equipment and systems first. Ideally, one should adopt a full liveaboard role for the process; move on board if you have not done so already! Energize everything, and start operating over an extended period; in other words, stress-test the boat systems at normal operating levels. The first aim is to dry things out. In most cases electrical systems require the power to be on and the generation of heat. The worst enemy is moisture finding its way into the windings of alternators, starters, and motors. Although windings insulation in equipment these days is high quality, it is worthwhile going through the exercise of expelling moisture. Be thorough and leave nothing to chance.

50.3   Damage Cause and Effect. What are the actual processes that cause the damage? The first major effect is the ingress of moisture and water, lowering electrical insulation resistance. When combined with salt residues, this drop in resistance level can be considerable, with tracking usually occurring between the terminals on connector and terminal strips. Gradual breakdowns can lead to short-circuit conditions between those terminals. The next major effect is corrosion, which causes mechanical components such as bearings, shafts, and brush gear to seize. With salt it can cause the breakdown of exposed copper conductors, that familiar characteristic blackening of copper. Motorized equipment, in particular water pumps and bilge pumps, will fail to run and will trip the circuit breaker after several seconds; the main cause is seized bearings or shafts. The last major effect is simply the accumulation of water in a fitting, causing corrosion and, in most cases, a short-circuit immediately after the power is applied. In general, the most common cause of electrical failures on boats is connection failures. Therefore, detailed inspection and verification that connections are in good electrical condition are essential to reliability during the remainder of the sailing season. Indiscriminate spraying of solvents, cleaners, or similar water-displacement fluids is not a substitute for proper cleaning and checking.

50.4   Engine Starting Systems. As engine starting problems are also a major cause of vessel problems, and form the basis of reliable charging, it is best to start at this point.

       a.   **Starter Motor Connections.** Check all the starter motor connections and retighten them. Disconnect the battery first so that you don't short things out. Put a ring spanner or socket on and actually check and tighten; don't just check it by hand and try to move the connection.

b.   **Starter Motor Terminals.** Clean moisture from the area around the starter motor terminals by wiping with a clean, damp rag. Electrical cleaner, WD-40, or CRC may be used after wiping clean to form a protective film.

c.   **Engine Start.** Operate the starter and turn over the engine three or four times for up to 10 seconds to generate heat in the starter. This has the effect of heating the windings from inside and drying windings and brush gear. In most cases this can be done with the throttle at zero to prevent any starting. Be careful not to overdo it, though, and create excessive heat. If you have larger engines with the starter motor brush-gear accessible, open the cover and pull back the brushes against the spring retainers, as they tend to jam. If it is black and dirty, use a vacuum cleaner to extract dust; wash only with spray electrical cleaner.

d.   **Main Negative Connection.** Check and tighten the main negative on the engine block to battery cable. At the end of the previous season, the connection may have become loose due to vibration of the engine. Moisture and any resultant corrosion may have deteriorated the connection. It is good practice to remove the connection, clean the mating surfaces, and reconnect.

e.   **Preheating.** Check all preheating element connections; clean moisture and any dirt from around the terminals that may cause tracking and lower heater efficiency.

50.5   **Batteries.** It is important not to make assumptions that batteries are serviceable. The most common cause of failure after the winter layup is batteries that have sulfated while remaining partially charged.

a.   **Terminals.** Remove, clean, and tighten all main battery connections.

b.   **Electrolyte Levels.** If you have flooded-cell batteries, check that all electrolyte levels are correct.

c.   **Battery Security.** Check that the batteries are secured or held down in place.

d.   **Battery Casing Tops.** Clean the battery case tops using a damp rag.

e.   **Isolation Switches**. Operate battery isolation switches several times when the engine is off, and "exercise" the switch contacts. Switches often develop high resistance across the contacts, and operation often alleviates the problem.

f.   **Battery Condition.** Where a battery is in any way suspect, in particular one that has obviously been sitting in a discharged state for several months, have the battery load tested to verify actual condition.

50.6 **Charging Systems.** Start the engine; run at medium rpms and check that the charging voltage is at 14V or, where a fast charge regulator is used, that voltages are correct. If the battery takes some time to rise to the nominal voltage, suspect that the battery has been sitting partially discharged for several months.

    a. **Rubber Drive Belts.** Check and retighten the alternator drive belt.

    b. **Terminals.** Check all the alternator connections and cable lugs; tighten or re-terminate as required.

    c. **Cable Terminations.** Check the security of the cable within the connector crimp. If loose or showing signs of fatigue, re-terminate it.

50.7 **Lighting Systems.** Perform the following checks.

    a. **Energization.** Switch on all internal lights; check and leave them on for at least 1 hour to warm and dry the fittings.

    b. **Operation.** Switch on all navigation lights and check that they are functioning; leave on for at least 1 hour to dry the fittings.

    c. **Lamp Bases.** Where possible, remove the lamp and refit; as the lamp socket spring may be seized or corroded, this action moves the spring and frees it up if it is seized. This is a surprisingly common failure mode.

50.8 **Marine Electronics.** The characteristic misting up of instrument displays is a typical sight after winter. The best remedy is to switch on the display lighting and leave it on for several hours. It is not a lot of heat, but it helps.

    a. **Antenna Connections.** Check that all antenna connections are secure. In particular, check VHF connectors at the mast base, or where they are made up the mast at the antenna.

    b. **Cable Condition.** Check that any exposed copper conductors have not deteriorated and that crimp connections are sound.

    c. **Antennae.** Check all antennae such as VHF, AIS, GPS, and NAVTEX for damage.

    d. **GPS.** Check that the GPS antenna is undamaged, and vertical if on a stern rail or pushpit-type support.

    e. **Autopilot.** Check the autopilot socket outlets for moisture. If caps or protective covers have not been properly screwed on, corrosion may have occurred around the small pins. Clean and dry them out before use.

    f. **Gas Detector.** Check gas detector operation. Sensor heads can become degraded by moisture and be ineffective. Turn on the detector and test after 1 hour. Ideally, sensor heads should be wrapped up for the winter. Use a disposable butane lighter to check them.

g. **Operation Check.** Operate all electronics equipment for several hours with lights on. The heating effect of the equipment power supplies and lamps should dry out any internal condensation that can cause malfunction on circuit boards.

h. **Ground Connections.** Check that all electronic device grounds are tight and in good condition.

50.9 **Windlass, Winches, Thrusters, and Toilets.** Perform the following checks.

a. **Terminals.** Check, clean, and tighten the windlass, powered winch, and thruster motor terminal connections.

b. **Operation Check.** Lower the anchor and operate the motor on load for two complete anchor retrievals. Warm up motor, subject to the duty cycle capability of the motor.

c. **Foot Switches.** Open and check the footswitches for water ingress. This is a very common failure mode and has been known to cause uncontrolled operation when shorted out with water. Check that contacts are in good condition; it may save you the task of hauling it in by hand in a failure.

d. **Thruster Load Test.** Operate the thruster under load for 3 minutes, the typical rating in most cases. This will help dry out moisture. Where possible, check brush gear and that the brushes are moving freely. This is important; subsequent arcing may cause motor commutator damage as well as poor operation.

e. **Electric Toilets.** Operate electric toilets several times to heat up the pump motor windings. Check that connections are in good condition and not corroded.

50.10 **Wiring Systems.** Inspect the following.

a. **Terminations.** Check all connections and re-terminate and seal if required. In particular, the bilge pump connections should be carefully inspected.

b. **Mast Lighting Circuits.** Check mast lighting junction circuits; this may be a junction box or terminal strip, or it may be a plug-and-socket arrangement. The typical fault is that screws are corroded; in many case the contact resistance deteriorates with increases in voltage drop and causes failure of the connection.

c. **DC Outlets.** Check spotlight socket outlets for moisture. Clean and dry if moisture is found.

d. **Bonding Connections.** Inspect all bonding and lightning system connections. Where any degradation has occurred, the connection should be re-terminated. For the systems to operate properly, they must have a low resistance

e.   **Circuit Breakers.** Operate and exercise all circuit breakers and switches several times.

f.   **Fuses.** Remove any installed fuses and inspect for deterioration. Replace the fuses if not in perfect condition.

50.11   **AC Power Systems.** Perform all of these important integrity tests.

a.   **GFCI/RCD Testing.** Test the proper operation of ground/earth tripping devices such as GFCIs and RCDs using the test button.

b.   **Shore Power Inlet.** The shore power inlet socket is the first area to examine. Inspect it and ensure that it is clean and dry. If you had the shore power cable plugged in all winter, also examine the plugs and sockets for signs of water and moisture ingress.

c.   **AC Wiring Check.** Arrange for all permanent vessel AC wiring to have a 500VDC insulation resistance (IR) (Megger or mega) test; this should be done every 12 months in any case. After 6 months laid up, the system can easily drop down below the mandatory and minimum 1 megohm (MΩ), making it unsafe. This test should include a test of the genset alternator windings.

d.   **Safety Grounds.** Check the integrity of all grounding connections and verify they are in good condition, but don't disconnect when the power is on.

e.   **AC Inverter.** Check that all AC inverter connections are in good condition.

f.   **Genset.** Service the engine and perform an oil change. Where it has a separate battery, carry out the battery checks.

g.   **Genset Operation.** Run the generator and load it up as much as possible.

50.12   **Equipment Expiry Dates.** Marine insurance companies have rejected a billion dollars in claims in the last decade based on latent or compliance deficiencies unrelated to the actual claim. This is a good time to audit all of the equipment and systems on board your boat for policy compliance. Check and verify all fire extinguishers are within the expiry date. The same applies to EPIRBs, flares, life jackets, updated electronic charts and anything else that has expiry dates or date limitations. Insurance coverage issues might also arise if you have retrofitted lithium-ion batteries, and whether you have documented proof that the installation was installed by an appropriately qualified professional. Every jurisdiction has differing laws on this, so play safe, check and document everything. Be meticulous with documentation and have a conversation with your insurer.

Technician Service Calls

51.1   About Calling Service Technicians. In the previous three editions, I researched and included a comprehensive list of service companies, including marine electricians, marine electronics technicians, and other qualified electrical experts, who came highly recommended. Marine electricians often have a merchant marine or naval background and will probably have many years of sea service behind them. They understand the environmental factors affecting marine electrical installations and are qualified to work on both AC and DC systems, as well as on most electronics. Beware of automotive electricians claiming to be marine electricians—they are not. There are some very good automotive electrical tradesmen doing marine work; go on recommendations if at all possible. Ask them if they own a boat! Use caution when domestic electricians claim to be marine electricians. There are many good technicians and tradespersons around who have good industrial electrical backgrounds and will do a good job. Look for ABYC certification. If you are getting AC work done, ask to see a license or some qualification, and snap an image of it for reference. Get references or check their backgrounds if possible. It's your life in the balance. In the United States many will be ABYC certified; check the accreditation. Given that we are well into the information age and you can search for a service person on your mobile phone or at home on your computer, the need for a more comprehensive list is now redundant. Many of those listed in earlier editions are no longer in business—or have sensibly gone sailing. I was unable to verify many of the previous listings; those noted in the tables below are the best I could curate.

51.2   Service Person Etiquette. I recall an episode when a boat arrived from a Pacific cruise. The skipper told me that his radar had been out for some months; he could only get the display partially working but no image. To really upset his day, I went to the stern-mounted scanner and simply switched on a local power switch. Imagine the reaction—he simply had forgotten to check it. On another occasion I visited a vessel where the skipper was cursing the new satellite communication system, as he couldn't get it to boot up. After a quick read of the operation manual, I asked him to type in a command line. I immediately observed his use of a back slash instead of a forward slash, as shown in the manual. Instant boot-up—and one very embarrassed person. Ask the following questions and consider the following points.

a.   **Did I operate the equipment properly?** Read the manual again and go back to basics. Only when you are sure that you have operated the equipment properly and it still doesn't work should you call a serviceperson.

b.   **Are all the plugs in and the power on?** It is amazing how many people forget to plug in an antenna or aerial or simply forget to put the power on. If the power is on at the breaker and not on at the equipment, double-check that the circuit connection on the back of the switchboard is not disconnected. Check that the equipment fuse has not ruptured. In other words, check the obvious.

c.   **What were the circumstances at time of fault?** Ask yourself what you were doing immediately prior to the fault. Many faults occur immediately after working on often-unrelated systems. The inadvertent disturbance of connections can and does occur regularly, so go and check.

d.   **Record the fault conditions.** Write down the fault and the situation when the unit failed. If a profile can be built up, it may point to other factors. This will not only assist the service person but also may assist you in resolving the problem and avoiding an expensive callout.

e.   **Keep a good technical file on board.** If possible, obtain copies of all technical manuals, as service people cannot carry or even get every manual. This will save you money if you can provide the information, as time will be saved. Make sure you have copies of circuit diagrams and anything that supports your devices.

f.   **Clean Up.** Clean up the area to be worked on. It is unfair to expect service people to work on filthy engines and dirty bilges. If you want or don't mind grime tracked through the boat, ignore this advice. A good serviceperson may simply decline to come back again.

g.   **Tool Kits.** Have a good tool kit ready. It is impossible to carry a complete tool set onto every boat. Technicians will greatly appreciate any assistance like this. Make sure your flashlights work, and make sure that where access to equipment is through lockers, they are emptied and clear. If technicians have to keep going back to their transport for tools, it is on your tab.

h.   **Refreshment Support.** Don't offer the technician a beer or coffee until the job is finished. Refreshments are appreciated, but they don't get the job done. Offer beverages *after* you get the invoice or pay. More hours will mean more dollars on your account.

i.   **Be Ready.** Make sure your crew is dressed for early-morning service calls. It isn't fun to be greeted by hungover or scantily-clad crewmembers when troubleshooting or tracing cable runs; the crew doesn't like it either.

j.   **Technician Collection.** If collecting a technician from ashore, make sure you both know where to meet. I have been left standing many times due to confusion about the pickup point. Please bail the dinghy first and have a dry towel for the technician to sit on. In most cases, the tools and spares will need some place dry as well.

k.   **Job Definition.** Write down the job you want done in a work order, and pay for it. Do not enter into verbal agreements. Define the scope of work or assistance. When the contracted work is complete, do not blame the service technician for all the other electrical problems that exist on your boat. Some people do this to avoid paying accounts. If you do not pay your account, I can assure you that word travels fast, and you may end up without help when you need it. If you want additional work done, document the request.

l. **Pay the Bill.** Do not sail off without settling your account. You may be arrested at the next port, have a writ nailed to your mast, lose your yacht to pay accounts, and end up in jail. I know these things can happen because I have had to do them. Technicians and electricians do not just get mad, they get even; they have families to feed and expenses to pay, just like you.

m. **Bill Shock.** There are quite a number of boating people who do not mind paying a $100 fee for a service call plus $90/hour for a washing machine mechanic to fix a $1,000 machine. But when it comes to paying a highly qualified marine electrical or electronic service technician to fix essential and very expensive equipment on $400,000 vessels, they think $100 an hour is exorbitant, even though only hourly rates (no travel time or service fee) are charged. What can I say?

51.3 Foreign Cruising Assistance. Electrical terminology varies throughout the world, and many have limited English skills. There are even variations between the various English-speaking countries. The following glossary gives some common translations for various devices in French and Spanish and is not exhaustive. If you are voyaging to Portuguese-speaking locations, Italy, Greece, and so on, there will also be communication challenges.

Table 51-1. Basic Three-Language Electrical Glossary

English	French	Spanish
Audible alarm	Avertisseur sonore	Bocina electronica
Alternator	Alternateur	Alternador
Alternator rating	Puissance de l'alternateur	Potencia del alternador
Alarm panel	Tableau des alarmes	Tarjeta instrumentos
Battery	Batterie	Bateria (Accumulador)
Bolt	Boulon	Perno, tornillo
Circuit breaker	Coupe-circuit or interrupteur	Fusible
Connection	Connexion, cablage	Conexion
Circuit diagram	Schema de cablage electrique	Eschema de connexiones
Current (electrical)	Courant	Corriente
Drive belt	Courroie de transmission	Correa de ventilador
Disconnect	Deconnecter, isoler	Desconectar
Electrician	Electricien	Electricista
Element	Element	Resistencia
Fault	Default	Defecto
Ignition switch	Contact (moteur)	Llave de contacto
Insulation	Isolement	Aislamiento

English	French	Spanish
Current level	Intensite (amps)	Intensidad
Fuse	Fusible	Fusible
Ground (earth)	Terre	Conectar con tierra
Lights	Feux	Luz
Light bulb	Ampoule electrique	Bombilla, foco
Lightning	Eclair	Relampago, rayo
Navigation lights	Feux de position	Luz de navigacio
Overheat	Surchauffe	Recalentarse
Oil pressure sensor	Sonde de pression d'huile	Sensore pressione olio
Pressure gauge	Manometer d'huile	Monometero de aceite
Preheating glow plugs	Bougies de prechauffage	Precalentamiento
Relay	Relais	Rele
Recharge	Recharge de batterie	Recargar
Short circuit	Court-circuit	Cortocircuito
Starter motor	Demarreur	Motor de arranque
Sensor	Capteur, sonde	Sensore
Switch	Bouton poussoir/interrupteur	Pulsador
Tachometer	Compte-tours	Tacometro
Temperature sensor	Sonde de temperature	Sensore tempertura
Transmitter	Emetteur	Transmisor
Voltmeter	Voltmeter	Voltimetro
Voltage	Tension de systeme	Tension del systema
Voltage drop	Chute de tension	Voltaje, bajar
Water pump	Pompe a eau	Bomba
Wire	Cable ou fil (electrique)	Alambre
Engine not starting	Le moteur ne demarre pas	El motor no arranca

Tool Kits and Spare Parts

52.1   Tool Kits and Spares Parts. There is no real answer to what an optimum tool kit should look like. A stainless-steel socket set, along with a set of wrenches and ring spanners of all the required sizes, is essential. Also, depending on your own equipment requirements, a simple torque wrench is worth having. I carry a bearing-puller set. A set of rechargeable battery–powered tools is high on the list. I have a simple 18V Ryobi skin with spare battery, and it is very versatile. This powers the drill, sander, a grinding and cutting wheel, a vacuum cleaner, and an emergency LED light. A set of screwdrivers is also paramount. Invest in tethered tools if you work up the mast. If a reasonable level of self-sufficiency is required, the following tools and equipment should be carried on board every vessel. The list should be used as a basic itemized checklist or the basis for your own kit and something to think about.

Table 52-1 Tools and Spares List

Recommended Tools	Consumables and Spares
Electrical Pliers	Electrical Tape
Long Nose Pliers	Self-bonding Tape
Side Cutting Pliers	Denso Tape
Cable Crimpers (Ratchet Type)	Heatshrink Tubing
Electrical Screwdriver Set	Nylon Cable Ties (3 sizes)
Jewelers' Screwdriver Set	Solder
Socket Wrench Set	Crimp Connector/Terminal Set
Ring Spanner Set	Nav Light Lamps (if not LED)
Adjustable Wrench Set	Anti-seize Lubricant
Open End Wrench Set	Electrical Cleaner CRC
Pipe Wrench	Dispersant (WD-40)
Multi-Grip Pliers	Deck Winch Maintenance Kit
Vise Grip Pliers	Water Pump Impeller × 2
Circlip Plier Set	Hacksaw Blades × 10
Oil Filter Wrench	Junior Hacksaw Blades × 10
Torque Wrench	Thread Tape × 2
Battery Powered Vacuum Cleaner	Silicon Sealant
Battery Powered Drill	Drill Bit Set
Battery Powered Sander	Emery cloth/sandpaper
Battery Powered Grinder	Grinding and Cutting Wheels

Recommended Tools	Consumables and Spares
Battery Powered Soldering Iron	Fuel Separator Filters × 2
Battery Powered Vacuum Cleaner	Oil Filters × 5
Vernier Calipers	Fuel Filters × 5
File, Medium-Small Half-Round	Alternator
File, Small Round	Alternator Drive Belt × 2
File Set, Rat Tail	Alternator Regulator
Mini File Set	Set of Fuses for Electronics
Hacksaw	Slow Blow Fuse for Windlass
Junior Hacksaw	Manual Bilge Pump Kit
Bolt Cutters	Water Pump Repair Kit
Multimeter	Potable Water Filters
Metal Cold Chisel Set	Toilet Repair Kit
Lump Hammer	Sealant Cartridges
Ball-peen/pein Hammer	Lube Oil for 2 engine changes
Punch Set	Grease Cartridges
Portable Bench Vise	Hose Clamp Selection
Drill Driver Hex Torx Bit Set	316 Stainless Screw Selection
Rubber Mallet	316 Stainless Bolt and Nut Selection
Wire Brush	Infrared Thermometer Gun
Grease Gun	O-ring Kit
Tube Cutter	
Hex (Allen) Key Set	
Tape Measure	

Service Directories

Table 53-1. Service Directory: Canada, US East Coast, Gulf of Mexico

Port	Company	Contact Numbers
Summerside, PE (Canada)	N S C Electronics	902-436-7521
Dartmouth, NS (Canada)	Mackay Communications	902-468-8480
Dartmouth, NS (Canada)	Atlantic Electronics	902-468-3628
Stonington, ME	Blackmore Electronics	207-367-2703
Searsport, ME	Hamilton Marine	207-548-6302
Portland, ME	Sawyer & Whitten Marine	207-879-4500
York, ME	Navtronics	207-363-1150
Gloucester, MA	Seatronics	978-281-0034
Harwich, MA	Dempsey Marine Electronics	508-360-1094
Fairhaven, MA	Chris' Electronics	508-994-8257
Newport, RI	Electra Yacht of Newport	401-338-0700
Jamestown, RI	Jamestown Electronics	401-423-2253
Hampton Bays, NY	Hampton Navigation	631-723-6915
Annapolis, MD	Mid-Atlantic Marine Electronics	410-919-9399
Annapolis, MD	PKYS Inc. Marine Electrics	410-280-2267
Cambridge, MD	Mid Shore Electronics	410-228-7335
Virginia Beach, VA	Marlin Marine Electronics	252-562-0600
Portsmouth, VA	J & L Marine Electronics	757-465-8323
Charleston, SC	Tidal Marine Electronics	843-763-8553
Savannah, GA	Coastal Marine Electronics	912-303-0042
Port St. Lucie, FL	Accurate Marine Electronics	305-432-8252
West Palm Beach, FL	Poseidon Marine Electronics	561-232-1413
Pompano Beach, FL	McLaughlin Marine Electronics	954-975-2112
Fort Lauderdale, FL	Yachtronics	954-763-1618
Fort Lauderdale, FL	Island Marine Electric	954-524-3177
Fort Lauderdale, FL	816 Marine Electric	202-344-5809
Key West, FL	Harbor Marine Electric	305-394-2825
Marathon, FL	Seamark Electronics	305-743-6633

Port	Company	Contact Numbers
Jacksonville, FL	A & G Marine Electronics	904-388-3690
Tavernier, FL	Keys Marine Electronics	305-852-4668
Fort Myers, FL	Felix Marine Electronics	239-936-6463
Naples, FL	Sea Tech Marine Electronics	239-430-1111
Sarasota, FL	SOS Marine Electronics	941-806-9395
Destin, FL	Redmond Marine Electronics	850-837-8092
Pensacola, FL	Georges Marine Electronics	850-456-4553
Panama City, FL	Marine Electric	850-234-6533
Orange Beach, AL	Xtreme Marine Electronics	251-981-1466
Biloxi, MS	Advanced Marine	228-374-6747
Paradis, LA	Wheelhouse Electronics	985-758-1010
Clear Lake Shores, TX	True North Marine	281-549-4300
Corpus Christi, TX	Lighthouse Technologies	361-510-2628
South Padre Island, TX	Boudreault Marine Electronics	512-203-9627

Table 53-2. Service Directory: Caribbean

Port	Company	Contact Numbers
St. Michael (Barbados)	Marine Power Solutions	246-435-8127
Kingston (Jamaica)	Mackay Marine	281-478-6245
Georgetown (Cayman Island)	Harbour House Marina	345-640-4700
Santo Domingo (Dom Rep)	Electromarine Services Alegra	829-926-4772
Santo Domingo (Dom Rep)	Marina Zarpar	809-523-5858
San Juan (Puerto Rico)	Caribbean Radio & Telephone	787-310-1024
Bayamon (Puerto Rico)	ATK Marine Service	939-276-9214
Cabo Rojo (Puerto Rico)	Twin Electronics	787-851-8825
Nanny Cay Tortola (BVI)	Cay Electronics	284-494-2400
Virgin Gorda (BVI)	Virgin Gorda Yacht Harbour	284-495-5500
St. Thomas (USVI)	Budget Marine VI Boatyard	340-779-2219
St. Thomas (USVI)	Tropi Com	340-775-4107
Anguilla	Necol NV	721-580-8148
New Guinea (St. Kitts)	St. Kitts Marine Works	869-662-8930

(continued)

621

Table 53-2. *Continued*

Port	Company	Contact Numbers
Pelican Key (Sint Maarten)	Necol NV	721-580-8148
English Harbour (Antigua)	The Signal Locker	268-460-1528
Pointe A Pitre (Guadeloupe)	Marinelectronic	+41 021-801-0246
Baie-Mahault (Guadeloupe)	SAD	059-051-0540
Pointe à Pitre (Guadeloupe)	Fred Marine	590-590-907-137
Le Marin (Martinique)	Diginav	+33 596-747662
Fort-de-France (Martinique)	Maximarine	800-828-5010
Cole Bay (Sint Maarten)	Electec	721-544-2051
Rodney Bay (St. Lucia)	MarinTek	758-450-0552
Rodney Bay (St. Lucia)	Regis Electronics	758-452-0205
St. George (Grenada)	Turbulence Ltd.	473-439-4495
Prickly Bay (Grenada)	Spice Island Marine	473-444-4342/4257
Carricou (Grenada)	Carricou Marine	473-443-6292
St. Vincent (Grenadines)	Barefoot Yacht Charters	784-456-9526
Barbados	Marine Power Solutions	246-435-8127
Chaguaramas (Trinidad)	Dockyard Electrics	868-634-4272
Chaguaramas (Trinidad)	Electropics (Peake Yacht Services)	868-634-2232
Chaguaramas (Trinidad)	Goodwood Marine	868-634-2203
Aruba (Netherlands Antilles)	Wind Creek Marina Oranjestad	+297 588-0260
Willemstaad (Curacao, NI)	Curacao Marine	599-9-465-8936
Santa Marta (Colombia)	Marina Santa Marta	+57 310-512-6883
Cartagena (Colombia)	Electronica Maritima	+57 566-33-909
Colon (Panama)	Shelter By Marina	507-433-3581
Panama	Centro Marino	575-663-3909
Bocos del Toro (Panama)	Bocas Marina	507-757-9800
Carmen (Mexico)	Marine Electronics Service	+52 938-382-5185
Cancun (Mexico)	Marina VV	+52 998-234-0100
La Ceiba (Honduras)	Eagle Marine	504-9995-2077
Placencia (Belize)	Thunderbirds Boatyard	510-634-1054
Golfito (Costa Rica)	Banana Bay Marina	506-277-502-55
San Jose (Costa Rica)	Promarina	506-221-562-22

Table 53-3. Service Directory: Canada, US West Coast, Mexico

Port	Company	Contact Numbers
Homer, AK	South Central Radar	907-235-8008
Anchorage, AK	Polar Marine	907-602-1020
Victoria, BC (Canada)	Anchor Marine Electric	250-386-8375
Victoria, BC (Canada)	All Marine Electric	250-217-2756
Port Hardy, BC (Canada)	Stryker Electronics	250-949-8022
Vancouver, BC (Canada)	Roton Industries	604-688-2325
Vancouver, BC (Canada)	Pacific Yacht Systems	604-284-5171
Vancouver, BC (Canada)	A-Sea Marine	604-338-9920
Vancouver, BC (Canada)	LeftMost Yacht Services	778-227-5389
Seattle, WA	Lunde Marine Electronics	206-789-3011
Seattle, WA	Hullux Marine Electrical	206-488-5311
Seattle, WA	Delta Marine Electrics	206-285-4934
Port Townsend, WA	West Coast Marine Electronics	206-604-5168
Gig Harbor, WA	Flecks Marine Service	253-229-6388
Clackamas, OR	Deep Creek Marine Elec	503-922-3259
San Francisco, CA	Cal-Marine Electronics	415-391-7550
San Francisco, CA	Farallon Marine	415-331-1924
San Francisco, CA	Fox Marine	510-868-5041
Petaluma, CA	Fred Fritz Electronics	707-762-9198
Alameda, CA	Reliable Marine Electronics	510-864-7141
Oakland, CA	Anyboat Marine Electronics	510-430-2660
Costa Mesa, CA	Seaside Marine Electronics	949-675-7866
Costa Mesa, CA	Alcom Marine Electronics	949-515-1727
Long Beach, CA	Long Beach Marine Electronics	563-594-8888
Long Beach, CA	Current Marine Electrical	562-850-3690
Marina del Rey, CA	Maritime Communications	310-821-4958
San Pedro, CA	Neptune Electronics	310-833-5291
National City, CA	Honor Marine Communications	619-233-7666
San Diego, CA	Shelter Island Marine Electronics	619-223-2182
San Diego, CA	Custom Marine Electronics	619-224-3646
San Diego, CA	Seanet Electronics	619-222-3407

(continued)

Table 53-3. *Continued*

Port	Company	Contact Numbers
San Diego, CA	Upgrade Marine	619-518-5685
San Diego, CA	Marks Marine Electronics	619-540-9875
San Diego, CA	San Diego Boat Electric	619-218-1018
Ensenada (Mexico)	Aquamar de Ensenada	01-646-174-0045
Ensenada (Mexico)	Niza Marine	52-1 (646) 174-2422
Puerto Vallarta (Mexico)	Imptronic SA	01 322-224-2534
Puerto Vallarta (Mexico)	Marine Vallarta	01 322-224-2534
Guaymas (Mexico)	Star Marine	01 800-830-3936
La Paz (Mexico)	Agencia Arjona	01 612-122-3333
La Paz (Mexico)	JWI Marine	01 612-139-5436

Table 53-4. Service Directory: UK and Ireland

Port	Company	Contact Numbers
South Kessock, Scotland	Gael Force Marine	+44 1463-229-400
Aberdeen, Scotland	Woodsons of Aberdeen	+44 1224-722-884
Edinburgh, Scotland	Blue V	+44 131-331-4546
Ayrshire, Scotland	Boat Electrics & Electronics	+44 1294-602-203
Ardrossan, Scotland	MB Marine Electrics & Electronics	+44 1294-602-003
Newcastle Upon Tyne	Storrar Marine	+44 191-2661037
Durham	Radio Holland	+44 142-983-9280
Douglas, Isle of Man	Bottom Line Ltd.	+44 800-246-1005
Pwllheli, Wales	Rowlands Marine Electronics	+44 1758-613-193
Lowestoft, East Suffolk	KM Electronics	+44 1502-569-079
Great Yarmouth, Norfolk	ElecTech Solutions	+44 1493-657-200
Harwich, Essex	PRS Communications Ltd.	+44 1255-240-523
Maldon, Essex	Mantsbrite	+44 1621-853-003
Rochester, Kent	Ships Electronic Services	+44 1634-295-500
Shoreham, West Sussex	Eurotek Marine	+44 1273-818-990
Itchenor, West Sussex	Greenham Regis	+44 1243-511-070
Portsmouth, Hampshire	MEI Ltd.	+44 239-232-6366
Lymington, Hampshire	Greenham Regis Marine	+44 1590-671-144
Cowes, Isle of Wight	D.G. Wroath	+44 1983-281-467

Port	Company	Contact Numbers
Southampton, Hampshire	Marine Electronic Systems	+44 2380-663-316
Southampton, Hampshire	A1 Marine Electronics	+44 7560-315-959
Hamble, Hampshire	Hudson Marine Electrics	+44 2380-455-129
Hamble, Hampshire	Diverse Performance Systems	+44 2380-453-399
Poole, Dorset	Seacraft Marine Ltd.	+44 7795-331-755
Portland, Dorset	Apollo Marine Systems	+44 1305-860-220
Weymouth, Dorset	JG Technologies	+44 1305-787-788
Portland, Dorset	Apollo Marine Systems Ltd.	+44 1305-860-220
Plymouth, Devon	PR Systems Marine Electronics	+44 1752-936-145
Dartmouth, Devon	AK Marine Electronics	+44 1803-833-300
Brixham, Devon	Tecmarine Ltd.	+44 1752-269-632
Topsham, Devon	NAUTEXE Ltd.	+44 7817-470-721
Falmouth, Cornwall	BT Marine Electronics	+44 1326-312-444
Dublin, Ireland	Marine Electrics	+353 1-230-0422
Killbegs, Ireland	Barry Electronics Ltd.	+353 74-973-1215
Cork, Ireland	CH Marine	+353 21-431-5700
Skibbereen, Ireland	Belco Marine	+353 28-36209

Table 53-5. Service Directory: Europe, Atlantic, Indian Ocean

Port	Company	Contact Numbers
Oslo (Norway)	Navigasjonsbutikken AS	+47 401-44410
Gothenburg (Sweden)	Vastkustens El Marin AB	+46 31-769-75-00
Gothenburg (Sweden)	AB Ramnav	+46 31-7758-700
Copenhagen (Denmark)	Tempo Marine Service	+45 23-64-22-00
Bremerhaven (Germany)	Roth Funkteknik & Yachtelect	+49 471-501-06-70
Emden (Germany)	Nordwest-Funk	+49 21-999060
Harlingen (Netherlands)	DTM Maritieme Service	+31 517-712-040
Ijmuiden (Netherlands)	WNL Marine Electronics	+31 255-534-755
Amsterdam (Netherlands)	JNAV	+31 642-444-433
Rotterdam (Netherlands)	CWN Marine	+31 107-500-550
Nieuwpoort (Belgium)	Nieuwpoort Marine	+32 585-85-800
Le Havre (France)	Radio Holland FR	+33 632-230-030

(continued)

625

Table 53-5. *Continued*

Port	Company	Contact Numbers
Roscoff (FR)	Navig'elec	+33 679-477-059
St. Helier (Jersey)	PC Boat Sales	+44 1534-737-537
St. Peter Port (Guernsey)	Radio & Electronic Services	+44 1481-249-114
St. Vaast La Hougue (France)	Marelec Electronics Navigation	+33 2-33-54-63-82
La Rochelle (France)	Pochon	+33 546-413-053
Hendaye (France)	Technic Marine Hendaye	+33 611-466-393
Anglet (France)	Espace Electronique Marine	+33 559-634-401
Leioa (Spain)	Electronautic	+34 662-933-662
Arteixo (Spain)	Stay Electronics	+34 638-987-085
Vigo (Spain)	Technonavalia Marine	+34 669-236-608
Vigo (Spain)	Electronica Ria de Vigo	+34 214-457-210
Cadiz (Spain)	Sandvik Marine Electronics	+34 956-099-101
Caravelos (Portugal)	Seatec	+351 214-457-210
Lisboa (Portugal)	Radio Holland Portugal	+351 213-976-087
Quarteira (Portugal)	EasyNav	+351 918-483-305
Madeira (Portugal)	Funchal Marina	+351 912-304-508
Madeira (Portugal)	Calheta Marina	+351 291-824-003
Las Palmas (Canary Islands)	Alisios Sailing Center	+34 928-233-171
Las Palmas (Canary Islands)	Servicios Electronicos	+34 928-248-150
Las Palmas (Canary Islands)	Distrimar	+34 928-469-617
Fuerteventura (Canary Islands)	Distrimar	+34 928-866-608
Ponta Delgada (Azores)	Mid Atlantic Yachts Services	+351 292-391-616
Sao Vicente (Cape Verde Islands)	boatCV — meio-do-atlantico	+238 230-0382
Walvis Bay (Namibia)	SMD Marine Electronics	+264 64-200-300
Walvis Bay (Namibia)	Mainmast Electronics	+264 64-205-999
Cape Town (South Africa)	Cape Marine Electronics	+27 21-421-5629
Durban (South Africa)	SEA Tech Marine Electronics	+27 31-206-0827
Antanarive (Madagascar)	Landis Madagascar	+261 20-22-251-51
Port Louis (Mauritius)	Seatronics	+230 5970-6131
Providence, Mahé Island (Seychelles)	Taylor Smith Naval Services	+248 438-4755
St. Gilles (Réunion Island, France)	Ralph Nautic Service	+262 262-42-07-21
Malé (Maldives)	Beach Marine	+960 330-00-72

Table 53-6. Service Directory: Mediterranean

Port	Company	Contact Numbers
Gibraltar	Electromed	+34 956-632-274
Valetta (Malta)	Yachting Partners	+356 699-494-685
Palma De Mallorca (Spain)	E-Touch Navcom	+34 627-362-256
Palma De Mallorca (Spain)	Navigation & Communication	+34 971-718-662
Palma De Mallorca (Spain)	Tallamar Electronics	+34 971-918-600
Mahon, Menorca (Spain)	Sailpower	+34 679-516-829
Ibiza (Spain)	Nautica Ereso	+34 971-314-122
Almeria (Spain)	Alamar Centro Nautico	+34 950-497-947
Malaga (Spain)	Navtronics Elec Naval	+34 629-521-511
Valencia (Spain)	RN Yacht Electric	+34 661-507-825
Denia-Alicante (Spain)	Denitel	+34 965-786-388
Girona (Spain)	Nortec Europe	+34 696-087-166
Barcelona (Spain)	Ventus Electronica Nautica	+34 636-132-066
Nice (France)	Nautor Villefranche	+33 493-766-007
Antibes (France)	ISEA Electronics	+33 60 736 2789
Saint Tropez (France)	Electronic Technic Marine	+33 (0)4-94-362-256
Portoferraio (Italy)	Elletromare Elba	+39 565-916-400
Palermo (Italy)	SE.A Assistance di Sergio Abatte	+39 349-291-3901
Messina (Italy)	Elletromar di lo Forte Rosali	+39 335-104-5636
Ostia Antica (Italy)	HMP Marine	+39 656-337-282
Naples (Italy)	Quality Yachting	+39 3473-723-816
Salerno (Italy)	Technovela	+39 08923-1396
Gallipoli (Italy)	Electronica Navale di De Rosa	+39 0833-202251
Bari (Italy)	E.G.C. Electronica Navale	+39 803-384-044
Rimini (Italy)	P.A.I.N.E. SNC di Renzi & Bianchi	+39 541-216-29
Venezia (Italy)	Navitronic	+39 339-39-63-550
Cagliari, Sardinia (Italy)	Rinaldi Nautica	+39 703-044-101
Rijeka (Croatia)	2bmarine	+385 99-213-0148
Split (Croatia)	MCI Servis D.O.O.	+385 913-948-383
Marina (Croatia)	Dark Blue Nautica	+385 919-505-309
Vodice (Croatia)	Wasi D.O.O.	+385 224-40-874
Dubrovnik (Croatia)	Wasi D.O.O.	+385 204-56-847

Table 53-6. *Continued*

Port	Company	Contact Numbers
Piraeus (Greece)	Aegean Electronics	+30 21-0412-3000
Piraeus (Greece)	Nova Electronics	+30 21-0412-1578
Piraeus (Greece)	Space Electronics	+30 21-0417-4614
Athens (Greece)	Hellas Marine Electronics	+30 21-0992-8128
Athens (Greece)	AMP Marine Electronics	+30 693-816-9733
Piraeus (Greece)	Gavrilis Marine Electronics	+30 694-823-9640
Leros—Dodecanese (Greece)	Agmar Marine	+30 697-697-4416
Corfu (Greece)	Electromar Marine Electronics	+30 6936-000-262
Rhodes (Greece)	Karras Marine Electronics	+30 6945-352-241
Bodrum (Turkey)	Motif Yachting	+90 252-316-1087
Istanbul (Turkey)	Morse Marine	+90 216-909-7853
Marmaris (Turkey)	TMS Technical Services	+90 252-422-0031
Limassol (Cyprus)	Cass Technava Ltd.	+357 2581-9921

Table 53-7. Service Directory: Pacific, New Zealand, Australia

Port	Company	Contact Numbers
Honolulu (Hawaii)	Oceantronics	808-522-5600
Honolulu (Hawaii)	Navtech Marine Electronics	808-848-7600
Honolulu (Hawaii)	PDF Inc.	808-847-1900
Honolulu (Hawaii)	Pacific Star Marine	619-997-8838
Kailua (Hawaii)	K & C Marine	808-294-0941
Papeete (French Polynesia)	Sin Tung Hing Marine	+689 549-454
Papeete (French Polynesia)	Assystem Polynesie	+689 425-962
Papeete (French Polynesia)	Marine Supplies SARL	+689-888-470
Nadi (Fiji)	Baobab Industries	+679 999-4780
Nadi (Fiji)	Yacht Help	+679 675-09112
Noumea (New Caledonia)	Alto Marine	+687 259-612
Noumea (New Caledonia)	Marine Corail	+687 254-460
Nuku Hiva (Marquesas)	Nuku-Hiva Yacht Services	+689 40-92-07-50
Galapagos (Ecuador)	Electronautica (Isla Santa Cruz)	+593 525-26-058
Opua (New Zealand)	Bay of Island Marine	+64 9402-7876

Port	Company	Contact Numbers
Whangarei (New Zealand)	Infracom Electronics	+64 9438-4644
Bay of Plenty (New Zealand)	Bay Marine Electronics	+64 7577-0250
Northland (New Zealand)	Waypoint Electronics	+64 2-749-6529
Tauranga (New Zealand)	Navtech	+64 2-744-22194
Auckland (New Zealand)	Mainsail Electronics	+64 2-179-1125
Auckland (New Zealand)	Beacon Marine	+64 9-360-0121
Rangiora (New Zealand)	Nautic Electronics	+64 3-327-3715
Christchurch (New Zealand)	Wright Technologies	+64 3-365-2541
Dunedin (New Zealand)	Navcom Electronics	+64 3-488-1644
Cairns, Queensland	Markwell Marine	+61 7-4030-0100
Cairns, Queensland	Brian Swinton Marine Elect	+61 7-4035-5311
Townsville, Queensland	Navcom	+61 7-4771-2422
Whitsundays, Queensland	Ion Marine Electrical	+61 403-872-274
Airlie Beach, Queensland	AB Marine Electronics	+61 7-4946-4070
Airlie Beach, Queensland	Twenty 16 Communications	+61 7-4946-5203
Mackay, Queensland	Pioneer Technologies	+61 7-4959-2066
Gladstone, Queensland	Modern Marine Electronics	+61 7-4976-9595
Burnett Heads, Queensland	Rampant Marine Electronics	+61 417-718-610
Buddina, Queensland	Industrial & Marine Electronics	+61 412-135-459
Brisbane, Queensland	Trymax Marine	+61 07-3245-3633
Brisbane, Manly Queensland	ADL Marine Electrics	+61 7-3348-3130
Gold Coast, Queensland	Odyssey Marine Electrical	+61 438-667-687
Sydney, New South Wales	Barrenjoey Marine Electrics	+61 2-9997-6822
Melbourne, Victoria	Goaty's Marine Electrical	+61 0418-327-403
Melbourne, Victoria	Offshore Marine Electronics	+61 3-9597-0528
Port Adelaide, South Australia	Quin Marine	+61 8-8440-2800
Hobart, Tasmania	Cordell Marine	+61 0409-162-742
Mandurah, Western Australia	Mandurah Marine Electrical	+61 0409-084-682
Perth, Western Australia	Maritime Electronic Services	+61 08-9335-2716
Darwin, Northern Territory	Wheelhouse Marine Electronics	+61 8-8981-2948

ACKNOWLEDGMENTS

Thanks to Grahame MacCleod, Paul Checkley, and Frank Penrose for the illustrations. Grateful thanks to the following friends, colleagues, and companies for their input, advice, and assistance during the preparation of the fourth edition and the previous three editions. I would like to acknowledge the use of source information and data, downloaded and derived from technical data sheets; user, installation, and maintenance manuals; and the many conversations with other marine system professionals who contributed advice and information.

Batteries: Aceleron, Exide, Grabat Energy, Lithionics, Mastervolt, Optima Batteries, Ocean Planet Energy (Firefly Battery Company), Oceanvolt, Rolls Battery Engineering, RELiON, Sonnenschein Batteries, Super-B Lithium Power, Trojan Battery Company.

Battery Charging: Mastervolt, NewMar, Victron Energie, Xantrex.

Alternative Energy Charging: Eclectic Engineering, DuPont, Hydromax, Hamilton Ferris, Marlec Engineering, MarineKinetix, Maxeon, Primus Windpower, Seamap, Superwind, TESUP, Sunware, Uniteck, Watt&Sea.

Alternator Charging: Adverc BM, Bosch, Balmar, Delphi, Hitachi, Lestek, Lucas CAV, Mastervolt, Motorola, Sterling Power, Tuff Stuff Silver Bullet, Valeo (Sev-Marchal & Paris-Rhone).

Battery Chargers: Cristec, Mastervolt, ProMariner, Victron Energie, Xantrex.

DC Systems Wiring: Ancor, Electra Cables, E-T-A, BEP, Blue Sea Systems, Index Marine, Littlefuse, NewMar Power, Perko, Pirelli.

Corrosion: Bridco, Canada Metal, M.G. Duff & Co., Swagelok, Unified Alloys.

Lighting: Aquasignal, Cantalupi, Dr. LED, Hella Marine, Imtra, Lumishore, Perko, Shadow Caster.

Lightning: Future Fibres, Marinco (Dynaplate), Newmar, Southern Spars, Sparcraft, Sertec.

Anchor Windlasses, Deck Winches, Furlers: Andersen, Bada, Facnor, Furlex, Five Oceans, Harken, Hutton-Arco, Lighthouse, Lofrans, Maxwell, Muir, Profurl, Quick Spa, Reckmann, Seldén Mast AB, Keel Servant.

Thrusters: Lewmar, MaxPower, Quantum Stabilizers, Quick Spa, Sleipner, VETUS, Wesmar.

DC Motors: ASEA, Printed Motor Works, SKF Marine.

Diesel Engines: Beta Marine, Bukh Diesel, Cummins, Caterpillar, Detroit Diesel Dahl, FloScan, Klinger-Thermoseal, Nanni Diesel, Mercedes, Parker Racor, Perkins, SeaSeal, Solé Diesel, Steyr Motors, T-ISS Safety Suppliers, Trident Marine Systems, Volvo Penta, VETUS, Yanmar, Westerbeke.

Engine Electrics and Instrumentation: Olympic Instruments, VDO Marine.

Electric Propulsion: Curtis Instruments, Integrel Solutions.

AC Power Systems: Merlin Gerin, Siemens, ABB.

AC Generators: Fischer Panda, Kohler, Mase, Northern Lights, Onan, Whisperpower.

AC Inverters: Xantrex.

Shore Power Systems: Bridgeport Magnetics, Guest Corporation, Hubbel, Marinco, Voltsafe.

Refrigeration and HVAC: Climma, Dometic, Dieselheat, Dupont, Dickenson Marine, Eberspacher, Force 10, Glacier Bay, HFL, International Thermal Research, Pompanette, VETUS, Wallas, Webasto Isotherm.

Water and Sewage Systems: Aqualarm, Attwood, Blakes and Taylors, Cleghorn Waring, Clamp-Aid, Envirovac, Electrosea, Flojet, Forespar, Flair-It, Gobius, Groco, Galleymaid, General Ecology, Jabsco, Jubilee Clips, Johnson Pumps, John Guest, Kuuma, MyCelx, Marco, Pentair Shurflo, PEXtite, Pur, Raritan, Rule, Sea-Fresh, Spectra Watermakers, Sea Recovery, Schenker, TruDesign, TMC, VacuFlush, VETUS, Sigmar, SeaLand, SPXFlow, SureCal, Seewater, Touch Sensor Technologies, Whale, Wavestream, Webasto Isotherm, Water Witch, Xylem, West Marine.

Hydraulic Power Systems: MTE Hydraulics, NAVTEC, Reckmann.

Electronic Charting: C-MAP, Raymarine, Simrad.

GPS: Garmin, Navico, NASA, Raymarine, Simrad.

AIS: Icom, Simrad, Navico, Raymarine, Weatherdock AG.

Instrument Systems: Airmar, Autonnic Research, Actisense, Brookes & Gatehouse, CZone, Cyclops Marine, FT Technologies, Maretron, Navico, NKE Marine Electronics, NASA Marine, Navpod, Ockum Instruments, Raymarine, Simrad, TackTick, Windex.

Electronics Interference: Dairyland, NewMar Power.

Autopilots: Jefa Steering Systems, Navico, Octopus Pumps, Raymarine, Simrad.

GMDSS: ACR Artex, Ocean Signal.

Navtex and Weatherfax: ICS Electronics, Furuno, Kantronics.

VHF Radio: AC Antennas, BBC, Icom, PSI Company, Raymarine, Shakespeare Marine, Standard Horizon, McGill Microwave Systems.

HF/SSB Radio: Icom, Kiss-SSB, SEA, Codan, Shakespeare Marine.

Ham Radio: ARRL, Dockside Radio, Icom, Radio Society of Great Britain, Sail Mail, Sea Mail, Xaxero, SCS (Spezielle Communications System), WinLink.

Satellite Communications: AST Space Mobile, Astranis, European Space Agency, Eutelsat, Globalstar, HughesNet, Heavens Above, Iridium, KVH, Kuiper, INMARSAT, Omnispace, One Web, Starlink, Thuraya, Telesat Lightspeed, Viasat.

Mobile Phones: Poynting, WhatsAPP, weBoost.

Radar: Furuno, Navico, Questus, Raymarine, Scanstrut.

Radar Reflectors: Echomax, Firdell.

Onboard Computing: Predictwind, Resolve-tec.

Entertainment Systems: Clarion Marine, Fusion, Glomex, James Loudspeaker, JL Audio, Poly Planar, Rockford Fosgate, Roswell Marine Audio, SiriusXM.

Sonar and Fish-finders: Airmar, Furuno, Garmin, Humminbird, Lowrance, Raymarine, ROFFS.

Safety and Security Systems: C-pod, Gost Global, Integrated Gas Technologies, Nauticalert, Sea-AI, Siren Marine, Yacht Sentinel, Yacht Protector.

Maintenance and Troubleshooting: Cortec Corp, Barden Bearings, Boeshield, FAG, Miller-Stephenson, SKF Marine, Timken, Ultra Safety Systems.

Publications: *Latitudes & Attitudes, Cruising World* magazine, *Maasmond Maritime* (Daily Newsclippings), *Ocean Navigator* magazine, *PassageMaker* magazine, *Practical Boat Owner, Yachting Monthly, Noonsite, Caribbean Compass, Sailing World, Marine Electronics Journal, Professional Mariner.*

Yacht Manufacturers: Hallberg-Rassy Yachts, Beneteau, Catalina Yachts, Hinckley Yachts, Arcona Yachts, Tartan Yachts, Island Packet Yachts, Nordhavn, Nautor Swan, Grand Banks, Pacific Seacraft, Sweden Yachts, X-yachts.

National Administrations: British Marine, International Marine Certification Institute (IMCI), Maritime and Coastguard Agency (MCA), National Oceanic and Atmospheric Administration (NOAA), Royal National Lifeboat Institution (RNLI), UK Hydrographic Office (UKHO), United States Coast Guard (USCG).

UK Associations and Clubs: Cruising Association (CA), Island Cruising NZ, Noonsite, Ocean Cruising Club (OCC), The Royal Cruising Club (RCC), Royal Yachting Association (RYA), Westerly Owners Association (WOA), World Cruising Club (ARC Rally).

BIBLIOGRAPHY

Admiralty List of Radio Signals, Volume 5. Taunton, UK: The Hydrographic Office.

American Boat and Yacht Council (ABYC).

Automotive Electrics and Electronics, 4th edition. Bosch.

Engine Monitoring on Yachts. Donat, Hans. VDO Marine.

Electrical Engineer's Reference Book, 15th edition. Say, M. G.

Electrical Safety Engineering, 3rd edition. W. Fordham Cooper.

GMDSS Handbook. London: International Maritime Organisation.

Handbook for Marine Radio Communications. Croydon, UK: Lloyds of London Press.

International Regulations for Preventing Collisions at Sea, 1972.

Lloyd's Rules for Yachts and Small Craft.

Metal Corrosion in Boats, 3rd edition. Warren, Nigel. Sheridan House.

Newnes Electrical Pocket Book, 22nd edition. E. A Reeves, editor.

The Oxford Companion to Ships and the Sea. Peter Kemp, editor.

The Rigging Handbook. Toss, Brion. Adlard Coles Nautical.

INDEX